# MATHEMATICS IN INDUSTRY 8



THE EUROPEAN CONSORTIUM
FOR MATHEMATICS IN INDUSTRY

*SUBSERIES*

A. Di Bucchianico
R.M.M. Mattheij
M.A. Peletier
*Editors*

# Progress in Industrial Mathematics at ECMI 2004

With 299 Figures, 44 in Color, and 35 Tables

Springer

*Editors*

A. Di Bucchianico
R.M.M. Mattheij
M.A. Peletier

Technische Universiteit Eindhoven
Department of Mathematics and
Computer Science
Postbus 513
5600 MB Eindhoven, The Netherlands

A.d.Bucchianico@tue.nl
R.M.M.Mattheij@tue.nl
M.A.Peletier@tue.nl

Mathematics Subject Classification (2000):
35-XX, 60-XX, 62-XX, 65-XX, 76-XX, 92-XX

ISBN 978-3-642-06633-7 e-ISBN 978-3-540-28073-6



Springer is a part of Springer Science+Business Media

springeronline.com


Printed in Germany



Cover design: *design & production* GmbH, Heidelberg
Printed on acid-free paper

# Preface

In the autumn of 1985 ESMI (European Symposium on Mathematics in Industry), the predecessor of ECMI, took place in Amsterdam. During that meeting the ideas were born that eventually lead to the foundation of ECMI as we know it now. Many successful meetings followed this 'ECMI-1985' and during this period ECMI became a brand name for Industrial Mathematics. The adulthood of ECMI is apparent from the many things it has achieved since then, as a truly European institution devoted to promote Industrial Mathematics in education and research. It took nearly 20 years to have another ECMI meeting, the 13-th, held in the Netherlands again, now in Eindhoven, June 2004. During the preparations for this meeting we were joined by the European Network for Business and Industrial Statistics (ENBIS), an organisation with objectives similar to those of ECMI. It enlarged the scope of the meeting and opened up a number of opportunities for further co-operation. For one thing, ECMI-people have less tradition in employing theory and methods from Stochastics. Yet new challenges in Science and Industry increasingly cross borders between traditional mathematical areas. Multidisciplinarity applies to Industrial Mathematics as a whole and in fact Industrial Mathematics is multidisciplinary par excellence.

The Technische Universiteit Eindhoven (TU/e) is a relatively young university. Although not large, it recently came out as second in ranking of European Universities of Technology (see Third European Report on S&T Indicators 2003). Also the city of Eindhoven looks rather young, despite the fact that it has an old history. This modern face of the city is probably typical for the spirit here and, for that matter, in the larger region. Also the greater Eindhoven region does well as it ranks among the top three regions in Europe regarding technological and industrial innovation. The theme of this conference, Industrial Mathematics, is aptly fitting in with this. Indeed, nowadays Mathematics is generally accepted as a Technology, playing a crucial role in many branches of industrial activity, for optimising both processes and products.

Since Industrial Mathematics is a vast and diverse area, each ECMI conference chooses a number of (application) themes to focus on. This time they were Aerospace, Electronic Industry, Chemical Technology, Life Sciences, Materials, Geophysics, Financial Mathematics and Water flow. The majority of the subjects of the talks were on these topics indeed. In particular the talks of the invited speakers were related to these main themes. They delivered excellent lectures, most of which are reported in these proceedings. In alphabetical order the speakers were Søren Bisgaard (Amherst, MA), Rainer Helmig (Stuttgart), John Hinch (Cambridge), John Hunt (London), Chris Rogers (Cambridge), Cord Rossow (Braunschweig), Fabrizio Ruggeri (Milano), Wim Schoenmakers (Leuven), Bernard Schrefler (Padova), and Michael Waterman (Los Angeles, CA). Moreover there was a plenary talk by Sabine Zaglmayr, the winner of the Wacker price for the best thesis on Industrial Mathematics.

Organizing a meeting like this is a multi-person undertaking. During the last three years a dedicated group of people has devoted much of their time to making this event a success, eventually growing to quite a large number of persons who were actively involved in the lubrication of it all at the meeting. We are very grateful for their help. Special mention should be made of the help we received from our university congress bureau and our CASA secretariat. It goes without saying, however, that the actual success of this meeting was due to the participants. The conference was attended by some 400 people, from all continents, who altogether gave over 300 talks. There were excellent contributions by the invited speakers, a large number of high quality minisymposia, and many interesting contributed talks. All speakers were invited to submit a contribution to these proceedings, which therefore record the majority of the talks. We are most grateful to the many reviewers who helped us in the refereeing process.

At this place we would also like to thank the companies and institutions that participated in the exhibition, which was conducive to providing a proper atmosphere. We are particularly indebted to the many sponsors who made it possible to keep the fees quite moderate and yet have a nice social programme and affordable catering. The Local Organising Committee deserves special thanks for the many smaller and larger things that they have done. In particular I am personally very indebted to my two co-editors, Sandro Di Bucchianico and Mark Peletier. Their continuous enthusiasm, constructive ideas, as well as their skills in technical editing have proven invaluable. On behalf of all three of us I trust that these proceedings will be useful for all those who are interested in the use and the usefulness of Mathematics in Industry.

*Bob Mattheij*
Eindhoven, February 2005

# Contents

## Part I Theme: Aerospace

## Part VII Theme: Financial Mathematics

# Part I

## Theme: Aerospace

# The MEGAFLOW Project – Numerical Flow Simulation for Aircraft

C.-C. Rossow, N. Kroll, and D. Schwamborn

[1] Deutsches Zentrum für Luft- und Raumfahrt e. V. (DLR) in the Helmholtz-Association
[2] Institute of Aerodynamics and Flow Technology D-38108 Braunschweig, Germany
cord.rossow@dlr.de, norbert.kroll@dlr.de

**Summary.** Some years ago the national CFD project MEGAFLOW was initiated in Germany, which combined many of the CFD development activities from DLR, universities and aircraft industry. Its goal was the development and validation of a dependable and efficient numerical tool for the aerodynamic simulation of complete aircraft which met the requirements of industrial implementations. The MEGAFLOW software system includes the block-structured Navier-Stokes code FLOWer and the unstructured Navier-Stokes code TAU. Both codes have reached a high level of maturity and they are intensively used by DLR and the German aerospace industry in the design process of new aircraft. Recently, the follow-on project MEGADESIGN was set up which focuses on the development and enhancement of efficient numerical methods for shape design and optimization. This paper highlights recent improvements and enhancements of the software. Its capability to predict viscous flows around complex industrial applications for transport aircraft design is demonstrated. First results concerning shape optimization are presented.

## 1 Introduction

Aerospace industry is increasingly relying on advanced numerical flow simulation tools in the early aircraft design phase. Today, computational fluid dynamics has matured to a point where it is widely accepted as an essential, complementary analysis tool to wind tunnel experiments and flight tests. Navier-Stokes methods have developed from specialized research techniques to practical engineering tools being used for a vast number of industrial problems on a routine basis [51]. Nevertheless, there is still a great need for improvement of numerical methods, because standards for simulation accuracy and efficiency are constantly rising in industrial applications. Moreover, it is crucial to reduce the response time for complex simulations, although the relevant geometries and underlying physical flow models are becoming increasingly complicated. In order to meet the requirements of German aircraft industry, the national project MEGAFLOW was initiated some years ago under the leadership of DLR [28, 29]. The main goal was to focus and direct development activities carried out in industry, DLR and universities towards industrial needs. The close collaboration between the partners led to the development and validation of a common aerodynamic simulation system providing both a structured and an unstructured prediction capability for complex applications. This software is still constantly updated to meet the requirements of industrial implementations.

In the first phase of the project the main emphasis was put on the improvement and enhancement of the block-structured grid generator MegaCads and the Navier-Stokes solver FLOWer. In a second phase the activities were focused on the development of the unstructured/hybrid Navier-Stokes solver TAU. Due to a comprehensive and cooperative validation effort and quality controlled software development processes both flow solvers have reached a high level of maturity and reliability. In addition to the MEGAFLOW initiative, considerable development and validation activities were carried out in several DLR internal and European projects which contributed to the enhancement of the flow solvers. The MEGAFLOW software is used in the German aeronautic industry and research organizations for a wide range of applications. Due to the use of common software, the process of transferring latest research and technology results into production codes has been considerably accelerated.

Recently, based on the MEGAFLOW network the national project MEGADESIGN (2004-2007) was set up [26]. Its main objective is to enhance and establish numerical shape optimization tools within industrial aircraft design processes. The project deals with several key issues including suitable techniques for geometry parameterization, meshing and mesh movement methods, efficiency and accuracy improvements of the flow solvers as well as flexible and efficient deterministic and stochastic based optimizers.

The present paper describes the features of the MEGAFLOW software and demonstrates its capability on the basis of several industrial relevant applications. Finally, the perspective and future requirements of CFD for industrial applications are shortly outlined.

## 2 MEGAFLOW software

The MEGAFLOW software offers flow prediction capabilities which are based on both block-structured and hybrid meshes. Details are given in [25].

### 2.1 Grid Generation

For the generation of block-structured grids the interactive system MegaCads has been developed. Specific features of the tool are the parametric construction of multi-block grids with arbitrary grid topology, generation of high-quality grids through advanced elliptic and parabolic grid generation techniques, construction of overlapping grids and batch functionality for efficient integration in an automatic optimization loop for aerodynamic shape design [12]. The limitation of MegaCads is the non automatic definition of the block topology which for rather complex configurations may result in a time consuming and labor intensive grid generation activity. Besides MegaCads, the commercial software package ICEM-HEXA and specialized in-house codes are used for specific applications.

In contrast to the block-structured approach, no major development activities have been devoted to the generation of unstructured meshes within the MEGAFLOW project. A strategic cooperation, however, has been established with the company CentaurSoft [3] which provides the hybrid grid generation package Centaur. The software consists of three major parts. An interactive program reads in the CAD data of the geometry under consideration, performs some CAD cleaning if necessary and sets up the grid generation process. In a second step the surface and volume grid are generated automatically. For viscous calculations a quasi-structured prismatic cell layer with a specified number of cells around the geometry surface ensures high resolution of boundary layer effects. In a third step grid adaptation may be used to locally refine grid resolution. During the cooperation the Centaur grid generation software has been substantially advanced for transport aircraft applications. Improvements are underway to include for example the generation of non isotropic elements and wake surfaces. Within the MEGADESIGN project the partner EADS-M is developing fully automatic hybrid grid generation software which is adapted to massively parallel distributed computers.

## 2.2 Flow Solvers

The main components of the MEGAFLOW software are the block-structured flow solver FLOWer and the unstructured hybrid flow solver TAU. Both codes solve the compressible three-dimensional Reynolds averaged Navier-Stokes equations for rigid bodies in arbitrary motion. The motion is taken into account by transformation of the governing equations. For the simulation of aero-elastic phenomena both codes have been extended to allow geometry and mesh deformation. In the following sections the specific features of the Navier-Stokes codes are briefly described.

### *Block-Structured Navier-Stokes Code FLOWer*

The FLOWer-Code is based on a finite-volume formulation on block-structured meshes using either the cell vertex or the cell-centered approach. For the approximation of the convective fluxes a central discretization scheme combined with scalar or matrix artificial viscosity and several upwind discretization schemes are available [27]. Integration in time is performed using explicit multistage time-stepping schemes. For steady calculations convergence is accelerated by implicit residual smoothing, local time stepping and multigrid. Preconditioning is used for low speed flows. For time accurate calculations an implicit time integration according to the dual time stepping approach is employed. The code is highly optimized for vector computers. Parallel computations are based on MPI [6].

A variety of turbulence models is implemented in FLOWer, ranging from simple algebraic eddy viscosity models over one- and two-equation models up to differential Reynolds stress models. The Wilcox k-$\omega$ model is the standard model in FLOWer which is used for all types of applications, while for transonic flow the linearized algebraic stress model LEA [42] and the nonlinear EARSM of Wallin [52] have shown to improve the prediction of shock locations. Furthermore, the SST model of Menter [36] is available for a better prediction of separating flows. All two-equation models can be combined with Kok's modification [23] for improved prediction of vortical flows. For supersonic flows different compressibility corrections are available. Recently, within the EU project FLOMANIA Reynolds stress models based on the Wilcox stress-$\omega$ model [53] and the so-called SSG/LRR-$\omega$ model, a combination of the Wilcox stress-$\omega$ and the Speziale-Sarkar-Gatski model [47], have been implemented into FLOWer [17]. Particularly the SSG/LRR-$\omega$ model has been applied to a wide variety of test cases, ranging from simple airfoils to complex aircraft configurations and from transonic to high-lift conditions. Generally improved predictions have been obtained, while the numerical behavior of the Reynolds stress models appeared to be as robust as that of two-equation models. Fig. 1 shows the predicted pressure and skin friction distribution obtained with the Wilcox k-$\omega$ and with the SSG/LRR-$\omega$ model for the Aerospatiale A airfoil at $M_\infty = 0.15,\ \alpha = 13.3°,\ Re = 2 \times 10^6$, demonstrating the improvement by Reynolds stress modeling.

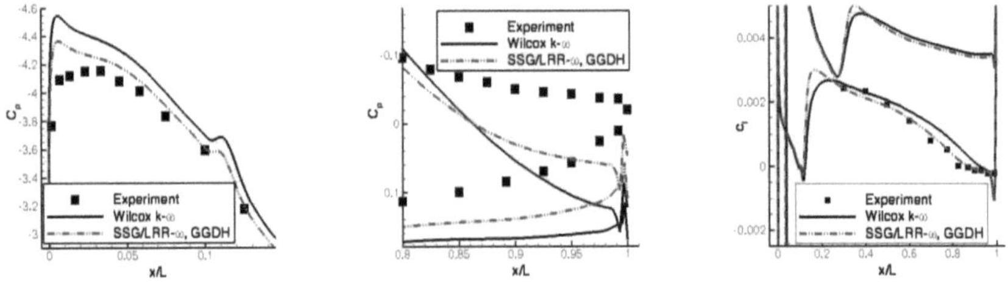


**Fig. 1.** Pressure distribution (near leading and trailing edge) and skin friction distribution for Aerospatiale A airfoil ( $M_\infty = 0.15,\ \alpha = 13.3°,\ Re = 2 \times 10^6$) calculated with the Reynolds stress turbulence model implemented in FLOWer.

Besides the modeling accuracy for turbulent flows, the numerical robustness of the respective transport equation turbulence models for complex applications has been a major issue. In FLOWer numerical stability has been enhanced by an implicit treatment of the turbulence equations and different limiting mechanisms that can be activated by the user. The convergence behavior of the FLOWer-Code for a rather complex application is demonstrated in Fig. 2(a). Results of a viscous computation for a helicopter fuselage are shown [32]. The rotor is modeled through a uniform actuator disc. The grid consists of 94 blocks and 7 million grid points. The residuals for density and turbulence quantities are reduced several orders of magnitude. In this low Mach number case the preconditioning technique has been employed.

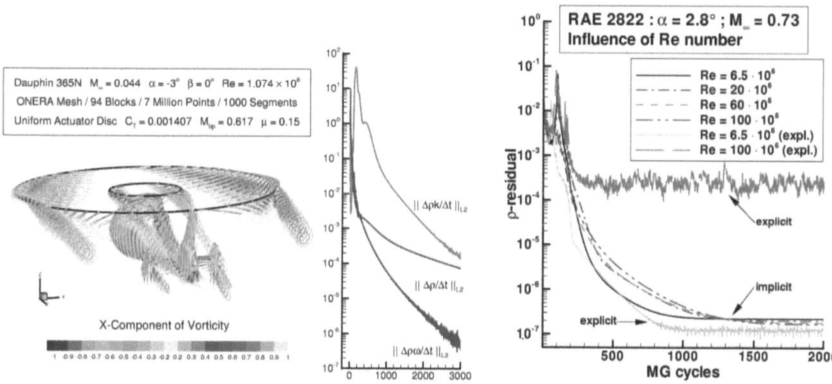


(a) Viscous calculation for Dauphin helicopter fuselage at $M_\infty = 0.044$, convergence behavior of mass and k-$\omega$ turbulence equations.

(b) Effect of Reynolds number on convergence for the RAE 2822 airfoil at $M_\infty = 0.73$, $\alpha = 2.8°$.

**Fig. 2.**

The fully implicit integration of the turbulence equations also ensures efficient calculations on highly stretched cells as they appear in high Reynolds number flows [18]. Fig. 2(b) shows the convergence history of FLOWer for the calculation of the viscous flow around the RAE 2822 airfoil at different Reynolds numbers. The advantage of the fully implicit method compared to the explicit multigrid scheme with point implicit treatment of source terms is evident.

FLOWer is able to perform transition prediction on airfoils and wings using a module consisting of a laminar boundary layer code and an $e^N$-database method based on linear stability theory [30]. Fig. 3 shows the predicted and measured force polars and transition locations of a subsonic laminar airfoil. This approach substantially improves the quality of predicted force coefficients. The experimentally determined transition points are reproduced with high accuracy. The transition prediction capability has been extended to 2D high-lift systems.

An important feature of FLOWer is the Chimera technique, which considerably enhances the flexibility of the block-structured approach [21, 45]. This technique mainly developed within the German/French helicopter project CHANCE [46] enables the generation of a grid around a complex configuration by decomposing the geometry into less complex components. Separate component grids are generated which overlap each other and which are em-

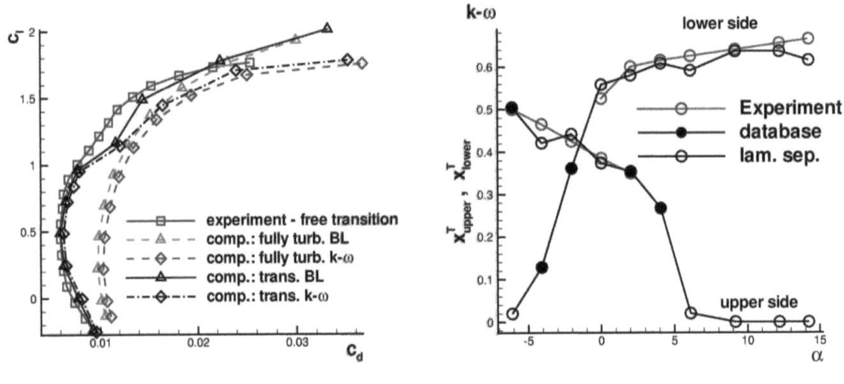


(a) (b)

**Fig. 3.** Transition prediction with $e^N$-database method for laminar Sommers airfoil at $M_\infty = 0.1$ and $Re = 4 \times 10^6$, (a) force polars calculated fully turbulent and with transition, (b) computed and measured transition locations.

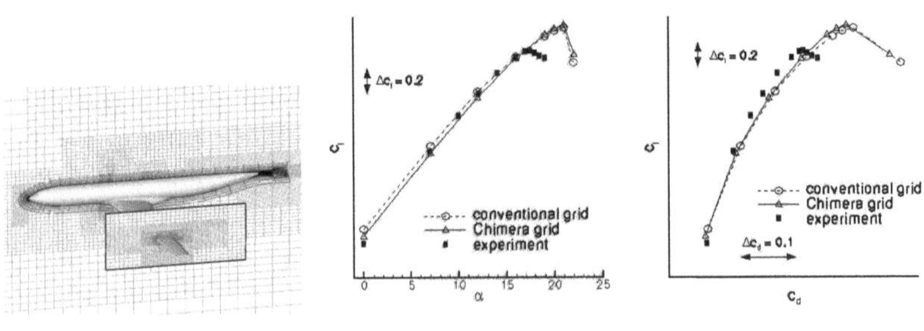


(a) (b) (c)

**Fig. 4.** Viscous computation about a 3D high-lift configuration using the Chimera technique of the block-structured FLOWer-Code, $M_\infty = 0.174$, $\alpha = 7°$ .

bedded in a Cartesian background grid that covers the whole computational domain. In combination with flexible meshes, the Chimera technique enables an efficient way to simulate bodies in relative motion. The communication from mesh to mesh is realized through interpolation in the overlapping area. The search for cells which are used for interpolation is performed using an alternating digital tree method. In the case when a mesh overlaps a body which lies inside another mesh, hole cutting procedures have to be used in order to exclude the invalid points from computation. Further simplification of the grid generation procedure is achieved by a fully automatic Cartesian grid generator. The grid generator places fine grids around the component grids and puts successively coarsened grids around the fine grids. Patched grid interfaces with hanging nodes are used at the interface between the grid blocks of the Cartesian mesh. In the vicinity of the configuration the Cartesian grid generator creates non isotropic cells which are adapted to the size of the cells in the component grids. This ensures accuracy in the overlap regions. The potential of the Chimera technique is demonstrated in Fig. 4 in case of the viscous calculation around a 3D high-lift configuration. Separate component grids have been generated for body, wing, flap and slat. The background grid has been produced with the automatic Cartesian grid generator. With this approach the time for grid generation has been considerably reduced. The whole grid consists of 4 million points in total. Fig. 4(b) and Fig. 4(c) show the distribution of lift versus angle of attack and lift versus drag, respectively. The results obtained on the Chimera grid are compared with computations carried out on a conventional block-structured grid and with experimental data. It can be seen that the computations on the different meshes agree very well and they are in quite good correlation to the experiments. Differences between computations and experiments occur at the angle of attack where lift breaks down.

### *Hybrid Navier-Stokes Code TAU*

The Navier-Stokes code TAU [19, 49] makes use of the advantages of unstructured grids. The mesh may consist of a combination of prismatic, pyramidal, tetrahedral and hexahedral cells and therefore combines the advantages of regular grids for the accurate resolution of viscous shear layers in the vicinity of walls with the flexibility of grid generation techniques for unstructured meshes. The use of a dual mesh makes the solver independent of the type of cells that the initial grid is composed of. Various spatial discretization schemes were implemented, including a central scheme with artificial dissipation and several upwind methods. The basic hybrid TAU-Code uses an explicit Runge-Kutta multistage scheme in combination with an explicit residual smoothing. In order to accelerate convergence, a multigrid procedure was developed based on the agglomeration of the control volumes of the dual grid for coarse grid computations.

In order to efficiently resolve detailed flow features, a grid adaptation algorithm for hybrid meshes based on local grid refinement and wall-normal mesh movement in semi-structured near-wall layers was implemented. This algorithm has been extended to allow also for de-refinement of earlier refined elements thus enabling the code to be used for unsteady time-accurate adaptation in unsteady flows. Fig. 5 gives a simple example of the process for viscous airfoil calculation. First a flow solution is calculated on a basic grid (a). After some refinement an adapted grid/solution is obtained (b). Changing the flow parameters and specifying e.g. that the number of mesh points should not increase any further, the de-refinement interacts with the refinement (c) and finally the new shock position is resolved (d).

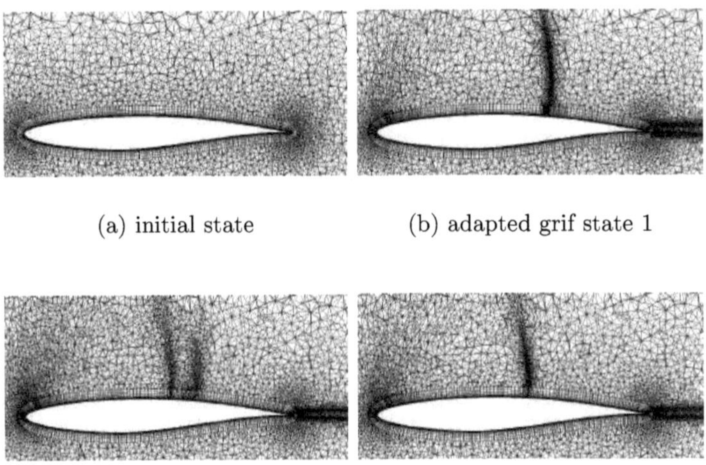

(a) initial state (b) adapted grif state 1

(c) intermediate state (d) sdapted grid state 2

**Fig. 5.** Demonstration of the dynamic mesh refinement and de-refinement capability of the TAU-Code.

With respect to unsteady calculations, the TAU-Code has been extended to simulate a rigid body in arbitrary motion and to allow grid deformation. In order to bypass the severe time-step restriction associated with explicit schemes, the implicit method based on the dual time stepping approach is used. For the calculation of low-speed flows, preconditioning of the compressible flow equations similar to the method used in FLOWer was implemented. One of the important features of the TAU-Code is its high efficiency on parallel computers. Parallelization is based on the message passing concept using the MPI library [6]. The code is further optimized either for cache or vector processors through specific edge coloring procedures.

The standard turbulence model in TAU is the Spalart-Allmaras model with Edwards modification, yielding highly satisfactory results for a wide range of applications while being numerically robust. Besides this model, a number of different k-$\omega$ models with and without compressibility corrections are available. Also nonlinear explicit algebraic Reynolds stress models (EARSM) and the linearized LEA model [42] have been integrated. Several rotation corrections for vortex dominated flows are available for the different models. Finally, there are options to perform detached eddy simulations (DES) based on the Spalart-Allmaras model [48] and so-called Extra-Large Eddy Simulations (XLES) [24].

The explicit character of the solution method severely restricts the CFL number which in turn often leads to slow convergence, especially in the case of large scale applications. In order to improve the performance and robustness of the TAU-Code, an approximately factored implicit scheme has been implemented [16]. The LU-SGS (Lower-Upper Symmetric Gauss-Seidel) scheme has been selected as a replacement for the Runge-Kutta scheme. In contrast to fully implicit schemes, this method has low memory requirements, low operation counts and can be parallelized with relative ease. Compared to the explicit Runge-Kutta method, the LU-SGS scheme is stable with almost no time step restrictions. An example of the performance improvement achieved is given in Fig. 6, where two convergence histories for viscous calculations on a delta wing are shown. The calculations were performed with multigrid on 16 processors of a Linux cluster. The figure shows the residual and the rolling moment against iteration count. In terms of iterations LU-SGS can be seen to converge approximately twice as fast as the Runge-Kutta scheme. Furthermore, one iteration of LU-SGS costs roughly 80% of one Runge-Kutta step. This results in a reduction of the overall calculation time by a factor of 2.5.

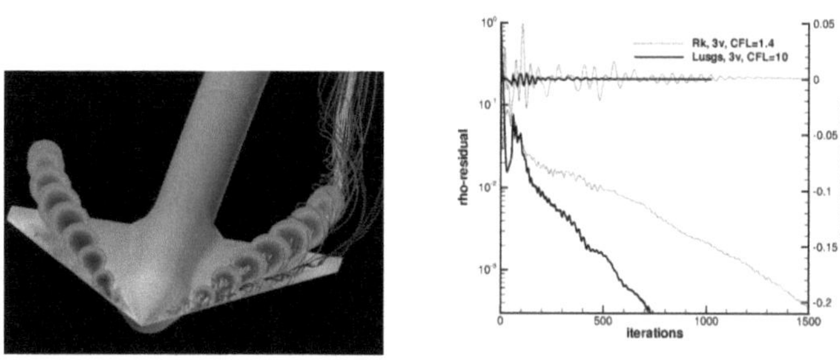


**Fig. 6.** Convergence behaviour of the hybrid TAU-Code for calculations of viscous flow around a delta wing at $M = 0.5$, $\alpha = 9°$. Comparison of the baseline Runge-Kutta scheme (RK) and the implicit LU-SGS scheme.

As the Chimera technique has been recognized as an important feature to efficiently simulate maneuvering aircraft, it has been also integrated into the TAU-Code [34]. In the context of hybrid meshes the overlapping grid technique allows an efficient handling of complex configurations with movable control surfaces (see Fig. 7). For the intergrid communication linear interpolation based on a finite element approach is used in case of tetrahedral mesh elements. For other types of elements (prisms, hexahedrons, pyramids) linear interpolation is performed by splitting the elements into tetrahedrons. Like in FLOWer, the search algorithm for donor cells is based on the alternating digital tree data structure. The current implementation of the Chimera technique can handle both steady and unsteady simulations for inviscid and viscous flows with multiple moving bodies. The technique is available in parallel mode. In Fig. 8 results of a viscous Chimera calculation for a delta wing with trailing edge flaps are shown [43]. The component mesh of the flap is designed to allow a flap deflection of ±15°. The comparison of calculated and measured surface pressure distributions at both 60% and 80% cord length shows good agreement.

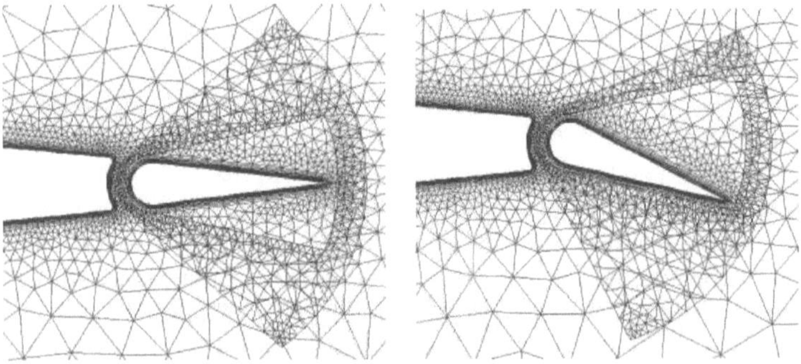

**Fig. 7.** Hybrid Chimera grid for delta wing with a movable flap.

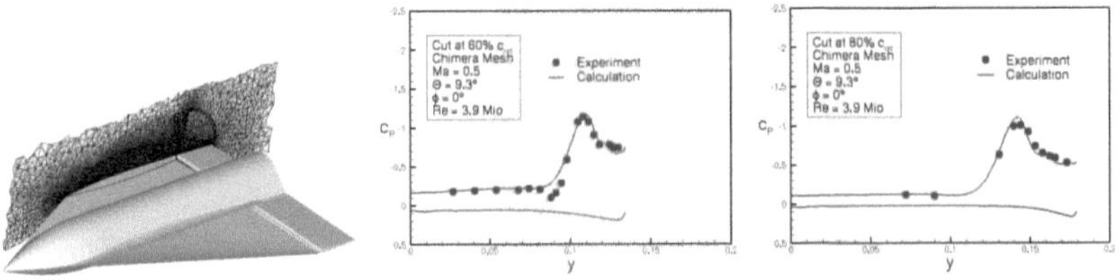


**Fig. 8.** Viscous computation of a delta wing with trailing edge flap using the Chimera option of the hybrid TAU-Code, surface pressure distributions for flap deflection angle $\theta = 0°$ at 60% and 80% cord.

## 3 Software validation

Software validation is a central and critical issue when providing reliable CFD tools for industrial applications. Among others, the verification and validation exercises should address consistency of the numerical methods, accuracy assessment for different critical application cases and sensitivity studies with respect to numerical and physical parameters. Best practice documentation is an essential part of the work. Over the last few years the MEGAFLOW software has been validated within various national and international projects for a wide range of configurations and flow conditions (see e.g. [25, 40]). This section shows sample results for a subsonic and transonic validation test case.

Flow prediction for a transport aircraft in high-lift configuration is still a challenging problem for CFD. The numerical simulation addresses both complex geometries and complex physical phenomena. The flow around a wing with deployed high-lift devices at high incidence is characterized by the existence of areas with separated flow and strong wake/boundary layer interaction. The capabilities of the MEGAFLOW software to simulate two- and three-dimensional high-lift transport aircraft configurations has been extensively validated within the European high-lift program EUROLIFT I [39]. One of the investigated test cases is the DLR-F11 wing/body/flap/slat configuration.

Fig. 9 highlights a comparison of lift and total drag results of the unstructured TAU-Code and the block-structured FLOWer-Code with experimental data from the Airbus LWST low speed wind tunnel in Bremen, Germany. Both, the block-structured grid generated by the DLR software MegaCads and the hybrid mesh generated by FOI contain about 3 million grid points to allow for a fair comparison of the methods.

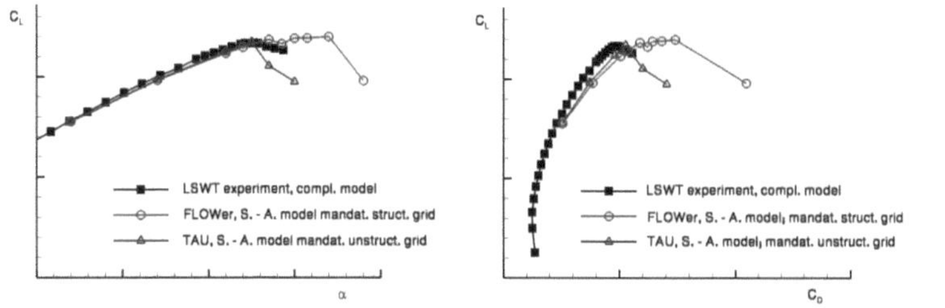


**Fig. 9.** Viscous computations for DLR-F11 high-lift configuration at $M_\infty = 0.18$, $Re = 1.4 \times 10^6$, lift as function of angle of attack and as function of drag.

Calculations for the start configuration at $M_\infty = 0.18$ and $Re = 1.4 \times 10^6$ were performed with FLOWer and TAU using the Spalart-Allmaras turbulence model with Edwards modification (SAE). In both cases preconditioning was used to speed up steady state convergence and to improve accuracy at the predominantly low speed conditions. In the linear range of the polar, the numerical results compare quite well with each other and with experimental data. At higher angle of attack differences occur between the TAU and FLOWer results. TAU predicts the lift break down at a lower angle of attack, which is in better agreement with the experimental results.

In the framework of the AIAA CFD Drag Prediction Workshop I [1], the accuracy of the MEGAFLOW software was assessed to predict aerodynamic forces and moments for the DLR-F4 wing-body configuration [38]. In Fig. 10 lift coefficient as function of drag and angle of attack for Case 2 ($M_\infty = 0.75$, $Re = 3 \times 10^6$) calculated with FLOWer and TAU are presented. These results were obtained using grids generated in-house at DLR. On request all calculations were performed fully turbulent. The FLOWer computations were carried out on a grid with 3.5 million points using central discretization with a mixed scalar and matrix dissipation operator and the $k/\omega$ -LEA turbulence model. The TAU results are based on an initial grid containing 1.7 million points which was adapted for each angle of attack yielding grids with 2.4 million points. In addition, an adaptation of the prismatic grid towards $y^+ = 1$ was done. Central discretization with standard settings of artificial dissipation was used. Turbulence was modeled with the one-equation model of Spalart-Allmaras. As can be seen from Fig. 10 the fully turbulent FLOWer computations over predict the measured drag curve by approximately 20 drag counts. Investigations have shown [38] that inclusion of transition in the calculation reduces the predicted drag by 14 drag counts, reducing the over prediction of drag to approximately 6 drag counts. The results of the unstructured fully-turbulent computations with TAU perfectly match with the experimental data. However, as for the structured computations, hybrid calculations with transition setting will reduce the predicted level of drag, in this case by approximately 10 drag-counts. Fig. 10 also shows the comparison of predicted and measured lift coefficient as a function of angle of attack. The values calculated by FLOWer agree very well with the experiment, whereas the results obtained with TAU over predict the lift almost in the whole range of angle of attack.

For the pitching moment (Fig. 11) the results obtained with FLOWer agree very well with experimental data. This is due to the fact that the surface pressure distribution predicted with the FLOWer-Code is in good agreement with the experiment. In case of the hybrid TAU-Code there are some discrepancies between the predicted and measured surface pressures resulting in a significant over prediction of the pitching moment. Further investigations [38] have shown that the improved results obtained with the FLOWer-Code are mainly attributed to a lower level of numerical dissipation (improved grid resolution and matrix dissipation) combined with the advanced 2-equation $k/\omega$ -LEA turbulence model.

Within the second AIAA drag prediction workshop [2] the hybrid TAU-Code was further assessed with respect to performance calculations for a wing/body/pylon/nacelle configuration at transonic flow conditions [11]. For this exercise the Spalart-Allmaras one-equation turbulence model was used.

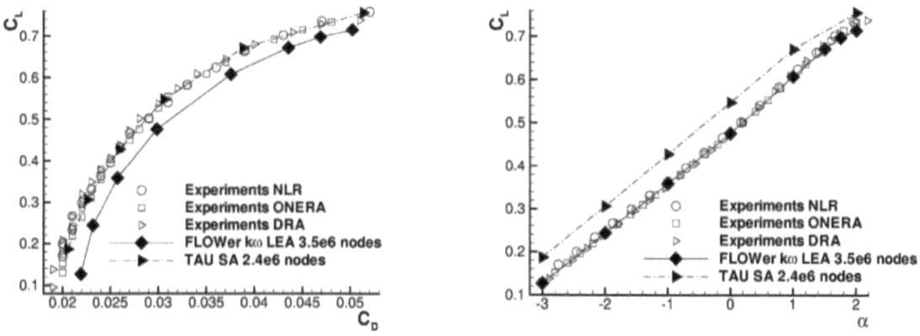


**Fig. 10.** Viscous calculations for DLR-F4 wing/body configuration (AIAA DPW I, case 2), $C_L(C_D), C_L(\alpha)$.

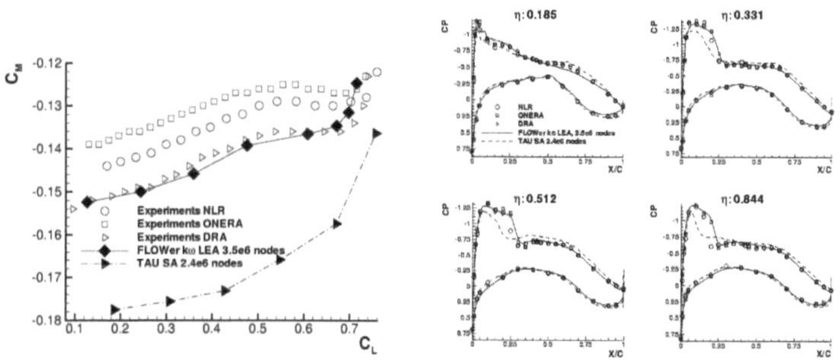


**Fig. 11.** Viscous calculations for DLR-F4 wing/body configuration (AIAA DPW I), $C_M(C_L)$ polar, surface pressure.

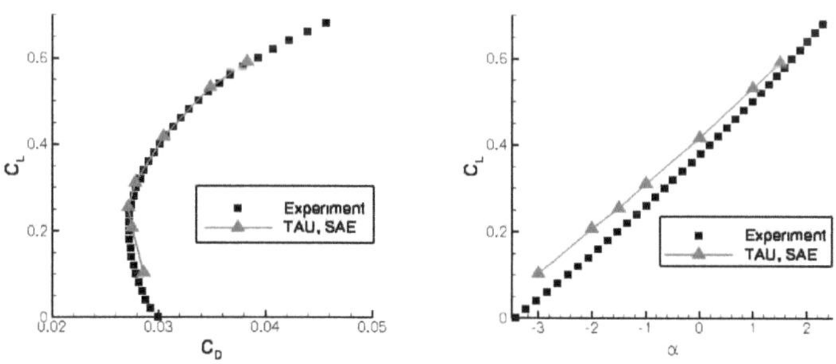


**Fig. 12.** TAU results for DLR-F6 wing/body/pylon/nacelle configuration (AIAA DPW II), $M_\infty = 0.75$

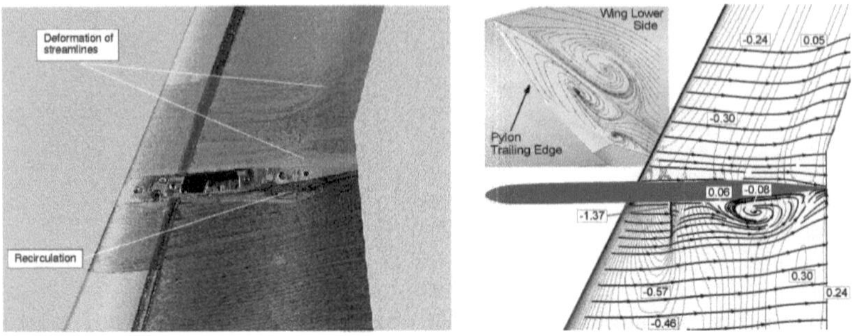


**Fig. 13.** Oil flow pictures (experiments) and streamlines (TAU results), DLR-F6 wing/body/pylon/nacelle configuration, wing lower and pylon inboard side, $M_\infty = 0.75$, $C_L = 0.5$.

The drag polar is predicted in good agreement with the experimental data while the lift is constantly over predicted (see Fig. 12). A detailed analysis of the flow features reveals that in principle all areas of flow separations on the investigated DLR F6 configuration are identified, however, compared with experiments the sizes of those areas are slightly under predicted (wing upper side) or over predicted (wing lower side). Fig. 13 compares measured and predicted flow features near the pylon inboard side at the wing lower side. This difference results in systematic deviations of the pressure distributions and pitching moments.

## 4 Industrial Applications

The MEGAFLOW software is intensively used at DLR and the German aircraft industry for many aerodynamic problems. Some typical large scale applications listed below demonstrate the capability of the software to support aircraft and helicopter design.

### *Civil transport aircraft at cruise conditions*

One key issue during the design of an enhanced civil aircraft is the efficient engine-airframe integration. Modern very high bypass ratio engines and the corresponding close coupling of engine and airframe may lead to substantial loss in lift and increased installation drag. At DLR, numerical and experimental studies have been devoted to estimate installation drag with respect to variations of engine concepts and the installation positions [13, 41]. For numerical investigations in this field both the block-structured FLOWer-Code and the hybrid TAU-Code have been used. Fig. 14 shows the hybrid grid in the symmetry plane for the DLR-F6 configuration [10]. The initial grid generated with Centaur consists of about 4.6 million nodes. Several solution based grid adaptation steps have been performed resulting in grids between 7.5 and 8.5 million nodes depending on the investigated engine concept. In Fig. 14 the lift as a function of the installation drag is plotted for three different positions of the CFM56 long duct nacelle ($M_\infty = 0.75$ and $Re = 3 \times 10^6$). The engines are represented by through-flow nacelles. Results predicted with the TAU-Code (symbols) and measured in the ONERA S2MA wind tunnel (lines) are shown. The agreement is very satisfactory demonstrating that the influence on installation drag due to varying engines locations or sizes can be accurately predicted by the TAU-Code [10].

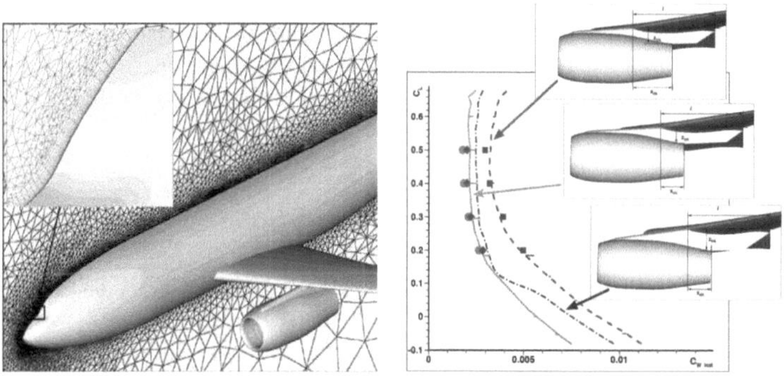

**Fig. 14.** Prediction of engine-airframe interference drag using the TAU-Code, left: hybrid grid for DLR-F6 configuration, right: lift as a function of installation drag for three different position of CFM56 engine, $M_\infty = 0.75, Re = 3 \times 10^6$, symbols: calculation, lines: experiment.

Viscous computations with the block-structured FLOWer-Code were performed for the DLR-ALVAST configuration with turbofan engines for the most interesting conditions 'Start of Cruise' (SOC), and 'Through Flow Nacelle' (TFN) representing a flight-idle power setting [41]. Computations were carried out at $M_\infty = 0.75$, $Re = 3 \times 10^6$ and with a constant lift coefficient of $C_L = 0.5$. Fig. 15 shows the impact of the power setting. Computed lines of constant Mach number in the engine symmetry plane are shown. The primary differences caused by the SOC thrust condition are the strong velocity increase in the jets up to supersonic speed and the resulting significant shear layers at the jet boundaries due to the larger velocity differences. Fig. 15 also shows corresponding computed and measured pressure distributions at the wing cross section $\eta = 33\%$ (inboard of nacelle). The most significant difference between the SOC and TFN condition is a lower pressure level for SOC in the mid chord area at the wing lower side. This influence is captured quite well by the numerical simulation.

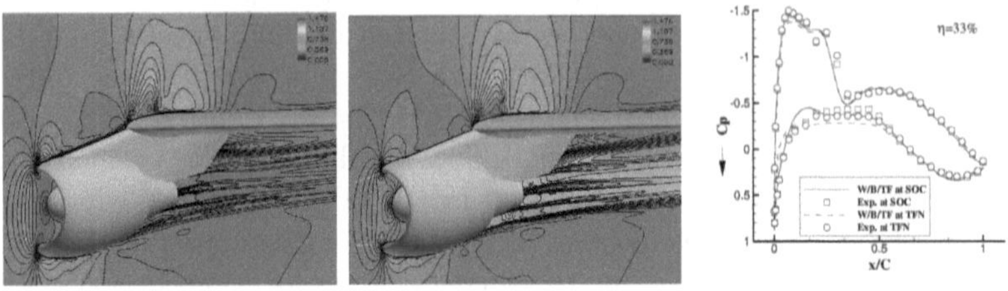


(a) TFN (b) SOC (c)

**Fig. 15.** Viscous calculation of DLR ALVAST configuration with FLOWer at $M_\infty = 0.75$, $C_L = 0.5$, influence of thrust condition of turbofan engine, (a) and (b) constant Mach number distribution for TFN and SOC, (c) surface pressure distribution at cross section $\eta = 33\%$.

### *Civil transport aircraft at high-lift conditions*

Based on thorough development and validation efforts of the hybrid unstructured approach employing both the Centaur grid generation software and the Navier-Stokes-Code TAU, complex high-lift flows become more and more accessible. As an example the flow around the DLR ALVAST model in high lift configuration equipped with two different engine concepts, the VHBR (Very High Bypass Ratio) and the UHBR (Ultra High Bypass Ratio) engine has been computed [35]. The numerical simulations are focused on complex flow phenomena arising from the engine installation at high-lift conditions. Special attention was paid to a possible reduction of the maximum lift angle resulting from dominant three-dimensional effects due to engine installation. Fig. 16 displays the surface pressure coefficient of the ALVAST high-lift configuration with installed VHBR and UHBR engine at an angle of attack of $\alpha = 12°$ in take-off conditions. The computations were performed on a hybrid grid with 10 million points generated by Centaur. In Fig. 17(a) the vortex shedding from the inboard side of the nacelle is shown. The vortex originates from the rolling-up of the shear layer and crosses the slat and the wing upper side. Using the computational data as input this vortex system could be identified with PIV visualization in a recent wind tunnel campaign. Fig. 17(a) also shows the impact of the two different engine concepts on the span wise lift distribution. For the VHBR concept the lift loss on the wing due to engine mounting is roughly compensated by the lift generated by the nacelle itself. For the UHBR concept the wing lift loss is slightly stronger than for the VHBR. Nevertheless, it is overcompensated by the higher lift carried by the large nacelle.

One key aspect of the development of a new transport aircraft is the design of a sophisticated and optimal high-lift system for take-off and landing conditions. A possibility to increase maximum lift is the usage of small delta wing like plates on the engine nacelles, the so-called nacelle strakes. These strakes generate vortices which run above the wing for high angles of attack. These vortices influence the wing and slat pressure distributions and shift the flow separations to higher angles of attack. At cruise flight conditions the strakes should not produce any significant additional drag. Previous investigations based on hybrid grid RANS solutions using the DLR TAU software have shown that for civil transport aircraft the influence of the nacelle strakes on lift and drag can be computed qualitatively [15]. In order to quantitatively predict the lift increment due to the strakes, care must be taken generating and adapting the grid with and without strakes. The idea has been to use the final adapted grid of the configuration with nacelle strakes and to fill the strakes with tetrahedral elements so that a nearly identical grid for the configuration with and without strakes can be build. The initial grid generation has been performed with Centaur. The element sizes have been controlled by several sources in the region where the strake vortices appear. The near wall region has been resolved by 25 layers of prismatic elements. The initial grid contains approximately 13.05 million points. The TAU grid adaptation has been used to insert additional points in areas of large gradients and to fulfill a $y^+$ of nearly one. The three times adapted grid contains approximately 16.71 million points. The filling of the strake volume has been performed using customized tools based on MegaCads [12] and the NETGEN [4] software. Fig. 17(b) shows the adapted grid in the vicinity of the nacelle strake. The filled strake volume is visible. The solutions have been calculated using the TAU-Code for the flow condition $M_\infty = 0.18$, $Re = 3$ million and a between 8° and 16° . Fig. 17(c) demonstrates the resolution of the strake vortex and an iso-vorticity plane for $\alpha = 10°$ . It has been shown that for this configuration a lift increase of $\Delta C_L \approx 0.1$ can be found both from the numerical calculations and the experiments although the absolute maximum lift values differ [14].

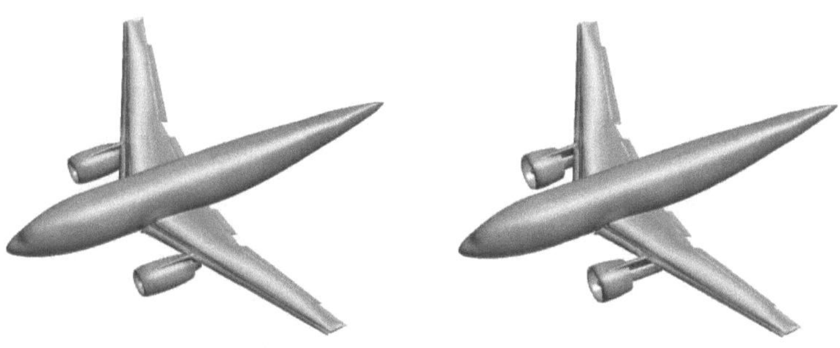

**Fig. 16.** Viscous simulation of the ALVAST high-lift configuration with VHBR (left) and UHBR (right) engine using the TAU-Code, surface pressure distribution, $M_\infty = 0.22$, $\alpha = 12°$, $Re = 2 \times 10^6$.

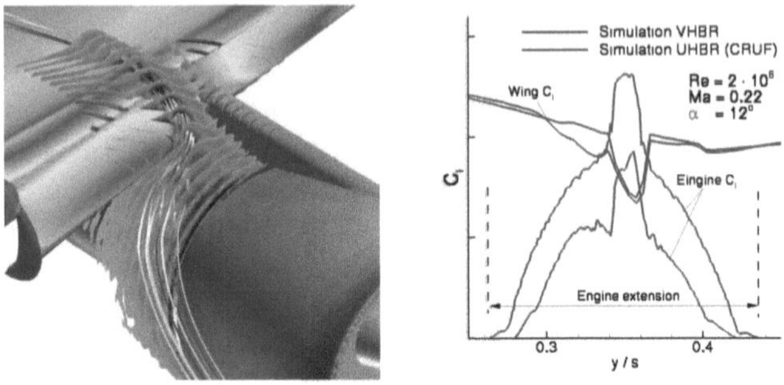


(a) Engine interference for ALVAST high-lift configuration with VHBR and UHBR engine $M_\infty = 0.22$, $\alpha = 12°, Re = 2 \times 10^6$ , left: nacelle vortex, right: lift distribution of wing and nacelle.

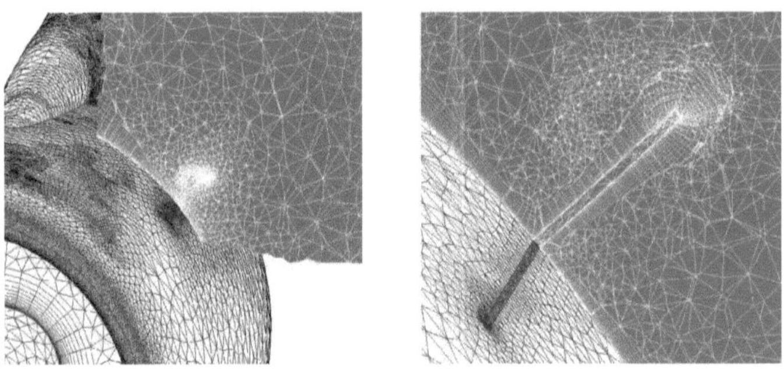

(b) Civil transport high-lift configuration with nacelle strakes, filled strake grid.

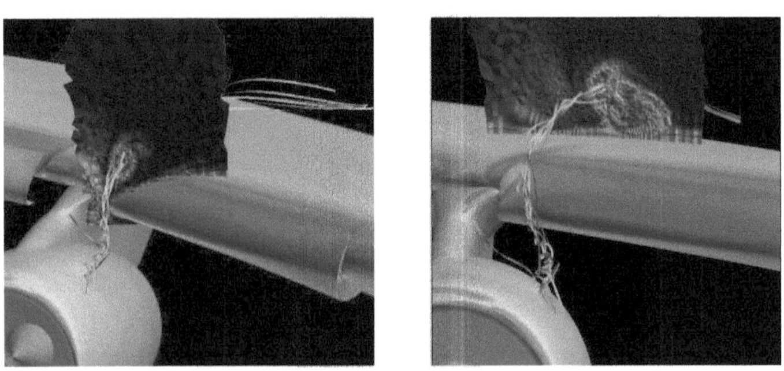

(c) Civil transport high-lift configuration with nacelle strakes, calculated streamlines and iso-vorticity cut planes.

**Fig. 17.**

### *Military aircraft*

Concerning military aircraft applications numerical simulations for the X-31 configuration have been carried out with the TAU-Code [5]. These computations show the capability of the TAU-Code to simulate complex delta wing configurations with rounded leading edges. Fig. 18(a) shows the numerically obtained 3D flow field over the X-31 configuration indicating the complexity of the vortex flow topology over the wing and the fuselage. Comparisons with experimental data show good agreement regarding the vortex topology. In Fig. 18(b) an oil flow picture of the X-31 clean wing from low speed experiments is shown in comparison to the corresponding CFD result. The angle of attack is $\alpha = 18°$ at a Reynolds number of 1.0 million. The attachment line of the strake vortex and the main wing vortex as well as the separation line of the main wing vortex near the leading edge is emphasized indicating that the flow topology from the calculation fits quite well with the experiment.

### *Helicopter*

At DLR large effort is devoted to the enhancement of the MEGAFLOW software for helicopter applications. The development and validation activities are carried out in the German/French project CHANCE [46]. They include performance prediction of the isolated rotor in hover and forward flight as well as the quasi-steady and time-accurate simulation of the complete helicopter including engines and main and tail rotor.

The aerodynamic assessment of helicopter main rotors requires a computational procedure with fluid-structure coupling including trim. The results which are presented here were obtained with a weak coupling (see [37]) between the RANS solver FLOWer and the comprehensive rotor simulation code S4 in which the blade structure is modeled as a beam. The test case is the four-bladed 7A-rotor with rectangular blades in high-speed forward flight ($M_\omega R = 0.64$, $M_\infty = 0.256$ with an advance ratio of $\mu = 0.4$). Fig. 18(c) presents the grid system used while Fig. 18(d) compares the measured with the predicted data. The overall agreement of the coupled solution (FLOWer/S4 coupling) with the experimental data is acceptable although the negative peak in normal force around 120 azimuth is not well computed. This phenomenon is subject of ongoing research. The results of the simplified blade element aerodynamic module of S4 are presented by dashed lines in Fig. 18(d). It is obvious that this simplified aerodynamic model is not able to capture the time dependent blade load history.

A quasi-steady computation of the flow-field around the Eurocopter EC-145 helicopter has been carried out [32, 31]. The effect of engines and rotors has been simulated by means of in-/outflow boundary conditions and by actuator discs respectively. As visualized in Fig. 19(a), the rotor downwash results in an asymmetrical flow pattern on the fuselage surface. The figure shows separation lines and singular points on the boot and tail boom. Moreover,

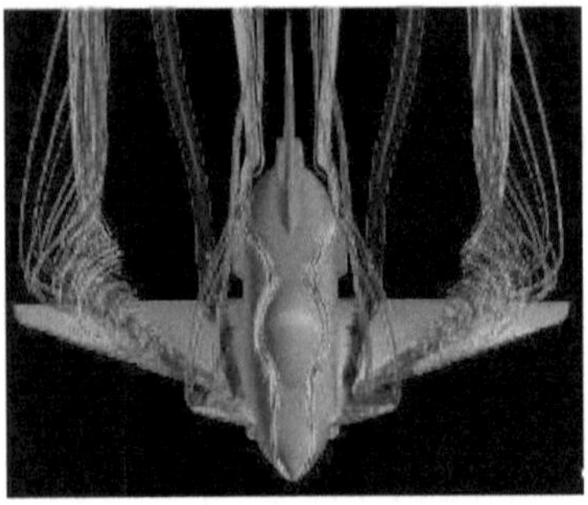

(a) 3D flow field of the X-31 configuration at 18° angle of attack, TAU-Code.

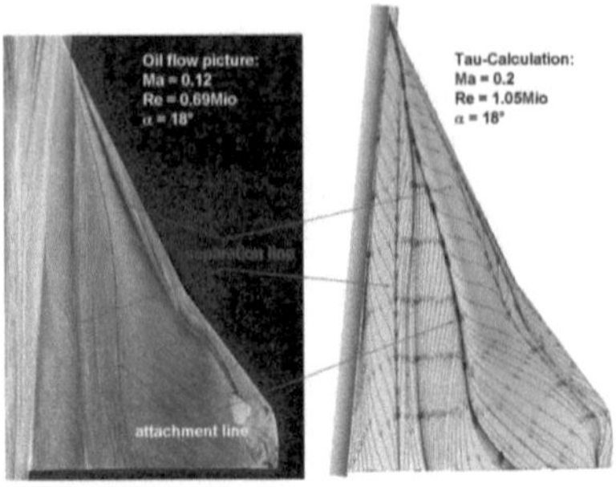


(b) X-31 clean wing, left: oil flow visualization, right: surface streamlines obtained with TAU-Code.

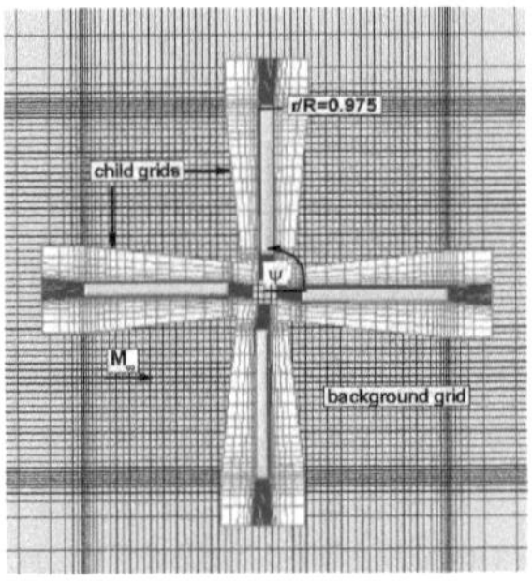


(c) Chimera grid system around 4-bladed 7A-rotor.

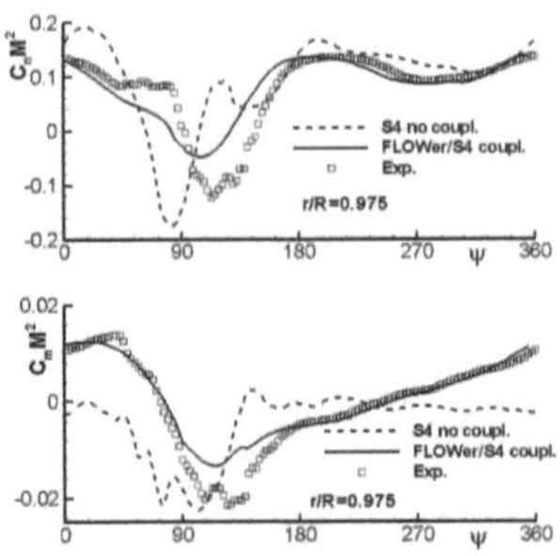


(d) Comparison of predicted and measured normal force and pitching moment coefficients versus azimuth for a high-speed forward flight test case of the 7A rotor.

**Fig. 18.**

the right vertical stabilizer experiences a much higher loading as the left one. In Fig. 19(b) the surface temperature distribution and a 3D-contour for temperature of $T = 60°C$ are depicted. Again the rotor downwash produces an asymmetrical temperature wake, which results in a single hot spot ($T = 60°C$) on the left horizontal stabilizer.

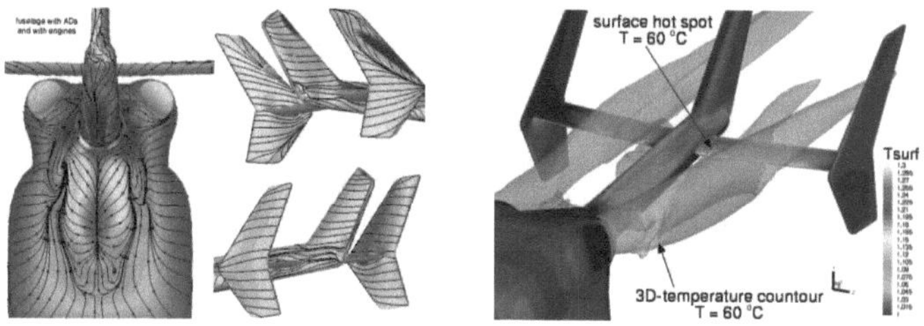


(a) (b)

**Fig. 19.** (a) $C_P$-distribution and friction lines on the EC145 fuselage, visualisation of separation areas on the boot and vertical stabilisers. (b) Temperature surface distribution and 3D-contour (T=60°C), visualisation of the impact of engine plumes on horizontal stabilisers.

# 5 Multidisciplinary simulations

The aerodynamic performance of large transport aircraft operating at transonic conditions is highly dependent on the deformation of their wings under aerodynamic loads. Hence accurate performance predictions require fluid/structure coupling in order to determine the aerodynamics of the configuration in aero-elastic equilibrium. Consequently, at DLR major effort is currently devoted to couple the flow solvers FLOWer and TAU with numerical methods simulating the structure. The activities include the development of efficient and robust grid deformation tools, accurate interpolation tools for transferring data between the fluid grid and the structure grid as well as the implementation of suitable interfaces between the flow solvers and the structural solvers. Concerning structure, both high-fidelity models (ANSYS, NASTRAN) and simplified models (beam model) are considered.

The importance of fluid/structure coupling is demonstrated in Fig. 20. Within the European project HiReTT Navier-Stokes calculations were performed for a wing-body configuration of a modern high speed transport type aircraft at $M_\infty = 0.85$ and $Re = 32.5 \times 10^6$. The block-structured FLOWer-Code was used on a grid with about 3.5 million points. The $k/\omega$ turbulence model was employed. Two types of calculations were carried out. On the one hand the aerodynamic behavior of the jig-shape was predicted. On the other hand the aero-elastic equilibrium was determined by a fluid/structure coupling. For this calculation the coupling procedure of the University of Aachen (Lehr und Forschungsgebiet fr Mechanik) was used [8]. It is based on the FLOWer-Code for the fluid and a beam model for the structure. From Fig. 20 it is obvious that good agreement with experimental data obtained in the ETW can only be achieved with the fluid/structure coupling.

The improvement of maneuverability and agility is a substantial requirement of modern fighter aircraft. Most of today's and probably future fighter aircraft will be delta wing configurations. The flow field of such configurations is dominated by vortices resulting from flow separation at the wings and the fuselage. The time lag between vortex position and state with respect to the on-flow conditions of the maneuvering aircraft can lead to significant phase shifts in the distribution of loads. Reliable results for the analysis of the flight properties can only be achieved by a combined non-linear integration of the unsteady aerodynamics, the flight motion and the elastic deformation of the aircraft structure.

Within the DLR internal project SikMa [5, 44] a multidisciplinary simulation tool for maneuvering aircraft is being developed and validated. The unstructured, time-accurate flow solver TAU is coupled with a computational module solving the flight-mechanic equations and a structural mechanics code determining the structural deformations. By use of an overlapping grid technique (Chimera), simulations of complex configurations with movable control surfaces are possible. Fig. 21 shows an example of a multidisciplinary simulation of coupled aerodynamics and flight-mechanics. In this simulation the delta wing is released at a roll angle of zero degree and a pitching angle of $\alpha = 9°$ while the trailing edge flaps are deflected to $\eta = \pm 5°$ , respectively. On the upper right side of the figure the pressure distribution is shown at a stage where the flaps are fully deflected. On the upper left side the corresponding pitching and rolling moment are depicted as a function of the roll angle. The time histories of the rolling angle and the flap deflection angle are shown at the bottom of Fig. 21.

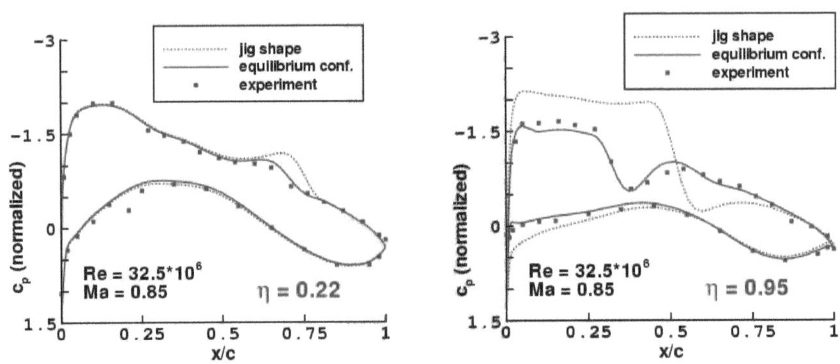


**Fig. 20.** $C_P$-distribution for different span wise sections for a wing/body configuration, numerical results obtained for pre-deformed geometry (dashed line) and with fluid/structure coupling (full line).

## 6 Numerical optimization

For aerodynamic shape optimization, FLOWer and TAU offer an inverse design mode which is based on the inverse formulation of the small perturbation method according to Takanashi [50]. The method has been extended to transonic flows [7] and is capable of designing airfoils, wings and nacelles in inviscid and viscous flows.

In the context of regional aircraft development various wing designs for transonic flow were performed at DLR with the inverse mode of the Navier-Stokes solver FLOWer. As design target suitable surface pressure distributions were specified subject to geometrical constraints and a given lift coefficient. Fig. 22(a) shows the comparison of drag rise between an early baseline wing and an improved wing as a function of Mach number. The reduction of drag in the higher Mach number range is clearly visible. The constraint with respect to the lift coefficient was satisfied.

The inverse design methodology coupled to the hybrid TAU-Code was also applied to the design of wing-mounted engine nacelles [55]. Fig. 22(b) shows results of the redesign of an installed nacelle. The aircraft geometry under consideration is the DLR ALVAST wing/body/pylon/nacelle config-

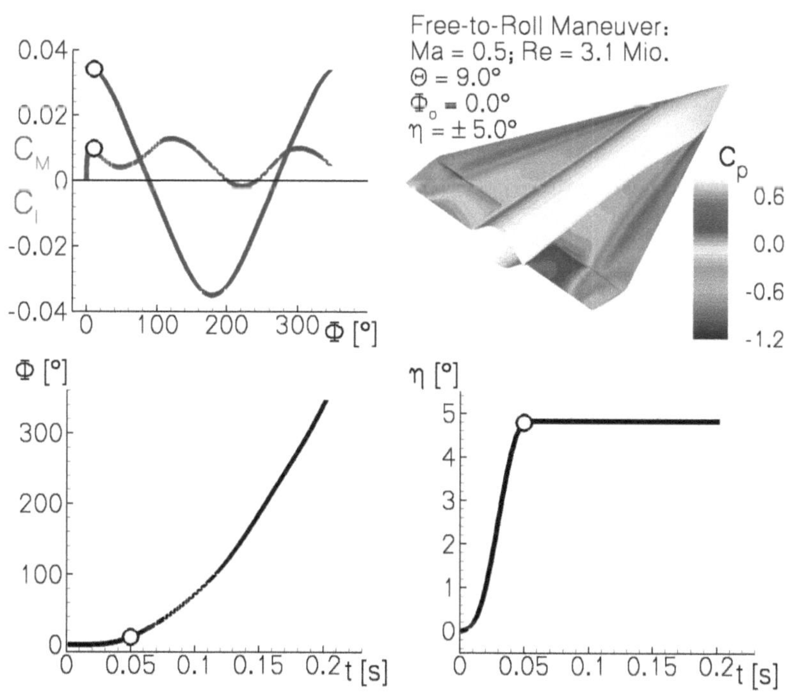


**Fig. 21.** Coupled aerodynamics and flight mechanics simulation for a rolling delta wing with trailing edge flaps using the TAU-Code.

uration equipped with a VHBR engine. The initial nacelle geometry is set up by the scaled profiles of the side section only. The prescribed nacelle target pressure distribution corresponds to the surface pressure distribution of the installed VHBR nacelle. The redesign was performed for inviscid flow at $M_\infty = 0.75$, $\alpha = 1.15°$ and the stream tube area ratio $\varepsilon_{FAN} = 0.96$. Fig. 22(b) shows surface pressure distributions and nacelle profiles in three circumferential sections. As can be seen, the prescribed pressure distributions are met in all three sections. This demonstrates that the inverse design methodology is capable of designing installed engine nacelles.

The inverse design method is very efficient; however it is restricted to a prescription of a target pressure distribution. A more general approach is the numerical optimization in which the shape, described by a set of design parameters, is determined by minimizing a suitable cost function subject to some constraints. At DLR high-lift system optimization is of major interest. Hence, the MEGAFLOW software has been coupled to various optimization strategies. As a demonstration results of a drag optimization for a 3-element airfoil in take-off configuration [54] are presented in Fig. 23. A limit in pitching moment has been prescribed as secondary constraint. In total 12 design variables are taken into account. These are slat and flap gap, overlap and deflection. In addition, the slat and flap cut-out contours are parameterized by three variables each. The optimization method is based on a deterministic SUBPLEX strategy. The Navier-Stokes FLOWer-Code is used to predict the flow field. The block-structured grid has about 80.000 grid points. In the left part of Fig. 23 the initial and optimized slat and flap contours are shown,

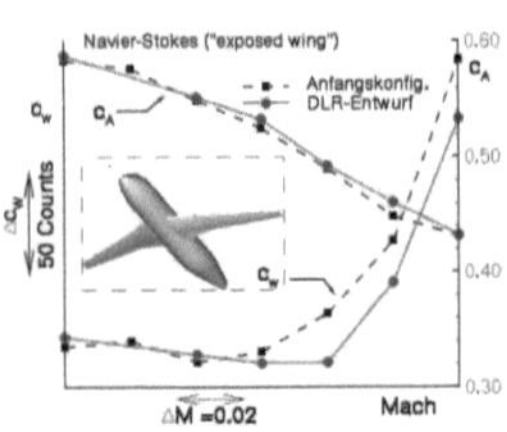


(a) Inverse wing design using FLOWer, drag rise lift as function of Mach number for baseline configuration and optimized configuration.

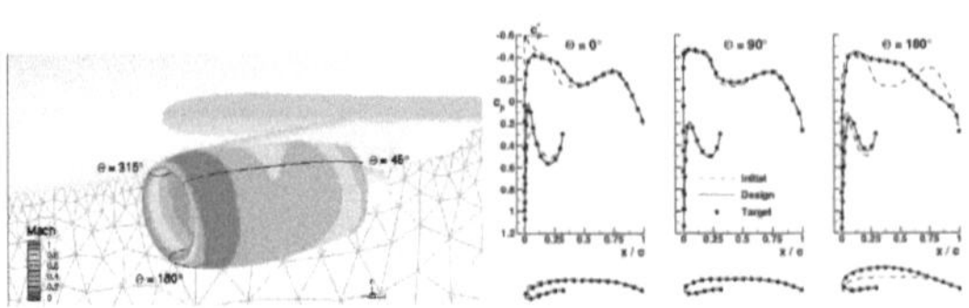

(b) Redesign of an installed nacelle using the TAU- Code, surface pressure distribution and nacelle profiles in three circumferential sections.

**Fig. 22.**

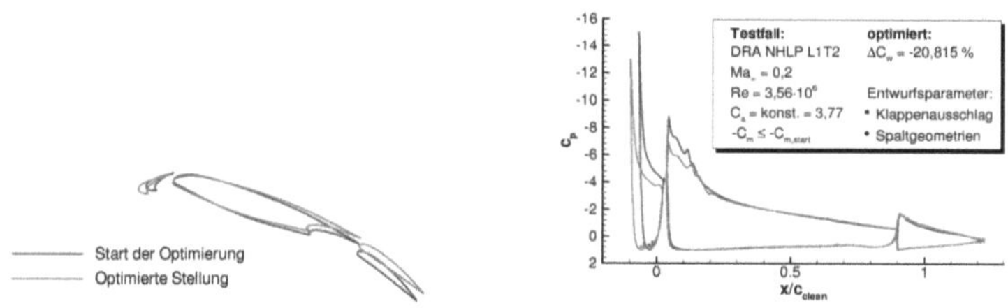


**Fig. 23.** Setting optimization of a 3-element airfoil using the FLOWer-Code.

in the right part the corresponding pressure distributions. The optimization affects the element chord, setting and deflection angle as well as the angle of attack. The optimization results in a decrease in total drag of 21%, while the maximum lift is slightly improved by 2%.

Because detailed aerodynamic shape optimizations still suffer from high computational costs, efficient optimization strategies are required. Regarding the deterministic methods, the adjoint approach is seen as a promising alternative to the classical finite difference approach (see e.g. [22]), since the computational cost does not depend on the number of design parameters. Accordingly, within the MEGAFLOW project an adjoint solver following the continuous adjoint formulation has been developed and widely validated for the block-structured flow solver FLOWer [20]. The adjoint solver can deal with the boundary conditions for drag, lift and pitching-moment sensitivities. The adjoint option of the FLOWer-Code has been validated for several 2D as well as 3D optimization problems controlled by the (adjoint) Euler equations. Within the ongoing MEGADESIGN project the robustness and efficiency of the adjoint solver will be further improved, especially for the Navier-Stokes equations. The adjoint solver implemented in FLOWer is currently transferred to the unstructured Navier-Stokes solver TAU.

To demonstrate the capability of the adjoint approach to handle many design parameters with low cost, the optimization of a supersonic transport wing/body configuration has been carried out [9]. The baseline geometry is based on the EUROSUP [33] geometry (Fig. 24), which is a supersonic commercial aircraft of 252 seats capacity, designed for a range of 5,500 nautical miles with supersonic cruise at Mach number $M_\infty = 2.0$. The optimization goal is to minimize the drag at a fixed lift coefficient of $C_L = 0.12$. The fuselage incidence is allowed to change in order to maintain the lift coefficient but it should not be greater than 4 degrees to the onset flow. In order to explore the full potential of the adjoint technique, no specific restrictions are set to define the parameterization. 74 design variables were used to change the twist, the thickness and the camber line at specific wing sections and 10 more design variables allowed changing the radial distribution of the fuselage. A minimum allowable value of the fuselage radius and a minimum wing thickness law were imposed in order to prevent unrealistic aircraft. After ge-

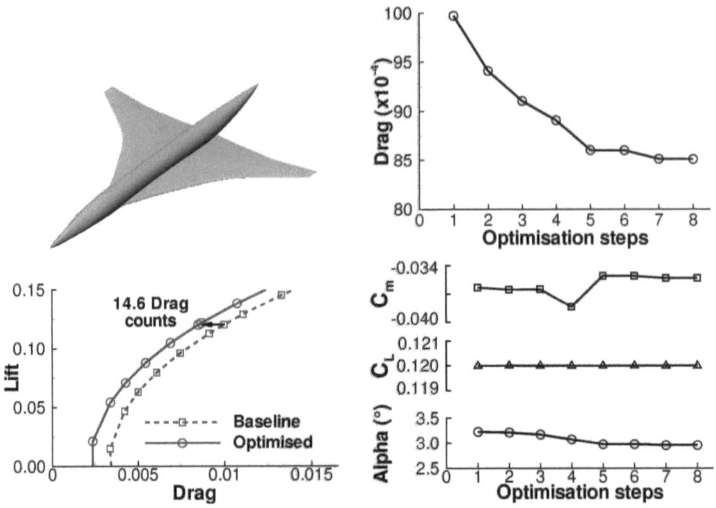


**Fig. 24.** Shape optimization of supersonic transport aircraft at $M_\infty = 2.0$ (drag minimization at constant lift).

ometrical modifications, the intersection of wing and fuselage is recalculated automatically by the DLR in-house grid generator MegaCads for each new configuration. At $M_\infty = 2.0$, the main aerodynamic effects are well predicted using the Euler equations. Therefore, the aerodynamic states are computed by FLOWer running in Euler mode. The constraint on the lift is handled using the target lift mode available in FLOWer which automatically adjusts the angle of attack to reach the desired lift. In the present optimization problem, the unique aerodynamic constraint is the lift, which is handled directly by FLOWer and the geometrical constraints are automatically fulfilled during the parameterization. Fig. 24 shows the evolution of the drag coefficient during the optimization, where an optimization step includes the evaluation of the gradient and the line search. About 8 optimization steps were necessary to achieve the optimum, which represents 54 aerodynamic computations and 8 adjoint flow evaluations. This approach is more than 11 times faster than using brute force optimization based on finite differences. The optimum configuration has 14.6 less drag counts than the baseline geometry. It can be seen in Fig. 24 that FLOWer keeps the lift constant during the complete optimization and the angle of attack decreases slightly by about 0.3 degrees. The pitching moment decreases by about 2.8%. It is interesting to analyze the evolution of the performance around the design point. The lower left picture of Fig. 24 shows the polar both for the baseline and the improved geometries. It can clearly be seen that there is an almost constant reduction of the drag for the whole polar of the optimized geometry and not only at the main design point ($C_L = 0.12$).

## 7 Conclusions and perspective

The main objective of the MEGAFLOW initiative was the development of a dependable, effective and quality controlled software package for the aerodynamic simulation of complete aircraft. Due to its high level of maturity, the MEGAFLOW software system is being used extensively throughout Germany for solving complex aerodynamic problems - especially in industrial development processes. However, since industry is still demanding more accurate and faster simulation tools, further development is required despite the high level of numerical flow simulation established today. Four major fields of further research activities may be identified:

The first field is the enhancement of numerical methods by new algorithms and solution strategies. Here, accuracy, robustness, and efficiency have to be addressed, while recognizing that these are contradicting requirements. In the design process of the aerospace industry with its severe time constraints, the difficult – with respect to required man-power usually unpredictable – set-up of highly accurate computations can not be tolerated. However, to establish numerical simulation during design, where decisions involving extreme economical risks have to be made, accuracy and reliability are crucial, which is why expensive wind tunnel testing is still indispensable. Furthermore, the efficiency of numerical methods has to be substantially improved. Relying solely on the progress of computational hardware is not an option, since over the last two decades the size of the problems to be simulated increased in parallel to or even faster than advancements in computer technology.

Second, the physical modeling of fluid flow needs further to be addressed. Despite long-time efforts, the current status of modeling of turbulence and transition is still inadequate for the highly complex flows to be simulated in aircraft design. Due to the immense computational effort required, the direct numerical simulation (DNS) or even Large Eddy Simulation (LES) of fluid flow will not be a practical alternative even for the next four or five decades. Therefore, reliable modeling of turbulence and transition will become decisive to bring numerical simulation as a routinely used tool into the aeronautical design process.

Third, the architecture of the simulation software is becoming more and more a strategic issue. On the one hand the software architecture must thoroughly exploit computational capabilities like parallelism, which requires a certain degree of dedication to a certain computational environment; on the other hand the software should be portable to different hardware arrangements. Furthermore, the software must be flexible with respect to coupling with other disciplines and integration into optimization strategies to allow the definition of an interdisciplinary simulation and optimization environment. At last, the software architecture must allow continuous upgrading for algorithmic and modeling improvements.

The last field to be addressed is validation. This requires on the one hand the thorough definition of suitable experiments by using most advanced measuring techniques. Especially for the envisaged simulation of unsteady flows with moving bodies and actuated control surfaces, corresponding experimental data are lacking. On the other hand, due to unavoidable effects such as grid dependency and limitations in physical modeling, the assessment of uncertainties in numerical simulation and a resulting statement of reliable applicability is becoming a major matter of future concern.

Development activities in the direction of the issues summarized above have been initiated in the now ongoing German CFD project MEGADESIGN, which is a follow-on project to the German MEGAFLOW initiative.

*Acknowledgement.* The authors would like to thank their colleagues of the DLR Institute of Aerodynamics and Flow Technology for providing the material presented in this paper. Thanks also to C. Braun from the University of Aachen, who provided the numerical results shown in Figure 27. Furthermore, the partial funding of the MEGAFLOW and MEGADESIGN project through the German Government in the framework of the aeronautical research program is gratefully acknowledged.

## References

1. $1^{st}$ AIAA CFD Drag Prediction Workshop. http://www.aiaa.org/tc/apa/dragpredworkshop/dpw.html.
2. $2^{nd}$AIAA CFD Drag Prediction Workshop. http://ad-www.larc.nasa.gov/tsab/cfdlarc/aiaa-dpw/.
3. CentaurSoft: http://www.centaursoft.com/.
4. NETGEN, http://www.hpfem.jku.at/netgen , 2004.
5. Schütte. A., G. Einarsson, B. Schöning, A. Raichle, W. Mönnich, Th. Alrutz, Neumann. J., and J. Heinnecke. Numerical Simulation of Maneuvering Combat Aircraft. In *STAB 2004, to appear in Notes on Numerical Fluid Mechanics and Multidisciplinary Design (NNFM)*. Springer Verlag, 2005.
6. P. Aumann, H. Barnewitz, H. Schwarten, K. Becker, R. Heinrich, B. Roll, M. Galle, N. Kroll, Th. Gerhold, and M. Schwamborn, D.and Franke. MEGAFLOW: Parallel Complete Aircraft CFD. *Parallel Computing*, 27:415–440, 2001.
7. W. Bartelheimer. An Improved Integral Equation Method for the Design of Transonic Airfoils and Wings. In *AIAA 95-1688*, 1995.
8. C. Braun, A. Bouche, and J. Ballmann. Numerical Study of the Influence of Dynamic Pressure and Deflected Ailerons on the Deformation of a High Speed Wing Model. In Krause. E., W. Jaeger, and M. Resch, editors, *High Performance Computing in Science and Engineering 2004*. Springer Verlag, 2004.
9. J. Brezillon and N.R. Gauger. 2D and 3D Aerodynamic Shape Optimization Using Adjoint Approach. *Aerosp. Sci. Technol.*, 8:715–727, 2004.
10. O. Brodersen. Drag Prediction of Engine-Airframe Interference Effects Using Unstructured Navier-Stokes Calculations. *Journal of Aircraft*, 39(6):927–935, 2002.
11. O. Brodersen, M. Rakowitz, S. Amant, Larrieu. P., D. Destarac, and M. Sutcliffe. Airbus, ONERA and DLR Results from the 2nd AIAA Drag Prediction Workshop. In *AIAA 2004-0391*, 2004.

12. O. Brodersen, A. Ronzheimer, R. Ziegler, T. Kunert, J. Wild, and M. Hepperle. Aerodynamic Applications Using Megacads. In M. Cross et al., editor, $6^{th}$ *International Conference on Numerical Grid Generation on Computational Field Simulations,ISGG*, pages 793–802, 1998.
13. O. Brodersen and A. Stürmer. Drag Prediction of Engine-Airframe Interference Effects Using Unstructured Nvier-Stokes Calculations. In *AIAA 2001-2414*, 2001.
14. O. Brodersen and J. Wild. DLR-IB 124-2004-18, 2004.
15. O. Brodersen, J. Wild, S. Melber-Wilkending, and L. Lekemark. In *DLR-IB 129-2003-34*, DLR Braunschweig, 2003.
16. R. Dwight. A Comparison of Implicit Algorithms for the Navier-Stokes Equations on Unstructured Grids. In *Proceedings of the ICCFD Conference*, Toronto, Canada, 2004.
17. B. Eisfeld. Numerical Simulation of Aerodynamic Problems with a Reynolds Stress Turbulence Model. In *STAB 2004, to appear in Notes on Numerical Fluid Mechanics and Multidisciplinary Design (NNFM)*. Springer Verlag, 2005.
18. J.K. Fassbender. Improved Robustness for Numerical Simulation of Turbulent Flows around Civil Transport Aircraft at Flight Reynolds Numbers. In *DLR-FB 2003-09*, 2003.
19. M. Galle. Ein Verfahren zur numerischen Simulation kompressibler , reibungsbehafteter strömungen auf hybriden netzen. In *DLR-FB 99-04*, 1999.
20. N.R. Gauger. Das Adjungiertenverfahren in der aerodynamischen Formoptimierung. In *DLR-FB 2003-05*, 2003.
21. R. Heinrich and N. Kalitzin. Numerical Simulation of Three-Dimensional Flows Using the Chimera Technique. *Notes on Numerical Fluid Mechanics*, 72:15–23, 1999.
22. A. Jameson, L. Martinelli, and N. Pierce. Optimum Aerodynamic Design Using the Navier-Stokes Equations. *Theoret. Comput. Fluid Dynamics*, 10:213–237, 1998.
23. J.C. Kok and F.J. Brandsma. Turbulence Model Based Vortical Flow Computations for a Sharp Edged Delta Wing in Transonic Flow Using the Full Navier-Stokes Equations. In *NLR-CR-2000-342*, 2000.
24. J.C. Kok, H.S. Dol, B. Oskam, and H. van der Ven. Extra-Large Eddy Simulation of Massively Separated Flows. In *AIAA-Paper, 2004-0264*, 2004.
25. N. Kroll and J.K. Fassbender(Eds.). *Notes on Numerical Fluid Mechanics and Multidisciplinary Design*, volume 89, chapter MEGAFLOW - Numerical Flow Simulation for Aircraft Design. Springer, 2005.
26. N. Kroll, Gauger. N. R., J. Brezillon, K. Becker, and V. Schulz. Ongoing Activities in Shape Optimization within the German Project MEGADESIGN. In *ECCOMAS 2004*, Jyväskylä, Finland, July 2004.
27. N. Kroll, R. Radespiel, and C.-C. Rossow. Accurate and Efficient Flow Solvers for 3D-Applications on Structured Meshes. In *AGARD R-807*, pages 4.1–4.59, 1995.
28. N. Kroll, C.C. Rossow, K. Becker, and F. Thiele. The MEGAFLOW project. *Aerosp. Sci. Technol.*, 4:223–237, 2000.
29. N. Kroll, C.C. Rossow, D. Schwamborn, K. Becker, and G. Heller. MEGAFLOW - A Numerical Flow Simulation Tool for Transport Aircraft Design. *ICAS*, 1.10.5, 2002.

30. A. Krumbein. Coupling of the DLR Navier-Stokes Solver FLOWer with an $e^N$-Database Method for Laminar-Turbulent Transition Prediction on Airfoils. *Notes on Numerical Fluid Mechanics*, 77:92–99, 2002.
31. F. Le Chuiton. Chimera simulation of a complete helicopter with rotors as actuator discs. In *STAB 2004, to appear in Notes on Numerical Fluid Mechanics and Multidisciplinary Design (NNFM).* Springer Verlag, 2005.
32. F. LeChuiton. Actuator Disc Modeling for Helicopter Rotors. *Aerosp. Sci. Technol*, 8:285–297, 2004.
33. D.A. Lovell. Aerodynamic Research to Support a Second Generation Supersonic Transport Aircraft - the EUROSUP Project. In *ECCOMAS 1998*, 1998.
34. A. Madrane, A. Raichle, and A. Stürmer. Parallel Implementation of a Dynamic Overset Unstructured Grid Approach. In *ECCOMAS 2004*, Jyväskylä, Finland, July 2004.
35. S. Melber. 3D RANS Simulations for High-Lift Transport Aircraft Configurations with Engines. *DLR-IB 124-2002/27*, 2002.
36. F.R. Menter. Two-Equation Eddy-Viscosity Turbulence Models for Engineering Applications. *AIAA Journal*, 32:1598–1605, 1994.
37. K. Pahlke and B. van der Wall. Chimera Simulations of Multibladed Rotors in High-Speed Forward Flight with Weak Fluid-Structure-Coupling. In *29th European Rotorcraft Forum*, page 63, Friedrichshafen, Germany, 2003.
38. M. Rakowitz, M. Sutcliffe, B. Eisfeld, D. Schwamborn, and J. Bleecke, H.and Fassbender. Structured and Unstructured Computations on the DLR-F4 Wing-Body Configuration. In *AIAA 2002-0837*, 2002.
39. R. Rudnik. Towards CFD Validation for 3D High Lift Flows - EUROLIFT. In *ECCOMAS 2001*, Swansea, United Kingdom, 2001.
40. R. Rudnik, S. Melber, A. Ronzheimer, and O. Brodersen. Three-Dimensional Navier-Stokes Simulations for Transport Aircraft High Lift Configurations. *Journal of Aircraft*, 38:895–903, 2001.
41. R. Rudnik, C.C. Rossow, and H. v. Geyr. Numerical Simulation of Engine/Airframe Integration for High-Bypass Engines. *Aerosp. Sci. and Technol.*, 6:31–42, 2002.
42. T. Rung, H. Lübcke, M. Franke, L. Xue, F. Thiele, and S. Fu. Assessment of Explicit Algebraic Stress Models in Transonic Flows. In *Proceedings of the $4^th$ Symposium on Engineering Turbulence Modeling and Measurements*, pages 659–668, France, 1999.
43. A. Schütte, G. Einarsson, A. Madrane, B. Schöning, W. Mönnich, and W.-R. Krüger. Numerical Simulation of Maneuvering Aircraft by CFD and Flight Mechanic coupling. In *RTO Symposium*, Paris, April 2002.
44. A. Schütte, G. Einarsson, B. Schöning, A. Madrane, W. Mönnich, and W. Krüger. Numerical Simulation of Manoeuvring Aircraft by Aerodynamic and Flight Mechanic Coupling. In *RTO AVT Symposium Paris*, 2005.
45. Th. Schwarz. Development of a Wall Treatment for Navier-Stokes Computations Using the Overset Grid Technique. In *26th European Rotorcraft Forum*, page 45, 2000.
46. J. Sidès, K. Pahlke, and M. Costes. Numerical Simulation of Flows around Helicopters at DLR and ONERA. *Aerosp. Sci. Technol.*, 5:35–53, 2001.
47. C.G. Speziale, S. Sarkar, and T.B. Gatski. Modeling the pressure-strain correlation of turbulence: an invariant dynamical systems approach. *Journal of Fluid Mechanics*, 227:245–272, 1991.

48. M. Strelets. Detached Eddy Simulation of massively separated flows. In *AIAA-Paper 2001-0879*, 2001.
49. Gerhold T., O. Friedrich, Evans J., and M. Galle. Calculation of Complex Three-Dimensional Configurations Employing the DLR-TAU Code. In *AIAA 97-0167*, 1997.
50. S. Takanashi. Iterative Three-Dimensional Transonic Wing Design Using Integral Equations. *Journal of Aircraft*, 22(8), 1985.
51. J.B. Vos, A. W. Rizzi, D. Darracq, and E. H. Hirschel. Navier-Stokes Solvers in European Aircraft Industry. *Progress in Aerospace Sciences*, 38:601–697, 2002.
52. S. Wallin and A.V. Johansson. An Explicit Algebraic Reynolds Stress Model for Incompressible and Compressible Turbulent Flows. *J. Fluid Mech.*, 403:89–132, 2000.
53. D.C. Wilcox. *Turbulence Modeling for CFD, DCW Industries*. CA, La Cañada, 1998.
54. J. Wild. Validation of Numerical Optimization of High-Lift Multi-Element Airfoils based on Navier-Stokes-Equations. In *AIAA 2002-2939*, 2002.
55. R. Wilhelm. An Inverse Design Method for Designing Isolated and Wing-Mounted Engine Nacelles. In *AIAA 2002-0104*.

# Gradient Computations for Optimal Design of Turbine Blades

K. Arens[1], P. Rentrop[1], and S.O. Stoll[2]

[1] TU München, Zentrum Mathematik arens@ma.tum.de
[2] TU Karlsruhe, IWRMM

**Summary.** The optimal profile of turbine blades is crucial for the efficiency of modern powerplants. The applied SQP algorithms are based on gradient information.

**Key words:** sensitivity method, adjoint method, turbine design

## 1 Introduction

In power plants the aerodynamic optimization of turbine blades is crucial for efficiency considerations. The profile of the turbine blade is described by Bézier polynomials, where the coefficients are used as design variables in a nonlinear optimization procedure. The fluid-mechanics are modeled by the 2D Euler equations. The gas flow through the blade row suffers from the occurrence of shock-waves. These shock-waves produce high losses of energy and therefore of efficiency. By optimizing the blade profile shock-waves can nearly be avoided or remarkably reduced in their strengths.

## 2 Model Problem

As a model problem the flow through a nozzle with region $\Omega$ as in Fig. 1 will be considered . The fluid dynamics are governed by the 2D Euler gas equations. With density $\rho$, momentum in $x$-direction $m = pu$, momentum in $y$-direction $n = pv$ and total energy $E$, the *conservative* variable vector $U$, the gas equations are written as a conservation law of hyperbolical type

$$\frac{\partial U}{\partial t} + \frac{\partial}{\partial x} F(U) + \frac{\partial}{\partial y} G(U) = 0 \tag{1}$$

$$U = \begin{pmatrix} \rho \\ m \\ n \\ E \end{pmatrix}, \; F(U) = \begin{pmatrix} m \\ mu + p \\ mv \\ (E+p)u \end{pmatrix}, \; G(U) = \begin{pmatrix} n \\ nu \\ nv + p \\ (E+p)v \end{pmatrix}. \tag{2}$$

For the present the design problem will be discussed for the stationary 2D Euler equation

$$F(U)_x + G(U)_y = 0 \tag{3}$$

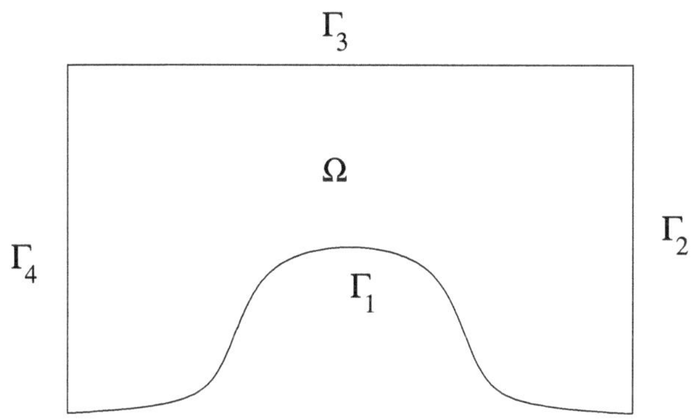


**Fig. 1.** Two dimensional model problem

At the inlet $\Gamma_4$ density $\rho$, inflow angle $\angle(u, v)$, velocity $v = 0$ and pressure $p$ are given. At the outlet $\Gamma_2$ pressure $p$ is prescribed. At the boundaries $\Gamma_1$ and $\Gamma_3$ $V^T n = un_x + vn_y = 0$ holds for the velocity in the normal direction $n$ of $\delta\Omega = \Gamma_1 \bigcup \ldots \bigcup \Gamma_4$. The upper wall is fixed whereas the lower wall $\Gamma_1$ should be optimized via

$$y(x) = \begin{cases} 0 & : \; -0.5 \le x < 0 \\ \sum_{i=1}^{4} \alpha_i b_i(x) & : \; 0 \le x < 1 \\ 0 & : \; 1 \le x < 1.5 \end{cases} \tag{4}$$

The coefficients $\alpha_i$, $i = 1, \ldots, 4$ are the design parameters. The functions $b_i$ are chosen as $b_i(x) = x^{i+1}(x-1)^2$. At $\Gamma_1$ a pressure distribution is prescribed as nominal pressure $p^d$. The objective is to find $\alpha_i$ such that the functional $I$ is minimized

$$I = \frac{1}{2} \int_{\Gamma_1} (p - p^d)^2 \, ds. \tag{5}$$

To use efficient optimization algorithms like Sequential Quadratic Programming (SQP) the gradient information has to be provided, see [9]

$$\frac{\partial I}{\partial \alpha_i} = \frac{\partial I}{\partial U} \frac{\partial U}{\partial \alpha_i}. \tag{6}$$

Whereas $\frac{\partial I}{\partial U}$ can be calculated analytically the sensitivities $\frac{\partial U}{\partial \alpha}$ must be calculated numerically. There are three different approaches to calculate the gradient information: i) by finite differences, ii) via the sensitivity equation, iii) by an adjoint method, see also [8, 7, 10].

# 3 Gradient Computation

## 3.1 Finite Differences

The approximation $\frac{\partial U}{\partial \alpha} \approx \frac{U(\alpha+\Delta\alpha)-U(\alpha)}{\Delta\alpha}$ by finite differences has several disadvantages. This method is too imprecise for our purpose if the mesh is not parameterized. Additionally for every design parameter a new mesh must be calculated.

## 3.2 Sensitivity Equation

Explicit differentiation of the Euler gas equations with respect to $\alpha$ results in the *sensitivity equation*, see [5, 3, 2, 6]

$$\frac{\partial \mathbf{s}}{\partial t} + \frac{\partial}{\partial x}(\frac{dF(U)}{dU}\mathbf{s}) + \frac{\partial}{\partial y}(\frac{dG(U)}{dU}\mathbf{s}) = 0 \tag{7}$$

with

$$\mathbf{s} = \frac{\partial U}{\partial \alpha} = \begin{pmatrix} \frac{\partial \rho}{\partial \alpha} \\ \frac{\partial m}{\partial \alpha} \\ \frac{\partial n}{\partial \alpha} \\ \frac{\partial E}{\partial \alpha} \end{pmatrix} = \begin{pmatrix} \rho_\alpha \\ m_\alpha \\ n_\alpha \\ E_\alpha \end{pmatrix}. \tag{8}$$

This conservation law has to be solved for every design parameter $\alpha$.

## 3.3 Adjoint Method

The principle of the adjoint method lies in solving a dual problem which leads to the same result as the original problem. To state the dual problem for the model problem (3), (4) a Lagrange formalism is implemented, see [10]. To achieve the full information for the adjoint equation the Euler equations and the boundary conditions are coupled to the functional $I$ via the Lagrange multipliers $\Lambda$ and $\mu$.

$$I = \frac{1}{2}\int_{\Gamma_1} (p-p^d)^2\, ds + \int_\Omega \Lambda^T (F(U)_x + G(U)_y)\, d\Omega + \int_{\Gamma_1} \mu V^T n\, ds. \tag{9}$$

Differentiation by $\alpha_i$ leads to the adjoint equation

$$-(\frac{\partial F}{\partial U})^T \Lambda_x - (\frac{\partial G}{\partial U})^T \Lambda_y = 0 \quad \text{in } \Omega, \tag{10}$$

and the boundary conditions

$$\Lambda^T(\frac{\partial F}{\partial U}n_x + \frac{\partial G}{\partial U}n_y)\frac{\partial U}{\partial \alpha_i} = 0 \quad \text{on } \Lambda_k, k = 2, 3, 4, \tag{11}$$

$$\Lambda^T(\frac{\partial F}{\partial U}n_x + \frac{\partial G}{\partial U}n_y)\frac{\partial U}{\partial \alpha_i} + \frac{\partial p}{\partial U}(p - p^d)\frac{\partial U}{\partial \alpha_i} + \mu n \frac{\partial V}{\partial U}\frac{\partial U}{\partial \alpha_i} = 0 \quad \text{on } \Gamma_1. \tag{12}$$

After solving the adjoint equation only once one receives

$$\frac{dI}{d\alpha_i} = \frac{1}{2}\int_{-1}^{1}(p - p^d)^2 \frac{db_i}{dx}\,dx + \int_{-1}^{1}\mu V n \frac{db_i}{dx}\,dx + \int_{\Gamma_1}\mu V \frac{\partial n}{\partial \alpha_i}\,ds. \tag{13}$$

In comparison to the sensitivity equation method where a system of differential equations has to be solved for every $\alpha_i$ a less costly scalar product has to be solved for every $\alpha_i$.

## 4 Optimal Turbine Blade

As the Adjoint method has not yet been implemented in the optimization algorithms of our industrial partner the following optimal design results were achieved by using the sensitivity equation approach and an adopted SQP solver [4]. Fig. 2 shows the starting profile and an optimal profile of a turbine blade. Fig. 3 was generated with TASCflow, see [1], it shows the pressure

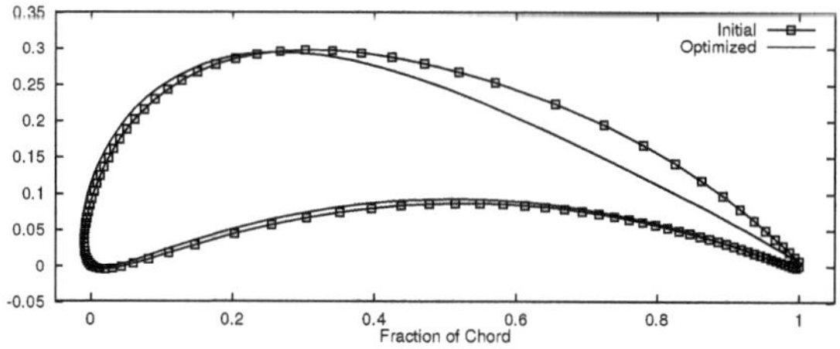


**Fig. 2.** Starting profile (dotted) and optimal profile

distribution before and after the optimization process. The darker the region the higher are the pressure values indicating shock regions. In Fig. 3 the optimal profile on the right shows a significant less pressure value (dark grey: initial profile, light grey: optimal profile).

### Acknowledgement

The authors are strongly indebted to Prof. Dr. A. Gilg and Dr. U. Wever from CT, Siemens Munich.

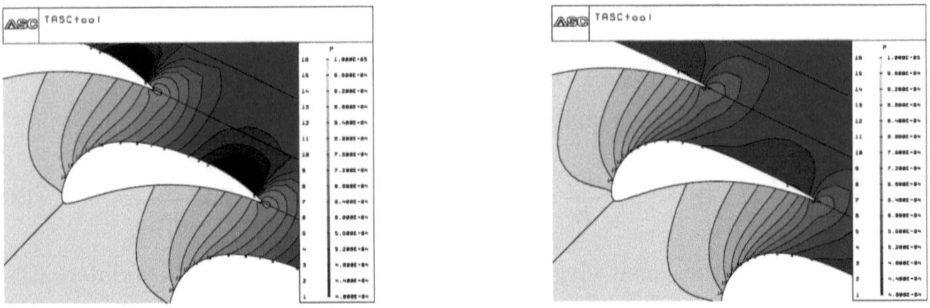


**Fig. 3.** Pressure distribution for starting and optimal profile

## References


1. *TASCflow: User Documentation.* Advanced Scientific Computing Ltd., Waterloo, Ontario, Canada, 1995.
2. J.R. Appel. *Sensitivity Calculations for Conservation Laws with Application to Discontinuous Fluid Flows.* PhD thesis, Virginia Polytechnic Institute and State University, 1997.
3. J.R. Appel and M.D. Gunzburger. Sensitivity calculations in flows with discontinuities. In *Proceedings 14th AIAA Applied Aerodynamics Conference New Orleans, LA.*, 1996.
4. R. Bell, U. Wever, and Q. Zheng. Profile optimization for turbine blades. *Surv. Math. Ind.*, 10(1):23–44, 2001.
5. J. Borggard and J. Burns. A sensitivity equation approach to shape optimization in fluid flows. In M. Gunzburger, editor, *Flow Control*, pages 49–78. Springer, 1995.
6. J. Borggard and J. Burns. A pde sensitivity equation method for optimal aerodynamic design. *J. Comput. Phys.*, 136(2):366–384, 1997.
7. M.B. Giles and N.A. Pierce. Analytic adjoint solutions for the quasi-one-dimensional euler equations. *J. Fluid Mech.*, 426:327–345, 2001.
8. A. Iollo, M.D. Salas, and S. Taasan. Shape optimization governed by the euler equations using an adjoint method. Technical Report 93-78, ICASE, 1993.
9. P. Spelucci. *Numerische Verfahren der nichtlinearen Optimierung.* Birkhäuser, 1993.
10. S.O. Stoll. *Das adjungierte Verfahren zur Geometrieoptimierung unter der Restriktion von hyperbolischen Erhaltungsgleichungen.* Number 365 in Fortschr.-Ber. VDI Reihe 20. VDI Verlag, 2003.

# Fast Numerical Computing for a Family of Smooth Trajectories in Fluids Flow

G. Argentini

Riello Group, via degli Alpini 1, 37045 Legnago (Verona), Italy
gianluca.argentini@riellogroup.com

**Summary.** In this work I present a technique of construction and fast evaluation of a family of cubic polynomials for analytic smoothing and graphical rendering of particles trajectories for flows in a generic geometry. The principal result of the work was implementation and test of a method for interpolation of 3D points by regular parametric curves, and fast and efficient evaluation of these functions for a good resolution of rendering. For this purpose I have used a parallel environment using a multiprocessor cluster architecture. The efficiency of the used method is good, mainly reducing the number of floating-points computations by caching the numerical values of some line-parameter's powers, and reducing the necessity of communication among processes. This work has been developed for the Research & Development Department of my company for planning advanced customized models of industrial burners.

**Key words:** computational fluid dynamics, cubic spline interpolation, parallel computing, parallel efficiency.

## 1 Introduction

Industrial and power burners have some particular requirements, as a customized study of the geometry for combustion head and combustion chamber for an optimal shape of the flame. Rapid prototyping for an accurate design of the correct geometry involves a numerical simulation of the gas or oil flows in the burner's components.

The necessity of an high graphic resolution requires a large amount of particles paths for tracing the streamlines of flow. Hence the numerical computation is memory and cpu very expensive for the used hardware environment. In a tipical simulation the number of paths to compute is some thousands, and the number of geometrical points to interpolate for each path is some thousands too. For the treatment of this large amount of data a parallel environment can be very useful.

## 2 Fitting trajectories with cubic polynomials

We suppose to have a dataset output from pre-processing and processing phases of a simulation, for example from numerical resolution of Navier-Stokes equations or from Cellular Automaton models [1]. We would a fast and flexible method to obtain from those data an accurate paths tracking of fluid particles with a smooth 3D visualization of trajectories, possibly with continuous slope and curvature. Our experience shows that Computational Fluid Dynamics packages have some limits in this post-processing phase, principally due to a rigid resolution of the initial mesh and to a small degree of parallelism.

Let **S** the number of 3D points for each trajectory and **M** the total number of trajectories from simulation dataset. We have tested that usual interpolation methods have some disadvantages for our aims: for example Bezier-like is not realistic in case of twisting or diverging speed-fields; Chebychev or Least-Squares-like are too rigid for a customized application; polynomial fitting is simple but often shows spurious effects as Runge phenomenon [6]. We have elaborated a *spline*-based technique.

We suppose $\mathbf{S} = 4\text{x}\mathbf{N}$. For every group of four points, the interpolation is obtained by three cubic polynomials imposing four analytical conditions: passage at $\mathbf{P}_k$ point, $1 \leq k \leq 3$; passage at $\mathbf{P}_{k+1}$ point; continuous slope and curvature at $\mathbf{P}_k$ point. For smooth rendering and for avoiding excessive twisting of trajectories, the cubics $\mathbf{u}_k$ are added to the Bezier curve $\mathbf{b}$ associated to the four points: $\mathbf{v}_k = \alpha\mathbf{b} + \beta\mathbf{u}_k$, $0 < \alpha,\ \beta < 1$ (Fig.1).

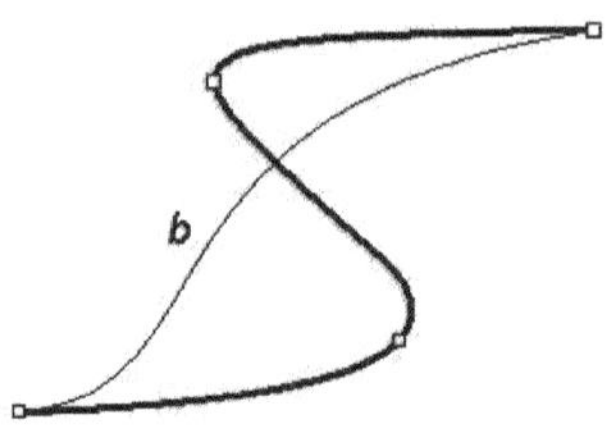


**Fig. 1.** Spline-based method with continuous slope and curvature; **b** is the Bezier curve interpolating the four points.

In our simulations we have chosen $\alpha = \beta = 0.5$ . Let $\mathbf{b} = As^3 + Bs^2 + Cs + D$, $0 \leq s \leq 1$, the Bezier curve of control points $\mathbf{P}_1$, ..., $\mathbf{P}_4$, and let $\mathbf{u}_k = at^3 + bt^2 + ct + d$, $0 \leq t \leq 1$, be the spline between two points. One can see that the coefficients of this spline can be computed by a matrix-vector product $\mathbf{coeff} = \mathbf{T} * \mathbf{p}$ where $\mathbf{coeff} = (a, b, c, d)$, $\mathbf{p} = (\mathbf{P}_{k+1}, \mathbf{P}_k, B, C, 1)$ and $\mathbf{T}$ is a $4 \times 5$ numerical matrix, constant for every groups of points and for every trajectory. If we define the $4\mathbf{M} \times 5\mathbf{M}$ *global matrix*

$$\mathbf{G} = \begin{pmatrix} \mathbf{T} & \mathbf{0} & \cdots & \mathbf{0} \\ \mathbf{0} & \mathbf{T} & \cdots & \mathbf{0} \\ \vdots & \vdots & \ddots & \vdots \\ \mathbf{0} & \mathbf{0} & \cdots & \mathbf{T} \end{pmatrix}$$

where $\mathbf{0}$ is a $4 \times 5$ zero-matrix, and define the vector $\mathbf{s} = (\mathbf{P}_{k+1}, \mathbf{P}_k, B_1, C_1, 1, ..., \mathbf{P}_{k+1}, \mathbf{P}_k, B_{\mathbf{M}}, C_{\mathbf{M}}, 1)$, one can compute for every two-points group the coefficients of cubic splines for all the $\mathbf{M}$ trajectories with the matrix-vector product $\mathbf{c} = \mathbf{G} * \mathbf{s}$. The matrix $\mathbf{G}$ is *sparse* with density equal at most to $1/\mathbf{M}$; if $\mathbf{M} = 1000$, the density of 0.001 is a very good value for obtain the benefits of sparsity methods, mainly in computational total time and memory allocation [2].

## 3 Computing splines

For computing the coefficients of all the splines involved in the simulation, the complexity analysis shows a total number of operations of order $\mathbf{M} * \mathbf{N}$. Using $\mathbf{P}$ computational processes on a multiprocessor environment, a useful method is the distribution of $\mathbf{M}/\mathbf{P}$ trajectories to every process. In this way every process receives $\mathbf{M}/\mathbf{P}$ rows of the matrix $\mathbf{G}$ for computing splines by matrix-vector multiply. In a first experiment (fall 2003), we have used the Linux cluster at CINECA, Bologna (Italy), equipped with Pentium III 1.133 GHz processors, and a software environment constituted by C programs and MPI libraries [3]. The use of such parallel routines has been useful only for startup of multi-processes and data distribution. Tests have shown a quasi-linear *speedup*, in the sense of parallelism, for all the values of $\mathbf{M}$ and $\mathbf{N}$ respect to the number $\mathbf{P}$ of used processes (Fig.2).

In a second experiment (winter 2003), we have used a multinode Windows 2000 cluster of our company, equipped with a total of 4 Intel Xeon 3.2 GHz processors and 4 GB Ram, and a parallel environment using MATLAB 6.5 scripts on distributed package's sessions on nodes. Tests have shown very high performances for splines computation using the internal algorithms of sparse matrix-vector multiply for the matrix $\mathbf{G}$.

## 4 Valuating splines

After the computation of splines, we have focused on their valuations on a suitable set of parameter's values. This set can be chosen large enough to obtain a fine sampling for an high graphic resolution. Consequently the amount of computation can be very huge, so that it is necessary an adequate method to valuate all the splines for all the trajectories.

Let $\mathbf{V}$+1 the number of ticks for each spline valuation with a uniform sampling; then the ticks are (0, $1/\mathbf{V}$, $2/\mathbf{V}$, . . ., ($\mathbf{V}$-1)/$\mathbf{V}$, 1). The values of

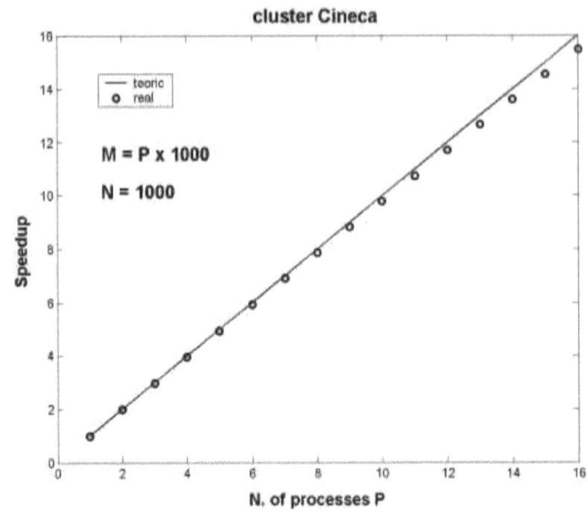


**Fig. 2.** Speedup registered with Linux cluster at Cineca.

splines parameter $t$ are (0, 1, 2, 3)-th degree powers of this array. The value of a cubic at $t_0$ can be view as a dot product:

$$at_0^3 + bt_0^2 + ct_0 + d = (a, b, c, d).(t_0^3, t_0^2, t_0, 1)$$

This fact permits to consider the constant $4 \times (\mathbf{V}+1)$ matrix

$$\mathbf{T} = \begin{pmatrix} 0 \; (1/\mathbf{V})^3 \cdots ((\mathbf{V}-1)/\mathbf{V})^3 \; 1 \\ 0 \; (1/\mathbf{V})^2 \cdots ((\mathbf{V}-1)/\mathbf{V})^2 \; 1 \\ 0 \; (1/\mathbf{V})^1 \cdots ((\mathbf{V}-1)/\mathbf{V})^1 \; 1 \\ 1 \quad 1 \quad \cdots \quad 1 \quad 1 \end{pmatrix}$$

We consider the $\mathbf{M} \times 4$ matrix

$$\mathbf{C} = \begin{pmatrix} a_1 & b_1 & c_1 & d_1 \\ a_2 & b_2 & c_2 & d_2 \\ \vdots & \vdots & \vdots & \vdots \\ a_{\mathbf{M}} & b_{\mathbf{M}} & c_{\mathbf{M}} & d_{\mathbf{M}} \end{pmatrix}$$

where each row contains the coefficients of a spline interpolating two points in a single trajectory. Then the $\mathbf{M} \times (\mathbf{V}+1)$ matrix product $\mathbf{E} = \mathbf{C} * \mathbf{T}$ contains in each row the values of a cubic between two data points, for all the $\mathbf{M}$ trajectories (*Eulerian view*). In a similar way on can consider a *Lagrangian view* for computing the values of all the cubics in a single trajectory. It can be easily shown that the total number of operations for computing all the values along each trajectory is of order $\mathbf{N} \times \mathbf{M} \times (\mathbf{V}+1)$.

## 5 Computing values of splines

For the computation of the values we have used the cluster of our company with multisessions of MATLAB package as parallel environment. It is fundamental for this step the improvement of performances due to the usage of LAPACK level 3 Blas routines incorporated in Matlab [4].

Another feature of this method is the fact that the matrix $\mathbf{T}$ is constant, hence it is computed only once, requires a small memory allocation so its values can be stored permanently in the cache. With $\mathbf{P}$, number of processes, divisor of $3\mathbf{N}$, total number of two-points groups, the method used has been the distribution of $3\mathbf{N}/\mathbf{P}$ matrices $\mathbf{C}$ to every process.

The performances of multiprocess products show a quite linear speedup respect the $\mathbf{P}$ variable and a total computation time of order $\mathbf{N}\times\mathbf{M}$; increasing the value of $\mathbf{M}$ or $\mathbf{N}$ for a better resolution, the time spent on computation doesn't change if the value of processes is increased (*Gustafson Law*) [5].

## 6 Conclusions

These techniques have supplied good results for improving performances of post-processing phase in CFD simulations. Further work is planned for implementing a *global matrix* product for the splines evaluation, with the purpose of using the sparse matrices benefits to reduce total execution time and memory allocation.

## References

1. G. Argentini. Message passing fluids: molecules as processes in parallel computational fluids. In J. Dongarra, D. Laforenza, and S. Orlando, editors, *Recent Advances in Parallel Virtual Machine and Message Passing Interface: 10th European PVM/MPI Users' Group Meeting, Venice, Italy, 2003*, volume 2840 of *LNCS*, pages 550–554, Berlin, 2003. Springer Verlag.
2. J.R. Gilbert, C.B. Moler, and R. Schreiber. Sparse matrices in MATLAB: design and implementation. *SIAM Journal of Matrix Analysis and Application*, 13(1):333–356, 1992.
3. M. Lanzarini. Science and Supercomputing at CINECA: 2003 Report. Technical report, CINECA, Bologna, 2004.
4. C.B. Moler. MATLAB incorporates LAPACK. Increasing the speed and capabilities of matrix computation. *MATLAB News & Notes*, Winter 2000, 2000.
5. P. Pacheco. *Parallel programming with MPI.* Morgan Kaufmann, San Francisco, 1997.
6. A. Quarteroni, R. Sacco, and F. Saleri. *Numerical Mathematics*, volume 37 of *Texts in Applied Mathematics.* Springer Verlag, Berlin, 2000.

# Optimal Control of an ISS-Based Robotic Manipulator with Path Constraints

S. Breun and R. Callies

Zentrum Mathematik M2, Technische Universität München
Boltzmannstr. 3, 85748 Garching, Germany
breun@ma.tum.de and callies@ma.tum.de

**Summary.** Optimal path-constrained trajectories of an ISS-based, three-link robot are investigated with a monorail as an additional fourth and prismatic joint. This results in a problem of optimal control for a multiple constrained nonlinear system of differential-algebraic equations. After transformation into minimum coordinates, the only remaining control is the acceleration of the end-effector along the prescribed trajectory, replacing four actuator torques/forces in the original formulation. The simpler structure is achieved at the price of introducing piecewise defined equations of motion, two highly nonlinear control constraints and two state constraints of first order. Switching points between partly linear and fully rotational motion are optimized. Solutions are presented including touch points of the state constraints with the two control constraints being active simultaneously. For the mathematical treatment of those problems, new interior point conditions are derived.

**Key words:** differential-algebraic control problem, robotic motion

## 1 Introduction

To reduce time-consuming extravehicular activities onboard of the International Space Station ISS, a promising approach is to substitute robotic manipulators for missing manpower. Important steps to maintain operational safety are monorails attached to the ISS structure and partly guiding the robot's motion, the spatial prescription of the end-effector trajectories and motion planning strategies that take into account the reduced accuracy of the linear motion compared to the rotational joints.

Optimal path-constrained trajectories of an ISS-based, three-link robot are investigated with a monorail as an additional fourth and prismatic joint. Operation of the highly accurate end-effector makes sense only while the low accuracy monorail motion stops. This results in a problem of optimal control for a multiple constrained nonlinear system of differential-algebraic equations. Switching points between partly linear and fully rotational motion are optimized simultaneously.

## 2 Optimal Control Problem

Fig. 1 gives a schematic sketch of the robot with three rotational joints and one prismatic joint. The $\alpha_i$ ($i = 1, 2, 3$) denote the angles of the rotational

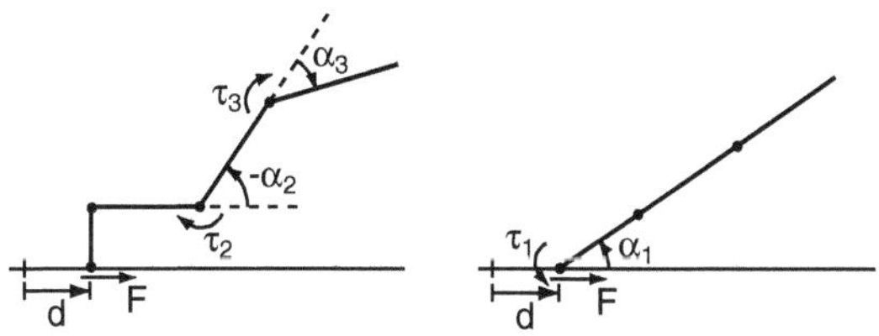


**Fig. 1.** Schematic side and top view of the investigated ISS-robot.

joints (with the respective actuator torques $\tau_i$) and $d$ the position of the prismatic joint (with the actuator force $F$). Only two configurations are admitted: Either the monorail is fixed and the three rotational joints are active for high-precision manipulations or the outer rotational link is fixed and the monorail is active for longer-distance motion. This results in piecewise defined equations of motion of the following type

$$\boldsymbol{M}(\boldsymbol{q})\ddot{\boldsymbol{q}} + \boldsymbol{k}(\boldsymbol{q}, \dot{\boldsymbol{q}}) = \boldsymbol{T} \tag{1}$$

with the state variables $\boldsymbol{q} = (d, \alpha_1, \alpha_2)^T$ or $\boldsymbol{q} = (\alpha_1, \alpha_2, \alpha_3)^T$ and the controls $\boldsymbol{T} = (F, \tau_1, \tau_2)^T$ or $\boldsymbol{T} = (\tau_1, \tau_2, \tau_3)^T$, depending on the active configuration. $\boldsymbol{M}(\boldsymbol{q})$ denotes the mass matrix and $\boldsymbol{k}(\boldsymbol{q}, \dot{\boldsymbol{q}})$ the vector of centrifugal and Coriolis forces. State variables at initial time and at final time $t_f$ are prescribed:

$$\boldsymbol{q}(0) = \boldsymbol{q}_0, \;\; \dot{\boldsymbol{q}}(0) = \dot{\boldsymbol{q}}_0, \;\; \boldsymbol{q}(t_f) = \boldsymbol{q}_f, \;\; \dot{\boldsymbol{q}}(t_f) = \dot{\boldsymbol{q}}_f \tag{2}$$

Control constraints read as follows:

$$|T_i| \leq T_{i0}, \quad i = 1, 2, 3 \tag{3}$$

To avoid damage of the ISS structure during motion, the end-effector has to follow a pescribed path $\tilde{\boldsymbol{r}}(s) \in \mathbb{R}^3$ with the path coordinate $s$. This results in equality constraints

$$\boldsymbol{r}(\boldsymbol{q}) = \tilde{\boldsymbol{r}}(s) \tag{4}$$

with the position of the end-effector $\boldsymbol{r}(\boldsymbol{q})$.

Minimum time solutions are looked for subject to the conditions (1)–(4). This yields a differential algebraic system of differential index 3.

## 3 Transformation into Minimum Coordinates

To avoid severe mathematical problems from the algebraic constraints (4), they are eleminated by transformation into minimum coordinates (cf. [1]). Differentiating (4) twice with respect to time and substituting into (1) yields

$$\boldsymbol{T} = \boldsymbol{b}(s)\ddot{s} + \boldsymbol{c}(s,\dot{s}) \tag{5}$$

$$\tilde{\boldsymbol{b}}(s) := \boldsymbol{M}(\boldsymbol{q}(s))\, \boldsymbol{r}_{\boldsymbol{q}}^{-1}(\boldsymbol{q}(s)), \quad \boldsymbol{b}(s) := \tilde{\boldsymbol{b}}(s)\, \tilde{\boldsymbol{r}}_s(s) \tag{6}$$

$$\boldsymbol{c}(s,\dot{s}) := \tilde{\boldsymbol{b}}(s) \left[\tilde{\boldsymbol{r}}_{ss}(s)\dot{s}^2 - \left(\frac{d}{dt}\boldsymbol{r}_{\boldsymbol{q}}(\boldsymbol{q}(s))\right) \dot{\boldsymbol{q}}(s,\dot{s})\right] + \boldsymbol{k}(\boldsymbol{q}(s), \dot{\boldsymbol{q}}(s,\dot{s})) \tag{7}$$

With the new and only control $u := \ddot{s}$ and the new state variables $x_1 = s$ and $x_2 = \dot{s}$, the new differential equations are linear and very simple. The difficulties are transformed into the constraints (3)

$$-T_{i0} \le b_i(s)\ddot{s} + c_i(s,\dot{s}) \le T_{i0}, \quad i = 1,2,3 \tag{8}$$

For $b_i(s) \neq 0$, this yields the mixed control and state multi-constraint

$$u_{\min}(s,\dot{s}) := \max_i \left(u_{\min,i}(s,\dot{s})\right) \le \ddot{s} \le \min_i \left(u_{\max,i}(s,\dot{s})\right) =: u_{\max}(s,\dot{s})$$

with

$$u_{\max,i}(s,\dot{s}) := \frac{\operatorname{sign}(b_i(s))\, T_{i0} - c_i(s,\dot{s})}{b_i(s)},$$

$$u_{\min,i}(s,\dot{s}) := \frac{-\operatorname{sign}(b_i(s))\, T_{i0} - c_i(s,\dot{s})}{b_i(s)}$$

For a regular parameterization of $\tilde{\boldsymbol{r}}(s)$, i.e. $|\tilde{\boldsymbol{r}}_s(s)| \neq 0$, there is at least one nonzero element $b_i(s)$ (cf. (6) with $\boldsymbol{M}$ and $\boldsymbol{r}_{\boldsymbol{q}}^{-1}$ regular). For $b_i(s) = 0$, (8) yields a pure state constraint: $|c_i(s,\dot{s})| \le T_{i0}$.

The result of the transformation is a well-structured optimal control problem:

$$I(u) = t_f \quad \longrightarrow \quad \min \tag{9}$$

subject to the equations of motion

$$\dot{x}_1 = x_2\,, \quad \dot{x}_2 = u\,, \tag{10}$$

the boundary conditions

$$x_1(0) = s_0\,, \;\; x_2(0) = \dot{s}_0\,, \;\; x_1(t_f) = s_f\,, \;\; x_2(t_f) = \dot{s}_f\,, \tag{11}$$

the control constraints

$$g_1(x_1,x_2,u) := u - u_{\max}(x_1,x_2) \le 0 \tag{12}$$

$$g_2(x_1,x_2,u) := u_{\min}(x_1,x_2) - u \le 0 \tag{13}$$

and the state constraints

$$h(x_1,x_2) := u_{\min}(x_1,x_2) - u_{\max}(x_1,x_2) \le 0 \tag{14}$$

$$\tilde{h}(x_1,x_2) := |c_i(x_1,x_2)| - T_{i0} \le 0 \quad \text{for } b_i(x_1) = 0,\, i = 1,2,3 \tag{15}$$

## 4 Optimal Control Theory

The optimal control problem (9)–(15) is transformed into a multi-point boundary value problem in a well-known manner (see e.g. [3], [2])

$$\dot{x}_1 = x_2,\ \ \dot{x}_2 = u^*,\ \ \dot{\lambda}_1 = -H_{x_1} = -\lambda_2 \frac{\partial u^*}{\partial x_1},\ \ \dot{\lambda}_2 = -H_{x_2} = -\lambda_1 - \lambda_2 \frac{\partial u^*}{\partial x_2}$$

with the adjoint variables $\lambda_1$, $\lambda_2$ and the Hamiltonian $H = \lambda_1 x_2 + \lambda_2 u$. The optimal control $u^*$ is given by

$$u^*(x_1, x_2) = \begin{cases} u_{\min}(x_1, x_2) & \text{if } \lambda_2 > 0 \\ u_{\max}(x_1, x_2) & \text{if } \lambda_2 < 0 \\ u_{\text{sing}}(x_1, x_2) & \text{if } \lambda_2 = 0 \quad \forall\, t \in [t_1, t_2] \subseteq [0, t_f] \end{cases}$$

At switching points $t_s$ of the control between $u_{\min}$ and $u_{\max}$, the condition

$$\lambda_2|_{t_s} = 0$$

holds together with the interior point conditions

$$x_1(t_s^-) = x_1(t_s^+),\ \ x_2(t_s^-) = x_2(t_s^+),\ \ \lambda_1(t_s^-) = \lambda_1(t_s^+),\ \ \lambda_2(t_s^-) = \lambda_2(t_s^+)$$

Another type of switching occurs at time $t_c$, if there is a change between the configurations with partly linear and fully rotational motion. For $t_c$ fully optimized, the interior point conditions derived from the generalized first order necessary conditions of optimal control theory read as follows

$$x_{1,2}(t_c^-) = x_{1,2}(t_c^+),\ \ \lambda_1(t_c^-) = \lambda_1(t_c^+),\ \ x_2(t_c) = 0,\ \ H|_{t_c^-} = H|_{t_c^+}$$

If (14) becomes active, then both control constraints (12), (13) become active too and the constraint qualification [3] doesn't hold any more

$$\text{rank} \begin{pmatrix} \frac{\partial g_1}{\partial u} & g_1 & 0 & 0 \\ \frac{\partial g_2}{\partial u} & 0 & g_2 & 0 \\ \frac{\partial h^1}{\partial u} & 0 & 0 & h \end{pmatrix} \neq 3\,, \qquad h^1(x_1, x_2, u) := \dot{h}(x_1, x_2) \tag{16}$$

*Classical optimal control theory for state constraints is not applicable here!*

To overcome this difficulty, we observe that $u$ is uniquely determined by

$$u = u_{\min}(x_1, x_2) = u_{\max}(x_1, x_2)$$

For a boundary arc on $[t_1, t_2] \subseteq [0, t_f]$, $t_1 < t_2$, $h^1(x_1, x_2, u) = 0$ serves as an additional constraint and completely determines $x_1, x_2$ together with $h(x_1, x_2) = 0$. Because $x_1, x_2$ also have to satisfy (10), in general $x_1, x_2$ are overdetermined. Thus, no boundary arcs are expected (and also not detected). General contact points, however, are possible, but only *touch points* are found. Application of the generalized first order necessary conditions leads to the numerically stable *interior point conditions* for a touch point at $t = t_t \in ]0, t_f[$ even in case of (16)

$$x_{1,2}(t_t^-) = x_{1,2}(t_t^+),\ \ h|_{t_t} = 0,\ \ h^1|_{t_t} = 0,\ \ H|_{t_t^-} = H|_{t_t^+} \tag{17}$$

# 5 Numerical Example

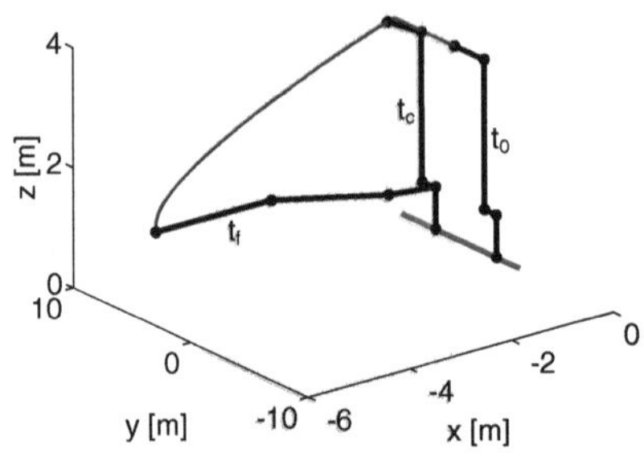


**Fig. 2.** Robotic manipulator and the prescribed path for the end-effector.

For the example system in Fig. 2 a high precision numerical solution of the transformed multi-point boundary value problem has been obtained by the new multiple shooting code JANUS [2]. Total time of motion is $t_f = 11.736\,s$, after $t_c = 6.0314\,s$ optimal switching from the partly linear to the fully rotational motion takes place. The optimal solution in Fig. 3 contains one touch

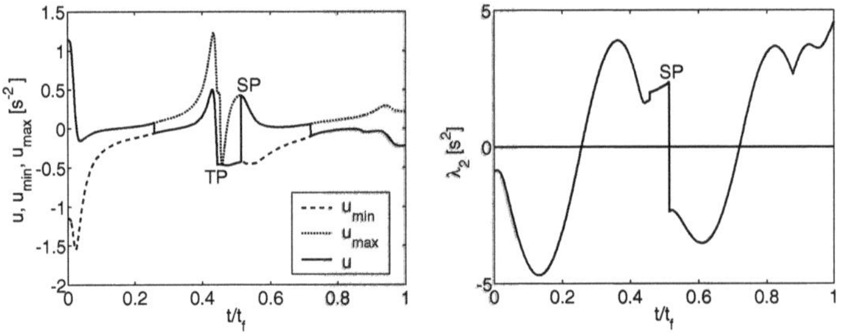


**Fig. 3.** The optimal control and the accompanying switching function.

point of the state constraint (TP) where (17) holds. The switching point of configurations (SP) is also a switching point of the control. An interesting detail: $\lambda_2$ is jumping across zero, not continuously passing through it. By backward transformation the actuator force and torques are easily obtained.

## References

1. J. E. Bobrow, S. Dubowsky, and J. S. Gibson. Time-Optimal Control of Robotic Manipulators Along Specified Paths. *Int. J. Robot. Res.*, 4(3):3–17, 1985.
2. R. Callies. Entwurfsoptimierung und optimale Steuerung. Habilitationsschrift, Technische Universität München, 2000.
3. R. F. Hartl, S. P. Sethi, and R. G. Vickson. A Survey of the Max. Principles for Optim. Control Probl. with State Constraints. *SIAM Rev.*, 37(2):181–218, 1995.

# Rigorous Analysis of Extremely Large Spherical Reflector Antennas: EM Case

E.D. Vinogradova[1], S.S. Vinogradov[2], and P.D. Smith[1]

[1] Department of Mathematics, Macquarie University, Sydney 2109, Australia.
evinogra@maths.mq.edu.au, pdsmith@maths.mq.edu.au
[2] School of Physics, University of Sydney, Sydney 2006, Australia.
svinogra@physics.usyd.edu.au

**Summary.** The transmitting spherical reflector antenna (SRA) has a well-known rigorous solution form as a second kind Fredholm system that is well conditioned when truncated to a finite system. The size of such systems for extremely large SRAs require specially designed highly efficient numerical algorithms to make their analysis feasible. Two significant features of the system are that its convolution format admits a computationally rapid implementation of the bi-conjugate gradient method, and at high frequencies, a certain decoupling occurs. These features allow an effective numerical treatment of apertures some thousands of wavelengths.

**Key words:** spherical reflector antenna, electromagnetics, method of regularisation, iterative methods

## 1 Introduction

Reflector antennas have long been studied by mostly high-frequency asymptotic techniques; however none predicts the spatial electromagnetic (EM) field distribution with uniform accuracy. On the other hand the lower frequency Method of Moments becomes computationally impractical for apertures exceeding about one hundred wavelengths.

When treated as a classical mixed-boundary value problem the antenna fields are expanded in spherical wave harmonics, and rigorous solutions for both acoustic and electromagnetic diffraction from the SRA may be derived, as in [3], Part 2. These rely heavily on the Method of Regularisation (MoR) outlined in [3], Part 1. In both acoustic and EM cases we obtained second kind Fredholm equations as infinite systems of linear algebraic equations (i.s.l.a.e.) to be solved for some modified Fourier coefficients. These systems can be solved very effectively with a truncation method. Rapid convergence with a proper choice of truncation number $N_{tr}$ delivers the Fourier coefficients with predictable accuracy. It was shown in [3], Part 2, that four-digit accuracy for the acoustic SRA of radius $a$ is obtained with $N_{tr} = [2\pi a/\lambda] + 12$ ($\lambda$ wavelength).

Regularisation of the corresponding EM problem for the SRA produces a coupled pair of i.s.l.a.e. for the electric and magnetic Fourier coefficients. Matrix element computation is based entirely on recurrence formulae, obviating the need for time-consuming numerical integration, and direct solution methods for the linear system makes it feasible to investigate relatively large SRAs, of aperture size $D$ up to $D/\lambda = 500$ in the acoustic case, and $D/\lambda = 250-300$ in the EM case (see [3], Part 2).

It is highly desirable to treat even larger SRAs in the quasi-optical region where $1000 \leq D/\lambda \leq 10000$. The successful acoustic treatment in [2] does not simply transfer to the EM case because of the strong imbalance in the magnitude of electric and magnetic Fourier coefficients that prevents their accurate calculation if the bi-conjugate gradient method (Bi-CG) is directly applied; at the very highest frequencies, convergence fails.

In this regime, we show that the original system approximately decouples so that primary beam diffraction can be described by a separate system for each polarization; the methodology of [2] then provides an effective treatment of the decoupled equations. We examine the error dependence on iteration number, and analyse its accuracy for huge SRAs.

## 2 The Decoupled System at High Frequencies

Let $\theta_0$ be the polar angle describing the SRA angular size; its aperture diameter and electrical size are $D = 2a\sin\theta_0$ and $D/\lambda = \pi^{-1}ka\sin\theta_0$, where $k = 2\pi/\lambda$. The GO-focal distance is $f = \frac{1}{2}a$ so $f/D = (4\sin\theta_0)^{-1}$.

The solution described in [3], Part 2, for the transmitting SRA, excited by a complex point Huygen's source (CPHS), incorporates the so-called polarisation constants arising from TE- and TM-wave coupling. Their elimination from the relevant equations ((4.180)-(4.183) of [3], Part 2) produces the following system to be solved for the unknown electric $\{X_n\}_{n=1}^{\infty}$ and magnetic $\{Y_n\}_{n=1}^{\infty}$ Fourier coefficients that are $O\left(n^{-1}\right)$ as $n \to \infty$: for $m = 1, 2, 3, \ldots$,

$$X_m - \sum_{n=1}^{\infty}(X_n\varepsilon_n + \alpha_n)\left(R_{nm} - \frac{\gamma_1}{\Delta}R_{n0}R_{0m}\right) = i\gamma_2\sum_{n=1}^{\infty}(Y_n\mu_n + \beta_n)\,Q_{n0}R_{0m}, \tag{1}$$

$$Y_m - \sum_{n=1}^{\infty}(Y_n\mu_n + \beta_n)\left(Q_{nm} + \frac{\gamma_3}{\Delta}Q_{n0}Q_{0m}\right) = i\gamma_4\sum_{n=1}^{\infty}(X_n\varepsilon_n + \alpha_n)\,R_{n0}Q_{0m}. \tag{2}$$

The remaining coefficients in (1) and (2) are the "incomplete scalar products"

$$\begin{Bmatrix} Q_{nm} \\ R_{nm} \end{Bmatrix} = \frac{1}{\pi}\left\{\frac{\sin(n-m)\theta_0}{n-m} \pm \frac{\sin(n+m+1)\theta_0}{n+m+1}\right\}; \tag{3}$$

the asymptotically small parameters $\varepsilon_n = 1 + 4ika\psi_n'(ka)\zeta_n'(ka)(2n+1)^{-1}$, $\mu_n = 1 - i(2n+1)\psi_n(ka)\zeta_n(ka)/ka$ that are $O\left(n^{-2}\right)$ as $n \to \infty$; the constants arising from elimination of the polarisation constants

$$\begin{aligned} \gamma_1 &= 4ka(1-Q_{00}) + Q_{00}/ka, \quad \gamma_2 = 4ka/\Delta, \\ \gamma_3 &= 4kaR_{00} + (1-R_{00})/ka, \quad \gamma_4 = (\Delta ka)^{-1}, \end{aligned} \tag{4}$$

where $\Delta = 4kaR_{00}(1-Q_{00}) - Q_{00}(1-R_{00})/ka$; and coefficients arising from the CPHS located at a complex point $r_s = d + ib$,

$$\alpha_n = 4ikat_n(kr_s)\zeta_n'(ka), \quad \beta_n = -i(2n+1)t_n(kr_s)\zeta_n(ka)/ka, \tag{5}$$

where $t_n(kr_s) = (i\psi_n'(kr_s) + \psi_n(kr_s))/kr_s$, and $\psi_n(z) = \sqrt{\pi z/2}J_{n+\frac{1}{2}}(z)$, $\zeta_n(z) = \sqrt{\pi z/2}H^{(1)}_{n+\frac{1}{2}}(z)$ are the spherical Bessel functions in Debye notation.

The computation of the matrix elements in (1), (2) is rapid if recurrence formulae are used, and matrix fill-time is reasonable when $D/\lambda \leq 300$. As $D/\lambda$ increases, matrix inversion becomes prohibitively expensive. Moreover, commonly available PCs do not possess the necessary memory capacity and speed to process matrix equations of extremely large size. An alternative that is time and memory efficient employs the Bi-Conjugate Gradient Method (Bi-CG) in which the matrix-vector multiplications of the iterative algorithm are effected by the FFT. This approach succeeded for the acoustic analogy of extremely large SRAs ($D/\lambda \leq 5000$) [2]. A similar attack on the coupled equations (1), (2) failed because of the vastly differing magnitudes of the electric and magnetic coefficients ($\|Y_n\| \ll \|X_n\|$).

Principal plane patterns for various values of $D/\lambda$ (up to 300) showed that their deviation in symmetry vanishes quite rapidly as $D/\lambda$ increases, indicating decoupling between the TE- and TM-waves. The terms on the right hand sides of (1), (2) containing $X_n$ or $Y_n$ provide a perturbation to the reduced system formed by their omission. A crude analytical estimate showing that the perturbation is proportional to $O\left((D/\lambda)^{-2}\right)$, in the sense of the norm estimate, was numerically confirmed for various values of $D/\lambda$ and $\theta_0$. Thus, at high values of $D/\lambda$, the equations (1) and (2) become, approximately,

$$X_m^d - \sum_{n=1}^{\infty} X_n^d \varepsilon_n R_{nm}^d = \sum_{n=1}^{\infty} \alpha_n R_{nm}^d, \tag{6}$$

$$Y_m^d - \sum_{n=1}^{\infty} Y_n^d \mu_n Q_{nm}^d = \sum_{n=1}^{\infty} \beta_n Q_{nm}^d, \tag{7}$$

where $m = 1, 2, 3, \ldots$, the superscript $d$ indicating the *decoupled system*, and

$$R_{nm}^d = R_{nm} - (R_{00})^{-1} R_{n0}R_{0m}; \; Q_{nm}^d = Q_{nm} + (1-Q_{00})^{-1} Q_{n0}Q_{0m}. \tag{8}$$

The difference between solutions to system (1), (2) and systems (6), (7), truncated to a finite order $N_{tr}$ for four-digit accuracy, is measured by

$$er_X = \left| \|X_n\| - \|X_n^d\| \right| / \|X_n\|, \; er_Y = \left| \|Y_n\| - \|Y_n^d\| \right| / \|Y_n\|. \quad (9)$$

Let us examine the SRA excited by a real Huygens source ($b = 0$) located at $r_s = d + ib$ with $d/a = 0.52$ and $\theta_0 = 35.13°$ (the Arecibo observatory SRA angle). The dependence of $er_X$, $er_Y$ on $D/\lambda$ is shown in Fig. 1(a). It confirms the crude analytical estimate of the scale of decoupling. Radiation patterns for various $f/D$ ratios and electrical sizes $D/\lambda$ were computed; for any $f/D$ value, even for small antennas (e.g., $D/\lambda = 20$, shown in Fig. 1(b)), the radiation patterns coincide graphically (to within 1%). As $D/\lambda$ increases, the difference rapidly diminishes below the error due to truncation itself.

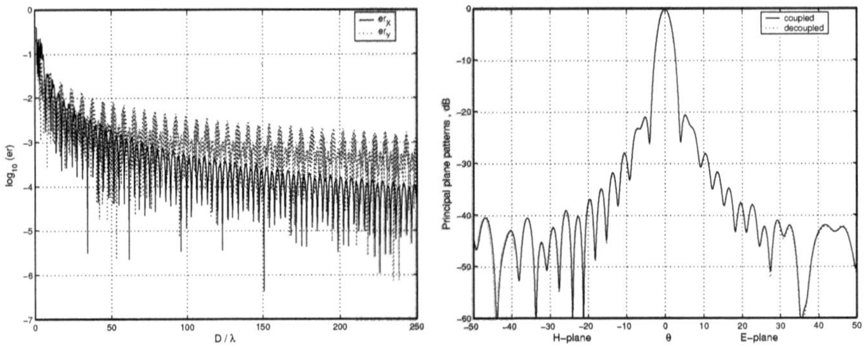


**Fig. 1.** (a) The error measures $er_X$, $er_Y$; (b) the principal plane patterns $D/\lambda = 20$.

## 3 Algorithm Performance on the Decoupled System

The decoupled system (6), (7) has much in common with that describing the acoustic case [2]; its form, as a discrete convolution, allows matrix-vector products, performed iteratively by the Bi-CG algorithm, to be very efficiently implemented with the discrete FFT. The matrix is not retained in memory; it is represented by a simple vector requiring far less than the prohibitively large storage needed by the full matrix. Thus computations with extremely high rank systems become feasible, as in [2]. The convergence of the iterative algorithm may be estimated using the difference of solutions $\mathbf{X}_{N_i}, \mathbf{Y}_{N_i}$ and $\mathbf{X}_{N_{i-1}}, \mathbf{Y}_{N_{i-1}}$ at steps $N_i$ and $N_{i-1}$ via

$$er_X(N_i) = \left| \|\mathbf{X}_{N_i} - \mathbf{X}_{N_{i-1}}\| \right| / \|\mathbf{X}_{N_i}\|, er_Y(N_i) = \left| \|\mathbf{Y}_{N_i} - \mathbf{Y}_{N_{i-1}}\| \right| / \|\mathbf{Y}_{N_i}\| \quad (10)$$

Figure 2(a) shows these error measures as a function of $N_i$ with $D/\lambda = 1500$, $d/a = 0.52$ and $\theta_0 = 35.13°$; $er_X(N_i)$ and $er_Y(N_i)$ fall below $10^{-8}$ when $N_i = 879$ and $N_i = 2699$, respectively. Figure 2(b) shows the comparable result for a huge SRA ($D/\lambda = 12000$) of matrix rank $N_{tr} = 2^{16} = 65536$.

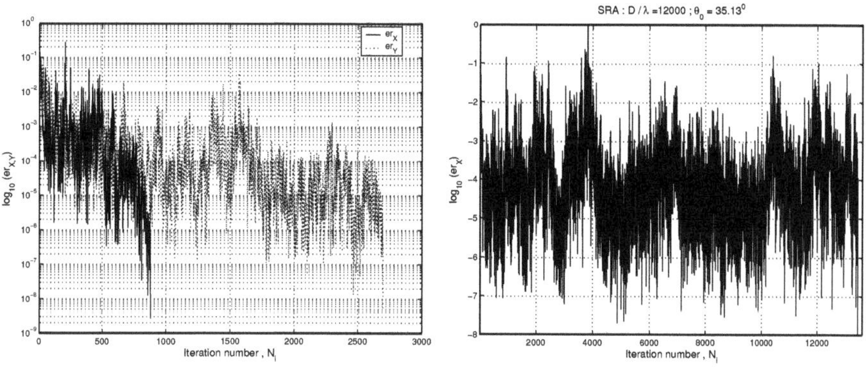


**Fig. 2.** The error measures $er_X(N_i)$, $er_Y(N_i)$ for $D/\lambda = 1500$ (a) and 12000 (b).

# 4 Conclusions

At high frequencies the equations describing the SRA are effectively decoupled, increasingly so as its electrical size grows. The decoupled approximation is valid within engineering accuracy when $D/\lambda \geq 20$. The analysis of large SRAs is effectively and efficiently performed by the Bi-CG algorithm in which matrix-vector multiplication is performed with discrete FFTs as illustrated with an example with $D/\lambda = 12000$. Our calculations were also validated against those of [1] on the performance of a sub-aperture of diameter 105 metres in the Arecibo reflector. This approach is also applicable to various problems solved by the MoR [3].

# References


1. A.W. Love. Radiation patterns and gain for a nominal aperture of 105 meters in the Arecibo Spherical Reflector. *IEEE Antennas Propagat. Mag.*, 43(1):20–30, 2001.
2. P.D. Smith, E.D. Vinogradova, and S.S. Vinogradov. Analysis of extremely large spherical reflector antennas. In *Proc. International Symposium on Electromagnetic Theory*, pages 730–732, Pisa, Italy, 2004.
3. S.S. Vinogradov, P.D. Smith, and E.D. Vinogradova. *Canonical Problems in Scattering and Potential Theory, Part 1: Canonical Structures in Potential Theory, Part 2: Acoustic and Electromagnetic Diffraction by Canonical Structures.* Chapman & Hall/CRC Press, Boca Raton FL, 2001, 2002.

# Part II

## Theme: Electronic Industry

# Simulation and Measurement of Interconnects and On-Chip Passives: Gauge Fields and Ghosts as Numerical Tools

Wim Schoenmaker[1], Peter Meuris[2], Erik Janssens[3], Michael Verschaeve[4], Ehrenfried Seebacher[5], Walter Pflanzl[6], Michele Stucchi[7], Bamal Mandeep[8], Karen Maex[9], and Wil Schilders[10]

1 MAGWEL wim.schoenmaker@magwel.com
2 MAGWEL peter.meuris@magwel.com
3 MAGWEL erik.janssens@magwel.com
4 MAGWEL michael.verschaeve@magwel.com
5 austriamicrosystems ehrenfried.seebacher@austriamicrosystems.com
6 austriamicrosystems walter.pflanzl@austriamicrosystems.com
7 IMEC michele.stucchi@imec.be
8 IMEC bamal.mandeep@imec.be
9 IMEC karen.maex@imec.be
10 PHILIPS wil.schilders@philips.com

**Summary.** This paper describes the present status of using lattice gauge and ghost field methods for the simulation of on-chip interconnects and integrated passive components at low and high frequencies. Test structures have been developed and characterized in order to confront the simulation techniques with experimental data. The solution method gives results that are in agreement with the measurements.

**Key words:** Interconnects, Integrated Passives, Characterization, Simulation, Ghost Fields

## 1 Introduction

With the further downscaling of deep-submicron CMOS devices a continuous increase in transistors switching rates is achieved. This allows for faster circuits and as a consequence, more powerful products become available to consumers. The downscaling not only has an impact on the speed of information processing as a results of fast switching times. Moreover, per unit chip area a much larger number of active devices is encountered. In other words: the transistor density has continuously increased over the years. This evolution was captured in the famous Moore's law [4] predicting that every 18 months the performance of integrated circuits will double. Derivations of Moore's law are that the cost per transistor will drop exponentially or that the clock frequency of the integrated circuits will grow exponentially. It should be emphasized that Moore's statement has been the driver behind the tremendous growth of the semiconductor industry, but it should also be stressed that Moore's law extrapolates an early observation, that at some instance will break down because physical laws will be violated or economical constraints will not release the required investments for the technology development and manufacturing. In the present work, we are primarily interested in the physical issues that will ultimately prevent us from sustaining in agreement with Moore's prediction. Actually, our approach is based on the very conventional attitude to keep in pace with the Moore's law. As a consequence, design methodologies that worked fine in the passed have to be upgraded to incorporate the new physical phenomena that come with cranking up the frequencies and densifying the active devices. Moreover, new interconnect technology is needed to guarantee that the gain in switching speed is not annihilated by interconnects that suffer from too much delay and loss. In order to achieve these goals a number of challenges need to be addressed.

### Which difficulties are to be expected?

The amount of difficulties coming with further downscaling of the integrated circuit is huge. The interested reader can find a detailed account in the annual revised International Technology Roadmap for Semiconductors [1]. Just to mention a view : printing the small structures on Silicon, will require further research in lithography. Keeping the source and drain well separated and at the same time reducing the channel lengths of the transistors, will require increasing control over the activation and diffusion of the dopants. The engineering of the channel will require several modifications in order to suppress the short-channel effect (SCE). The latter corresponds to lowering of the threshold voltage and results into a less clearer distinction between the on and off state of the transistor.

Apart from all the difficulties ("challenges") that one encounters inside the active devices or in-Silicon, there are also many issues to be dealt with for the interconnects or the on-Silicon architecture. The transistor densification requires that the interconnects have less spacing and cross talk becomes a serious issue. Not only do interconnects act as receivers for signals in neighboring runners, the currents in the runners are also re-distributed due to the presence of signals in neighboring lines. This is the proximity effect. These effects all occur as high-frequency. Of course, the well-known skin effect also plays an important role on wide ( $\sim 1\mu$m) interconnect at frequency of $\sim 20$ GHz, which represents the wire bandwidth necessary to allow a correct propagation of 1-2 GHz clock signals. Signal delay is an effect of major importance and the technological way to reduce it, is by reducing the resistance of the interconnects and to reduce the capacitances of the runners. The resistance can be lowered by choosing different metallic materials (Al $\rightarrow$ Cu) and the capacitances can be lowered by using dielectric materials with a lower permittivity (lowK materials). Here, a first hard limit is encountered from physics : the lowest permittivity that ever can be reached is $\varepsilon_r = 1$, being the permittivity of vacuum. The lowest values that are presently available are 1.7-2.3 and belong to porous materials that suffer from mechanical stability. Therefore it is not evident that these materials are suitable for use in the back-end processing.

The difficulties that we will address in this paper deal with design. Whereas at low frequencies it suffices to characterize the interconnect layout by its lumped resistance and lumped capacitance parameters, at high frequencies the full electromagnetic characterization is required. It is desired that the designer still has access to compact models that characterize the structures, the building of these compact models requires a full electromagnetic analysis in the frequency range of interest and in three spatial dimensions. Two-dimensional considerations are too restrictive since modern interconnect layouts are done in a multi-layer pattern. By inclusion of the frequency dependence, *i.e.* the physics of the electromagnetic fields, above mentioned effects are captured.

## 2 The Maxwell Equations and the Drift-Diffusion Equations

After having described the problem under consideration, we will give in this section the physical equations corresponding to it. As was stated above, we want to obtain compact models for given structures in three dimensions that describe their current-voltage characteristics accurately. These characteristics are the results of an interplay between electromagnetic fields and their sources being the charge and current densities. The latter are described by the Maxwell equations that are summarized below :

$$\nabla \cdot \mathbf{D} = \rho \tag{1}$$

$$\nabla \cdot \mathbf{B} = 0 \tag{2}$$

$$\nabla \times \mathbf{E} = -\frac{\partial \mathbf{B}}{\partial t} \tag{3}$$

$$\nabla \times \mathbf{H} = \mathbf{J} + \frac{\partial \mathbf{D}}{\partial t} \quad . \tag{4}$$

In here, $\mathbf{D}$, $\mathbf{E}$, $\mathbf{B}$, $\mathbf{H}$, $\mathbf{J}$ and $\rho$ denote the electrical induction, the electric field, the magnetic induction, the magnetic field, the current density and the charge density. The following constitutive equations relate the inductances to the field strengths :

$$\mathbf{B} = \mu \mathbf{H} \ , \qquad \mathbf{D} = \varepsilon \mathbf{E} \ . \tag{5}$$

The permittivity, $\varepsilon$, and permeability, $\mu$ and the constitutive equation that relates the current $\mathbf{J}$ to the electric field and the carrier densities, are determined by the medium under consideration. For a conductor the current $\mathbf{J}$ is given by Ohm's law :

$$\mathbf{J} = \sigma \mathbf{E} \ . \tag{6}$$

At metallic contact, we assume that the contact is such a high quality conductor, that the tangential electric field in the contact pad vanishes. In other words at the contact, we have

$$\mathbf{n} \times \mathbf{E}(\mathbf{x}, \mathbf{t}) = 0 \ , \qquad \mathbf{x} \in \text{contact} \ . \tag{7}$$

Furthermore, at the metallic contact pad it is assumed that the perpendicular component of the magnetic induction $\mathbf{B}$ vanishes :

$$\mathbf{B} \cdot \mathbf{n} = 0 \ , \qquad \mathbf{x} \in \text{contact} \ . \tag{8}$$

Outside the contact pads different boundary conditions can be explored. The precise implementation depends on the problem under consideration and the frequencies of interest. It should be stressed that the failure or success of a high-frequency simulation strongly depends on the proper choice and treatment of the boundary conditions. In the semiconducting regions, the current $\mathbf{J}$ consists of negatively and positively charged carrier currents obeying the current continuity equations.

$$\nabla \cdot \mathbf{J}_{\mathrm{n}} - q \frac{\partial n}{\partial t} = U(n,p) \ , \tag{9}$$

$$\nabla \cdot \mathbf{J}_{\mathrm{p}} + q \frac{\partial p}{\partial t} = -U(n,p) \ . \tag{10}$$

In here, the charge and current densities are

$$\rho = q(p - n + N_D - N_A) \ , \tag{11}$$

$$\mathbf{J}_{\mathrm{n}} = q\mu_{\mathrm{n}} n \mathbf{E} + kT\mu_{\mathrm{n}} \nabla n \ , \tag{12}$$

$$\mathbf{J}_{\mathrm{p}} = q\mu_{\mathrm{p}} p \mathbf{E} - kT\mu_{\mathrm{p}} \nabla p \tag{13}$$

and $U(n,p)$ is the generation/recombination rate of charged carriers. The current continuity equations provide the solution of the variables $n$ and $p$. Up to this point we have not faced the need for introducing the Fermi potentials as well as the Poisson potential. These variables enter the description through the boundary conditions. At ohmic contacts it is assumed that charge neutrality is valid and that the applied bias is equal to the Fermi potential. In particular, in the drift-diffusion model, the carrier densities are given in terms of the Poisson and Fermi potential as

$$p = n_i \exp \frac{q}{k_{\mathrm{B}} T} (\varphi_p - V) \ ,$$

$$n = n_i \exp \frac{q}{k_{\mathrm{B}} T} (V - \varphi_n) \ . \tag{14}$$

Using the charge neutrality, the contacts are characterized by $p - n + N_D - N_A = 0$ and $\varphi_p = \varphi_n = V_{app}$, where the latter is the applied voltage. Since the boundary conditions are formulated in terms of potentials, it makes sense to introduce the magnetic vector potential $\mathbf{A}$ next to the electric scalar potential $V$ in the following way : The magnetic induction $\mathbf{B}$ is given by

$$\mathbf{B} = \nabla \times \mathbf{A} \tag{15}$$

and using (3), the electric field is given by

$$\mathbf{E} = -\nabla V - \frac{\partial \mathbf{A}}{\partial t} \ . \tag{16}$$

The Maxwell equations are expressed in terms of the potential formulation as follows :

$$-\nabla \cdot \varepsilon \left( \nabla V + \frac{\partial \mathbf{A}}{\partial t} \right) = \rho \tag{17}$$

$$\nabla \times \nabla \times \mathbf{A} = \mu_0 \mathbf{J} - \mu_0 \varepsilon \frac{\partial}{\partial t} \left( \nabla V + \frac{\partial \mathbf{A}}{\partial t} \right) \ . \tag{18}$$

Since the operator $\nabla \times \nabla \times$ has no inverse, the vector potential is not uniquely defined and a gauge condition should be added. It is very appealing to use the Coulomb gauge since in this gauge the Poisson equation remains unaltered. In other words : the Poisson equation has no frequency-dependent terms. Thus we obtain :

$$\nabla \cdot (\varepsilon \mathbf{A}) = 0 \qquad \text{and} \qquad -\nabla \cdot (\varepsilon \nabla V) = \rho \tag{19}$$

From (16) it follows that $\mathbf{E}$ and $\mathbf{A}$ should be considered on equal level. As a consequence, since $\mathbf{E}$ is assigned to *links* of the computational grids, this should also be the case for the variables $\mathbf{A}$. This observation has far reaching consequences. In order to compute link variables as fundamental unknowns, the corresponding discretization should reflect this point. Setting up discretized equations for these variables amounts to assigning a pointer to every link in the grid.

## 3 Gauge Fields and Ghost Fields

The discretization of Ampére's equation (18) can be done by applying Stokes' theorem twice [3]. In Fig. 1, this method is illustrated. Each link of the grid provides one equation and one unknown, *i.e.* the projection of $\mathbf{A}$ along the link : $A_{ij} = \mathbf{A} \cdot \mathbf{e}_{ij}$, where $\mathbf{e}_{ij}$ is the unit vector pointing from node $i$ to node $j$. However, the singularity of the $\nabla \times \nabla \times$ operator pops up as redundancy in the system of equations, *i.e.* the equations are not independent. In fact, the discretization of the gauge condition by applying Gauss' theorem for each node results into an additional system of equations that just eliminates the

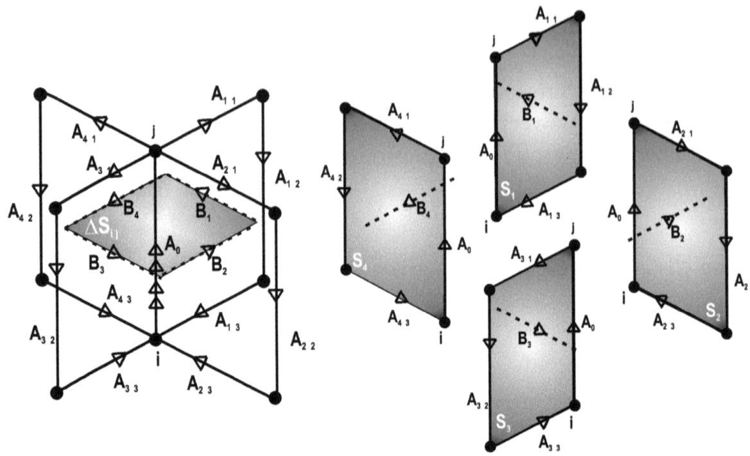

**Fig. 1.** Illustration of the discretization of $\nabla \times \nabla \times \mathbf{A}$ .

redundancy. Although we could proceed with the system of equations (17, 18, 19), there is a serious drawback in this formulation. The combined system of equations is not square, *i.e.* there are more equations for the link variables as $A_{ij}$ than there are unknowns. The mismatch in the counting of the number equations and the number of unknowns corresponds to the number of nodes in the grid [6]. The lack of having a square matrix corresponding to the system of equations for the the link variables obstructs the use of iterative linear solvers, that are nowadays the workhorses is simulation [5]. However, from the observation that the size of the mismatch is just equal to the number of nodes, suggests that we could recover a square formulation by adding additional degrees of freedom. To be more precise : for each node we need one additional unknown. This strategy lies at the heart of the solving techniques exploited by MAGWEL [2]. The feasibility of this approach was shown in [6, 7]. The additional collection of unknowns has been named a ghost field : $\chi$. Just as the quantum ghost particles that are indispensable to formulate the problem in a mathematical consistent and computable way, this ghost field has a classical (non-quantum) basis and is also indispensable to formulate the problem in an attractive numerical scheme. Whereas in the past, only analytical efforts shaped our language to describe a physical problem we now enter into an era in which the desire to address computational methods also contributes to the language of physics. After inclusion of the ghost field, the equations (17, 18, 19) become

$$-\nabla \cdot (\varepsilon \nabla V) = \rho \tag{20}$$

$$\nabla \times \nabla \times \mathbf{A} + \nabla \chi = \mu_0 \mathbf{J} - \mu_0 \varepsilon \frac{\partial}{\partial t} \left( \nabla V + \frac{\partial \mathbf{A}}{\partial t} \right) \tag{21}$$

$$\nabla^2 \chi + \nabla \cdot \mathbf{A} = 0 \tag{22}$$

Finally, after applying a small-signal analysis by setting all variables $X = X_0 + (X_R + iX_I)\exp(i\omega t)$ we finally arrive at the system of equations that may formally be given by equation (23) :

$$\begin{bmatrix} A(\omega) & B(\omega) \\ C(\omega) & D(\omega) \end{bmatrix} * \begin{bmatrix} \mathbf{X} \\ \mathbf{Y} \end{bmatrix} = \begin{bmatrix} \mathbf{F} \\ \mathbf{G} \end{bmatrix} . \tag{23}$$

In here, the vector **X** corresponds to the usual set of technology computer-aided design (TCAD) variables and the vector **Y** corresponds to the electro-*magnetic* (EM) extension to incorporate high-frequency effects, *i.e.*

$$\mathbf{X} = \begin{bmatrix} V \\ \varphi_p \\ \varphi_n \end{bmatrix} , \qquad \mathbf{Y} = \begin{bmatrix} \mathbf{A} \\ \chi \end{bmatrix} . \tag{24}$$

In particular the structure of the matrix becomes

$$\begin{bmatrix} A_0 + \omega A_1 & \omega B \\ C_0 + \omega C_1 & D_0 + \omega D_1 + \omega^2 D_2 \end{bmatrix} . \tag{25}$$

Starting from the formulation given above, a new simulation tool has been constructed that allows a detailed computation of the electromagnetic behavior of on-chip structures, taking into account the presence of the semiconducting substrates and the junctions therein. As can be observed from (25), we see that at low frequencies a decoupling occurs. The B-matrix gets small and it suffices to compute the solution **X** that can be inserted in the second equation for **Y**. The feed-back of **Y** on the solution for **X** is negligible. In Fig. 2, the convergence behavior is illustrated that is typically observed by iteratively solving the TCAD and the EM problem.

## Gauges and Ghosts : The History in a Nutshell

The history of gauge theories is a long and fascinating story. It is likely that the story has not reached its end. Reflecting back on electromagnetism, the first scientific formulation was done in terms of *forces*. The Coulomb law, experimentally verified by Cavendish, gives the forces acting between charges. Similarly, the law of Biot-Savart that lies at the heart of describing magnetic interactions is also expressed in terms of forces. A major breakthrough and change in perception was introduced by Faraday, who puts emphasis on the *fields*. In other words : whereas for forces it is always needed to have at least two particle participating in the description of electromagnetic interactions, the fields are modifications of the surrounding space of a single particle. The reality of such vacuum modifications have become even more acceptable after the discovery that light consists of electromagnetic waves. In order to compute the properties of electromagnetic fields it turned out to be quite convenient to introduce the scalar and vector potentials $V$ and **A**. Note that this incorporates a next level of abstraction : the potentials are not uniquely defined

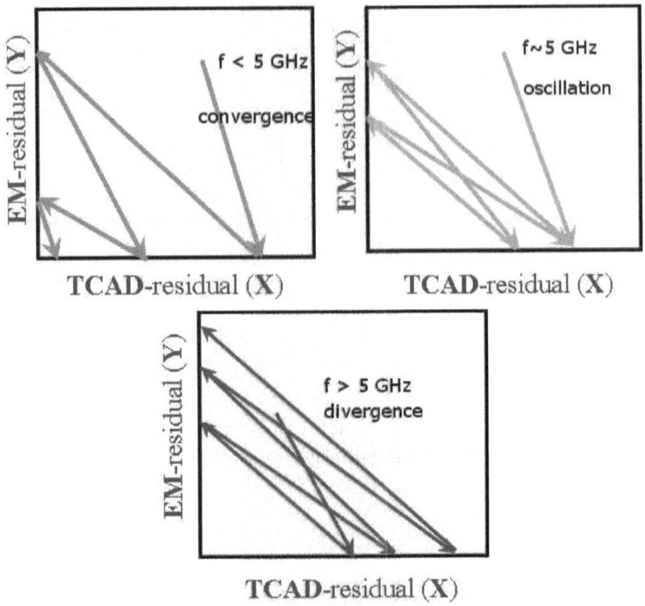


**Fig. 2.** Convergence behavior of the TCAD and EM residuals in different frequency ranges.

and equivalent descriptions exist by making the following substitutions or performing a gauge transformation :

$$V(\mathbf{x},t) \rightarrow V'(\mathbf{x},t) = V(\mathbf{x},t) + \frac{\partial \Lambda(\mathbf{x},t)}{\partial t}$$
$$\mathbf{A}(\mathbf{x},t) \rightarrow \mathbf{A}'(\mathbf{x},t) = \mathbf{A}(\mathbf{x},t) + \nabla \Lambda(\mathbf{x},t) \ . \tag{26}$$

It was Hermann Weyls [8] great contribution to observe that this non-uniquess is related to a symmetry principle : In each space-time point the local frame that determines the real and imaginary part of a quantum-mechanical wave function, may be arbitrary chosen. This means that the wave function can be multiplied with an arbitrary phase factor $g((\mathbf{x},t) = \exp \mathrm{i}\Lambda(\mathbf{x},\mathrm{t})$. This is an element of the unitary group $U(1)$. Weyl named this non-uniquenes "Maszstab" (gauge) invariance. In 1954, C.N. Yang and R. Mills proposed a model that generalizes the group $U1)$ to a non-abalian group $SU(2)$ The Weinberg-Salam model for weak interactions is (1967) is the first successful application of this idea. Soon the next success was recorded by applying the Yang-Mills concepts to the strong interactions that resulted into the theory of quantum-chromodynamics. Here the symmetry group is $SU(3)$. Nowadays, so-called gauge theories are the basis for our understanding of fundamental interactions.

The use of gauge theories was very much enforced by the aim to understand the elementary particles at the quantum level. However, the quantization of gauge theories was hampered precisely because of the underlying gauge invariance. A major achievement was realized by Fadeev and Popov, Feynman, 't Hooft and Veltman who were able to respect unitary principles by introducing a fictitious particle, a ghost particle that is only present inside closed loops of the scattering Feynman diagrams. The ghost particle paved the road towards a consistent quantization of gauge theories.

With the advent of computers, the ghost particle also can play an important role. This ghost field differs essentially from the quantum ghost field. Whereas the latter carries energy, albeit only inside virtual processes, *i.e.* inside quantum loops, the 'computational' ghost field is a zero-energy field. The field exists, *i.e.* is different from zero, while the computation is still iterating towards its solution. When arriving at the solution the computational ghost field fades out.

Ghost fields may appear a nuisance in modern physics. However, if one classifies ghost fields as indispensable computational but unphysical dynamical variables or degrees of freedom, they have been used for many years. An example is provided by the electromagnetic fields in the Lorentz gauge in free space. The Green function of this field is :

$$G(\mathbf{k}, \omega) = \frac{\delta^{\mu,\nu}}{|\mathbf{k}|^2 - \omega^2 + \mathrm{i}\varepsilon} , \tag{27}$$

where $\delta^{\mu,\nu}$ with $(\mu, \nu = 0, 1, 2, 3)$ is the Minkowski metric of space and time.

$$\delta^{\mu,\nu} = \begin{bmatrix} -1\,0\,0\,0 \\ 0\ \ 1\,0\,0 \\ 0\ \ 0\,1\,0 \\ 0\ \ 0\,0\,1 \end{bmatrix} \tag{28}$$

Let the electromagnetic wave propagate in the $z$- direction. There are two transversal components, $A_x$ and $A_y$. Actually the time component, that corresponds to the $\delta^{00} = -1$ weight of the Green function, is unphysical. This part cancels the longitudinal component, *i.e.* $\delta^{33} = 1$, that is also unphysical, such that only the two physical transversal components remain. One can fairly say that the fields $V$ and $A_z$ are ghost fields for the description of electromagnetic waves in free space. An application of these ideas in computational electromagnetism is provided by a transformation of the variables $(V, \mathbf{A}, \chi)$ to the variables $(V, \mathbf{E})$, by using $\mathbf{A} = \frac{\mathrm{i}}{\omega}(\mathbf{E} + \nabla V)$. In this formulation, $V$ can also be viewed as a ghost field. Note that a ghost field is still needed to regulate the singular operation $\nabla \times \nabla \times \mathbf{E}$.

## 4 Applications

In the section, we will present a number of applications. The examples illustrate how high-frequency effects are manifest in on-chip structures and how the environment consisting of the substrate and dielectric material will impact the results. The first example shows that substrate currents are induced by high-frequency signals in the runners above the substrate. The second example discusses a metal-insulator-metal capacitor. It is shown that the lumped element parameter, *i.e.* the capacitance varies as a function of the frequency. The third example shows how co-planar strip lines can be characterized starting from the Maxwell equations and the constitutive equations and deriving the lumped-element parameters. In the fourth example, a full analysis is done of a spiral inductor above a conductive substrate.

### Substrate Currents.

A U-shaped conductor is positioned above a conductive substrate as is illustrated in Fig. 3. The conductor is biased with a high-frequency AC signal. An alternating magnetic induction is injected in the substrate and Faraday's law implies that circular electric fields are generated in the substrate. Since the substrate is conductive, eddy currents will flow. Naively, one might expect to interrupt the flow of the eddy currents by putting insulating trenches in the substrate. In Fig. 3, a "+"-shaped trench is etched in the substrate. This will indeed have some effect, however as can be seen in the Fig. 3, the eddy currents are still present. This is because displacement currents will be induced in the trenches. This example illustrates that in order to characterize structures that are composed of dielectrics and conductive materials at high-frequencies, all terms in the Maxwell equations are needed for capturing the full physical picture.

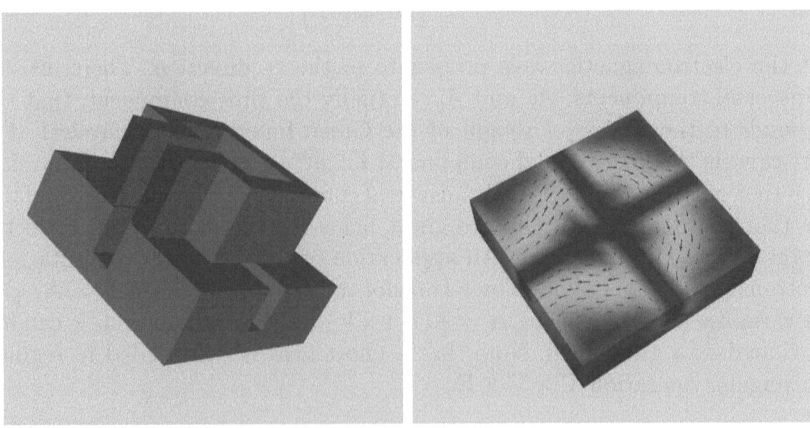

**Fig. 3.** 3D view of a U-turn structure above a conductive substrate (left). The induced substrate current is shown (right).

### Metal-Insulator-Metal Capacitor

In Fig. 4, a 3D view of the MIM capacitor is show. Above a substrate, a metallic layer is deposited. Next a thin layer or dielectric is deposited and next a second metallic plate is deposited. Above this plate a 'thick layer of dielectric material is deposited and the contact pad is attached. The contact pad is attached to the second metallic plate of the condenser by a grid of vias that are etched in the top dielectric layer. The vias are seen in Fig. 4. The structure was designed, processed and characterized by austriamicrosystems. In Fig. 5 the comparison is shown of the measured and the simulated capacitance as a function of the frequency. The plateau in the experimental data around 25 GHz is presently under study.

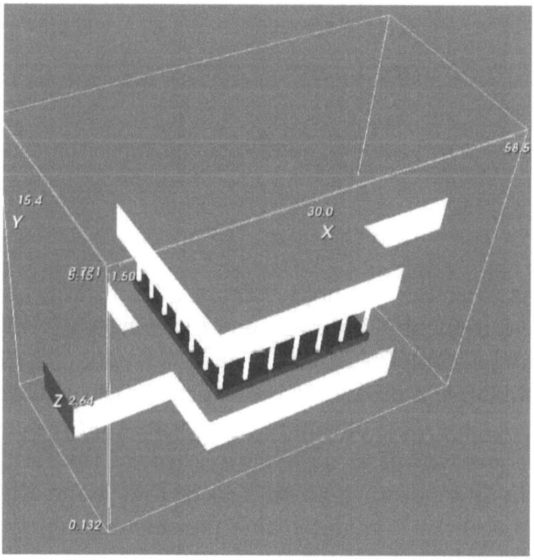


**Fig. 4.** 3D view of the Metal-Insulator-Metal (MIM) capacitor.

### Co-planar striplines

The layout of the simulated coplanar line is shown in Fig. 6. This structure represents a large on-chip wire, running between by two adjacent grounded conductors; the silicon substrate is also grounded, but most of the fields are supposed to be concentrated between the wire and the conductors. This structure is used to characterize the behavior of interconnect materials, namely insulators and conductors, at high frequency. Input for the simulation are material parameters such as the effective dielectric constant and loss tangent of the insulator, the resistivity of the conductor and the geometry of the structure, *i.e.* the wire width, length, spacing and the thickness of layers.

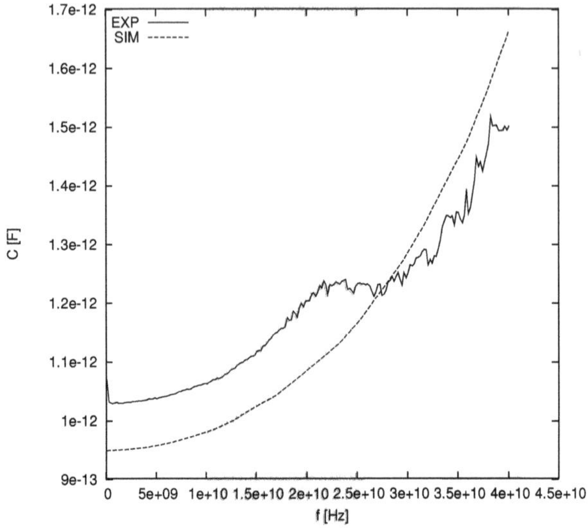


**Fig. 5.** Comparison of the measured and simulated capacitor of the MIM structure as a function of the frequency. The solid lines are measurements and dashed lines are simulations.

Simulations of the structure without contact pads have been compared with S-parameters measurements in a frequency range from 1GHz to 30GHz, after de-embedding the parasitics originating from the pads. At these frequencies, the skin effect and current crowding are present in the structure. Both effects are clearly shown in Fig. 7. The simulation shows that, when using all terms in the Maxwell equations, it is possible to capture the full physical picture and determine the line parameters. In Fig. 8, Fig. 9 and Fig. 10, a comparison is shown between the simulated line parameters and the measured line parameters. Even at high frequencies, there is a very good match between simulation and experiment. The line parameters R,L and C vs. frequency are very important for estimating and designing the signal propagation on-chip. The increase in resistance and the decrease in inductance is due both to the skin effect and to the current crowding : the skin effect increases R, the current crowding increases R and decreases L since there is a reduction of the size of the inductance loop made by the wire and the return path on the two conductors.

## Spiral inductor

One of the standard examples of on-chip passives is the design of a spiral conductor. The layout of the spiral is given in Fig. 11 spiral is realized in the 0.35 $\mu$m technology of The structure under study is depicted in Fig. 12. The design was carried out by austriamicrosystems in the framework of the

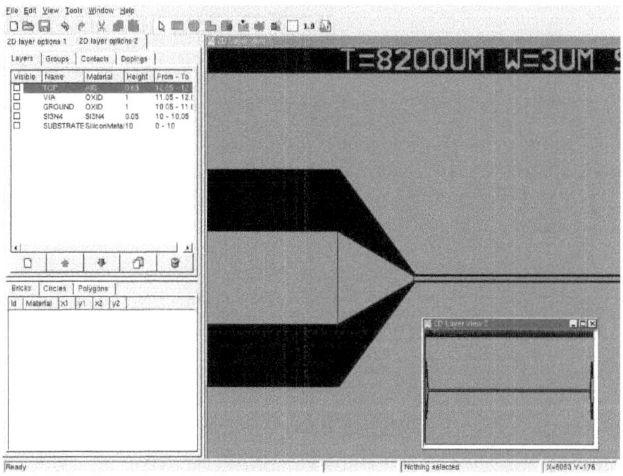


**Fig. 6.** The layout of the coplanar line under study.

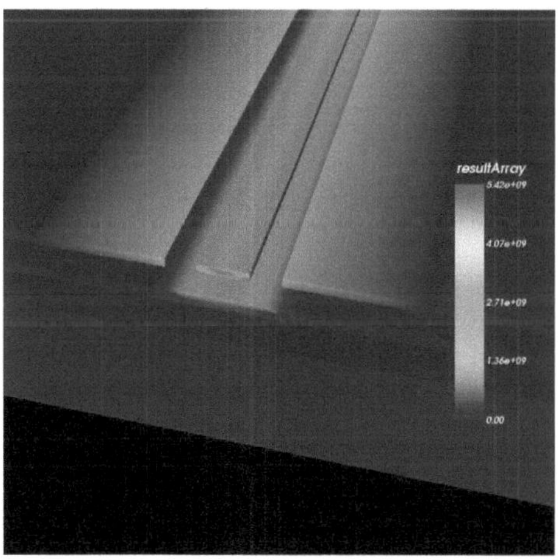


**Fig. 7.** The current distribution at 30GHz in the coplanar line under study.

CODESTAR project and the spiral was processed using the 0.35 $\mu$m technology of austriamicrosystems.

The S-parameters of the spiral inductor have been measured and the Q-factor was extracted from the data. The extracted and simulated results for the Q-factor are shown in Fig. 13. The simulation was carried out using a mesh of 44550 nodes, in a simulation domain of 1000 $\mu$m $\times$ 1000 $\mu$m $\times$ 307.56 $\mu$m, for 24 frequency points in a frequency range from DC to 23 GHz.

The simulation predicts the location of the resonance frequency. At this frequency the electric energy equals the magnetic energy and the structure's

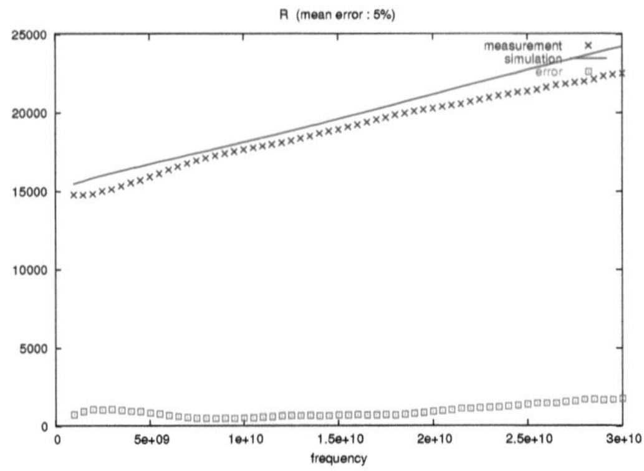


**Fig. 8.** The resistance of the coplanar line.
solid line = simulation, × = measurement, square = error

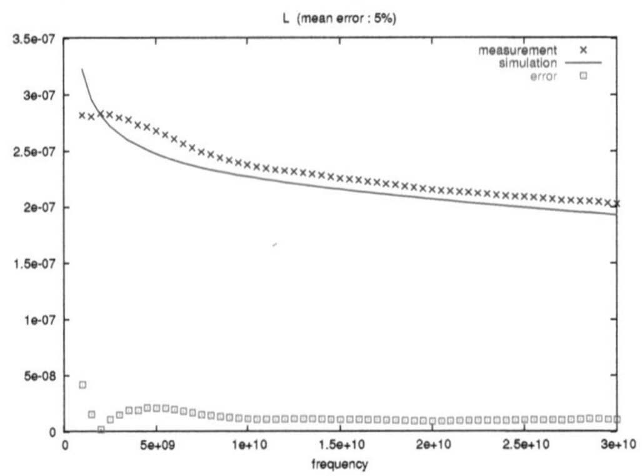


**Fig. 9.** The inductance of the coplanar line.
solid line = simulation, × = measurement, square = error

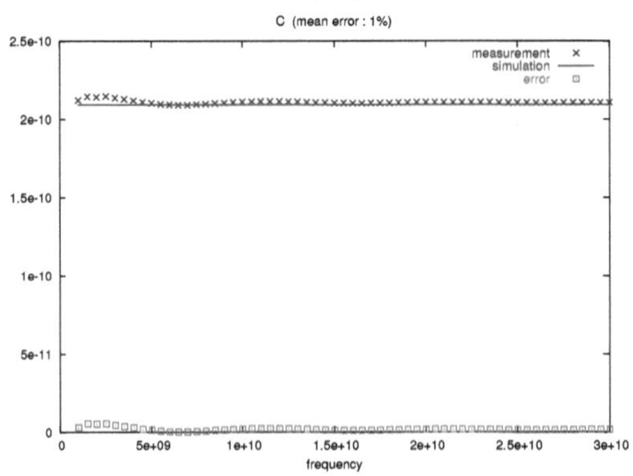


**Fig. 10.** The capacitance of the coplanar line.
solid lines = simulation, × = measurement, square = error

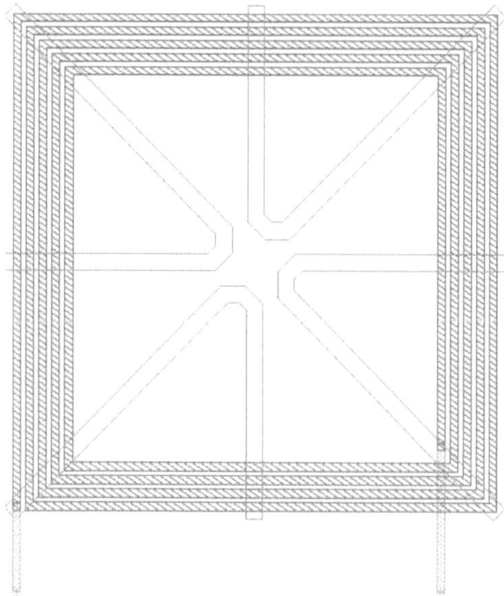

**Fig. 11.** The layout of the spiral inductor developed by austriamicrosystems.

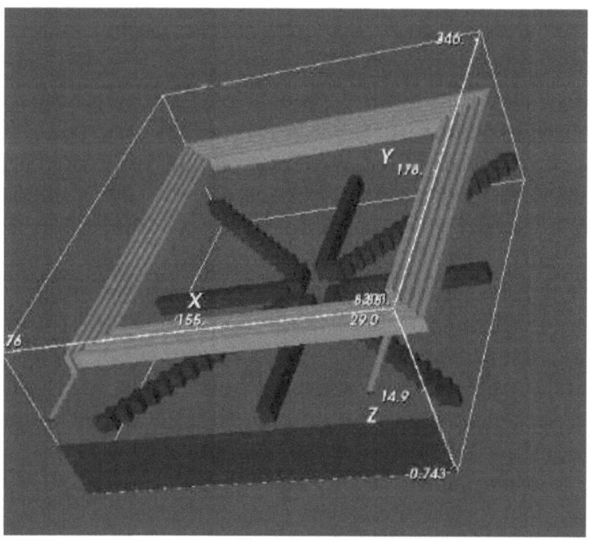


**Fig. 12.** The geometry of the spiral inductor under study.

behavior changes from inductive to capacitive and the resulting quality factor vanishes.

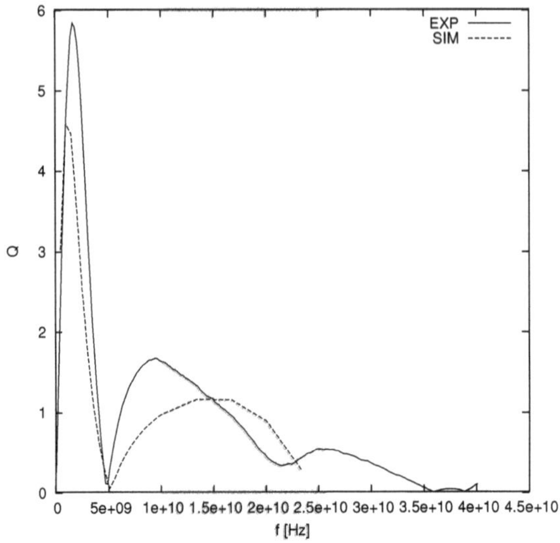


**Fig. 13.** The Q-factor as a function of frequency. The experimental results are indicated by solid line, while the simulation result is shown as a dashed line.

# 5 Conclusions

This paper presented an approach to handle on-chip electromagnetic effects. The underlying idea is that at high frequencies, the inductive parts of the electric fields are a substantial fraction of the total electric field. This is clearly illustrated for the skin effect : at the inner part of the runner where the skin effect is manifest, the inductive part of the electric field will compensate the external electric field with the result that the total electric field vanishes. As a consequence, no current is present. In general, at high frequencies the inductive effects will be present ubiquitously. The computation of these fields requires that the environment is faithfully included in the set up of the computation. In particular, the simulation of part of the integrated circuits puts severe demands on the structure editor. Whereas "toy" problems can still be edited using rather elementary building blocks, the industrial problems need much more powerful tools to prepare the simulation. Within the EUI project CODESTAR, a structure editor is developed that is capable of reading GDS files. The latter contain the mask information that is needed to print the on-chip structures. The geometrical data is very fine-detailed, since processing constraints are also taken into consideration. For example, the vias that connect the metal layers in Fig. 4, are drilled in a 0.35 micron technology. This via structure introduces about 10.000 additional nodes in the grid. Keeping in mind a rule of thumb that every node introduces 10 fundamental unknowns, the computational burden that induced by the vias is tremendous. For that reason mesh cleaning is a necessity and tools for that purpose have been developed. Once that an appropriate mesh has been found, the computation of the fields should be done. Now the boundary conditions come into the picture. For TCAD-alike problems, the boundary conditions for the vector potential can be chosen of the Dirichlet type. In the simulations that have been discussed in this paper, the boundary conditions for the vector potental was chosen to be $A_{ij} = 0$ for the link $ij$ being at the surface of the simulation domain and $\chi_i = 0$ for node $i$ being at the surface of the simulation domain. Such an approach requires a sufficient amount of space around the structures of interest in order to capture the electromagnetic energy. This free-field space adds additional grid nodes to the computational problem and hands-on experience has to be gathered to find sensible trade-off between accuracy and speed. Our method to discretize the vector potential by taking into account its geometrical meaning, *i.e.* by respecting the fact that it is a vector field, contrary to a scalar field, turns out to be beneficial in the sense that even rather crude meshes are able to grab the behavior of these fields under high-frequency biases. There is a lesson to be learned from this observation : The discretization of physical fields should be guided by the geometrical character of the field under consideration. Many software tools implicitly take this aspect already into account. However, there are also numerous tools that ignore the geometrical connection and assign vectors (one-forms) and surfaces (two-forms) to the nodes of the computational grid. Our observations discourage such implementations.

*Acknowledgement.* Part of the work was funded by the EU project CODESTAR -IST-2001-34058.

## References


1. The international technology roadmap for semiconductors. `http://itrs.org`.
2. `http://www.magwel.com`.
3. P. Meuris, W. Schoenmaker, and W. Magnus. Strategy for electromagnetic interconnect modeling. *IEEE Trans. on CAD of Integ. Syst. and Circ.*, 20:753–762, 2001.
4. G. Moore. Cramming more components onto integrated circuits. *Electronics*, 38, 1965.
5. O. Schenk, S. Roellin, and A. Gupta. The effect of unsymmetric matrix permuatioans and scaling in semiconductor device and circuit simulation. *IEEE Trans. on CAD of Integ. Syst. and Circ.*, 23, 2004.
6. W. Schoenmaker, W. Magnus, and P. Meuris. Ghost fields in classical gauge theories. *Phys. Rev. Lett.*, 81, 2002.
7. W. Schoenmaker and P. Meuris. Electromagnetic interconnect and passives modeling: Software implementation issues. *IEEE Trans. on CAD of Integ. Syst. and Circ.*, 21:534–543, 2002.
8. H. Weyl. Gravitation und Elektrizität. *Sitzungsberichte der Preussischen Akademie der Wissenschaften*, 26:465–480, 1918.

# Eigenvalue Problems in Surface Acoustic Wave Filter Simulations

S. Zaglmayr[1], J. Schöberl[2], and U. Langer[3]

[1] FWF-Start Project Y-192 "3D hp-Finite Elements", Johannes Kepler University Linz, Altenbergerstraße 69, 4040 Linz, Austria sz@jku.at
[2] Radon Institute for Computational and Applied Mathematics (RICAM), Altenbergerstraße 69, 4040 Linz, Austria joachim.schoeberl@oeaw.ac.at
[3] Institute of Computational Mathematics, Johannes Kepler University Linz, Altenbergerstraße 69, 4040 Linz, Austria ulanger@numa.uni-linz.ac.at

**Summary.** Surface acoustic wave filters are widely used for frequency filtering in telecommunications. These devices mainly consist of a piezoelectric substrate with periodically arranged electrodes on the surface. The periodic structure of the electrodes subdivides the frequency domain into stop-bands and pass-bands. This means only piezoelectric waves excited at frequencies belonging to the pass-band-region can pass the devices undamped.

The goal of the presented work is the numerical calculation of so-called "dispersion diagrams", the relation between excitation frequency and a complex propagation parameter. The latter describes damping factor and phase shift per electrode.

The mathematical model is governed by two main issues, the underlying periodic structure and the indefinite coupled field problem due to piezoelectric material equations. Applying Bloch-Floquet theory for infinite periodic geometries yields a unit-cell problem with quasi-periodic boundary conditions. We present two formulations for a frequency-dependent eigenvalue problem describing the dispersion relation.

Reducing the unit-cell problem only to unknowns on the periodic boundary results in a small-sized quadratic eigenvalue problem which is solved by QZ-methods. The second method leads to a large-scaled generalized non-hermitian eigenvalue problem which is solved by Arnoldi methods.

The effect of periodic perturbations in the underlying geometry is confirmed by numerical experiments. Moreover, we present simulations of high frequency SAW-filter structures as used in TV-sets and mobile phones.

**Key words:** piezoelectric effect, periodic structures, Bloch theory, eigenvalue problems.

# 1 Introduction

This work deals with mathematical modeling and numerical simulation of periodic piezoelectric Surface Acoustic Wave filters (briefly SAW-filters) and results in the computation of so-called "dispersion diagrams". We focus on surface acoustic wave devices used for frequency filtering in wireless communication such as standard components in TV-sets and cellular phones. However, there are many other application fields of SAW-devices as in radar and sensor technology and non-destructive measurement.

A SAW filter consists of a piezoelectric substrate onto whose surface electrode structures are evaporated. We want to concentrate on analyzing frequency filtering effects caused by the periodic arrangement of the electrodes. In practical periodic SAW-filters one arranges some hundreds up to some thousands of electrodes periodically in order to gain the so-called stop-band phenomena. The nature of periodic structures prohibits the propagation of SAWs excited in several frequency ranges. The frequency domain is classified into pass-bands, *i.e.* frequencies for which excited surface waves get through the periodic piezoelectric device, and stop-bands, *i.e.* frequencies which cannot pass trough. Therefore, the piezoelectric device can be used for frequency filtering.

A fundamental and recommendable introduction to acoustic field problems, various (surface) wave modes and piezoelectricity is provided by Auld in [3]. The numerical solution of piezoelectric systems via the finite element method is treated e.g. by Lerch in [16]. An overview of the historical development of SAW-devices is given in [19]. The principles of periodic SAW-devices are treated in some IEEE papers like [12], however, in most of them only pure-propagating modes are simulated.

The mathematical justification for the quasi-periodic field distribution is given by Bloch-Floquet theory, which analyzes the spectral properties of ordinary and partial differential operators on periodic structures. This theory was developed by Bloch for solving special problems in quantum mechanics, where one deals with periodic Schrödinger operators, and by Floquet for ordinary differential equations. A description by physicists can be found in [17] and in [2]. A functional analytic approach is provided by Simon and Reed [20]. The generalization to partial differential equations with periodic coefficients was done by Bensoussan, Lions and Papanicolaou in [6] for real and elliptic problems and by Kuchment [13], who applied the theory to scalar equations on photonic and acoustic band-gap devices in [4].

Bloch-Floquet theory states that the solution on periodic structures can be decomposed into quasi-periodic functions, so-called Bloch waves. Therefore the problem can be restricted to the unit-cell, *i.e.* the domain including one electrode. Successive arrangement of this unit-cell yields the original geometry. In order to describe the original periodic system, appropriate quasi-periodic boundary conditions have to be established.

The unit-cell problem turns out to be a coupled-field eigenvalue problem depending on either the frequency or the complex propagation constant. The numerical solution requires discretization, which is done by the Finite Element Method (FEM), and the application of an eigenvalue solver. We introduce step-by-step the mathematical tools for handling periodic structures, *i.e.* formulating and incorporating appropriate boundary conditions and corresponding discretization methods.

We begin with the scalar wave problem and establish three different solution methods for computing the dispersion diagram. All these methods result in non-hermitian eigenvalue problems of linear or quadratic form. Applying the established methods to periodic structures on piezoelectric problems is formally equivalent to the scalar wave model problem. However, the matrices get indefinite and worse conditioned due to piezoelectric properties, which requires special numerical treatment. Mathematical modeling results in two reasonable versions of frequency-dependent eigenvalue problems, one of quadratic form and the other one of generalized linear form. This requires special theory and numerics of algebraic eigenvalue problems.

In [5] a recommendable collection of state-of-the-art direct and iterative methods for large-scale eigenvalue problems is given. The book includes improved algorithms and implementational details. Tisseur [23] specializes on quadratic eigenvalue problems and Lehouqh [14] on Arnoldi and Implicit Restarted Arnoldi Methods (IRAM). A collection of structure-preserving methods is provided in [8].

The stated eigenvalue problems are solved numerically by our open-source high-order Finite Element solver NGSolve [22] in combination with the mesh generator Netgen [21]. For solving the occurring eigenvalue problems we link the software packages Lapack [1], providing direct methods, and Arpack [9], providing Implicit Restarted Arnoldi methods.

The main goal of this paper is the detailed derivation of a mathematical model for surface wave propagation in periodic piezoelectric structures including numerical solution methods and simulation of practical filter structures. The paper is organized as follows. We start with the technical details of surface acoustic wave filters including some first model assumptions, which are based on physical considerations, in Section 2. An introduction to piezoelectric equations is given in Section 3. To gain a detailed mathematical modeling we treat the two main subproblems separately, those are wave propagation in periodic media and the piezoelectric coupled field problem. In Section 4 we derive mathematical tools and solution strategies for the dispersion context of a scalar model problem with periodic coefficients. Section 5 starts with mathematical tools for the piezoelectric coupled field equations and results in combining the solution methods derived in 4 to piezoelectric equations. Numerical results are presented in Section 6. First, the effect of periodic perturbations in the underlying geometry is confirmed. Second, we present simulations of a high frequency SAW-filter structures as used in TV-sets or GSM-mobile phones.

# 2 Problem Description and First Model Assumptions

## 2.1 Surface Acoustic Qave Filters

We study a piezoelectric *surface acoustic wave* (*SAW, Rayleigh-wave*) device as used for frequency filtering in telecommunications. The main components of such devices are a piezoelectric substrate and two interdigital transducers (IDT) (see Fig. 1). Such an IDT is a comb of electrodes evaporated on the top surface of the piezoelectric crystal. Due to the underlying piezoelectric substrate an IDT transforms an alternating voltage into mechanical deformations. An acoustic wave can be excited. Vice versa, mechanical vibrations of the substrate evoke surface charges on the electrodes. An electric signal can be measured at the receiving IDT.

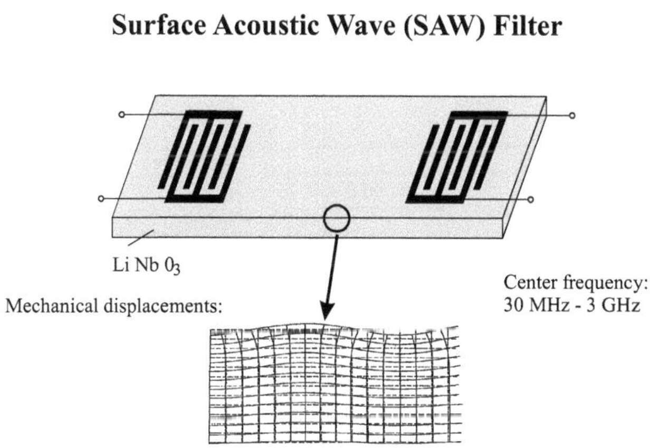


**Fig. 1.** Principal SAW filter consisting of piezoelectric substrate and input/output IDTs [15].

We focus on periodic SAW-filters where frequency filtering is achieved by periodic arrangement of electrodes on the surface of the piezoelectric substrate. If an acoustic wave propagates on the surface through the periodic structure, it is partially reflected at each electrode. Depending on the excitation frequency of the acoustic wave the reflected parts interfere constructively or not. If there is huge number of electrodes and the reflections interfere constructively, the wave propagation is prohibited, although the reflections at each electrode are very small. This effect occurs in whole frequency bands, so called band-gaps or stop-bands.

Numerical simulation of the full three-dimensional device is not reasonable. We already perform some model reduction on the geometric domain based on physical considerations: We denote the direction of periodicity by $(x)$, the surface normal direction by $(y)$ and their perpendicular direction by $(z)$. The dimensional extension of electrodes in $(z)$ – direction is huge in comparison to the periodicity. Moreover, we assume homogenous material topology in $(z)$ – direction. We are mainly interested in the propagation of Rayleigh-waves and their interaction with the periodic structure. These waves live near the surface, the amplitude decreases rapidly within depth and becomes negligibly small within the depth of a few wavelengths.

In general, surface waves are three-dimensional, but the relevant Rayleigh-waves depend only on the sagittal plane, *i.e.* the plane spanned by the direction of propagation and the surface normal. Thus, the mechanical and electric fields only depend on $x$ and $y$ coordinates. We can restrict the computational geometry to two dimensions.

In practical SAW devices the IDTs consist of some hundreds up to some thousands of electrodes. Therefore, extending the electrodes periodically to infinity is a suitable approximation.

We choose the infinite 2-dimensional domain which is periodic in the $x$-direction to model the piezoelectric substrate with a huge amount of periodically arranged electrodes. See Fig. 2.

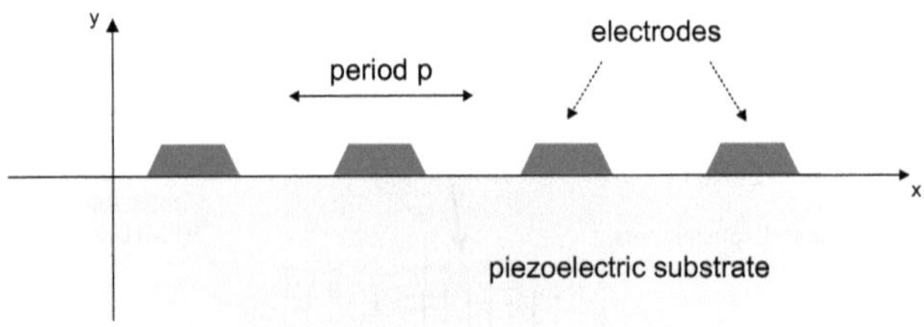


**Fig. 2.** 2D periodic geometry

## 2.2 Quasi-periodic Wave Propagation and the Dispersion Diagram

We will see that in periodic structures the mechanical deformation $u(x,t)$ and the electric potential $\Phi(x,t)$ of surface acoustic waves can be decomposed into quasi-periodic Bloch-waves of the form

$$u(x,t) = e^{i\omega t} e^{(\alpha+i\beta)x} u_p(x), \qquad \Phi(x,t) = e^{i\omega t} e^{(\alpha+i\beta)x} \Phi_p(x)$$

with the $p$-periodic functions $u_p$, $\Phi_p$. The wave-propagation can be described by the functional context between the frequency $\omega$ and the propagation parameter $\alpha + i\beta$, which is of great interest for engineers designing SAW-filters. The aim of this work is the full calculation of the dispersion diagram, which gives the relation between $\omega$, and the attenuation $\alpha$ and the phase shift $\beta$ in each periodic cell.

We can observe several wave modes in the dispersion diagram (see Fig. 3): Surface waves belonging to pass- and stop-bands, but also bulk waves which

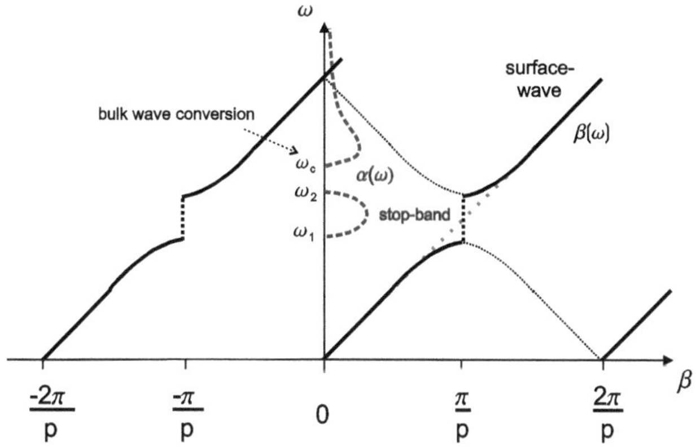


**Fig. 3.** Dispersion diagram: structure with periodical arranged electrodes

are volume waves. For surface wave propagation the frequency domain is classified in pass-bands and stop-bands as follows:

Wave propagation occurs below the lower stop-band edge $\omega_1$. Surface waves can pass the periodic structure undamped, *i.e.* they belong to the pass-band.

For stop-band frequencies $\omega \in (\omega_1, \omega_2)$ the wave reflections occurring at each electrode, interfere constructively. The wave gets exponentially damped $(\alpha \neq 0)$.

Above a certain frequencies $\omega_c$ also bulk waves are excited by IDTs. A small damping coefficient $\alpha$ is introduced, by the lack of energy into the material caused by bulk waves. This effect is called "bulk wave conversion" and can only be simulated if the model includes wave absorption of the material.

The dotted straight line in Fig. 3 shows the dispersion context in homogenous materials, where no stop-band effects occur since there are no interfering reflections.

## 3 The Piezoelectric Equations

Piezoelectric materials are characterized by the following two effects. The *direct piezoelectric effect* states that a mechanical deformation of a piezoelectric substrate evokes an electric field, which can be measured by charges on the surface. The effect is reversible: a piezoelectric crystal shrinks or stretches, if it is exposed to an electric field (*converse piezoelectric effect*). These phenomena result from special asymmetries occurring in some crystalline materials (e.g. in quartz by nature or in industrial produced ceramics). These effects cannot exist in isotropic media, *i.e.* piezoelectric materials are always anisotropic. To gain the piezoelectric equations we have to combine electrostatics and elastodynamics. We state the equations in the three-dimensional space.

*Elasticity*

For an impressed volume force density $f(x)$ the elastic equation of motions states that the mechanical displacement $u$ and the mechanical stresses $T$ are related as

$$\frac{\partial^2 u}{\partial t^2}(x,t) - \mathrm{div}_x T = f(x). \tag{1}$$

The elastic strains $S$ are defined by the geometrical properties

$$S = \frac{1}{2}(\nabla u + (\nabla u)^t). \tag{2}$$

*Electrostatics*

The electric field intensity $E$ can be expressed by an electric potential $\Phi$ as

$$E = -\nabla \Phi. \tag{3}$$

Piezoelectric materials are insulators, *i.e.* there are no free volume charges. Therefore electrostatics gives

$$-\mathrm{div} D = 0 \tag{4}$$

for the dielectric displacement vector $D$.

*Piezoelectric material laws*

We assume a linear piezoelectric coupling of elastic and electric fields, since nonlinear coupling terms are negligible small. Extending Hook's law and the electrostatic equation for the dielectric displacement by the direct or respectively the converse piezoelectric effect yields

$$\begin{aligned} T_{ij} &= c_{ijkl} S_{kl} - e_{kij} E_k, \\ D_i &= e_{ijk} S_{jk} + \varepsilon_{ik} E_k, \end{aligned} \tag{5}$$

where $c$ denotes the mechanical stiffness tensor, $\varepsilon$ the dielectric permittivity tensor, $e$ the piezoelectric coupling coefficient tensor.

We point out that the mechanical stiffness matrix and the permittivity matrix are symmetric. Since the direct and converse piezoelectric effect are symmetric, the coupling coefficients are equal for both effects. Due to symmetry considerations we can reduce the four material tensors: $c$ to a $6 \times 6$ symmetric matrix, $\varepsilon$ to a $3 \times 3$ symmetric matrix and $e$ to a $6 \times 3$ matrix. We refer the interested reader to [3] for more details on piezoelectric equations.

## 4 A Scalar Model Problem

To get a better insight into the problem of wave propagation in periodic media and to construct methods for the computation of dispersion diagrams we start with a scalar model problem. We consider the scalar wave equation with periodic coefficients. By the periodic arrangement of the cells $\Omega_k^p = [kp,(k+1)p] \times [0,H]$ we derive the strip $\Omega := \bigcup_{k=-\infty}^{\infty} \Omega_k^p$, which is periodic in $(x_1)$. This will be the underlying geometry modeling the infinite periodic domain (see Fig. 4). We search for general solutions $u(x,t)$ of the scalar wave equation

$$\begin{aligned} \tfrac{\partial^2 u}{\partial t^2}(x,t) - \mathrm{div}_x(a(x)\nabla_x u(x,t)) &= 0 \text{ on } \Omega, \\ a(x)\tfrac{\partial u}{\partial n}(x,t) &= 0 \text{ on } \Gamma_N, \\ u(x,t) &= 0 \text{ on } \Gamma_D. \end{aligned} \tag{6}$$

Since we are interested in the structure of the solution space, we state no initial conditions. The positive coefficient function $a$ describes the periodical properties of the material in $x_1$-direction, *i.e.*

$$a(x_1+p, x_2) = a(x_1, x_2) \qquad \forall\, (x_1,x_2) \in \Omega. \tag{7}$$

The classical formulation requires higher regularity on the coefficients and on the solutions. With regard to the weak formulation derived later we assume the periodic coefficient $a$ to be positive and piecewise constant. Moreover, the arrangement of $\Gamma_N$ and $\Gamma_D$ is assumed to coincide with the periodic nature of the domain, as shown in Fig. 4. Note that we state no radiation conditions in $x_1$-direction.

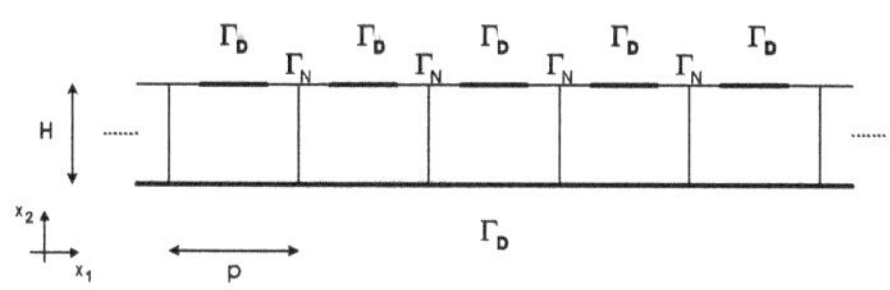


**Fig. 4.** Infinite periodical cluster 2D ($\Omega$)

We can separate the time-dependency by shifting the problem to the frequency domain. Therefore, we apply the *time-harmonic ansatz*

$$\hat{u}(x,t) = \hat{u}(x)\, e^{i\omega t}. \tag{8}$$

Form now on we suppress the hat-marker for the complex function $\hat{u}(x) \equiv u(x)$ and agree that to obtain physical results we have to consider the real parts afterwards. The wave-equation (6) transforms to the following eigenvalue problem with periodic-coefficient $a(.)$:

Find the complex-valued eigensolutions $u$ and eigenvalues $\omega \geq 0$ of

$$-\mathrm{div}(a(x)\nabla u(x)) \;=\; \omega^2 u(x) \quad \forall x \in \Omega. \tag{9}$$

### 4.1 Bloch's Theorem and the Quasi-Periodic Unit-Cell Problem

The problem above states an eigenvalue problem with periodic coefficients in an unbounded domain. Bloch-Floquet theory deals with the analysis of partial differential operators with periodic coefficients.

**Bloch theorem on the spectra of periodic operators**

*We assume the hermitian partial differential operator* $A : C^2(\Omega, \mathbb{C}) \to C(\Omega, \mathbb{C})$ *to be invariant w.r.t. translations* $T_p$ *of length* $p$ *in* $x_1$*-direction,* i.e.

$$T_p A = A T_p \quad \textit{with} \quad T_p : f(.,.) \to \quad f(. + p, .) \,.$$

*For every m-dimensional eigenspace* $\mathcal{E}_A(\lambda) := \{v | \, Av = \lambda v\}$*, there exists a set of* **Bloch waves** $(\varphi_i)_{1 \le i \le m}$ *spanning* $\mathcal{E}_A(\lambda)$*, i.e. satisfying*

$$A\varphi_j = \lambda \varphi_j \quad \textit{and} \ \exists \alpha_j, \beta_j \in \mathbb{R} : \ T_p \varphi_j = e^{(\alpha_j + i\beta_j)p} \varphi_j. \tag{10}$$

Lions [6] deals with elliptic operators, but restricts the solution space to the case $\alpha = 0$. The general case is treated in Kuchment [13].

Our problem requires the calculation of Bloch waves solving (9), which are assumed to be *quasi-periodic* in $x_1$-direction,

$$\exists \, \alpha, \beta \in \mathbb{R} \, \forall \, (x_1, x_2) \in \Omega \, : \, u(x_1, x_2) = u_p(x_1, x_2) e^{(\alpha + i\beta)x_1} \tag{11}$$

with $u_p$ being periodic, i.e. $u_p(x_1 + p, x_2) = u_p(x_1, x_2) \ \forall \, (x_1, x_2) \in \Omega$,
or equivalently

$$\exists \, \alpha, \beta \in \mathbb{R} \, \forall \, (x_1, x_2) \in \Omega \, : \, u(x_1 + p, x_2) = u(x_1, x_2) e^{i(\alpha + i\beta)p}. \tag{12}$$

Apparently quasi-periodic Bloch waves are fully described by

- a periodic function $u_p(\cdot)$,
- the complex propagation constants $\alpha + i\beta$.

Bloch's theorem justifies a reduction of the infinite problem to one single cell. We choose $\Omega_0^p$ and refer to the quasi-periodicity of the Bloch-waves by introducing quasi-periodic boundary conditions on the interfaces $\Gamma_L = \partial\Omega_{-1}^p \cap \partial\Omega_0^p$, $\Gamma_R = \partial\Omega_0^p \cap \partial\Omega_1^p$.

We state the **quasi-periodic unit-cell problem** in strong form as

$$-\text{div}(a \nabla u) = \omega^2 u \quad \text{in } \Omega_0^p \tag{13}$$

$$u = 0 \qquad \text{on } \Gamma_{0,D} \tag{14}$$

$$a \frac{\partial u}{\partial n} = 0 \qquad \text{on } \Gamma_{0,N} \tag{15}$$

$$\gamma \, u(x_1, x_2) = u(x_1 + p, x_2) \quad \text{for } (x_1, x_2) \in \Gamma_L \tag{16}$$

$$-\gamma \, a(x_1, x_2) \frac{\partial u}{\partial n_l}(x_1, x_2) = a(x_1 + p, x_2) \tfrac{\partial u}{\partial n_r}(x_1 + p, x_2) \quad \text{for } (x_1, x_2) \in \Gamma_L, \tag{17}$$

where $\gamma := e^{(\alpha + i\beta)p}$ and $n_l$,$n_r$ denote the outer normal vectors on $\Gamma_L$ and $\Gamma_R$, respectively.

### 4.2 The Mixed Variational Formulation

The variational formulation includes the real-valued or complex-valued Sobolev-spaces

$H^1(\Omega_0^p) := \{u \mid \int_{\Omega_0^p} |u|^2\,dx + \int_{\Omega_0^p} |\nabla u|^2\,dx < \infty\}$ and

$H^1_{0,D}(\Omega_0^p) := \{u \in H^1(\Omega_0^p) \,|u = 0 \text{ on } \Gamma_D\}$. The weak formulation of (13)-(15) in $H^1_{0,D}$ gives

$$\int_{\Omega_0^p} \nabla u \nabla v\,dx - \omega^2 \int_{\Omega_0^p} uv\,dx + \int_{\Gamma_L} a\frac{\partial u}{\partial n} v\,ds + \int_{\Gamma_R} a\frac{\partial u}{\partial n} v\,ds = 0,$$

where we assume $a \in L_\infty(\Omega_0^p)$. The incorporation of the quasi-periodic boundary conditions (16)–(17) is done by a mixed formulation. First, we identify $\Gamma_R$ and $\Gamma_L$ by a reference boundary $\Gamma$. Second, we define the trace-operators for the restriction of left and right boundary, but with respect to the reference boundary $\Gamma$, especially the superposition of the trace operator on $\Gamma_l$ or $\Gamma_r$ and the identification of the boundaries with $\Gamma$:

$$\begin{array}{rccc} tr_l : H^1(\Omega) & \rightarrow & H^{\frac{1}{2}}(\Gamma) \\ u & \mapsto & u_l \end{array}, \quad \begin{array}{rccc} tr_r : H^1(\Omega) & \rightarrow & H^{\frac{1}{2}}(\Gamma) \\ u & \mapsto & u_r \end{array}.$$

Third, by introducing a new unknown for the normal-derivative with respect to $\Gamma$

$$\lambda := a\frac{\partial u}{\partial n_l} \quad \in H^{-\frac{1}{2}}(\Gamma)$$

we can reformulate the weak formulation of (13)–(17) as non-symmetric **mixed variational formulation on the unit cell**:

Find $(u, \lambda)$ in $H^1(\Omega) \times H^{-\frac{1}{2}}(\Gamma)$ such that

$$\begin{aligned} \int_{\Omega_0^p} a\nabla u \nabla v\,dx - \omega^2 \int_{\Omega_0^p} uv\,dx + < tr_l v - \gamma\, tr_r v, \lambda > &= 0 \;\; \forall v \in H^1(\Omega), \\ < tr_r u - \gamma\, tr_l u, \mu > &= 0 \;\; \forall \mu \in H^{-\frac{1}{2}}(\Gamma). \end{aligned} \tag{18}$$

We used the duality product on $\Gamma$ denoted by $< .,. > := < .,. >_{H^{\frac{1}{2}}(\Gamma)\times H^{-\frac{1}{2}}(\Gamma)}$. For regular functions this coincides with the $L_2$-inner-product. The normal derivative $\lambda$ takes the role of a *Lagrange-parameter.*

### 4.3 The Frequency-Dependent Eigenvalue Problem

In the mixed variational problem (18) we are interested in possible solutions $(u, \lambda)$ in combination with the parameter-dependence on $\omega$ and $\gamma$. There are two possibilities to extract a parameter-dependent eigenvalue problem:

1. *Find all eigensolutions $(u, \lambda)$ of (18) with positive eigenvalues $\omega^2$ depending on the parameter $\gamma$.* If we want to calculate the whole dispersion context, the EVP has to be stated depending on a complex parameter $(\alpha + i\beta)$, i.e. two real parameters. This approach is suitable if we state the problem only for pass-bands, i.e. $\gamma = e^{i\beta}$.

2. *Find all eigensolutions* $(u, \lambda)$ *of (18) with eigenvalues* $\gamma \in \mathbb{C}$ *depending on the real-valued frequency* $\omega$. Since we are interested in general complex-propagation parameters $\alpha + i\beta$, we choose this frequency-dependent approach.

Defining the frequency-dependent bilinear form

$$k_\omega(u,v) := \int_{\Omega_0^p} a\nabla u \nabla v \, \mathrm{d}x - \omega^2 \int_{\Omega_0^p} uv \, \mathrm{d}x, \tag{19}$$

we get an **abstract version of the non-symmetric frequency-dependent eigenvalue-problem** for the quasi-periodic unit-cell problem:

Find eigenfunctions $(u, \lambda) \in H^1_{0,D}(\Omega_p^p) \times H^{-\frac{1}{2}}(\Gamma)$ referring to the eigenvalue $\gamma \in \mathbb{C}$

$$\begin{aligned} k_\omega(u,v) \quad & + < (tr_l - \gamma\, tr_r)v, \lambda > = 0 \;\forall v \in H^1_{0,D}(\Omega_0^p) \\ < (tr_r - \gamma tr_l)u, \mu > \quad & = 0 \;\forall \mu \in H^{-\frac{1}{2}}(\Gamma) \end{aligned} \tag{20}$$

dependent on the frequency $\omega \in \mathbb{R}^+$.

### 4.4 Galerkin-Discretization of the Frequency-Dependent EVP

We assume a Galerkin-discretization $V_h \subset H^1_{0,D}(\Omega_0^p)$ by $H^1$-conforming finite elements. The choice of a finite element base for $H^{-\frac{1}{2}}(\Gamma)$ is more challenging. If we consider a general discretization of the right and the left boundary we are faced with the discretization of the dual space for the Lagrange-multiplier. This can be done by Mortar finite elements as suggested in [7, 24].

If we use periodic meshes, in the sense that the left and the right boundary are discretized equivalently, we can avoid the assembling of the FE-space for the Lagrange-parameter and simply use nodal constraints on the boundary. In that case, the degrees of freedom are directly connected and so the discrete matrices of the trace-operators are simply identity matrices.

We define the discretized system matrix $K_\omega := [K_{\omega,jk}] = [k_\omega(\varphi_k, \varphi_j)]$ for an $H^1$-conforming finite-element base $\{\varphi_j\}$ spanning $V_h$. The FE-discretization of (20)

$$\begin{pmatrix} K_\omega & Tr_l^{\,t} \\ Tr_r & 0 \end{pmatrix} \begin{pmatrix} u_h \\ \lambda_h \end{pmatrix} = \gamma \begin{pmatrix} 0 & Tr_r^{\,t} \\ Tr_l & 0 \end{pmatrix} \begin{pmatrix} u_h \\ \lambda_h \end{pmatrix} \tag{21}$$

We classify the degrees of freedom corresponding to the left ($l$), the right ($r$) boundary, and the remaining ones ("inner" degrees of freedom, $i$). The dimensions $n_l = n_r$, $n_i$ are defined coinciding with this classification and $\dim(V_h) =: n = n_i + 2 \cdot n_l$.

Considering the sparsity and the symmetry of the FE-matrices we arrive at a **parameter-dependent discretized generalized eigenvalue-system** of the following structure:

$$\left(\begin{array}{ccc|c} K_{\omega,ii} & K_{\omega,li}^T & K_{\omega,ri}^T & 0 \\ K_{\omega,li} & K_{\omega,ll} & 0 & I \\ K_{\omega,ri} & 0 & K_{\omega,rr} & 0 \\ \hline 0 & 0 & I & 0 \end{array}\right) \begin{pmatrix} u_i \\ u_l \\ u_r \\ \lambda \end{pmatrix} = \gamma \left(\begin{array}{ccc|c} 0 & 0 & 0 & 0 \\ 0 & 0 & 0 & 0 \\ 0 & 0 & 0 & I \\ \hline 0 & I & 0 & 0 \end{array}\right) \begin{pmatrix} u_i \\ u_l \\ u_r \\ \lambda \end{pmatrix}. \tag{22}$$

*Remark 1.* The generalized eigenvalue problem $Ax = \gamma Bx$ defined in (22) has the following properties:

1. The right-hand side matrix $B$ has a large kernel ($\dim(\ker B) = n_i + n_l$), which corresponds to infinite eigenvalues. There are $n_i + n_l$ infinite and $2n_l$ finite eigenvalues.
2. The eigenvalue-problem is *symplectic*, i.e. if $\gamma$ is an finite non-zero eigenvalue then $\frac{1}{\gamma}$ is also an eigenvalue. One can exploit and preserve the special structure by using structure-preserving computational methods as proposed by Merhmann in [18].
3. Concerning dispersion diagrams we are mainly interested in eigenvalues $\gamma = e^{(\alpha+i\beta)p}$ near the unit-circle, i.e. $|\gamma| \approx 1$.

### 4.5 A Model Improvement by Absorbing Boundary Conditions

So far we have used standard boundary conditions on the bottom boundary of the cell. Since we are interested in surface effects, we do not want to simulate the whole thickness of the underlying substrate, we cut the domain a few wavelengths away from the surface. The assumption of Dirichlet or Neumann boundary conditions is not suitable, since these types of artificial boundary introduce unnatural reflections. Moreover, damping effects in surface waves, caused by bulk wave radiation effects, are only possible in models including wave absorption into the substrate. These reflections can be avoided or at least minimized by the choice of absorbing boundary conditions (ABCs).

**First order absorbing boundary conditions** are introduced by complex-valued frequency-dependent Robin boundary conditions of the form

$$n^T(a\nabla u) = i\omega u \text{ on } \Gamma_{bot}.$$

This condition is exact for plane waves in outer normal direction $n$, but still leads to partial reflections for general plane waves. This approach leads to the complex-symmetric bilinear form

$$k_\omega^{\mathrm{ABC}}(u,\ v) := \int_{\Omega_0^p} a\nabla u \nabla v \,\mathrm{d}x + i\omega c(u,\ v) - \omega^2 \int_{\Omega_0^p} uv \,\mathrm{d}x \tag{23}$$

with $c(u,v) := \int_{\Gamma_{bot}} uv\, ds$.

Quite recently **the method of perfectly matched layers (PML)** became very popular. We do not want to go into a detailed description of this method. In order to construct solution methods which can be also applied to PML boundaries, we only point out its effect on the structure of the corresponding bilinear form $k_\omega$. Technically one introduces an artificial boundary layer in which the coefficients of the underlying PDE are extended into the complex plane. On the infinite level PML perfectly absorbs plane waves in any arbitrary direction. On the discrete level the quality of absorption can be controlled by the choice of the FE-discretization. By this approach the bilinear form extends to

$$k_\omega^{\mathrm{PML}}(u,v) := \int_{\Omega_0^p} \tilde{a} \nabla u \nabla v \, \mathrm{d}x - \omega^2 \int_{\Omega_0^p} \tilde{\rho} u v \, \mathrm{d}x \tag{24}$$

with complex-valued parameters $\tilde{a}$ and $\tilde{\rho}$. This bilinear form is again *complex-symmetric.*

*Remark 2.* By the choice of the proposed ABCs, the system-matrix $K_\omega$ in the generalized algebraic EVP (22) gets complex-valued and is complex-symmetric.

### 4.6 Solution Strategies

In this section we want to construct two strategies for the solution of the generalized algebraic EVP (22) for complex-valued and complex-symmetric matrices $K_\omega$.

We state two reduced eigenvalue problems which still have the same finite spectrum as the initial system. This is achieved by reducing infinite eigenvalues referring from the large kernel of the right-hand-side matrix in (22).

#### The Inner-Node-Matrix Method

Substituting first $u_r = \gamma u_l$ and then $\lambda = -K_{\omega,li} u_i - K_{\omega,ll} u_l$ leads to a generalized non-hermitian linear eigenvalue problem of the form

$$\begin{pmatrix} K_{\omega,ii} & K_{\omega,il} \\ K_{\omega,ir}^T & 0 \end{pmatrix} \begin{pmatrix} u_i \\ u_l \end{pmatrix} = \gamma \begin{pmatrix} 0 & -K_{\omega,ir} \\ -K_{\omega,il}^T & -K_{\omega,ll} - K_{\omega,rr} \end{pmatrix} \begin{pmatrix} u_i \\ u_l \end{pmatrix}. \tag{25}$$

We point out that in above problem none of the two matrices is regular nor symmetric, but by spectral transformation coinciding with $\mu := \frac{1}{\gamma - 1}$ we get the following equivalent problem:

Find eigenvectors $\begin{pmatrix} u_i \\ u_l \end{pmatrix} \in \mathbb{C}^{n_i + n_l}$ w.r.t. the eigenvalues $\mu = \frac{1}{\gamma-1}$:

$$\begin{pmatrix} 0 & -K_{\omega,ir} \\ -K_{\omega,il}^T & -K_{\omega,ll} - K_{\omega,rr} \end{pmatrix} \begin{pmatrix} u_i \\ u_l \end{pmatrix} = \mu \begin{pmatrix} K_{\omega,ii} & K_{\omega,il} + K_{\omega,ir} \\ K_{\omega,il}^T & K_{\omega,ll} + K_{\omega,rr} \end{pmatrix} \begin{pmatrix} u_i \\ u_l \end{pmatrix}. \tag{26}$$

The right-hand-side matrix is obviously regular and complex-symmetric. Moreover, all involved matrices are sparse.

An implementation of the implicitly restarted Arnoldi-algorithm is provided by the software-package ARPACK. The package includes an iterative solver for generalized non-hermitian eigenvalue $Av = \gamma Bv$ which only requires matrix-vector products and the application of the inversion. Therefore, the sparsity of the FE-matrices can be exploited. In each frequency step we have to perform a Sparse-Cholesky factorization of $B$.

### The Schur-Complement Method

We start with the already reduced system stated in (25) and take the Schur-complement with respect to the inner degrees of freedom. Using the classification in inner, left, right degrees of freedom we state the Schur-complement of $K_\omega$ as

$$S := -\left(K_{\omega,li}, K_{\omega,ri}\right) K_{\omega,ii}^{-1}(K_{il} - \gamma K_{ir}) + \begin{pmatrix} K_{\omega,ll} & 0 \\ 0 & K_{\omega,rr} \end{pmatrix}. \tag{27}$$

Substituting $u_i$ by $u_i = -K\omega, ii^{-1}(K_{il} - \gamma K_{ir})u_l$ in (25) we result in the following **frequency-dependent quadratic eigenvalue problem**:

Find eigenpairs $(\gamma, u_l) \in \mathbb{C} \times \mathbb{C}^{n_l}$ such that

$$\gamma^2 S_{lr} u_l + \gamma(S_{ll} + S_{rr})u_l + S_{lr}^T u_l = 0. \tag{28}$$

In each frequency step we first calculate the inverse of the sparse and complex-symmetric matrix $K_{\omega,ii}^{-1}$ by a Sparse-Cholesky-factorization and assemble the Schur-complement. The quadratic eigenvalue problem is tackled by linearization to a double-sized generalized eigenvalue problem, which is solved by the QZ-method implemented in LAPACK.

## 5 Piezoelectric Equations and Periodic Structures

In this section we want to combine the three main modeling steps,

- the underlying piezoelectric equations, which lead to a coupled field problem of saddle-point structure (indefinite, but symmetric),
- absorbing boundary conditions for acoustic waves in piezoelectric media in order to enable wave absorption of the substrate,
- acoustic wave propagation in periodic structures and its solution strategies.

Due to the governing piezoelectric equations mathematical modeling, analysis and solution strategies get more technical. One has to overcome some problems due to the indefinite saddle-point structure of piezoelectric equations. But the quasi-periodic problem results in a formally equivalent eigenvalue problem, which can be solved numerically with the methods introduced above.

### 5.1 2-D Geometry and Anisotropic Materials

At first, we adopt the three-dimensional piezoelectric equations given in (1)–(5) to the fact that surface waves only depend on the sagittal plain. This was the justification to reduce the geometry to the plain spanned up by the direction of surface wave propagation ($x_1$) and the normal onto the surface ($x_2$). Due to the anisotropic properties of the material general surface waves can polarize (particle motion) outside the sagittal plane. Even though all field quantities only depend on the ($x_1$, $x_2$) – plane, a mechanical deformation in ($x_3$) – direction is possible. Therefore, the equations for the elastic strain (2) and the electric field (3) simplify to

$$S(u) = \frac{1}{2}\Big(\nabla_{(x_1,x_2,x_3)} u(x_1,x_2) + \big(\nabla_{(x_1,x_2,x_3)} u(x_1,x_2)\big)^t\Big), \tag{29}$$

$$E = \nabla_{(x_1,x_2,x_3)} \Phi(x_1,x_2) = \left(\frac{\partial \Phi}{\partial x_1}, \frac{\partial \Phi}{\partial x_2}, 0)^t\right). \tag{30}$$

From now on we denote the equations (1),(29),(3),(29),(4) as the *governing piezoelectric equations for the three-dimensional mechanical displacement* $u = (u_1,\ u_2,\ u_3)^t$ *and the scalar potential* $\Phi$.

### 5.2 The Underlying Infinite Periodic Piezoelectric Problem

*The cell-based periodic model geometry*

The periodic geometry $\Omega$ can be described in terms of successive arrangement of a unit-cell $\Omega_0^p$ (with $\mathrm{diam}_{x_1}(\Omega_0^p) = p$) of an analogous structure as shown in Fig. 5. We denote the translation of this cell parallel to the $x_1$-axis as the $k$-th cell $\Omega_k^p := \Omega_k^p := \{y = (k.p,\ 0) + x | x \in \Omega_0^p\}$ and achieve a representation of an infinite periodic strip $\Omega$ by $\Omega := \bigcup_{k=-\infty}^{\infty} \Omega_k^p$.

Each cell basically consists of a piezoelectric substrate $\Omega_{k,S}$ with one evaporated electrode $\Omega_{k,E}$; these two domains are disjoint but matching. In numerical computation we will choose the model geometry shown in Fig. 4.

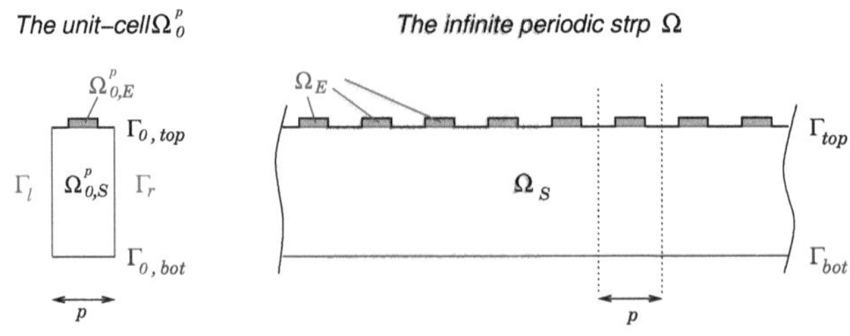


**Fig. 5.** Underlying cell-based periodic geometry

*The piezoelectric equations with periodic coefficients on $\Omega$*

Shifted to the frequency domain by a harmonic ansatz, the piezoelectric equations for the mechanical displacement $(u_1,\ u_2,\ u_3)$ and the scalar potential $\Phi$ are

$$\begin{aligned} -\mathrm{div}\Big(c : \big(\nabla u + (\nabla u)^t\big) + \varepsilon : \nabla\Phi\Big) &= \omega^2 \rho u \text{ in } \Omega, \\ -\mathrm{div}\Big(e^t : \big(\nabla u + (\nabla u)^t\big) - \varepsilon : \nabla\Phi\Big) &= 0 \qquad \text{in } \Omega, \end{aligned} \tag{31}$$

with underlying periodic structure $\Omega$ and periodic coefficient matrices $T_p\ c = c,\ T_p\ e = e,\ T_p\ \rho = \rho$. On the metallic electrodes $\Omega_E$ the piezoelectric coupling coefficient $e$ is set to zero. Concerning the boundary conditions we choose homogenous Dirichlet boundary condition for the potential on $\Gamma_E := \partial\Omega \cap \partial\Omega_E =: \Gamma_D$ in order to model short-circuited electrodes. The remaining top-surface boundary is assumed to be charge-free. Concerning the mechanical field the whole top-surface boundary is assumed to be stress-free. Therefore, the following boundary conditions are claimed for (31)

$$\begin{array}{lrl} \text{short-circuited electrodes} & \Phi = 0 & \text{on } \Gamma_E := \partial\Omega \cap \partial\Omega_E, \\ \text{stress-free:} & n^t.T = 0 & \text{on } \Gamma_{top}, \\ \text{charge-free:} & n^t.D = 0, & \text{on } \Gamma_{top} \backslash \Gamma_E. \end{array} \tag{32}$$

$$\text{Absorbing BCs on } \Gamma_{bot} \tag{33}$$

with normal stresses $n^t.T := n^t\big(c : \big(\nabla u + (\nabla u)^t\big) + e : \nabla\Phi\big)$ and normal charges $n^t.D := n^t\big(e^t : \big(\nabla u + (\nabla u)^t\big) - \varepsilon : \nabla\Phi\big)$.

Solving the periodic problem again requires the computation of Bloch waves. Therefore, it can be restricted to a piezoelectric unit-cell problem with quasi-periodic boundary conditions for mechanical and electric field quantities. Analogous to the scalar model, we begin with the mathematical tools required for the piezoelectric unit-cell problem with standard-boundary conditions. The incorporation of the quasi-periodicity will be done in the second step.

For sake of simplicity we assume the charge- and stress-free boundary conditions

$$n^t.T = 0, \quad n^t.D = 0 \quad \text{on } \Gamma_{bot} \tag{34}$$

on the bottom boundary in the first stage of modeling.

### 5.3 Piezoelectric Equations in Weak and Discretized Form

Restriction of the time-harmonic piezoelectric equations stated in (31),(32),(34) onto the unit-cell $\Omega_o^p$ and its weak formulations yields the following eigenvalue problem.

Find eigensolutions $(u, \Phi) \in \big[H^1(\Omega_0^p)\big]^3 \times H^1_{0,D}(\Omega_0^p)$ corresponding to the eigenvalues $\omega^2$ such that $\forall v \in H^1(\Omega_0^P)^3 := \big[H^1(\Omega_0^P)\big]^3$, $\forall \Psi \in H^1_{0,D}(\Omega_0^p)$

$$\begin{aligned} \int_{\Omega_0^p} (Bv)^T : c\,Bu \quad &+ \int_{\Omega_0^p} (S(v))^t : e^t \nabla \Phi \,\mathrm{d}x = \omega^2 \int_{\Omega_0^p} \rho\, v^t u \,\mathrm{d}x \\ \int_{\Omega_0^p} (\nabla \Psi)^t : e\,Bu \,\mathrm{d}x \; &- \int_{\Omega_0^p} (\nabla \Psi)^t \varepsilon \,\nabla \Phi \;\mathrm{d}x \;= 0 \end{aligned} \tag{35}$$

Due to the large kernel in the right hand side the eigenvalue problem is degenerated, which leads to infinite eigenvalues.

For the sake of simplicity we introduce the mechanical bilinear form $a_{uu}(u, v) := \int_{\Omega_0^p} S(v)^t : c\,S(u)\,\mathrm{d}x$, the piezoelectric coupling bilinear forms $a_{u\Phi}(u, \Phi) = a_{\Phi u}(\Phi, u) := \int_{\Omega_0^p} S^t(u) : e^t \nabla \Phi \,\mathrm{d}x$, the dielectric $a_{\Phi\Phi}(\Phi, \Psi) = \int_{\Omega_0^p} (\nabla \Psi)^t \varepsilon \,\nabla \Phi \,\mathrm{d}x$, and the mechanical mass bilinear form $m_{uu}(u, v) := \int_{\Omega_0^p} \rho\, v^t u \,\mathrm{d}x$.

*On the structure of piezoelectric discretized eigenvalue-problems*

The discretization of $H^1(\Omega_0^p)^3$ and $H^1(\Omega_0^p)$ with conforming finite elements yields a algebraic eigenvalue-problem of the special saddle-point structure

$$\begin{pmatrix} A_{uu} & A_{u\Phi} \\ A_{\Phi u} & -A_{\Phi\Phi} \end{pmatrix} \begin{pmatrix} u_h \\ \Phi_h \end{pmatrix} = \omega^2 \begin{pmatrix} M_{uu} & 0 \\ 0 & 0 \end{pmatrix} \begin{pmatrix} u_h \\ \Phi_h \end{pmatrix}. \tag{36}$$

The matrix blocks correspond to the mechanical, dielectric and piezoelectric bilinear forms. Therefore, the problem is symmetric since the sub-matrices satisfy $A_{uu} = A_{uu}^t, A_{\Phi\Phi} = A_{\Phi\Phi}^t, A_{\Phi u} = A_{u\Phi}^t, M_{uu} = M_{uu}^t$. The eigenvalue problem is degenerated. It possesses *$dim(\Phi_h)$ infinite eigenvalues.*

The Schur-complement with respect to the potential $\Phi_h$ yields

$$(A_{uu} + A_{u\Phi} A_{\Phi\Phi}^{-1} A_{\Phi u}) u_h = \omega^2 M_{uu} u_h,$$

which states a positive-definite eigenvalue problem. However, we will not pursuit this strategy, due to the computational costs for inverting $A_{\Phi\Phi}$.

### 5.4 The Quasi-Periodic Unit-Cell Problem

Due to Bloch's theorem (general version stated in [13]) we use Bloch waves

$$\begin{aligned} u(x_1, x_2) &= u_p(x_1, x_2) e^{(\alpha + i\beta) x_1} \text{ with } u_p \; p-\text{periodic in } x_1 \\ \Phi(x_1, x_2) &= \Phi_p(x_1, x_2) e^{(\alpha + i\beta) x_1} \text{ with } \Phi_p \; p-\text{periodic in } x_1 \end{aligned}$$

as ansatz for the eigenfunctions in the periodic piezoelectric eigenvalue problem (31)–(32) together with either (33) or (34).

Therefore, **the quasi-periodic unit-cell problem** is stated by (31),(32) and (33) or (34) restricted onto $\Omega_0^p$ together with the quasi-periodic boundary conditions

$$\begin{aligned} \gamma\, u(x_1, x_2) &= u(x_1 + p, x_2) && \text{for } (x_1, x_2) \in \Gamma_L, \\ -\gamma\, n^t \cdot T(x_1, x_2) &= n^t \cdot T(x_1 + p, x_2) && \text{for } (x_1, x_2) \in \Gamma_L, \\ \gamma\, \Phi(x_1, x_2) &= \Phi(x_1 + p, x_2) && \text{for } (x_1, x_2) \in \Gamma_L, \\ -\gamma\, n^t \cdot D(x_1, x_2) &= n^t \cdot D(x_1 + p, x_2) && \text{for } (x_1, x_2) \in \Gamma_L, \end{aligned} \tag{37}$$

with $\gamma := e^{(\alpha+i\beta)p}$.

Now, the solution strategy is formally equivalent to that presented for the scalar model problem. We interpret the quasi-periodic unit-cell problem as eigenvalue-problem for the propagation-constant $\gamma$ while depending on the frequency $\omega$.

We identify the quasi-periodic boundaries $\Gamma_l$ and $\Gamma_r$ with a reference boundary $\Gamma$. The corresponding trace operators $tr_l$ and $tr_r$ are defined as the composition of the standard $H^1$-trace operator onto $\Gamma_L$ and the boundary identification of $\Gamma_l$ or $\Gamma_r$ with $\Gamma$: $tr_l : H^1\Omega_p^0 \rightarrow H^{\frac{1}{2}}(\Gamma_l) \stackrel{id}{\rightarrow} H^{\frac{1}{2}}(\Gamma)$, and vice versa for $tr_r$. The trace-operator on the three-dimensional mechanical field $u$ in $H^1(\Omega_0^p)^3$ is defined component-wise as $tr_l u := (tr_l u_1,\ tr_l u_2,\ tr_l u_3)$ in $H^{\frac{1}{2}}(\Gamma)^3 := [H^{\frac{1}{2}}(\Gamma)]^3$ Furthermore, by introducing new unknowns for the normal fluxes on the left boundary with respect to $\Gamma$

$$\lambda := n^t \cdot T \in H^{-\frac{1}{2}}(\Gamma)^3 \quad \text{and} \quad \zeta := n^t \cdot D \in H^{-\frac{1}{2}}(\Gamma), \tag{38}$$

we result in the **frequency-dependent mixed variational formulation**:

Find eigensolutions $(u,\ \Phi,\ \lambda,\ \zeta)$ corresponding to eigenvalues $\gamma \in \mathbb{C}$ with $(u,\ \Phi,\ \lambda,\ \zeta) \in H^1(\Omega_0^p)^3 \times H^1_{0,D}(\Omega_0^p) \times H^{-\frac{1}{2}}(\Gamma)^3 \times H^{-\frac{1}{2}}(\Gamma)$ such that $\forall v \in H^1(\Omega_0^p)^3, \forall \Psi \in H^1_{0,D}(\Omega_0^p), \forall \mu \in H^{-\frac{1}{2}}(\Gamma)^3, \forall \nu \in H^{-\frac{1}{2}}(\Gamma)$

$$\begin{aligned}
&a_{uu}(u,v) + a_{u\Phi}(\Phi,v) - \omega^2 m(u,v) + < (tr_l - \gamma\, tr_r)v, \lambda > &= 0\\
&a_{\Phi u}(u,\Psi) - a_{\Phi\Phi}(\Phi,\Psi) \qquad\qquad\quad + < (tr_l - \gamma\, tr_r)\Psi, \zeta > &= 0\\
&< (\gamma\, tr_l - tr_r)u, \mu > &= 0\\
&< (\gamma\, tr_l - tr_r)\Phi, \nu > &= 0
\end{aligned} \tag{39}$$

is satisfied for given parameters $\omega^2$.
$< .,. >_{H^{\frac{1}{2}}(\Gamma)\times H^{-\frac{1}{2}}(\Gamma)}$, respectively.

Again, the introduced unknowns $\lambda,\ \zeta$ for the normal fluxes on $\Gamma_l$ with respect to $\Gamma$ take the role of **Lagrange-multipliers**.

To gain a compact formalism we agree on the abbreviations $\tilde{u} := (u, \Phi) \in H^1_{0,D_4}(\Omega_0^p)^4 := H^1(\Omega_0^p)^3 \times H^1_{0,D}(\Omega_0^p)$, and $\tilde{v} := (v, \Psi)$, and on the frequency-dependent piezoelectric bilinear form

$$\begin{aligned}
k^\omega(\tilde{u},\tilde{v}) &:= k^\omega\big(\ (u,\Phi),\ (v,\Psi)\ \big)\\
&:= a_{uu}(u,v) + a_{u\Phi}(\Phi,v) - \omega^2 m(u,v)\\
&\quad + a_{\Phi u}(u,\Psi) - a_{\Phi\Phi}(\Phi,\Psi).
\end{aligned} \tag{40}$$

An abstract version of the non-symmetric frequency-dependent eigenvalue problem for the quasi-periodic unit-cell problem can be stated.

Find eigensolutions $(\tilde{u}, \tilde{\lambda}) \in H^1_{0,D}(\Omega_0^p)^4 \times H^{-\frac{1}{2}}(\Gamma)^4$ corresponding to eigenvalues $\gamma \in \mathbb{C}$ such that

$$\begin{aligned} k_\omega(\tilde{u}, \tilde{v}) \qquad & + < \tilde{\lambda}, (tr_l - \gamma\, tr_r)\tilde{v} > = 0 \;\forall \tilde{v} \in H^1(\Omega_0^p)^4 \\ < (\gamma\, tr_l - tr_r)\tilde{u}, \tilde{\mu} > \qquad & = 0 \;\forall \tilde{\mu} \in H^{-\frac{1}{2}}(\Gamma)^4 \end{aligned} \tag{41}$$

is satisfied for given parameters $\omega^2$.
The duality-product $< .,. >$ refers to $< .,. >_{H^{\frac{1}{2}}(\Gamma)^4 \times H^{-\frac{1}{2}}(\Gamma)^4}$.

*Model extension to absorbing boundary conditions*

In case of piezoelectric equations absorbing boundary conditions are a bit challenging. The degeneration of the frequency-dependent eigenvalue problem causes some technical difficulties. However, we only state the formal characteristics of the extended system bilinear forms

$$k_\omega^{\mathrm{ABC}} := a\big((u,\Phi),(v,\Psi)\big) + i\,\omega\, c\big((u,\Phi),(v,\Psi)\big) - \omega^2 m\big((u,\Phi),(v,\Psi)\big), \tag{42}$$

$$k_\omega^{\mathrm{PML}} := \tilde{a}\big((u,\Phi),(v,\Psi)\big) - \omega^2\, \tilde{m}\big((u,\Phi),(v,\Psi)\big). \tag{43}$$

The absorbing bilinear form $c(.,.)$ is positive-definite. The complex-valued PML-bilinear forms $\tilde{a}(.,.)$ and $\tilde{m}(.,.)$ are complex-symmetric.

*The discretized eigenvalue problem*

Analogous to the scalar case, we assume matching meshes on the left and the right boundary. Therefore, a discretization of $H^{-\frac{1}{2}}(\Gamma)^4$ by Mortar-Elements can be avoided. We can use nodal constraints for the Lagrange-parameter and the discrete trace-operators corresponding to $tr_l$ and $tr_r$ simplify to identity matrices.

Discretization of $H^1_{0,D}(\Omega_0^p)$ for the frequency-dependent piezoelectric bilinear form is done in the way already described for (36). Galerkin-discretization of (41) leads to **parameter-dependent discretized generalized eigenvalue-system** (compare with (22))

$$\left(\begin{array}{ccc|c} K_{\omega,ii} & K_{\omega,li}^T & K_{\omega,ri}^T & 0 \\ K_{\omega,li} & K_{\omega,ll} & 0 & I \\ K_{\omega,ri} & 0 & K_{\omega,rr} & 0 \\ \hline 0 & 0 & I & 0 \end{array}\right) \begin{pmatrix} \tilde{u}_i \\ \tilde{u}_l \\ \tilde{u}_r \\ \lambda \end{pmatrix} = \gamma \left(\begin{array}{ccc|c} 0 & 0 & 0 & 0 \\ 0 & 0 & 0 & 0 \\ 0 & 0 & 0 & I \\ \hline 0 & I & 0 & 0 \end{array}\right) \begin{pmatrix} \tilde{u}_i \\ \tilde{u}_l \\ \tilde{u}_r \\ \lambda \end{pmatrix}, \tag{44}$$

where each $\tilde{v^i}$ refers to 4 degrees of freedom $(u_1^i, u_2^i, u_3^i, \Phi^i)$ and each classified $(i,l,r)$ matrix block is of the following (complex)-symmetric saddle point structure

$$K_{\omega,\alpha,\beta} = \begin{pmatrix} K_{\omega,\alpha,\beta,\,uu} & K_{\omega,\alpha,\beta,\,\Phi u}^T \\ K_{\omega,\alpha,\beta,\,\Phi u} & -K_{\omega,\alpha,\beta,\,\Phi\Phi} \end{pmatrix} \qquad \text{for } \alpha,\beta \in \{i,l,r\}. \tag{45}$$

Standard boundary conditions on the bottom leads to real-valued matrices, absorbing ones to complex-valued ones. We agreed on suppressing the $h$-subscript denoting the discrete level.

Due to the abstract formulation of the scalar and the piezoelectric eigenvalue problem we can apply the solution strategies of the scalar model, i.e. the Inner-Node-Matrix method or the Schur-Complement method in Subsection 4.6.

# 6 Numerical Results

In case of a 2-dimensional geometry the implementation of the Schur-Complement method together with a Sparse-Cholesky-Factorization is suitable. However, if one thinks about simulations on 3-dimensional geometries, one have to perform the Inner-Node-Matrix method.

The Schur-Complement method is implemented in the high order FE-Solver NGSolve [22] using an LAPACK eigenvalue solver (`zgeev, dggev`) [1].

## 6.1 The Scalar Model Problem

We use the scalar model problem to examine the specific influence of periodic perturbation on surface wave propagation. Therefore, we determine the dispersion context for 3 different problem types based on the geometry shown in Fig. 4:

1. Wave propagation in *homogenous media*, where we assume homogenous Neumann BCs on the top and the surface ($\Gamma_D = \emptyset$, $\Gamma_N := \Gamma_{top} \cup \Gamma_{bot}$) (see Fig. 6).
2. Wave propagation in *periodic media*, where periodic perturbations are simulated by periodically arranged homogenous Dirichlet- and Neumann-BCs on the top surface (see Fig. 4). The homogenous Dirichlet conditions is used as imitation of short-circuited electrodes, where a vanishing potential can be assumed (see Fig. 7).
3. Wave propagation in *periodic media* with first order *absorbing boundary conditions* on the bottom surface $\Gamma_{bot}$. The periodic structure is modeled as described in item 2. See Fig. 8.

In the three following dispersion diagrams complex propagation constants which belong to pass-bands are drawn in gray, those belonging to stop-bands in black. These diagrams include both bulk waves and surface waves. The classification can be performed by examining the corresponding eigenvectors.

In the homogenous case, there are no stop-bands. We gain pure imaginary propagation constants $i\beta$ corresponding to continuous pass-bands.

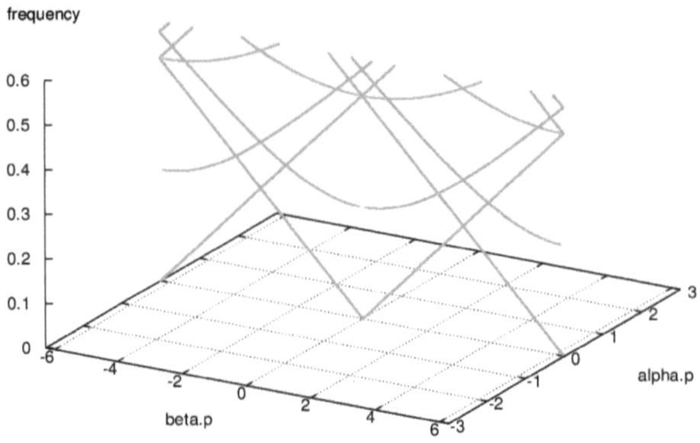


**Fig. 6.** Scalar model: dispersion relation in a homogenous structure.

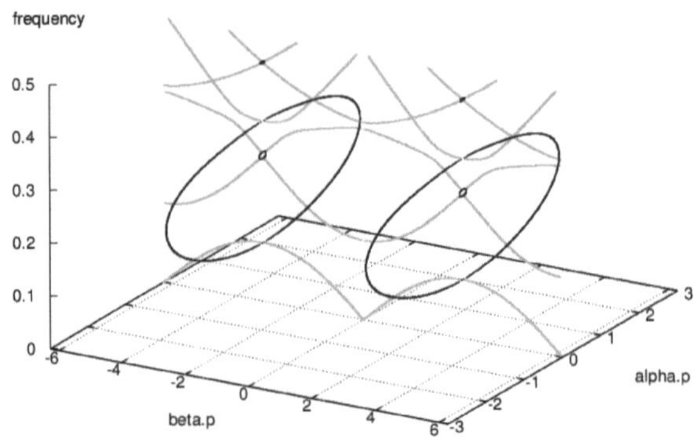


**Fig. 7.** Scalar model: Dispersion relation in a periodic structure.

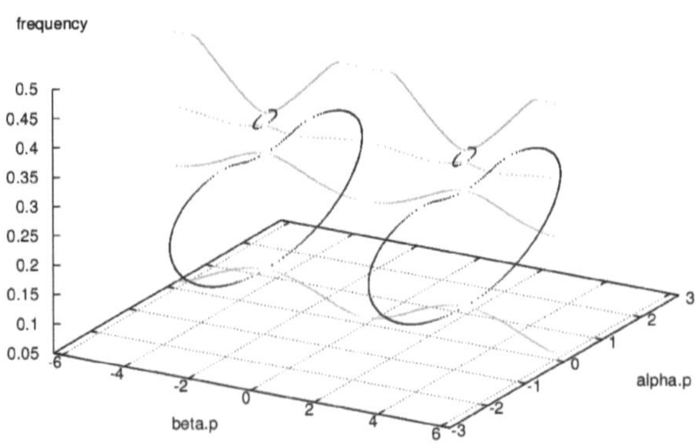


**Fig. 8.** Scalar model: Dispersion relation in a periodic structure with ABC.

## 6.2 Simulation of a Piezoelectric Periodic Structure

Since the dispersion diagram gives information on many parameters of wave propagation, which are used in other models and simulations, we want to determine the eigenvalues very accurately. Our main aspects are frequency domains where the dimension of the unit-cell is in the range of half the wavelength. Therefore, higher order polynomials should approximate these waves very accurately even for coarse meshes. However, the entering corners of the electrodes and the jumping coefficients cause singularities in the solution. These singularities cannot be resolved simply by increasing the polynomial order of the ansatz functions, but only by a special local mesh-refinement denoted as *hp-refinement*. Both methods consist of two main steps. First the computation of an inverse (SC-method) or respectively a Sparse-Cholesky decomposition. Second the solution of an eigensystem, here the decrease of degrees of freedom is very important for decreasing computational times.

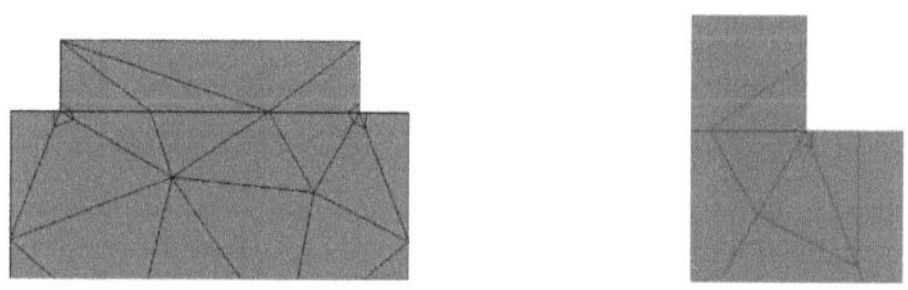

**Fig. 9.** Special local refinement at singularities

We simulate the dispersion context of a TV-filter structure as used in practice with Lithium-Niobate substrate and aluminum short-circuited electrodes. The topology is chosen as shown in Fig. 4. On the bottom first order absorbing boundary conditions are assumed. We used 52 elements of polynomials order $p = 4$ and an hp-refinement of 3 levels, which results in $4 \cdot 609$ degrees of freedom. Fig. 10 shows a two-dimensional plot of the dispersion context near the stop-band of the chosen filter structure. On the left the context between the frequency and the attenuation-constant $\alpha$ per cell is drawn, while on the

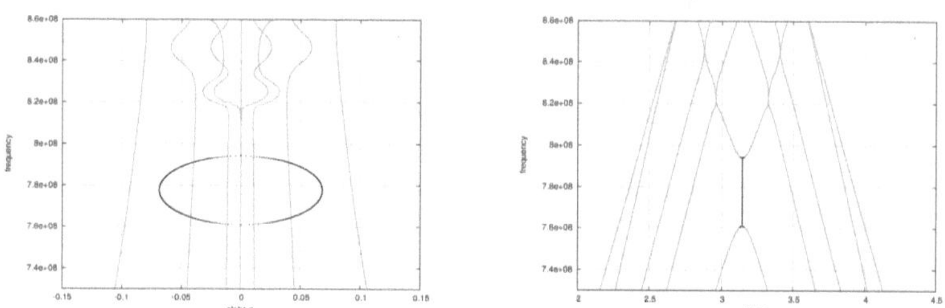


**Fig. 10.** Dispersion context of piezoelectric structure with periodic arranged electrodes.

right the context between frequency and phase shift $\beta$ in each cell. Above the upper stop-band edge we can observe an increased attenuation caused by bulk wave radiation, which is enabled by absorbing boundary conditions at the bottom and does not occur in simulations including only standard boundary conditions.

## 7 Conclusions

We gave a full detailed modeling for piezoelectric surface acoustic wave filters. We started with developing mathematical tools for periodic structures for the scalar wave equation. In order to reduce the computation domain while allowing wave absorption we introduced absorbing boundary conditions at the artificial bottom boundary. By an abstract formulation we achieved that the developed methods are directly applicable on the piezoelectric field equations.

With the Inner-Node-Matrix and the Schur-Complement method we provided and implemented two solution strategies. The Schur-Complement method is suitable for solving the dispersion context for the three dimensional piezoelectric equations with an underlying two dimensional geometry strategy. However, if one wants to extend the model to 3 dimensional geometries, iterative algorithms using only matrix-vector products are recommendable. This is provided by the Inner-Node-Matrix method.

Another possible model improvement would be gained by perfectly matched layers which allow an improved wave absorption into the material. We showed that the introduced methods are still applicable in such models.

The developed algorithms can be also applied to other problem fields including periodic structures like Maxwell's equations for simulating photonic crystals.

By numerical experiments we compared the dispersion diagrams of homogenous versus periodic structures and observed the classification of the frequency domain into pass- and stop-band in the later one. Finally, we simulated a piezoelectric structure as used for frequency filtering in common TV-sets.

*Acknowledgement.* This research has been supported by the Austrian Science Foundation - 'Fonds zur Förderung der wissenschaftlichen Forschung (FWF)' - under the project grants SFB 1306 and the START-Project Y-192. We would like to thank Epcos AG, Munich, for the interesting problem and the financial support, especially Dr. N. Finger, Dr. G. Kovacs, and Dr. K. Wagner. Moreover, we acknowledge the fruitfull research cooperation with Dr. M. Hofer and Professor R. Lerch of the Department of Sensor Technology at the University of Erlangen, Germany.

## References


1. E. Anderson, Z. Bai, C. Bischof, J. Demmel, J. Dongarra, J. Du Croz, A. Greenbaum, S. Hammarling, A. McKenney, S. Ostrouchov, and D. Sorensen. *LAPACK Users' Guide*. SIAM, Philadelphia, third edition, 1999.
2. Mermin Ashcroft. *Solid State Physics*. Holt-Sounders International, 1976.
3. B. Auld. *Acoustic Fields and Waves in Solids*, volume 1,2. Krieger, second edition, 1990.
4. W. Axmann and Kuchment P. An efficient finite element method for computing spectra of photonic and acoustic band-gap materials, scalar case. *Journal of Computational Physics*, 150:468–481, 1999.
5. Z. Bai, J. Demmel, J. Dongarra, A. Ruhe, and H. van der Vorst, editors. *Templates for the solution of Algebraic Eigenvalue Problems: A Practical Guide*. SIAM, Philadelphia, 2000.
6. A. Bensoussan, J.L. Lions, and G. Papanicolaou. High frequenca wave propagation in periodic structures. In J. L. Lions, G. Papanicolaou, and R.T. Rockafellar, editors, *Asymptotic Analysis for periodic structures*, Studies in Mathematics and its Applications, pages 614–626. North-Holland, 1978. Section 3, Spectral theroy of differential operators with periodic coefficients.
7. C. Bernardi, Y. Maday, and A. Patera. A new nonconforming approach to domain decomposition: the Mortar element method. pages 13–51, 1994.
8. B. Bunse-Gerstner, R. Byers, and V. Mehrmann. A chart on numerical methods for structured eigenvalue problems. *SIAM Journal of Matrix Analysis and its Applications*, 13:419–453, 1992.
9. F.M. Gomes and D.C. Sorensen. ARPACK++: A C++ implementation of ARPACK eigenvalue package. Technical report, Computational and Applied Mathematics, Rice University, 1997.
10. M. Hofer, N. Finger, S. Zaglmayr, J. Schöberl, G. Kovacs, U. Langer, and R. Lerch. Finite element calculation of the dispersion relations of infinitely extended saw structures , including bulk wave radiation. In Vittal S. Rao, editor, *Proceedings of SPIE's 9th Annual International Symposium on Smart Structures and Materials*, pages 472–483. SPIE, 2002.
11. M. Hofer, N. Finger, S. Zaglmayr, J. Schöberl, G. Kovacs, U. Langer, and R. Lerch. Finite element calculation of wave propagation and excitation in periodic piezoelectric systems. In *Proceedng of the Fifth World Congress on Computational Mechanics (WCCM V)*. Vienna University of Technology, Austria, 2002.
12. M. Koshiba, S. Mitobe, and M. Suzuki. Finite-element solution of periodic waveguides for acoustic waves. *IEEE Transactions on Ultrasonics, Ferroelectrics and Frequency Control*, 34(4), 1987.
13. P. Kuchment. *Floquet theory of partial differential equations*, volume 60 of *Operator Theory Advances and Applications*. Birkhaeuser Verlag, Basel and Boston, 1993.
14. R.B. Lehoucq, D.C. Sorensen, and C. Yang. Arpack users' guide: Solution of large scale eigenvalue problems with implicitly restarted Arnoldi methods. Technical report, Computational and Applied Mathematics, Rice University, 1997.
15. R. Lerch. Analyse hochfrequenter akustischer Felder in Oberflächenwellenfilter-Komponenten. *Archiv für elektronische Übertragungstechnik (AEU)*, 44(4):317–327, 1990.

16. R. Lerch. Simulation of Piezoelectric Devices by Two- and Three-Dimensional Finite Elements. *IEEE Transactions on Ultrasonics, Ferroelectrics and Frequency Control*, 37(3):233–247, 1990.
17. O. Madelung. *Grundlagen der Halbleiterphysik*, chapter 12. Folgerungen aus der Translationsinvarianz. Springer, Berlin, 1970.
18. V. Mehrmann and D. Watkins. Structure-preserving methods for computing eigenpairs of large sparse skew-Hamiltonian/Hamiltonian pencils. *SIAM J. Sci. Comput.*, 22(6):1905, 2001.
19. D.P. Morgan. History of SAW Devices. In *IEEE Frequency Control Symposium*, pages 439–460, 1998.
20. N. Reed and B. Simon. *Analysis of Operators*, volume 4 of *Methods of Modern Mathematical Physics*, chapter 13 Spectral Analysis. Academic Press, 1978.
21. J. Schöberl. NETGEN - an advancing front 2D/3D-mesh generator based on abstract rules. *Comput. Visual.Sci*, (1):41–52, 1997.
22. J. Schöberl. NGSolve. Online manual, 2003.
23. F. Tisseur and K. Meerbergen. The quadratic eigenvalue problem. *SIAM Review*, 43(2):235–286, 2001.
24. B. Wohlmuth. A Mortar finite element method using dual spaces for the Lagrange multiplier. *SIAM Journal on Numerical Analysis*, 38(3):989–1012, 2001.
25. S. Zaglmayr. Eigenvalue problems in surface acoustic wave filter simulations. Master's thesis, Institute of Computational Mathematics, Johannes Kepler University Linz, Austria, 2002.

# Diffraction Grating Theory with RCWA or the C Method

N.P. van der Aa[1]

Technische Universiteit Eindhoven, P.O. Box 513, 5600 MB Eindhoven, The Netherlands n.p.v.d.aa@tue.nl

**Summary.** Diffraction gratings are often used in optical metrology. When an electromagnetic wave is incident on a grating, the periodicity of the grating causes a multiplicity of diffraction orders. In many metrology applications one needs to know the diffraction efficiency of these orders. Since the period of a grating is often of the same order of magnitude as the wavelength, it is needed to solve Maxwell's equations rigorously in order to obtain these diffraction efficiencies. Two of those methods are the rigorous coupled-wave analysis (RCWA) and the C method.
In this paper a comparison is made between RCWA and the C method with respect to accuracy and speed. Restrictions are made to one-interface problems, which means that only two media are involved separated by one interface, and only gratings are considered with a periodicity in only one direction.

**Key words:** diffraction gratings, C method, RCWA.

## 1 Introduction

When the grating's period is of the same order of magnitude as the wavelength, rigorous methods are required to solve Maxwell's equations. At the time Jean Chandezon introduced his method [1, 2], another method called rigorous coupled-wave analysis (RCWA), was already widely used [3, 4]. The main question remains when one should use RCWA or the C method. Although both methods have a completely different approach for solving the grating problem, it is widely known, that solving eigenvalue problems is the most computationally expensive operation in both methods. That is why this paper concentrates on the computations of the eigenvalue problems to select a criterion for the usage of a certain method. Therefore, the differences between the methods will be discussed and, as an example, a sinusoidal grating is used to illustrate the criterion.

## 2 Mathematical problem

An infinitely long, one-dimensional grating with only one interface is shown in Fig. 1. One-dimensional implies that the grating is periodic, say with period $\Lambda$, in the $x$-direction and constant in the $y$-direction. The fact that the grating is assumed to be infinitely long, allows a restriction to only one period. The domain exists of two media, denoted by $\Omega_1$ (usually air) and $\Omega_2$ (dielectric or metal). The boundaries are denoted by $\Gamma_m$ for $m = 1, ..., 5$.

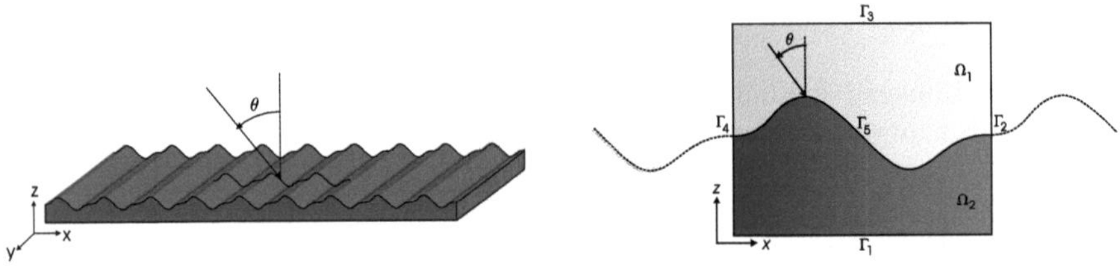


**Fig. 1. Left**: three dimensional representation of the diffraction grating; **right**: the domain of interest.

The media are assumed to be linear with respect to the electromagnetic fields, homogeneous, isotropic, time-invariant, dispersion-free, source-free and non-magnetic. The electromagnetic fields are assumed to be time-harmonic, which implies that the initialization phase is neglected. The incident field is either TE or TM polarized. All these assumptions reduce the local Maxwell equations to a generalized Helmholtz equation [2].

$$\nabla^2 F(x,z) + k^2 n^2(x,z) F(x,z) = 0, \tag{1}$$

where $F$ is either the electric field $E_y$ for TE polarized light or the magnetic field $H_y$ for the TM case. The parameter $n = \sqrt{\varepsilon(x,z)\mu_0}$ is the refractive index and $k = \omega\sqrt{\varepsilon_0\mu_0}$ is the wave number.

On boundaries $\Gamma_1$ and $\Gamma_3$, the outgoing wave condition holds, which means that the fields have to be finite for $z \to \pm\infty$. The restriction to only one period gives a pseudo-periodic boundary condition at $\Gamma_2$ and $\Gamma_4$ by invoking the Floquet-Bloch theorem.

$$F(x,z) = F(x+\Lambda, z)\exp(i\sin\theta), \quad 0 \le x < \Lambda, \quad -\infty < z < \infty. \tag{2}$$

The last boundary is $\Gamma_5$ and on this interface, the tangential components of the electromagnetic fields are continuous.

## 3 Solution methods

From general grating theory it is known that above and below the grating grooves, the Rayleigh expansion holds as a solution of the field:

$$F(x,z) = \sum_{m=-\infty}^{\infty} A_m \exp(ik_{xm}x + ik_{zm}z), \tag{3}$$

where the $A_m$ are the reflection coefficients in the upper half-space or the transmission coefficients in the lower half-space and $k_{xm}$ and $k_{zm}$ are known coefficients. This Rayleigh expansion is a direct consequence of the outgoing wave condition, the pseudo-periodic boundary condition and the Helmholtz equation that holds in the upper and lower half-space. The reason the Rayleigh expansion does not hold inside the grating grooves is that the complex permittivity is not a constant, but a function of $x$ and $z$. This leads to an eigenvalue problem in both methods. The details of the methods are discussed separately.

- **RCWA**
  By eliminating the $z$-dependency of the complex permittivity, it is possible to write the solution inside the grooves as a Fourier expansion, since only a dependency on the periodic coordinate $x$ is present. The way RCWA accomplishes this, is by slicing up the grating domain such that inside each slice, the permittivity only depends on $x$. At the boundaries between two slices, the tangential components of the electromagnetic fields are continuous. In this way, the unknown reflection and transmission coefficients of the upper and lower half-space can be connected to each other and determined. However, introducing the Fourier expansion in the Helmholtz equation gives an eigenvalue problem of size $2N+1$ for both TE and TM polarization for every slice.

- **C method**
  The C method uses a completely different approach. The method uses the idea that if the grating interface were flat, the Rayleigh expansions would be valid for the entire domain, except at the interface. The C method ensures the grating interface to be flat by introducing a new coordinate system. A restriction of the method is that the interface can be described by a function of $x$, i.e. $z = a(x)$. There are parametric descriptions, but that is only for stability purposes. The coordinate transformation is given by

  $$u = x, \quad v = y, \quad w = z - a(x). \tag{4}$$

  The periodicity is preserved in the coordinate $u$ and the grating interface is now described by a flat line given by $w = 0$. However, in the generalized Rayleigh expansion, a new unknown turns up. By substituting this expansion into the transformed Helmholtz equation, an eigenvalue system has to be solved for each **medium**, but since TE and TM polarization cannot be separated this time, the size is $4N+2$.

Figure 2 illustrates how the two methods handle the mathematical model obtained in Section 2. The main differences between the C method and RCWA are:

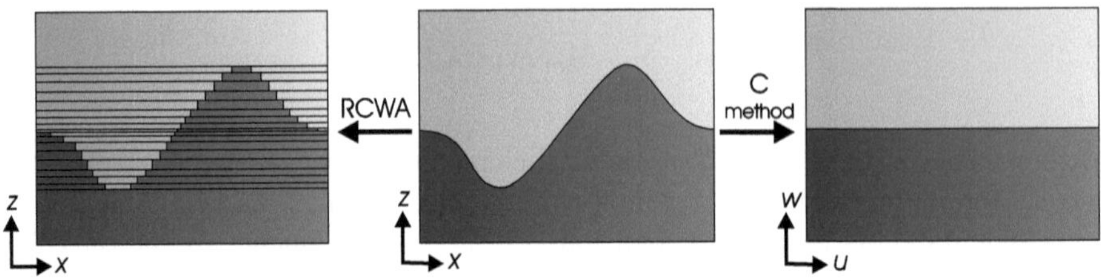


**Fig. 2.** Schematic representation of the way the RCWA and C method remove the $z$ dependency.

- There is one eigenvalue problem per **layer** of size $2N + 1$ for RCWA vs. one per **medium** of size $4N + 2$ for the C method.
- RCWA solves one eigenvalue system for each polarization state, while the C method solves one eigenvalue system for both TE and TM polarization simultaneously.
- RCWA approximates the grating interface, while the C method does not.
- RCWA can handle all types of diffraction gratings, including overhanging gratings, while the C method is restricted to interfaces which can be described by a function of the periodicity coordinate.

A general eigenvalue system of size $p \times p$ takes $O(p^3)$ flops. For the C method only two eigenvalue systems have to be solved of twice the size of the eigenvalue systems obtained with RCWA, but when RCWA uses $q$ layers it also has $q$ eigenvalue systems. Altogether, this implies that RCWA may have 8 times more layers than the number of media for the C method to have an equal number of computations.

## 4 Results

To show the results, test case 2 from [2] has been used. It concerns a sinusoidal grating with a period equal to twice the wavelength. The refractive index of the upper medium is 1 (air), while the one of the lower medium is 1.5 (dielectric). The amplitude of the sine equals the size of the wavelength.

Figure 3 shows the results of RCWA for several values of $N$ and several numbers of layers $q$. It can be seen that it is not the number of harmonics $N$ that determines the diffraction efficiency mostly, but the number of layers $q$. To have the relative difference between RCWA and the C method below 1%, RCWA already needs 15 to 20 layers, while for 0.1% 50 to 80 layers are necessary. It should be noticed that the layer thickness has been chosen equidistant.

To conclude, this paper shows that the number of layers needed to approximate the grating to obtain an accurate (defined by user) result, is the most important criterium and not the number of harmonics. Secondly, for general

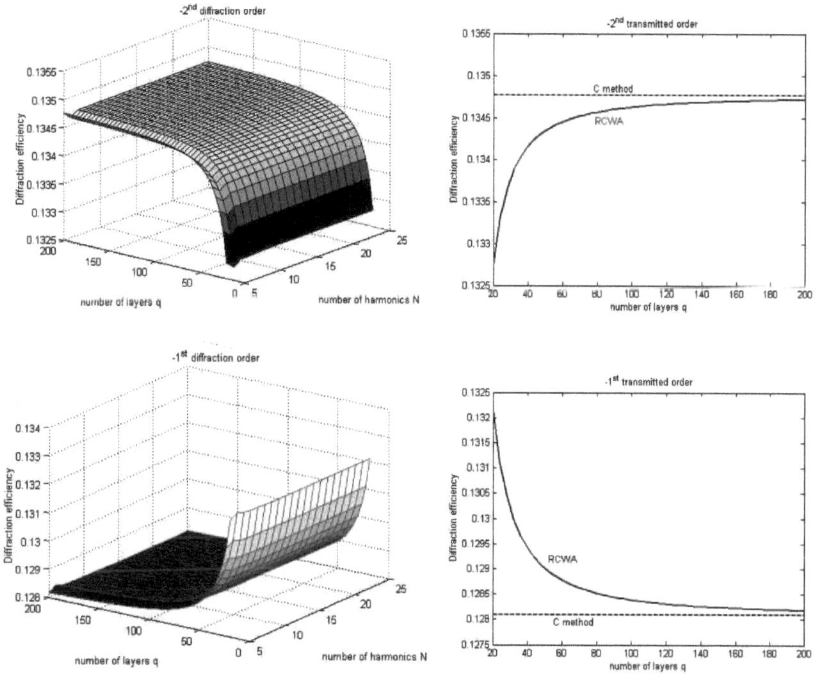


**Fig. 3.** Diffraction efficiencies of the -2$^{\text{nd}}$ and -1$^{\text{th}}$ diffraction order as a function of the number of layers and the number of harmonics (*left*) and if $N = 14$ a comparison with the C method.

grating profiles the C method will obtain the answer with less computational efforts if RCWA uses more than 10 layers to approximate the grating profile.

## References


1. J. Chandezon *et al.* A new theoretical method for diffraction gratings and its numerical application. *Journal of Optics*, 11(4):235–241, 1980.
2. Lifeng Li *et al.* Rigorous and efficient grating-analysis method made easy for optical engineers. *Applied Optics*, 38(2):304–313, 1999.
3. M. G. Moharam *et al.* Formulation for stable and efficient implementation of the rigorous coupled-wave analysis of binary gratings. *J. Opt. Soc. Am. A*, 12(5):1068–1076, May 1995.
4. Mark van Kraaij. Comparison of the rigorous coupled-wave analysis and multiple shooting. *Proc. 13th European Conference on Mathematics for Industry, ECMI 2004, Eindhoven, The Netherlands.*, 2004.

# Relocation of Electric Field Domains and Switching Scenarios in Superlattices

L.L. Bonilla, G. Dell'Acqua, and R. Escobedo

Universidad Carlos III de Madrid, Leganés, Spain.
bonilla@ing.uc3m.es

**Summary.** A numerical study of domain wall relocation during slow voltage switching is presented for doped semiconductor superlattices. Unusual relocation scenarios are found and interpreted according to previous theory.

**Key words:** superlattices, relocation of domains, voltage switching

## 1 Introduction

Semiconductor superlattices are essential ingredients in fast nanoscale oscillators, quantum cascade lasers and infrared detectors. Quantum cascade lasers are used to monitor environmental pollution in gas emissions, to analyze breath in hospitals and in many other industrial applications. A semiconductor superlattice (SL) is formed by growing a large number of periods with each period consisting of two layers, which are semiconductors with different energy gaps but having similar lattice constants, such as GaAs and AlAs. The conduction band edge of an infinitely long ideal SL is modulated so that it looks like a one-dimensional (1D) crystal consisting of a periodic succession of a quantum well (GaAs) and a barrier (AlAs). Vertical charge transport in a SL subject to strong electric fields exhibits many interesting features, and it is realized experimentally by placing a doped SL of finite length in the central part of a diode (forming a $n^+$-$n$-$n^+$ structure) with contacts at its ends. In this paper, we study the relocation of electric field domains which appear in a strongly doped, dc voltage biased SL when the voltage is switched between two different values. Our model consists of a system of spatially discrete drift-diffusion equations (DDE) for the electric field and current, an algebraic constraint representing voltage bias, initial and boundary conditions [2]. By numerically solving this model, we find that the current through the SL exhibits very different patterns involving several mechanisms for relocating electric field domains, depending on the way the voltage is switched.

## 2 The Sequential Tunnelling Model

We use the discrete drift-diffusion model described in the review paper [2]. It consists of the following Poisson and charge continuity equations:

$$F_i - F_{i-1} = \frac{e}{\varepsilon}(n_i - N_D^w), \tag{1}$$

$$\frac{\mathrm{d}n_i}{\mathrm{d}t} = J_{i-1\to i} - J_{i\to i+1}, \tag{2}$$

for the average electric field $-F_i$ and the two-dimensional (2D) electron density $n_i$ at the $i$th SL period (which starts at the right end of the $(i-1)$th barrier and finishes at the right end of the $i$th barrier), with $i = 1, \ldots, N$. Here $N_D^w$, $\varepsilon$, $-e$ and $eJ_{i\to i+1}$ are the 2D doping density at the $i$th well, the average permittivity, the electron charge and the tunnelling current density across the $i$th barrier, respectively. The SL period is $l = d + w$, where $d$ and $w$ are the barrier and well widths, respectively. Time-differencing (1) and inserting the result in (2), we obtain the following form of Ampere's law:

$$\frac{\varepsilon}{e}\frac{\mathrm{d}F_i}{\mathrm{d}t} + J_{i\to i+1} = J(t). \tag{3}$$

The space-independent unknown function $eJ(t)$ is the total current density through the SL. Quantum mechanical calculations show that the *constitutive relation* for the tunnelling current density $eJ_{i\to i+1}$ is [2]

$$J_{i\to i+1} = \frac{n_i v(F_i)}{l} - D(F_i)\frac{n_{i+1} - n_i}{l^2}. \tag{4}$$

The nonlinear smooth functions of electric field, $v(F)$ and $D(F)$ have dimensions of velocity and diffusivity, respectively, and their explicit expressions can be found in Appendix A of [2]. It is important to mention that the drift velocity $v(F)$ has a first local maximum at $(F_M, v_M)$, ($F_M$ and $v_M$ are both positive), it is positive for positive $F$, and $v(0) = 0$. Fig. 1 shows $v/v_M$ as a function of $F/F_M$. $D(F) > 0$ for non-negative $F$.

Substituting (1) and (4) in (3), we find the DDEs:

$$\frac{\mathrm{d}F_i}{\mathrm{d}t} + v(F_i)\,\frac{F_i - F_{i-1}}{l} - D(F_i)\,\frac{F_{i+1} - 2F_i + F_{i-1}}{l^2} = \frac{e}{\varepsilon}\left[J - \frac{N_D^w v(F_i)}{l}\right], \tag{5}$$

with $i = 1, \ldots, N$. Bias and boundary conditions are

$$F_0 = F_{N+1} = \rho_c J, \quad \frac{1}{N}\sum_{i=1}^{N} F_i = \frac{V(t)}{Nl}, \tag{6}$$

in which $\rho_c > 0$ and $V(t) > 0$ are the resistivity of the contacts and the voltage, respectively. To analyze this model, it is convenient to render all equations dimensionless. We adopt $F_M$, $N_D^w$, $v_M$, $v_M l$, $eN_D^w v_M/l$ and $\varepsilon F_M l/(eN_D^w v_M)$ as

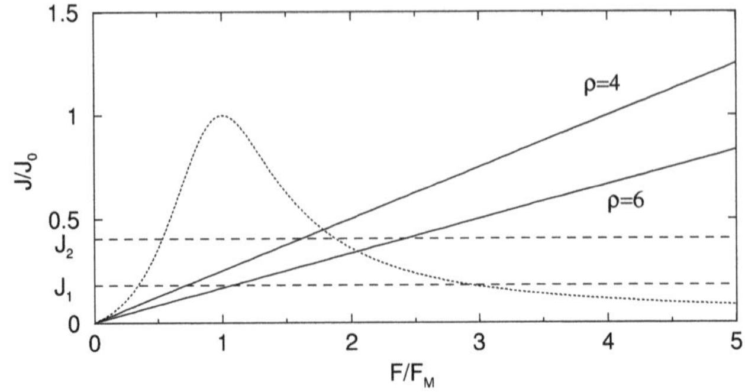


**Fig. 1.** Dimensionless current density (or drift velocity) versus $F/F_M$ and boundary conditions $E = \rho J$ for $\rho = 4$ and $\rho = 6$. Stable stationary solutions are found for $J_1 < J < J_2$ (dashed lines). Current density unit is $eJ_0 = eN_D^w v_M/l = 2.88\,\mathrm{A\,cm^{-2}}$.

units of $F_i$, $n_i$, $v(F)$, $D(F)$, $eJ$ and $t$, respectively. Typical SL parameters as in [1] are $b = 4\,\mathrm{nm}$, $w = 9\,\mathrm{nm}$, $F_M = 6.92\,\mathrm{kV\,cm^{-1}}$, $N_D^w = 1.5 \times 10^{11}\mathrm{cm^{-2}}$, $v_M = 156\,\mathrm{cm\,s^{-1}}$, $v_M l = 2.03 \times 10^{-4}\,\mathrm{cm^2\,s^{-1}}$ and $eJ_0 = eN_D^w v_M/l = 2.88\,\mathrm{A\,cm^{-2}}$. For a circular sample with a diameter of 120 $\mu$m, the units of current and time are 0.326 mA and 2.76 ns, respectively. The nondimensional equations of the model are:

$$\frac{\mathrm{d}E_i}{\mathrm{d}t} + v(E_i)\,\frac{E_i - E_{i-1}}{\nu} - D(E_i)\,\frac{E_{i+1} - 2E_i + E_{i-1}}{\nu} = J - v(E_i), \quad (7)$$

$$\frac{1}{N}\sum_{i=1}^{N} E_i = \Phi, \quad E_0 = E_{N+1} = \rho J. \quad (8)$$

Here we have used the same symbol for dimensional and dimensionless quantities except for the electric field ($F$ dimensional, $E$ dimensionless). The parameters $\nu = eN_D^w/(\varepsilon F_M)$, $\rho = \rho_c e v_M N_D^w/(lF_M)$, and $\Phi = V/(F_M N l)$ are dimensionless doping density, contact resistivity and average electric field (bias), respectively. For the above mentioned 9/4 SL, $\nu \simeq 3$. The contact resistivity $\rho$ will be selected in certain ranges to be specified below and the variation of $\Phi$ will be explained in the next Section.

## 3 Switching Scenarios

Numerical solution of the equations (7) - (8) with different initial field profiles shows that (for constant $\Phi$) the stable field profiles $\{E_i\}$ are time-independent, step-like and increasing with $i$: typically they consist of two flat regions called electric field domains separated by an abrupt transition region called a domain wall or charge monopole [2]. The current–voltage diagram for these stable solutions is depicted in Fig. 2.

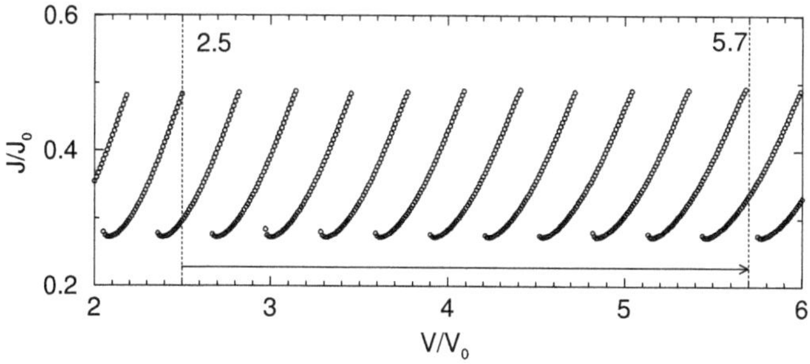


**Fig. 2.** Current–voltage diagram showing stationary branches and applied voltage step. The unit of voltage is 0.36 V.

The electric field profiles of each branch of solutions in Fig. 2 differ in the location of their domain wall: counting branches in the direction of increasing voltage, the profiles of the $j$th branch have their domain wall located in the $(N-j+1)$th SL period. Notice that for certain values of the voltage, several branches with different current are possible (multistability). If we switch the voltage from a value $V_{ini}$ corresponding to one branch to a final value $V_{fin} = V_{ini}+\Delta V$ corresponding to different branches, the domain wall has to relocate in a different SL period. During switching, $V(t) = V_{ini}+\dot{V}t$, with $\dot{V} = \Delta V/\Delta t$, and $\Delta t$ is the ramping time. We now study what happens during switching for different values of $\Delta t$.

The case of very small $\Delta t$ (nanoseconds) and small $\Delta V$ (spanning two branches) was studied theoretically in [1]. In this paper, we consider much larger values of $\Delta t$ and $\Delta V$, and observe several new phenomena. If $\Delta t$ is in the range of microseconds and $\rho$ is appropriate, the current oscillates with time, as shown in Fig. 3 for $\Delta V$ as depicted in Fig. 2. For a fixed doping density $\nu$ there are two currents $J_1$ and $J_2$ (marked in Fig. 1) such that a domain wall in an infinitely long SL remains stationary if $J_1 < J < J_2$, and it moves to the right (resp. left) if $J < J_1$ (resp. $J > J_2$) [3].

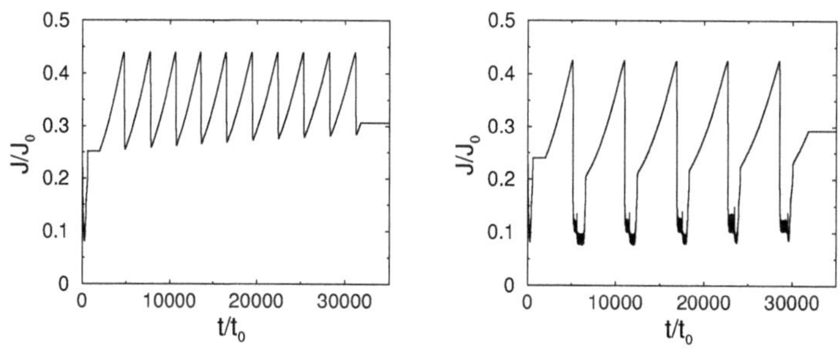


(a) (b)

**Fig. 3.** Current density for (a) $\rho = 4$ and (b) $\rho = 6$. The unit of time is 2.76 ns.

Let $\rho_j$, $j = 1, 2$, be such that $v(\rho_j J_j) = J_j$. During voltage switching, the current changes as in Fig. 3(a) if $\rho < \rho_1$, and as in Fig. 3(b) if $\rho_1 < \rho < \rho_2$. If $\rho > \rho_2$, there are no stationary solutions and the current oscillates periodically in time, as in the Gunn effect [2].

Our simulations show that the electric field profile corresponding to Fig. 3(a) consists of slow change of a step-like profile during the finite time intervals in which the current increases followed by a rapid motion of the domain wall, one SL period to the left, when the current is near its local maximum value (which is larger than $J_2$). The domain wall motion is followed by a drop in the current. This situation lasts until the end of switching and the number of maxima of the current is equal to the number of branches skipped during voltage switching. For $\rho_1 < \rho < \rho_2$, Fig. 3(b) shows that the number of current maxima during switching is half the (even) number of branches skipped during voltage switching. Near each current maximum (with $J > J_2$), the domain wall traverses one SL period to the left. Immediately afterwards, a pulse of the electric field is created at the injecting left contact and travels to the end of the SL accompanied by the motion of the old domain wall to the right. This is the tripole-dipole scenario discovered in [1] and characterized by a succession of double peaks of the current followed by a succession of single peaks. The novelty is that voltage switching continues after the domain wall has arrived at a stable location and the same process continues until $t = \Delta t$. Then the number of current peaks larger than $J_2$ is half that in Fig. 3(a). If $\Delta t$ is smaller than a critical value, there is only one large current peak during switching followed by the double peaks - single peak succession typical of only one tripole-dipole scenario.

*Acknowledgement.* We thank B. Birnir, O. Sánchez and J. Soler for fruitful discusions. This work has been supported by the MCyT grant BFM2002-04127-C02, and by the European Union under grant HPRN-CT-2002-00282. R.E. has been supported by a postdoctoral grant awarded by the Autonomous Region of Madrid.

## References

1. A. Amann, A. Wacker, L.L. Bonilla, and E. Schöll. Dynamic scenarios of multistable switching in semiconductor superlattices. *Phys. Rev. E*, 63(066207):066207–1–066207–8, 2001.
2. L.L. Bonilla. Theory of nonlinear charge transport, wave propagation and self-oscillations in semiconductor superlattices. *J. Phys.: Cond. Matter*, 14:R341–R381, 2002.
3. A. Carpio, L.L. Bonilla, A. Wacker, and E. Schöll. Wavefronts may move upstream in semiconductor superlattices. *Phys. Rev. E*, 61:4866–4876, 2000.

# Quantum Kinetic and Drift-Diffusion Equations for Semiconductor Superlattices

L.L. Bonilla and R. Escobedo

Universidad Carlos III de Madrid, Madrid, Spain.
bonilla@ing.uc3m.es

**Summary.** A nonlocal (quantum) drift-diffusion equation for the electric field and the electron density is derived from a Wigner-Poisson equation modelling quantum vertical transport in strongly coupled semiconductor superlattices, by using a consistent Chapman-Enskog procedure. Numerical solutions for a device consisting of a n-doped superlattice placed in a $n^+$-$n$-$n^+$ diode under a constant voltage bias are presented and compared with those obtained by using a semiclassical approximation.

**Key words:** Superlattices, Chapman-Enskog, quantum drift-diffusion equation.

Industrial uses of semiconductor superlattices (SLs) include fast nanoscale oscillators, terahertz and infrared detectors and quantum cascade lasers. The Wigner-Poisson system for 1D electron transport in the lowest miniband of a strongly coupled SL is:

$$\frac{\partial f}{\partial t} + \frac{i}{\hbar}\left[\mathcal{E}\left(k + \frac{1}{2i}\frac{\partial}{\partial x}\right) - \mathcal{E}\left(k - \frac{1}{2i}\frac{\partial}{\partial x}\right)\right] f$$

$$+\frac{ie}{\hbar}\left[W\left(x + \frac{1}{2i}\frac{\partial}{\partial k}, t\right) - W\left(x - \frac{1}{2i}\frac{\partial}{\partial k}, t\right)\right] f = Q[f], \qquad (1)$$

$$\varepsilon \frac{\partial^2 W}{\partial x^2} = \frac{e}{l}(n - N_D), \qquad (2)$$

$$n = \frac{l}{2\pi}\int_{-\pi/l}^{\pi/l} f(x,k,t)dk = \frac{l}{2\pi}\int_{-\pi/l}^{\pi/l} f^{FD}(k;n)dk, \qquad (3)$$

$$f^{FD}(k;n) = \frac{m^* k_B T}{\pi\hbar^2}\ln\left[1 + \exp\left(\frac{\mu - \mathcal{E}(k)}{k_B T}\right)\right]. \qquad (4)$$

Here $f$, $n$, $N_D$, $\mathcal{E}(k)$, $l$, $k_B$, $T$, $W$, $\varepsilon$, $m^*$ and $e > 0$ are the one-particle Wigner function, the 2D electron density, the 2D doping density, the miniband dispersion relation, the SL period, the Boltzmann constant, the lattice temperature, the electric potential, the SL permittivity, the effective mass of the electron, and minus the electron charge, respectively. The left-hand side of (1) can be straightforwardly derived from the Schrödinger-Poisson equation for the wave function in the miniband using the definition of the 1D Wigner function: $f(x,k,t) = \sum_{j=-\infty}^{\infty} \int \overline{\psi(x+jl/2,y,z,t)}\psi(x-jl/2,y,z,t)e^{ijkl}dx_\perp$ $[(\psi(x,x_\perp,t) = \sum_{q,q_\perp} a(q,q_\perp,t)\varphi_q(x)e^{iq_\perp \cdot x_\perp}$, $x_\perp = (y,z)$, is a superposition of the Bloch states corresponding to the miniband$]$. The collision term $-Q[f]$ in (1) is the sum of $\nu_e\left(f - f^{FD}\right)$, which represents energy relaxation towards a 1D effective Fermi-Dirac (FD) distribution $f^{FD}(k;n)$ (local equilibrium), and $\nu_i[f(x,k,t) - f(x,-k,t)]/2$, which accounts for impurity elastic collisions [4]. For simplicity, the collision frequencies $\nu_e$ and $\nu_i$ are fixed constants. Exact and FD distribution functions have the same electron density, thereby preserving charge continuity as in the Bhatnagar-Gross-Krook (BGK) collision models [2]. Then the chemical potential $\mu$ depends on $n$ and is found by inverting the exact relation (3).

It is convenient to derive the charge continuity equation and a nonlocal Ampère's law for the current density. The Wigner function $f$ is periodic in $k$; its Fourier expansion is $\sum_{j=-\infty}^{\infty} f_j(x,t)\, e^{ijkl}$. Defining $F = \partial W/\partial x$ (*minus* the electric field) and the average $\langle F\rangle_j(x,t) = \frac{1}{jl}\int_{-jl/2}^{jl/2} F(x+s,t)\,ds$, it is possible to obtain the following equivalent form of the Wigner equation

$$\frac{\partial f}{\partial t} + \sum_{j=-\infty}^{\infty} \frac{ijl}{\hbar}\, e^{ijkl}\left(\mathcal{E}_j\, \frac{\partial}{\partial x}\langle f\rangle_j + e\,\langle F\rangle_j\, f_j\right) = Q[f], \tag{5}$$

where $\mathcal{E}(k) = \Delta\,(1-\cos kl)/2$ is the tight-binding dispersion relation ($\Delta$ is the miniband width) and $v(k) = \frac{\Delta l}{2\hbar}\sin kl$ is the miniband group velocity. Integrating this equation over $k$ yields the charge continuity equation $\frac{\partial n}{\partial t} + \frac{\partial}{\partial x}\sum_{j=1}^{\infty}\frac{2jl}{\hbar}\left\langle \mathrm{Im}(\mathcal{E}_{-j}f_j)\right\rangle_j = 0$, from which we can eliminate the electron density by using the Poisson equation and integrating over $x$, thereby obtaining the nonlocal Ampère's law for the total current density $J(t)$:

$$\varepsilon\,\frac{\partial F}{\partial t} + \frac{2e}{\hbar}\sum_{j=1}^{\infty} j\langle \mathrm{Im}(\mathcal{E}_{-j}f_j)\rangle_j = J(t). \tag{6}$$

To derive the QDDE, we shall assume that the electric field contribution in (5) is comparable to the collision terms and that they dominate the other terms (*the hyperbolic limit*) [4]. Let $v_M$ and $F_M$ be the electron velocity and field positive values at which the (zeroth order) drift velocity reaches its maximum. In this limit, the time $t_0$ it takes an electron with speed $v_M$ to traverse a distance $x_0 = \varepsilon F_M l/(eN_D)$, over which the field variation is of order $F_M$, is much longer than the mean free time between collisions, $\nu_e^{-1} \sim \hbar/(eF_M l) = t_1$. We therefore define the *small parameter* $\epsilon = t_1/t_0 = \hbar v_M N_D/(\varepsilon F_M^2 l^2)$ and formally multiply the first two terms on the left side of (1) or (5) by $\epsilon$ [4]. After obtaining the number of desired terms, we set $\epsilon = 1$. The solution of (5) for $\epsilon = 0$ is calculated in terms of its Fourier coefficients as

$$f^{(0)}(k;F) = \sum_{j=-\infty}^{\infty} \frac{(1 - ij\mathcal{F}/\tau_e)\, f_j^{FD}}{1 + j^2\mathcal{F}^2}\, e^{ijkl}, \tag{7}$$

where $\mathcal{F} = \langle F \rangle_1 / F_M$, $F_M = \frac{\hbar}{el}\sqrt{\nu_e(\nu_e + \nu_i)}$ and $\tau_e = \sqrt{(\nu_e + \nu_i)/\nu_e}$.

The Chapman-Enskog *Ansatz* for the Wigner function is [4]:

$$f(x, k, t; \epsilon) = f^{(0)}(k;F) + \sum_{m=1}^{\infty} f^{(m)}(k;F)\, \epsilon^m, \tag{8}$$

$$\varepsilon \frac{\partial F}{\partial t} + \sum_{m=0}^{\infty} J^{(m)}(F)\, \epsilon^m = J(t). \tag{9}$$

The coefficients $f^{(m)}(k;F)$ depend on the 'slow variables' $x$ and $t$ only through their dependence on the electric field and the electron density. The electric field obeys a reduced evolution equation (9) in which the functionals $J^{(m)}(F)$ are chosen so that the $f^{(m)}(k;F)$ are bounded and $2\pi/l$-periodic in $k$. Differentiating the Ampère's law (9) with respect to $x$, we obtain the charge continuity equation. Moreover the condition, $\int_{-\pi/l}^{\pi/l} f^{(m)}(k;n)\, dk = 2\pi\, f_0^{(m)}/l = 0$, $m \geq 1$, ensures that $f^{(m)}$, $m \geq 1$, do not contain contributions proportional to the zero-order term $f^{(0)}$. Inserting (8) and (9) into (5), we find the hierarchy:

$$\mathcal{L}f^{(1)} = -\left( \frac{\partial f^{(0)}}{\partial t} + \sum_{j=-\infty}^{\infty} \frac{ijl\mathcal{E}_j e^{ijkl}}{\hbar} \frac{\partial}{\partial x} \langle f^{(0)} \rangle_j \right)\Bigg|_0 \tag{10}$$

$$\mathcal{L}f^{(2)} = -\left( \frac{\partial f^{(1)}}{\partial t} + \sum_{j=-\infty}^{\infty} \frac{ijl\mathcal{E}_j e^{ijkl}}{\hbar} \frac{\partial}{\partial x} \langle f^{(1)} \rangle_j \right)\Bigg|_0 - \frac{\partial}{\partial t} f^{(0)}\Big|_1, \tag{11}$$

and so on, where $\mathcal{L}u(k) \equiv ie\hbar^{-1} \sum_{-\infty}^{\infty} jl \langle F \rangle_j u_j e^{ijkl} + (\nu_e + \nu_i/2) u(k) + \nu_i u(-k)/2$, and the subscripts 0 and 1 in the right hand side of these equations mean that $\varepsilon\, \partial F/\partial t$ is replaced by $J - J^{(0)}(F)$ and by $-J^{(1)}(F)$, respectively.

The solvability conditions for the linear hierarchy of equations yield $J^{(m)} = \frac{2e}{\hbar} \sum_{j=1}^{\infty} j \langle \mathrm{Im}(\mathcal{E}_{-j} f_j^{(m)}) \rangle_j$, which can also be obtained by insertion of (8) in (6). In the tight-binding dispersion relation case, the leading order of the Ampère's law (9) is

$$\varepsilon \frac{\partial F}{\partial t} + \frac{ev_M}{l} \langle n\mathcal{M}V(\mathcal{F}) \rangle_1 = J(t), \tag{12}$$

$$V(\mathcal{F}) = \frac{2\mathcal{F}}{1 + \mathcal{F}^2}, \quad v_M = \frac{\Delta l\, \mathcal{I}_1(M)}{4\hbar\tau_e \mathcal{I}_0(M)}, \quad \mathcal{M}\left(\frac{n}{N_D}\right) = \frac{\mathcal{I}_1(\tilde{\mu})\, \mathcal{I}_0(M)}{\mathcal{I}_1(M)\, \mathcal{I}_0(\tilde{\mu})}, \tag{13}$$

$$\mathcal{I}_m(s) = \int_{-\pi}^{\pi} \cos(mk)\, \ln\left(1 + e^{s - \delta + \delta\cos k}\right) dk, \tag{14}$$

provided $\delta = \Delta/(2k_BT)$ and $\tilde{\mu} \equiv \mu/(k_BT)$. Here $M$ (calculated graphically in Fig. 1 of [4]) is the value of the dimensionless chemical potential $\tilde{\mu}$ at which (3) holds with $n = N_D$. The drift velocity $v_M V(\mathcal{F})$ has the Esaki-Tsu form with a peak velocity that becomes $v_M \approx \Delta l I_1(\delta)/[4\hbar\tau_e I_0(\delta)]$ in the Boltzmann limit [5] ($I_n(\delta)$ is the modified Bessel function of the $n$th order).

To find the first-order correction in (9), we first solve (10) and find $J^{(m)}$ for $m = 1$. The calculation yields the first correction to (12) (here $'$ means differentiation with respect to $n$) [4]

$$\varepsilon \frac{\partial F}{\partial t} + \frac{ev_M}{l} \mathcal{N}\left(F, \frac{\partial F}{\partial x}\right) = \varepsilon \left\langle D\left(F, \frac{\partial F}{\partial x}, \frac{\partial^2 F}{\partial x^2}\right)\right\rangle_1 + \langle A\rangle_1 J(t), \quad (15)$$

$$A = 1 + \frac{2ev_M}{\varepsilon F_M l(\nu_e + \nu_i)} \frac{1 - (1 + 2\tau_e^2)\mathcal{F}^2}{(1 + \mathcal{F}^2)^3} n\mathcal{M}, \quad (16)$$

$$\mathcal{N} = \langle nV\mathcal{M}\rangle_1 + \langle (A-1)\langle\langle nV\mathcal{M}\rangle_1\rangle_1\rangle_1 - \frac{\Delta l\tau_e}{F_M\hbar(\nu_e + \nu_i)} \left\langle \frac{B}{1 + \mathcal{F}^2}\right\rangle_1, \quad (17)$$

$$D = \frac{\Delta^2 l^2}{8\hbar^2(\nu_e + \nu_i)(1 + \mathcal{F}^2)} \left(\frac{\partial^2 \langle F\rangle_1}{\partial x^2} - \frac{4\hbar v_M \tau_e C}{\Delta l}\right) \quad (18)$$

$$B = \left\langle \frac{4\mathcal{F}_2 n\mathcal{M}_2}{(1 + 4\mathcal{F}_2^2)^2} \frac{\partial\langle F\rangle_2}{\partial x}\right\rangle_1 + \mathcal{F}\left\langle \frac{n\mathcal{M}_2(1 - 4\mathcal{F}_2^2)}{(1 + 4\mathcal{F}_2^2)^2} \frac{\partial\langle F\rangle_2}{\partial x}\right\rangle_1,$$
$$- \frac{4\hbar v_M(1 + \tau_e^2)\mathcal{F}(n\mathcal{M})'}{\Delta l\tau_e(1 + \mathcal{F}^2)} \left\langle n\mathcal{M}\frac{1 - \mathcal{F}^2}{(1 + \mathcal{F}^2)^2} \frac{\partial\langle F\rangle_1}{\partial x}\right\rangle_1, \quad (19)$$

$$C = \left\langle \frac{(n\mathcal{M}_2)'}{1 + 4\mathcal{F}_2^2} \frac{\partial^2 F}{\partial x^2}\right\rangle_1 - 2\mathcal{F}\left\langle \frac{(n\mathcal{M}_2)'\mathcal{F}_2}{1 + 4\mathcal{F}_2^2} \frac{\partial^2 F}{\partial x^2}\right\rangle_1,$$
$$+ \frac{8\hbar v_M(1 + \tau_e^2)(n\mathcal{M})'\mathcal{F}}{\Delta l\tau_e(1 + \mathcal{F}^2)} \left\langle \frac{(n\mathcal{M})'\mathcal{F}}{1 + \mathcal{F}^2} \frac{\partial^2 F}{\partial x^2}\right\rangle_1. \quad (20)$$

Here $\mathcal{M}_2(n/N_D) \equiv \mathcal{I}_2(\tilde{\mu})\mathcal{I}_0(M)/[\mathcal{I}_1(M)\mathcal{I}_0(\tilde{\mu})]$ and $\mathcal{F}_2 \equiv \langle F\rangle_2/F_M$. If the electric field and the electron density do not change appreciably over two SL periods, $\langle F\rangle_j \approx F$, the spatial averages can be ignored, and the *nonlocal* QDDE (15) becomes the *local* generalized DDE (GDDE) obtained from the semiclassical theory [4]. The boundary conditions for the QDDE (15) (which contains triple spatial averages) need to be specified on the intervals $[-2l, 0]$ and $[Nl, Nl + 2l]$, not just at the points $x = 0$ and $x = Nl$, as in the case of the parabolic GDDE. Similarly, the initial condition has to be defined on the extended interval $[-2l, Nl + 2l]$.

Fig. 1 shows the evolution of the current during the self-sustained oscillations that appear when the QDDE (15) and (2) are solved for boundary conditions $\varepsilon\partial F/\partial t + \sigma F = J$ at each point of the intervals $[-2l, 0]$ and $[Nl, Nl + 2l]$ and appropriate dc voltage bias. The contact conductivity $\sigma$ is selected so that $\sigma F$ intersects $eN_D v_M V(F/F_M)/l$ on its decreasing branch, as in the theory of the Gunn effect [1]. Parameter values correspond to a 157-period 3.64 nm GaAs/0.93 nm AlAs SL at 5K, with $N_D = 4.57 \times 10^{10}$ cm$^{-2}$, $\nu_i = 2\nu_e = 18 \times 10^{12}$ Hz under a dc voltage bias of 1.62 V. Cathode and anode contact conductivities are 2.5 and 0.62 $\Omega^{-1}$cm$^{-1}$, respectively.

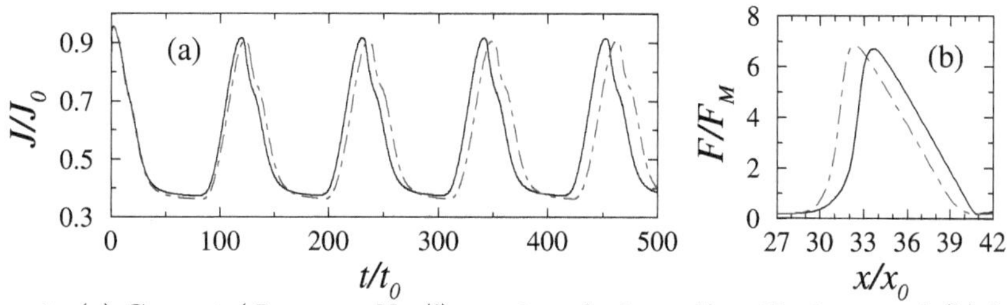


**Fig. 1.** (a) Current ($J_0 = ev_M N_D/l$) vs. time during self-oscillations, and (b) fully developed dipole wave. Solid line: QDDE, dashed line: GDDE. Parameter values: $x_0 = 16$ nm, $t_0 = 0.24$ ps, $J_0 = 1.10 \times 10^5$ A/cm$^2$.

We observe that the field profile of the dipole wave during self-oscillations is sharper in the case of the GDDE than in the case of the QDDE. The local spatial averages appearing in the QDDE have a smoothing effect on the sharp gradients of the electric field. This smoothing effect produces rounder and smaller dipole waves in the QDDE, as compared to the same solution for the GDDE. The equal-area rule as in the theory of the Gunn effect hints that smaller waves are faster [3], resulting in a slightly larger frequency for the self-oscillations in the QDDE (37.6 GHz) than in the case of the GDDE (36.8 GHz).

*Acknowledgement.* This work has been supported by the MCyT grant BFM2002-04127-C02-01, and by the European Union under grant HPRN-CT-2002-00282. R.E. has been supported by a postdoctoral grant awarded by the Consejería de Educación of the Autonomous Region of Madrid.

## References

1. L. L. Bonilla and. Theory of nonlinear charge transport, wave propagation, and self-oscillations in semiconductor superlattices. *J. Phys.: Condens. Matter*, 14(R341–R381), 2002.
2. P.L. Bhatnagar, E.P. Gross, and M. Krook. A model for collision processes in gases. i. Small amplitude processes in charged and neutral one-component systems. *Phys. Rev.*, 94:511–525, 1954.
3. V.L. Bonch-Bruevich, I.P. Zvyagin, and A.G. Mironov. *Domain electrical instabilities in semiconductors.* Consultants Bureau, New York, 1975.
4. L.L. Bonilla, R. Escobedo, and Á. Perales. Generalized drift-diffusion model for miniband superlattices. *Phys. Rev. B*, 68(241304):241304–1–241304–4, 2003.
5. A.A. Ignatov and V.I. Shashkin. Bloch oscillations of electrons and instability of space-charge waves in superconductor superlattices. *Sov Phys. JETP*, 66:526–530, 1987. [Zh. Eksp. Teor. Fiz. **93**, 935 (1987)].

# Model Order Reduction of Nonlinear Dynamical Systems

C. Brennan[1], M. Condon[1], and R. Ivanov[1,2]

[1] School of Electronic Engineering, Dublin City University, Dublin 9, IRELAND
{brennanc; condonm; ivanovr}@eeng.dcu.ie
[2] Institute for Nuclear Research and Nuclear Energy, 72 Tzarigradsko chaussee, 1784 Sofia, BULGARIA

**Summary.** Some propositions for approximation of the controllability and observability gramians for nonlinear systems are presented. This enables a balancing type model reduction to be performed for nonlinear systems in the same manner as for linear systems.

**Key words:** gramians, controllability, observability, Lyapunov equations

## 1 Introduction

Nonlinear systems arise in all aspects of engineering - aeronautics, chemical and processing industries, high-speed electronics and so on. The complexity of modern systems is such that their simulation may involve the solution of several thousands of non-linear coupled ordinary (ODE) or partial differential equations (PDE). This can prove computationally arduous both in terms of speed and memory requirements even with state-of-the-art workstations. To this end, model reduction techniques are of paramount importance in that they permit repetitive and iterative simulation for both design and optimization purposes to proceed in a reasonable time-frame.

Typically, all but one of the continuous variables of the PDE are discretised and the system can be written in the form of a (very large size) system of ODE:

$$\dot{x} = f(x) + Bu(t) \tag{1}$$

$$y = h(x(t)) \tag{2}$$

where $f : \mathfrak{R}^n \to \mathfrak{R}^n$ and $h : \mathfrak{R}^n \to \mathfrak{R}^q$ are nonlinear functions, $u(t) \in \mathfrak{R}^p$ is regarded as an input to the system and $y(t) \in \mathfrak{R}^q$ is an output. The goal in model reduction is to replace the $n$–dimensional system (1)–(2) with a system of much smaller dimension $k \ll n$, such that the input-output behavior of the reduced order system satisfactorily represents the behavior of the full system.

Now while model reduction techniques for linear systems abound, general user-friendly robust reduction techniques for nonlinear systems remain to be found. The existing methods for model order reduction of nonlinear systems can be classified as follows: A. Empirical methods. These methods use experimental or simulation data for the input-output behavior of the system [3]; B. Volterra methods. These are based on polynomial expansion of the nonlinear function $f(x)$ and taking into account the first several terms of the expansion [4, 2]; C. Trajectory approximation methods. Included here are the so called Proper orthogonal decomposition and Piecewise polynomial approximation.

One approach for linear reduction familiar, in particular, to control engineers is balanced truncation. It involves the determination of two specific matrices, namely the controllability and observability gramian matrices, which when balanced enable the important states in a system to be determined and those of lesser importance for the input-output mapping to be eliminated. For nonlinear systems, the computation of exact gramians is impractical so empirical or approximate constant gramians are required in lieu of the exact solutions. This enables a balancing type model reduction process to then proceed for nonlinear systems in the same manner as for linear systems. In what follows a new approach for the determination of approximate empirical gramians will be discussed.

## 2 Linear time-varying systems

The gramians for Linear time-varying systems (LTVS) provide a motivation for constructions suitable for nonlinear systems. For a LTVS

$$\dot{x}(t) = A(t)x(t) + B(t)u(t), \qquad y(t) = C(t)x(t) \tag{3}$$

the fundamental solution is defined as the solution of:

$$\dot{\Theta}(t) = A(t)\Theta(t), \qquad \Theta(0) = I \tag{4}$$

where $I$ is the corresponding identity matrix. For example, if $A$ is a constant matrix, (as for the linear time invariant system – LTIS) then one simply recovers the very well known solution $\Theta(t) = \exp(At)$. The Controllability and the Observability gramians are [5]:

$$P = \int_0^\infty \Theta^{-1}(-\tau)B(-\tau)B^T(-\tau)\Theta^{-1T}(-\tau)d\tau \tag{5}$$

$$Q = \int_0^\infty \Theta^T(\tau)C^T(\tau)C(\tau)\Theta(\tau)d\tau \tag{6}$$

Strictly speaking, the gramians for LTVS must depend on $t$ as shown in [5]. However, for the purposes of model reduction, constant gramians are preferred and the constant versions (5) and (6) are used as approximations.

The expressions in (5) and (6) are generalisations of the gramians for LTIS where $\Theta(t) = \exp(At)$:

$$P = \int_0^\infty e^{A\tau} B B^T e^{A^T\tau} d\tau, \qquad Q = \int_0^\infty e^{A^T\tau} C^T C e^{A\tau} d\tau. \tag{7}$$

## 3 Nonlinear systems

One approach for construction of empirical gramians is outlined in the paper [3]. However, instead of considering delta-inputs, it is more natural to analyze the system in the vicinity of an equilibrium point when $u(t) = 0$ . Consider the vicinity of an isolated asymptotically stable equilibrium point (steady–state solution) which is supposed to be a constant solution and is chosen for simplicity at $x = 0$, i.e. $f(t, 0) \equiv 0$, [1].

In what follows, it is proposed to make use of an approximation for the most natural object – the fundamental solution $\Theta$ of (1) that would generalize the $\exp(At)$ term for linear systems ($f(x) = Ax$). This is reasonable since the projection Krylov spaces for linear systems are generated by their fundamental solution $\exp(At)$. The constructions would, in general, depend on $\Theta$ for negative times which is unavoidable. For linear systems, of course, there is a simplification since $\left(e^{A(-t)}\right)^{-1} \equiv e^{At}$ so this does not present a limitation but in general, $\Theta^{-1}(-t) \neq \Theta(t)$ , cf. (5).

Let $x^{ilm}(t)$ be the solution of (1) with $u \equiv 0$ and with initial condition $x^{ilm}(0) = c_m T_l e_i$. It is assumed that this initial condition does not take the system outside the region of attraction of the equilibrium point $x = 0$. Then the 'state-space average' of the 'nonlinear' fundamental solution may be defined as:

$$\langle \Theta(t) \rangle = \frac{1}{rs} \sum_{m=1}^{s} \sum_{l=1}^{r} \sum_{i=1}^{n} \frac{1}{c_m} x^{ilm}(t) e_i^T T_l^T \tag{8}$$

where $\mathbf{M} \equiv \{c_1, c_2, \ldots, c_s\}$ is a set of $s$ positive constants, $\mathbf{T}^n \equiv \{T_1, T_2, \ldots, T_r\}$ – a set of $r$ orthogonal $n \times n$ matrices and $\mathbf{E}^n \equiv \{e_1, e_2, \ldots, e_n\}$ a set of standard unit vectors in $\Re^n$. The purpose of using these sets is an attempt to ensure that the entire region of feasible values of initial inputs/states is covered and probed. The set $\mathbf{E}^n$ defines the standard directions and the set $\mathbf{T}^n$ defines 'rotations' of these directions. The set $\mathbf{M}$ introduces different scales for each direction of the initial states/inputs.

The following constructions of empirical controllability and observability gramians are now suggested:

**Definition 1.** *For the system in (1) – (2), the empirical controllability gramian is defined as:*

$$\tilde{P} = \int_0^\infty \langle \Theta(-\tau) \rangle^{-1} B(-\tau) B^T(-\tau) \langle \Theta(-\tau) \rangle^{-1T} d\tau \tag{9}$$

*where $\langle \Theta(t) \rangle$ is as described in (8).*

Of course, this construction requires that $\langle\Theta(-\tau)\rangle$ is invertible for all $\tau \geq 0$. (9) is obviously a generalisation of (5).

**Definition 2.** *For the system in (1) – (2) the empirical observability gramian is defined as:*

$$\tilde{Q} = \int_0^\infty z^T(\tau)z(\tau)d\tau \tag{10}$$

*where $z(\tau) \in \Re^n$ is given by:*

$$z(t) = \frac{1}{rs}\sum_{i,l,m}\frac{1}{c_m}y^{ilm}(t)e_i^T T_l^T$$

*and $y^{ilm}(t)$ is the output which corresponds to an initial state $x^{ilm}(0) = c_m T_l e_i$ and a zero source term.*

Both gramians (9) and (10) when applied to LTVS (or LTIS) thus result in the usual gramians i.e. (5) and (6).

## 4 Illustrative numerical example

The ladder of Fig. 1 represents a heat flow model [6]. The voltage at the $m$-th node represents the temperature on a rod at a distance, proportional to $m$ (i.e. the distance is being discretized). The (input) voltage at node 1 represents the heat source. The nonlinearities represent the conductivity dependence on the temperature. The output is taken as the average voltage at all nodes, representing the average temperature of the rod. The choice of different parameters of the circuit represents different spatial or environment conditions [6]. The nonlinear resistor introduces quadratic nonlinearity at each node $i_{\text{nl}}(v) = gv^2$ for $v > 0$. Varying $g$ can change the magnitude of the nonlinearity. The condition $v > 0$ is achieved by taking the input current $u(t) > 0$. The other parameters are $C = r = 1$.

The example enables confirmation of the effectiveness of the reduction based on the proposed empirical gramians – Fig. 2.

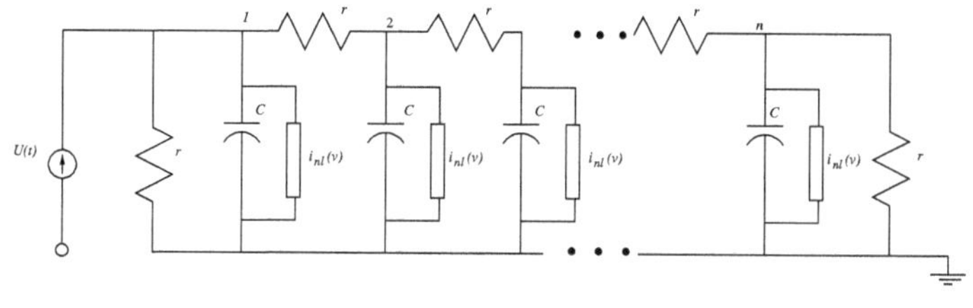

**Fig. 1.** Circuit with quadratic nonlinearity

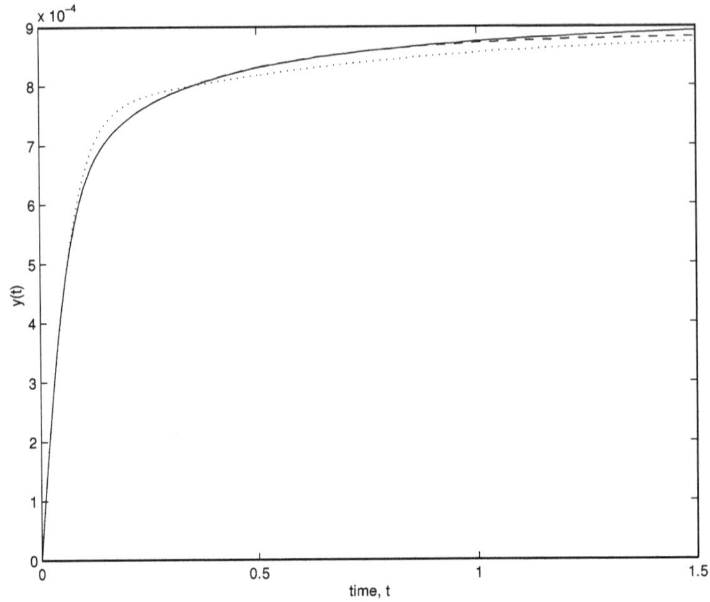


**Fig. 2.** Comparison between output from nonlinear model ($n = 100$, $g = 10$) and reduced order $k = 3$ models – step input: (a) solid line-original nonlinear model; (b) dashed line – reduced order model, reduction based on the novel empirical gramians (9) and (10); (c) dotted line – reduced model, with gramians based only on the linear part of the system.

## References


1. M. Condon and R. Ivanov. Empirical balanced truncation of nonlinear systems. *Journal of Nonlinear Science*, 2004. To appear.
2. M. Condon and R. Ivanov. Nonlinear systems – algebraic gramians and model reduction. *COMPEL Journal*, 24(1), 2005. To appear.
3. S. Lall, J.E. Marsden, and S. Glavaski. A subspace approach to balanced truncation for model reduction of nonlinear control systems. *International Journal of Robust and Nonlinear Control*, 12:519–535, 2002.
4. J.R. Phillips. Projection-based approaches for model reduction of weakly nonlinear, time-varying systems. *IEEE Transactions on computer-aided design of integrated circuits and systems*, 22(2):171–187, 2003.
5. P. Van Dooren. Gramian based model reduction of large-scale dynamical systems. In D.F. Griffiths and G.A. Watson, editors, *Numerical Analysis 1999*, Research Notes in Mathematics Series, pages 231–247. CRC Press LLC, 2000.
6. T. Veijola and L. Costa. Combined electrical and thermal circuit simulation using APLAC. In *Circuit Theory Laboratory Report Series, No. CT-34*. Helsinki University of Technology, 1998.

# Electrolyte Flow and Temperature Calculations in Finite Cylinder Caused by Alternating Current

A. Buikis and H. Kalis

Institute of Mathematics of Latvian Academy of Sciences and University of Latvia, Akadēmijas laukums 1, Riga, LV-1524, Latvia, buikis@latnet.lv, kalis@lanet.lv

**Summary.** The distribution of electromagnetic fields, forces and source term of temperature induced by an alternating axially-symmetric system of electric current in a cylinder of a finite length with 6 electrodes has been investigated and calculated in [2, 1].

In this paper the three-phase alternating current with phase shift 120 degree is fed to every of 9 discreate circular conductors-electrodes, which are placed on the internal wall of the cylinder. The motion of electrolyte and temperature distribution in a cylinder has been calculated in dependence of the arrangement of electrodes .

**Key words:** Magneto-hydrodynamic flow, temperature, electrolyte, cylinder.

## 1 Introduction

In many technological applications it is important to mix and heat an electroconductive liquid, using various magnetic fields. One of the modern areas of applications developed during last years is effective use of electrical energy produced by alternating current in production of heat energy. This process is ecologically clean. Devices based on this principle are developed during last ten years. This work presents the mathematical model of one of such devices. It is a finite cylinder with 9 metal coils-electrods positioned on its inner surface with a fixed distance one from the other. By connecting those coils to three–phase alternating current, they irradiate energy. If this cylinder is placed, for example, in a house heating system together with a small electromotor which rotates water in the entry of cylinder and pumps water through it, we obtain an effective, compact and ecologically clean house heating device.

## 2 Mathematical Model

In this work we considere a finite cylinder $\tilde{\Omega} = \{(r, z) : 0 < r < a,\ 0 < z < Z\}$ with 9 metal coils-electrodes $L_i = \{(r, z),\ r = a,\ z = z_i\},\ 0 < z_i < Z,\ i = \overline{1, 9}$, positioned on its inner surface with a fixed distance one from the other. Alternating current with density $j_i = j_0 \cos(\tilde{\omega} t + (i-1)\theta)$, is fed to every of 9 discrete circular conductors.

Here $j_0$ is the amplitude, $\tilde{\omega} = 2\pi f$, $f$ are the angular frequency and frequency of the alternating current, $\theta$ is the phase (usually $\theta = 120^0$, $f =$ 50Hz) and $t$ is the time ($j_0 \approx 10^6 \frac{\mathrm{A}}{\mathrm{m}^2}$).

The current creates in the weakly conductive liquid-electrolyte axial $F_z$ and radial $F_r$ components of the electromagnetic force (Lorentz' force).

For calculating the electromagnetic fields, the averaging method over the time interval $2\pi/\tilde{\omega} = 1/f$ is used. The averaged values of force $< F_r >$, $< F_z >$ give rise to a liquid (electrolyte) motion.

At the inlet of the cylinder we have a uniform velocity $U_0 \approx 0.1 \frac{\mathrm{m}}{\mathrm{s}}$, but the swirl velocity is taken as the induced by the rigid body rotation with angular velocity $\Omega_0 \approx 4s^{-1}$.

The liquid have following parameters:
kinematic viscosity $\nu \approx 10^{-5} \frac{\mathrm{m}^2}{\mathrm{s}}$, density $\rho \approx 1000 \frac{\mathrm{kg}}{\mathrm{m}^3}$, the electric conductivity $\sigma \approx 100 \Omega^{-1}\mathrm{m}^{-1}$, the specific heat capacity $c \approx 4000 \frac{\mathrm{J}}{\mathrm{kg.K}}$, the heat conductivity $\lambda \approx 0.6 \frac{\mathrm{W}}{\mathrm{m.K}}$ and the heat exchange coefficient $\alpha \approx 12 \frac{\mathrm{W}}{\mathrm{m}^2.\mathrm{K}}$.
The radius $a$ of the cylinder is 0.05m, the length $Z$ of the cylinder is 0.35m.

The axially-symmetric stationary Navier-Stokes equations for vorticity function $\omega$, hydrodynamic-stream function $\psi$ and circulation $W$ in the cylindrical coodinates ($r$, $\varphi$, $z$) are used in the non-dimensional form with swirl number $\Gamma$ , Reynolds number $Re$ and Taylor number $Te$ [1].

These equations were put in the dimensionless form scaling all the lengths to $r_0 = a$ (the inlet radius of the tube), the axial velocity $v_z$ to $U_0$( the uniform inlet axial velocity), swirl velocity to $W_0 = r_0 V_0$, the azimuthal velocity $v_\varphi$ to $V_0 = r_0 \Omega_0$, the vorticity $\omega$ to $\omega_0 = U_0/r_0^2$ and stream function $\psi$ to $\psi_0 = U_0 r_0^2$.

In the part of the inlet ($z = 0,\ 0 \le r < r_1$) the axial streams are assumed to have a uniform velocity $U_0 : W = \omega = 0, \qquad \psi = 0.5r^2$;
in the other part of the inlet ($z = 0,\ r_1 \le r \le 1$) the swirl velocity profile is induced by the rigid body rotation with the angular velocity $\Omega_0 : W = r^2,\ \omega = 0,\ \psi = 0.5(r_1^2 + \beta(r^2 - r_1^2))$,
where $\beta \approx 0.1$ is the velocity ratio of the coaxial free stream velocity to axial jet velocity $U_0$.

The steady heat transport equation with source terms $j_\varphi$ and with constant properties ([1] heat convection is neglected) contain the heat sources parameter $K_T$, Biot number $Bi$, Prandtl number $Pr$, Peclet $Pe$ and Eckert $Ec$numbers. The dimensionless temperature $T = \frac{\tilde{T} - T_a}{T_b - T_a}$, where $T_a = 293$K, $T_b = 353$K are external and permissible temperatures, $\tilde{T}$ is dimensional temperature.

## 3 The Finite-Difference Approximations and Numerical Results.

The presence of large parameters of first order derivatives $(\Gamma, Re, Pe)$ in the systems of differential equations couses additional numerical difficulties for the application of the general finite-difference methods . Thus special monotonous approximations are constructed [3], using functions of matrix and the exponential functions $s(x) = x/(\exp(x)-1) > 0,\ s'(x) = \frac{ds}{dx} < 0 (s(0) = 1,\ s'(0) = -0.5)$ with Patankar approximations [4]. We consider an uniform grid with steps $h_1$, $h_2$ in the $r$, $z$ directions. The coresponding finite difference scheme [2] is calculated with the under relaxation method.

As the basis for the calculations of 9 circular conductors $L_i$ are chosen, which are arranged in the axial direction at the points $z_i = 0.2i,\ i = \overline{1,9}$. The results of numerical experiments was obtained in the case of $h_1 = h_2 = 0.1,\ r_1 = 0.5,\ \Gamma \in [0,8],\ Re = 500,\ K_T = 1.2,\ Pr = 67,\ Pe = 10^3,\ Ec = 10^{-8},\ l = Z/a = 3$.

The values of averaged forces $< F_z >$, $< F_r >$, curl of forces $< f^{\varphi} >$ and maximal value of the source function $< j_{\varphi}^2 >$ depending of the arrangement of 9 conductors by numbering $nj = [123456789]$ are in the Table 1.

**Table 1.** The extremal values of averaged forces and curl of forces

| $Nr.$ | $nj$ | $< F_z >$ | $< F_r >$ | $< f^{\varphi} >$ | $< j_{\varphi}^2 >$ |
|---|---|---|---|---|---|
| 1 | [123456789] | [-17.7;1.10] | [-11.9;11.9] | [-1.4;122] | 2.49 |
| 2 | [135792468] | [-1.10;17.7] | [-11.9;11.9] | [-122;0.6] | 2.49 |
| 3 | [147258369] | [-69.0;3.50] | [-50.5;50.5] | [-49.;200] | 12.9 |
| 4 | [761835924] | [-51.1;40.0] | [-36.8,40.6] | [-247;132] | 12.1 |
| 5 | [531642789] | [-28.3;47.5] | [-21.8;26.3] | [-193;195] | 9.70 |
| 6 | [478591623] | [-31.4;15.4] | [-20.3;20.3] | [-59.;188] | 8.83 |
| 7 | [963852741] | [-3.50;69.0] | [-50.5;50.5] | [-200;49.] | 12.9 |
| 8 | [258147369] | [-69.0;3.50] | [-50.5;50.5] | [-49.;200] | 12.9 |
| 9 | [369147258] | [-69.0;3.50] | [-50.5;50.5] | [-49.;200] | 12.9 |

We obtain for the dimensionless values of $\psi_{max}$, $W_{max}$, $\omega \in [\omega_{min}, \omega_{max}]$ $T_{max}$, $T_{av} = \frac{1}{l}\int_0^l \int_0^1 rT(r,z)\mathrm{d}r\mathrm{d}z$ depending of different connection of electrods and of the parameter $\Gamma$ by $Te = 0.1$ following selected results:

1. Conductors are series connected, one after another $nj = [123456789]$ (see in the Table 1, $Nr.1$):
$\Gamma = 0,\ \psi_{max} = 0.20,\ \omega \in [-1.1, 6.9],\ T_{max} = 0.052,\ T_{av} = 0.027$, – on the cylinder surface by the electrode developed small vortex, induced by the Lorentz force;

2. Conductors are connected to each other skipping two of them, the ends of 3 wires are in the begin of electrodes $nj = [147258369]$ (see in the Table 1,

$Nr.3$):
a) $\Gamma = 0$, $\psi_{max} = 0.38$, $\omega \in [-9.0, 70]$, $T_{max} = 0.50$, $T_{av} = 0.43$, Fig. 1 shows big vortex by the last electrod, induced by Lorentz force, Fig. 3 shows the distribution of temperature;
b) $\Gamma = 8$, $\psi_{max} = 0.29$, $\omega \in [-16, 80]$, $W_{max} = 0.72$, $T_{max} = 0.53$, $T_{av} = 0.45$, Fig. 2 shows that this vortex decreases and at the inlet of the cylinder the vortex-breakdown from the swirling jets develops([2]), the distribution of the temperature is more uniform;

3. The first 6 conductors are connected to each other skipping one of them, but the last 3 are series connected $nj = [531642789]$ (see in the Table 1, $Nr.5$): $\Gamma = 0$, $\psi \in [-0.13, 0.18]$, $\omega \in [-35, 7]$, $T_{max} = 0.12$, $T_{av} = 0.10$, Fig. 4 shows big vortex by the first electrods, induced by Lorentz force, but the temperature is small;

4. The ends of 3 wires are in the end of electrods, $nj = [963852741]$ (see in the Table 1, $Nr.7$):
a) $\Gamma = 0$, $\psi \in [-0.32, 0.16]$, $\omega \in [-107, 32]$, $T_{max} = 0.26$, $T_{av} = 0.20$, – we can see big vortex by the first electrods (similar with case 3.);
b) $\Gamma = 1$, $\psi \in [-0.83, 0.17]$, $\omega \in [-108, 414]$, $W_{max} = 0.52$, $T_{max} = 0.24$, $T_{av} = 0.18$, – the vortex by the electrods decreases and gets up.

## 4 Conclusion

1. The results of the numerical experiments with 9 circular conductors reported here had give some new physical conclusions on the flow behavior and distribution of temperature in the cylinder.
2. The averaged values of the electric field, electromagnetic forces, the azimuthal component of the curl of forces' vector and the heat source are calculated for different arrangement of the electrodes.
3. Using monotone finite-difference schemes for calculations, the average in the time axially-symmetric motion of electrolyte and the temperature distribution in a cylinder have been obtained:
   1) the vortex formation inside the cylinder ;
   2) the distribution of temperature depending of arrangement of the electrods (the maximal dimensionless temperature for the conductors connected to each other skiping two of them ($nj = [147258369]$) is 10 times higher than in the case when the conductors are connected in series).

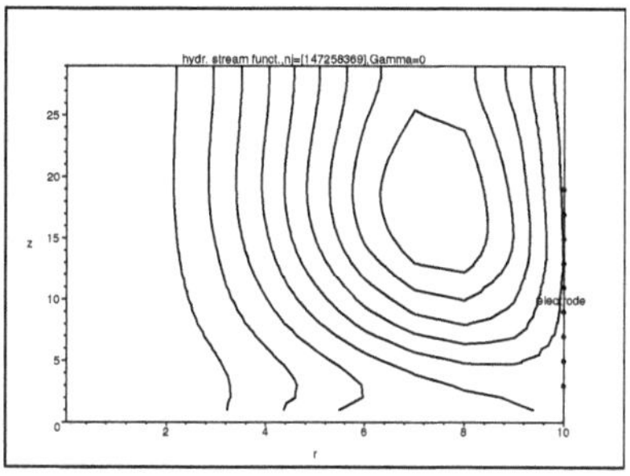


**Fig. 1.** The stream functions $nj$ = [147258369], $\psi \in (0.00, 0.38)$, $\Gamma = 0$

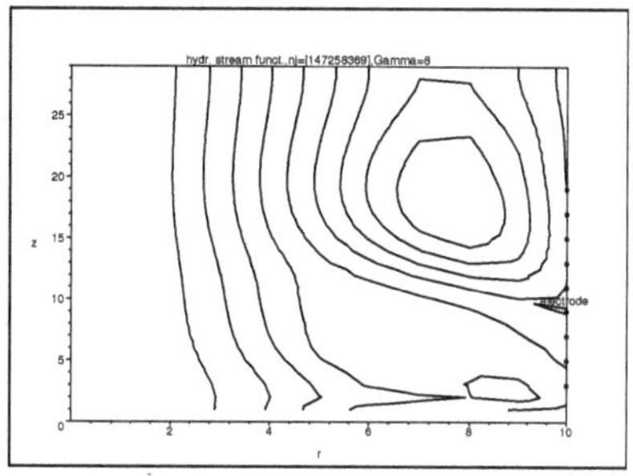


**Fig. 2.** The stream functions $nj$ = [147258369], $\psi \in (0.00, 0.29)$, $\Gamma = 8$

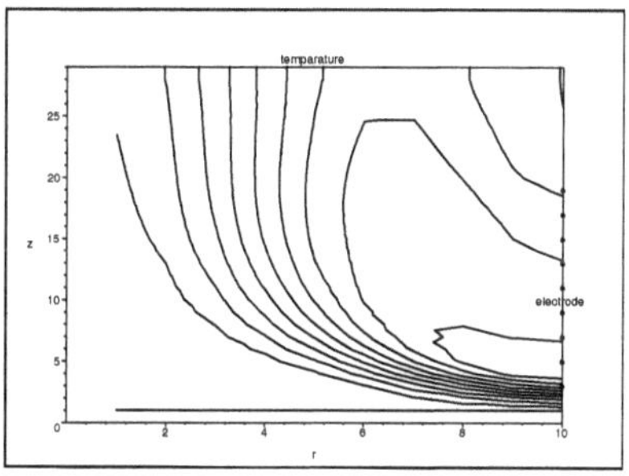


**Fig. 3.** Temperature $nj$ = [147258369], $T_{max} = 0.50$, $\Gamma = 0$

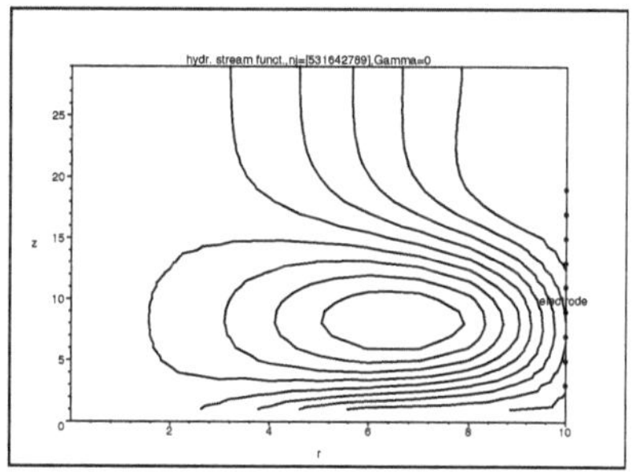


**Fig. 4.** The stream functions $nj$ = [531642789], $\psi \in (-0.13, 0.18)$, $\Gamma = 0$

# References

1. A. Buikis, R. Čiegis, and A.D. Fitt, editors. *Creation of temperature field in finite cylinder by alternated electromagnetic force.*, volume 5 of *Mathematics in Industry*. Progress in Industrial Mathematics at ECMI 2002. 12th Conference of the European Consortium for Mathematics in Industry., Springer, 2004.
2. A. Buikis and H. Kalis. Flow and temperature calculation of electrolyte for a finite cylinder in the alternating field of finite number circular wires. *Magnetohydrodynamics*, 40:77–90, 2004.
3. H. Kalis. Special computational methods for solution of MHD problems (in russian). *Magnetohydrodynamics*, 30(2):144–155, 1994.
4. S. Patankar. *Calculation of the heat transfer and fluid flow problems(in Russian).* Moscow, 1984.

# Numerical Simulation of the Problem Arising in the Gyrotron Theory

J. Cepitis[1], O. Dumbrajs[2], H. Kalis[1], and A. Reinfelds[1]

[1] Institute of Mathematics Latvian Academy of Sciences and University of Latvia, Akadēmijas laukums 1, LV-1524 Rīga, Latvia lzalumi@latnet.lv
[2] Helsinki University of Technology, FIN-02150 Espoo, Finland dumbrajs@csc.fi

**Summary.** Numerical aspects for solving of certain problem arising in gyrotron theory are discussed. Particularly, finite-difference schemes using quasistationarization and method of lines were applied and the relevant results analyzed.

**Key words:** Finite-difference schemes, method of lines, gyrotrons, quasisteady solutions, Robin's boundary conditions, spectral problems.

## 1 Introduction

Gyrotrons are microwave sources whose operation is based on the stimulated cyclotron radiation of electrons oscillating in a static magnetic field. Gyrotron oscillators can have a wide application, including radars, advanced communication systems, technological processes, atmospheric sensing, ozone conservation, artificial ionospheric mirror, extra-high-resolution spectroscopy, etc. However, the main application of powerful gyrotrons is electron cyclotron resonance plasma heating in tokamaks and stellarators and the noninductive current drive in tokamaks. Extensive literature exists on various aspects of these devices, see [3].

Competition between the amplitudes of nonstationary gyrotron oscillations $f_s$ and the complex transverse momentum of electrons $p$ in different modes ($s = 1, \dots, m$) can be described by the following nonlinear system of equations [1, 4]

$$\begin{cases} \dfrac{\partial p}{\partial x} + i(|p|^2 - 1)p = i \sum\limits_s f_s \exp(i\Delta_s x + \Psi_s), \\ \dfrac{\partial^2 f_s}{\partial x^2} - i\dfrac{\partial f_s}{\partial t} + \delta_s f_s = I_s \dfrac{1}{4\pi^2} \displaystyle\int_0^{2\pi} \int_0^{2\pi} p \exp[-i(\Delta_s x + \Psi_s)]\, d\Theta_0\, d\Phi. \end{cases}$$

Here $i$ is the imaginary unit, $x \in [0, L]$ and $t \geq 0$ is the normalized axial and time coordinates, for every number of mode $s$ the corresponding $\Delta_s$ is the frequency mismatch, $\delta_s$ describes variation of the critical frequencies, $I_s$ is the dimensionless current, $\Psi_s$ is the phase of the mode.

This system of equations has to be supplemented by the standard initial conditions $p(0) = \exp(i\Theta_0)$,

$$f_s(0,x) = f_{s0}(x), \tag{1}$$

where $f_{s0}(x)$ are given complex functions, and by the boundary conditions for the field at the entrance and at the exit to the interaction space in the cavity of gyrotron

$$f_s(t,0) = 0, \qquad \frac{\partial f_s(t,L)}{\partial x} = -i\gamma_s f_s(t,L), \tag{2}$$

where $\gamma_s$ are positive parameters which describe the wave number at the resonators exit.

The word combination "competition between the amplitudes of oscillations in different modes" is used in the sense that most of these modes vanishes in direction of axis $x$ which is located along the length of the gyrotron cavity.

The boundary condition of Robin's type at the resonators exit causes additional difficulties for the numerical simulation, particularly for the constructing of difference schemes. Moreover, these schemes require many hours of computing time for every single set of gyrotron operating parameters and this severely restricts the parameters space which could be investigated. It was discovered that the results of the computations depend in a nontrivial manner on the chosen spatial and temporal step-lengths. In particular it was noticed that the temporal step-length cannot be taken too small, otherwise unstable and unpredictable results were obtained.

In order to study peculiarities of used numerical methods it is appropriate to investigate the single mode case having also special restrictions to the variable $p$.

If $|p|^2 = 1 + C$, ($C = \text{const} \le 0$), and in the single-mode approximation ($m = 1$ and subscripts in designations are omitted) the nonlinear system of gyrotron equations reduces to the following form:

$$i\frac{\partial f}{\partial t} = \frac{\partial^2 f}{\partial x^2} + \delta f - iIe^{-i\Delta_* x}\int_0^x f(t,\xi)e^{i\Delta_*\xi}\,d\xi, \tag{3}$$

where $\Delta_* = \Delta + C$, with corresponding to (1) and (2) initial and boundary conditions.

Using the designation $y(t,x) = \int_0^x f(t,\xi)e^{i\Delta_*\xi}\,d\xi$ and the transformation $\bar{f} = e^{i\Delta_* x} f$ the integro-differential equation (3) can be written in the form of the system of two partial differential equations

$$\begin{cases} \dfrac{\partial \bar{f}}{\partial t} = -i\left(\dfrac{\partial^2 \bar{f}}{\partial x^2} - 2i\Delta_*\dfrac{\partial \bar{f}}{\partial x} + (\delta - \Delta_*^2)\bar{f}\right) - Iy(t,x) \\ \dfrac{\partial y}{\partial x} = \bar{f}(t,x), \end{cases} \tag{4}$$

with the following conditions:

$$\bar{f}(t,0)=0,\quad \frac{\partial \bar{f}(t,L)}{\partial x}=-i(\gamma-\Delta_*)\bar{f}(t,L),$$
$$\bar{f}(0,x)=e^{i\Delta_* x}f_0(x). \tag{5}$$

The numerical simulation of the problem (4), (5) in the case $I=0$ and $\delta=$ *const* was investigated in the article [2]. Now we are presenting observations for the general case.

## 2 Numerical Simulation

### 2.1 Quasistationarization

We represent the quasisteady solution of the problem (4), (5) in the form

$$\bar{f}(t,x)=g(x)\exp(i\alpha t),\quad y(t,x)=w(x)\exp(i\alpha t), \tag{6}$$

where $\alpha$ is a complex number $\alpha=\alpha_1+i\alpha_2$ ($\alpha_2$ is a temporal damping factor: if $\alpha_2>0$, the solution of (4) decreases in time, if $\alpha_2<0$, this solution increases, and for $\alpha_2=0$ this solution is oscillating in time).

Substituting the (6) into equation (4) and conditions (5), we obtain the Sturm-Liouville problem for the system of two ordinary differential equations

$$\begin{cases} g''(x)-2i\Delta_* g'(x)+\tilde{\lambda}g(x)-Iiw(x)=0 \\ w'(x)=g(x) \\ g(0)=w(0)=0,\quad g'(L)=-i(\gamma-\Delta_*)g(L), \end{cases} \tag{7}$$

where $\tilde{\lambda}=\alpha+\delta-\Delta_*^2$ is complex eigenvalue.

If $\delta=$ const then for the nontrivial solution of problem (7) we obtain a transcendental complex equation for calculating the eigenvalue $\tilde{\lambda}$

$$\mu_3\kappa_3(\mu_2-\mu_1)+\mu_2\kappa_2(\mu_1-\mu_3)+\mu_1\kappa_1(\mu_3-\mu_2)=0, \tag{8}$$

where $\kappa_j=(\mu_j-i(\Delta_*-\gamma))\exp(\mu_j L)$, $\mu_j, j=1,2,3$, are the three roots of the following complex cubic equation:

$$\mu^3-2i\Delta_*\mu^2+\tilde{\lambda}\mu-Ii=0,$$

and number the roots of the equation (8) $\lambda^{(k)}$, $k=1,2,\dots$ by increasing their real parts. It is seen that the parameter $\delta$ affects only the values of $\alpha_1$, i.e., determines mainly the spatial, but not the temporal behavior of the function $f$.

The corresponding complex eigenfunctions $g^{(k)}(x)$ can be obtained from the problem (7) and the quasisteady solution for every eigenvalue $\lambda^{(k)}=\alpha^{(k)}-\delta+\Delta_*^2$ in the form

$$f^{(k)}(t,x) = \exp(-i\Delta_* x) g^{(k)}(x) \exp(i\alpha^{(k)} t).$$

From $\alpha_2^{(k)} = Im(\lambda^{(k)}) > 0$ follows that asymptotically the solutions tends to zero, if $t$ tends to $\infty$. In practice $\delta$ is not a constant, but has the form

$$\delta(x) = \left(\sinh\left(\frac{L}{2}\right)\right)^{-\frac{1}{2}} \sinh\left(x - \frac{L}{2}\right).$$

### 2.2 Method of Lines

For approximation of the derivatives we considered:

- Uniform grid with the grid points

$$x_j = jh, \qquad j = \overline{0,n}, \tag{9}$$

  where $h = L/n$ is the space step;
- Nonuniform grid with the grid points as the roots of the Chebyshev polynomials of the second kind

$$x_j = 0.5L(1 - \cos(\pi j/n)), \quad j = \overline{0,n}. \tag{10}$$

Considering only the spatial discretization (the variable $x$ is discretized and the variable $t$ is continuous) and using the matrix $D, D^2$ of derivatives in the grids (9),(10) (see, [2]) we obtain the system of ordinary differential equations in the following matrix-vector form:

$$\frac{df_h}{dt} = Gf_h, \quad f_h(0) = f_{0h}, \tag{11}$$

where $G = B - IE^-(D^{-1}E^+)$, $\quad B = -i(D^2 + \delta E)$, $E^\pm = diag(\exp(\pm i\Delta_* x_h))$.
The solution $f_h = f_h(t)$ of the problem (11) can be obtained in two forms:

- Using matrix-exponent function

$$f_h(t) = \exp(Gt) f_{0h};$$

- Using the spectral decomposition of matrix $G = RD_0R^{-1}$, where $D_0$ is the diagonal eigenvalues matrix, $R$ is the eigenvectors matrix with corresponding eigenvectors in the columns of the matrix

$$f_h(t) = R\exp(D_0 t)(R^{-1} f_{0h}). \tag{12}$$

It were appeared that for any $n$ first $n_1 < n$ eigenvalues well coincide with the roots of the equation (8), but some last $n_2 = n - n_1$ eigenvalues were "parasitic" (see the example in [2]). These $n_2$ eigenvalues substantially differ from the precise ones and can cause inaccuracy for the calculated value of $f_h(t)$. Consequently they must be excluded from the evaluation (12). For this purpose we used the spectral decomposition method for the solving the system (11) in one of the following forms

$$f_h(t) = R_* \exp(D_0^* t)(R^{-1} f_{0h})_*,$$

or

$$f_h(t) = R_* \exp(D_0^* t)(R_*' R_*)^{-1} R_*' f_{0h},$$

where $R_*$ is the matrix $R$ without the last $n_2$ columns, $R_*'$ is the transpose matrix of $R_*$, $D_0^*$ is the quadratic matrix of $D_0$ without the last $n_2$ columns and rows and $(R^{-1} f_{0h})_*$ is the vector-column $(R^{-1} f_{0h})$ without the last $n_2$ rows.

## 3 Conclusions

Numerical experiments show that for uniform grid the two level approximation in the time the time step $\tau$ must be small. Therefore for the finite-difference scheme the number of grid points increased ($n \geq 200$) and large computer time is required. So, more suitable is to use the nonuniform grid and algorithm of method of lines, which is modified using the spectral decomposition method without some last inaccurate eigenvalues.

Our observations make possible more efficiently to numerically solve the general system of equations for arbitrary number $m$ of gyrotron oscillation modes.

*Acknowledgement.* This work is partly financially supported by the project No FU05-CT-2003-0083 of the European program EUROATOM.

## References

1. M.I. Airila and O. Dumbrajs. Generalized gyrotron theory with inclusion of adiabatic electron trapping in the pressence of a depressed collection. *Phys. Plasmas*, 8:1358–1362, 2001.
2. O. Dumbrajs, H. Kalis, and A. Reinfelds. Numerical solution of single-mode gyrotron equation. *Mathematical Modelling and Analysis*, 9:25–38, 2004.
3. C.J. Edgcombe, editor. *Gyrotron Oscillators: Their Principles and Practice.* Taylor and Francis, Ltd. London, 1993.
4. N.A. Zavolsky, G.S. Nusinovich, and A.B. Pavelyev. Stability of single-mode oscillations and nonstationary processes in gyrotrons with very large low quality factor resonators. In *Gyrotrons. Academy of Sciences of USSR, Gorky*, pages 84–112, Russian, 1989.

# A Deterministic Multicell Solution to the Coupled Boltzmann-Poisson System Simulating the Transients of a 2D-Silicon MESFET

C. Ertler, F. Schürrer[1], and O. Muscato[2]

[1] Institute of Theoretical and Computational Physics, Graz University of Technology, Graz, Austria
ertler@itp.tu-graz.ac.at, schuerrer@itp.tu-graz.ac.at
[2] Dipartimento di Matematica e Informatica, Università di Catania, Catania, Italy
muscato@dmi.unict.it

**Summary.** A deterministic solution method for the coupled Boltzmann-Poisson system regarding spatially two-dimensional problems is presented. The method is based on a discontinuous piecewise polynomial approximation of the carrier distribution function. The conduction band of silicon is modelled by a non-parabolic six-valley model. In particular, we applied the multicell method to simulate the transients of a silicon MESFET. The results are compared to Monte Carlo simulations.

**Key words:** Electron transport, semiconductors, kinetic theory.

## 1 Introduction

According to the shrinking size of modern semiconductor devices, a detailed kinetic description of the occurring transport processes becomes indispensable. This is achieved on semiclassical grounds by the coupled Boltzmann-Poisson system. Solving this coupled nonlinear system is a difficult, challenging task owing to its complicated mathematical structure.

The stochastic Monte Carlo (MC) technique is widely used, because of its direct physical interpretation and its ease to include various physical effects. However, this technique suffers from the disadvantages of statistically noisy results and an inefficient simulation of transients. Therefore, the development of alternative, deterministic methods, which are accompanied with less computational burden, has become an active field of research.

In 2003, Carillo et al. [1] were successful in applying a high-order shock-capturing scheme (WENO) as proposed in [4] to the Boltzmann transport equation (BTE). This non-oscillatory upwind finite difference scheme prevents so-called shocks of the carrier distribution function, which can appear according to the hyperbolic nature of the BTE. Their simulation results of a one-dimensional $n^+$-n-$n^+$-silicon diode are in excellent agreement with MC-calculations but with the advantage of saving CPU time and a noise-free resolution. Recently, we were able to reproduce their results by using an even faster deterministic multicell method, which is based on a discontinuous piecewise polynomial approximation of the carrier distribution function [2]. In this paper, we extend the multicell method to spatially two-dimensional problems.

The paper is organized as follows. The underlying physical model of silicon is shortly described in Sect. 2. We sketch the multicell method in Sect. 3 and finally apply it to a two-dimensional short channel silicon MESFET. The numerical results are compared to Monte Carlo data obtained with the Damocles code [3].

## 2 Physical Assumptions

We consider a non-parabolic multivalley model for the conduction band of silicon by considering six equivalent valleys lying around the six energy minima along the $\Delta$-directions of the first Brillouin zone. In the principal axis system of the ellipsoidal shaped isoenergetic surfaces, the dispersion relation regarding the electron energy $\varepsilon$ reads

$$\gamma = \varepsilon(1+\alpha\varepsilon) = \frac{\hbar^2}{2m_0}k^{*2}. \tag{1}$$

Here, we introduced the nonparabolicity parameter $\alpha$, the free electron mass $m_0$ and the starred wave vector $k_i^* = (m_0/m_i)^{1/2}k_i$, $i = 1,2,3$ with $m_1 = m_2 = m_\mathrm{t}$ and $m_3 = m_\mathrm{l}$ denoting the transversal and longitudinal effective mass of the electrons, respectively.

The transport of the electrons in each single valley under the impact of an electric field $\mathbf{E}$ is governed by the BTE for the corresponding space-, momentum- and time-dependent electron distribution function $f^\alpha(\mathbf{r},\mathbf{k},t)$:

$$\frac{\partial f^\alpha}{\partial t} + \mathbf{v}\frac{\partial f^\alpha}{\partial \mathbf{r}} - \frac{e\mathbf{E}}{\hbar}\cdot\frac{\partial f^\alpha}{\partial \mathbf{k}} = C[f^\alpha] + C_{\mathrm{i.v.}}[f^\alpha, f^\beta] \quad \alpha,\beta = 1,\ldots,6\ . \tag{2}$$

Here, $e$ is the elementary charge, $\mathbf{v}$ denotes the group velocity of the electrons, $C[f^i]$ and $C_{\mathrm{i.v.}}[f^\alpha, f^\beta]$ labels the collision operators according to intra- and intervalley scattering processes. The electric potential $\Phi(\mathbf{r},t)$ is determined by the Poisson equation

$$\Delta_\mathbf{r}\Phi(\mathbf{r},t) = \frac{e}{\varepsilon_0\kappa_s}\left[n(\mathbf{r},t) - n_\mathrm{d}(\mathbf{r})\right], \tag{3}$$

where $\varepsilon_0$ denotes the vacuum permittivity and $\kappa_s$ is the dielectric constant of the considered semiconductor. The particle densities of the electrons and donors are represented by $n$ and $n_\mathrm{d}$, respectively. The system of BTE's couples nonlinearly to the Poisson equation via $\mathbf{E} = -\nabla_\mathbf{r}\Phi$ and

$$n(\mathbf{r},t) = \frac{1}{4\pi^3} \int_{1.\mathrm{BZ}} \mathrm{d}\mathbf{k}\, f(\mathbf{k},\mathbf{r},t), \tag{4}$$

where the integration is extended over the first Brillouin zone (1.BZ).

We assume the electrons to interact only with phonons and neglect electron-electron and impurity scattering. The phonon gas is considered to remain in permanent equilibrium at the lattice temperature $T$. By considering a non-degenerate electron gas, the collision operator typically reads

$$C[f] = \frac{V}{8\pi^3}\left(\int P(\mathbf{k}',\mathbf{k}) f(\mathbf{k}')\mathrm{d}^3k' - \int P(\mathbf{k},\mathbf{k}') f(\mathbf{k})\mathrm{d}^3k'\right), \tag{5}$$

where $P(\mathbf{k}',\mathbf{k})$ represents the transition rate from state $\mathbf{k}'$ to $\mathbf{k}$ and $V$ denotes the volume of the crystal. Isotropic transition rates for acoustic and optical phonon scattering are obtained by applying the deformation potential approximation as given in [5].

## 3 The Multicell Method for Spatially Two-Dimensional Problems

The mathematical structure of the collision integrals [5] suggests to express the BTE in spherical coordinates of the starred wave vector $\mathbf{k}^*$. Moreover, to ensure that the particle conservation is established by the derived model equations, we introduce a distribution function weighted by the density of states:

$$\psi(\mathbf{k},t) = \gamma^{1/2}(1+2\alpha\varepsilon) f(\mathbf{k},t). \tag{6}$$

The method is based on a partition of the whole phase space into tiny cells. The energy $\varepsilon$, the polar angular variable $\theta = \cos\vartheta$ and the azimuthal angular variable $\varphi$ are equidistantly discretized as follows:

$$\begin{aligned} \varepsilon_\nu &= \nu\Delta_\varepsilon, \quad \nu = 0,\dots,N_\varepsilon, \quad \Delta_\varepsilon = \varepsilon_{max}/N_\varepsilon \\ \theta_\mu &= -1 + \mu\Delta_\theta, \quad \mu = 0,\dots,N_\theta, \quad \Delta_\theta = 2/N_\theta \\ \varphi_\sigma &= \sigma\Delta_\varphi, \quad \sigma = 0,\dots,N_\varphi, \quad \Delta_\varphi = 2\pi/N_\varphi \\ B_\nu &= [\varepsilon_{\nu-1},\varepsilon_\nu], \ B_\mu = [\theta_{\mu-1},\theta_\mu], \ B_\sigma = [\varphi_{\sigma-1},\varphi_\sigma]. \end{aligned}$$

Here, we introduced a maximum value $\varepsilon_{max}$ for the energy. In the case of the spatial $x$ and $y$ variables we use a non-uniform grid $B_\lambda = [x_{\lambda-1}, x_\lambda]$, $\Delta_x^\lambda = x_\lambda - x_{\lambda-1}$, $\lambda = 1,\dots,N_x$ and $B_\eta = [y_{\eta-1}, y_\eta]$, $\Delta_y^\eta = y_\eta - y_{\eta-1}$, $\eta = 1,\dots,N_y$, which allows us to refine the grid resolution in spatial regions, where rapid variations are expected, *e.g.*, at junctions. For a compact presentation, we introduce the vector $\pi = (\varepsilon,\theta,\varphi,x,y)$, the multi-indices $\gamma = (\nu,\mu,\sigma,\lambda,\eta)$ and $N = (N_\varepsilon, N_\theta, N_\varphi, N_x, N_y)$. Next, we assume the distribution function to be constant within a cell $B_\gamma = B_{\nu\mu\sigma}^{\lambda\eta} = B_\nu \times B_\mu \times B_\sigma \times B_\lambda \times B_\eta$:

$$\psi^\alpha(\pi,t) \approx \sum_{\gamma=1}^{N} \psi_\gamma^\alpha(t)\chi_{B_\gamma}(\pi) \tag{7}$$

with the characteristic function $\chi_{B_\gamma}(\pi)$ and the time-dependent unknowns $\psi_\gamma^\alpha(t)$.

By using this *Ansatz* (7) and integrating the resulting BTE over each single cell, a linear system of ordinary differential equations (ODE's) is established. In order to perform the numerical evaluation of the collision operators, we assumed the phonon energies to be integer multiples of the energy discretization length $\Delta_\varepsilon$. According to the definition (6), each equation can be physically interpreted as a particle continuity equation of a certain cell. However, when integrating the advection terms (left hand side of the BTE), the problem arises to evaluate the discontinuous function (7) at the boundaries of the cells $B_\gamma$. The correct way of determinating it in the appearing boundary terms is indicated by the characteristics of the advection terms, which is equivalent to an application of an upwind scheme. A simple forward Euler scheme turned out to be sufficiently accurate for performing the time integration of the resulting system of linear ODEs. The Poisson equation is solved self-consistently at each time step by applying a finite element Galerkin approach.

## 4 Numerical Results

We simulate a two-dimensional silicon MESFET shaped as indicated in Fig. 1 with the applied potentials at source $V_s = 0$ V, gate $V_g = -0.8$ V and drain $V_d = 1$ V. The donor densities are chosen as $n = 10^{17}$ cm$^{-3}$ and $n^+ = 3 \times 10^{17}$ cm$^{-3}$. Figure 2 shows the distribution of the electron density and the mean energy per electron in the MESFET at steady state. Finally, a comparison between deterministic and MC results for the particle density for certain cut lines of the MESFET is presented in Fig. 3.

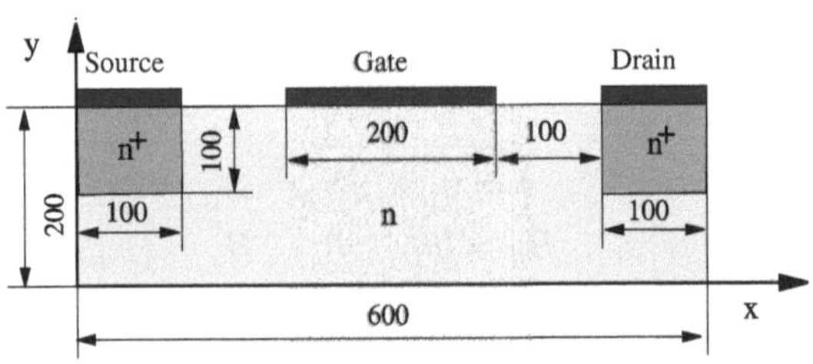


**Fig. 1.** Schematic illustration of the simulated Si-MESFET. Lengths are given in nm.

*Acknowledgement.* This work was supported by the Fonds zur Förderung der wissenschaftlichen Forschung, Vienna, under contract number P14699-TPH.

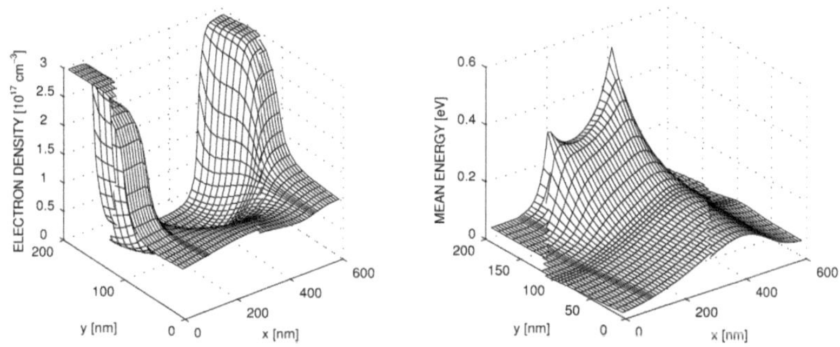


**Fig. 2.** Steady state electron density and mean energy versus position in the Si-MESFET.

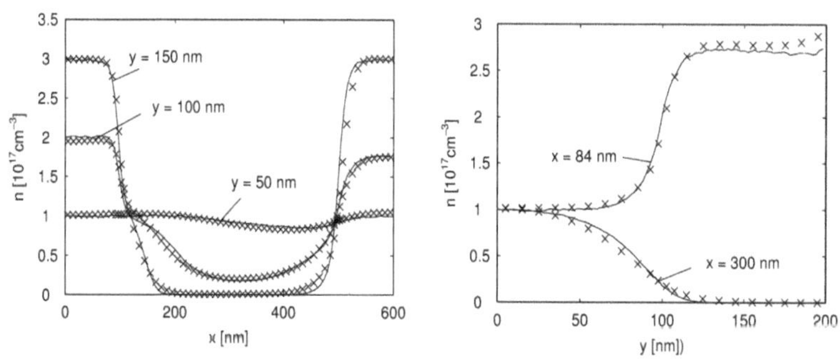


**Fig. 3.** The stationary-state electron density $n$ for certain cut lines of the Si-MESFET. (–): MC data; (×): multicell method.

# References


1. J. A. Carrillo, I. M. Gamba, A. Majorana, and C. W. Shu. A WENO-solver for the transients of Boltzmann-Poisson system for semiconductor devices: performance and comparisons with Monte Carlo methods. *J. Comput. Phys.*, 184:498–525, 2003.
2. C. Ertler and F. Schürrer. A multicell solution for the transients of a 1D $n^+$-n-$n^+$-silicon diode with respect to different crystallographic directions. *J. Gen. Math. A*, under review, 2004.
3. S. E. Laux, M. V. Fischetti, and D. J. Frank. Monte carlo analysis of semiconductor devices: the damocles program. *IBM J. Res. Develop.*, 34:466–494, 1990.
4. X. D. Liu, S. Osher, and T. Chan. Weighted essentially non-oscillatory schemes. *J. Comput. Phys.*, 115:200–212, 1994.
5. M. S. Lundstrom. *Fundamentals of Carrier Transport.* Cambridge University Press, Cambridge, 2000.

# Some Remarks on the Vector Fitting Iteration

W. Hendrickx[1], D. Deschrijver, and T. Dhaene

University of Antwerp, Department of Mathematics and Computer Science, Middelheimlaan 1, B-2020 Antwerpen, Belgium, wouter.hendrickx@ua.ac.be

**Summary.** *Vector Fitting* (VF) is an iterative technique to construct rational approximations based on multiple frequency domain samples, introduced by Gustavsen and Semlyen [1, 3]. VF is nowadays widely investigated and used in the *Power Systems* and *Microwave Engineering* communities. Numerical experiments show that VF has favorable convergence properties. However, so far, no theoretical proof for its convergence, or conditions to guarantee convergence, have been published. This paper gives a description of a general iterative Least-Squares framework for rational approximation and shows that VF fits into this framework.

**Key words:** Vector Fitting, rational interpolation, System Identification, Least-Squares.

## 1 Introduction

In *System Theory*, it is common practice to approximate the frequency domain response of a *Linear Time-Invariant system* (LTI system) by a rational pole-zero function. Finding such an approximation is inherently a difficult problem due to the non-linearity of the approximant. To remove the non-linearity, the denominator is often fixed at some well-chosen polynomial or the system is linearized in some way. Of course, this can degrade the quality of the approximant, or can even make accurate approximation impossible.

VF consists of an iterative pole relocation scheme. In each iteration step a linear Least-Squares (LS) problem is solved, to come up with more accurate approximations of numerator and denominator. New estimates of the poles are based on the approximations of the previous iteration.

In this contribution we position the VF technique in a broader LS rational approximation framework. This way, we want to facilitate further exploration of the theoretical properties of the VF technique. Furthermore, we offer some insight into the initial choice of pole locations of the VF algorithm. For completeness, we note that the iteration treated in this paper is related to the Sanathanan and Koerner iteration [2].

## 2 An iterative scheme for solving rational LS problems

Suppose that we are trying to approximate a function $H$ by a model of the form

$$\tilde{H}(s) = \frac{\sum_{i=1}^{N} \alpha_i f_i(s)}{\sum_{j=1}^{D} \beta_j g_j(s)} =: \frac{p(s,\alpha)}{q(s,\beta)} \tag{1}$$

where the $f_i$ and $g_j$ are fixed basis functions for the nominator and denominator respectively. Furthermore, $\alpha_i$ and $\beta_j$ are unknown coefficients. To resolve the ambiguity in the definition, it is possible to choose $\alpha_N = 1$ for example. The $p$ and $q$ serve as an abbreviation, $\alpha$ and $\beta$ are shorthands for the tuples $(\alpha_1, \ldots, \alpha_N)$ and $(\beta_1, \ldots, \beta_D)$, respectively.

Now suppose we have sampled $H$ at certain points $(s_k)_{k=1}^{n}$. Our goal is to approximate $H$ by a function of the form $\tilde{H}$ in an LS sense:

$$\operatorname{argmin}_{\alpha,\beta} \sum_{k=1}^{n} \left| H(s_k) - \tilde{H}(s_k) \right|^2 \tag{2}$$

The problem with this formulation is that both numerator and denominator contain unknown variables $\alpha_i$ and $\beta_j$, so basic techniques for solving LS problems do not apply.

It is tempting to rewrite the LS problem as

$$\operatorname{argmin}_{\alpha,\beta} \sum_{k=1}^{n} \left| \sum_{i=1}^{N} \alpha_i f_i(s_k) - H(s_k) \sum_{j=1}^{D} \beta_j g_j(s_k) \right|^2 \tag{3}$$

which is a simple linear LS problem of the form $\operatorname{argmin}_x \|Ax - b\|_{l^2}$. Unfortunately this formulation is not equivalent with problem (2). Rewriting (2) gives:

$$\operatorname{argmin}_{\alpha,\beta} \sum_{k=1}^{n} \frac{1}{|q(s_k,\beta)|^2} \; |p(s_k,\alpha) - H(s_k)\, q(s_k,\beta)|^2 \tag{4}$$

which resembles (3), except for the weighting factor $\dfrac{1}{|q(s_k,\beta)|^2}$.

The following iterative scheme can be applied: Start by setting $|q(s,\beta^{(0)})| = 1$. Calculate the sequences $\alpha^{(t)}$ and $\beta^{(t)}$ by iteratively solving

$$\operatorname{argmin}_{\alpha^{(t)},\beta^{(t)}} \sum_{k=1}^{n} \frac{1}{\left|q(s_k,\beta^{(t-1)})\right|^2} \; \left|p(s_k,\alpha^{(t)}) - H(s_k)\, q(s_k,\beta^{(t)})\right|^2 \tag{5}$$

(which is a basic LS problem in $\alpha^{(t)}$ and $\beta^{(t)}$) for $t = 1, 2, \ldots$ Note that the weighting factor is approximated by the denominator from the last iteration.

## 3 The Vector Fitting methodology

In this section we will repeat the classical formulation of the VF methodology. Suppose we want to approximate the function $f : \mathbb{C} \to \mathbb{C}$ by a rational function and that $f$ is known at a fixed set of sample points $(s_k)_{k=1}^n$. Now take an arbitrary function $\sigma : \mathbb{C} \to \mathbb{C}$ and assume that both $\sigma(s)f(s)$ and $\sigma(s)$ can be approximated by rational functions using *the same set of poles* $(\overline{a}_i)_{i=1}^D$ (and linear and constant terms). Formally, we have:

$$\begin{pmatrix} f(s)\,\sigma(s) \\ \\ \sigma(s) \end{pmatrix} = \begin{pmatrix} \sum_{i=1}^D \dfrac{c_i}{s-\overline{a}_i} + e + h\,s \\ \\ \sum_{i=1}^D \dfrac{\tilde{c}_i}{s-\overline{a}_i} + 1 \end{pmatrix} \tag{6}$$

We can now multiply the second row by $f(s)$ and evaluate the system in each of the samples $s_k$. If we assume the poles of $\overline{a}_i$ are fixed beforehand, we get a system of linear equations in the unknowns $(c_i, \tilde{c}_i, e, h)$ by equating the first row with the second row. This system is overdetermined if a lot of samples are available. In that case it can be solved using classical LS techniques.

We now proceed by writing both $\sigma(s)\,f(s)$ and $\sigma(s)$ in function of their zeros and poles:

$$\begin{pmatrix} f(s)\,\sigma(s) \\ \sigma(s) \end{pmatrix} = \begin{pmatrix} \dfrac{\prod_{i=1}^{D+1}(s-z_i)}{\prod_{i=1}^{D}(s-\overline{a}_i)} \\ \dfrac{\prod_{j=1}^{D}(s-\tilde{z}_j)}{\prod_{i=1}^{D}(s-\overline{a}_i)} \end{pmatrix} \tag{7}$$

Dividing the first row by the second, we get an approximation for $f$ of the form:

$$f(s) = \frac{\prod_{i=1}^{D+1}(s-z_i)}{\prod_{j=1}^{D}(s-\tilde{z}_j)} \tag{8}$$

Note that the zeroes of $\sigma$ became the poles of our approximation.

The above procedure can be applied in an iterative fashion: the poles found in the last iteration can be inserted in equation (7) as *guesses* for the actual poles $(\overline{a}_i)_{i=1}^D$. Eventually, we want this procedure to converge. By this we mean that the guessed poles $\overline{a}_i$ become close enough to the real poles of $f$. In that case $\sigma$ will be approximately 1 and we have found an approximation for $f$.

One problem that remains is the choice of the initial pole locations. This problem will be addressed in Section 5.

## 4 How VF fits in

Fix $D$ (the degree of the denominator), fix $(a_i)_{i=1}^D$ (the *starting poles*) and set $N = D + 2$. Now choose the following basis for the first iteration:

$$f_i(s) = g_i(s) = \frac{1}{s - a_i} \quad \text{for } i = 1, \ldots, D$$
$$f_{D+1}(s) = g_{D+1}(s) = 1 \quad f_{D+2}(s) = s \tag{9}$$

First note the following:

$$\text{span}\,\langle f_1, \ldots, f_{D+2} \rangle = \frac{\mathbb{C}_{D+1}[s]}{\prod_{i=1}^{D}(s - a_i)} \tag{10}$$

$$\text{span}\,\langle g_1, \ldots, g_{D+1} \rangle = \frac{\mathbb{C}_{D}[s]}{\prod_{i=1}^{D}(s - a_i)} \tag{11}$$

where $\mathbb{C}_k[s]$ denotes the polynomials in $s$ of degree equal or less than $k$. If a polynomial $p(s) \in \mathbb{C}_k[s]$, it's always possible to factor $p$ completely, i.e. $p(s) = \prod_{i=1}^{k}(s - x_i)$ for certain $x_i \in \mathbb{C}$.

Now using the basis functions specified in (9) we can proceed with the iterative method proposed in Section 2. The first iteration produces two sets of coefficients $\alpha_i^{(1)}$ and $\beta_i^{(1)}$. Using (10) and (11) we can rewrite $p$ and $q$ as

$$p(s, \alpha^{(1)}) = \frac{\prod_{i=1}^{D+1}(s - z_i^{(p,1)})}{\prod_{i=1}^{D}(s - a_i)} \qquad q(s, \beta^{(1)}) = \frac{\prod_{j=1}^{D}(s - z_j^{(q,1)})}{\prod_{i=1}^{D}(s - a_i)} \tag{12}$$

The second iteration (using $q(s, \beta^{(1)})$ as a weighting factor, as in Sect. 2) produces $p(s, \alpha^{(2)})$ and $q(s, \beta^{(2)})$. (12) can be applied to these new functions (just replace the 1's by 2's). At first sight this does not really resemble the vector fitting methodology. Rewriting the defining equation (5) of the iterative scheme shows the following:

$$\sum_{k=1}^{n} \left| \frac{\prod_{i=1}^{D}(s_k - a_i)}{\prod_{j=1}^{D}(s_k - z_j^{(q,1)})} \right|^2 \left| \frac{\prod_{i=1}^{D+1}(s_k - z_i^{(p,2)})}{\prod_{i=1}^{D}(s_k - a_i)} - H(s_k) \frac{\prod_{j=1}^{D}(s_k - z_j^{(q,2)})}{\prod_{i=1}^{D}(s_k - a_i)} \right|^2 \tag{13}$$

which simplifies to

$$\sum_{k=1}^{n} \left| \frac{\prod_{i=1}^{D+1}(s_k - z_i^{(p,2)})}{\prod_{j=1}^{D}(s_k - z_j^{(q,1)})} - H(s_k) \frac{\prod_{j=1}^{D}(s_k - z_j^{(q,2)})}{\prod_{j=1}^{D}(s_k - z_j^{(q,1)})} \right|^2 \tag{14}$$

Using (10) and (11) with $a_i$ replaced by $z_j^{(q,1)}$, we see that the LS problem we solve in the second iteration is exactly that solved in the vector fitting technique:

$$\sum_{k=1}^{n} \left| \sum_{i=1}^{D} \frac{c_i}{(s_k - z_i^{(q,1)})} + e + h\, s_k - H(s_k) \sum_{j=1}^{D} \frac{d_j}{(s_k - z_j^{(q,1)})} - H(s_k)\, \gamma \right|^2$$

where $\gamma$ is chosen 1.

## 5 Initial pole placement

In order to get a system of linear equations that is not too badly conditioned, it is important to choose the initial poles at *good locations.* As all samples lie on the complex axis, choosing poles too far to the left in the complex plane makes the real part of the poles dominate the matrix entries

$$\frac{1}{s_k - \overline{a}_i} = \frac{1}{j(\omega_k - \mathcal{I}\overline{a}_i) - \mathcal{R}\overline{a}_i} \approx \frac{-1}{\mathcal{R}\overline{a}_i}$$

which renders all entries equally small.

Ideally, one would like to put the initial poles close to some of the sample points. Doing so makes some of the elements in the linear systems matrix significantly larger than all the other elements. This improves conditioning of the system. Of course the limiting case, where we let the poles coincide with some sample points, produces a matrix with some elements infinitely large and all others zero. In that case all information would be lost.

Therefore, we suggest to place poles on a line, parallel and close to the imaginary axis in order to get good conditioning. Originally, pole placement on a line through the origin was suggested [1]. To our experience this gives similar results.

The VF methodology also introduces the flipping of the poles around the imaginary axis between each two iterations in order to obtain a model which has all its poles in the left half-plane. In the context of system identification, this means that the modeled system is stable. Flipping a pole to the left half plane is equivalent to multiplying the approximant by the all-pass function

$$F(s) = \frac{s - p}{s - (-\mathcal{R}p + \mathcal{I}p)} \qquad |F(j\omega)|^2 = \frac{|\omega - \mathcal{I}p|^2 + |\mathcal{R}p|^2}{|\omega - \mathcal{I}p|^2 + |-\mathcal{R}p|^2} = 1$$

where $p$ is a pole. This means that the amplitude of the system remains the same, only the phase changes.

## References

1. Bjorn Gustavsen and Adam Semlyen. Rational approximation of frequency domain responses by Vector Fitting. *IEEE Transactions on Power Delivery*, 14(3):1052–1061, 1999.
2. C. K. Sanathanan and J. Koerner. Transfer function synthesis as a ratio of two complex polynomials. *IEEE trans. Automat. Contr.*, 8:56–58, 1963.
3. A. Semlyen and B. Gustavsen. Vector Fitting by pole relocation for the state equation approximation of nonrational transfer matrices. *Circuits and systems: Analog and Digital signal processing*, 19(6):549–566, 2000.

# Krylov Subspace Methods in the Electronic Industry

P. Heres[1] and W. Schilders[2]

[1] Technische Universiteit Eindhoven, Department of Mathematics and Computer Science, p.j.heres@tue.nl
[2] Philips Research Laboratories, Eindhoven

**Summary.** Krylov subspace methods are well-known for their nice properties, but they have to be implemented with care. In this article the mathematical consequences encountered during implementation of Krylov subspace methods in an existing layout-simulator are discussed. Briefly, the representation in a circuit is visited and two methods to avoid parts of the redundancy are drawn.

**Key words:** Model order reduction, Krylov subspace methods, orthogonalisation, circuit simulation.

## 1 Introduction

Wireless applications are gaining interest in the electronic industry nowadays. Integration plays a more and more important role in the design of these applications. Technologies like SoC and RF-SiP are needed to meet the demands set by the consumer market. All this makes an accurate and fast modelling of the electromagnetic (EM) effects of passive electronic structures needed.

The EM analysis of arbitrary shaped layouts can be calculated with existing tools. One specific example of such a tool was the drive for our research. The Boundary Element Models initially generated by this tool can be simply too large to be handled. Several reduction methods can be applied to make the treatment of these models feasible.

In stead of the already implemented reduction method, Krylov subspace methods, like the methods presented in [4] and [5], were proposed to be implemented in the layout simulator. These methods were chosen because of their well-known properties with respect to preservation of stability and passivity. In this article mathematical consequences encountered during implementation are discussed.

## 2 Equation setting

We consider the following set of equations:

$$\begin{bmatrix} \mathbf{C} & \mathbf{0} \\ \mathbf{0} & -\mathbf{L} \end{bmatrix} \frac{d}{dt} \begin{bmatrix} \mathbf{v} \\ \mathbf{i} \end{bmatrix} + \begin{bmatrix} \mathbf{G} & \mathbf{P}^T \\ \mathbf{P} & -\mathbf{R} \end{bmatrix} \begin{bmatrix} \mathbf{v} \\ \mathbf{i} \end{bmatrix} = \begin{bmatrix} \mathbf{B} \\ \mathbf{0} \end{bmatrix} \mathbf{u} \tag{1}$$

In this system the values for the capacitive elements are in the matrix $\mathbf{C} \in \mathbb{R}^{n \times n}$, the inductive values are in $\mathbf{L} \in \mathbb{R}^{m \times m}$. The matrices $\mathbf{G} \in \mathbb{R}^{n \times n}$ and $\mathbf{R} \in \mathbb{R}^{m \times m}$ represent the resistive values. $\mathbf{P} \in \mathbb{R}^{m \times n}$ is an incidence matrix consisting of 1's, $-1$'s and 0's. We denote by $\mathbf{u}$ the input signal. The state space vector consists of voltages $\mathbf{v}$ and currents $\mathbf{i}$. In this way the system represents an RCL-circuit. Despite the precise formulation in (1), the methods mentioned in this paper are generally applicable to systems of this form:

$$\begin{aligned} \mathbf{C}\frac{d}{dt}\mathbf{x}(t) &= -\mathbf{G}\mathbf{x}(t) + \mathbf{B}_i\mathbf{u}(t) \\ \mathbf{y}(t) &= \mathbf{B}_o^T\mathbf{x}(t). \end{aligned} \tag{2}$$

These are not specific for circuits.

The latter system is a Linear Time Invariant system. Because the matrix $\mathbf{C}$ can be singular, this can be a Differential Algebraic Equation (DAE). A common way to solve these systems is to transform them to the frequency domain with a Laplace transform:

$$\begin{aligned} (\mathbf{G} + s\mathbf{C})\mathbf{X}(s) &= \mathbf{B}_i\mathbf{U}(s) \\ \mathbf{Y}(s) &= \mathbf{B}_o^T\mathbf{X}(s); \end{aligned} \tag{3}$$

After elimination of the state space vector $\mathbf{X}(s)$ a transfer function is obtained:

$$\mathbf{H}(s) = \mathbf{B}_o^T(\mathbf{G} + s\mathbf{C})^{-1}\mathbf{B}_i \tag{4}$$

This function gives a direct relation between the input and the output of the system and is therefore representative for the behaviour of the system in frequency domain. If the system has more than one inputs and outputs, the transfer function is a matrix representing the transfers from one port to the other. Typically, one tries to approximate the behaviour of this transfer function.

## 3 Model Order Reduction

The aim of Model Order Reduction is to capture the essential features of a large model into a much smaller approximation. Thus, the large system is replaced by a smaller approximation, with the same amount of input signals, *i.e.* ports in terms of a circuit and a comparable behaviour.

The idea behind Krylov subspace methods is to generate a (basis for a) Krylov space. A Krylov space is defined as:

$$\mathcal{K}_n(\mathbf{b}, \mathbf{A}) = [\mathbf{b}, \mathbf{A}\mathbf{b}, \ldots \mathbf{A}^n\mathbf{b}] \tag{5}$$

Next an orthonormal basis of this small space is calculated and the system matrices are projected onto this basis. Due to space limitations for the publication, we refer the reader to [3] for the issues induced by the orthogonalisation of the Krylov space.

If a system has more than one port, then $\mathbf{B}$ is a matrix; the number of the columns in $\mathbf{B}$ is equal to the number of ports, say $p$. In that case the Krylov space consists of blocks: with every iteration a block of $p$ columns is added to the Krylov space. This makes our approximation $p$ columns and rows larger.

Well-known Krylov subspace methods in chronological order of publication are PVL [2], PRIMA [5] and Laguerre-SVD [4]. PVL and PRIMA make use of the fact that the transfer function can be written as:

$$\mathbf{H}(s) = \mathbf{B}_o^T(\mathbf{G} + s\mathbf{C})^{-1}\mathbf{B}_i = \mathbf{B}_o^T(\mathbf{I} - (s - s_0)\mathbf{A})^{-1}\mathbf{R} \tag{6}$$

with $\mathbf{A} = -(\mathbf{G} + s_0\mathbf{C})^{-1}\mathbf{C}$ and $\mathbf{R} = (\mathbf{G} + s_0\mathbf{C})^{-1}\mathbf{B}_i$. With this formulation a Krylov space is generated, which represents the moments of the transfer function:

$$\mathcal{K}_q(\mathbf{R}, \mathbf{A}) = [\mathbf{R}, \mathbf{A}\mathbf{R}, \ldots \mathbf{A}^q\mathbf{R}] \tag{7}$$

Laguerre-SVD is based on the fact that the transfer function can be expanded into scaled Laguerre functions in frequency domain:

$$\begin{aligned} \mathbf{H}(s) &= \mathbf{L}^T(\mathbf{G} + s\mathbf{C})^{-1}\mathbf{B} = \\ &\frac{2\alpha}{s+\alpha}\mathbf{L}^T \sum_{n=0}^{\infty} \left((\mathbf{G} + \alpha\mathbf{C})^{-1}(\mathbf{G} - \alpha\mathbf{C})\right)^n (\mathbf{G} + \alpha\mathbf{C})^{-1}\mathbf{B}_i \left(\frac{s-\alpha}{s+\alpha}\right)^n \end{aligned} \tag{8}$$

From this expansion very naturally a definition for a Krylov subspace arises. The starting vector is then $\mathbf{R} = (\mathbf{G} + \alpha\mathbf{C})^{-1}\mathbf{B}_i$ and the generating matrix $\mathbf{A} = (\mathbf{G} + \alpha\mathbf{C})^{-1}(\mathbf{G} - \alpha\mathbf{C})$. Then the Krylov space is defined as in (7).

Advantages of Krylov subspace methods are that they are very generally applicable, because $\mathbf{C}$ and $\mathbf{G}$ do not need to be regular. Furthermore, they are relatively cheap. Because it can be proven that the moments in the moment expansion of the transfer function are preserved, the methods are accurate. For PRIMA and Laguerre-SVD it is proven that stability and passivity of the system are preserved during reduction. Especially this last property is important in the implementation of Model Order Reduction methods into the layout simulator. PVL convergences faster than PRIMA, but stability can be lost in this methods. Therefore, in this setting PVL is left out of consideration.

In spite of these advantages, there are some severe disadvantages known for Krylov subspace methods. First of all no error bound is known in general. For PVL a bound is known and published in [1]. In PRIMA and SVD-Laguerre it is not known when to stop. Hence, easily an unnecessarily large approximation is generated.

## 4 Validation of results

Because the original model can be represented as a circuit, an RCL-circuit, and because the EM layout simulator uses a representation of its reduced model in a circuit, it is desired to represent our reduced model in terms of a circuit. This circuit representation enables us to use the speed of existing circuit simulators, in evaluating the behaviour of the reduced model.

In the original model, see (1) the state space vector consisted of voltages and currents. After projecting the system onto a smaller space, these voltages and currents are mixed and therefore the physical meaning of the reduced model is lost. Hence, it is not possible to represent the system without making use of controlled sources or controlled components. Nevertheless, we obtained a circuit representing the reduced model and this representation is tested and compared to the output of the layout simulator. The results in frequency domain can be made as accurate as wanted, together with the increasing size of the reduced system. More important is that the results for a transient analysis is stable. This was not the case for the existing reduction method, which gave a good approximation in frequency domain, but could be instable in time domain.

## 5 Redundancy

Next to the already mentioned disadvantages of Krylov subspace methods, there is another drawback to Krylov subspace methods. Because they do not carefully choose the needed information, a lot of information is incorporated in the smaller model which is not needed for a good approximation. So, even if we stopped the iterative process in time, the models are redundant. In our research we found two ways to avoid parts of this redundancy, without too much computational expenses.

The first proposal is a deflation of converged columns. Sometimes it can happen that a column is generated which already existed in the space. At that moment we want to stop iterating with this direction and want to be able to proceed with the other columns in the block. This convergence should be treated with care, because if we violate the basic property of Krylov spaces, the small approximation can become really cumbersome. In the Block Arnoldi Algorithm, used to generate the Block Krylov space a specialized QR, *i.e.* a rank-revealing QR step is substituted. In this way smaller approximations with the same transfer function can be generated.

Our second proposal is to remove insignifant poles, via an eigendecomposition of the reduced system. Because the reduced system is small, a full eigendecomposition can be calculated cheaply:

$$\mathbf{CV} = \mathbf{GV\Lambda}. \tag{9}$$

Here the diagonal matrix $\mathbf{\Lambda} \in \mathbb{C}^{q\times q}$ consists of the eigenvalues, where $q$ is the size of the reduced system. The associated eigenvectors are in $\mathbf{V} \in \mathbb{C}^{q\times q}$. Once this decomposition is obtained, the transfer function can be written in a pole-residue expansion:

$$\mathbf{H}(s) = c + \sum_{j=1}^{q} \frac{r_j}{s - p_j}, \tag{10}$$

with $r$ and $p \in \mathbb{C}$.

We saw that in this sum there are terms which do not contribute to the transfer function. This can be either because $r_j$ is very small or $p_j$ is very large. These poles are removed, which comes down to removing the associated columns from $\mathbf{V}$. Complex poles are always removed in conjugate pairs. Naxt a real basis is generated for the eigenvector matrix. This is finally used to project our reduced system on.

## 6 Conclusions

In this article we presented the mathematical challenges of implementing Krylov subspace methods in an existing layout simulator. We showed that Krylov subspace methods are efficient for the given examples, but have to implemented with care. Several adjustments can be implemented to the existing methods, to make them more efficient. There is an obvious need for realization. Realization enables the application of Model Order Reduction in time domain simulations of the EM behaviour.

## References

1. Z. Bai, R.D. Slone, W.T. Smith, and Q. Ye. Error Bound for Reduced System Model by Padé Approximation via the Lanczos Process. *IEEE Trans. CAD-IC and Syst.*, 18(2):133–141, February 1999.
2. Peter Feldmann and Roland W. Freund. Efficient Linear Circuit Analysis by Padé Approximation via the Lanczos Process. *IEEE Trans. Computer-Aided Design*, 14:137–158, 1993.
3. P.J. Heres and W.H.A. Schilders. Deflation of Converged Columns in Krylov Subspace Methods. *To be published*, 2004.
4. Luc Knockaert and Daniel De Zutter. Passive Reduced Order Multiport Modeling: The Padé-Arnoldi-SVD Connection. *Int. J. Electronics and Communications*, 53:254–260, 1999.
5. A. Odabasioglu and M. Celik. PRIMA: Passive Reduced-order Interconnect Macromodeling Algorithm. *IEEE. Trans. Computer-Aided Design*, 17(8):645–654, August 1998.

# On Nonlinear Iteration Methods for DC Analysis of Industrial Circuits

M. Honkala, J. Roos, and V. Karanko

Circuit Theory Laboratory, Helsinki University of Technology, P.O.Box 3000, FI-02015 HUT, Finland. {mikko,janne,ville}@ct.hut.fi

**Summary.** Several iterative methods have been tested in nonlinear DC analysis of industrial electronic circuits.

**Key words:** Nonlinear equations, iterative methods, trust region.

## 1 Introduction

Modern electronic circuits are typically large, consisting of thousands of transistors and other components. For the simulation of these large, nonlinear circuits, efficient iteration methods are needed. Choosing the nonlinear iteration method for a circuit simulator, we have to take into account the special properties of both the circuit equations and the circuit simulator, in our case, APLAC (www.aplac.com).

The DC analysis of APLAC is based on the (modified) Newton–Raphson (NR) method. It has fast local convergence, but, especially in cases where the initial guess for the nonlinear iteration is poor, NR iteration may diverge or the convergence may be extremely slow. Usually, other methods with strong convergence properties (*e.g.*, homotopy methods) are slow, while faster methods (*e.g.*, methods that approximate the inverse of the Jacobian matrix) have convergence problems. Therefore, one has to compromise between speed and reliable convergence.

Our goal is to find a method or a combination of methods that converges robustly enough for badly scaled DC analysis and is also reasonably fast.

In APLAC, transistors and other nonlinear components are modeled such that the current functions and their first derivatives are available. Therefore, the Jacobian matrix and the gradient are easy to obtain, but, *e.g.*, the construction of the Hessian matrix would need expensive numerical computation. In addition, the Jacobian matrices are sparse and often nearly singular.

We concentrate on some trust region and tensor methods [1, 7], which should be efficient in the case of nearly singular Jacobian matrices and do not need the computation of Hessian matrices. We also improve the convergence of the methods using a non-monotone strategy and compare their efficiency to NR and some conjugate gradient (CG) methods. All the methods have been implemented in the in-house development version of APLAC using the Matlab C-function libraries. Simulations with real-life circuits are presented.

## 2 Equation formulation

The nonlinear circuit equations for DC analysis can be written in the algebraic form

$$\mathbf{f}(\mathbf{x}) = \mathbf{0}, \tag{1}$$

where $\mathbf{x}$ is the vector of the unknown voltages and currents. Function values and derivatives can be directly obtained from the model equations. If we define the objective function as

$$F = \frac{1}{2}\|\mathbf{f}(\mathbf{x})\|_2^2 = \frac{1}{2}\mathbf{f}(\mathbf{x})^{\mathrm{T}}\mathbf{f}(\mathbf{x}), \tag{2}$$

the gradient is

$$\mathbf{g} = \nabla F = \mathbf{J}^{\mathrm{T}}\mathbf{f}(\mathbf{x}), \tag{3}$$

where $\mathbf{J}$ is the Jacobian matrix. The Hessian matrix

$$\mathbf{H} = \nabla^2 F = \frac{\partial \mathbf{g}}{\partial \mathbf{x}} = \frac{\partial\left(\mathbf{J}^{\mathrm{T}}\right)}{\partial \mathbf{x}}\mathbf{f}(\mathbf{x}) + \mathbf{J}^{\mathrm{T}}\mathbf{J} \tag{4}$$

is possible to obtain, but it would need expensive numerical computation and, therefore, methods using the Hessian are omitted.

In this paper we study damped iterations

$$\mathbf{x}_{k+1} = \mathbf{x}_k + \lambda_k \boldsymbol{\Delta}\mathbf{x}_k, \tag{5}$$

with the damping factor $\lambda_k$, $0 < \lambda_k \leq 1$. The update $\boldsymbol{\Delta}\mathbf{x}_k$ is

$$\boldsymbol{\Delta}\mathbf{x}_k^{\mathrm{NR}} = -\mathbf{J}_k^{-1}\mathbf{f}_k, \tag{6}$$

$$\boldsymbol{\Delta}\mathbf{x}_k^{\mathrm{SD}} = -\mathbf{J}_k^{T}\mathbf{f}_k = -\mathbf{g}_k, \tag{7}$$

$$\boldsymbol{\Delta}\mathbf{x}_k^{\mathrm{CG}} = -\mathbf{g}_k + \beta_{k-1}\boldsymbol{\Delta}\mathbf{x}_{k-1}, \tag{8}$$

for NR, steepest-descent (SD), and CG methods, respectively. In CG, formulas for $\beta$ are called Fletcher–Reeves, Polak–Ribière, and Dai–Yuan formulas and are given as [2, 5]

$$\beta_{k-1}^{\mathrm{FR}} = \frac{\|\mathbf{g}_k\|^2}{\|\mathbf{g}_{k-1}\|^2}, \quad \beta_{k-1}^{\mathrm{PR}} = \frac{\mathbf{g}_k^T\mathbf{y}_{k-1}}{\|\mathbf{g}_{k-1}\|^2}, \quad \beta_{k-1}^{\mathrm{DY}} = \frac{\|\mathbf{g}_k\|^2}{\mathbf{d}_{k-1}^T\mathbf{y}_{k-1}}, \tag{9}$$

respectively, where $\mathbf{y}_{k-1} = \mathbf{g}_k - \mathbf{g}_{k-1}$.

## 3 Line-search methods

For global convergence, a line search is performed in the step direction, i.e., adjust $\lambda_k$ such that

$$\|\mathbf{f}(\mathbf{x}_k + \lambda_k \mathbf{\Delta x}_k)\| < \|\mathbf{f}(\mathbf{x}_k)\| . \tag{10}$$

## 4 Trust-region methods

A trust region $B = \{\mathbf{x} \mid \|\mathbf{x} - \mathbf{x}_k\| \leq \delta\}$, where $\delta$ is the trust-region radius, is the region where the linear or quadratic model $\mathbf{m}(\mathbf{x})$ is assumed to approximate $\mathbf{f}(\mathbf{x})$. In the trust-region methods, the iteration step, $\mathbf{\Delta x}_k$, is obtained by minimizing the model within the trust region:

$$\min_{\|\mathbf{\Delta x}\| \leq \delta} \mathbf{m}(\mathbf{x}_k + \mathbf{\Delta x}). \tag{11}$$

The trust-region radius, $\delta$, is adaptively adjusted during the iteration.

## 5 Non-monotone strategy

The monotone-decrease requirement (10) is sometimes too strict. It may produce unnecessary line searches or trust-region reductions and, thus, slow down the iteration. This can be avoided using a non-monotone strategy, which has been effectively used within iterative line-search methods [4] and trust-region methods [3].

The idea of a non-monotone strategy is very simple. Instead of demanding the function norm to be smaller than the norm in the previous iteration, it is required to stay below the maximum of $M + 1$ earlier norms, i.e.,

$$\|\mathbf{f}(\mathbf{x}_{k+1})\| < \max_{0 \leq j \leq m(k)} \|\mathbf{f}(\mathbf{x}_{k-j})\|. \tag{12}$$

where $m(0) = 0$ and $m(k) = \min\,[m(k-1)+1, M]$. This loosens the too strict decrease conditions, but ensures that the function norm is reduced within $M$ iterations and that the iteration does not diverge.

## 6 Dog-leg method

Dog leg (DL) [6] is a trust-region method that combines the NR and SD methods. If the NR step is inside the trust region, it is accepted as a trial step. Otherwise, the point that minimizes the objective function in the direction of SD, the Cauchy point, is computed. If the Cauchy point is outside the trust region, a damped SD step to the trust-region boundary is taken. When the Cauchy point is inside the trust region, a step is taken to the trust region boundary between the Cauchy point and the NR point.

The non-monotone strategy can be applied to this method, too.

## 7 Tensor methods

Tensor methods with line search were presented in [7]. In [1], tensor methods with 2D trust-region methods were introduced.

In these methods, the quadratic model is

$$\mathbf{m}(\mathbf{x} + \mathbf{\Delta x}) = \mathbf{f}(\mathbf{x}_k) + \mathbf{J}_k\mathbf{\Delta x} + 1/2\mathbf{T}_k\mathbf{\Delta x}\mathbf{\Delta x}, \quad (13)$$

where $\mathbf{T}_k$ is the tensor obtained from interpolating past function values. Although a quadratic model is used, there is no need for the Hessian matrix. The iteration update $\mathbf{\Delta x}$ is found by minimizing $\|\mathbf{m}(\mathbf{x} + \mathbf{\Delta x})\|$.

The tensor methods with line-search and 2D trust-region methods were implemented according to [1]. The non-monotone strategy was applied to these methods, too.

## 8 Results

NR, DL, and tensor with line-search and 2D trust-region methods, were implemented with monotone and non-monotone strategies, but the three CG methods with monotone line search only. All the methods were implemented in APLAC using the Matlab C-library functions. Simulations were performed with relevant industrial and benchmark circuits (Table 1). The DC analyses with CG methods did not converge or stopped at the maximum number of iterations. The results of NR and DL iterations are presented in Table 2 and of tensor methods in Table 3. In Tables 2 and 3, "nc" stands for no convergence and "max" for maximum number of iterations used. The DL method seemed to be the best.

**Table 1.** Test circuits.

| Cir. | nodes | BJTs | MOSFETs |
|---|---|---|---|
| 1 | 53 | – | 74 |
| 2 | 117 | 17 | 42 |
| 3 | 124 | 8 | 14 |
| 4 | 177 | 41 | – |
| 5 | 475 | – | 88 |
| 6 | 475 | 104 | 41 |
| 7 | 518 | 148 | – |
| 8 | 721 | 96 | – |
| 9 | 2200 | 254 | 179 |

**Table 2.** Simulation results with NR and DL methods.

| | Newton–Raphson | | | | Dog Leg | | | |
|---|---|---|---|---|---|---|---|---|
| | monotone | | non-monotone | | monotone | | non-monotone | |
| Cir. | iter. | CPU/s | iter. | CPU/s | iter. | CPU/s | iter. | CPU/s |
| 1 | 240 | 25.1 | 12 | 2.1 | max | – | 27 | 3.0 |
| 2 | max | – | max | – | 30 | 3.3 | 68 | 5.7 |
| 3 | 10 | 1.8 | 21 | 2.3 | 17 | 2.1 | 14 | 2.0 |
| 4 | max | – | nc | – | 242 | 17.8 | 75 | 6.7 |
| 5 | max | – | max | – | 161 | 30.8 | 75 | 16.7 |
| 6 | nc | – | nc | – | 80 | 23.3 | 121 | 34.5 |
| 7 | nc | – | 185 | 89.0 | nc | – | nc | – |
| 8 | 38 | 15.0 | 47 | 17.0 | 65 | 20.3 | 36 | 12.4 |
| 9 | nc | – | nc | – | 285 | 323.0 | 266 | 303.0 |

**Table 3.** Simulation results with tensor methods.

| | Line Search | | | | 2D Trust Region | | | |
|---|---|---|---|---|---|---|---|---|
| | monotone | | non-monotone | | monotone | | non-monotone | |
| Cir. | iter. | CPU/s | iter. | CPU/s | iter. | CPU/s | iter. | CPU/s |
| 1 | max | – | 21 | 2.6 | 24 | 3.3 | 159 | 21.8 |
| 2 | max | – | max | – | max | – | 98 | 11.8 |
| 3 | 23 | 2.6 | nc | – | 26 | 3.2 | max | – |
| 4 | max | – | nc | – | max | – | max | – |
| 5 | nc | – | nc | – | max | – | nc | – |
| 6 | nc | – | nc | – | nc | – | max | – |
| 7 | nc | – | max | – | nc | – | max | – |
| 8 | 54 | 21.7 | 52 | 21.2 | max | – | max | – |
| 9 | nc | – | nc | – | 185 | 236.0 | 112 | 142.0 |

## References


1. A. Bouaricha and R.B. Schnabel. TENSOLVE: A software package for solving systems of nonlinear equations and nonlinear least squares problems using tensor methods. Preprint MCS-P463-0894, Mathematics snd Computer Science Division, Argonne National Laboratory, Argonne, IL, 1994.
2. Y.H. Dai and Y. Yuan. A nonlinear conjugate gradient method with a strong global convergence property. *SIAM Journal on Optimization*, 10:177–182, 1999.
3. N.Y. Deng, Y. Xiao, and F.J. Zhou. Nonmonotonic trust region algorithm. *Journal of optimization theory and applications*, 76(2):259–285, 1993.
4. L. Grippo, F. Lampariello, and S. Lucidi. A nonmonotone line search technique for Newton's method. *SIAM J. Numer. Anal.*, 23(4):707–716, August 1986.
5. C.T. Kelley. *Iterative Methods for Optimization.* SIAM, Philadelphia, 1999.
6. M.J.D. Powell. A new algorithm for unconstrained optimization. In J.B. Rosen, O.L. Mangasarian, and K. Ritter, editors, *Nonlinear Programming*, pages 31–65. Academic Press, New York, 1970.
7. R.B. Schnabel and P.D. Frank. Tensor methods for nonlinear equations. *Siam Journal on Numerical Analysis*, 21(5):815–843, October 1984.

# Implementing Efficient Array Traversing for FDTD-lumped Element Cosimulation

L. R. de Jussilainen Costa

Circuit Theory Laboratory, Helsinki University of Technology, P.O. Box 3000, FI-02015 HUT, Finland. luis@ct.hut.fi

**Summary.** The efficient implementation of the FDTD algorithm in C, particularly the data types and nested loops required, is discussed. The different constructs were run on four computer platforms indicating significant performance improvement with proper implementation. The extent of the improvement depends on the data type, compiler and computer used.

**Key words:** FDTD implementation; nested loops; optimal loops; array types.

## 1 Introduction

The finite difference time-domain (FDTD) method for solving Maxwell's equations discretises the six field components in space and time and uses difference equations to simulate the electric and magnetic fields in the time domain [6]. For cosimulation with a lumped element (LE) circuit simulator like Aplac[1],the current density is divided into two parts: the conduction current density in the dielectric medium and the current introduced by the LE circuit into a given region in the medium [5]. The field and material parameter values are stored in arrays whose position in the array represents the location in space at a given time. These values are updated using values from the previous time point. Hence, the FDTD algorithm entails accessing and updating floating point numbers in several three dimensional (3D) arrays.

In the following, the implementation of two array types, a 3D array and a one dimensional array referred to here as vector, and their traversal in two different ways are discussed. The program, written in C [3], emulates the FDTD algorithm. The times taken to traverse the array types in the two ways by the optimally compiled program are compared. Traversing is performed in a manner natural to the language and in another more efficient way. It turns out that the execution speed of the program is compiler-dependent, and judicious programming [1] can improve execution speed significantly.

[1]See www.aplac.com.

## 2 Implementing the Data Types and Array Traversing

In this discussion, double precision numbers are updated by the FDTD algorithm but the conclusions are true for single precision, too. The 3D array and vector can be allocated statically, whence the number of grid points in the $x$, $y$ and $z$ directions must be known at compile time, or, preferably, dynamically with the `malloc()` function where this data is required only at run time [4, pp. 945–946]. A vector can be used instead of the 3D array in Fig. 3 by arranging the data, for example, as illustrated in Fig. 1.

| $0,0,0$ | $1,0,0$ | $\cdots$ | $X,0,0$ | $0,1,0$ | $1,1,0$ | $\cdots$ | $X,Y,0$ | $0,0,1$ | $\cdots$ | $i,j,k$ | $\cdots$ | $X,Y,Z$ |
|---|---|---|---|---|---|---|---|---|---|---|---|---|

**Fig. 1.** A vectorised 3D array.

As in the 3D array, the coordinate $(i, j, k)$ specifies a position in the vector of length $XYZ$ as well, but now the position is explicitly calculated as

$$\text{location} = XY \cdot k + X \cdot j + i = X(Y \cdot k + j) + i. \tag{1}$$

The statement `*(*(*(ar3d+k)+j)+i)` accesses an element in the 3D array `ar3d` with three additions and dereferences, while for the vector, from (1), two additions and multiplications suffice. So, it seems the vector is more efficient.

The ultimate criterion for choosing an array type for programming the FDTD algorithm depends on how efficiently the array is traversed. Another desirable feature is code readability. This discussion is limited to traversing the entire 3D computational space. The efficiency of the two nested loop implementations discussed below for the two data types are compared in Section 3.

The standard code to traverse a 3D array and a vector from point $(i_s, j_s, k_s)$ to point $(i_e, j_e, k_e)$ is given in Fig. 2. Assigning three auxilliary pointers, `dep = ar3d`, `row = *dep` and `col = *row`, as shown in Fig. 3, speeds up traversing. This scheme maximises memory access in unit strides resulting in better performance since the next array (memory) location is simply obtained by incrementing the current position value by one.

```
for (k=ks; k<=ke; k++) {
 for (j=js; j<=je; j++) {
  for (i=is; i<=ie; i++) {
   *(*(*(ar3d+k)+j)+i) = 1.0;}}}
```

```
for (k=ks; k<=ke; k++) {
 for (j=js; j<=je; j++) {
  for (i=is; i<=ie; i++) {
   *(ar1d+X*(Y*k+j)+i) = 1.0;}}}
```

**Fig. 2.** Standard code fragments to traverse a 3D array (left) and a vector (right).

Moving computations from the inner to the outer loops or altogether outside (a technique called frequency reduction [2, Section 12-5.2]) results in further speed up. For the vector, rearranging (1) to allow for frequency reduction and replacing the operators =, * and + with the computationally cheaper binary

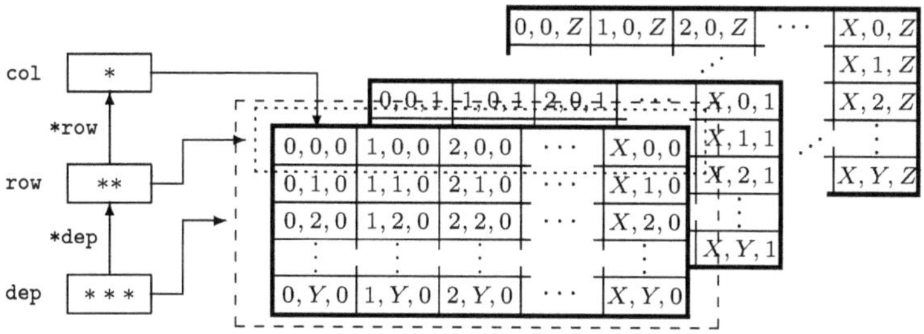


**Fig. 3.** Pointers to the first column and row in the first matrix of a 3D array. The numbers in the boxes are the position coordinates in the array.

operator += results in an efficient method for traversing the array. The refined code fragments for the two array types are given in Fig. 4.

The loop structure for the two data types is similar so that macros implementing either one form or the other can be defined, as shown in Fig. 5. Writing the program in this macro language allows the development of easily maintainable FDTD software that is portable on several platforms and always having the optimum data type for efficient nested loop traversing.

```
double ***dep, **row, *col;
 ...
dep = ar3d + ks;
for (k=ks; k<=ke; k++) {
 row = *dep + js;
 col = *row + is;
 dep++;
 for (j=0; j<=je; j++) {
  for (i=0; i<=ie; i++) {
   *col++ = 1.0;}
  row++;
  col = *row + is;}}
```

```
int adi, adj, adk; double *ptr;
 ...
adi = X - ie - 1 + is;
adj = (Y - je - 1 + js)*X;
adk = (Y*ks + js)*X + is;
ptr = ar1d + adk - adj;
for (k=ks; k<=ke; k++) {
 ptr += adj;
 for (j=js; j<=je; j++) {
  for (i=is; i<=ie, i++) {
   *ptr++ = 1.0;}
  ptr += adi;}}
```

**Fig. 4.** Refined code for traversing a 3D array (left) and a vector (right).

## 3 Comparison of the Two Data Types

Memory overhead, larger for the 3D array than the vector, is negligible compared to the field and material parameter array requirements. Traversing speeds for the two array types using the standard and refined loop implementations were compared by compiling the code on the four platforms with the compilers in Table 1. As this work aims to implement an FDTD field simulator as part of Aplac, the compilation options were those used for Aplac.

The arrays were allocated statically and dynamically and were traversed alone and as members of a structure. The starting time $t_0$ and the time after $10^5$ traversals $t_1$ were recorded using C's standard library function `clock()` to find $(t_1 - t_0)/$`CLOCKS_PER_SEC`. The resulting times are given in Table 2(a).

Time differences in the performance of the two data types are not big enough to be definitive, but it is clear the refined loops are faster. The run

**Table 1.** The platforms, compilers and optimisation options used.

| Processor | Operating System |
|---|---|
| Digital AlphaPC 164LX 533 MHz | Digital UNIX OSF1 V4.0D (Rev. 878) |
| HP D270/2 (2 × PA RISC 2.0 D270) | HP-UX B.10.20 A 9000/871 |
| Sun Ultra 2 UPA/SBus (2 × SUNW, Ultra-SPARC 200 MHz) | SunOS Release 5.5.1 [UNIX(R) System V Release 4.0] |
| AMD Athlon(tm) 800 MHz Processor | Linux version 2.2.16-22 |

| Compiler | Compilation command and options | Platform |
|---|---|---|
| DEC C V5.6-071 | `cc -std -O5 -Olimit 3000` | Alpha |
| HP C Preproc. A.10.32.17<br>HP C Compiler A.10.32.17<br>HP Linker ld B.10.33 | `cc -Aa -Onolimits` | HP-UX |
| WorkShop Compilers C 4.2 | `cc -fast -O4 -xtarget=ultra2/2200` | Sun |
| Gnu CC gcc ver. 2.96 [2] | `gcc -ansi -O` | Linux |

[2] Test code compilation with gcc versions 2.96 and 3.3.1 with -O3 optimisation gave faster execution, but the general result was the same. However, gcc 2.96 -O3 causes Aplac to malfunction.

times of the stand-alone arrays were significantly faster than those for the arrays in the structure, implying that operation -> is expensive. Also, the larger the array dimension in the innermost loop the faster the array is traversed.

A second test was run, now traversing seven arrays in the refined loops only, performing the following calclulation emulating an FDTD update equation:

```
*p=0.9*(*p)+0.8*(*(q++)*(*(r++)-*(s++))-*(t++)*(*(u++)-*(v++)));
```

p, q, r, s, t, u and v are pointers to an element in seven different arrays, either 3D or a vector. This test result is given in Table 2(b), which indicates that the vector is the obvious choice in the Alpha computer and, though the difference is not as dramatic, in the Sun and Linux computers. In the HP-UX computer, however, the choice is the 3D array. This unexpected result indicates that the compiler (and probably the platform, too) has a marked bearing on the traversing efficiency and so also on the choice of the array type.

```
#define DeclVars double ***dep,**row,*ptr
#define UseArray(a) double ***(a)
#define SetPtrTo(a) dep=(a)+ks
#define IncrKPtr row=*dep+js;ptr=*row+is;\
        dep++
#define IncrJPtr row++;ptr=*row+is
```

```
#define DeclVars int adi,adj,adk;double *ptr
#define UseArray(a) double *(a)
#define SetPtrTo(a) adi=X-ie-1+is;\
        adj=(Y-je-1+js)*X;adk=(y*ks+js)*X+is;\
        ptr=(a)+adk-adj
#define IncrKPtr ptr+=adi
#define IncrJPtr ptr+=adj
```

```
DeclVars; UseArray(arr);
...
SetPtrTo(arr);
for (k=ks; k<=ke; k++) {
  IncrKPtr;
  for (j=js; j<=je; j++) {
    for (i=is; i<=ie; i++) {
      *ptr++ = 1.0;}
    IncrJPtr;}}
```

**Fig. 5.** Macros to implement looping for the 3D array (top-left) and vector (top-right), and the resulting macro code for nested looping (bottom-centre).

**Table 2.** (a) Time taken to traverse one array $10^5$ times. 'Std.' refers to the standard and 'Ref.' to the refined nested loops. (b) Time taken to traverse seven arrays as part of a data structure, i.e., `s->a`, $10^5$ times using the refined nested loop.

| No. of arrays: 1 | Array size: 50 × 20 × 10 | | | | | | | |
|---|---|---|---|---|---|---|---|---|
| Array | Static | | Dynamic | | Static | | Dynamic | |
| | Std | Ref. | Std | Ref. | Std | Ref. | Std | Ref. |
| | Alpha | | | | HP-UX | | | |
| ***ar3d | 3.17 | - | 4.53 | 4.08 | 6.47 | - | 15.98 | 6.58 |
| *ar1d | 3.38 | 2.58 | 3.33 | 2.57 | 6.40 | 6.77 | 6.41 | 6.85 |
| s->ar3d | - | - | 4.65 | 4.33 | - | - | 20.64 | 6.51 |
| s->ar1d | - | - | 3.25 | 2.65 | - | - | 12.25 | 6.70 |
| | Sun | | | | Linux | | | |
| ***ar3d | 7.73 | - | 26.19 | 8.70 | 4.83 | - | 13.58 | 4.81 |
| *ar1d | 7.20 | 7.29 | 6.81 | 8.75 | 6.75 | 4.70 | 6.07 | 4.70 |
| s->ar3d | - | - | 36.44 | 7.85 | - | - | 16.69 | 5.80 |
| s->ar1d | - | - | 21.48 | 7.70 | - | - | 10.23 | 4.85 |

(a)

| Array size | Alpha | | HP-UX | |
|---|---|---|---|---|
| | 1D | 3D | 1D | 3D |
| 50×20×10 | 76.96 | 112.68 | 2522.73 | 569.42 |
| 50×10×20 | 74.56 | 110.98 | 2482.34 | 253.81 |
| 20×50×10 | 72.36 | 114.25 | 2100.40 | 182.17 |
| 20×10×50 | 75.18 | 116.75 | 2078.75 | 660.35 |
| 10×50×20 | 79.01 | 138.28 | 2559.74 | 546.42 |
| 10×20×50 | 76.00 | 132.66 | 2336.87 | 673.79 |
| | Sun | | Linux | |
| 50×20×10 | 172.92 | 179.67 | 245.19 | 252.01 |
| 50×10×20 | 172.38 | 181.06 | 250.19 | 257.17 |
| 20×50×10 | 172.14 | 183.55 | 246.84 | 271.48 |
| 20×10×50 | 173.79 | 184.60 | 247.59 | 271.49 |
| 10×50×20 | 175.75 | 191.56 | 252.38 | 303.79 |
| 10×20×50 | 175.51 | 192.35 | 247.17 | 303.84 |

(b)

## 4 Conclusions

A 3D array and a vector, implemented in C, were traversed in two differently realised nested loops with the FDTD algorithm in mind. Traversing times for the two array types, indicating program efficiency, show that the array type choice is compiler and computer platform dependent. Although the compiler optimises the program for speed, manipulating the loops using compiler programming techniques results in more efficient code. A macro language may be used to program the algorithm allowing development of easily maintainable code having the optimum data and program structure on several platforms. Programming in other languages will probably give similar timing results.

## References


1. K. Dowd and C. Severance. *High Performance Computing.* Second Edition, O'Reilly & Associates, Inc., Cambridge, 1998.
2. Tremblay J-P. and Sorenson P.G. *The Theory and Practice of Compiler Writing.* McGraw-Hill Book Company, Singapore, 1985.
3. B.W. Kernighan and D.M. Ritchie. *The C Programming Language.* Second edition, Prentice Hall PTR, New Jersey, 1988.
4. W.H. Press, S.A. Teukolsky, W.T. Vettering, and B.P. Flannery. *Numerical Recipes in C: The Art of Scientific Computing.* Second Edition, Cambridge University Press, New York, 1992.
5. W. Sui. *Time-Domain Computer Analysis of Nonlinear Hybrid Systems.* CRC Press, Boca Raton, 2002.
6. K.S. Yee. Numerical solution of initial boundary value problems involving maxwell's equations in isotropic media. *IEEE Trans. on Antennas and Propagation,* AP-14:302–307, 1966.

# Thermal Modeling of Bottle Glass Pressing

P. Kagan and R.M.M. Mattheij

CASA, Technische Universiteit Eindhoven, The Netherlands

**Summary.** Finite element approximation in space and Crank-Nicolson approximation in time are used to model incompressible creeping flow of molten glass with temperature dependent viscosity. Iso-P triangle elements and second degree approximation of temperature and velocity fields are applied. Localized thermal behavior is captured with adaptively refined unstructured mesh.

**Key words:** Glass pressing, thermal modeling, finite elements.

## 1 Introduction

In bottle manufacturing, a gob of molten glass is formed into a parison in a plunger and mold machine. Most of the used analysis techniques are based on experimentally acquired knowledge [1]. On the other hand, mathematical modeling can prove a decisive factor in production optimization, *e.g.*, [2, 3]. Therefore, this article reports on the study aiming at acquiring further insight into thermomechanical aspects of glass forming by means of numerically simulating the involved processes.

## 2 Physical model

*Thermodynamics* The glass density $\rho$ and specific heat $c_p$ are assumed constant based on the experimental data [4] and [6], respectively. Furthermore, the concept of "effective" heat conduction [6] is used to model heat transfer within the molten glass. That is, $\boldsymbol{q} = -\kappa \, \mathrm{grad}\,(T)$, where $T$ is the absolute temperature of the glass. Moreover, heat convection within the glass and the viscous dissipation are neglected due to the creeping nature of the glass flow. Finally, boundary heat transfer is modeled according to $\boldsymbol{q} \cdot \boldsymbol{n} = \alpha \left[T - \bar{T}\right]$, where $\boldsymbol{n}$ is the unit vector normal to the boundary, $\bar{T}$ is the temperature of the boundary and $\alpha$ is the convection coefficient.

*Fluid dynamics* Based on extensive experimental data analyzed in [1] glass is modeled as an isotropic incompressible Newtonian viscous fluid. The Cauchy stress tensor $\boldsymbol{T}$ is $\boldsymbol{T} = -p\boldsymbol{I} + 2\mu\boldsymbol{D}'$, where $p$ is the pressure, $\boldsymbol{I}$ is the unit second order tensor and $\boldsymbol{D}'$ is the deviatoric part of the velocity gradient. The dynamic viscosity $\mu$ depends on the temperature according to the experimentally established Fulcher relation $\log \frac{\mu}{\mu_0} = -a + b\frac{10^3}{T - T_0}$, where the numerical values of the constants depend on the glass composition. Furthermore, dimensions of a typical parison (length $\sim 0.1m$) and duration of a typical forming cycle (time $\sim 1\mathrm{s}$) suggest that the glass flow is creeping, so inertial and convective accelerations are neglected. Nonpenetration and full slip boundary conditions are assumed along the mold and the plunger, and material surface kinematics is assumed along the free surface. Plunger kinematics is prescribed, that is, linearly diminishing plunger velocity is imposed. Finally, attention is focused on the axisymmetric problem described with reference to the cylindrical coordinates $(r, z)$ along the unit vectors $(\boldsymbol{e}_r, \boldsymbol{e}_z)$, respectively.

*Equations of motion* In what follows $P$ is the interior of the computational domain, $\partial P$ is the boundary of $P$, $\partial P^p$, $\partial P^f$, $\partial P^m$ and $\partial P^s$ are parts of $\partial P$ corresponding to plunger, free surface, mold and symmetry axis, respectively. The glass behavior is described by the solution of

$$\rho\dot{e} + \operatorname{div}(\boldsymbol{q}) = 0 \qquad @\ P, t > 0, \tag{1a}$$

$$\operatorname{div}(\boldsymbol{v}) = \boldsymbol{D} \cdot \boldsymbol{I} = 0 \qquad @\ P, \tag{1b}$$

$$\operatorname{div}(\boldsymbol{T}) = 0 \qquad @\ P, \tag{1c}$$

satisfying the thermal initial and boundary conditions

$$T = T^0 \qquad @\ t = 0, \tag{2a}$$

$$-\kappa \operatorname{grad}(T) \cdot \boldsymbol{n} = \alpha\left[T - \bar{T}\right] \qquad @\ \partial P^p, \partial P^f, \partial P^m, \tag{2b}$$

$$-\kappa \operatorname{grad}(T) \cdot \boldsymbol{n} = 0 \qquad @\ \partial P^s, \tag{2c}$$

and the flow boundary conditions

$$[\boldsymbol{v} - \bar{\boldsymbol{v}}] \cdot \boldsymbol{n} = 0, \qquad \boldsymbol{T} \cdot [\boldsymbol{t} \otimes \boldsymbol{n}] = 0 \quad @\ \partial P^p, \partial P^m, \partial P^s, \tag{3a}$$

$$\boldsymbol{T} \cdot [\boldsymbol{n} \otimes \boldsymbol{n}] = p_{ext}, \quad \boldsymbol{T} \cdot [\boldsymbol{t} \otimes \boldsymbol{n}] = 0 \quad @\ \partial P^f. \tag{3b}$$

In these expressions $\bar{\boldsymbol{v}}$ is the velocity of the boundary and $\boldsymbol{n}$ and $\boldsymbol{t}$ are the unit vectors normal and tangent to the corresponding components of $\partial P$, respectively.

*Free surface kinematics* The position of the free surface $\partial P^f$ at time $t$ is implicitly represented by $f = \bar{f}(r, z, t) = z - \bar{\eta}(r, t) = 0$. The material surface assumption together with a neglected convective term enforce the motion of the free surface in accordance with the solution of

$$\dot{f} = -\frac{\partial \eta}{\partial t} + z = 0 \qquad @\ t > 0, \tag{4}$$

$$\eta = 0 \qquad @\ t = 0. \tag{5}$$

## 3 Finite element model

*Thermal problem* The space finite element approximation of Equation (1a) is developed by means of Galerkin weighted residual formulation with six node triangle shape functions $N_j(r,z)$:

$$\boldsymbol{M}\dot{\boldsymbol{T}} + \boldsymbol{S}\boldsymbol{T} = \boldsymbol{F} \quad @\, t > 0, \tag{6a}$$

$$M_{ij} = \sum_e \int_{P^e} \rho c_p N_i N_j \, dv^e, \tag{6b}$$

$$S_{ij} = -\sum_e \int_{P^e} \text{grad}(N_i) \cdot [-\kappa \text{grad}(N_j)] \, dv^e + \sum_{e \in \partial P} \int_{\partial P^e} N_i \alpha N_j \, da^e, \tag{6c}$$

$$F_i = -\sum_{e \in \partial P} \int_{\partial P^e} N_i \alpha \bar{T} \, da^e, \tag{6d}$$

$$\boldsymbol{T} = \boldsymbol{T}^{\mathbf{0}} \quad @\, t = 0. \tag{6e}$$

In these expressions, $P^e$ is a generic finite element domain and $\partial P^e$ is its boundary. Equations (6a) describe an ordinary initial value problem that is conveniently solved by Crank-Nicolson finite difference scheme [5] with time step $\Delta t$:

$$\left[\boldsymbol{M} + \frac{1}{2}\Delta t \boldsymbol{S}\right] \boldsymbol{T}^{n+1} = -\Delta t \boldsymbol{F}^{n} + \left[\boldsymbol{M} - \frac{1}{2}\Delta t \boldsymbol{S}\right] \boldsymbol{T}^{n}. \tag{7}$$

Finally, boundary and initial conditions are imposed and the resulting algebraic linear equations are solved leading to the nodal values $\boldsymbol{T}^{n+1}$ of the glass temperature at time $t^{n+1}$.

*Flow problem* Following the guidelines of Babuška-Brezzi condition [7] the pressure $p$ is approximated in terms of three-node triangular shape functions $N_j^p(r,z)$, while the velocity components $v_r$ and $v_z$ are approximated in terms of six node triangle shape functions $N_j^v(r,z)$. The mixed finite element approximation of (1b)– (1c) is

$$\boldsymbol{A}\boldsymbol{X} = \boldsymbol{0}, \quad \boldsymbol{A}_{ij} = \begin{bmatrix} \boldsymbol{S}_{ij}^{vv} & \boldsymbol{S}_{ij}^{vp} \\ \boldsymbol{S}_{ij}^{pv} & \boldsymbol{0} \end{bmatrix}, \boldsymbol{X}_j = \begin{bmatrix} \boldsymbol{V}_j \\ P_j \end{bmatrix}, \boldsymbol{V}_j = \begin{bmatrix} V_j^r \\ V_j^z \end{bmatrix}, \tag{8a}$$

$$\boldsymbol{S}_{ij}^{vv} = \sum_e \int_{P^e} - \begin{bmatrix} N_{i,r}^v & 0 & \frac{1}{r}N_i^v & N_{i,z}^v \\ 0 & N_{i,z}^v & 0 & N_{i,r}^v \end{bmatrix} \mu \begin{bmatrix} 2N_{j,r}^v & 0 \\ 0 & 2N_{j,z}^v \\ \frac{2}{r}N_j^v & 0 \\ N_{j,z}^v & N_{j,r}^v \end{bmatrix} dv^e, \tag{8b}$$

$$\boldsymbol{S}_{ij}^{vp} = \sum_e \int_{P^e} - N_i^v \begin{bmatrix} N_{j,r}^p \\ N_{j,z}^p \end{bmatrix} dv^e, \tag{8c}$$

$$\boldsymbol{S}_{ij}^{pv} = \sum_e \int_{P^e} N_i^p \left[N_{j,r}^v + \tfrac{1}{r}N_j^v \; N_{j,z}^v\right] dv^e. \tag{8d}$$

In these expressions, *e.g.*, $N_{i,r}$ designates $\frac{\partial N_i}{\partial r}$, etc. Finally, boundary conditions are imposed and the resulting algebraic linear equations are solved leading to the nodal values of the glass velocity and pressure. Rank of the coefficient matrix must be monitored carefully while solving (8a) since the described flow finite element fails the patch test when too many velocity degrees of freedom are constrained [7].

*Mesh deformation problem* Thermomechanical behaviour of the glass during the pressing involves large displacements of the material. Consequently, the Lagrangian finite element mesh used for solving the thermomechanical problem undergoes significant distortion and most certainly becomes invalid unless special care is taken. At present, complete reconstruction of the mesh is used followed by projecting the temperature from deformed mesh onto new mesh by means of Taylor series truncated after the linear term.

## 4 Results

The initial geometry of mold and plunger used at this stage of the study are depicted in metric units on the left of Fig. 1. The finite element mesh was adaptively refined toward the external boundaries of the computational domain in order to resolve sharp changes of the temperature and the viscosity. Consequent subplots of this picture present the calculated temperature and the radial and axial velocities, respectively. The mesh and the calculated temperature at the half pressing time $t = 0.6\,\mathrm{s}$ and at the end of the pressing $t = 1.2\mathrm{s}$ are presented on the left and the right of Fig. 2, respectively.

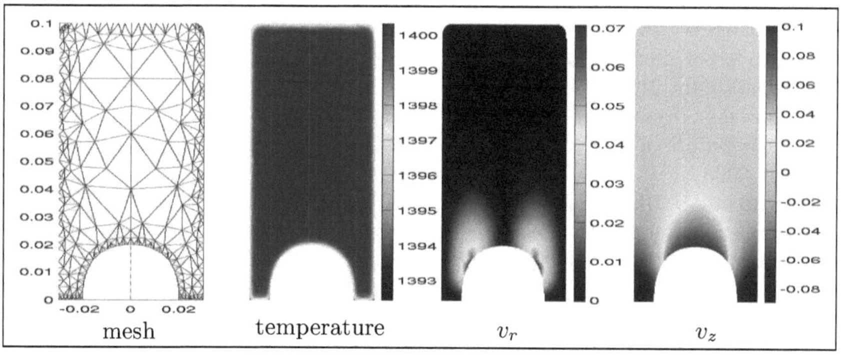


**Fig. 1.** Mesh and velocity components at $t = 0$, temperature at $t = 0.03\,\mathrm{s}$

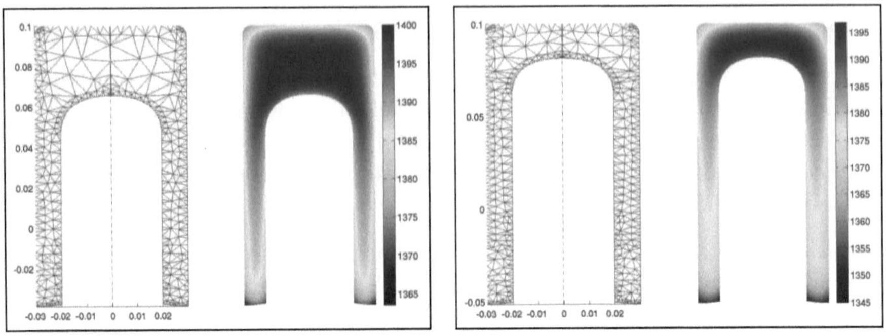

**Fig. 2.** Mesh and temperature: left $t = 0.6$s, right $t = 1.2$s

# 5 Conclusions

The results presented in this article extend our insight into the behavior of the molten glass during the pressing stage of bottle manufacturing. Two major observations are obvious already at this early stage of the study. First, the glass temperature is significantly affected by the heat convection across the rigid boundaries and the free surface. Secondly, the changes of the glass viscosity due to the thermal inhomogeneity considerably affect the glass flow pattern. From the conceptual point of view, our analysis shows real potential for the purposes of process optimization.

# References


1. C.L. Babcock. *Silicate Glass Technology Methods.* Wiley, New York, 1977.
2. J.M.A. Cesar de Sa. Numerical modeling of glass forming processes. *Engineering Computations*, 3(4):266–275, 1986.
3. K. Laevsky. *Pressing of glass in bottle and jar manufacturing: numerical analysis and computation.* PhD thesis, Eindhoven University of Technology, The Netherlands, 2003.
4. I. Sawai and S. Inoue. Spezifisches gewicht der ternären gläser $CaO-Na_2O-SiO$ bei hoher temperatur. *Journal of the Society of Chemical Industry of Japan, Transactions*, 43(2):47B–49B, 1940.
5. G. Strang and G.J. Fix. *An Analysis of the Finite Element Methods.* Prentice-Hall, Englewood Cliffs, N.J., 1973.
6. A.F. Van Zee and C.L. Babcock. A method for the measurement of thermal diffusivity of molten glass. *Journal of American Ceramic Society*, 34(8):244–250, 1951.
7. O.C. Zienkiewicz and R.L. Taylor. *The Finite Element Method.* Butterworth Heinmann, 5 edition, 2000.

# Simulation of Pulsed Signals in MPDAE-Modelled SC-Circuits

S. Knorr[1] and U. Feldmann[2]

[1] Bergische Universität Wuppertal, FB C, Gaußstr. 20, D-42119 Wuppertal
knorr@math.uni-wuppertal.de
[2] Infineon Technologies AG, Balanstr. 73, D-81541 Munich
uwe.feldmann@infineon.com

**Summary.** The simulation of circuits including signals with widely separated time scales can easily become very time-consuming. To avoid this, a multidimensional signal model was developed. The resulting system of network equations can be solved very efficiently by a method of characteristics. We investigate the applicability of this method to circuits including digital signal structures. Moreover, systems given in linear-implicit form are solved using the multidimensional approach.

## 1 Introduction

Signals with widely separated time scales often arise in radio frequency application. To describe such signals more efficiently, a multidimensional model has been developed, which transfers the circuit's differential-algebraic equations (DAE) to a multirate system of partial differential-algebraic equations (MPDAE). A specially tailored method of characteristics has already been sucessfully used to solve MPDAE-modelled network equations governed by semi-explicit DAEs including harmonic signals [4].

Now, we want to apply the method of characteristics to MPDAE-modelled switched capacitor (SC) circuits. In those circuits, transistors are driven by high frequency pulses, which are characterized by a digital signal structure.

In the first test example of a switched capacitor filter, the applicability of the method to the non-harmonic, strongly nonlinear signals is investigated. The second circuit of the Miller integrator serves to simulate network equations, which are given in linear-implicit form.

## 2 Switched capacitor filter

The first test example is the switched capacitor filter depicted in Fig. 1. A sinusoidal input signal charges the first capacitor driven by the pulse $p_a$ and

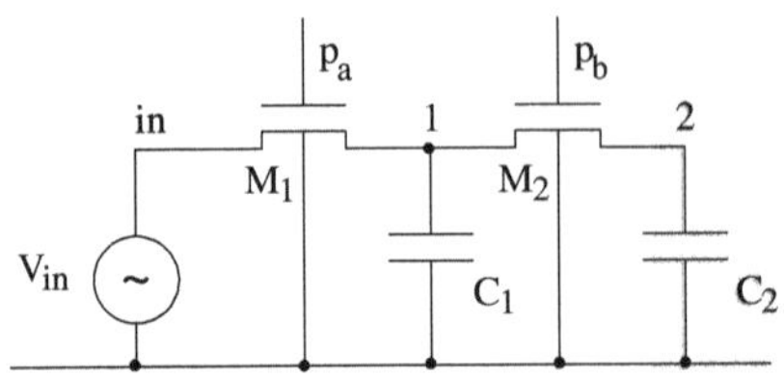


**Fig. 1.** Switched capacitor filter

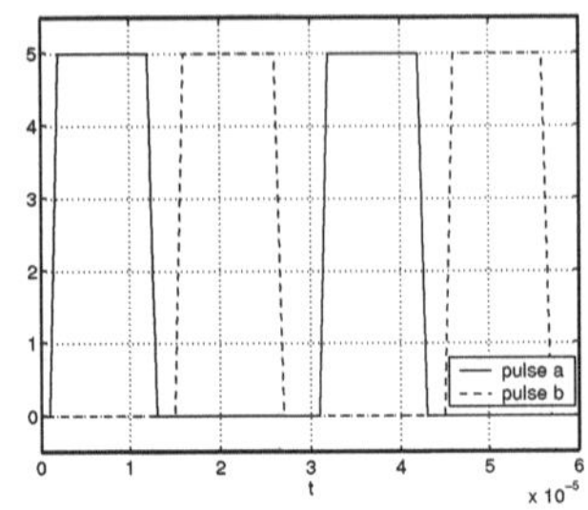


**Fig. 2.** Pulses $p_a$ and $p_b$

this charge is transmitted to the second capacitor driven by the pulse $p_b$. The transistors work as switches and the pulses have to be complementary to each other as shown in Fig. 2. The equations for the two nodes are in ODE form given by

$$-I_{DS}(u_{p_a}, u_{in}, u_1, 0) + I_{DS}(u_{p_b}, u_1, u_2, 0) + C_1 \dot{u}_1$$
$$+ CGSO \cdot W \cdot \frac{d(u_1 - u_{p_a})}{dt} + CGDO \cdot W \cdot \frac{d(u_1 - u_{p_b})}{dt} = 0 \qquad (1)$$
$$-I_{DS}(u_{p_b}, u_1, u_2, 0) + C_2 \dot{u}_2 + CGSO \cdot W \cdot \frac{d(u_2 - u_{p_b})}{dt} = 0 \qquad (2)$$

with overlap capacitances $CGSO$, $CGDO$ and transistor width $W$. For the drain to source current $I_{DS}(u_{gate}, u_{drain}, u_{source}, u_{bulk})$ of the MOS-transistors $M_1$ and $M_2$, a level-1 model by Stichman-Hodges is used [1].

The pulses work at the fast time scale $T_2 = 3 \cdot 10^{-5}$ s, whereas the sinusoidal input $V_{in}$ oscillates with $T_1 = 10^{-3}$ s. To describe these widely separated time scales more efficiently, a multidimensional signal model is applied. A detailed description of this modelling approach can be found in [2].

## 3 Multidimensional approach

To decouple the different time scales of the switched capacitor circuit, a corresponding variable is assigned to each of them. For two different time scales this approach generalizes a two-tone signal $s(t)$ to a so-called multivariate funtion (MVF) $\hat{s}(t_1, t_2)$, for example

$$s(t) = \sin\left(\frac{2\pi}{T_1} t\right) \sin^2\left(\frac{\pi}{T_2} t\right) \quad \rightsquigarrow \quad \hat{s}(t_1, t_2) = \sin\left(\frac{2\pi}{T_1} t_1\right) \sin^2\left(\frac{\pi}{T_2} t_2\right).$$

The original signal can always be reconstructed by $s(t) = \hat{s}(t, t)$.

Applying this multidimensional signal model to the SC-filter circuit transfers the network-ODE (1)+(2) to a multirate partial differential equation (MPDE):

$$(C_1 + CGSO \cdot W + CGDO \cdot W)\left(\frac{\partial u_1(t_1,t_2)}{\partial t_1} + \frac{\partial u_1(t_1,t_2)}{\partial t_2}\right)$$
$$= I_{DS}(u_{p_a}(t_2), u_{in}(t_1), u_1(t_1,t_2), 0) - I_{DS}(u_{p_b}(t_2), u_1(t_1,t_2), u_2(t_1,t_2), 0)$$
$$+ CGSO \cdot W \cdot \frac{du_{p_a}(t_2)}{dt_2} + CGDO \cdot W \cdot \frac{du_{p_b}(t_2)}{dt_2} \tag{3}$$

$$(C_2 + CGSO \cdot W)\left(\frac{\partial u_2(t_1,t_2)}{\partial t_1} + \frac{\partial u_2(t_1,t_2)}{\partial t_2}\right)$$
$$= I_{DS}(u_{p_b}(t_2), u_1(t_1,t_2), u_2(t_1,t_2), 0) + CGSO \cdot W \cdot \frac{du_{p_b}(t_2)}{dt_2}\,. \tag{4}$$

As the PDE is of hyperbolic type, we are able to apply the method of characteristics described in [4]. The ODEs arising in the characteristic system of the MPDE are solved via discretization along the characteristic curves, which are straight lines in the direction of the diagonal. Boundary conditions are given by the periodicity of the MVFs. The simulation results for node 2, which coincide with solutions generated by MATLAB-routines, are shown in Fig. 3.

Thus, the application of the method of characteristics to network equations including digital signal structures works successfully. In the following, we investigate a system given in a linear-implicit form.

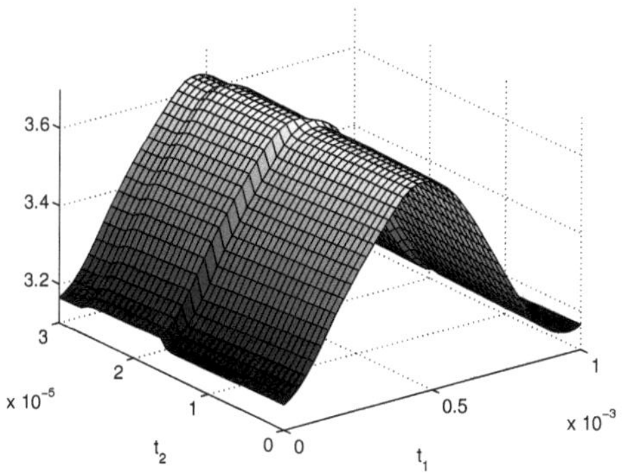


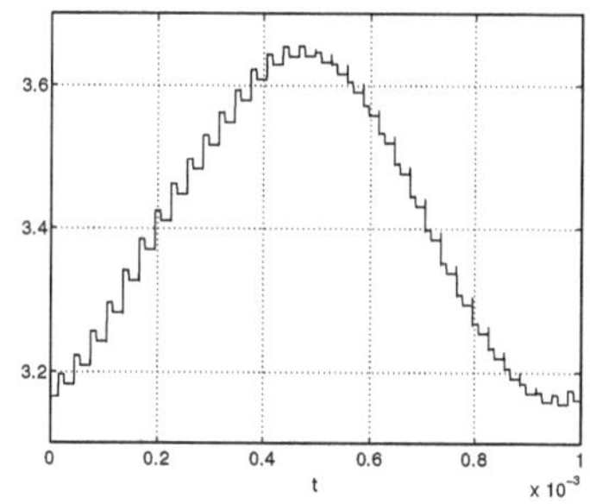


**Fig. 3.** MPDE-solution (left), reconstructed ODE-solution (right)

# 4 Miller integrator

The Miller integrator in Fig. 4 produces the negative integral of the input signal at node 3. The sinusoidal input with period $T_1 = 10^{-5}$ s is sampled periodically with $T_2 = 25 \cdot 10^{-9}$ s. Pulses $p_a$ and $p_b$ have a similar behaviour as above (see Fig. 2).

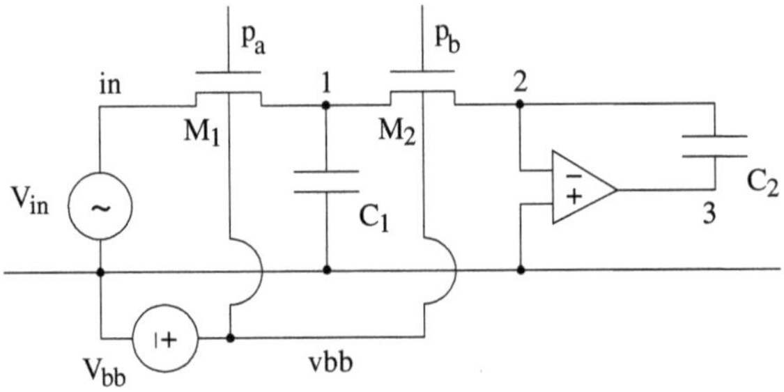


**Fig. 4.** Miller integrator

How the index of the network equations may depend on the value of technical circuit parameters is investigated in [3]. In our example, the network equations are an index-1 DAE-system. Again, bivariate functions are introduced for all state variables and sources, which leads to multirate partial differential-algebraic equations (MPDAE) given in a linear-implicit form:

$$C_1 \cdot \left( \frac{\partial u_1(t_1,t_2)}{\partial t_1} + \frac{\partial u_1(t_1,t_2)}{\partial t_2} \right) = I_{DS}(u_{p_a}(t_2), u_{in}(t_1), u_1(t_1,t_2), vbb) - I_{DS}(u_{p_b}(t_2), u_1(t_1,t_2), u_2(t_1,t_2), vbb) \quad (5)$$

$$C_2 \cdot \left( \frac{\partial \left[u_2(t_1,t_2) - u_3(t_1,t_2)\right]}{\partial t_1} + \frac{\partial \left[u_2(t_1,t_2) - u_3(t_1,t_2)\right]}{\partial t_2} \right) = I_{DS}(u_{p_b}(t_2), u_1(t_1,t_2), u_2(t_1,t_2), vbb) \quad (6)$$

$$0 = u_3(t_1,t_2) + 1000 \cdot u_2(t_1,t_2) \quad (7)$$

with a negative substrate bias voltage $vbb$.

Again, the method of characteristics described in the previous section was used to solve the system. Also for this example, the one-dimensional solution reconstructed from the MPDAE-solution coincides well with a corresponding MATLAB-solution of the original network equations. Figure 5 shows the simulation results for node 1. Thus, equations given in linear-implicit form can also be solved via the multidimensional approach.

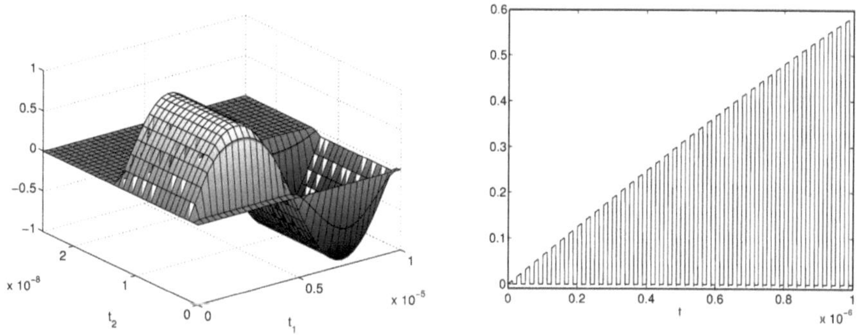


**Fig. 5.** MPDAE-solution (left), zoom into DAE-solution (right)

# 5 Conclusions

The approach via characteristic systems was successfully applied to MPDAE-modelled pulsed signals in switched capacitor circuits. Not only harmonic but also digital-like signals can be simulated using the described method of characteristics. In addition, network equations given in linear-implicit form can be solved as well as explicit ones. In any case, the efficiency of the multidimensional approach and of the specially tailored method can be exploited.

*Acknowledgement.* This work has been supported within the federal BMBF project with the grant number 03GUNAVN. The authors are indebted to Michael Günther and Roland Pulch for helpful discussions.

# References

1. P. Antognetti and G. Massobrio. *Semiconductor device modeling with SPICE.* McGraw-Hill, New York, 1987.
2. A. Bartel, M. Günther, R. Pulch, and P. Rentrop. Numerical techniques for different time scales in electric circuit simulation. In M. Breuer, F. Durst, and C. Zenger, editors, *High Performance Scientific and Engineering Computing*, Lecture Notes in Computational Science and Engineering, pages 343–360. Springer, 2002.
3. M. Günther and U. Feldmann. CAD-based electric-circuit modeling in industry II: Impact of circuit configurations and parameters. *Surv. Math. Ind.*, 8:131–157, 1999.
4. R. Pulch and M. Günther. A method of characteristics for solving multirate partial differential equations in radio frequency application. *Appl. Numer. Math.*, 42:397–409, 2002.

# A More Efficient Rigorous Coupled-Wave Analysis Algorithm

M.G.M.M. van Kraaij[1] and J.M.L. Maubach[2]

[1] Technische Universiteit Eindhoven (TU/e), Den Dolech 2, Eindhoven, The Netherlands. M.G.M.M.v.Kraaij@tue.nl
[2] Technische Universiteit Eindhoven (TU/e), Den Dolech 2, Eindhoven, The Netherlands. J.M.L.Maubach@tue.nl

**Summary.** We present a modification of a well-known mathematical model based on the Rigorous Coupled-Wave Analysis (RCWA) that can be used to solve optical diffraction problems on periodic structures (both 1-D and 2-D gratings with approximated layer-structure). The algorithm calculates the reflected and transmitted field which in turn determine the diffraction efficiencies for all reflected and transmitted orders.

Results created with a Matlab implementation of the modified RCWA algorithm (MSolver) show excellent overlap with other published and measured data.

**Key words:** Rigorous Coupled-Wave Analysis, RCWA, Diffraction grating

## 1 Introduction

Lithography often uses gratings for various metrology tasks such as alignment, overlay metrology and CD metrology. With the tightening requirements on metrology accuracy it becomes increasingly more important to understand the behaviour of the grating in the metrology application using a rigorous mathematical diffraction model.

In order to understand the complexity of the grating problem, it is necessary to realize that nowadays gratings have complex profiles and consist of all kinds of different materials. Real-life lithography does not produce symmetric profiles with sinusoidal, rectangular or trapezoidal grooves for very high and very low groove frequencies. Moreover in the visible region and for shorter wavelengths the finite conductivity complicates the grating response and requires more complex mathematical models.

The RCWA algorithm is often used because of its good convergence and relatively simple implementation. The algorithm uses a layered structure to approximate the grating profile but for the material properties no approximations are used.

This paper presents a modified version of the RCWA algorithm which solves optical diffraction problems better for the case of highly conducting materials. For a full derivation of the discrete equations from Maxwell's equations, see [1, 2]. The modification in Sect. 3 from equation (6) on (based on material from [3]) makes the RCWA algorithm converge faster. Presented numerical results show that the modified method converges much faster, especially for metallic gratings.

## 2 The model

First consider the diffraction problem in Fig. 1 which leads to Maxwell's equations which are the basis for the RCWA algorithm. This is the standard model with the standard assumptions: A linearly polarized electromagnetic field with angle $\psi$ is obliquely incident at an arbitrary angle of incidence $\theta$ and at an azimuthal angle $\varphi$ upon a dielectric or lossy grating. The grating is assumed to be infinitely long in the periodic $x$-direction with a period $\Lambda$. The grating grooves along the $y$-direction are also assumed to be infinitely long. In the example below only two different media are present with refraction indices $n_I$ and $n_{II}$.

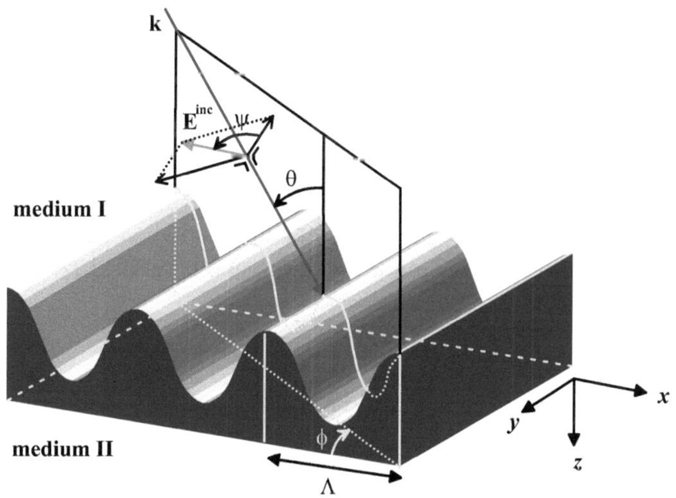


**Fig. 1.** One-dimensional periodic grating in $\mathbb{R}^3$

Fig. 2 shows how a general grating is approximated by a multilayered grating. Note that all calculations can be restricted to only one period. In each layer the material constants only depend on the horizontal $x$-coordinate and are independent of the vertical $z$-coordinate. Furthermore the different media are assumed to be homogeneous, linear and isotropic.

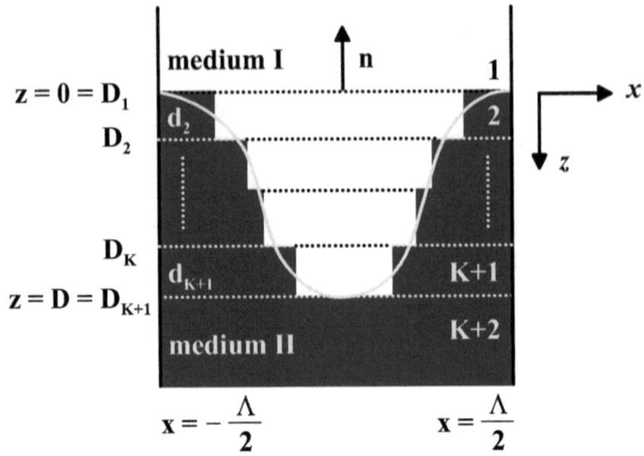


**Fig. 2.** Layered one-dimensional grating in $[-\frac{\Lambda}{2}, \frac{\Lambda}{2}] \times (-\infty, \infty)$

# 3 The equations and boundary conditions

Assuming that no primary or external sources are present and considering only time-harmonic field quantities, Maxwell's equations applied to the discrete model in Fig. 2 and the constitutive relations can be reduced[3] to the following equations for each grating layer $i$:

$$\frac{\partial}{\partial z} E_{i,x}(x,z) = -j\omega\mu_0 H_{i,y}(x,z) + \frac{\partial}{\partial x} E_{i,z}(x,z), \tag{1a}$$

$$\frac{\partial}{\partial z} H_{i,y}(x,z) = -j\omega\tilde{\varepsilon}_i(x) E_{i,x}(x,z), \tag{1b}$$

$$\frac{\partial}{\partial x} H_{i,y}(x,z) = j\omega\tilde{\varepsilon}_i(x) E_{i,z}(x,z). \tag{1c}$$

All the electric field components $E_i$ can be eliminated so that for each layer only one equation for the $y$-component of the magnetic field $H_i$ remains:

$$\frac{\partial^2}{\partial z^2} H_{i,y}(x,z) = -k_0^2 \frac{\tilde{\varepsilon}_i(x)}{\varepsilon_0} H_{i,y}(x,z) - \frac{\tilde{\varepsilon}_i(x)}{\varepsilon_0} \frac{\partial}{\partial x}\left(\frac{\varepsilon_0}{\tilde{\varepsilon}_i(x)} \frac{\partial}{\partial x} H_{i,y}(x,z)\right). \tag{2}$$

On the left and right of the domain the pseudo-periodic boundary conditions are applied and above and below the grating Rayleigh's radiation condition is used. Finally at the layer interfaces the continuity of the tangential electromagnetic field components are preserved. This can be reformulated in the following set of equations for the magnetic field:

$$H_{i,y}(x, D_i) = H_{i+1,y}(x, D_i), \tag{3a}$$

$$\frac{1}{\tilde{\varepsilon}_i(x)} \frac{\partial}{\partial z} H_{i,y}(x, D_i) = \frac{1}{\tilde{\varepsilon}_{i+1}(x)} \frac{\partial}{\partial z} H_{i+1,y}(x, D_i). \tag{3b}$$

[3]Only TM polarized incident light in a planar diffraction case is considered here. For more details on other diffraction cases see [4]

The complex permittivity $\tilde{\varepsilon}_i(x)$ within each layer is expanded in a Fourier series. The incident magnetic field $H_y^{inc}$ is assumed to be one plane wave and the magnetic field expansions for each grating layer are given by equation (4):

$$H_y^{inc}(x,z) = n_1 \left(\frac{\varepsilon_0}{\mu_0}\right)^{\frac{1}{2}} e^{-jk_0 n_1 (x \sin\theta + z\cos\theta)}, \tag{4a}$$

$$H_{1,y}(x,z) = n_1 \left(\frac{\varepsilon_0}{\mu_0}\right)^{\frac{1}{2}} \sum_n R_n e^{-j(k_{xn}x - k_{1,zn}z)} + H_y^{inc}(x,z), \tag{4b}$$

$$H_{i,y}(x,z) = n_1 \left(\frac{\varepsilon_0}{\mu_0}\right)^{\frac{1}{2}} \sum_n U_{i,n}(z) e^{-jk_{xn}x}, \tag{4c}$$

$$H_{K+2,y}(x,z) = n_1 \left(\frac{\varepsilon_0}{\mu_0}\right)^{\frac{1}{2}} \sum_n T_n e^{-j(k_{xn}x + k_{K+2,zn}(z-D))}. \tag{4d}$$

Note that these expansions already satisfy the pseudo-periodic boundary condition and Rayleigh's radiation condition. Substituting these expansions into (2) and truncating the equations results in:

$$\frac{d^2}{dz'^2} \boldsymbol{U}_i(z') = \mathbf{E}_i \left(\mathbf{K}_x \mathbf{P}_i \mathbf{K}_x - \mathbf{I}\right) \boldsymbol{U}_i(z'). \tag{5}$$

However equation (5) does not uniformly preserve the continuity of the appropriate field components across the discontinuities in one layer of the complex permittivity function. This is caused by the way in which the Fourier series are used in the truncated equations. We propose to use the truncated equations:

$$\frac{d^2}{dz'^2} \boldsymbol{U}_i(z') = \mathbf{P}_i^{-1} \left(\mathbf{K}_x \mathbf{E}_i^{-1} \mathbf{K}_x - \mathbf{I}\right) \boldsymbol{U}_i(z'). \tag{6}$$

This proposal is based on [3] which suggests that these truncations are better when there are discontinuities in one layer of the permittivity function. Equation (6) is not derived from (2) but from the basis equations (1) after multiplying (1b) with $1/\tilde{\varepsilon}_i(x)$. So instead of first eliminating the electric field components, substituting the expansions and truncating the equations, we now start with substituting the expansions in (1), truncating the equations and then eliminating the electric field components.

Finally the complex reflected and transmitted field amplitudes are determined by calculating eigenvalues and eigenvectors of equation (6) and using the boundary conditions (3) at the layer interfaces. An enhanced transmittance matrix approach is used to calculate the reflected and transmitted field amplitudes in a stable way [2].

## 4 Numerical results

In this example a simple binary gold grating with TM polarized incident light is used. All the important grating parameters can be found in Fig. 3. The convergence of the original RCWA algorithm is compared with the modified RCWA algorithm in Fig. 4. Here the diffraction efficiency of the 0th reflected order versus the total number of orders retained in the expansions is plotted. Note that the diffraction efficiency is just the amount of energy relative to the incident field. Clearly the modified RCWA algorithm converges much faster.

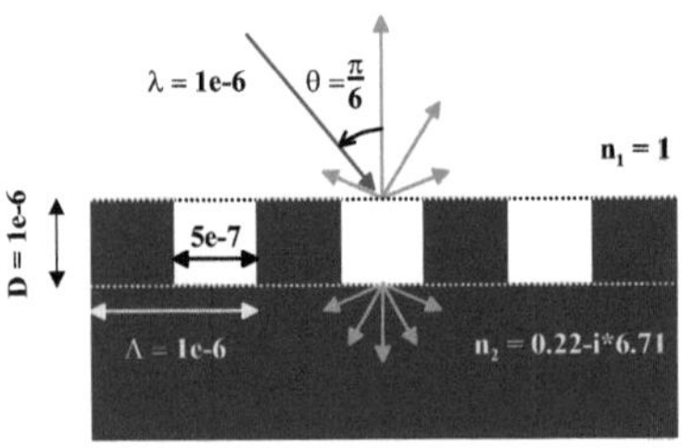


**Fig. 3.** Grating parameters

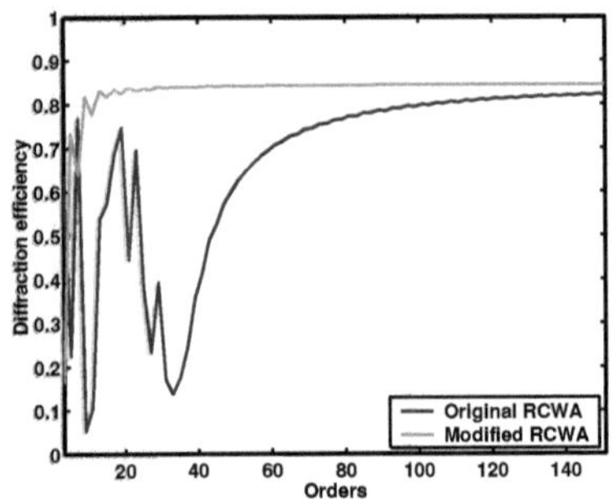


**Fig. 4.** Convergence 0th order

## 5 Conclusions

The original RCWA algorithm is worked out in detail for all diffraction cases and is extended with Fourier factorization rules from [3] which also improved convergence for the $C$-method. The modified RCWA algorithm now also performs well for TM polarized light on metallic gratings.

A Matlab implementation of the modified RCWA algorithm (MSolver) shows good overlap with other published and measured data.

## References

1. Max Born and Emil Wolf. *Principles of Optics*, chapter 13: Optics of Metals. Pergamon Press, 6th edition, 1980.
2. M. G. Moharam, Drew A. Pommet, and Eric B. Grann. Stable implementation of the rigorous coupled-wave analysis for surface-relief gratings: enhanced transmittance matrix approach. *J. Opt. Soc. Am. A*, 12(5):1077–1086, May 1995.
3. Lifeng Li. Use of Fourier series in the analysis of discontinuous periodic structures. *J. Opt. Soc. Am. A*, 13(9):1870–1876, September 1996.
4. M.G.M.M.v.Kraaij. A more Rigorous Coupled-Wave Analysis (MSolver). Master's thesis, Technische Universiteit Eindhoven, 2004.

# Iterative Solution Approaches for the Piezoelectric Forward Problem

M. Mohr*

Department of Sensor Technology, University of Erlangen-Nuremberg, Germany
Marcus.Mohr@lse.eei.uni-erlangen.de

**Summary.** One of the fields of engineering science in which numerical simulation is playing a role of increasing importance is the design of piezoelectric transducers. Efficient techniques to solve the forward problem of computing the mechanical displacements and electric potential for a given configuration play a crucial role in the design itself, but also in the related problem of identifying the correct material parameters. In this paper we consider the iterative solution of linear systems arising from a Finite-Element discretisation of the piezoelectric forward problem with the Generalised Minimal Residual method in combination with incomplete LU decomposition and inexact block diagonal preconditioning.

**Key words:** symmetric indefinite, piezoelectricity, iterative solver, GMRES, ILU, inexact block preconditioner.

## 1 Introduction

The coupling between the mechanical and electrical field in piezoelectric materials makes them very interesting for the design of sensors and actuators. Piezoelectric transducers can be found in a multitude of applications ranging from ultrasound devices in medical imaging and industrial cleaning over force and acceleration sensors to surface acoustic wave filters and micro-pumps to name only a few.

Design and optimisation of such actuators and sensors today rely to a large part on numerical simulation. Key issues here are of course efficiency and accuracy. The latter is influenced not only by the precision of the discretisation and solution process, but also by the precision of the material parameters entering the model. Thus, simulation based parameter identification is a topic of increasing relevance.

*This work has been supported by the Deutsche Forschungsgemeinschaft under Grant number Ka 1778/1–1.

Both the simulation of piezoelectric materials, as well as the inverse problem of parameter identification rely on the ability to efficiently solve the related forward problem, *i.e.*, to determine for given material parameters and boundary conditions the resulting electric field and mechanical displacements. The discretisation of this problem by means of the Finite-Element method (FEM) leads to a linear system with a symmetric but indefinite matrix. In this paper we consider the iterative solution of this system with the Generalised Minimal Residual (GMRES) method and report on results with ILU(k) and inexact block diagonal preconditioning.

## 2 Mathematical Model

In the case of a piezoelectric material the connection between the electric potential $\Phi$ and the mechanical displacements $\mathbf{d}$ is given by a system of four partial differential equations

$$\begin{aligned} \varrho\frac{\partial^2\mathbf{d}}{\partial t^2} - \mathcal{B}^T\left(c^E\mathcal{B}\mathbf{d} + e^T\operatorname{grad}\varphi\right) &= 0 \\ -\operatorname{div}\left(e\mathcal{B}\mathbf{d} - \varepsilon^S\operatorname{grad}\varphi\right) &= 0 \ . \end{aligned} \tag{1}$$

In (1) $c^E$, $\varepsilon^S$ and $e$ denote the tensors of elasticity, dielectricity and piezoelectric coupling. $\mathcal{B}$ represents a first order differential operator that is the transpose of the divergence of a dyadic. System (1) can be discretised with a standard Finite-Element approach using nodal Ansatz functions. For details see *e.g.*, [4]. The resulting linear system for the stationary part of the problem takes the form

$$\begin{pmatrix} M & P \\ P^T & -E \end{pmatrix}\begin{pmatrix} \mathbf{u} \\ \varphi \end{pmatrix} = \begin{pmatrix} \mathbf{f_m} \\ \mathbf{f_e} \end{pmatrix} , \tag{2}$$

where $M$ represents the mechanics block, $-E$ the electrostatics block and $P$ the piezoelectric coupling part. The vectors $\varphi$ and $\mathbf{u}$ contain the expansion coefficients of the discrete electric potential and mechanical displacement in terms of the Finite-Element basis. The vectors $\mathbf{f_m}$ and $\mathbf{f_e}$ arise from the boundary conditions for the mechanical and electrical unknowns. It is known that a standard FEM discretisation of the mechanical and electrical part of the problem will lead to symmetric positive definite matrices $M$ and $E$. Thus, from (2) we can expect the problem matrix to be symmetric, but indefinite. This can also be seen from the example spectrum plotted in Fig. 2.

## 3 Iterative Solution

We consider the iterative solution of (2) with the help of the GMRES method. Starting from an initial guess $x^{(0)}$ the method computes in each step a new iterate by $x^{(k)} = x^{(0)} + Q_k v^{(k)}$. Here $Q_k \in \mathbb{R}^{n\times k}$ is a matrix whose columns form an orthonormal basis of the Krylov subspace $K^k(A, r^{(0)}) = \text{span}\{r^{(0)}, Ar^{(0)}, A^2r^{(0)}, \ldots, A^{k-1}r^{(0)}\}$ and $v^{(k)} \in \mathbb{R}^k$ is chosen such that the norm of the new residual $\|b - Ax^{(k)}\|_2$ is minimal.

The advantage of GMRES is its stability, *i.e.*, it is guaranteed to converge. There is a grain of salt to this, however. The Arnoldi process used for computing the basis $Q_k$ requires storing all previous base vectors and the orthonormalisation becomes increasingly costly with each step. Thus, one often re-starts the method after $m$ iterations discarding the old Krylov subspace. This is denoted as GMRES(m). For more details on GMRES see *e.g.*, [5].

In this paper we employ right-preconditioning to improve convergence speed, *i.e.*, we replace the linear system $Ax = b$ by a preconditioned system $\tilde{A}y = b$ with $\tilde{A} = AP^{-1}$ and $x = P^{-1}y$, and consider the following two methods for preconditioning. The first approach is an ILU(k) preconditioner. Here one chooses $P = LU$, where $L$ and $U$ are the lower/upper triangular factors of an incomplete LU decomposition $A = LU - R$. The notion ILU(k) indicates that the level of fill-in allowed in the factors $L$ and $U$ is determined by its "distance" from the original sparsity pattern, for more details see *e.g.*, [1]. Computation of the decomposition was performed with the Euclid method from the hypre library, see *e.g.*, [3].

Our second approach consists in using an inexact block diagonal preconditioner (BDP). We set

$$P^{-1} = \begin{pmatrix} \tilde{M}^{-1} & 0 \\ 0 & -\tilde{E}^{-1} \end{pmatrix} . \tag{3}$$

Here $\tilde{M} = LU$ comes from an ILU(k) decomposition of $M$ alone, while $\tilde{E}^{-1}$ is derived implicitly by solving a linear system with $E$ by one cycle of algebraic multigrid, see *e.g.*, [2].

## 4 Numerical Experiments

For our numerical experiments we consider as test problem the unit cube $\Omega = (0,1)^3$ with an electrode of fixed unit potential on top. The Dirichlet boundary conditions are given by

$$\begin{aligned} u_x, u_y, u_z, \varphi &= 0 \text{ for } z = 0 \\ u_x &= 0 \text{ for } y = 0 \\ u_y &= 0 \text{ for } x = 0 \\ \varphi &= 1 \text{ for } z = 1 \text{ and } 0 \leqslant x, y \leqslant 1/2 \ . \end{aligned} \tag{4}$$

The remaining boundary conditions are of homogeneous Neumann type. We assume the cube to consist of the lead-zirconate-titanate ceramic PZT-4. Its material tensor consists of ten free parameters

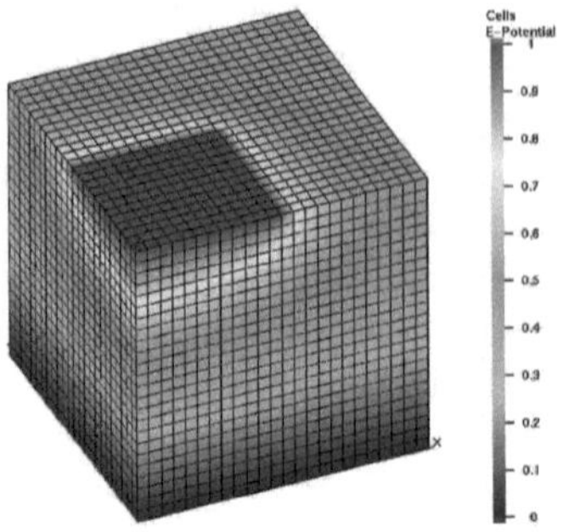


**Fig. 1.** Electric potential distribution in the unit cube for boundary conditions (4).

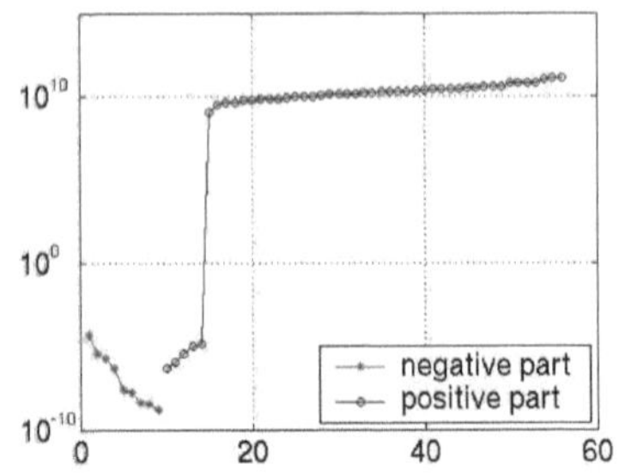


**Fig. 2.** Logarithmic plot of magnitude of eigenvalues of the system matrix from (2) for test problem with 8 elements.

$$\left(\begin{array}{cccccc|ccc}
c_{11}^E & c_{12}^E & c_{13}^E & \cdot & \cdot & \cdot & \cdot & \cdot & e_{31} \\
c_{12}^E & c_{11}^E & c_{13}^E & \cdot & \cdot & \cdot & \cdot & \cdot & e_{31} \\
c_{13}^E & c_{13}^E & c_{33}^E & \cdot & \cdot & \cdot & \cdot & \cdot & e_{33} \\
\cdot & \cdot & \cdot & c_{44}^E & \cdot & \cdot & \cdot & e_{15} & \cdot \\
\cdot & \cdot & \cdot & \cdot & c_{44}^E & \cdot & e_{15} & \cdot & \cdot \\
\cdot & \cdot & \cdot & \cdot & \cdot & \frac{1}{2}(c_{11}^E - c_{12}^E) & \cdot & \cdot & \cdot \\
\hline
\cdot & \cdot & \cdot & \cdot & e_{15} & \cdot & \varepsilon_{11} & \cdot & \cdot \\
\cdot & \cdot & \cdot & e_{15} & \cdot & \cdot & \cdot & \varepsilon_{11} & \cdot \\
e_{31} & e_{31} & e_{33} & \cdot & \cdot & \cdot & \cdot & \cdot & \varepsilon_{33}
\end{array}\right) \quad (5)$$

which are approximately on the order of $10^{10}$ ($c_{kj}^E$), $10^{-10}$ ($\varepsilon_{ii}$) and $10^0$ ($e_{kj}$). This leads to a large difference in the magnitude of the matrix entries and a bad conditioning, as can be seen in Fig. 2.

In order to test the convergence of the solver we use a discretisation with $24^3$ hexahedral elements and tri-linear basis functions. This leads to a problem with about 60.000 unknowns. We start from $x^{(0)} = 0$ and stop once $\|b - Ax^{(k)}\|_2 / \|b\|_2 < 10^{-6}$. The convergence speed for full, *i.e.*, un-restarted, GMRES with both approaches for preconditioning is plotted in Fig. 3. For comparison we also give the convergence rate for unpreconditioned GMRES. We note that due to the bad conditioning some form of preconditioning is definitely necessary. Figures 4 and 5 give estimates for the costs and storage requirements of the solution process. The ILU(k) preconditioner achieves a fast convergence especially for larger values of $k$, however, these induce significantly higher memory costs. Restarting does not significantly improve this situation. The inexact block diagonal approach on the other hand is not as good as the ILU preconditioner for small $k$, but for larger ones its arithmetic costs are nearly as small, while its memory costs are considerably smaller.

For the future we plan to evaluate the two approaches on larger problems and compare them against sparse direct solvers and iterative methods based on a Schur complement.

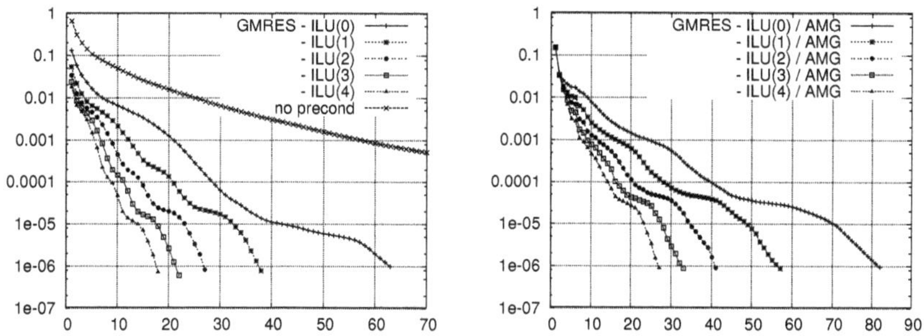


**Fig. 3.** Convergence behaviour of full GMRES with ILU(k) (left) and inexact block diagonal preconditioning (right).

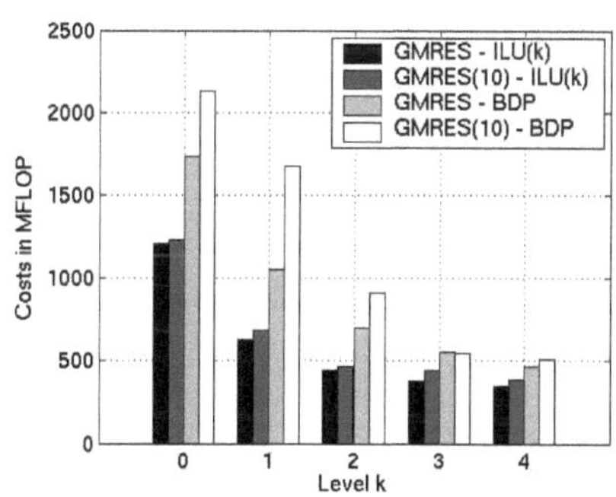


**Fig. 4.** Approximate costs to solve the linear system in arithmetic operations.

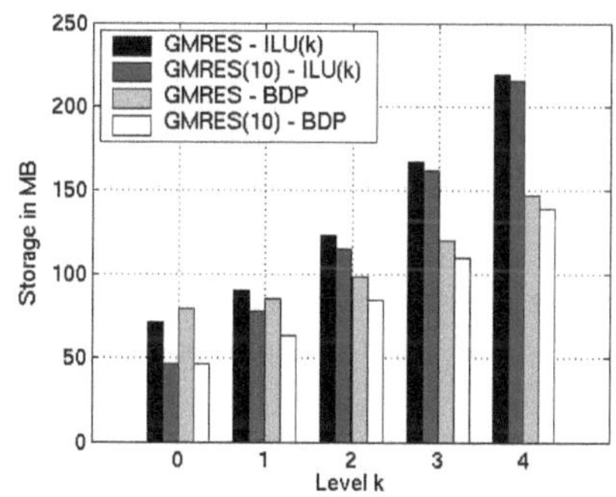


**Fig. 5.** Estimate of storage requirements for solution approach.

# References


1. R. Barrett, M. Berry, T. F. Chan, J. Demmel, J. Donato, J. J. Dongarra, V. Eijkhout, R. Pozo, C. Romine, and H. van der Vorst. *Templates for the Solution of Linear Systems: Building Blocks for Iterative Methods.* SIAM, 1994.
2. W. L. Briggs, Van E. Henson, and S. F. McCormick. *A Multigrid Tutorial.* SIAM, 2nd edition, 2000.
3. R. Falgout and U. Yang. Hypre: a Library of High Performance Preconditioners. In P. Sloot, C. Tan, J. Dongarra, and A. Hoekstra, editors, *Computational Science - ICCS 2002, Part III*, volume 2331 of *Lecture Notes in Computer Science*, pages 632–641. Springer, 2002. Also available as Lawrence Livermore National Laboratory technical report UCRL-JC-146175.
4. M. Kaltenbacher. *Numerical Simulation of Mechatronic Sensors and Actuators.* Springer Verlag, 2004.
5. H. A. van der Vorst. *Iterative Krylov Methods for Large Linear Systems.* Number 13 in Cambridge Monographs on Applied and Computational Mathematics. Cambridge University Press, 2003.

# Hydrodynamic Modeling of an Ultra-Thin Base Silicon Bipolar Transistor

O. Muscato

Dipartimento di Matematica e Informatica, Università di Catania, Catania, Italy
muscato@dmi.unict.it

**Summary.** Transport phenomena in a submicron *npn* silicon bipolar junction transistor are described by using an extended hydrodynamic model for the electrons, combined with a solution of the drift-diffusion model for the holes. Under suitable scaling assumptions, the above model reduces to the energy transport model, or to the Navier-Stokes-Fourier model, in which all the transport coefficients are now explicitly determined. The validity of the constitutive equations is investigated by using Monte Carlo simulations.

**Key words:** Electron transport, semiconductors, kinetic theory.

## 1 Introduction

Along with the reduction of horizontal dimensions, the vertical profile of silicon bipolar junction transistors (hereafter BJT) has been scaled aggressively over the last decade. Since in a *npn* BJT the minority carriers in the $(p)$ base are electrons, if the base thickness is order of electron mean-path $l_e$ ($\simeq$ 10-20 nm), the electron transport is quasi-ballistic, *i.e.*, only few collisions take place across the base region. In this case hot electrons phenomena happen in the device, which must be controlled by the CAD designer. In this regime, the standard drift-diffusion equations (hereafter DDE) is not able to simulate these devices because they do not include the carriers energy as a dynamic variable. In order to set up a judicious transport model, we observe that the massive holes are confined in the $(p)$ base in local thermal equilibrium, being minority carriers in all the device. For this reason, in the quasi ballistic regime, a transport model can be constructed by taking a solution of the DDE for the holes and an extended hydrodynamic model (EHM) for the electrons, in which more moments of the distribution function are considered, in order to capture the hot electron effects in the device. An extended hydrodynamic model for unipolar devices, formed by thirteen balance equations for the physical unknowns density, momentum, temperature, stress deviator, heat flux has been introduced [1, 6], whose main peculiarity is to be *physics-based*, *i.e.*, free of any fitting parameters. The above model, under suitable scaling hypothesis, reduces to the energy transport model (ETM), and to the Navier-Stokes-Fourier model (NSF) in which now all the transport coefficients are determined. In this paper we want to check, by means of Monte Carlo (MC) simulations, obtained with the `Damocles` code [3], the validity of the above models for the description of a realistic ultra-thin base *npn* 2D bipolar junction transistor (BJT) operating in quasi ballistic regime.

## 2 The Extended Hydrodynamic Model

Balance equations for one spherical parabolic conduction band, are usually obtained by taking suitable moments of the Boltzmann transport equation (BTE) [1]. A system of 13 equations in the 13 unknown moments $n$ (density), $V_i$ (velocity), $T$ (temperature), $\sigma_{<ij>}$ (stress deviator), $q_i$ (heta flux) is obtained providing constitutive equations for the high-order fluxes and the production terms are given.

The closure problem can be tackled with the help of the variational method known as maximum entropy principle, which allows the determination of the non-equilibrium distribution function, and consequently, of the constitutive relations. This system is named extended hydrodynamic model (EHM). The details of this procedure can be found in [1, 6]. Since the production terms are the moments over the collisional operator of the BTE, the collision mechanisms must be defined. We assume that the electrons interact with phonons (optical and acoustic) [2] and neglect the electron-electron and impurity scatterings.

## 3 Limit Models

From the EHM, under suitable scaling hypothesis, we can obtain some well known limit models, where now the transport coefficients are completely determined. A generalization of the phenomenological constitutive equations of Navier-Stokes-Fourier can be obtained by applying the *Maxwellian iteration technique* in which, the stress deviator $\sigma_{<ij>}$ and the heat flux $q_i$ appearing in the balance equations, can be expressed in terms of the variables $\{n, V_i, T\}$. One obtains for the first iterate [6]

$$q_i = -\kappa \frac{\partial T}{\partial x_i} - \frac{\alpha}{\beta} n k_B T V_i, \quad \sigma_{<ij>} = 0, \tag{1}$$

where $\kappa$ is the thermal conductivity, $\alpha$ and $\beta$ are average collisions rates. The equation (1) is a generalization of the Fourier law for heat conduction with an extra convective term. The second iterate gives a generalization of the usual Navier-Stokes law for the stress deviator

$$\sigma_{<ij>} = -2\mu\frac{\partial V_{<i}}{\partial x_{j>}} - \frac{4}{5}\frac{1}{\gamma}\frac{\partial q_{<i}}{\partial x_{j>}} - \frac{16}{15}\frac{\xi}{\gamma}\hat{\mathbf{L}}_{<ij>} + \frac{16}{15}\frac{1}{\gamma}\sum_{\eta=1}^{6}\mathcal{A}_\eta\sum_{r=1}^{4}\mathbf{L}^{2r+1}_{<ij>}\left[(N_\eta+1)H^+_{2r+1}+N_\eta H^-_{2r+1}\right] \qquad (2)$$

where $\mu = nk_BT/\gamma$ is the shear viscosity, and now all the transport coefficients are explicitly evaluated. Another macroscopic model, simpler respect to the hydrodynamic model but more accurate than the drift-diffusion one, is the energy transport model, which is based on the balance equations for the density and the average energy $W$[1]. If one considers a time scale such that the energy $W$ is not yet relaxed to its equilibrium value then, from the EHM, one obtains an Energy Transport Model (ETM) [5]. The constitutive equations for the velocity and energy-flux $S_i$ are of the form

$$V_i = D_{11}(W)\frac{\partial \log n}{\partial x_i} + D_{12}(W)\frac{\partial W}{\partial x_i} + D_{13}(W)\frac{\partial \varphi}{\partial x_i} \qquad (3)$$

$$S_i = D_{21}(W)\frac{\partial \log n}{\partial x_i} + D_{22}(W)\frac{\partial W}{\partial x_i} + D_{23}(W)\frac{\partial \varphi}{\partial x_i}, \qquad (4)$$

where the diffusion coefficients $D_{ij}$ are now exactly determined.

## 4 Numerical Results

In order to validate the above models, we consider the 2D silicon *npn* BJT structure shown in Fig. 1, operating in quasi ballistic regime (see [4] for the details). The device is at room temperature ($T_L$=300 K) and operates in the direct region with $V_e = 0$ V, $V_c$=2.5 V, and $V_b$=0.9 V. The closure relations for the high-order fluxes, and the production terms have been checked successfully in [4].

In Fig. 2 we report the constitutive relations for the heat flux equation (1), and for the stress deviator equation (2). We note that the heat flux is not well verified in the base-collector junction, where the electric field exhibits very high values, and inside the collector. The heat flux constitutive equation has peak values greater than the corresponding MC data, up to one order of magnitude. This phenomena can be justified by the fact that the Fourier law has been obtained by a linearization of the balance equations, which leads to constitutive equations valid for small gradients. Regarding to the stress deviator, we have some agreement in the emitter and in the last part of the collector, but the behaviour is completely different in the base-collector junction. For the ETM, we report in Fig. 3 with (ooo) the constitutive relations for the velocity and the energy flux given by equations (3), (4), in which the

[1] The average energy is related to the temperature $T$ by the following relation: $W = \frac{3}{2}nk_BT + \frac{1}{2}nm^\star V^2$

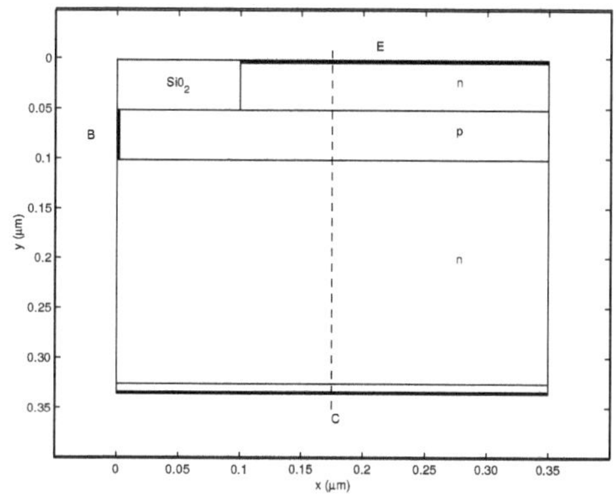


**Fig. 1.** Cross-section of the *npn* BJT used in the simulation. The electrical contacts are marked with thick lines. The different doping regions of the device are labelled by $n$ and $p$. The dashed line is the cross-section at $x = 0.175\ \mu$m.

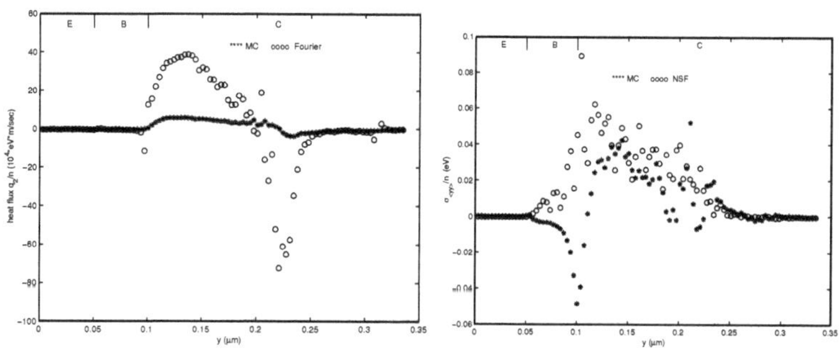


**Fig. 2.** On the left: spatial profile for the heat flux $q_2/n$ obtained by MC simulations (***) at cross section $x = 0.175\ \mu$m. With (ooo) we plot the Fourier equation (1) in which the values of the MC moments have been substitute. On the right the same figure for the stress deviator $\sigma_{<22>}/n$, with the Navier-Stokes equation (2).

MC moments have been substituted. In the same figure we plot with (***) the same quantities obtained by MC simulations, and those obtained by using the simulator DESSIS, with (xxx). DESSIS is a TCAD (by ISE) simulator, based on a standard DDE or an ETM with adjustable transport coefficients (the default parameters for silicon have been used in our simulations ). We notice that the constitutive equations data (ooo) are noisy: in fact if we substitute the MC moments $\{n, W\}$, and their derivatives into equations $(3) and (4)$, an extra numerical error is generated. Qualitatively we can say that the constitutive equations and the DESSIS data differ with respect to the MC data. All the velocities underestimate the MC data, whereas the energy fluxes overestimate the MC data. In conclusions we proved that a generalized Navier-Stokes-Fourier model was not able to reproduce correctly the heat flux near the base-collector junction where the electric field exhibits very high values, and inside

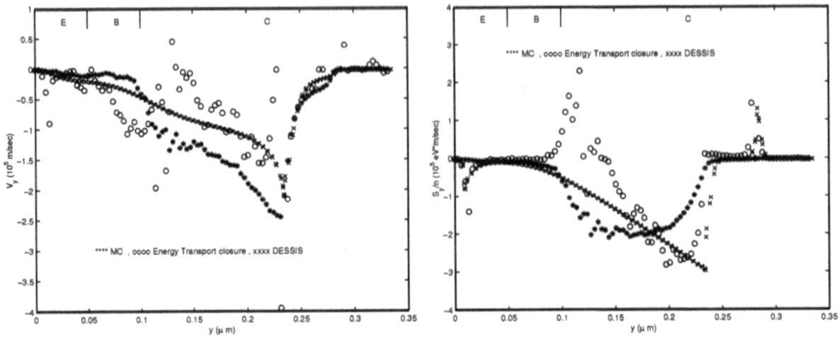


**Fig. 3.** On the left: spatial profile for the velocity $V_y$ obtained by MC simulations (***) at cross section x=0.175 $\mu$m. With (ooo) we plot the energy transport constitutive equation for the velocity (3); with (xxx) we plot the same quantity obtained with the energy transport simulator in DESSIS. On the right: the same figure for the energy flux $S_y/n$, with the energy transport constitutive equation for the energy flux (4).

the collector, where hot electrons effects can be relevant. Also the constitutive equations for an ETM, deduced from the EHM, fail to reproduce the MC data. The proposed EHM for the electrons, coupled with a solution of the DDE for holes, seems to be the best candidate for describing the quasi ballistic transport in a submicron BJT, as proved numerically in [4].

*Acknowledgement.* This work has been supported by COFIN 2002 , MURST 60%, European Community's Human Potential Programme under contract HPRN-CT-2002-00282, [HYKE].

## References

1. A.M. Anile and M. Trovato. *Phys. Lett. A*, 230:387–395, 1997.
2. C. Jacoboni and L. Reggiani. The Monte Carlo method for the solution of charge transport in semiconductors with applications to covalent materials. *Rev. Mod. Phys.*, 55:645–705, 1983.
3. S.E. Laux, M.V. Fischetti, and D.J. Frank. *IBM J. Res. Develop.*, pages 466–494, 1990.
4. O. Muscato. Hydrodynamic transport models for an ultra-thin base Si bipolar transistor. *J. Appl. Physics*, 96(2):1219–1229, 2004.
5. V. Romano. *Math. Meth. Appl. Sci.*, 24:439–471, 2001.
6. M. Trovato and L. Reggiani. *J. Appl. Phys.*, 85(8):4050–4065, 1999.

# Warped MPDAE Models with Continuous Phase Conditions

R. Pulch

Bergische Universität Wuppertal, Department of Mathematics and Natural Sciences, Chair of Applied Mathematics and Numerical Analysis, Gaußstr. 20, D–42119 Wuppertal, Germany
pulch@math.uni-wuppertal.de

**Summary.** In radio frequency (RF) application, electric circuits often exhibit multitone signals, where time scales differ by several orders of magnitude. Thus circuit simulation by means of transient analysis becomes inefficient. A multivariate model yields an alternative strategy considering amplitude as well as frequency modulation. Consequently, a warped multirate partial differential algebraic equation (MPDAE) has to be solved using periodic boundary conditions. Thereby, the determination of a local frequency function is crucial for the efficiency of the model. For this purpose, two special choices of continuous phase conditions are applied as additional boundary conditions. Numerical simulations show that these continuous phase conditions identify local frequency functions, which are physically reasonable.

**Key words:** multirate partial differential algebraic equation, phase condition, circuit simulation, frequency modulation, radio frequency.

## 1 Introduction

Numerical simulation of electric circuits rests upon a network approach, which yields systems of differential algebraic equations (DAEs), see [2]. In RF application, generated signals often exhibit widely separated time scales. For example, a slow oscillation may vary the amplitude of a carrier wave. Therefore a transient integration of the DAE system becomes costly, since the fastest rate restricts the step size.

A signal model using multivariate functions (MVFs) decouples the time scales and thus provides an alternative strategy. Consequently, Brachtendorf et al. [1] introduced a multirate partial differential algebraic equation (MPDAE), which allows the simulation of amplitude modulated signals in forced oscillators. If the circuit also includes autonomous time scales, frequency modulation may result, too. Narayan and Roychowdhury [4] generalised the model into a warped MPDAE for this case. Accordingly, a time-dependent local frequency function arises, which influences essentially the signal representation. However, an appropriate choice of the local frequency is unknown at the beginning.

We use continuous phase conditions to determine the local frequency function by the behaviour of corresponding MVFs. Thereby, the idea is to control the phase in slice planes of the MVF. This strategy yields additional boundary conditions for the warped MPDAE system in time domain. We apply this technique to a forced Van der Pol oscillator.

## 2 Multivariate Signal Model

To illustrate the multidimensional signal model, we consider a simple multitone oscillation

$$x(t) = \left[1 + \alpha \sin\left(\tfrac{2\pi}{T_1} t\right)\right] \sin\left(\tfrac{2\pi}{T_2} t + \beta \sin\left(\tfrac{2\pi}{T_1} t\right)\right) \tag{1}$$

for parameters $0 < \alpha < 1$, $\beta > 0$. If $T_1 \gg T_2$ holds, then a high-frequency oscillation arises, where amplitude as well as frequency is modulated by a slow oscillation. Hence we need many time steps to resolve this signal accurately. Alternatively, an own variable is introduced for each separate time scale, which yields directly the biperiodic function

$$\hat{x}_1(t_1, t_2) = \left[1 + \alpha \sin\left(\tfrac{2\pi}{T_1} t_1\right)\right] \sin\left(2\pi t_2 + \beta \sin\left(\tfrac{2\pi}{T_1} t_1\right)\right), \tag{2}$$

where the second period is transformed to 1. We can completely reconstruct the original signal via $x(t) = \hat{x}_1(t, t/T_2)$. This representation (2) is called a MVF of the multitone signal (1). Unfortunately, the MVF (2) exhibits many oscillations in the rectangle $[0, T_1[\times[0, 1[$ for large parameters $\beta$. Thus we include only the amplitude modulation part in a MVF, *i.e.*,

$$\hat{x}_2(t_1, t_2) = \left[1 + \alpha \sin\left(\tfrac{2\pi}{T_1} t_1\right)\right] \sin\left(2\pi t_2\right). \tag{3}$$

Now the function features a simple behaviour in $[0, T_1[\times[0, 1[$. Therefore we can represent this MVF with sufficient accuracy using relatively few grid points. The frequency modulation part is modelled by a separate function

$$\Psi(t) = \frac{t}{T_2} + \frac{\beta}{2\pi} \sin\left(\tfrac{2\pi}{T_1} t\right). \tag{4}$$

Now we are able to reconstruct the signal (1) applying $x(t) = \hat{x}_2(t, \Psi(t))$. The derivative $\nu := \Psi'$, which is a $T_1$-periodic time-dependent function, can be seen as a local frequency of the signal. Thus we obtain an efficient representation by means of this model.

Using the inappropriate MVF (2), the reconstruction formula indicates a local frequency $\nu \equiv 1/T_2$. It follows that the choice of a local frequency function is not unique and critical for the efficiency of the MVF model.

## 3 Warped MPDAE System

In general, an electric circuit is modelled by a DAE system of the form

$$\frac{\mathrm{d}\mathbf{q}(\mathbf{x})}{\mathrm{d}t} = \mathbf{f}(\mathbf{x}) + \mathbf{b}(t) \qquad \left(\mathbf{x}(t), \mathbf{b}(t), \mathbf{q}(\mathbf{x}), \mathbf{f}(\mathbf{x}) \in \mathbb{R}^k\right), \tag{5}$$

where $\mathbf{x}$ denotes unknown voltages and currents. The input signals $\mathbf{b}$ shall be $T_1$-periodic. We assume that $\mathbf{x}$ is a multitone signal of the discussed type. Applying the multivariate model, the DAE changes into the MPDAE

$$\frac{\partial \mathbf{q}(\hat{\mathbf{x}})}{\partial t_1} + \nu(t_1)\frac{\partial \mathbf{q}(\hat{\mathbf{x}})}{\partial t_2} = \mathbf{f}(\hat{\mathbf{x}}) + \mathbf{b}(t_1) \qquad \left(\hat{\mathbf{x}}(t_1, t_2) \in \mathbb{R}^k,\ \nu(t_1) \in \mathbb{R}\right) \tag{6}$$

with the MVF $\hat{\mathbf{x}}$ of $\mathbf{x}$. It follows that a $(T_1, 1)$-periodic MPDAE solution yields multitone DAE solution via $\mathbf{x}(t) = \hat{\mathbf{x}}(t, \int_0^t \nu(\tau)d\tau)$. Thereby, the $T_1$-periodic local frequency $\nu$ is a priori unknown and thus the system (6) is underdetermined. Houben [3] proposed minimum conditions, which reduce oscillatory behaviour in MVFs, to fix this function.

Alternatively, we try to control the phase in each slice plane of the MVF for constant $t_1$. A unifying effect shall produce simple MVF representations. Since the local frequency is a scalar function, we consider just a single component of the MVF $\hat{\mathbf{x}} = (\hat{x}^1, \ldots, \hat{x}^k)^T$, for example the first one. Now feasible choices for continuous phase conditions are

$$\hat{x}^1(t_1, 0) = \eta \quad (\eta \in \mathbb{R}) \quad \text{for all } t_1 \tag{7}$$

or

$$\left.\frac{\partial \hat{x}^1}{\partial t_2}\right|_{t_2=0} = 0 \quad \text{for all } t_1. \tag{8}$$

Consequently, we add either (7) or (8) to the biperiodic boundary conditions in a time domain method. Thus the resulting technique is cheaper in comparison to a minimisation procedure. The existence of MVFs satisfying one of the phase conditions can be motivated by transformations of MPDAE solutions.

## 4 Numerical Simulation

As benchmark, we consider a forced Van der Pol oscillator of the form

$$\begin{aligned} \dot{x} &= y \\ \dot{y} &= -10(x^2 - 1)y + (2\pi z)^2 x \\ 0 &= z - \left[1 + \tfrac{1}{2}\sin\left(2\pi 10^{-3} t\right)\right], \end{aligned} \tag{9}$$

which represents a DAE system of index 1. A multitone solution arises and we employ the warped MPDAE model. Numerical solutions are obtained by a time domain technique, which is based on characteristic curves, see [5]. Let $\nu_a$

and $\nu_b$ be the local frequencies, which are caused by the phase conditions (7) and (8), respectively. Figure 1 illustrates these functions, which are nearly the same ($|\nu_a - \nu_b| < 10^{-3}$). Since the frequencies respond to the input, they are physically reasonable. The corresponding MVFs $\hat{x}$ and $\hat{y}$ are shown in Fig. 2. The solutions belonging to the two phase conditions differ mainly by a translation in $t_2$-direction, which reflects that (6) is autonomous in the variable $t_2$. Although $\hat{x}$ exhibits nearly constant amplitude, $\hat{y}$ includes amplitude modulation. Finally, Fig. 3 displays the reconstructed DAE solution $x$ together with a reference solution of (9). We observe a phase shift in later cycles. Nevertheless, the other signal properties coincide at any time.

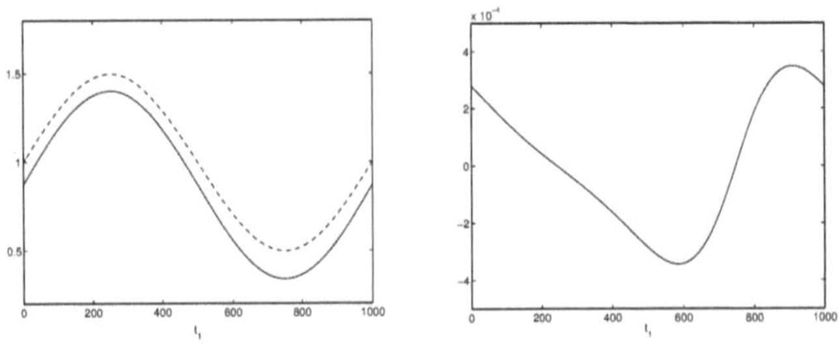


**Fig. 1.** Local frequency $\nu_a$ (solid line) together with input signal (dashed line) (left) and difference of local frequencies $\nu_a - \nu_b$ (right).

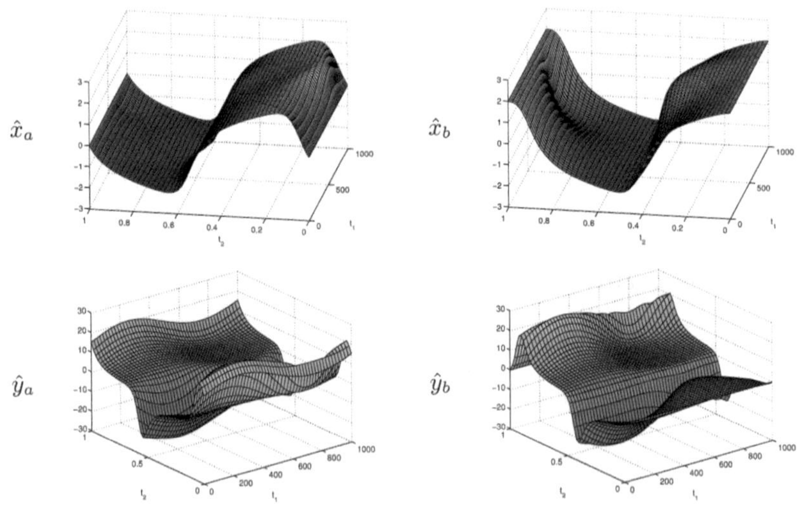


**Fig. 2.** MPDAE solutions using phase condition (7) (left) and (8) (right).

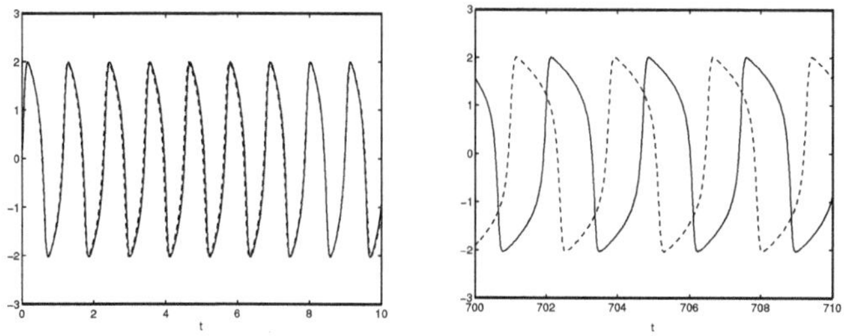


**Fig. 3.** DAE solution $x$ integrated by trapezoidal rule (solid line) and interpolated by MPDAE solution (dashed line) in time intervals $[0, 10]$ (left) and $[700, 710]$ (right).

# 5 Conclusions

A multivariate model for analysing oscillators, which produce amplitude as well as frequency modulated signals, has been presented. The arising MPDAE system demands the identification of an appropriate local frequency function. Numerical simulations demonstrate that continuous phase conditions are able to determine physically reasonable local frequencies. Thus corresponding MVFs exhibit a simple structure and the model becomes efficient. Underlying existence theorems using the phase conditions still have to be researched.

*Acknowledgement.* This work has been supported within the federal BOMB project No. 03GUNAVN. The author thanks M. Günther and S. Knorr for helpful discussions.

# References


1. H.G. Brachtendorf, G. Welsch, R. Laur, and A. Bunse-Gerstner. Numerical steady state analysis of electronic circuits driven by multi-tone signals. *Electrical Engineering*, 79:103–112, 1996.
2. M. Günther and U. Feldmann. CAD based electric circuit modeling in industry I: mathematical structure and index of network equations. *Surv. Math. Ind.*, 8:97–129, 1999.
3. S.H.M.J. Houben. Simulating multi-tone free-running oscillators with optimal sweep following. In W.H.A. Schilders, E.J.W. Ter Maten, and S.H.M.J. Houben, editors, *Scientific Computing in Electrical Engineering*, Mathematics in Industry, pages 240–247. Springer, 2004.
4. O. Narayan and J. Roychowdhury. Analyzing oscillators using multitime PDEs. *IEEE Trans. CAS I*, 50:894–903, 2003.
5. R. Pulch. Multi time scale differential equations for simulating frequency modulated signals. to appear in: Appl. Numer. Math., 2004.

# Exact Closure Relations for the Maximum Entropy Moment System in Semiconductor Using Kane's Dispersion Relation

M. Junk[1] and V. Romano[2]

[1] FB Mathematik, Universität des Saarlandes, Postfach 151150, 66041 Saarbrücken, Germany `junk@num.uni-sb.de`
[2] Dipartimento di Matematica e Informatica, Università di Catania, viale A. Doria 6 - 95125 Catania, Italy `romano@dmi.unict.it`

**Summary.** The maximum entropy moment systems of the Boltzmann equation is only solvable with physically unrealistic restrictions on the choice of the macroscopic variables. We show that no such difficulties appear in the semiconductor case if Kane's dispersion relation is used for the energy band of electrons. As an application the 5-moment model is discussed.

**Key words:** maximum entropy moment closure, semiconductor Boltzmann equation, Kane's dispersion relation.

## 1 The Maximum Entropy Moment Systems for Electrons in Semiconductors

In a semi classical approximation, a kinetic description of electrons in a semiconductor is given by a transport equation for the one particle distribution function $f(t, \boldsymbol{x}, \boldsymbol{k})$, which represents the probability of finding an electron at time $t$ in an elementary volume $d\boldsymbol{x}d\boldsymbol{k}$, around position $\boldsymbol{x}$ and with crystal momentum $\boldsymbol{k}$,

$$\frac{\partial f}{\partial t} + v_i(\boldsymbol{k})\frac{\partial f}{\partial x_i} - \frac{e}{\hbar}E_i\frac{\partial f}{\partial k_i} = \mathcal{C}[f]. \tag{1}$$

Here $e$ is the absolute value of the electron charge, $\boldsymbol{k}$ represents the crystal momentum of the electron and $\boldsymbol{E}$ is the electric field which is related to the electron distribution by Poisson's equation: $\boldsymbol{E} = -\nabla\varphi, \quad \varepsilon\Delta\varphi = -e(N_D - N_A - n)$, where $\varphi$ is the electric potential, $\varepsilon$ is the permittivity of the semiconductor, $N_D$ and $N_A$ are respectively the donor and acceptor density, and $n$ is the electron density. The latter is related to $f$ by $n = \int_B f d\boldsymbol{k}$, $B$ being the first Brillouin zone. The right hand side $\mathcal{C}[f]$ in (1) is the collision operator, which takes into account scattering of the electrons with acoustical and optical phonons and with impurities. The electron velocity $\boldsymbol{v}(\boldsymbol{k})$ depends on the electron energy $\mathcal{E}$ by the relation $\boldsymbol{v}(\boldsymbol{k}) = \frac{1}{\hbar}\nabla_{\boldsymbol{k}}\mathcal{E}$. In general, the expression of $\mathcal{E}$ (the so called band structure) depends on the material and is very complicated. A rough approximation is given by the *parabolic band* while a more refined model is given by *Kane's dispersion relation* which takes into account the non-parabolicity at high energies

$$\mathcal{E}(\boldsymbol{k}) = \frac{1}{1+\sqrt{1+2\frac{\alpha}{m^*}\hbar^2|\boldsymbol{k}|^2}}\frac{\hbar^2|\boldsymbol{k}|^2}{m^*} = \sqrt{\frac{1}{4\alpha^2}+\frac{\hbar^2|\boldsymbol{k}|^2}{2\alpha m^*}}-\frac{1}{2\alpha}, \qquad \boldsymbol{k}\in\mathbb{R}^3 \quad (2)$$

where $\alpha$ is the non-parabolicity parameter. The corresponding electron velocity is $\boldsymbol{v}(\boldsymbol{k}) = \frac{1}{\sqrt{1+\frac{2\alpha}{m^*}\hbar^2|\boldsymbol{k}|^2}}\frac{\hbar}{m^*}\boldsymbol{k}$. In the mathematical modelling of electron transport in semiconductors the Kane dispersion relation is considered as one of the best analytical approximation to the real energy band.

Besides the electron density $n$, other physically relevant quantities are the average electron velocity, energy and energy-flux

$$\boldsymbol{u} = \frac{1}{n}\int_{\mathbb{R}^3}\boldsymbol{v}(\boldsymbol{k})f\,d\boldsymbol{k}, \quad W = \frac{1}{n}\int_{\mathbb{R}^3}\mathcal{E}(\boldsymbol{k})f\,d\boldsymbol{k}, \quad \boldsymbol{S} = \frac{1}{n}\int_{\mathbb{R}^3}\boldsymbol{v}(\boldsymbol{k})\mathcal{E}(\boldsymbol{k})f\,d\boldsymbol{k}.$$

To generalize this observation, we introduce general weight functions $a_i : \mathbb{R}^3 \mapsto \mathbb{R}$ and the corresponding moments $\rho_i = \langle f, a_i\rangle$, $i = 1,\ldots,m$ where $\langle\cdot,\cdot\rangle$ denotes $\boldsymbol{k}$ integration. We split the vector of weight functions $\boldsymbol{a}$ into two subgroups. The first $m_1$ components of $\boldsymbol{a}$ are chosen as $(P_1(\boldsymbol{v}(\boldsymbol{k})),\ldots,P_{m_1}(\boldsymbol{v}(\boldsymbol{k})))$ where $P_1,\ldots,P_{m_1}$ are linearly independent polynomials with $P_1(\boldsymbol{v}) = 1$, and the remaining $m_2$ components give rise to energy moments $(\mathcal{E}(\boldsymbol{k})Q_1(\boldsymbol{v}(\boldsymbol{k})), \ldots, \mathcal{E}(\boldsymbol{k})Q_{m_2}(\boldsymbol{v}(\boldsymbol{k})))$ where, again, $Q_1,\ldots,Q_{m_2}$ are linearly independent polynomials and $Q_1(\boldsymbol{v}) = 1$.

Multiplying (1) with weight functions $\boldsymbol{a} = (a_1,\ldots,a_m)^T$ and integrating over $\boldsymbol{k}$, we obtain equations for the moments

$$\frac{\partial\boldsymbol{\rho}}{\partial t} + \frac{\partial}{\partial x_j}\langle f, v_j\boldsymbol{a}\rangle = \langle \mathcal{C}[f] + \gamma\boldsymbol{E}\cdot\nabla_{\boldsymbol{k}}f, \boldsymbol{a}\rangle, \qquad \gamma = e/\hbar. \quad (3)$$

The system would be closed if the particle distribution could be expressed in terms of the moment vector $\boldsymbol{\rho}$ as $f(t,\boldsymbol{x},\boldsymbol{k}) = F(\boldsymbol{\rho}(t,\boldsymbol{x}),\boldsymbol{k})$ A method to obtain such a relationship is the maximum entropy approach where $F(\boldsymbol{\rho},\boldsymbol{k})$ is taken as solution of the problem

$$\text{maximize } H(f) = -\langle f, \log f - 1\rangle \text{ with } f \geq 0 \text{ and } \langle f, \boldsymbol{a}\rangle = \boldsymbol{\rho} \quad (4)$$

It is important to remark that the maximum entropy distribution represents, in a statistical sense [3], the least biased estimator of the exact distribution $f$ on the base of the knowledge of a finite number of moments of $f$.

For general $a_i$, the formal solution of (4) is obtained with the method of Lagrange multipliers. We introduce the Lagrange functional $L(f,\boldsymbol{\lambda}) :=$

$H(f) - \boldsymbol{\lambda} \cdot (\boldsymbol{\rho} - \langle f, \boldsymbol{a} \rangle)$ where $\boldsymbol{\lambda}$ is the vector of Lagrange multipliers. The necessary condition that all directional derivatives vanish in the maximum $f_{\boldsymbol{\lambda}}$ leads to $f_{\boldsymbol{\lambda}} = \exp(\boldsymbol{\lambda} \cdot \boldsymbol{a})$. Finally, the Lagrange multipliers $\boldsymbol{\lambda}$ are chosen in such a way (if possible) that the moment constraints $\boldsymbol{\rho} = \langle f_{\boldsymbol{\lambda}}, \boldsymbol{a} \rangle$ are satisfied which gives rise to a function $\boldsymbol{\lambda} = \boldsymbol{\lambda}(\boldsymbol{\rho})$. We then introduce $F(\boldsymbol{\rho}, \boldsymbol{k}) = f_{\boldsymbol{\lambda}(\boldsymbol{\rho})}(\boldsymbol{k})$.

Depending on the choice of weight functions $a_i$, it can happen that problem (4) is not always solvable, [4, 5, 2, 7].

## 2 Solvability of the Maximum Entropy Problem

In order to state our main result, we first reformulate (4). For notational convenience, we measure $\mathcal{E}$, $\boldsymbol{k}, \boldsymbol{v}$ in units $1/(2\alpha)$, $\sqrt{m^*/(2\alpha\hbar^2)}$, and $1/\sqrt{2\alpha m^*}$ which leads to $\mathcal{E}(\boldsymbol{k}) = \sqrt{1 + |\boldsymbol{k}|^2} - 1, \quad \boldsymbol{v}(\boldsymbol{k}) = \frac{\boldsymbol{k}}{\sqrt{1+|\boldsymbol{k}|^2}}$. Note that for large $\boldsymbol{k}$, $\boldsymbol{v}(\boldsymbol{k})$ is bounded and $\mathcal{E}(\boldsymbol{k})$ grows only linearly due to the estimates

$$|\boldsymbol{v}(\boldsymbol{k})| < 1, \qquad |\boldsymbol{k}| - 1 \leq \mathcal{E}(\boldsymbol{k}) \leq 2|\boldsymbol{k}| + 1. \tag{5}$$

Based on $\mathcal{E}$ and $\boldsymbol{v}$ and two sets $\{P_1, \dots, P_{m_1}\}$, $\{Q_1, \dots, Q_{m_2}\}$ of linearly independent polynomials with $P_1 = Q_1 = 1$, we define the weight functions as

$$\boldsymbol{a} = (P_1(\boldsymbol{v}), \dots, P_{m_1}(\boldsymbol{v}), \mathcal{E}Q_1(\boldsymbol{v}), \dots, \mathcal{E}Q_{m_2}(\boldsymbol{v}))^T. \tag{6}$$

Since the assumption of a three-dimensional $\boldsymbol{k}$-space is not relevant for our argument, we assume $\boldsymbol{k} \in \mathbb{R}^d$. The moment set related to the weights $a_i$ is generated by the functions in $\mathcal{F} = \{f \geq 0 : f \not\equiv 0, |\boldsymbol{a}|f \in \mathbb{L}^1(\mathbb{R}^d)\}$. The corresponding moments are collected in $\mathcal{M} = \{\langle f, \boldsymbol{a} \rangle : f \in \mathcal{F}\}$. Using this notation and the definition of the entropy functional $H(f) = -\langle f, \log f - 1 \rangle$, we can restate (4) as

$$\text{maximize } H(f) \text{ subject to } f \in \mathcal{F} \text{ and } \langle f, \boldsymbol{a} \rangle = \boldsymbol{\rho} \tag{7}$$

Our main result is

**Theorem 1.** *The maximum entropy moment problem* (7) *is uniquely solvable for any* $\boldsymbol{\rho}$ *inside the open, convex cone* $\mathcal{M}$. *The solution is an exponential density* $\exp(\boldsymbol{\lambda} \cdot \boldsymbol{a})$ *for some* $\boldsymbol{\lambda} \in \mathbb{R}^m$ *depending on* $\boldsymbol{\rho}$.

To give an idea of the proof, first we observe that, up to normalization, every $f \in \mathcal{F}$ can be viewed as a probability density.

Moreover, if $P$ and $R$ are probability measures on the Borel sets $\mathcal{B}$ on $\mathbb{R}^d$, such that $P$ has a density with respect to $R$, *i.e.*, $P(A) = \int_A p_R \, dR$ with $A \in \mathcal{B}$, the relative entropy (or I-divergence) is defined as $I(P||R) = \int p_R \log p_R \, dR$.

Reformulating the maximum entropy problem in terms of relative entropy, one can get the proof of the main theorem by using a results of Csiszár [1] for measurable spaces. Here we skip all the technical details (the interested reader is referred to [6]).

# 3 The Euler-Poisson Model

As an example of application we analyze the Euler-Poisson model in the case of Kane's dispersion relation. It is based on the same moments employed in ideal gas dynamics, that is density $n$, average velocity $\boldsymbol{u}$ and average energy $W$

$$\frac{\partial n}{\partial t}+\frac{\partial(nu^i)}{\partial x^i}=0,\quad \frac{\partial(nu^i)}{\partial t}+\frac{\partial(nU^{ij})}{\partial x^j}=-enE_jH^{ij}+nC_u^i, \tag{8}$$

$$\frac{\partial(nW)}{\partial t}+\frac{\partial(nS^j)}{\partial x^j}=-enu_kE^k+nC_W, \tag{9}$$

where $U^{ij}=\frac{1}{n}\int_{\mathbb{R}^3}fv^iv^j d\boldsymbol{k}$, $H^{ij}=\frac{1}{n}\int_{\mathbb{R}^3}\frac{1}{\hbar}f\frac{\partial v_i}{\partial k_j}d\boldsymbol{k}$, $C_u^i=\frac{1}{n}\int_{\mathbb{R}^3}\mathcal{C}[f]v^i d\boldsymbol{k}$, $C_W=\frac{1}{n}\int_{\mathbb{R}^3}\mathcal{C}[f]\mathcal{E}(k)d\mathbf{k}$.

For the 5-moment case the weight function vector is $\boldsymbol{a}=(1,\boldsymbol{v},\mathcal{E})$ and the corresponding Lagrange multipliers are given by the vector $\boldsymbol{\lambda}=-\left(\lambda,\boldsymbol{\lambda}^v,\lambda^W\right)$. The MEP distribution function reads $f_{\boldsymbol{\lambda}}=\exp\left(-\lambda-\lambda_i^v v^i-\lambda^W\mathcal{E}\right)$ and one has the straightforward characterization of the cone $\Lambda$ (which is obviously convex and open) $\Lambda=\left\{\boldsymbol{\lambda}=-\left(\lambda,\boldsymbol{\lambda}^v,\lambda^W\right):\ \boldsymbol{\lambda}\in\mathbb{R}^5,\lambda^W>0\right\}$. By writing $d\boldsymbol{k}=\frac{m^*}{\hbar^3}\sqrt{2m^*\mathcal{E}(1+\alpha\mathcal{E})}(1+2\alpha\mathcal{E})d\mathcal{E}\,d\Omega$ with $d\Omega$ elementary solid angle, the explicit relation between the Lagrange multipliers and the macroscopic variables are given by

$$u_3=\frac{1}{d_0}\int_0^\infty v(\mathcal{E})e^{-\lambda^W\mathcal{E}}\sqrt{\mathcal{E}(1+\alpha\mathcal{E})}(1+2\alpha\mathcal{E})\left[\frac{\sinh z}{z^2}-\frac{\cosh z}{z}\right]d\mathcal{E} \tag{10}$$

$$W=\frac{1}{d_0}\int_0^\infty \mathcal{E}e^{-\lambda^W\mathcal{E}}\sqrt{\mathcal{E}(1+\alpha\mathcal{E})}(1+2\alpha\mathcal{E})\ \frac{\sinh z}{z}\ d\mathcal{E},\quad z=\lambda_3^v\ v(\mathcal{E}) \tag{11}$$

$$n=\pi\frac{(2m^*)^{3/2}}{\hbar^3}\ e^{-\lambda}d_0,\ d_0=\int_0^\infty e^{-\lambda^W\mathcal{E}}\sqrt{\mathcal{E}(1+\alpha\mathcal{E})}(1+2\alpha\mathcal{E})\ \frac{\sinh z}{z}\ d\mathcal{E}$$

It is relevant only to study the dependence of $\lambda_3^v$ and $\lambda^W$ on $u_3$ and $W$ because $\lambda$ plays only the role of a normalization factor. We want to investigate whether the moment cone, that is the set of moment for which the MEP distribution there exists, is sufficiently large for concrete applications. To this aim we have numerically checked the invertibility of the rectangle $\left\{(W,u_3)\in[0.04,0.35]\times\left[-1.2\times10^5,1.2\times10^5\right]\right\}$ under the mapping $(u_3,W)\mapsto(\lambda_3^v,\lambda^W)$ implicitly defined by the relations (10)-(11). $W$ is expressed in $eV$, $u_3$ in m/sec, $\lambda_3^v/\sqrt{m^*}$ in $1/\sqrt{eV}$ and $\lambda^W$ in $1/eV$. The numerical analysis (see figure) shows that the moment cone contains the above rectangle and therefore it is sufficiently wide to enclose the relevant physical region of velocity and energy.

## Acknowledgements

The author V.R. acknowledges that this work has been partially supported by M.I.U.R. (P.R.I.N. 2004 *Problemi matematici delle teorie cinetiche*) and by P.R.A. (ex fondi 60%).

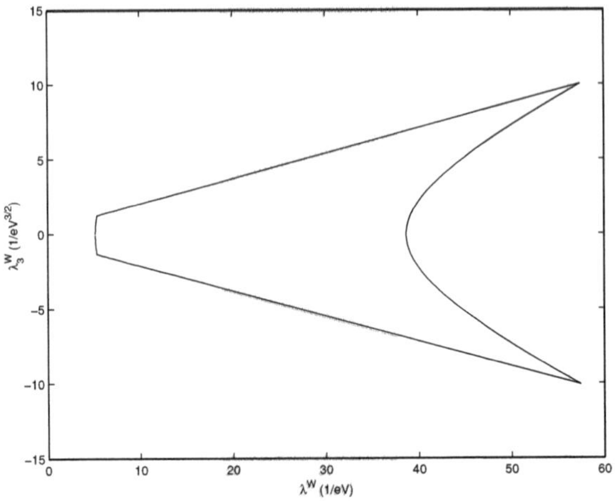


**Fig. 1.** image of the rectangle $\{(\lambda_3^v/\sqrt{m^*}, \lambda^W) \in [-10, 10] \times [1, 65]\}$ under the mapping $(\lambda_3^v, \lambda^W) \mapsto (u_3, W)$ defined by the relations (10)-($11_2$).

## References


1. I. Csiszár. I-divergence geometry of probabaility distributions and minimization problems. *Ann. of Prob.*, 3:146–158, 1975.
2. W. Dreyer, M. Junk, and M. Kunik. On the approximation of the Fokker-Planck equation by moment systems. *Nonlinearity*, pages 881–906, 2001.
3. E. T. Jaynes. Information theory and statistical mechanics. *Phys. Rev.*, 106:620–630, 1957.
4. M. Junk. Domain of definition of levermore's five-moment system. *J. Stat. Phys.*, 93:1143–1167, 1998.
5. M. Junk. Maximum entropy for reduced moment problems. *Models Methods Appl. Sci.*, 10:1001–1025, 2000.
6. M. Junk and V. Romano. Maximum entropy moment system of the semiconductor Boltzmann equation using Kane's dispersion relation. *Accepted for publication in Continuum Mech. Thermodyn.*, 2004).
7. M. Junk and A. Unterreiter. Maximum entropy moment systems and Galilean invariance. *Continuum Mech. Thermodyn.*, 14:563–576, 2002.

# Reduced Order Models for Eigenvalue Problems

J. Rommes[1]

Mathematical Institute Utrecht University
P.O.Box 80.010, 3508 TA Utrecht
The Netherlands
http://www.math.uu.nl/people/rommes
rommes@math.uu.nl

**Summary.** Two main approaches are known for the reduced order modelling of linear time-invariant systems: Krylov subspace based and SVD based approximation methods. Krylov subspace based methods have large scale applicability, but do not have a global error bound. SVD based methods do have a global error bound, but require full space matrix computations and hence have limited large scale applicability. In this paper features and short-comings of both types of methods will be addressed. Furthermore, ideas for improvements will be discussed and the possible application of Jacobi-Davidson style methods such as JDQR and JDQZ for model reduction will be considered.

**Key words:** reduced order models, eigenvalue problems.

## 1 Introduction

Dynamical systems and control systems arise from, for instance, partial differential equations and electrical circuits. Simulation and controller design for large scale systems can become extremely expensive in storage requirements and computations. A way to reduce these costs is to use reduced order models (ROMs), which preserve key characteristics of the original system, but are significantly smaller in dimension. This paper will summarize the two main approaches for reduced order modelling, Krylov subspace based and SVD based approximation methods, together with features and shortcomings. Furthermore, ideas for improvements will be discussed and the possible application of Jacobi-Davidson style methods such as JDQR and JDQZ for model reduction will be considered.

In Section 2, the reduced order modelling problem will be stated. In Section 3, existing ROM methods will be described. Section 4 explores the application of Jacobi-Davidson style methods to ROM problems and concludes.

## 2 Reduced Order Modelling Problem

In this paper linear time-invariant (LTI) systems will be considered:

$$\begin{cases} C\frac{dx(t)}{dt} &= Gx(t) + Bu(t) \\ y(t) &= Lx(t) + Mu(t) \end{cases} \tag{1}$$

where $x(t) \in \mathbb{R}^N$ is the state vector, $u(t) \in \mathbb{R}^m$ is the input function and $y(t) \in \mathbb{R}^p$ the output. The matrices $C, G \in \mathbb{R}^{N\times N}$ are system matrices. The matrices $B \in \mathbb{R}^{N\times m}$, $L \in \mathbb{R}^{p\times N}$ and $M \in \mathbb{R}^{p\times m}$ are distribution matrices. The number of state variables (the order of the system) is denoted by $N$. The number of input and output variables are denoted by $m$ and $p$ respectively.

The problem is to find an approximating system, the reduced order model:

$$\begin{cases} \tilde{C}\frac{d\tilde{x}(t)}{dt} &= \tilde{G}\tilde{x}(t) + \tilde{B}u(t) \\ \tilde{y}(t) &= \tilde{L}\tilde{x}(t) + \tilde{M}u(t) \end{cases} \tag{2}$$

with $\tilde{x}(t) \in \mathbb{R}^n$, $u(t) \in \mathbb{R}^m$ $\tilde{y}(t) \in \mathbb{R}^p$, $\tilde{C}, \tilde{G} \in \mathbb{R}^{n\times n}$, $\tilde{B} \in \mathbb{R}^{n\times m}$, $\tilde{L} \in \mathbb{R}^{p\times n}$ and $\tilde{M} \in \mathbb{R}^{p\times m}$. Note that the number of inputs and outputs is the same as for the original system, and that the input itself is not changed. The ROM should satisfy the following requirements (see also [1]):

- The order of the system is strongly reduced: $n \ll N$.
- The approximation error $||y - \tilde{y}||$, in appropriate norm, must be small.
- Important properties such as passivity and stability are preserved.
- The procedure for computing the ROM must be computationally efficient and numerically stable, and ideally has an automatic convergence test.

Physical realizability of the ROM may be an additional requirement.

## 3 Reduced Order Modelling Methods

SVD based and Krylov subspace based approximation methods will be described briefly in this section. From now on, SISO ($m = p = 1$) systems are considered. For an extended overview, see [1].

*SVD based methods*

Let $C = I$ and $M = 0$ and $G$ be stable in (2). The singular values of the Hankel operator

$$\mathcal{H} : u(t) \mapsto \int_{-\infty}^{0} Le^{G(t-\tau)}Bu(\tau)d\tau \quad t \geq 0,$$

are equal to the square roots of the eigenvalues of a product of two symmetric positive-definite matrices $P, Q \in \mathbb{R}^{N\times N}$ (the Gramians) [5]. Hence, the Hankel singular values $\sigma_i^H = \lambda_i^{1/2}(PQ), i = 1 \dots N$ form a discrete set. These Hankel singular values play a role for dynamical systems similar to the role singular values play for matrices. A reduced order model is constructed by truncating the original system in such a way that the $n$ largest Hankel singular values are preserved. In order to do this, first the matrices $P$ and $Q$ must be solved from two Lyapunov equations

$$GP + PG^T + BB^T = 0, \qquad G^TQ + QG + L^TL = 0 \tag{3}$$

Then, in order to truncate, a balancing transformation that diagonalizes both $P$ and $Q$ must be computed. This balancing transformation is computed using the standard SVD.

SVD based methods have the disadvantage that dense matrix computations are required to solve the two Lyapunov equations, so that large scale application is not attractive: the number of operations is $O(N^3)$. However, stability and passivity are preserved for the SVD-based methods, and moreover, there is a global error bound [5]:

$$||H(s) - H_n(s)||_{L_\infty} = \sup_\omega \sigma_{\max}(H(i\omega) - H_n(i\omega)) \leq 2 \sum_{i=n+1}^{N} \sigma_i^H$$

For more details about the balancing transformation and the Hankel operator, the reader is referred to [5].

*Krylov subspace based methods*

The transfer function of a linear dynamical system (2) is defined as the Laplace transform of the impulse response (for simplicity $M = 0$):

$$H(s) = L(sC - G)^{-1}B = L(I - (s - s_0)A)^{-1}R$$

with $A = -(G + s_0C)^{-1}C$ and $R = (G + s_0C)^{-1}B$. Note that it is assumed that $sC - G$ is a regular pencil and that $s_0$ is chosen such that $G + s_0C$ is nonsingular. A series expansion of $H$ around $s_0$ is

$$H(s) = \sum_{i=0}^{\infty} m_i(s - s_0)^i, \qquad m_i = LA^iR.$$

Krylov subspace based methods construct a ROM by matching moments $m_i$:

$$\tilde{H}(s) = \sum_{i=0}^{\infty} \tilde{m}_i(s - s_0)^i, \qquad \tilde{m}_i = \tilde{L}\tilde{A}^i\tilde{R},$$

with $m_i = \tilde{m}_i,\ i = 0, \dots, n$ for appropriate $n \ll N$.

Explicit computation of the moments $m_i = LA^iR$ is numerically unstable: within a few iterations the vector $A^iR$ will approximate the eigenvector corresponding to the dominant eigenvalue. Krylov subspace methods circumvent this difficulty by constructing an orthonormal basis for the Krylov subspace $\mathcal{K}^n(A, R) = \text{span}\{R, AR, \ldots, A^{n-1}R\}$.

Let the columns of $V_n \in \mathbb{R}^{N \times n}$ form an orthonormal basis for $\mathcal{K}^n(A, R)$, *i.e.*, $V_n^T V_n = I$ and $\text{span}(V_n) = \mathcal{K}^n(A, R)$. The ROM is now constructed by the state transformation $x \to V_n\tilde{x}$:

$$\begin{cases} (V_n^T C V_n)\frac{d\tilde{x}(t)}{dt} = (V_n^T G V_n)\tilde{x}(t) + (V_n^T B)u(t) \\ \tilde{y}(t) \qquad\qquad\quad = (LV_n)\tilde{x}(t) \end{cases}$$

Two well-known methods based on Krylov subspaces are Padé via Lanczos (PVL) [3] and PRIMA [7]. PVL exploits the connection between the Bi-Lanczos process and Padé approximation. After $n$ iterations of the process (two matrix-vector products per iteration), $2n$ moments are matched, while limited storage of iteration vectors is needed. However, the process may suffer from breakdowns and does not guarantee preservation of stability and passivity (see [2] for remedies). PVL does not provide the reduced system matrices.

PRIMA constructs an orthonormal basis for the Krylov subspace $\mathcal{K}^n(A, R)$ using Arnoldi iterations. After $n$ iterations of the PRIMA process, $n$ moments are matched. Storage and orthogonalization of the iteration vectors is needed, but the procedure is numerically stable and preserves stability and passivity. The reduced system matrices can easily be obtained from the PRIMA process.

## 4 New Research Directions

A method that combines the large scale applicability and stability of Krylov subspace methods with the global error bound and stability/passivity preservation of the SVD based methods may be fruitful. A first attempt to such a method, approximate balanced reduction, has been made in [9], but has not yet produced a robust and efficient method.

One may also consider Jacobi-Davidson methods [8]. The Jacobi-Davidson method is an efficient method for computing eigenpair approximations $(\theta_i, u_i)$ near a specific target for $Ax = \lambda x$, provided a good preconditioner is available for the correction equation. This correction equation

$$(I - u_i u_i^*)(A - \theta_i I)(I - u_i u_i^*)t = -(Au_i - \theta u_i) \tag{4}$$

has to be solved to modest accuracy every iteration of the Jacobi-Davidson process. Note that the pair $(\theta_i, u_i)$ is selected as a Ritz-pair and changes every iteration. The search space for $u_i$ of the Jacobi-Davidson process is extended orthogonally with $t$.

Now suppose that the transfer function is of the form $H(s) = L(sI - A)^{-1}R$. If the transfer function is computed exactly, then $(sI - A)$ has to be inverted for a range of values of $s$. Since this is not feasible for large systems, reduced order models are needed. With $s$ replaced by $\theta_i$, the operator $(sI - A)$ is equal to $-(A - \theta_i I)$. This suggests that the search space built during the

Jacobi-Davidson process may contain useful information for approximate inversion. The idea is to start Jacobi-Davidson processes for several targets and to combine the relevant parts of the search space into a new space. This new space can be used to construct a reduced order model, similar to the Krylov subspace based methods. Because the transfer function is evaluated for values $s = i\omega$, the dominant eigenvalues are the eigenvalues closest to the imaginary axis. The dominance of the actual contribution to the transfer function also depends on the component of $R$ in the direction of the corresponding eigenvector. This information has to be taken into account in the Jacobi-Davidson process.

An advantage of the Jacobi-Davidson approach is that matrix inversions are avoided. PRIMA and PVL need the $LU$-decomposition of $G + s_0C$ for this, while the JDQZ method [4] for generalized eigenproblems works with $G$ and $C$ directly. The JDQZ method computes a partial generalized Schur form for the generalized eigenproblem $Gx = \lambda Cx$. Similarly to the original Jacobi-Davidson method, the JDQZ method may be used for transfer functions of the form $H(s) = L(sC - G)^{-1}R$, where $C$ is allowed to be singular. Because of singularity of $C$, there may be eigenvalues at infinity, which are not of practical interest. Harmonic Petrov values can be used to avoid computing these eigenvalues. Another possibility is to use purification techniques [6]. An open issue is the construction of good preconditioners, which are of vital importance to the convergence of the Jacobi-Davidson process.

## References


1. A.C. Antoulas and D.C. Sorensen. Approximation of large scale dynamical systems: an overview. *Int. J. Appl. Math. Comput. Sci.*, 11(5):1093–1121, 2001.
2. Z. Bai and R.W. Freund. A partial Padé-via-Lanczos method for reduced-order modeling. Num. Anal. Manu. 99-3-20, Bell Laboratories, December 2000.
3. P. Feldmann and R.W. Freund. Efficient linear circuit analysis by Padé approximation via the Lanczos process. *IEEE Trans. CAD*, 14:639–649, 1995.
4. D.R. Fokkema, G. L.G. Sleijpen, and H.A. van der Vorst. Jacobi-Davidson style QR and QZ algorithms for the reduction of matrix pencils. *SIAM J. Sc. Comp.*, 20(1):94–125, 1998.
5. K. Glover. All optimal Hankel-norm approximations of linear multivariable systems and their $L^\infty$-error bounds. *Int. J. Control*, 39:1115–1193, 1984.
6. K. Meerbergen and A. Spence. Implicitly Restarted Arnoldi with Purification for the Shift-Invert Transformation. *Math. Comp.*, 66(218):667–689, 1997.
7. A. Odabasioglu and M. Celik. PRIMA: Passive Reduced-order Interconnect Macromodeling Algorithm. *IEEE Transactions on Computer-Aided Design of Integrated Circuits and Systems*, 17(8):645–654, 1998.
8. G.L.G. Sleijpen and H.A. van der Vorst. A Jacobi-Davidson Iteration Method for Linear Eigenvalue Problems. *SIAM J. Matrix Anal. Appl.*, 17(2):401–425, 1996.
9. D.C. Sorensen and A.C. Antoulas. The Sylvester equation and approximate balanced reduction. *Lin. Alg. Appl.*, (351–352):671–700, 2002.

# DRK Methods for Time-Domain Oscillator Simulation

M.F. Sevat[1], S.H.M.J. Houben[2], and E.J.W. ter Maten[3]

[1] Philips Research Laboratories, Eindhoven marcel.sevat@philips.com
[2] Magma Design-Automation, Eindhoven
[3] Philips Research Laboratories and Eindhoven University of Technology

**Summary.** This paper presents a new Runge-Kutta type integration method that is well-suited for time-domain simulation of oscillators. A unique property of the new method is that its damping characteristics can be controlled by a continuous parameter.

**Key words:** DRK method, time domain, oscillator simulation.

## 1 Introduction

In the case of weakly non-linear circuit behaviour, oscillators can be simulated in the frequency domain, using *e.g*,. the Harmonic Balance method. In the case of strongly non-linear circuit behaviour, they are to be simulated in the time domain, using *e.g.*, the BDF-methods or the Trapezoidal Rule (TR). If the start-up behaviour of an oscillator is to be observed, a time domain method is even mandatory. However, the BDF methods exert a considerable damping on an oscillatory solution of the circuit equations. The TR method, when used on oscillators, does not exert any damping at all for all frequencies (which is also not wanted). To remedy this situation, an integration method would be preferred that has some damping to avoid numerical instability, but still so small that its effect on the oscillation can be neglected. In the next sections, DRK methods will be investigated as potential candidates for such methods.

## 2 DRK methods

We apply a Diagonal Runge-Kutta (DRK) method to a general DAE of the form

$$\boldsymbol{g}(t, \dot{\boldsymbol{x}}, \boldsymbol{x}) = \boldsymbol{0}. \tag{1}$$

Given a step size $h$ and an initial value $\boldsymbol{x}_0$, the DRK method computes a sequence $\{\boldsymbol{x}_n\}$, where $\boldsymbol{x}_n$ is an approximation to the solution at $t = nh$. Given $(a_{ii})$ and $(b_i)$, $\boldsymbol{x}_{n+1}$ is computed from $\boldsymbol{x}_n$ as follows

$$\boldsymbol{g}\left(t_n + ha_{ii}, \boldsymbol{X}_n^{(i)}, \boldsymbol{x}_n + ha_{ii}\boldsymbol{X}_n^{(i)}\right) = \boldsymbol{0}, \tag{2a}$$

$$\boldsymbol{x}_{n+1} = \boldsymbol{x}_n + h\sum_{i=1}^{s} b_i \boldsymbol{X}_n^{(i)}. \tag{2b}$$

The quantities $\boldsymbol{X}_n^{(i)}$ are called the stages of the DRK method, and $s$ is called the stage count. Note that (2a) constitutes an Implicit Euler step with step-size $ha_{ii}$. So in essence, a DRK method is a linear combination of Implicit Euler steps.

### 2.1 Order conditions

For the ODE $\dot{x} = f(x)$, $f \in C^\infty(\mathbb{R}, \mathbb{R})$ we have $g(t, \dot{x}, x) = \dot{x} - f(x) = 0$ , leading to the following DRK procedure:

$$X_n^{(i)} = f(x_n + ha_{ii}X_n^{(i)}), \quad \text{for} \quad i = 1 \ldots s, \tag{3a}$$

$$x_{n+1} = x_n + h\sum_{i=1}^{s} b_i X_n^{(i)}. \tag{3b}$$

The order conditions up to order $k$ are now found by equating, for arbitrary $f$, the following terms at the point $h = 0$ (see [1]):

$$\frac{d^j x_{n+1}}{dh^j} = \frac{d^j x(t_n + h)}{dh^j}, \quad \text{for} \quad j = 0 \ldots k. \tag{4}$$

In [3] it is shown that this is only possible up to order 2. Then, the following order conditions should be satisfied:

$$\sum_{i=1}^{s} b_i = 1, \ \sum_{i=1}^{s} b_i a_{ii} = \frac{1}{2}. \tag{5}$$

The fact that DRK methods are limited to such low orders, appears to make them quite unappealing. In contrast to the common approach in Runge-Kutta theory, as presented in *e.g.*, [1], we do not aim for maximising of the order of the method. Rather we balance the desire for a high order against the goal of obtaining a method that does not damp out oscillations.

### 2.2 Stability conditions

To study stability we apply the DRK-methods to the Dahlquist test equation

$$\dot{x} = \lambda x, \quad \lambda \in \mathbb{C}.$$

Let $x_{n+1}$ be computed from $x_n$ with the DRK method using step size $h > 0$. Then

$$\zeta(h\lambda) := \frac{x_{n+1}}{x_n} = 1 + \sum_{i=1}^{s} \frac{b_i h\lambda}{1 - a_{ii}h\lambda} \tag{6}$$

defines the amplification factor. For a DRK method to be usable as an integration method for oscillator problems, $\zeta(h\lambda)$ should satisfy the following conditions:

$$|\zeta(j\omega)| \lesssim 1\,, \quad \omega \in \mathbb{R}, \tag{7a}$$

$$|\zeta(z)| < 1 \text{ for } \Re(z) < 0, \tag{7b}$$

$$\lim_{\Re(z)\to-\infty} \zeta(z) = 0, \tag{7c}$$

It can be shown that once condition (7a) is satisfied, and $\zeta(z)$ is analytic in the left half of the complex plane, then also condition (7b) is satisfied (see [2]). Assuming for the moment that condition (7a) is satisfied, then to also satisfy condition (7b) the poles of $\zeta(z)$ need to be in the right half of the complex plane. This leads to the following restriction on the coefficients $a_{ii}$:

$$a_{ii} > 0 \quad \text{for } i = 1, \ldots, s. \tag{8}$$

Note that even with this restriction satisfied, we still need to check on any proposed set of coefficients whether (7a) is satisfied. Applying condition (7c) to (6) leads to:

$$\sum_{i=1}^{s} \frac{b_i}{a_{ii}} = 1, \tag{9}$$

which embodies another restriction on the DRK coefficients.

## 3 Two-stage Example

To have a DRK method suitable for oscillator simulation, the coefficients $a_{ii}$ and $b_i$ should satisfy order and stability conditions as derived in the preceding sections. For two stages already solutions with one degree of freedom exist. For this particular case, the set of equations to be solved is:

$$b_1 + b_2 = 1, \quad b_1 a_{11} + b_2 a_{22} = \frac{1}{2}, \quad \frac{b_1}{a_{11}} + \frac{b_2}{a_{22}} = 1, \tag{10a}$$

$$a_{11} > 0, \quad a_{22} > 0. \tag{10b}$$

In the sequel, we denote $\gamma := a_{22}$ as the degree of freedom. We then find the following solution to (10):

$$b_1 = \frac{2\gamma^2 - 3\gamma + 1}{2\gamma^2 - 4\gamma + 1}, \quad b_2 = \frac{-\gamma}{2\gamma^2 - 4\gamma + 1}, \quad a_{11} = \frac{2\gamma - 1}{2\gamma - 2}, \quad a_{22} = \gamma, \tag{11a}$$

$$\gamma \in (0, \frac{1}{2}) \cup (1, \infty), \quad \gamma \neq \frac{1}{2 \pm \sqrt{2}}. \tag{11b}$$

It remains to be checked if this set of coefficients satisfies the condition (7a). To that end, we investigate

$$\zeta(j\omega) = \frac{b_1/a_{11}}{1 - ja_{11}\omega} + \frac{b_2/a_{22}}{1 - ja_{22}\omega}, \tag{12}$$

from which we find (see [3])

$$|\zeta(j\omega)|^2 = 1 - \frac{\omega^4\gamma^2(1-2\gamma)^2}{(1+\gamma^2\omega^2)[4(1-\gamma)^2 + \omega^2(1-2\gamma)^2]}. \tag{13}$$

Hence, $|\zeta(j\omega)| \leq 1$, which implies condition (7a). Note that by making $\gamma$ small enough, $|\zeta(j\omega)|$ can be brought arbitrarily close to 1. For $\gamma \to 0$ the method approaches the midpoint rule.

The stability diagrams for various values of $\gamma$ are shown in Fig. 1. This figure clearly illustrates the fact that $\gamma$ can be used to control the amount of damping on the imaginary axis. Furthermore, it shows that by sufficiently decreasing the value of $\gamma$ the amount of damping can be brought as close to zero as we require.

## 4 Alternative Formulation

Using the transformation $\widetilde{\boldsymbol{X}}_n^{(i)} = \boldsymbol{x}_n + ha_{ii}\boldsymbol{X}_n^{(i)}$ for $i = 1, \ldots, s$ and assuming that (9) holds, we obtain the following alternative formulation of the DRK method:

$$\boldsymbol{g}\left(t_n + ha_{ii}, \frac{\widetilde{\boldsymbol{X}}_n^{(i)} - \boldsymbol{x}_n}{ha_{ii}}, \widetilde{\boldsymbol{X}}_n^{(i)}\right) = \boldsymbol{0}, \tag{14a}$$

$$\boldsymbol{x}_{n+1} = \sum_{i=1}^{s} \frac{b_i}{a_{ii}} \widetilde{\boldsymbol{X}}_n^{(i)}, \tag{14b}$$

The numerical robustness of this alternative formulation is better than the one of the standard formulation, as it avoids the summation of relatively small quantities to the current approximation in the update equation (see (2b)). It thereby circumvents the unnecessary loss of accuracy. For the two-stage case, considered in the previous section, the coefficients (11) satisfy condition (9). So (14) holds for this case, with the following expressions for its coefficients:

$$\frac{b_1}{a_{11}} = \frac{2(\gamma-1)^2}{2\gamma^2 - 4\gamma + 1}, \quad \frac{b_2}{a_{22}} = \frac{-1}{2\gamma^2 - 4\gamma + 1}, \tag{15}$$

with the coefficients $a_{11}$, $a_{22}$ and the restrictions on $\gamma$ the same as in (11).

## 5 Conclusions

We developed a Diagonal Runge-Kutta algorithm that is particularly suited for transient simulation of oscillators, in the sense that it does not damp out any oscillation present in the solution of the circuit equations. In fact it has been shown that its damping characteristics can be controlled by a dedicated parameter. The new algorithm allows designers to better simulate oscillators, or to detect unwanted oscillation earlier than would be the case with standard integration methods.

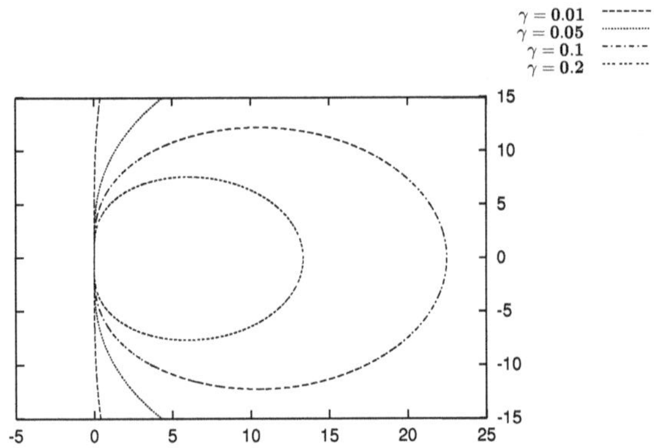


**Fig. 1.** Stability diagram of the DRK method in the complex $h\lambda$-plane.

## References

1. E. E. Hairer, S.P. Nø rsett, and G. Wanner. *Solving Ordinary Differential Equations I, Non-stiff problems.* Springer, Berlin, 2nd edition, 1993.
2. E. Hairer and G. Wanner. *Solving Ordinary Differential Equations II, Stiff and Differential Algebraic problems.* Springer, Berlin, 2nd edition, 1996.
3. S.H.M.J. Houben. *Circuits in Motion.* PhD thesis, Eindhoven University of Technology, 2003.

# Digital Linear Control Theory Applied To Automatic Stepsize Control In Electrical Circuit Simulation

A. Verhoeven[1], T.G.J. Beelen[2], M.L.J. Hautus[1], and E.J.W. ter Maten[3]

[1] Technische Universiteit Eindhoven averhoev@win.tue.nl
[2] Philips Research Laboratories
[3] Technische Universiteit Eindhoven and Philips Research Laboratories

**Summary.** Adaptive stepsize control is used to control the local errors of the numerical solution. For optimization purposes smoother stepsize controllers are wanted, such that the errors and stepsizes also behave smoothly. We consider approaches from digital linear control theory applied to multistep BDF-methods.

**Key words:** adaptive stepsize control, BDF-methods, digital linear control.

## 1 Introduction to error control

Transient simulation of electrical circuits is done by integration of the following implicit Differential-Algebraic Equation

$$\frac{d}{dt}\left[\mathbf{q}(t,\mathbf{x})\right] + \mathbf{j}(t,\mathbf{x}) = \mathbf{0}, \quad \mathbf{j}(0,\mathbf{x}(0)) = \mathbf{0}, \tag{1}$$

where $\mathbf{q},\mathbf{j} : \mathbb{R} \times \mathbb{R}^n \to \mathbb{R}^n$ are nonlinear functions that represent the charges and currents in the circuit, while $\mathbf{x}$ is the state vector. Because the BDF multistep-methods are the default methods used by analog circuit simulators, we will concentrate on these methods. While Runge-Kutta methods often contain an embedded reference method to estimate the local error, for the $k$-step BDF-method this can be done by means of the prediction $\hat{\mathbf{q}}_n$ which is based on the extrapolation of the previous $k+1$ values of $\mathbf{q}$. For the time-grid $\{t_i, i = 0, \ldots, N\}$ with time steps $h_i = t_i - t_{i-1}$ we obtain the estimate

$$\hat{r}_n = \frac{h_n}{t_n - t_{n-k-1}} \|\mathbf{q}(t_n, \mathbf{x}_n) - \hat{\mathbf{q}}_n\|. \tag{2}$$

If this estimate $\hat{r}_n$ is larger than a given tolerance level TOL, the current step is rejected. Otherwise, the solution $\mathbf{x}_n$ is accepted and the next numerical solution can be computed at a new timepoint.

The following stepsize controller is very commonly used for integration methods of order $p$:

$$h_n = \left(\frac{\varepsilon}{\hat{r}_{n-1}}\right)^{\frac{1}{p+1}} h_{n-1}, \tag{3}$$

where $\varepsilon = \theta$ TOL. It is based on the assumption that the error estimate satisfies the model

$$\hat{r}_n = \hat{\varphi}_n h_n^{p+1}, \tag{4}$$

where $\varphi_n$ is an unknown variable which is independent of $h_n$. This model is a good description for onestep methods and also a first-order approximation for the BDF-methods.

However, it appears that the controller from (3) may produce rather irregular error and stepsize sequences, which will decrease the effectiveness in optimization.

## 2 Control-Theoretic Approach to Stepsize Control

It is possible to use control-theoretic techniques for error control. In [2] this idea has been applied to onestep methods where we have the simple model (4). The logarithmic version of the onestep error model is

$$\log \hat{r}_n = (p+1)\log h_n + \log \hat{\varphi}_n. \tag{5}$$

Indeed, this implies that the sequence $\log \hat{r} = \{\log \hat{r}_n\}_{n\in\mathbb{N}}$ can be viewed as the output of a digital (*i.e.*, discrete) linear control system, where $\log h = \{\log h_n\}_{n\in\mathbb{N}}$ is the input signal and $\log \hat{\varphi} = \{\log \hat{\varphi}_n\}_{n\in\mathbb{N}}$ is an unknown output disturbance. In general, one can denote all finite linear models for $\log \hat{r}$ by

$$\log \hat{r} = G(q)\log h + \log \hat{\varphi}, \tag{6}$$

where $q$ is the shift-operator, with $q(\log h_n) = \log h_{n+1}$ and $G(q)$ being a rational function of $q$:

$$G(q) = \frac{L(q)}{K(q)} = \frac{\lambda_0 q^M + \cdots + \lambda_M}{q^M + \kappa_1 q^{M-1} + \cdots + \kappa_M}. \tag{7}$$

For the one-step model (5), we have $G(q) = p+1$. The input $\log h$ is computed on base of the previous values of the output $\log \hat{r}$ and the reference $\log \varepsilon$.

All linear controllers can be denoted by

$$\log h = C(q)(\log \varepsilon - \log \hat{r}), \tag{8}$$

where $C(q)$ is a rational function of $q$:

$$C(q) = \frac{B(q)}{A(q)} = \frac{\beta_0 q^{N-1} + \cdots + \beta_{N-1}}{q^N + \alpha_1 q^{N-1} + \cdots + \alpha_N}. \tag{9}$$

For the controller (3) we have that $C(q) = \frac{1}{p+1}\frac{1}{q-1}$.

## 3 Derivation of Process Model for BDF-Methods

Unfortunately, for the multistep BDF-methods, it is not possible to derive a linear model of the form of (6). In this case, we have the following nonlinear model for $\log \hat{r}$

$$\log \hat{r}_n = 2\log h_n + \log(h_{n-1}+h_n) + \cdots + \log(h_{n-p+1}+\cdots+h_n) + \log \hat{\varphi}_n - \log p!. \tag{10}$$

Note that $\log \hat{r}_n$ also depends on the previous stepsizes, because it is a multistep method. In [4] it is tried to approximate this model by the previous model for onestep methods. If the stepsizes only have small variations, also linearization can be used [1]. In [3] it is proved that the linearized model is equal to

$$\log \hat{r}_n = \sum_{k=0}^{p-1} (\gamma_p - \gamma_k) \log h_{n-k} + \log \hat{\varphi}_n, \quad \gamma_0 = -1, \gamma_k = \sum_{m=1}^{k} \frac{1}{m}. \tag{11}$$

This model can also be cast in the form of (6), where

$$G(q) = \frac{(1+\gamma_p)q^{p-1} + (\gamma_p - \gamma_1)q^{p-2} + \cdots + (\gamma_p - \gamma_{p-1})}{q^{p-1}}. \tag{12}$$

## 4 Design of Finite Order Digital Linear Stepsize Controller

Consider the error model in (6), which is controlled by the linear controller (8). It is assumed that $G(q)$ is already be known, while $C(q)$ still must be designed. Now, the closed loop dynamics are described by the following equations:

$$\begin{cases} \log h = U_r(q) \log \varepsilon + U_w(q) \log \hat{\varphi}, \\ \log \hat{r} = Y_r(q) \log \varepsilon + Y_w(q) \log \hat{\varphi}. \end{cases} \tag{13}$$

where by (7), (9) the transfer functions satisfy

$$\begin{aligned} U_r(q) &= \tfrac{B(q)K(q)}{A(q)K(q)+B(q)L(q)}, \; U_w(q) = \tfrac{-B(q)K(q)}{A(q)K(q)+B(q)L(q)}, \\ Y_r(q) &= \tfrac{B(q)L(q)}{A(q)K(q)+B(q)L(q)}, \; Y_w(q) = \tfrac{A(q)K(q)}{A(q)K(q)+B(q)L(q)}. \end{aligned} \tag{14}$$

Thus, the poles of the system are determined by the $N+M$ roots of the characteristic equation

$$R(q) \equiv A(q)K(q) + B(q)L(q) = 0.$$

If the poles lay inside the complex unity circle, the closed loop system is stable. Suitable choices are $R(q) = (q-r)^{N+M}$ or $R(q) = q^{N+M} - r^{N+M}$ for $r \in [0,1)$ [3]. Assume that $A(q), B(q)$ can be factorized like $A(q) = (q-1)^{p_A}(q+1)^{p_R}\tilde{A}(q)$ and $B(q) = (q+1)^{p_F}\tilde{B}(q)$. Then, the order of adaptivity is equal to $p_A$, while the stepsize and error filter orders are $p_R$ and $p_F$ [2]. The coefficients of $\tilde{A}, \tilde{B}$ can be computed from

$$(q-1)^{p_A}(q+1)^{p_R}\tilde{A}(q)K(q) + (q+1)^{p_F}\tilde{B}(q)L(q) = R(q). \tag{15}$$

## 5 Numerical Experiments

Consider the circuit which corresponding equations are given by:

$$\begin{aligned} C\dot{V}_1 + \tfrac{1}{R}V_1 - \sin(\omega_1 t) - \tfrac{1}{R_1}(V_2 - V_1) &= 0, \\ \tfrac{1}{R_1}(V_2 - V_1) - i_E &= 0, \\ V_2 - V_3 &= 0, \\ i_E - \tfrac{1}{R_2}(V_4 - V_3) &= 0, \\ C\dot{V}_4 + \tfrac{1}{R}V_4 - \sin(\omega_2 t) - \tfrac{1}{R_2}(V_3 - V_4) &= 0. \end{aligned}$$

| Parameters | Value |
|---|---|
| $\omega_1$ | $\frac{5}{2}\pi \cdot 10^3$ |
| $\omega_2$ | $\frac{1}{4}\pi \cdot 10^3$ |
| $R$ | 10 |
| $C$ | $10^{-3}$ |
| $R_1$ | 1 |
| $R_2$ | 1 |

A transient simulation along $[0, 0.08]$ is computed by a circuit simulator, while several stepsize controllers are used. Because the theory only holds for fixed integration order, the integration order is kept fixed at $p = 3$. By default, the simulator uses the controller (3) with a buffer such that the stepsize remains constant for small variations (case 1). This control action is removed for the other cases, because it destroys the characteristic behaviour of the designed controller. For all controllers we have $R(q) = (q - r)^{N+M}$. The smoothness of the stepsize and error sequence is quantified by the number $s(x) = \sqrt{\sum_{m=1}^{N}(x_m - x_{m-1})^2}/\|x\|_2$. Table 1 shows the results of the several testcases. For this circuit case 4 produces the smoothest results. Note at the decline of the number of rejections for the cases 1, 2 and 6 with $p_A = 1$ and $p_F = p_R = 0$. Figure 1 shows the results for cases 1 and 4.

**Table 1.** Numerical results.

| Case | $N$ | $M$ | $p_A$ | $p_F$ | $p_R$ | $r$ | # stepsizes | # rejections | # Newton iterations | $s(\log \hat{r})$ | $s(\log h)$ |
|---|---|---|---|---|---|---|---|---|---|---|---|
| 1 | 1 | 0 | 1 | 0 | 0 | 0 | 1258 | 222 | 1480 | 1.17 | 0.57 |
| 2 | 1 | 0 | 1 | 0 | 0 | $\frac{1}{2}$ | 1277 | 198 | 1475 | 0.82 | 0.38 |
| 3 | 2 | 0 | 2 | 0 | 0 | $\frac{1}{2}$ | 990 | 609 | 1599 | 1.14 | 0.83 |
| 4 | 2 | 0 | 1 | 1 | 0 | $\frac{1}{2}$ | 1053 | 0 | 1053 | 0.57 | 0.10 |
| 5 | 2 | 0 | 1 | 0 | 1 | $\frac{1}{2}$ | 1198 | 0 | 1198 | 0.75 | 0.22 |
| 6 | 3 | 2 | 1 | 0 | 0 | $\frac{1}{2}$ | 1015 | 0 | 1015 | 1.01 | 0.32 |

An important question is whether the new designed controllers also have a better performance if variable order is used. For many tested cases it was possible to get smoother results for a slightly increased or even decreased computational effort [3].

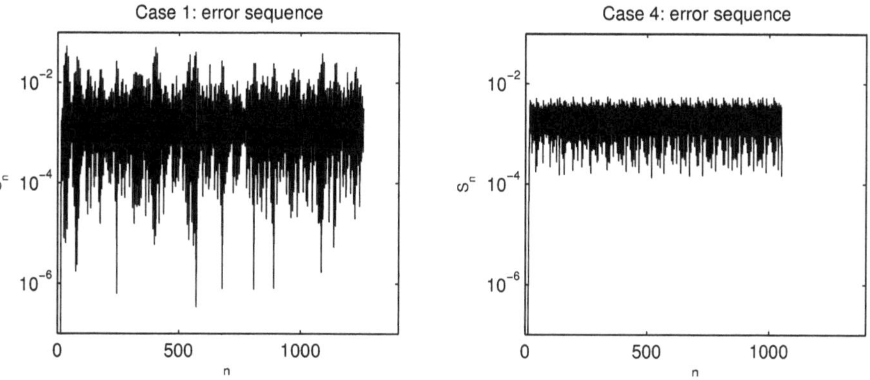


**Fig. 1.** Results of error sequences for cases 1 and 4 in Table 1.

# 6 Conclusions

If the error model is linear, control theory can indeed improve the smoothness of the results. For multistep BDF-methods applied to smooth problems, where the stepsizes have small variations, the linearized model works well. For more stiff problems it is better to use the one-step model.

The process model depends on the integration order. The designed controllers are also applicable to variable integration order. From the experiments it turns out that it is not attractive to use higher order adaptive controllers, while filtering can be attractive.

# References


1. A. Sjö. *Analysis of Computational Algorithms for Linear Multistep Methods.* PhD thesis, Lund University, Lund, Sweden, 1999.
2. G. Söderlind. Digital filters in adaptive time-stepping. *ACM Tr. on Math. Softw.*, V(N):1–24, 2002.
3. A. Verhoeven. Automatic control for adaptive time stepping in electrical circuit simulation. 2004.
4. M. Zhuang and W. Mathis. Research on stepsize control in the BDF method for solving differential-algebraic equations. *IEEE Proceedings*, pages 229–232, 1994.

# Part III

## Theme: Chemical Technology

# On the Dynamics of a Bunsen Flame

M.L. Bondar[1] and J.H.M. ten Thije Boonkkamp[2]

Technische Universiteit Eindhoven
Department of Mathematics and Computer Science,
PO Box 513 MB Eindhoven
mbondar@win.tue.nl[1] tenthije@win.tue.nl[2]

**Summary.** The stabilization of a Bunsen flame above the burner rim is simulated using the method of characteristics. Oscillations of the flame front and of its area due to flow oscillations are computed.

**Key words:** Bunsen flame, kinematic condition, method of characteristics.

## 1 Introduction

Combustion devices based on premixed flames have low emissions of pollutant gas. In the particular case of a Bunsen burner, the natural gas is premixed with air before combustion, giving a conical shaped flame that can be cooled such that the emission of pollutants is low. However, the sensitivity of Bunsen flames to flow oscillations leads to unwanted effects such as combustion instabilities and noise. Understanding and prediction of noise production is an important task in designing efficient, noise free combustion devices. This requires a transfer function which correlates the oscillations in the flow velocity and in the heat release rate. Given that the flow velocity fluctuations affect the flame by changing its area, which is proportional with the heat release rate, the first step in determining the transfer function consists of computing the instantaneous area of the flame. Thus, the determination of the location and of the shape of the flame front is required.

## 2 Flame front dynamics

The complete simulation of combustion dynamics is difficult and requires powerful computer resources. However, the main features of the flame response to flow oscillations can be captured with a reduced model based on the following assumptions. First, we assume that the flame is a surface which separates the burnt from the unburnt gas. This surface is referred to as the flame front. Next, we assume that the flame front moves at constant velocity, the laminar burning velocity $S_L$, in the direction normal to its surface towards the unburnt gas and, that it does not influence the flow, which is prescribed. Since the Bunsen flame is conical, we assume that the flame and the flame oscillations are axisymmetric. The flame front can be described as the level set of some combustion variable $G$, *i.e.*, $G(r,z,t) = G_0$ where $r$ and $z$ are the radial and axial coordinates, respectively, and $t$ is the time. Assuming the flame front is nowhere vertical, $z$ can be expressed as a function of $r$ and $t$, *i.e.*, $z = z(r,t)$. The motion of the flame front can be described by the kinematic relation,

$$\frac{\partial z(r,t)}{\partial t} + u\frac{\partial z(r,t)}{\partial r} - v + S_L\sqrt{\left(\frac{\partial z(r,t)}{\partial r}\right)^2 + 1} = 0, \tag{1}$$

where $u$ and $v$ are the radial and axial components of the gas velocity, respectively (see *e.g.* [1]). Solving (1) allows us to investigate the shape and the area of the flame.

The analytical solution of (1) in the case of a Poiseuille flow, which is a reasonable approximation for a flow in a duct, is described below.

## 3 Solution in the case of a Poiseuille flow

The flow is approximated with a Poiseuille flow, *i.e.*,

$$u(r,z,t) = 0, \qquad v(r,z,t) = v_0\left(1 - \left(\frac{r}{R}\right)^2\right), \tag{2}$$

where $v_0$, $(v_0 > S_L)$ and $R$ denote the maximum velocity of the flow and the duct radius, respectively. Introducing the notation $p := \partial z/\partial r$ and $q := \partial z/\partial t$, we obtain from (1) the following canonical form,

$$F(r,t,z,p,q) := q - v_0\left(1 - \left(\frac{r}{R}\right)^2\right) + S_L\sqrt{p^2+1} = 0. \tag{3}$$

Equation (3) is a nonlinear first order PDE. Given the initial conditions $z(r,0) = Z_0(r)$, $p(r,0) = Z_0'(r)$, it is possible to find an analytical solution for (3) by using the method of characteristics ([2]). This method reduces (3) to a system of five coupled ODEs. They hold along the characteristics, parametrized by $s$, and take initial values on the initial line $t = 0$, parametrized by $\sigma$. The unknowns of the system are $t(s,\sigma)$, $r(s,\sigma)$, $p(s,\sigma)$, $q(s,\sigma)$ and $z(s,\sigma)$. After a first evaluation the system reduces to a system of only three equations because $t(s,\sigma) = s$ and $q$ can be decoupled.

The following dimensionless variables are introduced, $r^* := r/R$, $t^* := t/\tau$, $\sigma^* := \sigma/R$, $z^* := z/R$, $\tau := R/S_L$ and $\hat{v} := v_0/S_L$. The scaled ODE system (we omitted the $*$) reads

$$\frac{\mathrm{d}r}{\mathrm{d}t} = \frac{p}{\sqrt{p^2+1}}, \qquad r(0,\sigma) = \sigma, \tag{4a}$$

$$\frac{\mathrm{d}z}{\mathrm{d}t} = \hat{v}(1-r^2) - \frac{1}{\sqrt{p^2+1}}, \qquad z(0,\sigma) = Z_0(\sigma), \tag{4b}$$

$$\frac{\mathrm{d}p}{\mathrm{d}t} = -2\hat{v}r, \qquad p(0,\sigma) = Z_0'(\sigma). \tag{4c}$$

The formal solution procedure for system (4) is as follows. First, from (4a) and (4c) we find the expression for $p(r,\sigma)$ along the characteristics. By substitution in (4a) we find the location of the characteristics in an implicit form $t = t(r,\sigma)$. From (4a) and (4b) we find the axial displacement along the characteristics, $z(r,\sigma)$. Finally we invert the implicit relation $t = t(r,\sigma)$ to find $\sigma = \sigma(r,t)$ and replace it in the expression for $z(r,\sigma)$ to find the axial displacement $z(r,t)$ of the flame front. For $t(r,\sigma)$ and $z(r,\sigma)$ the following expressions have been found, respectively,

$$t(r,\sigma) = -\int_\sigma^r \frac{c(\sigma) - \hat{v}x^2}{\sqrt{\left(c(\sigma) - \hat{v}x^2\right)^2 - 1}}\mathrm{d}x, \tag{5}$$

$$z(r,\sigma) = Z_0(\sigma) + \hat{v}t(r,\sigma) + \int_\sigma^r \frac{1 + \hat{v}x^2(c(\sigma) - \hat{v}x^2)}{\sqrt{\left(c(\sigma) - \hat{v}x^2\right)^2 - 1}}\mathrm{d}x, \tag{6}$$

where $c(\sigma) = \sqrt{1 + Z_0'(\sigma)^2} + \hat{v}\sigma^2 \geq 1$. These integrals can not be evaluated analytically, instead they can be formulated in terms of elliptic integrals, see [3]. To find the axial displacement $z(r,t)$, $\sigma$ as a function of $r$ and $t$ is needed. This is possible only if the Jacobian, $J(r,\sigma) = \partial t(r,\sigma)/\partial\sigma \neq 0$. For the initial condition $Z_0(r) = 0$, $Z_0'(r) = 0$ this is indeed the case. Moreover, this condition guaranties that no cusps in the flame front are created.

Since the velocity has a parabolic profile and vanishes at the burner rim, there is a region where the laminar burning velocity is bigger than the gas velocity. In this region the flame is pushed into the tube which contradicts with the movement of the real flame. To simulate the stabilization of the flame above the burner rim we apply the following procedure only on the domain where the gas velocity is bigger than the laminar burning velocity, *i.e.*, $0 \leq r \leq \delta$. Here $\delta$ is such that $\hat{v}(1-\delta^2) = 1$, *i.e.*, $\delta = \sqrt{1 - \hat{v}^{-1}}$. Since the inversion of (5) cannot be performed analytically, we use the following numerical approach. Let us introduce uniform discretizations for the space and time domains, *i.e.* $r_j = j\Delta r$, $t_i = i\Delta t$, with the grid size $\Delta r = \delta/M$ and the time step $\Delta t = 1/N$, $i = 1,\dots,N$, $j = 1,\dots,M$. For a given $r_j$ and a given $t_i$ we find the corresponding $\sigma_{i,j}$ by solving (5) through the secant method (the corresponding $\sigma_{i,j} > r_j$, for $i = 1,\dots,N$, $j = 1,\dots,M$). The axial displacement of the flame corresponding to the point $r_j$ at the moment $t_i$, $z(r_j,t_i)$ is then $z(r_j,\sigma_{i,j})$. For values of $\sigma > \delta$ the corresponding axial diplacement is negative. If for an $r_k$ the corresponding $\sigma_{i,k} > \delta$ we compute the axial displacement as $z(r_k,t_i) = z(r_k,\delta)$. This condition implies the fixation of the flame above the burner. Indeed, $z(\delta, t_i) = z(r_M, \sigma_{i,M})$ and since $\delta = r_M < \sigma_{i,M}$, from the previous condition we conclude that $z(\delta, t_i) = z(\delta, \delta) = 0$ for $i = 1, \ldots, N$. According to the present analytical model the flame reaches a stationary position after starting from an initial flat profile. The stationary position of the flame is numerically identical whith the steady solution of (1), which validates our results. An illustration of the stationary flame given by the system (4) is presented in Fig. 1.(a).

## 4 Flame response to flow perturbations

After the flame reaches its stationary position a perturbation is imposed in the gas velocity, of the form

$$v_0\left(1 - \left(\frac{r}{R}\right)^2\right)\varepsilon \sin \omega t, \tag{7}$$

where $\varepsilon$ and $\omega$ are the relative amplitude and the frequency of the velocity perturbation, respectively. Scaling the variables, the equation for the axial displacement of the flame front becomes

$$\frac{\partial z}{\partial t} - \hat{v}(1 - r^2)(1 + \varepsilon \sin \hat{\omega} t) + \sqrt{\left(\frac{\partial z}{\partial r}\right)^2 + 1} = 0, \tag{8}$$

where $\hat{\omega} := \omega R / S_L$. Assuming the perturbations in the gas velocity of small amplitude, the solution $z(r, t)$ of (8) can be expanded in an asymptotic expansion ([2]). Substituting the asymptotic expansion into (8) and collecting the terms of the same order leads to a system of equations. The small amplitude perturbation allows us to consider only the leading and the first order equations of the system. The leading order solution can be replaced by the steady solution of the (1) and the first order equation is a linear advection equation which is solved by using the upwind scheme. Combining the solutions of the leading and of the first order equations we arrive at an analytical numerical description of the flame front. The perturbation of the flame around the stationary position of the flame front can be seen in Fig. 1.(b), (c), (d).

The area of the flame can be computed easily by evaluating the formula

$$A(t) = 2\pi \int_0^{\delta} r \sqrt{\left(\frac{\partial z}{\partial r}\right)^2 + 1}\, \mathrm{d}r, \tag{9}$$

using the trapezoid approximation. The area oscillates at the same frequency as the velocity perturbation, but a phase difference exists between the two oscillations, see Fig. 2. The noise production can be evaluated by computing the phase difference between the two oscillations and the amplitude response (amplitude of area oscillations/amplitude of velocity oscillations) as function of frequency (see [4]).

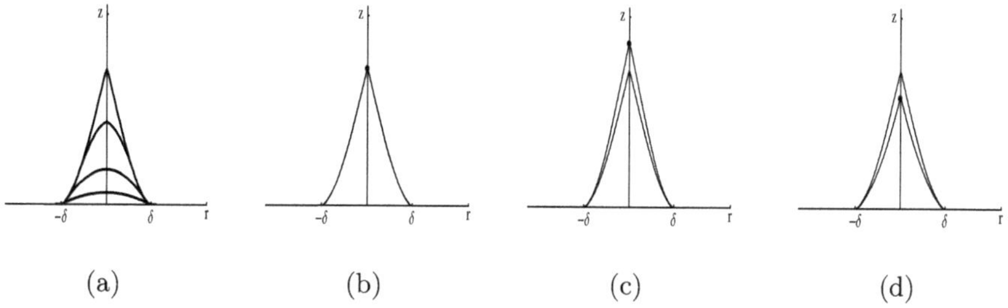


**Fig. 1.** Flame front dynamics. (a) Bunsen flame reaching its stationary position. (b), (c), (d) Oscillation of the flame front (solid line with bullets) around the stationary position (solid line). Here $\hat{v}$=5, $\hat{\omega} = 4$, $\varepsilon = 0.1$ (b) $t = 0$, (c) $t = 0.672$, (d) $t = 1.472$.

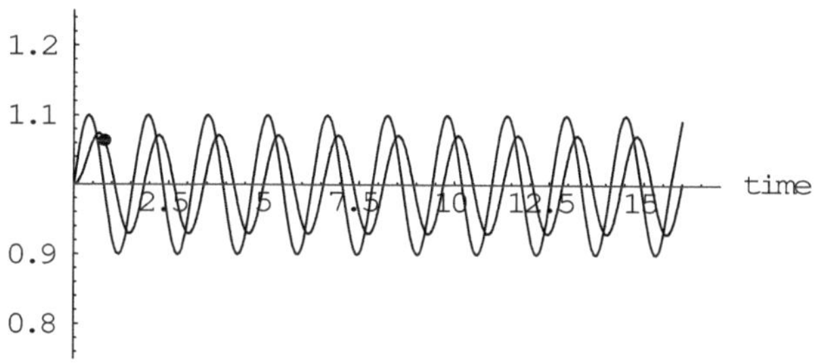


**Fig. 2.** Variation in time of the area of the flame (solid line with bullets) and of the velocity of the flow in the center of the duct (solid line), (normalized with the initial values). Here $\hat{v}$=5, $\hat{\omega} = 4$, $\varepsilon = 0.1$.

## References


1. M. Fleifil, A.M. Annaswamy, Z.A. Ghoneim, and A.F. Ghoneim. Response of a laminar premixed flame to flow oscillations; a kinematic model and thermoacoustic instability results. *Combust. Flame*, 106(4):487–510, 1996.
2. J. Kevorkian. In *Partial Differential Equations. Analytical Solution Techniques*, pages 358–369, 716–725. Chapman & Hall, New York-London, 1993.
3. L.M. Milne-Thomson. Elliptic integrals. In M. Abramowitz and I.A. Stegun, editors, *Handbook of Mathematical Functions, with Formulas, Graphs, and Mathematical Tables*, pages 587–626. Dover Publications, USA, 1974.
4. A. Putnam and W. Dennis. Burner oscillations of the Gauze-tone type. *J. Acoust. Soc. Am.*, 29(5):716–725, 1954.

# Index Analysis for Singular PDE Models of Fuel Cells

K. Chudej[1]

Lehrstuhl für Ingenieurmathematik, Universität Bayreuth, D-95440 Bayreuth
kurt.chudej@uni-bayreuth.de

**Summary.** A generalized definition is given for the time index and a new prototype example is introduced, which serves as a general case for the computation of the time index for a hierarchy of molten carbonate fuel cell models, including a 2D model. The time indices are computed by a new approach using linear integral equations.

**Key words:** Singular PDE, partial differential-algebraic equations, PDAE, time index, fuel cell, MCFC, integral equation.

## 1 Time Index: Definition and Prototype Example

Consider a singular PDE: Find $u(t, z)$ with time $t \in [0, T]$ and spatial coordinate $z \in \Omega = [0, 1]^d$, $d \in \{1, 2, 3\}$ s.t.

$$Au_t + \Psi(u, u_z, u_{zz}, t, z) = 0 \text{ on } [0, T] \times \Omega \quad (\text{if } d = 1) \tag{1}$$

($A$ is a given constant matrix, e.g. $A = \text{diag}(I, O)$) and initial conditions

$$A[u(0, z) - g(z)] = 0 \tag{2}$$

and nonlinear boundary conditions

$$h(u(t, 0), u(t, 1), u_z(t, 0), u_z(t, 1)) = 0 \quad (\text{if } d = 1). \tag{3}$$

For $d \geq 2$ Eq. (1) reads

$$Au_t + \Psi(u, u_{z_1}, \ldots, u_{z_d}, u_{z_1 z_1}, \ldots, u_{z_d z_d}, t, z) = 0. \tag{4}$$

We always assume in the following that the singular PDE is of parabolic-hyperbolic type, that the nonlinear boundary conditions (especially for the hyperbolic coordinates of $u$) are suitably posed and that the singular PDE has a solution.

The following definition is a generalization of [3] and [1]:

**Definition 1.** *If the matrix $A$ is regular, the (differential) time index is defined to be $\nu_t := 0$. Otherwise the (differential) time index $\nu_t$ is the smallest number of times, the singular PDE must be differentiated with respect to time $t$ in order to determine* $u_t = \varphi(t, z, u, \underbrace{u_z, u_{zz}, \dots}_{\text{spatial derivatives only}})$ *with a continuous function $\varphi$.*

The knowledge of the time index is important for the choice of a suitable numerical solution method.

*Prototype Example 1.* $d = 1$. Find scalar functions $u(t,z)$, $w(t,z)$, $v(t,z) > 0$, $\bar{w}(t,z)$, $\bar{v}(t,z) > 0$ s.t. the singular PDE (with positive constant $\lambda$)

$$u_t = \lambda u_{zz} + \psi_1(u, w, v) \tag{5a}$$
$$w_t = -v w_z + \psi_2(u, w, v) \tag{5b}$$
$$0 = v_z + \psi_3(u, w, v) \qquad \text{on } [0,T] \times \Omega \tag{5c}$$
$$\bar{w}_t = \bar{v}\bar{w}_z + \psi_4(u, \bar{w}, \bar{v}) \tag{5d}$$
$$0 = \bar{v}_z + \psi_5(u, \bar{w}, \bar{v}) \tag{5e}$$

with initial conditions $u(0,z) = g_1(z)$, $w(0,z) = g_2(z)$, $\bar{w}(0,z) = g_4(z)$ and boundary conditions $u_z(t,0) = 0$, $u_z(t,1) = 0$, and $w(t,0) = w_{\text{left}}(t)$, $v(t,0) = v_{\text{left}}(t)$, $\bar{w}(t,1) = \bar{w}_{\text{right}}(t)$, $\bar{v}(t,1) = \bar{v}_{\text{right}}(t)$.

Computation of the time index $\nu_t$: Partial differentiation of Eq. (5c) with respect to time yields

$$0 = v_{zt} + \frac{\partial \psi_3}{\partial v} v_t + \frac{\partial \psi_3}{\partial u} u_t + \frac{\partial \psi_3}{\partial w} w_t. \tag{6}$$

Plugging in the r.h.s. of Eqs. (5a, 5b) yields

$$0 = [\frac{\partial}{\partial z} + \frac{\partial \psi_3}{\partial v}][v_t] + \alpha(u, w, v, u_{zz}, w_z). \tag{7}$$

By a similar computation one gets

$$0 = [\frac{\partial}{\partial z} + \frac{\partial \psi_5}{\partial \bar{v}}][\bar{v}_t] + \bar{\alpha}(u, \bar{w}, \bar{v}, u_{zz}, \bar{w}_z). \tag{8}$$

If $\frac{\partial \psi_3}{\partial v} = 0$ and $\frac{\partial \psi_5}{\partial \bar{v}} = 0$ then

$$v_t(t,z) = v_{\text{left}}(t) - \int_0^z \alpha(u, w, v, u_{zz}, w_z)|_{(t,\tilde{z})} \, \mathrm{d}\tilde{z}, \tag{9}$$

$$\bar{v}_t(t,z) = \bar{v}_{\text{right}}(t) + \int_z^1 \bar{\alpha}(u, \bar{w}, \bar{v}, u_{zz}, \bar{w}_z)|_{(t,\tilde{z})} \, \mathrm{d}\tilde{z}, \tag{10}$$

therefore time index $\nu_t = 1$.

Otherwise integration yields linear Volterra integral equations of the second kind

$$0 = v_t(t,z) + \int_0^z k(t,\tilde{z}) v_t(t,\tilde{z})\, \mathrm{d}\tilde{z} + \beta(u,w,v,u_{zz},w_z)|_{(t,z)} \tag{11}$$

with $k(t,\tilde{z}) = \frac{\partial \psi_3}{\partial v}$, $\beta(u,w,v,u_{zz},w_z)|_{(t,z)} = \int_0^z \alpha(u,w,v,u_{zz},w_z)|_{(t,\tilde{z})}\, \mathrm{d}\tilde{z} - v_{\mathrm{left}}(t)$ for the unknown function $b(z) = v_t(t,z)$ and

$$0 = \bar{v}_t(t,z) + \int_1^z \bar{k}(t,\tilde{z}) \bar{v}_t(t,\tilde{z})\, \mathrm{d}\tilde{z} + \bar{\beta}(u,\bar{w},\bar{v},u_{zz},\bar{w}_z)|_{(t,z)} \tag{12}$$

with $\bar{k}(t,\tilde{z}) = \frac{\partial \psi_5}{\partial \bar{v}}$, $\bar{\beta}(u,\bar{w},\bar{v},u_{zz},\bar{w}_z)|_{(t,z)} = \int_1^z \bar{\alpha}(u,\bar{w},\bar{v},u_{zz},\bar{w}_z)|_{(t,\tilde{z})}\, \mathrm{d}\tilde{z} - \bar{v}_{\mathrm{right}}(t)$ for the unknown function $\bar{b}(z) = \bar{v}_t(t,z)$.

Both linear Volterra integral equations (11, 12) of the second kind can be solved uniquely (and depend continuously on $t$, $k$, $\bar{k}$, $\beta$, $\bar{\beta}$) for $b(z)$ and $\bar{b}(z)$, therefore time index $\nu_t = 1$.

The result still holds for vector functions $w$ and $\bar{w}$.

## 2 Time Index of Dynamic Fuel Cell Models

Three application examples of 1D respectively 2D families of dynamic models of molten carbonat fuel cells (MCFCs) from [2, 1, 5], fit into the setting of the prototype example 1.

*Example 2.* Simple 1D family of dynamic models of MCFCs: Find temperatures $\theta_s, \theta_a, \theta_c$, molar fractions $x_{a,j}, x_{c,j}$ ($j \in \{1,\ldots,7\}$) and molar flows $g_a, g_c$ (with $v_a := g_a\theta_a$ and $v_c := g_c\theta_c$) for fixed potential differences $\Phi_a, \Phi_e, \Phi_c$ s.t. to the singular PDE

$$\frac{\partial \theta_s}{\partial t} = \lambda \frac{\partial^2 \theta_s}{\partial z^2} + \varphi_1(\theta_s,\theta_a,\theta_c,x_a,x_c,\Phi_a,\Phi_e,\Phi_c)\,, \tag{13a}$$

$$\frac{\partial \theta_a}{\partial t} = -v_a \frac{\partial \theta_a}{\partial z} + \varphi_2(\theta_s,\theta_a,x_a,\Phi_a)\,, \tag{13b}$$

$$\frac{\partial x_{a,j}}{\partial t} = -v_a \frac{\partial x_{a,j}}{\partial z} + \varphi_{3,j}(\theta_s,\theta_a,x_a,\Phi_a)\,, \quad j = 1,\ldots,7\,, \tag{13c}$$

$$0 = \frac{\partial v_a}{\partial z} + \varphi_4(\theta_s,\theta_a,x_a,\Phi_a)\,, \tag{13d}$$

$$\frac{\partial \theta_c}{\partial t} = v_c \frac{\partial \theta_c}{\partial z} + \varphi_5(\theta_s,\theta_c,x_c,\Phi_c)\,, \tag{13e}$$

$$\frac{\partial x_{c,j}}{\partial t} = v_c \frac{\partial x_{c,j}}{\partial z} + \varphi_{6,j}(\theta_s,\theta_c,x_c,\Phi_c)\,, \quad j = 1,\ldots,7\,, \tag{13f}$$

$$0 = \frac{\partial v_c}{\partial z} + \varphi_7(\theta_s,\theta_c,x_c,\Phi_c)\,, \tag{13g}$$

boundary conditions

$$\frac{\partial \theta_s}{\partial z}(t,0) = \frac{\partial \theta_s}{\partial z}(t,1) = 0 \tag{14a}$$
$$\theta_a(t,0) = \theta_{a,\mathrm{in}}(t), \quad g_a(t,0) = g_{a,\mathrm{in}}(t) \tag{14b}$$
$$x_{a,j}(t,0) = x_{a,j,\mathrm{in}}(t)\,, \quad j = 1,\dots,7\,, \tag{14c}$$
$$\theta_c(t,1) = \theta_{c,\mathrm{in}}(g_a(t,1), x_a(t,1), \theta_a(t,1), g_a(t,0), x_a(t,0))\,, \tag{14d}$$
$$x_{c,j}(t,1) = x_{c,j,\mathrm{in}}(g_a(t,1), x_a(t,1), g_a(t,0), x_a(t,0))\,,\ j = 1,\dots,7\,, \tag{14e}$$
$$g_c(t,1) = g_{c,\mathrm{in}}(g_a(t,1), x_a(t,1), g_a(t,0), x_a(t,0))\,, \tag{14f}$$

and initial conditions

$$\theta_s(0,z) = \theta_{s,0}(z), \ \ \theta_a(0,z) = \theta_{a,0}(z), \ \ \theta_c(0,z) = \theta_{c,0}(z), \tag{15a}$$
$$x_{a,j}(0,z) = x_{a,j,0}(z), \ \ x_{c,j}(0,z) = x_{c,j,0}(z)\,, \quad j = 1,\dots,7\,. \tag{15b}$$

Although the boundary conditions are slightly more complicated compared to example 1, the result $\nu_t = 1$ still holds [1].

*Example 3.* A more detailed 1D family of dynamic models of MCFCs: Find additionally potential differences $\Phi_a$, $\Phi_e$, $\Phi_c$ and cell voltage $V(t)$ s.t. to a singular PDE consisting of (13a-13g, 14a-14f, 15a-15b ) and

$$\frac{\partial \Phi_a}{\partial t} = \big(i - i_a(\theta_s, x_a, \Phi_a)\big)/c_a\,, \tag{16a}$$
$$\frac{\partial \Phi_e}{\partial t} = -\big(i - i_e(\Phi_e)\big)/c_e\,, \tag{16b}$$
$$\frac{\partial \Phi_c}{\partial t} = -\big(i - i_c(\theta_s, x_c, \Phi_c)\big)/c_c\,, \tag{16c}$$
$$\frac{\mathrm{d}V}{\mathrm{d}t} = \Big(\int_0^1 i(t,\bar z)\,\mathrm{d}\bar z - I_{\mathrm{cell}}(t)\Big)/c_v\,, \tag{16d}$$
$$0 = -\Phi_a(t,z) + \Phi_e(t,z) + \Phi_c(t,z) - V(t) \tag{16e}$$

and initial conditions

$$\Phi_a(0,z) = \Phi_{a,0}(z),\ \Phi_e(0,z) = \Phi_{e,0}(z),\ \Phi_c(0,z) = \Phi_{c,0}(z),\ V(0) = V_0. \tag{17}$$

Twice partial differentiating (16e) with respect to $t$ and substituting the r.h.s. of (16a–16d) each time is necessary to get an equation where $\frac{\partial i}{\partial t}$ appears:

$$\beta\,\frac{\partial i}{\partial t}(t,z) + \int_0^1 \frac{\partial i}{\partial t}(t,\bar z)\,\mathrm{d}\bar z + \psi(\theta_{s,zz}, \theta_{a,z}, \dots, \Phi_{c,z}, \theta_s, \dots, \Phi_c, \frac{\mathrm{d}I_{\mathrm{cell}}}{\mathrm{d}t}) = 0 \tag{18}$$

with a suitable function $\psi$ and $\beta := \frac{c_v}{c_a} + \frac{c_v}{c_e} + \frac{c_v}{c_c}$. Considering $t$ as a parameter, this is a linear Fredholm integral equation of second kind for the function $\xi(z) := \frac{\partial i}{\partial t}(z,t)$. The associated homogeneous integral equation

$$\beta\,\xi(z) + \int_0^1 \xi(\bar z)\,\mathrm{d}\bar z = 0 \tag{19}$$

has the unique solution $\xi(z) \equiv 0$, since $\beta \neq -1$ for the given data. Therefore Eq. (18) is uniquely and continuously solvable for $\frac{\partial i}{\partial t}$. As a result we obtain time index $\nu_t = 2$ [1].

*Example 4.* Simple 2D family of dynamic models of MCFCs: Find functions $u(t,z)$ (temperature of the solid), $v(t,z) > 0$, $\bar{v}(t,z) > 0$ (temperature times molar flow of the gas in the anode and cathode gas channel) and $w(t,z)$, $\bar{w}(t,z)$ (temperature and molar fractions of the gas in the anode and cathode gas channel) s.t. the singular PDE (with positive constant $\lambda$)

$$\begin{aligned} u_t &= \lambda \Delta u + \psi_1(u,w,v) && && (20a)\\ w_t &= -v w_{z_1} + \psi_2(u,w,v) && && (20b)\\ 0 &= v_{z_1} + \psi_3(u,w,v) && \text{on } [0,T] \times [0,1]^2 && (20c)\\ \bar{w}_t &= \bar{v} \bar{w}_{z_2} + \psi_4(u,\bar{w},\bar{v}) && && (20d)\\ 0 &= \bar{v}_{z_2} + \psi_5(u,\bar{w},\bar{v}) && && (20e) \end{aligned}$$

with initial conditions $u(0,z) = g_1(z)$, $w(0,z) = g_2(z)$, $\bar{w}(0,z) = g_4(z)$ and boundary conditions $\frac{\partial u}{\partial n}|_{\partial\Omega} = 0$, and $w(t,0,z_2) = w_{\text{west}}(t,z_2)$, $v(t,0,z_2) = v_{\text{west}}(t,z_2)$, $\bar{w}(t,z_1,1) = \bar{w}_{\text{north}}(t,z_1)$, $\bar{v}(t,z_1,1) = \bar{v}_{\text{north}}(t,z_1)$.

A slight change in the notation of example 1 yields time index $\nu_t = 1$.

The perturbation index of a linearized version of this PDAE is computed in [4].

*Acknowledgement.* This work is supported by the German Federal Ministry of Education and Research (BMBF) within the project Optimierte Prozessführung von Brennstoffzellensystemen mit Methoden der Nichtlinearen Dynamik.

## References

1. K. Chudej, P. Heidebrecht, V. Petzet, S. Scherdel, Schittkowski, K., H.J. Pesch, and K. Sundmacher. Index analysis and numerical solution of a large scale nonlinear PDAE system describing the dynamical behaviour of molten carbonate fuel cells. *Z. Angew. Math. Mech.*, accepted for publication (2004).
2. P. Heidebrecht. Modelling, analysis and optimisation of a molten carbonate fuel cell with direct internal reforming (DIR-MCFC). Dissertation, Otto-von-Guericke-Universität Magdeburg, Magdeburg, 2004.
3. W. Lucht and K. Debrabant. On quasi-linear PDAEs with convection: Applications, indices, numerical solution. *Applied Numerical Mathematics*, 42:297–314, 2002.
4. J. Rang and K. Chudej. A perturbation index for a singular PDE model of a fuel cell. Report, Technische Universität Clausthal, 2004.
5. K. Sternberg, K. Chudej, and H.J. Pesch. *Molten Carbonate Fuel Cell: Simulation and Optimization of a Partial Differential-Algebraic Dynamical System.*

# On the Modeling of the Phase Separation of a Gelling Polymeric Mixture

F.A. Coutelieris, G.A.A.V. Haagh, W.G.M. Agterof, and J.J.M. Janssen

Unilever Food Research Center, Oliver van Noortlan 120, 3130 AC, Vlaardingen, The Netherlands `jo.janssen@unilever.com`

**Summary.** The gelation of polymer mixtures under constant cooling rate has been found to be an attractive product structuring mechanism for the food industry. As applications become wider, a predictive method for the process is warranted. To this end, we apply the so-called '$S_\gamma$ concept' in a CFD module for the modeling for microstructure formation of gelling mixtures, where moments of the particle size distribution are evaluated using the local flow conditions as obtained from CFD simulations for the processes considered. The major driving force for these processes is the competition between phase separation, gelation and hydrodynamic phenomena such as break-up and coalescence. Based on theoretical investigations, analytical expressions for the source terms representing the hydrodynamics (break up and coalescence of the droplets) as well as the gelation process were produced. Constitutive models are developed to incorporate the effects of phase separation and gelation on the rheology of the phases. The simulations for different cooling rates clarified the inter-relationships between the competitive mechanisms by depicting the time interval of the domination of each.

## 1 Introduction

Modeling of phase separation is usually based on the Flory-Huggins theory while the kinetics of phase separation is often reasoned from Cahn-Hilliard theory [6, 3]. Actually, there is a lack of models for gelling systems. This study aims at the description of the phase separation of polymer mixtures in an inhomogeneous flow what is the situation under practical conditions. The approach we take is the so-called '$S_\gamma$ concept' where an arbitrary number of moments is used to describe the drop size distribution [8, 2, 9]. The essence of the method is that the evolution of the moments of a distribution can be analyzed using a transport equation consisting of a convective term, which can be coupled to the local flow characteristics through the source terms. Since no experimental data are available for gelling systems under inhomogeneous flow, the results and the relative discussion could be actually considered as a demonstration of the abilities that our analysis presents.

## 2 Theory

The domain (particle) size distribution can be described by a collection of moments of the distribution as

$$S_\gamma = n \int_0^\infty d^\gamma \, P(d) \, \mathrm{d}d, \tag{1}$$

where $n$ is the total number density and $P(d)$ is the size distribution of the droplets. In accordance to experimental observations [1], the droplet size often often follows log-normal distribution. The main advantage of the $S_\gamma$ function is that they satisfy the transport equations [8]:

$$\frac{\partial S_\gamma}{\partial t} + \underline{u} \cdot \nabla S_\gamma = s_i \tag{2}$$

where $\underline{u}$ is the local velocity vector in the process equipment, which can be obtained from the CFD flow calculations and $s_i$ denotes the source terms which represent the change in the particle size distribution as a result of local phenomena such as break-up, coalescence and particle growth as a consequence of the phase separation. Thus, the source term of (2) can be expressed as

$$s_i = s_{br} + s_{cl} + s_{gr} \tag{3}$$

where $s_{br}$, $s_{cl}$ and $s_{gr}$ are the respective source terms that can be modeled explicitly. The break-up source term is given as in [2],

$$s_{br} = \int_0^\infty \left[ \frac{d^\gamma}{\tau_{br}(d)} \left( N_f(d)^{\frac{3-\gamma}{3}} - 1 \right) \right] n \, P(d) \, \mathrm{d}d \quad \text{for } d > d_{cr}, \tag{4}$$

where $d_{cr}$ denotes a critical diameter as determined by the critical capillary number for laminar flow. These relationships depend on the viscosity ratio and the flow type, as has been discussed extensively [7].

By considering the change in $S_\gamma$ due to a single coalescence event, $\Delta S_\gamma^{cl}(d, d')$, between two droplets of diameters $d$ and $d'$, respectively, and using the Smoluchowski collision rate, the coalescence source term is given as ([2, 4]):

$$s_{cl} = \left( 2^{\gamma/3} - 2 \right) \left( \frac{6\varphi}{\pi} \right)^2 k_{coll} \, u_{rel}(d_{eq}) \, P_{coal}(d_{eq}) \, d_{eq}^{\gamma-4} \tag{5}$$

By assuming that the phase separation inside the binodal is due to spinodal decomposition, leaving nucleation and growth out of consideration and that the temperature decrement is the driving force for phase separation and gelation, the growth source term can be written as:

$$s_{gr} = d^{\gamma-3} \exp(\frac{\gamma^2 \hat{\sigma}^2}{2}) \frac{6}{\pi} \frac{d\varphi}{dt} - \gamma \left( H_0 t^{-\frac{2}{3}} \right) S_{\gamma-1} \tag{6}$$

Based on the well-known Flory model [6], the gelation kinetics were modeled accordingly to [5]. In brief, the reversible cross-link model reduces to a reversible dimerisation of individual cross-links depending on the functionality of each polymer, $f$, on the fraction of sites that can form cross-links, $\chi$, as well as on the initial number of moles of polymer per unit volume, $N$. The kinetics of the gel fraction can be described as a binary chemical reaction:

$$\frac{d\chi}{dt} = k_f N f (1 - \chi)^2 - k_b \chi \tag{7}$$

where $k_f$ is the forward rate for cross-link formation and $k_b$ is the backward rate for dissociation.

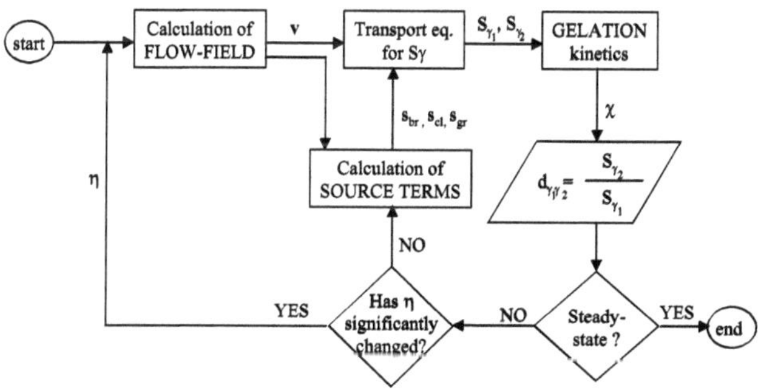


**Fig. 1.** Flowchart of the simulation algorithm

## 3 Results and Discussion

The flowchart of the simulation algorithm is shown in Fig. 1. The mixture gelatin-dextran-water is used as atypical example of the polymeric mixture of interest. To overcome the complications of inhomogeneous flow conditions in this initial study, a simple shear flow between two parallel plates of distance $0.1m$ has been considered. Initially, the flow (of velocity 0.1m/sec) is isothermal at 45°C and a homogeneous shear flow is present at a rate of shear of 1 s-1. At the same time, both the upper and the lower wall are cooled at a specified cooling rate to 20°C. The gelling system has been modeled for two different cooling rates: a high (30°C/min) and a low one (3°C/min). Since the temperature decrement is the driving force for the phase separation and gelation processes, the cooling rate influences significantly the time scale of the processes, and, therefore, on the composition of the mixture and the corresponding rheological properties.

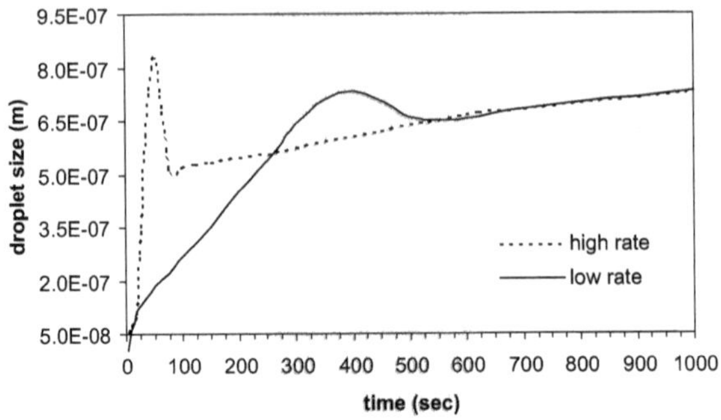


**Fig. 2.** Droplets' size ($d_{32}$) distribution for high and low cooling rates

The time evolution of the droplet size is presented in Fig. 2. The effect of the cooling rate on the mixture's microstructure is clear at the initial time period where a rapid increment of the droplets' size can be observed. Local maxima followed by local minima were observed for both the high and the cooling rates due to the competition between the hydrodynamic phenomena and the gelation. Finally, the droplets' size increases monotonically after the local minimum. This dynamical behavior of the system is clarified in Fig. 3a, where the relative significance of the source terms is shown for the case of high cooling rate. The source terms have been normalized by the summation of them at each time step, in order to be directly comparable. For the fast cooling rate, one can observe that, during the first 50 sec, the coalescence dominates and, therefore, the droplets' size increases presenting its local maximum at the same time. As the cooling effect terminates at $t = 50$ sec, the break up becomes significantly competitive to the coalescence for a period of about 20 sec and, finally, it dominates up to $t = 75$ sec, where the local minimum of the droplets size is presented. Then, grow-up starts to become competitive and it dominates after $t = 120$ sec, where a monotonic decrement of the domain size is presented. The competition between phase separation and gelation is responsible for the progressive weakening of the break up process, corresponding directly to a more smooth increment of the domain size. The above mechanism is further evaluated by the observations for the low cooling rate (Fig. 3b).

## 4 Conclusion

In the present work, the $S_\gamma$ concept has been applied to predict the microstructure formation in gelling polymer mixtures, which is governed by the competition between phase separation, gelation and hydrodynamic phenomena such as break-up and coalescence. The phase separation has been modeled

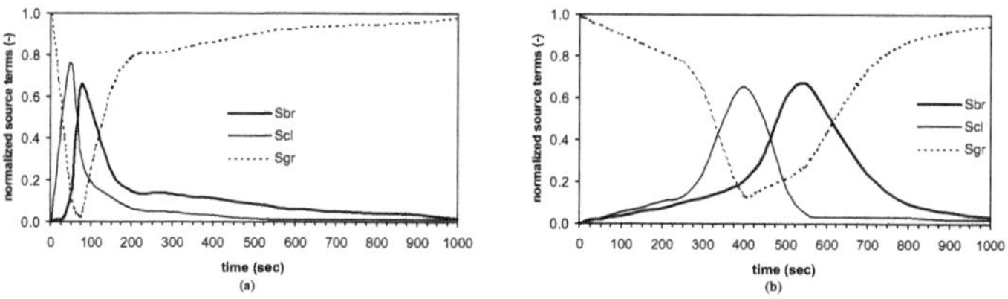


**Fig. 3.** Relative significance of source terms for high (a) and low (b) cooling rates

as spinodal decomposition, which affects the drop size distribution through the growth of domain sizes during spinodal decomposition and the increase in the volume fraction of the dispersed phase. Gelation has been modeled by using a reversible gelation kinetics description. Both phase separation and gelation significantly affect on the droplets' size, which is also influenced by the quench rate. Simulations for a shear flow that is cooled at two specified cooling rates have been carried out to demonstrate the competition between phase separation, gelation, and hydrodynamics (break-up and coalescence) in gelling two-phase systems.

## References

1. *Encyclopedia of Emulsion Technology*, volume 1. Marcel Dekker, 1983.
2. W.G.M. Agterof, G.E.J. Vaessen, G.A.A.V Haagh, J.K. Klahn, and J.J.M. Janssen. Prediction of emulsion particle sizes using a computational fluid dynamics approach. *Colloid Surface B*, 31:141–148, 2003.
3. J. W. Cahn and J. E. Hilliard. Free energy of a nonuniform system. i. interfacial free energy. *J. Chem. Phys.*, 28:258–267, 1958.
4. A.K. Chesters. The modeling of coalescence processes in fluid-liquid dispersions. *Chem. Eng.*, 69:259–270, 1991.
5. A.H. Clark. Biopolymer gelation—comparison of reversible and irreversible cross-link descriptions. *Polymer Gels Networks*, 1:139–158, 1993.
6. P.J. Flory. *Principles of Polymer Chemistry*. Cornell University Press, 1953.
7. H.P. Grace. Dispersion phenomena in high-viscosity immiscible fluid systems and application of static mixers as dispersion devices in such systems. *Chem. Eng. Commun*, 14(3-6):225–277, 1982.
8. A.M. Kamp, A.K. Chesters, C. Colin, and J. Fabre. Bubble coalescence in turbulent flows: A mechanistic model for turbulence-induced coalescence applied to microgravity bubbly pipe flow. *Int. J. Multiphase Flow*, 27:1363–1396, 2001.
9. J.K. Klahn, J.J.M. Janssen, G.E.J. Vaessen, Swart R. de, and W.G.M. Agterof. On the escape process during phase inversion of an emulsion. *Colloid Surface A*, 210:167–181, 2002.

# Iso-Surface Analysis of a Turbulent Diffusion Flame

B.J. Geurts[1,2]

[1] Mathematical Sciences, Faculty EEMCS, University of Twente, P.O. Box 217, 7500 AE Enschede, The Netherlands b.j.geurts@utwente.nl
[2] Fluid Dynamics Laboratory, Department of Applied Physics, Eindhoven University of Technology, P.O. Box 513, 5300 MB Eindhoven, The Netherlands

**Summary.** We analyze the evolution of a diffusion flame in a turbulent mixing layer. The location of the flame-center is defined by the "stoichiometric" interface. Geometrical properties such as its surface-area, wrinkling and curvature are characterized using an accurate numerical level-set quadrature method. This allows to quantify flame-properties as well as turbulence modulation effects due to coupling between combustion and turbulent transport. We determine the active flame-region which is responsible for the main part of the chemical processing in the flame.

**Key words:** Turbulence, combustion, iso-surface analysis, flame properties.

## 1 Introduction

In various combustion processes turbulent diffusion flames arise. These are characterized by a thin, distorted and lively evolving region where the conditions for combustion, such as presence of chemical species at appropriate concentration and temperature, are fulfilled. We will consider combustion in a turbulent mixing layer with stylized chemical reaction process. This model can be treated in full detail and provides an impression of the dominant turbulence modulation that arises from the coupling between the fluid-flow and the chemical reaction equations. Important global flame-properties will be quantified in detail by applying a new iso-surface quadrature method.

The central region of a diffusion flame may be visualized by monitoring the so-called "stoichiometric" interface. In a turbulent flow this interface develops into a complex, highly wrinkled surface. Fundamental properties such as the flame's surface-area and its wrinkling can be appreciated roughly by visual inspection. However, in order to become meaningful, a quantitative method of analysis is required. In this paper we will apply a new method for numerical integration over complex level-sets and show that an accurate impression of these properties and trends associated with variations in physical parameters can be obtained. Access to these fundamental aspects can be used as underpinning of theoretical and modeling studies aimed at better understanding of the combustion process or to allow more complex flames to be simulated.

The organization of this paper is as follows. In Section 2 we will introduce the model used to describe a diffusion flame in a temporal mixing layer. Section 3 is devoted to a description and application of the method of iso-surface analysis of the flame-center and the determination of the active flame region is discussed together with some concluding remarks.

## 2 Diffusion flame in a mixing layer

In this section we will introduce the mathematical model describing the flame problem studied in this paper. Subsequently, we will introduce the temporal mixing layer [3] and visualize the evolution of the flame.

The computational model is composed of the compressible flow equations for ideal gases, coupled to a system of advection-diffusion-reaction equations. The dimensionless equations can be expressed as

$$\partial_t \rho + \partial_j(\rho u_j) = 0 \tag{1}$$
$$\partial_t(\rho u_i) + \partial_j(\rho u_i u_j) + \partial_i p - \partial_j \sigma_{ij} = 0 \quad ; \quad i = 1, \ldots, 3 \tag{2}$$
$$\partial_t e + \partial_j((e+p)u_j) - \partial_j(\sigma_{ij} u_i) + \partial_j q_j - h_k \omega_k = 0 \tag{3}$$
$$\partial_t(\rho c_k) + \partial_j(\rho c_k u_j) - \partial_j(\pi_{kj}) - \omega_k = 0 \quad ; \quad k = 1, \ldots, N_s \tag{4}$$

where $\rho$ denotes the fluid mass-density, $u_i$ the velocity in the $x_i$ direction, $e$ the total energy density, $c_k$ the $k$-th chemical species concentration and $N_s$ the number of species respectively. In order to close this system of equations, additional constitutive equations need to be provided. The viscous fluxes are specified by $\sigma_{ij} = S_{ij}/Re$ and $\pi_{kj} = \partial_j c_k/(ReSc)$ with rate of strain tensor given by $S_{ij} = \partial_i u_j + \partial_j u_i - (2/3)\delta_{ij}\partial_k u_k$. The Reynolds ($Re$) and Schmidt ($Sc$) numbers characterize the strength of the viscous fluxes relative to the nonlinear convective contributions. The equation of state specifies the pressure $p$ through $e = p/(\gamma-1) + \rho u_i u_i/2$ where $\gamma \approx 7/5$. Finally, the heat flux vector $q_j = -\partial_j T/\{(\gamma-1)RePrM^2\}$ where $Pr$ is the Prandtl number, $M$ the Mach number and the temperature $T$ follows from the ideal gas law $\rho T = \gamma M^2 p$. We will use $Re = 50$, $M = 0.2$, $Pr = 1$ and consider different values for $Sc$ in the sequel.

The source terms in the species and energy equation represent the chemical processes that take place. The chemical reactions are characterized by reaction rates $\omega_k$ and the heat released in these reactions is given by $h_k\omega_k$ in which $h_k$ is the specific enthalpy associated with species $k$. The chemical reaction rate $\omega_k$ is assumed to be determined by an Arrhenius law. We will consider a single reaction in which fuel $F$ reacts with oxidizer $O$ to yield product $P$: $F + O \rightarrow P$. For this particular reaction we may express the reaction-rates as [1]

$$\omega_1 = -(\rho c_F)(\rho c_O) Da_O \exp(-\frac{Ze}{T}) \; ; \; \omega_2 = \alpha\omega_1 \; ; \; \omega_3 = -\omega_1(1+\alpha) \quad (5)$$

where we introduced the Zeldovich number $Ze$, put $Da_O = Da/W_O$ with $Da$ the Damköhler number and $W_i$ the molecular weight of species $i$. In addition, $\alpha = W_O/W_F$ and use was made of $W_P = W_F + W_O$. The source term for the energy equation may be written as

$$h_j\omega_j = h_1\omega_1 + \alpha h_2\omega_1 - (1+\alpha)h_3\omega_1 = \omega_1(h_1 + \alpha h_2 - (1+\alpha)h_3) = Q\omega_1 \quad (6)$$

where $Q$ will be referred to as the effective standard enthalpy of formation. In total this model requires four additional parameters: $Da_O$, $Ze$, $Q$ and $\alpha$. We will assume $\alpha = 1$, $Ze = 1$, $Da_O = 1$ and study variations in $Q$.

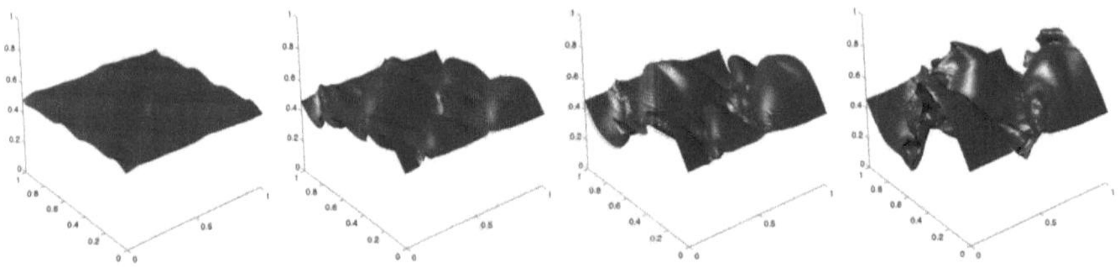

**Fig. 1.** Evolving stoichiometric surface $c_F - c_O = 0$ in a turbulent mixing layer. The snapshots are taken at $t = 15, 35, 55, 75$ (from left to right).

The consequences of combustion on turbulence may be illustrated with the canonical flow in a temporal mixing layer. In this flow two parallel fluid streams with different velocities merge and rapidly mix [3]. Initially, we consider the upper stream to contain fuel ($c_F = 1, c_O = 0$) and the lower stream to contain oxidizer ($c_O = 1, c_F = 0$). We adopt explicit Runge-Kutta time-stepping and finite volume discretization. The 'center' of the flame is defined through the "stoichiometric surface" $c_F - c_0 = 0$ as shown in Fig. 1.

## 3 Iso-surface analysis of turbulent flame properties

To quantify basic properties of an evolving diffusion flame we concentrate on "global" variables, such as the flame-area or wrinkling. The global variable corresponding to a density function $f$ and a level-set $S(a,t)$ is defined as

$$I_f(a,t) = \int_{S(a,t)} dA \; f(\mathbf{x},t) = \int_V d\mathbf{x} \; \delta(F(\mathbf{x},t) - a)|\nabla F(\mathbf{x},t)| f(\mathbf{x},t) \quad (7)$$

where $V$ is a fixed and arbitrary volume which encloses the level-set $S(a,t)$ defined as the set where $F(\mathbf{x},t) = a$ for a "level-function" $F$. The formulation in (7) was used as the basis of the numerical quadrature method in [2].

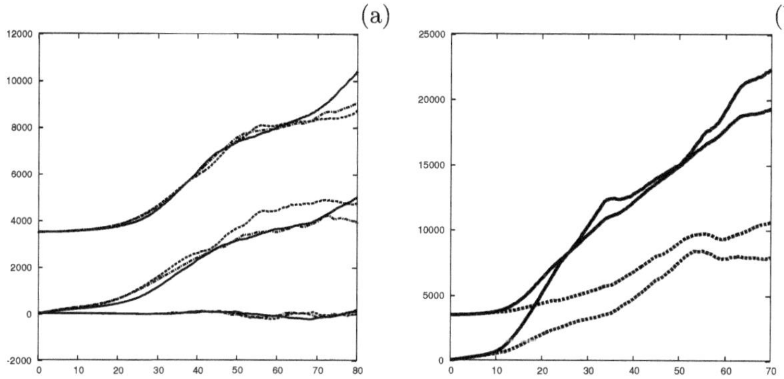


**Fig. 2.** Evolution of $A$ (top), $W$ (middle) and $C$ (bottom) for the flame in Fig. 1 at $32^3$ (solid), $64^3$ (dashed) and $96^3$ (dash-dot) with $Sc = 10$ and $Q = -1$ (a). Heat-release variations (b): solid: $Q = 0$, dashed: $Q = -100$ at $Sc = 50$.

For a diffusion flame the level-set function $F = c_F - c_O$. To determine the surface-area $A$ of the flame we adopt $f_A = 1$. A measure for the global "curvature" $C$ is obtained using as density

$$f_C(\mathbf{x}, t) = \nabla \cdot \mathbf{n} \;\; ; \;\; \mathbf{n}(\mathbf{x}, t) = \frac{\nabla F(\mathbf{x}, t)}{|\nabla F(\mathbf{x}, t)|} \tag{8}$$

where $\mathbf{n}$ denotes the unit normal on the flame surface. The 'wrinkling' $W$ is obtained using $f_W(\mathbf{x}, t) = |\nabla \cdot \mathbf{n}|$. In Fig. 2(a) we show estimates for $A$, $C$ and $W$ obtained at different resolutions. Already at a resolution of $96^3$ an acceptable accuracy is obtained which was further confirmed by results on finer grids. In Fig. 2(b) we varied the heat release parameter $Q$ and Schmidt number $Sc$. Evidently, a strong heat release induces a significant reduction in the area and wrinkling of the flame.

Motivated by the interpretation of the stoichiometric surface, we may introduce a "thick" active flame region around this surface, defined by $S(a) = \{\mathbf{x} \in R^3 | \; |c_F - c_O| \leq a\}$ in which the parameter $a$ is referred to as the stoichiometric interval width. The fuel processing-rate $\Gamma_F$ associated with $S(a)$ is

$$\Gamma_F(a, t) = \int_{-a}^{a} ds \int_{c_F - c_O = s} dA \; \omega_F(\mathbf{x}, t) \tag{9}$$

The processing rate $\Gamma_F$ arises mainly from nearby iso-surfaces $c_F - c_O = s$ where $s$ runs from $-a$ to $a$. When $a$ increases $\Gamma_F$ increases as well with a maximum at $a = 1$. This allows to define the $\varepsilon$-flame-region by $\Gamma_F(a, t)/\Gamma_F(1, t) = \varepsilon$ from which $a(\varepsilon, t)$ may be solved.

In Fig. 3 we collected the evolution of the stoichiometric interval width $a$. After the transitional stages a fairly constant value of $a$ defines the active flame region. The corresponding physical space region increases with time, as shown in Fig. 4.

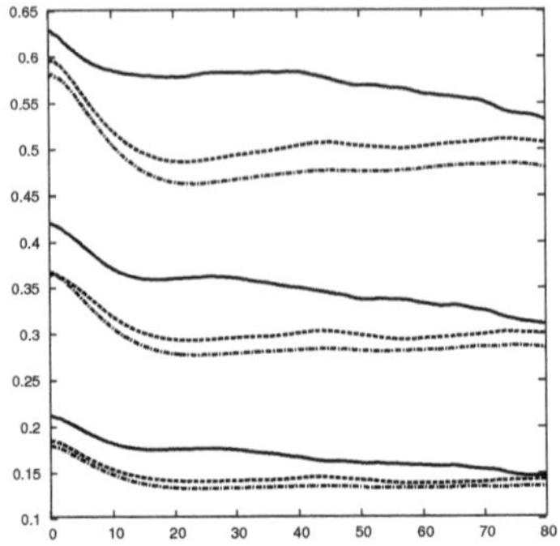


**Fig. 3.** Effective flame region corresponding to 75 % (top), 50 % (middle) and 25 % (bottom) of the total processing rate on $32^3$ (solid), $64^3$ (dashed) and $96^3$ (dash-dot).

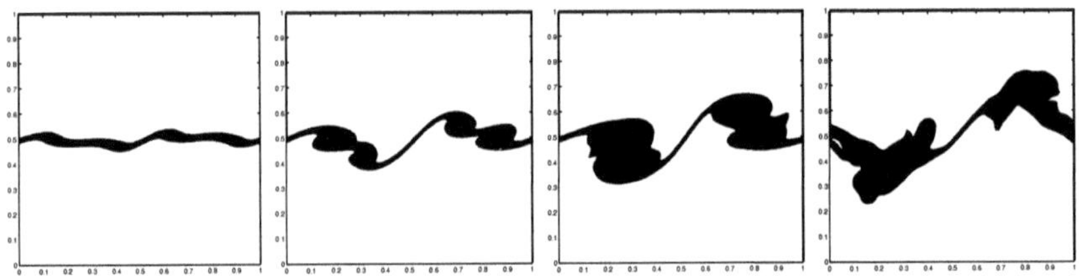

**Fig. 4.** Active flame region corresponding to 50% of the total combustion. A characteristic slice is shown at $t = 15, 35, 55, 75$ (from left to right).

In summary, we introduced a simple combustion model and studied a turbulent diffusion flame in a temporal mixing layer. The coupling between the combustion and the turbulent transport induces a significant modulation of the turbulent flow properties, e.g., characterized by a strongly reduced spreading rate of the mixing layer. Using a new method for integration over geometrically complex evolving level-sets, basic properties such as flame-area, wrinkling, curvature and active flame region were quantified.

*Acknowledgement.* The author would like to acknowledge fruitful discussions with Philip de Goey, Rob Bastiaans and Jeroen van Oijen of the Combustion Technology Group at Eindhoven University of Technology.

## References

1. R.J.M. Bastiaans, L.M.T. Somers, and H.C. de Lange. *DNS of non-premixed combustion in a compressible mixing layer.* R.T. Edwards Publishers, Philadelphia, 2001.
2. B.J. Geurts. Mixing efficiency in turbulent shear layers. *J. of Turb.*, 2:17, 2001.
3. A.W. Vreman, B.J. Geurts, and H. Kuerten. Large-eddy simulation of the turbulent mixing layer. *J. Fluid Mech.*, 339:357–390, 1997.

# A Simplified Model for Non–Isothermal Crystallization of Polymers

T. Götz[1] and J. Struckmeier[2]

[1] Department of Mathematics, University of Kaiserslautern, Erwin–Schrödinger–Str. 48, D–67663 Kaiserslautern, Germany. goetz@mathematik.uni-kl.de

[2] Department of Mathematics, University of Hamburg, Bundesstr. 55, D–20146 Hamburg, Germany. struckmeier@math.uni-hamburg.de

**Summary.** Recently, Burger and Capasso [$M^3AS$ **11** (2001) 1029–1053] derived a coupled system of partial differential equations to describe non–isothermal crystallization of polymers. The system is based on a spatial averaging of the underlying stochastic birth–and–growth process describing the nucleation and growth of single crystals. Using an appropriate scaling of the original system, we derive a simplified model which only consists of a reaction–diffusion equation with memory for the underlying temperature, such that the degree of crystallization can be explicitly given by a time integration of the temperature–dependent growth and nucleation rate. Numerical simulations indicate that the reduced model shows at least qualitatively the same behavior like the original model.

**Key words:** Crystallization of polymers, scaling properties, temperature equation with memory.

## 1 Introduction

The control and optimization of polymer crystallization in industrial applications requires an appropriate understanding of the physical–chemical phenomena occurring during the process. Hence, the mathematical modeling and simulation is an important task and there exists a large variety of different approaches in this direction. A recent overview on such models can be found in [2].

In the present work we are concerned with a pure deterministic model for non–isothermal polymer crystallization recently proposed by Burger and Capasso in [1]. The model consists of a system of partial differential equations for the crystalline volume fraction $\xi = \xi(x, t)$, the mean free surface distributions $v = v(x, t)$ and $w = w(x, t)$ of crystals as well as the underlying temperature field $\theta = \theta(x, t)$ in a two–dimensional domain.

In dimensionless form the system reads as

$$\dot{\theta} = \kappa \, \Delta\theta + L\dot{\xi} \tag{1a}$$
$$\dot{\xi} = \gamma \, (1-\xi) \, a(\theta) \, v \tag{1b}$$
$$\dot{v} = \lambda \, \mathrm{div} \, (a(\theta) \, w) + \delta \, a(\theta) \, b(\theta) \tag{1c}$$
$$\dot{w} = \lambda \, \nabla(a(\theta) \, v) \tag{1d}$$

subjected to the initial

$$\theta(x, 0) = \theta_0(x) \tag{2a}$$
$$\xi(x, 0) = 0 \tag{2b}$$
$$v(x, 0) = 0 \tag{2c}$$
$$w(x, 0) = 0 \tag{2d}$$

and boundary conditions

$$-\frac{\partial\theta}{\partial n} = \beta \, (\theta - \theta_{\text{out}}) \tag{3a}$$
$$v = -w^T \cdot n \, . \tag{3b}$$

Inspecting typical values of the parameters for isotactic polypropylene, we obtain the following scalings: $L = \mathcal{O}(1)$, $\kappa/\lambda = \mathcal{O}(1)$, $\gamma/\delta = \mathcal{O}(1)$ and $\kappa/\delta = \varepsilon \ll 1$.

Choosing the time scale of the nucleation process, i.e. $\delta = 1$ and introducing the function $u(x,t) = -\ln(1 - \xi(x,t))$, we obtain the system

$$\dot{\theta} = \varepsilon \Delta\theta + \gamma L e^{-u} a(\theta) v \tag{4a}$$
$$\dot{u} = \gamma a(\theta) v \tag{4b}$$
$$\dot{v} = \varepsilon \, \mathrm{div} \, (a(\theta) w) + a(\theta) \, b(\theta) \tag{4c}$$
$$\dot{w} = \varepsilon \, \nabla(a(\theta) v) \, . \tag{4d}$$

On the other hand, we could also consider the time scale of the diffusion, which means $\kappa = 1$. However, in industrial and technological applications one is usually more interested in the effects related to the nucleation and increase of crystallinity rather than in the mere diffusion process. Therefore we will focus in our subsequent discussion on the nucleation time scale.

## 2 Temperature Equation with Memory

If we define $\Phi(x,t) = (v(x,t), \ w_1(x,t), \ w_2(x,t))^T$ we may formulate (4c) and (4d) as a quasilinear first order hyperbolic system, see also [1]. Assuming $a'(\theta) = \mathcal{O}(1)$ and $\nabla\theta = \mathcal{O}(1)$ we can solve (4c) and (4d) using the method of characteristics and obtain

$$v(x,t) = \int_0^t a(\theta)b(\theta)\,ds + \mathcal{O}(\varepsilon^2) \qquad \text{and} \qquad w(x,t) = \mathcal{O}(\varepsilon)\,. \tag{5}$$

Integrating (4b) together with the initial condition (2b) and substituting the results into (4a) yields (up to higher order terms in $\varepsilon$) the following temperature equation with memory

$$\dot{\theta} = \varepsilon \Delta\theta + \gamma L e^{-u} a(\theta) \int_0^t a(\theta)b(\theta)\,ds \tag{6}$$

where $u = u(x,t)$ is now defined by

$$u(x,t) = \gamma \int_0^t a(\theta) \left( \int_0^s a(\theta)b(\theta)\,d\tau \right) ds\,. \tag{7}$$

Hence, for $\varepsilon \ll 1$, the initial–boundary value problem given by the system (1)–(3) may be substituted by the single reaction–diffusion equation (6) with memory together with (7) and initial and boundary conditions (2a) and (3a), respectively.

Similar models have also been derived by Kolmogoroff and Avrami, see [3] and references therein.

## 3 Numerical Results

As numerical scheme to solve (6) we use a first order explicit time integration together with a standard 5–point stencil for the Laplace operator. To ensure stability of the scheme and non–oscillating modes in the numerical approximates we should satisfy the condition $\varepsilon^2 k/h^2 < 1/8$, where $k$ and $h$ denote the step size in time and space, respectively. Because we apply a first order time integration the time step is chosen much smaller than given by the condition above in order to obtain a sufficiently accurate integration of the source term in the temperature equation.

For the growth and nucleation rates we use $a(\theta) = b(\theta) = \exp\left[-\kappa(\theta - \theta_{\text{ref}})\right]$ which describes the temperature dependence of the growth and nucleation rate observed in experiments at least qualitatively, see [4].

We perform two different simulations on the rectangle $[0,1] \times [0,2]$: in the first one we use the constant cooling temperature $\theta_{\text{out}} = 0$, in the second one the cooling temperature is $\theta_{\text{out}} = 0$ on the left and upper as well as $\theta_{\text{out}} = 1$ on the right and lower boundary of the rectangle. The initial temperature is homogeneous on the rectangle, *i.e.* $\theta_0(x) = 2$, and we use the parameters $L = 1/3$, $\beta = 10$, $\kappa = 3$ and $\theta_{\text{ref}} = 1/2$. The step size in space is given by $h = 1/40$, the time step equal to $k = 0.16$ and, finally, $\varepsilon = 10^{-4}$.

Figures 1 and 2 show the temperature and crystalline volume fraction for the two different profiles of the cooling temperature mentioned above. In both cases one observes a sharp front in the crystalline volume fraction moving in

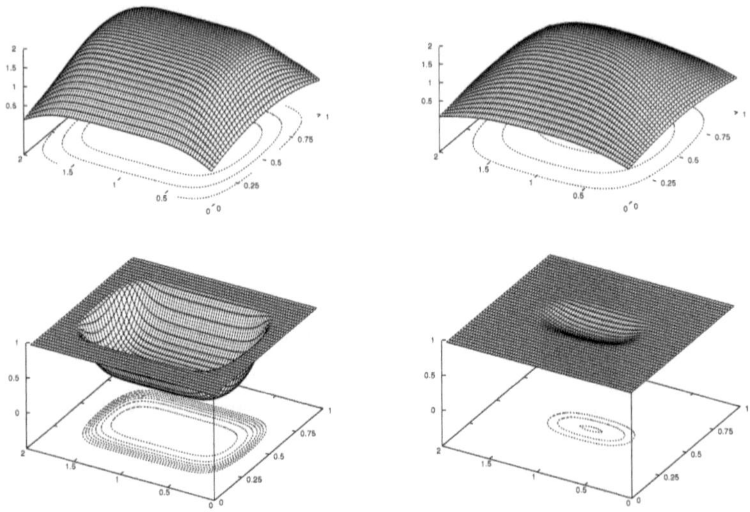

**Fig. 1.** Temperature (up) and crystalline volume fraction (down) for $t = 400$, 800.

time from the boundary into the interior domain. Like in [1] we do not observe such a moving front in the temperature fields, which indicates that the cooling at the boundary dominates the effect of latent heat during the crystallization process and confirms the validity of our asymptotic induced model reduction.

The influence of a non–uniform cooling temperature along the boundary is clearly indicated comparing the results shown in Figs. 1 and 2. Whereas the results in Fig. 1 seem to be completely symmetric with respect to the line $y = 1$, the non–uniform cooling temperature yields a shift of the higher crystalline volume fraction toward the lower cooling temperature at two boundary segments.

# 4 Conclusion

In the previous sections we discussed a deterministic model to describe the crystallization process of polymers. Referring to a recent work of Burger and Capasso we reconsidered the scaling properties of the model. Our main result is the reduction of the original model to a single reaction–diffusion equation with memory for the underlying temperature field. In the reduced model the crystalline volume fraction is obtained by integrating growth and nucleation rate over the temperature history. Our numerical results showed that the reduced model shows at least qualitatively the same behavior like the original model.

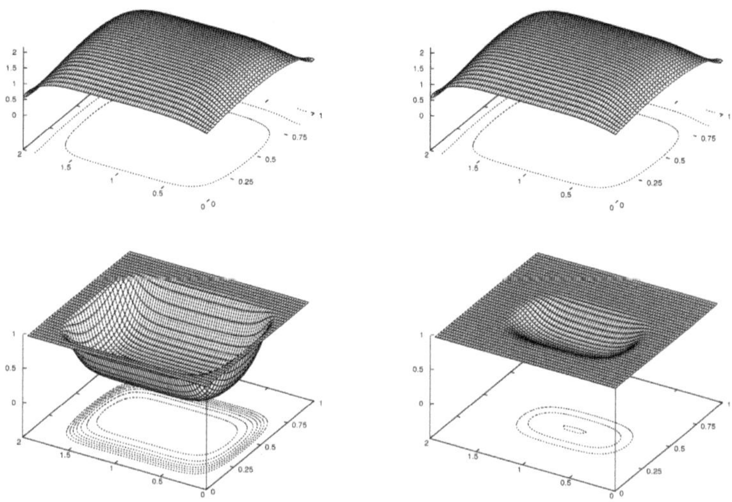

**Fig. 2.** Temperature (up) and crystalline volume fraction (down) for $t = 400$, 800.

A major goal in the mathematical modeling of polymer crystallization is the computation of an optimal control, in our case the cooling temperature along the boundary of the spatial domain, such that the crystallization is as uniform as possible. Here one may use our simplified model which may reduce the theoretical as well as numerical work when applying optimization strategies for partial differential equations. Results on the optimal control problem are currently under investigation.

*Acknowledgement.* The authors would like to thank Vincenzo Capasso for useful discussions on the subject.

## References

1. M. Burger and V. Capasso. Mathematical modelling and simulation of non–isothermal crystallization of polymers. *Math. Models Meth. Appl. Science*, 11(6):1029–1053, 2001.
2. V. Capasso. *Mathematical Modelling for Polymer Processing.* Springer, Berlin, 2002.
3. G. Eder. Mathematical modelling of crystallization processes as occuring in polymer processing. *Nonlinear Analysis*, 30(6):3807–3815, 1997.
4. E. Ratajski and H. Janeschitz-Kriegl. How to determine high growth speeds in polymer crystallization. *Colloid Polym. Sci.*, 274:938–951, 1996.

# Numerical Simulation of Cylindrical Induction Heating Furnaces

A. Bermúdez, D. Gómez, M. C. Muñiz, and P. Salgado

Depto. de Matemática Aplicada. Universidade de Santiago de Compostela. 15782 Santiago de Compostela. Spain.
mabermud@usc.es, malola@usc.es, mcarmen@usc.es, mpilar@usc.es

**Summary.** The aim of this work is to introduce and numerically solve an axisymmetric mathematical model for thermoelectrical simulation of an induction heating furnace.

**Key words:** Numerical methods, finite elements, induction heating, eddy current, phase change.

## 1 Introduction

The induction heating technique is widely used in the metallurgical industry in an important number of applications, such us metal smelting or purification systems. In general, an induction heating system consists of one or several inductors and metallic workpieces to be heated. The inductors are supplied with alternating current which induces eddy currents and heats the workpiece by means of Joule effect. The overall process is very complex and involves different physical phenomena: electromagnetic, thermal with change of phase and hydrodynamic in the liquid metal. Thus, numerical simulation represents an important tool to optimize the design of induction furnaces and understand their behavior. Indeed, we can find several publications devoted to numerically solving some of the previous problems (see [2] and references therein). From the mathematical point of view, it is needed to solve a coupled non linear system of partial differential equations which arises from a thermal-magneto-hydrodynamic problem.

In this paper, we will focus our attention on the thermoelectrical simulation of a cylindrical induction heating furnace. In particular, we will solve the heat transfer equation in transient state coupled with an eddy current problem. The main contribution with respect to previous thermoelectrical models ([2]) consists in introducing the change of state of the thermal problem, which leads to a nonlinearity in the heat equation.

The outline of the paper is as follows. By assuming cylindrical symmetry, we start describing the thermoelectrical model in a radial section of the domain. Finally, we propose a finite element method for its numerical solution and we present numerical results obtained for an industrial furnace used in silicon purification with our two-dimensional code.

## 2 Mathematical modelling

The induction coil of the furnace is replaced by $m$ cylindrical rings in order to consider the problem in an axisymmetric setting (see Fig. 1). We shall denote by $\Omega_0$ the workpiece to be heated, $\Omega_1$, $\Omega_2$, ..., $\Omega_m$ the turns of the coil and $\Omega_a$ the air around the conductors. This notation refers, in fact, to any radial section of these sets.

### 2.1 The electromagnetic submodel

The electromagnetic model is the so-called *eddy current* model which is obtained from the Maxwell's equations under the assumptions of low-frequency, harmonic regime and no charge density,

$$\mathbf{curl}\,\mathbf{H} = \mathbf{J}, \tag{1}$$

$$i\omega\mathbf{B} + \mathbf{curl}\,\mathbf{E} = \mathbf{0}, \tag{2}$$

$$\operatorname{div}\mathbf{B} = 0, \tag{3}$$

$$\operatorname{div}\mathbf{D} = 0, \tag{4}$$

where, $\mathbf{H}$, $\mathbf{D}$, $\mathbf{J}$, $\mathbf{B}$ and $\mathbf{E}$ are the complex amplitudes associated with the magnetic field, the electric displacement, the current density, the magnetic induction and the electric field, respectively; $\omega$ is the angular frequency. The system above is completed with the constitutive relations $\mathbf{B} = \mu\mathbf{H}$ and $\mathbf{D} = \varepsilon\mathbf{E}$, where $\mu$ is the magnetic permeability and $\varepsilon$ is the electric permittivity. We also need the Ohm's law which sets that $\mathbf{J} = \sigma\mathbf{E}$ inside conductors (where $\sigma$ is the electric conductivity) and $\mathbf{J} = 0$ in air.

We are interested in solving these equations by using a cylindrical coordinate system $(r, \theta, z)$ with the $z-$axis coinciding with the axis of the cylinder. From now on, we assume cylindrical symmetry, i.e. we suppose that none of the fields depends on the angular variable $\theta$. We further assume that the current density field has non-zero component only in the tangential direction $\mathbf{e}_\theta$,

$$\mathbf{J}(r, \theta, z) = J_\theta(r, z)\,\mathbf{e}_\theta. \tag{5}$$

From (3) we deduce that $\mathbf{B} = \mathbf{curl}\,\mathbf{A}$, where $\mathbf{A}$ is the called magnetic vector potential which we choose to be divergence-free. From (1) and (5), we have,

$$\mathbf{A}(r, \theta, z) = A_\theta(r, z)\mathbf{e}_\theta.$$

Then, by using the expressions of **curl** and div operators in cylindrical coordinates, we can write the eddy current problem in terms of $A_\theta$, as follows:

$$-\left(\frac{\partial}{\partial r}(\frac{1}{\mu r}\frac{\partial(rA_\theta)}{\partial r}+\frac{\partial}{\partial z}(\frac{1}{\mu}\frac{\partial A_\theta}{\partial z})\right)+i\omega\sigma A_\theta=\frac{\sigma C_k}{r}\text{ in }\Omega_k,\quad k=0,\ldots,m.$$

$$-\left(\frac{\partial}{\partial r}(\frac{1}{\mu r}\frac{\partial(rA_\theta)}{\partial r}+\frac{\partial}{\partial z}(\frac{1}{\mu}\frac{\partial A_\theta}{\partial z})\right)=0\text{ in }\Omega_a.$$

We notice that $C_k\in\mathbb{C}$, $k=0,\ldots,m$ are actually unknown constants. We refer the reader to [1] for details about the computation of the constants $C_k$ and a detailed deduction of the above equations.

In order to apply a finite element method to numerically solve the previous problem, we shall consider a rectangular box in the $(r,z)$-plane enclosing the induction heating system, and large enough for the magnetic field to be small at the boundaries of the box (see Fig. 1). Thus, the electromagnetic computational domain is $\Omega=\Omega_a\cup\Omega_0\cup\Omega_1\cup\cdots\cup\Omega_m$. The natural symmetry

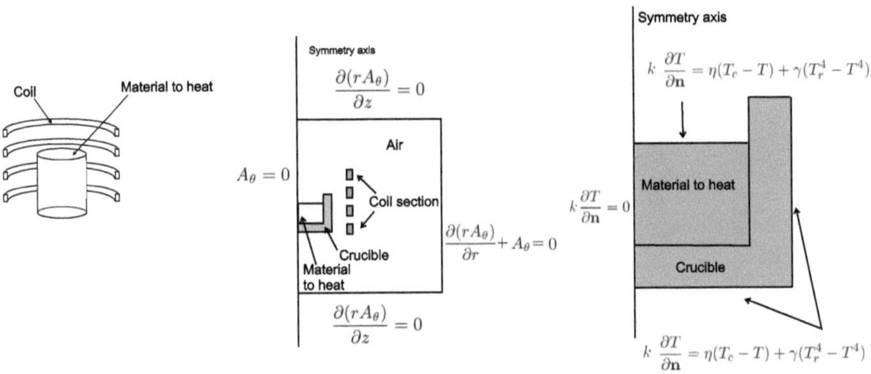


**Fig. 1.** Sketch of the workpiece and the coil. Boundary conditions of the thermo-electrical problem.

condition along the revolution axis $\Gamma_{\mathrm{d}}^{\mathrm{e}}$ is $A_\theta=0$ (see Fig. 1). On the lines which are perpendicular to this axis we impose $\partial(rA_\theta)/\partial z=0$ and following [2], on the boundary $\Gamma_{\mathrm{R}}^{\mathrm{e}}$ of the box which is parallel to the symmetry axis, we impose the Robin condition

$$\frac{\partial(rA_\theta)}{\partial r}+A_\theta=0.$$

## 2.2 The thermal submodel

The above model must be coupled with the heat equation in order to study the thermal effects in the workpiece. The thermal model is obtained from the heat transfer equation in transient state with change of phase since in general, the metal in the crucible can change of phase during the heating process. Hence, by assuming cylindrical symmetry, we have the equation,

$$\dot{e} - \frac{1}{r}\frac{\partial}{\partial r}\left(rk(r,z,T)\frac{\partial T}{\partial r}\right) - \frac{\partial}{\partial z}\left(k(r,z,T)\frac{\partial T}{\partial z}\right) = \frac{|J_\theta|^2}{2\sigma} \tag{6}$$

where $T$ is the temperature, $t$ is the time, $k$ is the thermal conductivity and $\dot{e} =: (\partial e/\partial t) + \mathbf{v}(\mathbf{x},t)\cdot\mathbf{grad}\, e$ denotes the material time derivative of enthalpy. We remark that the enthalpy density $e$ is expressed as a function of temperature by means of a *multivalued* function which depends on different physical parameters (see [1]). In this paper, we will assume that the velocity of the liquid metal $\mathbf{v}$ is null.

The computational domain of the thermal model, $\Omega_0$, is a radial section of the workpiece. The boundary of $\Omega_0$ splits into two parts: the symmetry axis and the rest, denoted by $\Gamma_{\mathrm{R}}^t$. We consider the following boundary conditions:

$$k(\mathbf{x},T)\frac{\partial T}{\partial \mathbf{n}} = 0 \qquad \text{on the symmetry axis,}$$

$$k(\mathbf{x},T)\frac{\partial T}{\partial \mathbf{n}} = \eta(T_c - T) + \gamma(T_r^4 - T^4) \qquad \text{on } \Gamma_{\mathrm{R}}^t,$$

where $\eta$ is the coefficient of convective heat transfer, $T_c$ and $T_r$ are the external convection and radiation temperatures, coefficient $\gamma$ is the product of emissivity by Stefan-Boltzman constant and $\mathbf{n}$ is the outward unit normal vector to the boundary.

## 3 Numerical solution

To integrate the equation (6) in time, we use a one-step implicit scheme. We denote by $\Delta_t = t^{n+1} - t^n$ the time step and by $X^n(r,z)$ the spatial position occupied at time $t^n$ by the material point which is at position $(r,z)$ at time $t^{n+1}$. Thus, at each time step we have to solve the following weak problems:

*(WTP) For each $n = 0,1,\ldots$, find a function $T^{n+1}$ such that*

$$\int_{\Omega_0} \frac{1}{\Delta t} e^{n+1} W r dr dz \;+\; \int_{\Omega_0} k(r,z,T^{n+1})\,\mathbf{grad}\,T^{n+1}\cdot\mathbf{grad}\,W\, r dr dz =$$
$$\int_{\Gamma_{\mathrm{R}}^t} (\eta(T_c - T^{n+1}) + \gamma(T_r^4 - (T^{n+1})^4)) W r\, d\Gamma \;+\; \int_{\Omega_0} \frac{1}{\Delta t} e^n \circ X^n\, W r dr dz +$$
$$\int_{\Omega_0} \frac{1}{2\sigma(r,z,T^{n+1})} |J_\theta^{n+1}|^2\, W r dr dz, \qquad \textit{for all test function } W.$$

*(WEP) Find a complex function $A_\theta^{n+1}$ satisfying $A_\theta^{n+1} = 0$ on $\Gamma_{\mathrm{d}}^e$ and*

$$\int_\Omega \left(\frac{1}{\mu(T^{n+1})r}\frac{\partial(rA_\theta^{n+1})}{\partial r}\frac{1}{r}\frac{\partial(r\bar{G})}{\partial r} + \frac{1}{\mu(T^{n+1})}\frac{\partial A_\theta^{n+1}}{\partial z}\frac{\partial \bar{G}}{\partial z}\right) r dr dz$$
$$+ \int_\Omega \mathrm{i}\omega\sigma(T^{n+1}) A_\theta^{n+1}\bar{G}\, r dr dz + \int_{\Gamma_{\mathrm{R}}^e} \frac{1}{\mu(T^{n+1})} A_\theta^{n+1}\bar{G}\, d\Gamma$$
$$= \sum_{k=1}^m \int_{\Omega_k} \sigma(T^{n+1})\, C_k \bar{G}\, dr\, dz, \qquad \textit{for all test function } G \textit{ null on } \Gamma_{\mathrm{d}}^e.$$

For the spatial discretization of problems ($WTP$) and ($WEP$) we consider continuous piecewise linear finite element spaces associated with triangular meshes of the domain for both fields $A_\theta$ and $T$. Notice that, at each time step, a coupled nonlinear system must be solved because of the heat source term is the Joule effect and physical parameters depend on temperature. We refer the reader to [1] to a detailed description of the iterative algorithms used to deal with the different non-linearities present in the problem.

Finally, we present some numerical results corresponding to the simulation of an industrial furnace used for silicon purification. We consider that the piece to be heated is silicon powder contained in a graphite crucible and both initially at 30 $^{\mathrm{o}}$C. Figure 2 shows the modulus of the current density and the temperature in silicon and graphite after 6 minutes, when the stationary state has been reached. We refer the reader to [1] for see the geometrical and physical data as well as other numerical results.

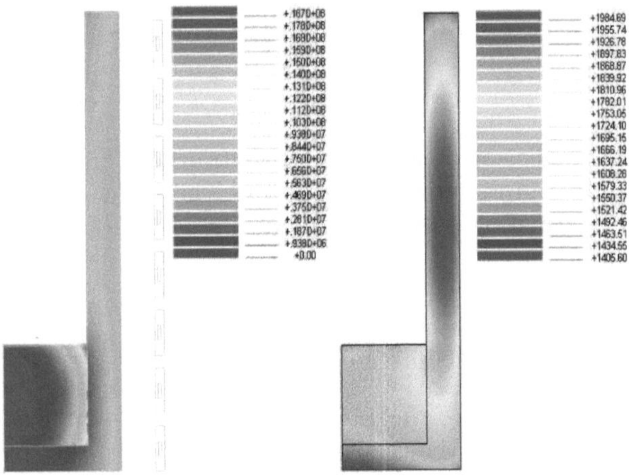


**Fig. 2.** Modulus of current density (left) and temperature (right) in the workpiece.

## References


1. A. Bermúdez, D. Gómez, M.C. Muñiz, and P. Salgado. Transient numerical simulation of a thermoelectrical problem in cylindrical induction heating furnaces. Submitted.
2. C. Chaboudez, S. Clain, R. Glardon, D. Mari, and M. Rappaz, J.and Swierkosz. Numerical modeling in induction heating for axisymmetric geometries. *IEEE Trans. Magn.*, 33:739–745, 1997.

# Thermal Radiation Effect on Thermal Explosion in a Gas Containing Evaporating Fuel Droplets.

I. Goldfarb[1], V. Gol'dshtein[1], D. Katz[1], and S. Sazhin[2]

[1] Department of Mathematics, Ben Gurion University of the Negev, Israel
goldfarb@math.bgu.ac.il, vladimir@math.bgu.ac.il, katsda@math.bgu.ac.il
[2] School of Engineering, Faculty of Science and Engineering, The University of Brighton, U.K. S.Sazhin@bton.ac.uk

**Summary.** Thermal explosion of diesel fuel droplets in the presence of thermal radiation is studied. The process is presented in terms of the dynamics of a multi-scale and singularly perturbed system, which is analyzed using the geometrical version of the Method of Integral Manifolds. Analytical estimates of the total ignition delay times in two limiting cases are obtained. The influence of the thermal radiation on the heat transfer and ignition delay time are clarified.

**Key words:** Diesel fuel, Combustion, Thermal explosion, Thermal radiation.

## 1 Introduction

The problem of thermal explosion in a gas containing fuel droplets is widely studied ([5, 9, 4]). In most cases these studies have been based on the applying computational fluid dynamics (CFD) packages ([4, 10]). An alternative approach to the problem is based on analytical studies of equations describing the limiting cases. The present study is focused on the latter approach. The zero-order approximation of the geometric version of the asymptotic Method of Integral Manifolds (MIM), developed in ([3, 6]), is used to study this process and to obtain analytical expressions for the total ignition delay, where appropriate. The effects of thermal radiation is taken into account, since the influence of the latter on heat transfer between droplets and gas can be noticeable in diesel engines. The classification of the possible thermal behavior of the system is suggested, and the impact of thermal radiation on the thermal explosion dynamics is clarified. Since thermal ignition delays seem to be typical for most explosive two-phase system, due to the heat exchange between the phases, we singled them out, as important characteristics of the model.

## 2 Physical model

We consider the ignition of spray as an explosion problem, where the droplets are regarded as the source of endothermicity. The main physical processes incorporated in the model are the evaporation of droplets, thermal radiation and highly exothermic oxidation reaction. The medium is modelled as a spatially homogeneous mixture of an optically thick, combustible gas with a mono-dispersed spray of evaporating fuel droplets. With a view to application of the results to diesel engines, we assume that gas pressure is constant. The system is assumed to be adiabatic.

The heat flow from gas to droplets is assumed to consist of two parts: convection and radiation heat flows. We assume that the thermal conductivity of the liquid phase is much greater than that of the gaseous phase, and the volume fraction of the liquid phase is much less than that of the gaseous phase. Thus, the heat transfer coefficient of the mixture is controlled by the thermal properties of the gaseous component. It is assumed that the burning process takes place in the gaseous phase only, and it is described by the first order exothermic reaction. Droplets velocities are neglected in the analysis. Therefore Nusselt (Nu) and Sherwood (Sh) numbers, describing the heat and mass transfer processes accordingly, are taken to be equal to 2. These assumptions allow us to describe the system by the following system of equations::

$$c_{pg}\rho_g\alpha_g\frac{dT_g}{dt} = \frac{\rho_g Q_f}{C_{fs0}}W - 4\pi R_d^2 n_d G\left(T_g, T_s\right) \tag{1}$$

$$W \equiv \alpha_g C_f A \exp\left(-\frac{E}{R_u T_g}\right)$$

$$\frac{d\left(R_d^3\right)}{dt} = -\frac{3R_d^2}{L\rho_L}G\left(T_g, T_s\right) \tag{2}$$

$$\alpha_g\frac{dC_f}{dt} = -W + \frac{4\pi R_d^2 n_d C_{fs0}}{L\rho_g}G\left(T_g, T_s\right) \tag{3}$$

$$G\left(T_g, T_s\right) = Lh_m\mu_f\left(C_{fs} - C_f\right) \tag{4}$$

$$G\left(T_g, T_s\right) \equiv \left(h_c\left(T_g - T_s\right) + k_1\sigma\left(T_g^4 - T_s^4\right)\right)$$

$$k_1 = R_d^b\left(7\cdot 10^4 - 20\cdot T_g\right),$$

where $c$ is the specific heat capacity, $\rho$ is the density, $\alpha$ is the dimensionless volumetric phase content, $Q$ is the specific combustion energy, $C_{fs0}$ is the initial value of the fuel vapor molar concentration near the surface of the droplets, $W$ is the chemical reaction term, $R_d$ is the radius of the droplets, $n$ is the number of droplets per unit volume, $T$ is the temperature, $L$ is the latent heat of evaporation, $\mu$ is the molar mass, $h_c$ and $h_m$ are convective heat transfer coefficients describing heat and mass fluxes respectively, $\sigma$ is the Stefan-Boltzmann constant. The subscript $g$ refers to gas mixture; $f$ –

combustible gas component of the mixture; $p$ – constant pressure; $s$ – to the saturation line; 0 – the initial state; $c$ – convection; $m$ – mass transfer.

The initial conditions are the following: $T_s|_{t=0} = T_{s0}$, $T_g|_{t=0} = T_{g0}$, $R_d|_{t=0} = R_{d0}$, $C_f|_{t=0} = C_{f0}$, $C_{fs0} = P_{s0}\left(R_u T_{s0}\right)^{-1}$, $R_u$ is the universal gas constant.

Following Semenov [8] we have introduced following dimensionless variables:

$$\theta_g = \frac{E}{R_u T_{g0}} \frac{T_g - T_{g0}}{T_{g0}}, \quad \theta_s = \frac{E}{R_u T_{g0}} \frac{T_s - T_{g0}}{T_{g0}}, \quad r = \frac{R_d}{R_{d0}}, \quad \eta = \frac{C_f}{C_{fs0}}. \tag{5}$$

Appropriate initial conditions are: $\theta_g|_{\tau=0} = 0$, $\theta_s|_{\tau=0} = \theta_{s0}$, $r_d|_{\tau=0} = r_{d0} = 1$, $\eta|_{\tau=0} = \eta_0$, and $\tau = t \cdot t_{react}^{-1}$.

After the substitution of the variables (5) into Equations (1)-(4) and the application of Frank-Kamenetskii's simplification [1] followed by the use of the appropriate energy integral, we present the initial system of equations in the following form:

$$\gamma \frac{d\theta_g}{d\tau} = \eta\left(r\right) \exp\left(\theta_g\right) - \varepsilon_1\left(\theta_g - \theta_s\right)\left(r + \varepsilon_3 r^{2+b}\right) \tag{6}$$

$$\frac{1}{\varepsilon_2} \frac{dr^3}{d\tau} = -\varepsilon_1\left(\theta_g - \theta_s\right)\left(r + \varepsilon_3 r^{2+b}\right) \tag{7}$$

$$\left(\sigma_1^* + \sigma_2^*\right)\left(\theta_g - \theta_s\right) + \frac{T_{g0}}{T_{s0}}\eta = \exp\left(\left(\frac{T_{y0} b^*}{T_{s0}}\right)\theta_s + \left(\frac{T_{g0}}{T_{s0}} - 1\right)\frac{L\mu_f}{T_{g0} R_u}\right) \tag{8}$$

For the problem under consideration the parameters $\gamma$ and $\varepsilon_2^{-1}$ are expected to be small. Therefore, the set of equations (6)-(8) represents singularly perturbed system of ODEs, and the geometric version of the Method of Integral Manifolds (GVMIM) ([3, 6]) is expected to be applicable.

Following the GVMIM, every phase trajectory of the system can be subdivided into the fast and slow parts. The fast parts are characterized by the high rate of change of one of the system variables while the others keep their initial values. At the slow parts the variables are almost constant compared with the fast parts. These parts can be identified with the integral manifolds, the location of which represents a separated problem. The general theory of integral manifolds states that the zero-order approximation of such manifold, termed as the slow surface, lies in the $\gamma$-vicinity of its exact place. It is a curve in the two-dimension case ([7, 2]).

The points dividing the slow curve into stable and unstable parts are called turning or stationary points. The slow curve has horizontal tangent at these points. The trajectory approaches the stable part of the slow curve and starts moving along it. The 'motion' along the stable part can take place until a turning point of the system is approached.

### 2.1 Fast gas temperature: $\varepsilon_2\gamma \ll 1$

In this case the slow curve of system (6)-(8) is presented by the following equation:

$$\Omega(\theta_s, r) = \eta(r)\exp\theta_g(\theta_s, r) - \varepsilon_1(\theta_g(\theta_s, r) - \theta_s)\left(r + \varepsilon_3 r^{2+b}\right) = 0, \qquad (9)$$

where the function $\theta_g(\theta_s, r)$ can be obtained from the Clapeyron-Clausius law (8), and $\eta(r)$ is determined by the energy integral of the system.

According to the definition given in the previous section, the turning points $(\theta_{T\Omega}, r_{T\Omega})$ of the slow curve are determined by the equation:

$$\Omega(\theta_{T\Omega}, r_{T\Omega}) = \frac{\partial\Omega}{\partial\theta_s}(\theta_{T\Omega}, r_{T\Omega}) = 0.$$

Equation (7) can be written in the following form:

$$\frac{dr^3}{d\tau} = -\varepsilon_1\varepsilon_2(\theta_g(\theta_s, r) - \theta_s)\left(r + \varepsilon_3 r^{2+b}\right) \qquad (10)$$

From Equation (9) it follows that:

$$\eta(r)\exp\theta_g(\theta_s, r) = \varepsilon_1(\theta_g(\theta_s, r) - \theta_s)\left(r + \varepsilon_3 r^{2+b}\right) \qquad (11)$$

Substitution of Equation (11) into Equation (10) and the following integration gives the following expression for the dimensionless ignition delay time:

$$\tau_{delay} = -\int_{r_{d0}}^{r_{T\Omega}} \frac{1}{(\delta-1)} \frac{d\eta(r)}{\eta(r)\exp(\theta_g(\theta_s, r))} \qquad (12)$$

### 2.2 Fast droplet radius: $\varepsilon_2\gamma \gg 1$

In this case the equation (7) is used to determine appropriate slow curve. Based on the physical background of the problem we can expect that the difference $(\theta_g(\theta_s, r) - \theta_s)$ is positive. Hence, the slow curve the slow curve is determined by the equation $r = 0$.

Remembering that in this case the dimensionless droplet temperature $\theta_s$ does not change, i.e. $\theta_s \equiv \theta_{s0}$, the integration of Equation (7) gives the following expression for the dimensionless ignition delay time:

$$\tau_{delay} = -\int_{r_{d0}}^{0} \frac{3r^2 dr}{\varepsilon_1\varepsilon_2(\theta_g(\theta_{s0}, r) - \theta_{s0})(r + \varepsilon_3 r^{2+b})} \qquad (13)$$

## 3 Conclusions

Heating, evaporation and ignition of diesel fuel droplets is studied. The geometric version of the Method of Integral Manifolds is used to investigate the impact of thermal radiation on the heat transfer between the droplets and gas. Analytical expressions for the delay time are obtained in two limiting cases.

The dependence of the system behavior on the values of its parameters is studied. The main type of the system behavior, the thermal explosion with delay, is investigated. Asymptotic analysis of the problem leads to two distinct scenarios: fast gas temperature and fast droplet radius. The analysis of the model allows us to conclude that in the parametric regions, where the first of the above-mentioned scenarios takes place, thermal radiation increases the delay time. Additionally, the delay time increases with the increase of the droplets number and of the dimensionless parameter $\varepsilon_1$, and decreases with the increase of the droplet radius and of the dimensionless parameter $\varepsilon_3$. A similar study has been performed in the parametric regions, where the second scenario takes place. In this case, thermal radiation decreases the delay time. Additionally, the delay time decreases with the increase of the droplets number density and the dimensionless parameters $\varepsilon_1$ and $\varepsilon_3$. It increases with the increase of the droplet radius.

## References


1. Frank-Kamenetskii D A. *Diffusion and Heat Exchange in Chemical Kinetics.* Plenum, New York, 2 edition, 1969.
2. Strygin B B and Sobolev V A. *Decomposition of Motions by the Integral Manifolds Method.* Nauka, Moscow, 1988.
3. Babushok V I and Gol'dshtein V M. Structure of the thermal explosion limit. *Combust.Flame*, 72:211–216, 1988.
4. Aggrawal S K. A review of spray ignition phenomena: present status and future research. *Prog. Energy Combust. Sci.*, 24:565–600, 1998.
5. Kuo K K. *Principles of Combustion.* Wiley, New York, 1986.
6. Gol'dshtein V M and Sobolev V A. Singularity theory and some problems of functional analysis. *AMS Translations*, 153:73–92, 1992.
7. Fenichel N. Geometric singular perturbation theory for ordinary differential equations. *J. Diff. Eq.*, 31:53–98, 1979.
8. Semenov N N. Zur theorie des verbrennungsprozesses. *Z. Phys. Chem*, 48:571–581, 1928.
9. Stone R. *Introduction to internal Combustion Engines.* MacMillan, London, 1992.
10. Sazhin S S Feng G Heikal M R Goldfarb I Goldshtein V and Kuzmenko G. Thermal ignition analysis of a monodisperse spray with radiation. *Combustion and Flame*, 124:684–701, 2001.

# Local Defect Correction for Laminar Flame Simulation

M. Graziadei[1] and J.H.M. ten Thije Boonkkamp[2]

Technische Universiteit Eindhoven, Department of Mathematics and Computer Science [1]m.graziadei@tue.nl [2]tenthije@win.tue.nl

**Summary.** An outline of the Local Defect Correction (LDC) method is given. The method is combined with a procedure to construct an orthogonal, curvilinear fine grid and it is applied to the thermo-diffusive model for laminar flames.

**Key words:** local defect correction, orthogonal curvilinear grid, laminar flames, thermo-diffusive model

## 1 Introduction

Boundary value problems (BVPs) can have solutions which exhibit very rapid variations in a relatively small part of the domain. Moreover, these so-called *high activity regions* are often of irregular shape. This certainly applies to laminar flames, where the solution varies very rapidly in the flame front, a thin region separating the burnt and unburnt gas mixture. A numerical solution method for such BVPs requires a grid that is very fine in the vicinity of the high activity region. One way to deal with such problems is to use the Local Defect Correction (LDC) method. Roughly speaking, the method combines a global coarse grid solution with a local fine grid solution, to improve the accuracy. Because of its irregular shape, a curvilinear fine grid is an obvious choice to cover the high activity region.

We have organised our paper as follows. In the next section we give a brief outline of the LDC method. Then, in Section 3, we describe a procedure to construct an orthogonal, curvilinear grid, and finally, in Section 4, we apply the method to the thermo-diffusive model for laminar flames.

## 2 An outline of LDC

In this section we present a brief outline of LDC; a more detailed discussion can be found in e.g. [5, 1, 4].

Consider the BVP

$$\mathcal{L}[u] = f, \quad \mathbf{x} \in \Omega, \tag{1a}$$
$$\mathcal{B}[u] = g, \quad \mathbf{x} \in \partial\Omega, \tag{1b}$$

where $\Omega \subset R^d\,(d = 2, 3)$ is a simply connected domain, $\mathcal{L}$ a linear elliptic operator and $\mathcal{B}$ a boundary operator, either of Dirichlet or Neumann type. Let us define a discretisation of (1) on a uniform (coarse) grid of size $H$, denoted by $\Omega_H$, covering $\Omega$, i.e.

$$\mathcal{L}_H[u_H] = f_H, \tag{2}$$

where the right-hand side $f_H$ contains both the source term $f$ and the contribution of the boundary conditions.

Suppose, $u$ changes very rapidly in a small, irregular subdomain $\Omega' \subset \Omega$. In this high-activity region, the grid size $H$ is definitely too large to capture the behaviour of $u$, so we formulate a new discrete BVP on $\Omega'$ by covering it with an orthogonal, curvilinear grid of characteristic grid size $h$. The discrete problem reads

$$\mathcal{L}_h[u_h] = f_h - \mathcal{G}_\Gamma^h \mathcal{P}^{h,H}[u^H|_\Gamma], \tag{3}$$

where the second term in the right-hand side of (3) represents the interpolation of the coarse grid solution $u^H$ on the interface $\Gamma = \partial\Omega' \backslash \partial\Omega$. The operator $\mathcal{P}^{h,H}$ is an interpolation operator, $\mathcal{L}_h$ is the discretisation of (1a), reformulated in the curvilinear coordinates defined in $\Omega'$, and $\mathcal{G}_\Gamma^h$ is the part operating on grid points on $\Gamma$. The numerical approximations on both grids can be combined into the composite grid solution $w_{H,h}$ as follows

$$w_{H,h}(\mathbf{x}) := \begin{cases} u_H(\mathbf{x}) & \text{if } \mathbf{x} \in \Omega_H \backslash \Omega'_H, \\ \mathcal{R}^{H,h}[u_h](\mathbf{x}) & \text{if } \mathbf{x} \in \Omega'_H, \end{cases} \tag{4}$$

where $\mathcal{R}^{H,h}$ is the restriction operator from the local coarse grid $\Omega'_H$ to the fine grid $\Omega'_h$.

From $w_{H,h}$ we can compute the following approximation $\tilde{d}_H$ of the local discretisation error of (2) on $\Omega'_H$

$$\tilde{d}_H(\mathbf{x}) = \begin{cases} 0 & \text{if } \mathbf{x} \in \Omega_H \backslash \Omega'_H, \\ (\mathcal{L}_H[w_{H,h}] - f_H)(\mathbf{x}) & \text{if } \mathbf{x} \in \Omega'_H. \end{cases} \tag{5}$$

Once $\tilde{d}_H$ has been computed, we can add it to the right hand side of (2), resulting in the equation

$$\mathcal{L}_H[\tilde{u}_H] = f_H + \tilde{d}_H. \tag{6}$$

Solving (6), we expect to get a better numerical approximation of (1). The procedure above can be repeated several times, giving rise to an iterative method. Convergence of this method is very fast, usually only one or two iterations are required [1].

# 3 Constructing an orthogonal curvilinear grid

The procedure in this section is based on [3]. The starting point is a non-orthogonal coordinate system: one family of coordinate lines will be kept, the other one will be transformed into a set of lines orthogonal to the first set. We will restrict ourselves to 2D domains. Consider the position vector $\mathbf{x} = \mathbf{x}(\xi, \eta)$ in a non-orthogonal coordinate system. Suppose, we want to keep the $\eta$-lines and we want to construct a family of coordinate lines, the $\zeta$-lines say, perpendicular to them. To do so, we introduce a function $k(\xi, \eta)$, being constant along the new $\zeta$-lines. Since the covariant base vector $\partial\mathbf{x}/\partial\xi$ is tangent to the $\eta$-lines, the orthogonality condition between the $\zeta$- and the $\eta$-lines can be formulated as

$$\nabla k \times \frac{\partial \mathbf{x}}{\partial \xi} = \mathbf{0}. \tag{7}$$

Substituting $\nabla k$, expressed in $(\xi, \eta)$-coordinates, into (7), we get the hyperbolic equation

$$\frac{\partial k}{\partial \eta} + F(\xi, \eta)\frac{\partial k}{\partial \xi} = 0, \quad F(\xi, \eta) := \left(\frac{\partial \mathbf{x}}{\partial \xi}, \frac{\partial \mathbf{x}}{\partial \eta}\right) \Big/ \left|\frac{\partial \mathbf{x}}{\partial \xi}\right|^2. \tag{8}$$

We have scaled the $(\xi, \eta)$-coordinates such that the corresponding grid sizes are equal to 1. For the discetisation of (8) we use central differences centred around the point $(\xi, \eta + \frac{1}{2})$.

The procedure to solve (8) is briefly as follows; see Fig. 1. We compute the function $\xi^*(\xi, \eta)$ that is the value of $\xi$ to which the point $(\xi, \eta)$ must be displaced along an $\eta$-line to get a trajectory orthogonal to it. Suppose we know $\xi^*(\xi, \eta)$ and we want to determine $\xi^*(\xi, \eta+1)$, subject to the initial condition $\xi^*(\xi, 1) = \xi$. We set

$$k(\xi, \eta + 1) = \xi, \tag{9}$$

and solve the discretisation of (8) for $k(\xi, \eta)$ by a backward step. After this we know $k$ and $\xi^*(\xi, \eta)$ on the $(\xi, \eta)$-points (in fact, $\xi^*(\xi, \eta) = \xi$) and $\xi^*(\xi, \eta)$ on the $(\zeta, \eta)$-points from the previous solution step. Then, $k(\xi^*, \eta)$ at the $(\zeta, \eta)$-points can be computed by a four-point Lagrangian interpolation. Furthermore $k$ is constant on the orthogonal trajectories $(k(\zeta, \eta) = k(\zeta, \eta + 1))$ and, because of (9), $k = \xi$ on the $(\eta + 1)$-line. From this we get

$$\xi^*(\xi, \eta + 1) = k(\zeta, \eta + 1) = k(\zeta, \eta) = k(\xi^*(\xi, \eta), \eta).$$

From the Cartesian coordinates of the $(\xi, \eta + 1)$-points and the values of $k(\zeta, \eta + 1)$ and $k(\xi, \eta + 1)$, we can get get the Cartesian coordinates of the $(\zeta, \eta)$-points by inverse interpolation.

We have applied the above procedure to a non-orthogonal grid system, where the $\eta$-lines are smoothed level curves of the course grid solution of (2). An advantage of this approach is that we do not need much grid points along the $\eta$-lines, since the solution is virtually constant along these lines.

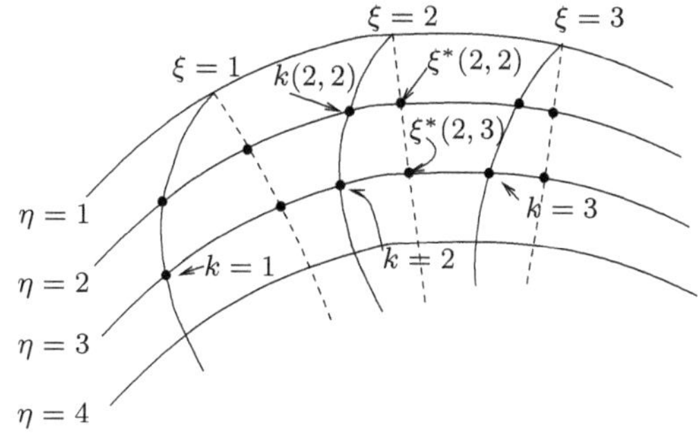


**Fig. 1.** Coordinate systems $(\xi, \eta)$ (solid lines) and $(\zeta, \eta)$ (dashed lines).

## 4 The thermo-diffusive model for laminar flames

Consider a laminar flame, propagating in an infinitely long tube. The thermo-diffusive model is a simplification of the conservation laws describing laminar flames. The main assumptions/approximations underlying the model are: the isobaric and constant density approximations and one-step chemistry [6]. In [2], it has been shown that this model allows for a travelling wave solution. If we furthermore assume a unit Lewis number, the propagation of the flame is governed by the following BVP

$$-\nabla^2 T + \big(V_0 + V\cos(\pi y/2L)\big)\tfrac{\partial T}{\partial x} = \omega(T), \quad x \in \mathbb{R},\ 0 < y < L, \tag{10a}$$

$$T(-\infty, y) = 0, \quad T(\infty, y) = 1, \quad \tfrac{\partial T}{\partial y}(x, 0) = \tfrac{\partial T}{\partial y}(x, L) = 0, \tag{10b}$$

where $T$ is a dimensionless temperature, $V_0$ the (unknown) velocity of the travelling wave and $V\cos(\pi y/2L)$ the velocity of the gas flow. The source term in the right-hand side of (10a) is given by

$$\omega(T) = \tfrac{1}{2}\beta^2(1-T)\exp\Big(-\tfrac{\beta(1-T)}{1-\alpha(1-T)}\Big), \tag{11}$$

with $\alpha$ and $\beta$ nondimensional parameters.

An expression for the speed $V_0$ can be obtained by integrating (10a) over the whole computational domain $\Omega$. This way we find

$$V_0 = \frac{1}{L}\iint_{\Omega} \omega \,\mathrm{d}S - \frac{2V}{\pi}. \tag{12}$$

Depending on the value of $V$, it is possible that $V_0 + V\cos(\pi y/2L) < 0$ in some part of the domain, resulting in an inversion of the flow, which is thus directed from the burnt towards the unburnt gases.

We have solved the BVP (10) on the finite domain $\Omega = (-6.1, 6.1) \times (0, 4)$, for the parameter values $\alpha = 0.83$, $\beta = 10$ and $V = 3$. After one LDC iteration the velocity $V_0$ converges to the value $-0.4292$. Figure 2 shows the coarse and fine grids used and Fig. 3 shows the computed temperature field.

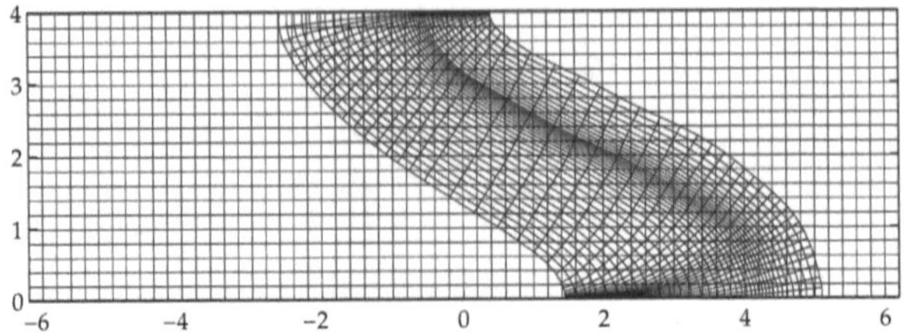


**Fig. 2.** Coarse and fine grid.

**Fig. 3.** Dimensionless temperature.

## References

1. M.J.H. Anthonissen. *Local Defect Correction Techniques: Analysis and Application to Combustion.* PhD thesis, Eindhoven, 2001.
2. H. Berestycki, B. Larrouturou, and P.L. Lions. Multi-dimensional travelling-wave solutions of a flame propagation model. *Arch. Rat. Mech. Anal.*, 111:33–49, 1990.
3. C.W. Davies. An initial value approach to the production of discrete orthogonal coordinates. *J. Comp. Phys.*, 59:164–178, 1981.
4. M. Graziadei, R.M.M. Mattheij, and J.H.M. Ten Thije Boonkkamp. Local defect correction with slanting grids. *Numer. Methods Partial Differential Eq.*, 20:1–17, 2004.
5. W. Hackbusch. Local defect correction and domain decomposition techniques. In K. Böhmer and H. J. Stetter, editors, *Defect Correction Methods. Theory and Applications, Computing, Suppl. 5*, pages 89–113, Wien, New York, 1984. Springer.
6. B. Larrouturou. Adaptive numerical simulation of premixed flame propagation. In T.J. Chung, editor, *Numerical Modeling in Combustion.* Taylor & Francis, Washington DC, 1993.

# Development of a Hierarchical Model Family for Molten Carbonate Fuel Cells with Direct Internal Reforming (DIR-MCFC)

P. Heidebrecht[1] and K. Sundmacher[1,2]

[1] Otto-von-Guericke-University Magdeburg, Process Systems Engineering, Universitätsplatz 1, 39106 Magdeburg, Germany
[2] Max-Planck-Institute for Dynamics of Complex Technical Systems, Sandtorstraße 1, 39106 Magdeburg, Germany

**Summary.** This contribution deals with the mathematical modelling of a high temperature molten carbonate fuel cell (MCFC) and serves as a basis for the following three contributions of this mini-symposium. After a motivation and a short introduction into the working principle of the MCFC, the most important equations of the model are presented. This model is applied for optimisation purposes and as a basis for the derivation of reduced models specifically designed for different tasks.

## Introduction

The molten carbonate fuel cell (MCFC) consumes fuel gases containing hydrogen, carbon monoxide and light hydro-carbons to produce electric energy. In addition to its high electric efficiency, its operating temperature of about 600°C makes the MCFC a suitable candidate for the coupled production of electricity and heat in stationary applications. Due to its insensitivity with respect to carbon monoxide, it is very flexible with respect to fuels giving the MCFC a high flexibility concerning its applications.

In Germany, the company MTU CFC Solutions has developed a 250 kWe MCFC system, called "Hotmodule" [1, 2]. Its electric system efficiency of about 50% is unsurpassed by conventional systems in this power class, and it has proven its feasibility and reliability in more than 20 successful field trial plants in Germany, Europe, Japan and the US. Currently, the project is heading towards commercialisation, with the planned start of a series production of economically competitive systems in 2006.

Although the Hotmodule has proven its reliability, there is still some optimisation potential left. As the Hotmodule is a high temperature fuel cell, the temperature distribution inside the stack is of crucial importance to the system performance. Too high temperatures cause material damage, while too low temperatures make the system inefficient. Unfortunately, information on this vital state from measurements is incomplete and difficult to get, so other ways of achieving reliable data are required, for example by using a state observer. Furthermore, the operating conditions for steady state operation and for load changes are determined empirically. This is mainly because the states inside the cell are not well known and thus large safety factors are applied. Consequently, the system is not operated in its optimal point and load changes are performed only in small steps. Finally, the optimal reforming concept is not clear. Three different methods of producing hydrogen from methane exist: external, indirect internal and direct internal reforming. Which combination of these three offers the best system performance is yet unknown. Thus, system design tools are needed.

All of these aims require mathematical models of strongly differing detail level and complexity. For example, the optimisation of input conditions requires a detailed steady state model, which must be quick to solve nevertheless, the design of an observer requires a dynamic model with only a few equations, and system design tools demand steady state models that are as simple as possible. Consequently, the number of models required to solve these questions is high.

To reduce the modelling effort, one single reference model is derived, from which all other required models are obtained either by physical simplifications or by mathematical reduction methods. The advantages of this hierarchical modelling strategy are clear: the overall modelling effort is reduced and the individual model variants are comparable to each other. Results obtained from one model can be transferred to another one more easily, and system parameters in different models have identical physical meanings.

In the following, we will shortly outline the basic working principle of the MCFC and give a short introduction into the model equations we apply.

## MCFC Working Principle

The working principle of the Molten Carbonate Fuel Cell is illustrated in Fig. 1.. It consists of two porous metallic electrodes, and a liquid electrolyte between them, which in the case of the MCFC is molten carbonate. Above and below the electrodes channels are located, through which gaseous reaction educts and products are transported. The anode channel is fed with a preheated mixture of desulphurised natural gas, that is mainly methane and steam. Methane is converted in the reforming process to a hydrogen-rich gas mixture at the reformer catalyst which is placed inside the anode channel:

$$CH_4 + 2H_2O \Longleftrightarrow CO_2 + 4H_2$$

Then hydrogen enters the porous anode electrode, where it meets the carbonate ions from the electrolyte melt. There these substances react to produce carbon dioxide, water and two free electrons according to the following oxidation reaction:

$$H_2 + CO_3^{2-} \Longleftrightarrow H_2O + CO_2 + 2e^-.$$

At typical MCFC temperatures, the conversion of the reforming process is severely limited by its chemical equilibrium. This limitation is overcome by the continuous removal of hydrogen by the electrochemical oxidation reaction. In addition to the direct mass integration between the reforming process and the oxidation, the endothermic reforming process receives its required heat from the exothermic oxidation reaction. Thus these two are coupled twofold.

The anode exhaust gas is mixed with air and the unoxidised components are fully oxidised in a catalytic combustion chamber. Because air is fed in excess, the exhaust gas of the combustion chamber still contains a significant amount of oxygen. This gas is then fed to the cathode channel where the electrochemical reduction reaction takes place. There, new carbonate ions are produced from carbon dioxide, oxygen and two electrons:

$$2e^- + {}^1\!/_2 O_2 + CO_2 \Longleftrightarrow CO_3^{2-}$$

The carbonate ions are transported towards the anode electrode through the electrolyte. The cathode exhaust gas leaves the system.

With this, a source of electrons is available at the anode electrode and electrons are consumed at the cathode electrode, so both electrodes can be electrically connected via any electric load and thus the cell serves as an electric energy supply device.

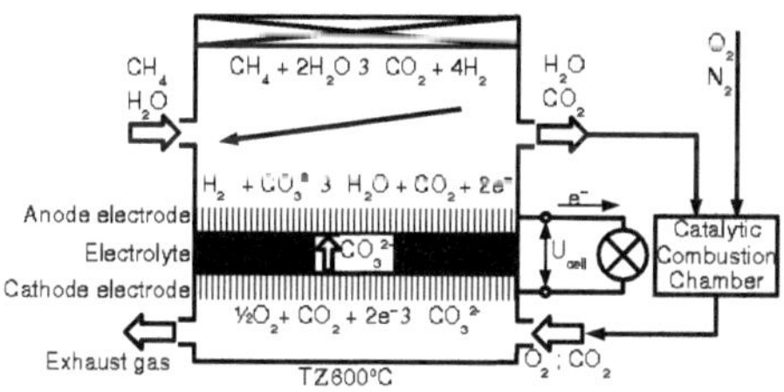


Fig. 1. Working principle of the MCFC with direct internal reforming.

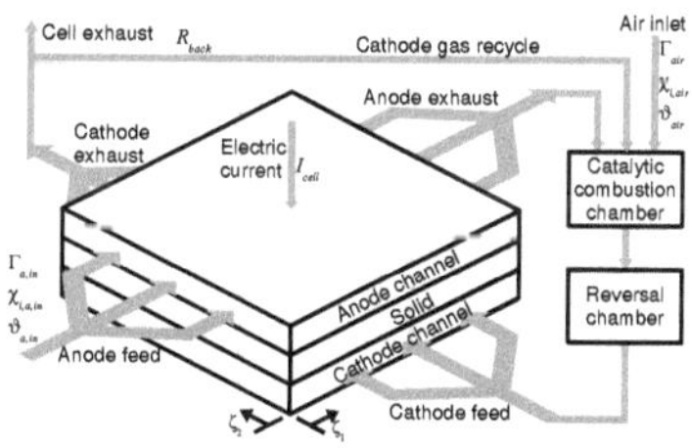


Fig. 2. Compartments and fluxes of the reference model consisting of anode and cathode gas layer, a solid phase, a catalytic combustion chamber and a reversal chamber. Also indicated are the input parameters of the system.

## The Reference Model

The reference model basically describes a spatially two-dimensional single MCFC in cross-flow configuration at galvanostatic operating mode (see

Fig. 2.). In addition to the cell, a catalytic combustion chamber is considered. The model is based on the balances of mass, charges and energy and fulfills the corresponding laws of conservation. It can be used to simulate the composition and flow of the gas inside the anode and cathode channels, predicts temperature distributions in both gas phases and in the solid parts of the cell and calculates the cell voltage and current density distribution. It is completely formulated in terms of dimensionless parameter groups [3, 4].

The model contains equations of different types, all of them have non-linear source terms:

- several hyperbolic partial differential equations (PDEs), describing the concentrations and temperatures in the gas phases,

$$\frac{\partial \chi}{\partial \tau} = -\gamma \frac{\partial \chi}{\partial \zeta} + \sigma_{\chi} (\chi, \vartheta, ...) \tag{1}$$

$$\frac{\partial \vartheta}{\partial \tau} = -\gamma \frac{\partial \vartheta}{\partial \zeta} + \sigma_{\vartheta} (\chi, \vartheta, ...) \tag{2}$$

- one parabolic PDE, describing the temperature in the solid cell parts,

$$\frac{\partial \vartheta_s}{\partial \tau} = \frac{\partial^2 \vartheta}{\partial \zeta^2} + \sigma_{\vartheta s} (\chi, \vartheta_s, \varphi) \tag{3}$$

- two ordinary differential equations (ODEs) with respect to the spatial coordinate, describing the gas flow,

$$0 = -\frac{\partial (\gamma \vartheta)}{\partial \zeta} + \sigma_{\gamma} (\chi, \vartheta, ...) \tag{4}$$

- three ODEs with respect to time, describing the changes of the electric potential,

$$\frac{\partial \varphi}{\partial \tau} = i - i_a (\chi, \vartheta_s, \varphi) \tag{5}$$

- one integral equation, defining the total cell current as the integral of the cell current density

$$\int_{\zeta} i(\zeta) \, d\zeta = I_{cell} \tag{6}$$

The model is completed by a number of implicit and explicit algebraic equations, and a set of initial and boundary conditions.

## Simulation Results and Model Applications

The presented model is able to simulate transient, spatially distributed concentration and temperature profiles, gas flows, current density distribution

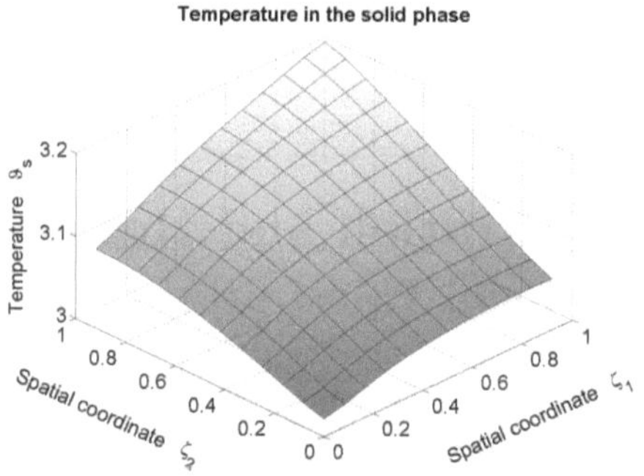


**Fig. 3.** Steady state temperature profile inside the cell's solid parts. The temperature range displayed here corresponds to a range from 894 K (621°C) to 954 K (681°C). Note that the anode gas flows along the $\zeta_1$-coordinate, while the cathode gas flows along $\zeta_2$.

and the cell voltage. As an example, Fig. 3. shows the simulated steady state temperature profile of the cell under certain operating conditions.

With this, the model includes all details being necessary to solve the open questions raised in the introduction. On the one hand it can be directly used for optimisation of input conditions and system design, and on the other hand it serves as a basis for the derivation of reduced models suitable for purposes like conceptual system design and control design. As an example, the reduction of this model by mathematical means for the design of an observer is presented by Mangold et al. [5].

## References

1. *www.mtu-cfc-solutions.de.*
2. M. Bischof and G. Huppman. *Journal of Power Sources*, 105:216–221, 2002.
3. P. Heidebrecht and K. Sundmacher. Dynamic modelling and simulation of a counter current molten carbonate fuel cell (mcfc) with internal reforming. *Fuel Cells*, 3-4:166–180, 2002.
4. P. Heidebrecht and K. Sundmacher. Molten carbonate fuel cell (mcfc) with internal reforming: model-based analysis of cell dynamics. *Chem. Eng. Sci.*, 58:1029–1036, 2003.
5. M. Mangold, M. Sheng, P. Heidebrecht, A. Kienle, and K. Sundmacher. Development of physical models for the process control of a molten carbonate fuel cell system. *Chem. Eng. Sci.*, 2004.

# Modelling of Filtration and Regeneration Processes in Diesel Particulate Traps

U. Janoske[1], T. Deuschle[2], and M. Piesche[2]

[1] University of Cooperative Education, Lohrtalweg 10, 74821 Mosbach, Germany janoske@ba-mosbach.de

[2] Institute of Mechanical Process Engineering, University of Stuttgart, Böblinger Strasse 72, 70199 Stuttgart, Germany deuschle@imvt.uni-stuttgart.de

**Summary.** The reduction of exhaust particulate emissions from diesel vehicles is a great upcoming challenge. As a result of their harmful effects, new legislation on diesel vehicles has been introduced throughout the world specifying low emission-levels. Today, the use of diesel particulate filter (DPF) in addition to engine modifications is the favoured method to fulfil these criteria. The principle of a DPF is based on the accumulation of particles in the alternating open and closed channels of the filter. The pressure drop over the DPF increases with time. This increase is associated with the rise of fuel consumption. For this reason, the deposited filter cake must be occasionally regenerated. To minimise complex and expensive investigations on test benches, a mathematical model has been developed describing the loading and regeneration behaviour of a DPF. The model is integrated in a commercial CFD-Code using user-defined subroutines (UDS). The CFD-Code was used for the calculation of the fluid flow and the particle tracks of different kinds of particles (e.g. soot, additives) in a two-dimensional model of the DPF. Thus, the axial and radial structure of the deposited particles on the filter can be determined. In the UDS models are implemented to calculate the pressure loss, the separation efficiency and the regeneration behaviour. Comparing the simulation results with the results gained experimentally, it can be seen that both sets of data concur. Further development concerning the implementation of a subroutine to describe the long-term behaviour and transport of the deposited particles will be carried out.

**Key words:** Numerical simulation, CFD, diesel particulate trap, filtration, regeneration

## 1 Introduction

Low fuel consumption combined with an excellent performance characterise modern passenger cars with direct injecting diesel engines. Conversely, the particle emissions from such engines are assumed to be a significant health hazard. According to the information currently available, diesel particle filters

combined with engine modifications are the favoured method to lower these emissions to the level of petrol engines. The principles upon that most filter systems are based nowadays are essentially the same: Particles are collected in a ceramic filter with a large filter surface. In this study a Silicon-Carbide (SiC) wall-flow filter with a honeycomb structure is used, whereby the channels are mutually locked in a checkerboard fashion. Due to the accumulation of particles in and on the channel walls the pressure drop over the DPF increases with time. This increase is associated with a rise in fuel consumption. For this reason, the deposited filter cake must be occasionally regenerated. Since the exhaust gas temperature under normal operation conditions is not sufficient for filter regeneration, the filters must be regenerated either using catalytic additives lowering the reaction ignition temperature or alternatively by increasing the temperature in the filter e.g. using methods of post-injection of fuel or electrical heaters [2, 1]. Another method for regenerating a DPF is the continuous regeneration (CRT), where nitrogen dioxide is used to oxidise the deposited soot in the filter.
A goal of this study is to set up a computational model that will allow the description of the loading and the regeneration behaviour in a DPF including transport phenomena of the deposits in the filter channel. In this article the examination is restricted to the loading, filtration and regeneration processes.

## 2 Simulation model

In order to limit the computational effort the loading and regeneration behaviour will only be performed at one channel of the DPF like shown in Fig. 1. Only one inlet channel and one outlet channel of the DPF are investigated. Therefore, symmetrical flow conditions within the channels are assumed, so that the modelling of only half of the channel is sufficient. Since non-stationary filtration processes with a variable height of the particle surface cannot be simulated using commercial CFD-codes, the computation is accomplished by combining a commercial CFD-code with self-defined program routines for a two-dimensional computational grid. The following describes the procedure of the simulation, as outlined in Fig. 2, in greater detail. After the computation of the flow and temperature field by the CFD-code, the particle tracks are calculated using an Euler-Lagrange approach. These information are the input data for the following user defined subroutines. Knowing the position of the separated particles on the ceramic wall and on the deposits already formed, the surface layer height and the flow resistance over the DPF can be computed. Regenerating the DPF leads to a decrease of the deposited mass and the pressure drop. After adjusting the computational grid according to the changed surface layer geometry the steps specified above are accomplished again. This procedure is repeated, until the desired period of operation or an inadmissible exhaust gas pressure drop is reached. Theses computations are performed under quasi-stationary conditions, which means that within one

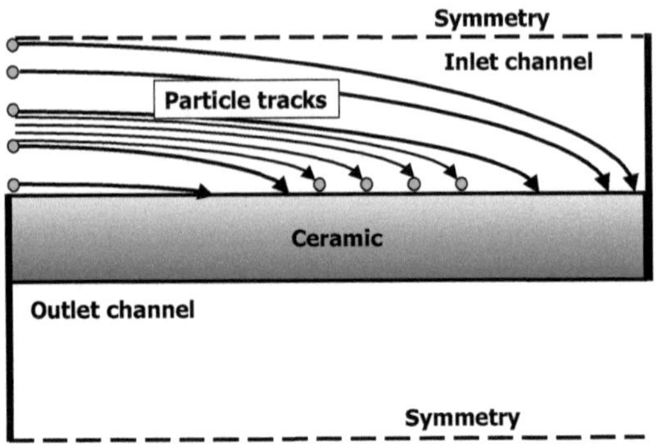


**Fig. 1.** Simulation model

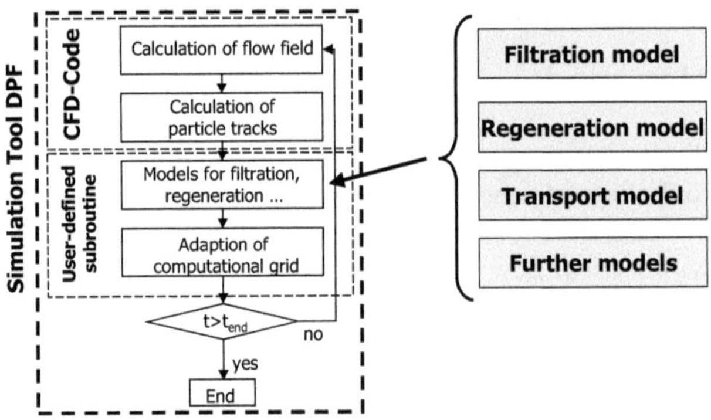


**Fig. 2.** Procedure of simulation for the characterisation of a DPF

time step the flow conditions are assumed to be constant. The steps done during the simulation of the soot loading are represented in Fig. 3. After computing the particle tracks with the Euler-Lagrange model, the sites of deposit on the ceramics and the surface layer are known. For reasons of limited computation time only a limited number of particle tracks can be computed. The laminar flow characteristics in the filter channel allow an interpolation between the computed particle deposition sites to produce deposition sites of fictitious particles. A reduction of the number of computed particle tracks can be accomplished leading to a reduced calculation time and a faster formation of the surface deposit layer. Limited storage capacities require further an effective use of storage resources. Since particle size distributions with several distribution classes are used for the exhaust, the reduction of the particle size distribution is performed by a so-called RRSB distribution (Rosin-Rammler-Sperling-Bennett). Thus, the distribution is described by the maximum and the minimum particle diameter and two parameters describing the RRSB. By

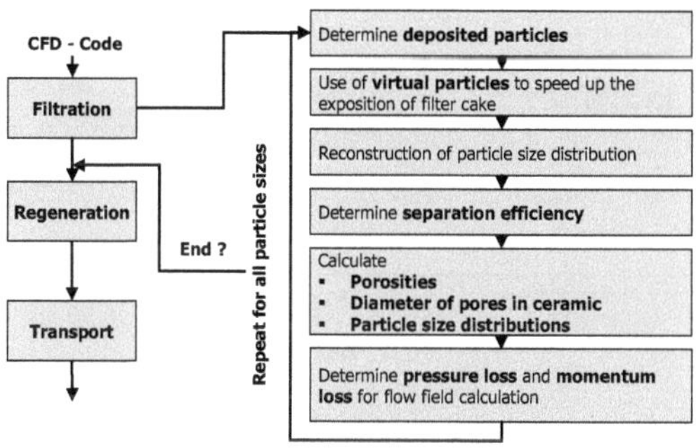


**Fig. 3.** Simulation of the loading behaviour

computing the particle tracks the position in axial direction of the filter channel is specified. The radial position of the deposited particle in the surface layer and in the ceramic material has to be determined separately.

## 3 Results

Comparing specific computed values with experimental results validated the filtration model. The experiments were performed in cooperation with the FKFS at the University of Stuttgart on a engine test bench (5 cylinder engine, 2.7 litres). As DPF, an Ibiden SiC filter of the dimensions 5.66“ x 6“ was used. The experiments were accomplished using sulphur-free fuel (< 10 ppm) with added iron or cerium-based additives.

Fig. 4 shows the pressure drop over the DPF as a function of the loading time for 3 different operating points for loading the DPF. The loads as well as the revolutions per minute were varied. The comparison between computation results and experimental data shows a good agreement.

## 4 Conclusion and Outlook

As shown, the developed computational model is able to simulate filter loading for experiments performed on engine test benches. The successful simulation of

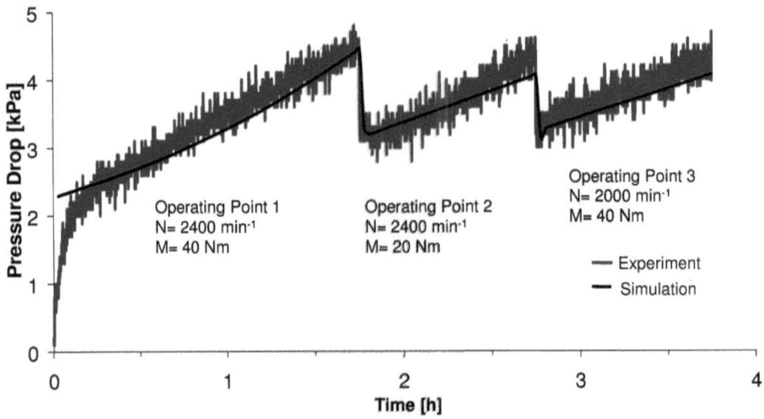


**Fig. 4.** Computation of a engine test bench cycle - Pressure drop over the DPF vs. loading time

the observed, time-dependent pressure drop confirms the assumptions used. The validation of the deposit layer height along the filter channel and the regeneration events are currently done. In the near future, models describing ash-transport and ageing processes are to be implemented into the program.

## References

1. A. G. Konstandopoulos, M. Kostoglou, E. Skaperdas, E. Papaioannou, D. Zarvalis, and E. Kladopoulou. Fundamental studies of diesel particulate filters: Transient loading, regeneration and aging. Technical Report SAE Technical Paper Series 2000-01-1016, 2000.
2. A.G. Konstandopoulos and J.H. Johnson. Wall-flow diesel particulate filters—their pressure drop and collection efficiency. Technical Report SAE Technical Paper Series 890405, 1989.

# Modelling the Shelf Life of Packaged Olive Oil Stored at Various Conditions

F.A. Coutelieris[1] and A. Kanavouras[2]

[1] Unilever Research and Development, Oliver van Noortlaan, 3130 AT, Vlaardingen, The Netherlands
[2] Corresponding author: Unilever Europe, Spreads and Cooking Products Category, Nassaukade 3, 3071 JL, Rotterdam, The Netherlands.
antonis.kanavouras@Unilever.com

**Summary.** A model was applied on experimental data to study the mass transport of oxygen diffusing through the oil phase and the packaging materials as well as the oxidation reactions. A nonlinear system was numerically solved for various combinations of materials, temperatures, and light availability, by adopting a typical Newton method, in conjunction with a multi-step up-winding finite differences scheme. The probability of the packaged olive oil not to reach the end of its shelf life ($P_{safe}$) and its time evolution, was in very good agreement with the experimental data. $P_{safe}$ was proposed as a reduction indicator for shelf life predictions at "real-life" conditions. Exposure to light at any pattern could significantly stimulate the oxidative degradations, only assisted by elevated temperatures and presence of oxygen. Plastic containers showed particularly higher protective role when oil was stored at light, while glass was the most protective material when oil was stored at dark.

## 1 Introduction

The type of material (plastics, glass, tin), the storage conditions (light, temperature) and the storage period can significantly influence the quality of olive oil [6]. In addition to the comprehensive experimental work on the oxidation of olive oil, [2] proposed a model based on the development of hydroperoxides as a function of both time and location in the package for a quick estimation of the product's response. [3] and [4] presented a two-dimensional model for the oxidation process of olive oil packaged in plastic bottles but without considering the diffusion of the flavor compounds in the oil phase and specific oxidative reactions. Furthermore no further refinement in terms of storage conditions, i.e. temperature and light was made. [8] presented an experimentally-based descriptive model for the shelf life of packaged olive, limited to chemical processes occurring inside the oil mass with the inadequacy of not incorporating the mass transport of the most oxidation-characteristic compounds due to diffusion as well as the interactions of the packaging materials with the flavor compounds.

## 2 Experimental

Extra virgin olive oil was placed in 500 mL PET, 500 mL PVC (Novapack, Co., Paris, IL, USA) or 500 mL glass bottles (Fisher Scientific Co., New Jersey, USA). The properties of the packaging materials were previously evaluated [8]. Half of the bottles were stored in dark and the other half were exposed to fluorescent light (four 40 W fluorescent light bulbs were placed at 30 cm above the bottles), all in controlled environment chambers at 15, 30 or 40°C. Separation and identification of hexanal was according to the previously developed methodology [8]. Statistical analysis was performed using commercial software (SAS® Proprietary Software Release 8.2, TS2M0, SAS Institute Inc., Cary, NC, USA).

## 3 Theory

In order to explain the oxidation process, a representative model for the evolution of hyperoxide in the packaged olive oil, based on the main chemical reactions:

$$O_2 \xrightarrow[h\nu]{k_a} O_3^- \tag{1}$$

$$RH + O_3^- \xrightarrow{k_b} ROOH$$

$$RH + O_2 \xrightarrow{k_c} ROOH \tag{2}$$

related to the oxidative degradation inside the oil phase, was applied. By assuming that the oil is quiescent, all the hyperoxide (ROOH) taking place in the above-mentioned reactions is finally transformed to hexanal, which could also be sorbed by the polymeric packaging materials and as quasi-steady state for the intermediate product $O_3^-$, the mass transport can be described by the following set of differential equations:

$$\frac{\partial C_{O_2}}{\partial t} = D_{O_2,mix}\frac{\partial^2 C_{O_2}}{\partial x^2} - \xi k_a C_{O_2} - k_c C_{O_2} C_{RH} \tag{3}$$

$$\frac{\partial C_{RH}}{\partial t} = -\xi k_a C_{O_2} - k_c C_{O_2} C_{RH} \tag{4}$$

$$\frac{\partial C_{hexanal}}{\partial t} = D_{hexanal,mix}\frac{\partial^2 C_{hexanal}}{\partial x^2} + \xi k_a C_{O_2} + k_c C_{O_2} C_{RH} \tag{5}$$

$$\frac{\partial C_{O_2}}{\partial t} = D_{O_2,wall}\frac{\partial^2 C_{O_2}}{\partial x^2}, \tag{6}$$

$$\frac{\partial C_{hexanal}}{\partial t} = D_{hexanal,wall}\frac{\partial^2 C_{hexanal}}{\partial x^2} \tag{7}$$

along with the appropriate boundary conditions in the oil-packaging interface.

The boundary value problem described above was discretized in space and time using a non-uniform finite-difference scheme [9]. A numerical algorithm, that involves a typical Newton method for non-linear systems [1] in conjunction with the finite differences scheme, was modified and adopted to handle non-linearity. The system was solved numerically with precision of order of $10^{15}$ for a range of storage temperatures (15°C, 30°C and 40°C), for various packaging materials (glass, PET, PVC) and light conditions (light, dark). The values for the parameters were taken from the relative literature [10, 5, 7, 3, 8]. When necessary, numerical interpolation or extrapolation was applied on the experimentally measured values.

Based on the concentration profiles of hexanal, the probability for the olive oil to reach the end of its shelf life during a certain time period is analogous to the ratio of the areas below and above an arbitrarily defined quality threshold. In other words, the probability of the oil to reach its self-life during the time period $[t_1, t_2]$ is analogous to the ratio of the relative areas, which on the other hand can be expressed by integrals of the spatially averaged hexanal concentration. Thus, we can define the probability, $P_{safe}$, for the oil not reaching the end of its shelf life period during the same time period $[t_1, t_2]$, as:

$$P_{safe} = 1 - \frac{\int_{t_1}^{t_2} \langle C_{hexanal} \rangle (t) dt}{\int_{0}^{t_2} \langle C_{hexanal} \rangle (t) dt} \tag{8}$$

where $t_1$ is the time when concentration reaches one critical value, considered as an upper limit for the quality acceptance. The brackets denote spatial averaging, and the upper edge of the integrals, $t_2$, has been set to 12 or 24 months in this study.

# 4 Result and Discussion

The experimentally measured values for hexanal [8] were used for the validation of the mathematical model. The agreement between model predictions and experimental data (discrete points) can be considered as sufficient since the averaged relative difference varies from 5.6% to 32.8% according to relative calculations (Fig. 1).

Figure 2 shows the time evolution of $P_{safe}$ for oil stored for 24 months at temperatures of 15°C, 30°C and 40°C for every 12 hours alternating light. In addition, the time evolution of $P_{safe}$ for oil stored at temperatures of 15, 30 and 40°C alternating every 4 months, and under continuous dark or light are also shown. The probability $P_{safe}$ after 24 months decreased significantly with the temperature increment, independently on the light conditions and

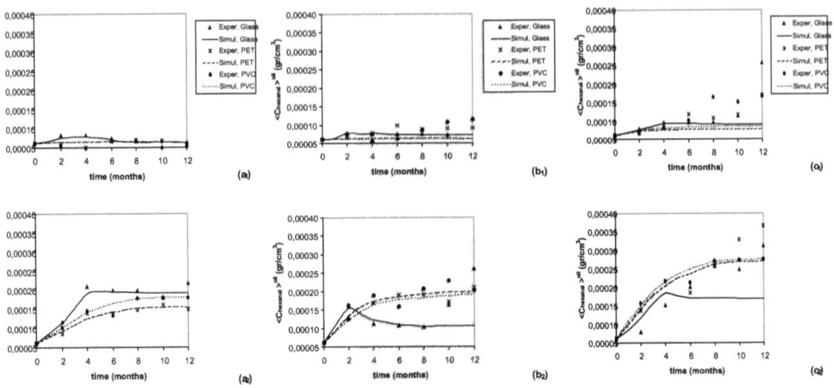


**Fig. 1.** : Time evolution of the spatially averaged hexanal concentration in the oil phase, $< C_{hexanal} >^{oil}$, for various packaging materials at 15°C ($a_1$, $a_2$), 30°C ($b_1$, $b_2$) and 40°C ($c_1$, $c_2$). Subscripts indicate the light conditions (1=dark, 2=light). Comparison of the experimental measurements (discrete points) with the simulations (solid lines

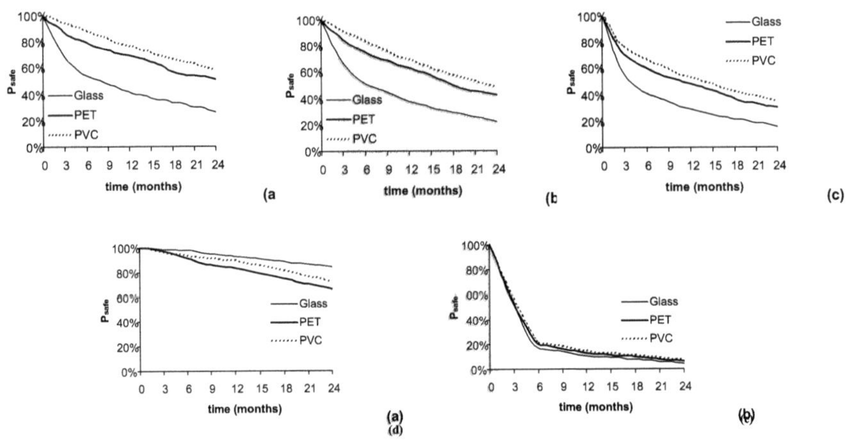


**Fig. 2.** Time evolution of $P_{safe}$ for oil stored for 24 months at temperatures of 15°C (a), 30°C (b) and 40°C (c) and daily alteration of light and dark and for oil stored at temperatures of 15, 30 and 40°C, alternating every 4 months, and under continuous dark (d) and continuous light (e).

the material. Elevated temperatures (40°C) combined with light, revealed an initial highly stimulated oxidation for oil stored in all packaging materials.

## 5 Conclusion

A satisfactory agreement of the model to the experimental results was shown through the low values of their relative differences ($< 33\%$ for any combination of storage conditions). By joining this accurate model with the $P_{safe}$ factor, enough evidence was obtained to support the benefits of storing the olive oil under continuous dark and low temperatures. For packaged olive samples stored at light, for the same storage temperature glass was a significantly less protective material, while PVC showed a higher protective role, although not that different to PET, most probably due to its higher oxygen diffusivity. The alternating presence of light had clearly reduced the probability of the oil to reach the end of its shelf life, compared to continuous light exposure.

## References

1. R. L. Burden and J.D. Faires. *Numerical analysis.* PWS-KENT Boston, 1989.
2. M. Dekker, M. Kramer, M.and van Beest, and Luning P. Modeling oxidative quality changes in several packaging concepts. In *Proceedings of the* 13$^{th}$ *IA-PARI conference on packaging CRC Press LLC*, volume 1, pages 297–303, 2002.
3. M. A. Del Nobile, M.L. Ambrosino, and Sacchi R. Design of plastic bottles for packaging of virgin olive oil. *J. Food Sci*, 68:170–175, 2003.
4. M. A. Del Nobile, M.L. Ambrosino, and Sacchi R. Influence of packaging geometry and material properties on the oxidation kinetics of bottled virgin olive oil. *J. Food Eng.*, 57:189–197, 2003.
5. A.E. Feigenbaum, V.J. Ducruet, S. Delpal, N. Wolff, J.P. Gabel, and J.C. Wittmann. Food and packaging interactions - penetration of fatty food simulants into rigid poly(vinyl chloride). *J. Agr. Food Chem*, 39:1927–1932, 1991.
6. F.R. Gutierrez, C.G. Herrera, and G-Q. Gutierez. Estudio de la cinética de evolutión de los índices de calidad del aceite de oliva virgen durante su conservatión en envases comerciales. *Grasas y Aceites*, 39:245–253, 1988.
7. P. Hernandez-Müňoz, R. Catala, and R. Gavara. Effect of sorbed oil on food aroma loss through packaging materials. *J. Agr. Food Chem*, 47:4370–4374, 1999.
8. A. Kanavouras, P. Hernandez-Müňoz, F. Coutelieris, and S. Selke. Oxidation derived flavor compounds as quality indicators for packaged olive oil. *J. Am. Oil Chem. Soc.*, acc., 2003.
9. W.H. Press, B.P. Flanner, S.A. Teukolsky, and T. Vetterling, W. Numerical recipes. Technical report, Cambridge University Press, 1986.
10. K. Toi. Diffusion and sorption of gases in poly(ethylene terephthalate). *J. Polymer Sci.*, 11:1839–1929, 1973.

# Nonlinear Model Reduction of a Dynamic Two-dimensional Molten Carbonate Fuel Cell Model

M. Mangold and Min Sheng

Max-Planck-Institut für Dynamik komplexer technischer Systeme,
Sandtorstraße 1, 39106 Magdeburg, Germany, mangold@mpi-magdeburg.mpg.de

**Summary.** A reduced nonlinear model of a planar molten carbonate fuel cell is presented. The model is derived from a spatially distributed dynamic model of the cell by applying the Karhunen Loève Galerkin procedure. The reduced model is of considerably lower order than the original one and requires much less computation time. The comparison between the two models shows that the reduced model can describe the dynamic of the temperature field with sufficient accuracy and has good extrapolation qualities with respect to changes in the model parameters.

**Key words:** molten carbonate fuel cell, dynamic simulation, reduced model, spatially distributed system, Karhunen-Loève decomposition, state observer.

## 1 Introduction

The molten carbonate fuel cell (MCFC) is a high-temperature fuel cell operated at 600°C - 700°C. Due to its high operation temperature, the MCFC offers advantages for the co-generation of heat and electricity. Currently, the development and operation of MCFCs as of other high temperature fuel cells is mainly based on experimental and empirical knowledge. However, model based process control and process design strategies can lead to a much better use of the fuel cells' capacities and increase the efficiency of the system, if suitable dynamic process models are available. Only very few detailed dynamic MCFC models have been published [2, 3, 4]. Those models consist of systems of algebraic and nonlinear partial differential equations in several space coordinates and are too complex for many process control purposes. The purpose of this contribution is to derive a reduced nonlinear dynamic model of a MCFC by applying the Karhunen-Loève-Galerkin method to the reference model. The reduced model is validated in test simulations by comparison with the reference model.

## 2 Spatially Distributed Reference Model of the MCFC

In this work, a planar cross-flow MCFC with direct internal reforming is considered. The model of the process used here is based on the following assumptions:

- Spatial gradients of the concentrations and of the temperature are considered in two space coordinates in the direction of the gas flows.
- As the dynamics of the temperature equations is much slower than that of the mass and charge balances, only the temperature equations is considered to be dynamic, the mass and charge balances are assumed to be at steady state.
- Temperature differences between the gases and the solid parts are neglected. This leads to a pseudo-homogeneous energy balance.
- The electrochemical reactions on anode and on cathode side are described by Butler-Volmer kinetics.

An energy balance of the system leads to a partial differential equation for the temperature that has the following structure:

$$
\begin{aligned}
0 = (\rho c_P) d \frac{\partial T}{\partial t} - \frac{c_P^A \dot{n}^A}{L_y} \frac{\partial T}{\partial y} - \frac{c_P^C \dot{n}^C}{L_z} \frac{\partial T}{\partial z} + \lambda d \left( \frac{\partial^2 T}{\partial y^2} + \frac{\partial^2 T}{\partial z^2} \right) \\
+ i^A \left( \frac{(-\Delta h_R^A)}{I\!RT} - \Delta\Phi^A \right) + i^C \left( \frac{(\Delta h_R^C)}{I\!RT} - \Delta\Phi^C \right) + \frac{I_{Cell}}{L_y L_z} \Delta\Phi^M \\
+ (-\Delta h_R^R) r_R + (-\Delta h_R^W) r_w =: Res(T)
\end{aligned}
\tag{1}
$$

The first line in (1) contains terms caused by convective and dispersive heat transport. The terms in the second and in the third line are nonlinear sources caused by the chemical and electrochemical processes inside the fuel cell. The evaluation of those terms requires the solution of mass and charge balances which complete the reference model. A detailed description of the model is given in [7].

## 3 Derivation of the Reduced MCFC Model

For model reduction of parabolic partial differential equations (PDEs) like (1), orthogonal projection methods have become a frequently used technique [5, 1]. The basic idea is to represent the unknown variable, *e.g.*, the temperature $T$, by an infinite sum of products of time dependent amplitude functions $T_i(t)$ and orthonormal, space dependent basis functions $\varphi_i(y, z)$:

$$
T(t, y, z) = \sum_{i=1}^{\infty} T_i(t) \varphi_i(y, z), \tag{2}
$$

A model reduction is achieved by approximating the infinite sum by a finite series

$$\tilde{T}(t,y,z) := \sum_{i=1}^{N^T} \tilde{T}_i(t)\varphi_i(y,z). \tag{3}$$

Clearly, the approximation (3) will not solve the PDE (1) exactly, but a nonzero residual will remain. In order to get conditions for the time functions $\tilde{T}_i(t)$, the projections of the residual *Res* in (1) onto the basis functions are required to vanish, *i.e.*,

$$\int_y^{L_y}\int_z^{L_z} Res(\tilde{T})\ \varphi_i(y,z) \stackrel{!}{=} 0, \qquad i = 1,\ldots,N^T. \tag{4}$$

This approach approximates the PDE (1) to $N^T$ ordinary differential equations. It reduces the infinite dimensional system to a finite dimensional one with $N^T$ dimensions, which is much easier to solve numerically.

For the model reduction of the MCFC model, not only the profile of the temperature, but also those of the molar fractions in the anode and in the cathode gas channels, as well as the profiles of the total molar flow rates have to be approximated by basis functions. The resulting reduced model is a low-order differential algebraic system of differential index one. The quality of the reduced model, *i.e.*, its deviation from the original model, mainly depends on two factors. The first one is the number of terms considered in the approximations of the spatial profiles. The second is the choice of the basis functions. A good approximation of the complete model by a low order reduced model is achievable, if suitable problem-specific basis functions are chosen. In this contribution, basis functions are derived numerically by applying the Karhunen-Loève decomposition method. The Karhunen-Loève decomposition (K-L decomposition) was originally developed for the description of stochastic data [6]. For the solution of partial differential equations, the K-L decomposition method can be used to generate basis functions for the Galerkin procedure [9] from simulation results with the original model taken at discrete time points, so-called snapshots. The K-L decomposition extracts the most typical or characteristic structure from these snapshots in the form of empirical eigenfunctions $\varphi_i(y,z)$. As shown in [9], the basis functions can be expressed as:

$$\varphi_i(y,z) = \sum_{j=1}^{N} {\alpha_j^i}^T \cdot v_j(y,z) \tag{5}$$

In the above equation, $N$ is the number of time points, for which simulation data are available; $v_j(y,z)$ denotes a snapshot taken at time point $j$; $\alpha_j^i$ is the $j$-th component of an eigenvector $\boldsymbol{\alpha}^i \in I\!R^N$ given by:

$$\boldsymbol{C}^M \boldsymbol{\alpha}^i = \lambda_i \boldsymbol{\alpha}^i, \tag{6}$$

where $\boldsymbol{C}^M$ is a symmetric $N \times N$ matrix whose elements are defined as

$$\boldsymbol{C}_{ij}^M = \frac{1}{N} \int_{y=0}^{L_y} \int_{z=0}^{L_z} v_i(y,z) v_j(y,z) dy dz. \tag{7}$$

The eigenvalue $\lambda_i$ may be interpreted as a measure of how well an eigenfunction $\varphi_i$ is able to approximate the time average of the snapshots [9]. In this sense, the eigenfunction $\varphi_1$ corresponding to the largest eigenvalue $\lambda_1$ is the most typical structure of the snapshots.

In order to determine suitable basis functions for the MCFC model, the response of the complete model to an increase of the cell current and to a subsequent decrease to the original value is computed numerically by using the method of lines. For the temperature profile, between 1 and 5 basis functions are chosen. For the other variables of the reduced model, two basis functions for each gas are found to be sufficient.

## 4 Validation of the Reduced Model

Test simulations are made in order to validate the reduced model by comparison with the original model. Special emphasis is laid on the extrapolation qualities of the reduced model. An example is shown in Fig. 1. It is found that already the approximation of the temperature profile by a single basis function leads to a quite satisfactory behaviour of the reduced model. The temperature error becomes very small, if 5 basis functions are used for the temperature. In all simulations shown in Fig. 1, the cell voltage of the reduced model matches the result of the complete model nearly perfectly. The K-L de-

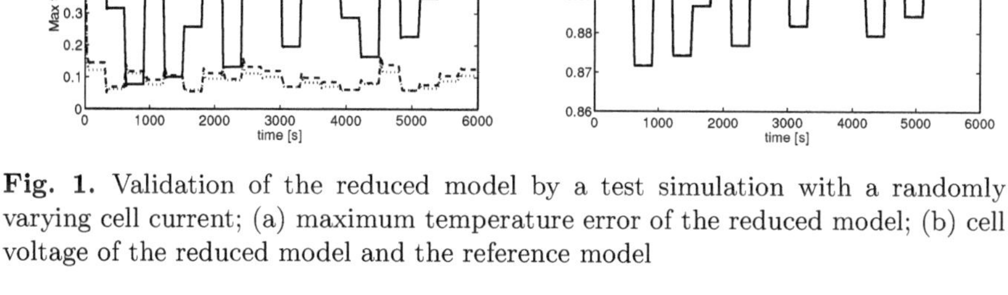


**Fig. 1.** Validation of the reduced model by a test simulation with a randomly varying cell current; (a) maximum temperature error of the reduced model; (b) cell voltage of the reduced model and the reference model

composition technique leads to a considerable reduction in terms of the order of the system as well as in terms of the computation time. After a spatial discretization, the complete model consists of about 12,000 equations. The simulation shown in Fig. 1 requires about 43,000s of CPU time on a PC with the complete model. In comparison, the reduced model consists of 25 equations, if 5 temperature basis functions are used. Its numerical solution takes about 380s of CPU time on the same PC. The decrease of the computational time achieved by the model reduction is not quite as strong as the decrease of the order of the system. The reason is that the evaluation of the reduced model equations is more complicated as it requires a numerical quadrature.

## 5 Conclusions

A reduced model of an MCFC is obtained by applying the Karhunen-Loève Galerkin method to a two-dimensional spatially distributed cross-flow model of the cell. The basic idea of the method is to approximate the profiles of the spatially distributed variables by basis functions obtained from test simulations with a detailed reference model. For the MCFC model considered here, this technique proves to be successful. The reduced model produces results that are very close to those of the original model, but it reduces the computation time by a factor of more than 100. Due to its properties, the reduced model is suitable for applications in the field of model based process control. An example is the state and parameter estimator described in [8].

*Acknowledgement.* This work is supported by the German Federal Ministry of Education and Research under contract No. 03C0345B.

## References

1. J.A. Atwell and B.B. King. Proper orthogonal decomposition for reduced basis feedback controllers for parabolic equations. *Math. Comp. Modelling*, 33:1–19, 2001.
2. W. He and Q. Chen. Three-dimensional simulation of a molten carbonate fuel cell stack under transient conditions. *J. of Power Sources*, 73:182–192, 1998.
3. P. Heidebrecht and K. Sundmacher. Dynamic modeling and simulation of a countercurrent molten carbonate fuel cell (MCFC) with internal reforming. *Fuel Cells*, 2:166–180, 2002.
4. P. Heidebrecht and K. Sundmacher. Molten carbonate fuel cells (MCFC) with internal reforming: Model-based analysis of cell dynamics. *Chem. Engng. Sci.*, 58:1029–1036, 2003.
5. K. Hoo and D. Zheng. Low-order control-relevant models for a class of distributed parameter systems. *Chem. Engng. Sci.*, 56:6683–6710, 2001.
6. M. Loève. *Probability Theory.* Van Nostrand, Princeton, NJ, 1955.

7. M. Mangold and M. M. Sheng. Nonlinear model reduction of a two-dimensional MCFC model with internal reforming. *Fuel Cells*, 4:68–77, 2004.
8. M. Mangold, M. Sheng, P. Heidebrecht, and K. Sundmacher. Development of physical models for the process control of a molten carbonate fuel cell system. *Chem. Engng. Sci.*, 59:4847–4852, 2004.
9. H.M. Park and D.H. Cho. The use of the Karhunen-Loève decomposition for the modeling of distributed parameter systems. *Chem. Engng. Sci.*, 51:81–98, 1996.

# Liquid/Solid Phase Change with Convection and Deformations: 2D Case

D. Mansutti, R. Raffo, and R. Santi

Istituto per le Applicazioni del Calcolo "M. Picone" (C.N.R.),
Viale del Policlinico, 137 – 00161 Roma (I)
d.mansutti@iac.cnr.it

**Summary.** We present the results of the numerical simulation of the first stages of the melting from a side of a gallium slab by adding to the heat transfer and to the melt flow the description of the effects of the deformations of the solid phase. The experiment by Gau and Viskanta in [4] has been considered.

**Key words:** Phase change, Deformations, Convection, Potential Functions

## 1 Introduction

In recent experiments of melting of pure metals [5], it seemed that the melt convection flow might be responsible of the structural variations in the solid portion close to the phase interface. However the physical quantities observed are very small and may be easily spoiled by experimental errors. In other words the nature of such processes indicate that mathematical and numerical assessments are necessary. Actually, mathematical model for liquid/solid phase transitions it has been built for the description of the dynamics of the liquid and solid (velocity field of the liquid and deformation field of the solid), the heat transport phenomena and the evolution of the phase front [1]. This model has already provided results in excellent agreement with the analytical solution [7] in the case of the solidification of a water layer.
Here, we approach the two-dimensional case. At this scope we have reformulated the mathematical model with the use of potential functions that allow to meet more easily the incompressibility constraint. As starting numerical test we have simulated the initial time steps of the experiment of the melting of a pure gallium slab described by Gau and Viskanta in [4]. This experiment is particularly suitable in order to identify the effects of the solid deformations in addition to those due to melt convection as it has been numerically simulated, not including the solid dynamics, by many specialists in the literature of the recent past years (included by one of the authors [2]). The new formulation

of the model is sketched in Sect. 2 and the numerical results are shown and discussed in Sect. 3.

## 2 Governing Equations and Reformulation

The equations governing the evolution of a continuum sample undergoing liquid/solid (L/S) phase transition have to describe the conservation laws of momentum, of energy and of mass; for their structure we refer to the books on classical mechanics (*e.g.*, [8]). A set of equations for each single phase is obtained together with the *jump conditions* for the balance of momentum, energy and mass across the phase interface. In the jump condition for the energy conservation law (the so-called *Stefan condition* [3]), most important is the contribution due to the release or the adsorption of latent heat corresponding respectively to the solidification or the melting processes.

In our model the liquid and the solid are described as an *incompressible viscous fluid* and an *isotropic linearly elastic material* [8]. This choice allows to keep average the level of difficulty of the final system of equations to be solved. Obviously, for the solid, a more appropriate model would be one describing correctly the specific material symmetry but, here, we aim to provide a first insight to the effects of the mechanical response of the solid within the transition process. Adding simplifying assumptions are: i) the density of the liquid and of the solid may be assumed equal, ii) the radiating heat is negligible, iii) liquid and solid interfacing particles do not slip over each other, iv) the material coefficients of the two phases may be assumed to be constant. With $t$ the time and $(x, y)$ the space cartesian coordinates, let us call $D_F$, $D_S$ and $\Gamma(t)$ the domains occupied respectively by the melt, by the solid and by the phase interface. Introducing the Boussinesq and Fourier approximations, the governing equations of the melt flow, holding in $D_F$, result:

$$\rho \frac{d\mathbf{v}}{dt} = -\nabla p + \mu_F \nabla^2 \mathbf{v} - \rho[1 - \alpha_F (T_F - T_p)]\mathbf{g} \tag{1}$$

$$\nabla \cdot \mathbf{v} = 0 \tag{2}$$

$$\rho c_F \frac{dT_F}{dt} = k_F \nabla^2 T_F + \mu_F \big[(\frac{\partial u}{\partial x})^2 + \frac{1}{2}(\frac{\partial u}{\partial y} + \frac{\partial v}{\partial x})^2 + (\frac{\partial v}{\partial y})^2\big] \tag{3}$$

where $\mathbf{v} = (u, v)$, $p$, $T_F$ and $\rho$ denote the velocity, the pressure, the temperature of the melt and the density of the sample, respectively. These equations are coupled with the following ones for the solid holding in $D_S$:

$$\rho \frac{\partial^2 \mathbf{U}}{\partial t^2} = \mu_S \nabla^2 \mathbf{U} - \rho[1 - \alpha_S (T_S - T_p)]\mathbf{g} \tag{4}$$

$$\nabla \cdot \mathbf{U} = 0 \tag{5}$$

$$\rho c_S \frac{\partial T_S}{\partial t} = k_S \nabla^2 T_S, \tag{6}$$

where $\mathbf{U}$ and $T_S$ denote the displacement and the temperature of the solid, respectively. The symbols $c$, $k$ and $\alpha$ with the appropriate subscript ($F$ or $S$) indicate respectively the heat capacity, the conductivity and the thermal expansion coefficients for the liquid and the solid, whereas $\mu_F$, $\lambda$ and $\mu_S$ are the viscosity coefficient of the melt and the Lamé constants of the solid. The jump conditions, that hold in $\Gamma(t)$, appear:

$$\mathbf{v}_F = \mathbf{v}_S \tag{7}$$

$$(-p\mathbf{I} + \mu_F(\nabla \mathbf{v}_F + (\nabla \mathbf{v}_F)^T)) \cdot \hat{\mathbf{n}} = \lambda \nabla \cdot \mathbf{U}_S \mathbf{I} + \mu_S(\nabla \mathbf{U} + (\nabla \mathbf{U})^T) \cdot \hat{\mathbf{n}} \tag{8}$$

$$-\rho \Lambda(\mathbf{v}_F \cdot \hat{\mathbf{n}} - u_{\mathbf{n}}) - k_S \nabla T_S \cdot \hat{\mathbf{n}} + k_F \nabla T_F \cdot \hat{\mathbf{n}} = 0 \tag{9}$$

where $\hat{\mathbf{n}}$ denotes the normal unitary vector on $\Gamma(t)$ and $\Lambda$ the latent heat. The set of equations (1) – (9) is completed by initial and boundary values according to the specific test case.

By observing that the vectors $\mathbf{v}$ and $\mathbf{U}$ are both required to be solenoidal, we have reformulated the above model on the basis of the Helmoltz-Hodge decomposition in order to meet more accurately and easily such constrain. This procedure is well known and experimented in fluid dynamics and leads to the scalar potential / stream function / vorticity formulation [6]. We propose to extend this approach also to the treatment of the solid by introducing the new unknowns, $\varphi$, $\chi$ and $\iota$ linked to $\mathbf{U}$ by these relations:

$$\mathbf{U} = (-\frac{\partial \chi}{\partial y} + \frac{\partial \varphi}{\partial x}, \frac{\partial \chi}{\partial x} + \frac{\partial \varphi}{\partial y}) \qquad \iota = \frac{\partial U_y}{\partial x} - \frac{\partial U_x}{\partial y}. \tag{10}$$

Accordingly, we transform the equation (4) by applying the curl operator. In doing so, (5) reduces to a simple Poisson equation for $\varphi$ and $\iota$ results solution of a scalar equation (instead of the vector equation (4) defining $\mathbf{U}$).

## 3 Numerical Test and Conclusions

We solved this model by a finite difference method based on a time Euler scheme and centered second order space schemes; front-fixing was used to handle the moving boundary. We adopted the initial and boundary values corresponding to the Gau and Viskanta experiment [4], describing the melting of a rectangular pure gallium slab heated on a vertical side; the other vertical side is lightly undercooled and the horizontal ones are insulated. Here, we show shots of the simulation of the first time instants obtained on a space grid $60 \times 10$ both in the solid and in the fluid domain. The maximum allowed time step was $\Delta t = 10^{-7}$. In Figs. 1 and 2 we plot the streamlines in the melt and the displacement vector field in the solid at time $t = 16$ sec and $t = 56$ sec, respectively. In Fig. 3 the profiles of the velocity components of the melt at three values of $y$ are sketched. Compared to those in absence of deformations [2], streamlines indicate an intensification of the upper mechanisms leading to faster melting than on the bottom. The melt flow is still multicellular. The non-zero velocities of the melt at the phase boundary are significant. The displacement field of the solid, essentially null far from the phase boundary, is the kinematical response of the solid to the pulling action of the melt. According to the hyperbolic nature of the momentum equation for the solid, displacements become oscillatory with time due to the unsteady excitation induced by the melt at the phase boundary. The (lengthy!) simulation of the remaining time interval of the experiment (up to 19 mins.) is still under process and, together with a mesh refinement analysis will be object of a future paper.

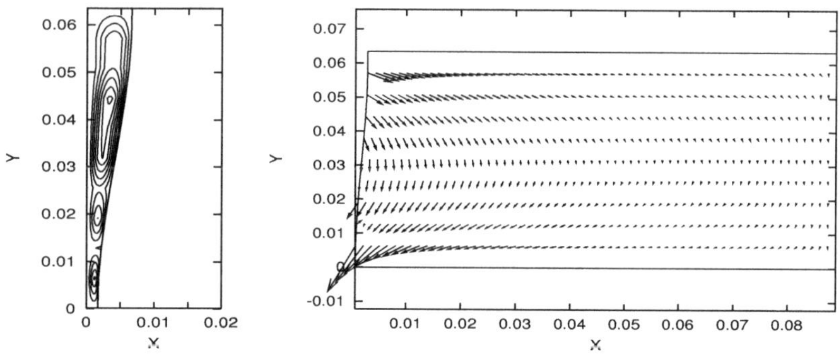


**Fig. 1.** Streamlines (melt) and displacement vector field (solid) at $t = 16$ sec.

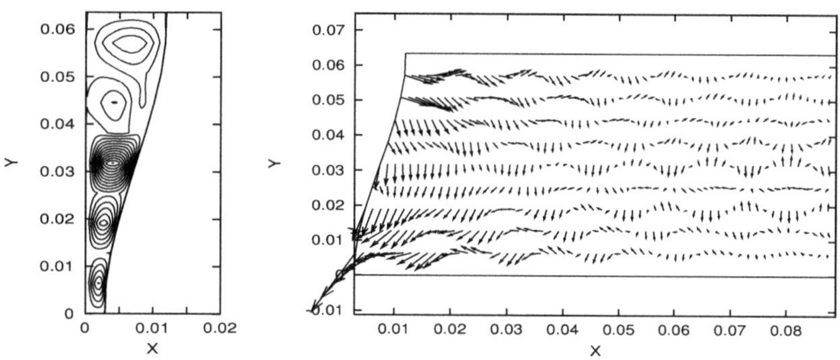


**Fig. 2.** Streamlines (melt) and displacement vector field (solid) at $t = 56$ sec.

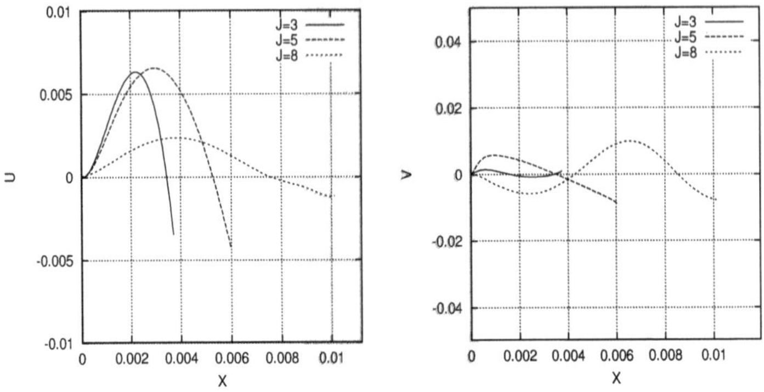


**Fig. 3.** The velocity components of the melt versus $x$ at three different levels at $t = 56$ sec.

*Acknowledgement.* D. Mansutti gratefully acknowledges Prof. K.R. Rajagopal for suggesting her to approach this model and for the useful hints provided. This work has been developed within a project funded by Agenzia Spaziale Italiana (contract n. ARS-99.39) and within the research team on Environmental Processes at I.A.C., Rome (C.N.R.).

## References

1. F. Baldoni. *Thermomechanics of Solidification.* Pittsburgh University Press, Pittsburgh, 1997.
2. M.M. Cerimele, D. Mansutti, and F. Pistella. Numerical modeling of liquid/solid phase transition - analysis of a gallium melting test. *Computers and Fluids*, 31:437–451, 2002.
3. J. Crank. *Free and moving boundary problems.* Oxford Science Publication, Oxford, 1984.
4. C. Gau and R. Viskanta. Melting and solidification of a pure metal on a vertical wall. *Transaction of the ASME*, 108:174–181, 1986.
5. P. Gondi, R. Montanari, E. Evangelista, and G. Buroni. X-ray study of structures of liquid metals with controlled convective motions. *Microgravity Quarterly*, 7(4):155–173, 1997.
6. G.J. Hirasaki and J.D. Hellums. Boundary conditions on the vector and scalar potentials in viscous three-dimensional hydrodynamics. *Quarterly of Applied Mathematics*, 28(2):163–178, 1970.
7. D. Mansutti, F. Baldoni, and K.R. Rajagopal. On the influence of the deformation of the forming solid in the solidification of a semi-infinte water-layer of fluid. *Mathematical Models & Methods in Applied Sciences*, 11(2):367–386, 2001.
8. J.C. Slattery. *Momentum, energy and mass transfer in continua.* McGraw-Hill, New York, 1972.

# Mathematical Modelling of Mass Transport Equations in Fixed-Bed Absorbers

A. Pérez-Foguet and A. Huerta

Laboratorio de Cálculo Numérico, Dept. de Matemática Aplicada III, E. T. S. Ingenieros de Caminos, Canales y Puertos, Universitat Politècnica de Catalunya, c/ Jordi Girona 1-3, Barcelona 08034, http://www-lacan.upc.es
agusti.perez@upc.es

**Summary.** This work presents a dimensionless analysis of mass transport equations in fixed-bed absorbers. Focus is centered in isothermal and incompressible problems, with special attention to nonlinear adsorption and desorption processes that take place at absorbent particles. The general differential–algebraic equation system is expressed in dimensionless form, and the model is particularized into four different formulations. The model is analyzed and used to simulate a standard industrial test efficiently. Formulations are selected depending on the relative importance of the different physical phenomena involved in each part of test.

**Key words:** dimensionless analysis, activated carbon filter, convection – diffusion – reaction equation, numerical simulation, mass conservation, adaptive modelling.

## 1 Introduction

Modelling adsorption and desorption in fixed–bed absorbers is of high interest in several industrial applications and consequently it has been widely studied (see [1, 2, 3] among many others). In the automotive sector, activated carbon filters called *canisters* are used to reduce the emission of hydrocarbons (HC) from the fuel tank. The production of these filters requires the verification of several quality and efficiency indicators. One of them is provided by the Working Capacity test (WC). This test measures the mass of butane that can be adsorbed by a canister at a prescribed loading – unloading sequence (imposed flows of a butane mixture and clean air respectively).

The transport of HC along the filter and the adsorption – desorption in the activated carbon particles requires accurate description of two spatial scales: the macro scale, for the canister itself with characteristic lengths of decimeters, and the micro scale (two orders of magnitude smaller, in the order of millimeters) for the activated carbon pellets. A model with two phases for each scale is considered here. The conservation mass of a single-solute is imposed at both scales. The spatial coordinates are averaged in a representative elementary volume [4] and equations of both scales are expressed in dimensionless form. After that, the micro scale equation is transformed into a system of two ordinary differential equations by imposing a spatial discretization of its weak form, which is defined in assumed spherical particles and incorporates appropriate boundary conditions (see [1] for a description of the related Homogeneous Surface Diffusion Model). The resulting algebraic – differential equation system is integrated numerically by an stabilized fractional step method in realistic finite element simulations.

## 2 Dimensionless model

Dimensionless form of mass balance equations is expressed in terms of the following variables: $\boldsymbol{x} = \boldsymbol{x}'/L$, $\mathrm{t} = V\mathrm{t}'/L$ and $\boldsymbol{v} = \boldsymbol{v}'/V$, where $L$ and $V$ are reference values for length and velocity, and $\boldsymbol{x}'$, $\mathrm{t}'$ and $\boldsymbol{v}'$ represent dimensional spatial coordinates, time and interparticle velocity respectively. Mass transport equations depend on the unknowns $\mathrm{c}(\boldsymbol{x},\mathrm{t}) = \mathrm{c}'(\boldsymbol{x}',\mathrm{t}')/c_{\mathrm{ref}}$, $\bar{\mathrm{q}}(\boldsymbol{x},\mathrm{t}) = \bar{\mathrm{q}}'(\boldsymbol{x}',\mathrm{t}')/q_{\mathrm{ref}}$, and $\mathrm{q}_{\mathrm{R}}(\boldsymbol{x},\mathrm{t}) = \mathrm{q}'_{\mathrm{R}}(\boldsymbol{x}',\mathrm{t}')/q_{\mathrm{ref}}$, where $q_{\mathrm{ref}}$ and $c_{\mathrm{ref}}$ are reference values, and $\mathrm{c}'$, $\bar{\mathrm{q}}'$ and $\mathrm{q}'_{\mathrm{R}}$ are equal to the interparticle concentration, the mean value (in the particle) of the mass adsorbed by unit of clean carbon mass, and the same magnitude but in the external surface of the particles.

On one hand, transport in the macro scale is given by the following dimensionless equation:

$$\frac{\partial \mathrm{c}}{\partial \mathrm{t}} + \boldsymbol{v}\cdot\boldsymbol{\nabla}\mathrm{c} = \boldsymbol{\nabla}\cdot\left(\frac{\boldsymbol{\nabla}\mathrm{c}}{P_{\mathrm{e}}}\right) - \left(\frac{S_{\mathrm{t}}}{B_{\mathrm{i}}E_{\mathrm{d}}} + \mathrm{r}_{\varepsilon_{\mathrm{p}}}\frac{\partial \mathrm{L}(\bar{\mathrm{q}})}{\partial \bar{\mathrm{q}}}\right)\frac{\partial \bar{\mathrm{q}}}{\partial \mathrm{t}} \tag{1}$$

where $\boldsymbol{\nabla}$ is the gradient operator with respect to $\boldsymbol{x}$, and the following dimensionless numbers are used: Peclet, $P_{\mathrm{e}}$, Biot, $B_{\mathrm{i}}$, and Staton, $S_{\mathrm{t}}$. The surface diffusion modulus is denoted by $E_{\mathrm{d}}$, the porosity ratio by $\mathrm{r}_{\varepsilon_{\mathrm{p}}}$, and the dimensionless Freundlich isotherm by $\mathrm{L}(\bar{\mathrm{q}})$. The parameters are defined as

$$\begin{gathered} P_{\mathrm{e}} = \frac{VL}{D}, \quad B_{\mathrm{i}} = \frac{k_{\mathrm{f}}c_{\mathrm{ref}}R}{D_{\mathrm{s}}q_{\mathrm{ref}}\rho_{\mathrm{s}}(1-\varepsilon_{\mathrm{p}})}, \quad S_{\mathrm{t}} = \frac{k_{\mathrm{f}}L(1-\varepsilon_{\mathrm{f}})}{VR\varepsilon_{\mathrm{f}}}, \quad E_{\mathrm{d}} = \frac{LD_{\mathrm{s}}}{VR^2}, \\ \mathrm{r}_{\varepsilon_{\mathrm{p}}} = \frac{1-\varepsilon_{\mathrm{f}}}{\varepsilon_{\mathrm{f}}}\varepsilon_{\mathrm{p}}, \quad \mathrm{L}(\bar{\mathrm{q}}) = \frac{1}{A^{1/n}}\bar{\mathrm{q}}'^{1/n}, \quad A = A'\frac{c_{\mathrm{ref}}^n}{q_{\mathrm{ref}}}, \end{gathered} \tag{2}$$

where $D$ is fluid diffusion, $k_{\mathrm{f}}$ the film mass transfer coefficient, $R$ the particle radius, $D_{\mathrm{s}}$ the surface diffusion, $\rho_{\mathrm{s}}$ the clean carbon density, $\varepsilon_{\mathrm{f}}$ and $\varepsilon_{\mathrm{p}}$ the interparticle and the intraparticle porosities, and $A'$ and $n$ the isotherm coefficients.

On the other hand, the dimensionless transport equations at the micro scale are found from the spatial discretization of the surface diffusion equation at particle level complemented with Robin boundary conditions:

$$\begin{aligned}\frac{\partial \bar{q}}{\partial t} &= 3\, B_i\, E_d\, (c - L(q_R)) \\ \frac{\partial q_R}{\partial t} &= 10\, B_i\, E_d\, (c - L(q_R)) + 35\, E_d\, (\bar{q} - q_R)\,.\end{aligned} \tag{3}$$

Note that diffusion of the intraparticle fluid phase is neglected, both to keep the model simple and because it is not relevant in gas absorbtion modelling [3]. The formulation of the model corresponding to equations (1) and (3) is called the *three Variables Formulation* (3VF). A first simplification of the model follows from $S_t/(B_i E_d) >> r_{\varepsilon_p} \frac{\partial L(\bar{q})}{\partial \bar{q}}$, which is verified with reference values given in table 1. However, note that the reaction term in equation (1) remains nonlinear due to the coupling with equations (3). A linear model is found only with linear isotherms (i.e., $n = 1$).

Apart from this obvious simplification three options can be developed, first assuming that diffusion inside particles, $D_s$, is large enough to consider $\bar{q} = q_R$. This case is referred as the first *two Variables Formulation* (2VF–A), and its reaction term depends on the film mass transfer coefficient, $k_f$, but it does not on $D_s$. The second approach in characterized by the hypothesis $c = L(q_R)$, which corresponds to the assumption that $k_f$ is large enough to consider Dirichlet conditions at the particle external surface. This formulation will be referred as 2VF–B and it depends on $S_t/B_i$ and therefore on $D_s$, but it does not on $k_f$. Finally the third formulation imposes simultaneously the hypothesis of 2VF–A and 2VF–B: $\bar{q} = q_R = L^{-1}(c)$. The model is then independent of $k_f$ and $D_s$. It will be referred as 1VF because it only depends on one variable. In this case, two equivalent formulations can be used, one in terms of c and the other in terms of $\bar{q}$. Both can be further simplified using the isotherm relationship presented in equation (2), assuming that $P_e$ is large and the following inequalities:

$$\frac{S_t}{B_i E_d} >> (1 + r_{\varepsilon_p}) \left( \frac{\partial L^{-1}(c)}{\partial c} \right)^{-1} \quad \text{and} \quad \frac{S_t}{B_i E_d} >> (1 + r_{\varepsilon_p}) \frac{\partial L(\bar{q})}{\partial \bar{q}}\,, \tag{4}$$

which are true for reference values, see table 1. Then, typical nonlinear first–order hyperbolic equations are obtained.

The solutions of these equations may present shocks when a high value of the unknowns precedes lower ones along the characteristics. In a one dimensional problem, with zero initial conditions and a boundary condition equal to $c^{in}$, or equivalently $\bar{q}^{in} = A(c^{in})^n$, chocks with the following velocities are found for each 1VF formulation:

$$v_{sh}^{c} = \frac{B_i E_d}{S_t} \frac{(c^{in})^{1-n}}{A\,n(2-n)} \quad \text{and} \quad v_{sh}^{\bar{q}} = \frac{B_i E_d}{S_t} \frac{(\bar{q}^{in})^{\frac{1}{n}-1}}{A^{\frac{1}{n}}}\,. \tag{5}$$

Note that these two velocities are in general not equal (except for $n = 1$). In order to determine which formulation is preferable, a global mass balance criterium is used: *In a one-dimensional problem, for any of time, the accumulated flow–in through the boundary should be equal to the mass inside the domain,*

**Table 1.** Dimensionless parameters for loading (left) and unloading (right)

| $P_e$ | $E_d$ | $A$ | $n$ | $r_{\varepsilon_p}$ | $P_e$ | $E_d$ | $A$ | $n$ | $r_{\varepsilon_p}$ |
|---|---|---|---|---|---|---|---|---|---|
| $10^5$ | 2.1 | 0.8 | 0.31 | 1.36 | $10^7$ | 0.021 | 1.04 | 0.31 | 1.36 |
| $S_t$ | $B_i$ | $S_t/B_i$ | $B_iE_d$ | $S_t/(B_iE_d)$ | $S_t$ | $B_i$ | $S_t/B_i$ | $B_iE_d$ | $S_t/(B_iE_d)$ |
| 50.3 | 0.083 | 603.3 | 0.17 | 292 | 0.51 | 0.084 | 6.03 | 0.002 | 292 |

*i.e. between the boundary and the position of the shock front.* This condition can be expressed as $v_{sh}\,\mathcal{M}(\bar{q}^{in}) = L(\bar{q}^{in})$, where $\mathcal{M}(\bar{q}) = S_t/(B_iE_d)\bar{q}$ is the approximation to the mass by unit of volume in dimensionless form under hypothesis of equations (4); the general expression is given by $\mathcal{M}(c, \bar{q}) = c + r_{\varepsilon_p} L(\bar{q}) + S_t/(B_iE_d)\bar{q}$. The c–based 1VF formulation verifies the condition only for $n = 1$, instead, the $\bar{q}$–based verifies the global mass balance for all $n$. Thus, the adequate 1VF is the $\bar{q}$–based formulation.

## 2.1 Dimensionless analysis

In this subsection, three relevant considerations about the dimensionless structure of the model are highlighted. First, note that hypothesis of equations (4), which have been used for 1VF analysis, allows to simplify also 2VF and 3VF formulations (although, as commented before, they remain nonlinear except for $n = 1$). Moreover, as $S_t/(B_iE_d)$ indicates the relative importance of the mass adsorbed with respect to the mass present in the interparticle fluid fase, hypothesis of equations (4) are expected to apply to all usual absorbent media. Recall that this ratio of dimensionless numbers is independent of reference velocity, $V$.

Second, the Biot number, $B_i$, which is also independent of $V$, indicates the relative importance of $k_f$ with respect to $D_s$. Previous works, see [1], present the following classification: high dependence on $k_f$ for $B_i << 1$, high dependence on $D_s$ for $B_i >> 100$, and relevance of both effects for $B_i \in [0; 100]$. In the model presented here, first case, $B_i << 1$, corresponds to 2VF–A formulation, which is governed by $S_t$, $B_iE_d$ and the other parameters of equations (2). Second case, $B_i >> 100$, corresponds to 2VF–B formulation, which depends on $S_t/B_i$, $E_d$ and the other parameters. And third case, $B_i \in [0; 100]$, to the general 3VF, which depend on all dimensionless parameters defined in equations (2).

Finally, note that the influence of reference velocity is restricted to $P_e$, $S_t$ and $E_d$. Low velocities imply small $P_e$ and large $S_t$ and $E_d$. As large $S_t$ and $E_d$ are also found with large $k_f$ and $D_s$, the 1VF formulation is expected to be the most appropriated for low velocities. Note that in this case the model depends basically on $S_t/(B_iE_d)$, and on $P_e$, for velocities in the order of $L/D$ or smaller.

## 3 Application: Working Capacity test

The model presented in previous section has been used to simulate the WC test. Each regime of the test (loading and unloading) is modelled with a different formulation. The representative dimensionless numbers for each regime are summarized in table 1. Note that, on one hand, $S_t/(B_i E_d)$ verifies hypothesis of equations (4), thus simplified formulations can be considered. Moreover, note that in both cases $B_i << 1$ thus 2VF–A or 1VF are expected. And finally, remark that 1VF is more likely to be appropriate for loading than 2VF–A, because of the low velocities, at least compared with those of unloading. On the other hand, $P_e^{-1}$ is much smaller than one (and therefore than $S_t/(B_i E_d)$), therefore interparticle diffusion is expected to be not relevant in any of both regimens, except, locally, in regions with very low velocities. It has been verified that the results obtained with calibrated parameters and real 3D–canisters, and using the same formulation throughout the domain (1VF for loading and 2VF–A for unloading) are satisfactory.

## 4 Conclusions

A mathematical model for transport and adsorption – desorption of hydrocarbons in activated carbon has been presented and analyzed. The model has been apply to the numerical simulation of the Working Capacity test with canisters (automotive filters for hydrocarbons in gas phase). Different formulations of the model are proposed for each part of the test, obtaining realistic results with complex tridimensional geometries.

*Acknowledgement.* The partial financial support of Expert Components S.A. and the Spanish government (Ministerio de Ciencia y Tecnología, grand number REN2001-0925-C03-01/CLI) are gratefully acknowledged.

## References

1. S. Baup, C. Jaffre, D. Wolbert, and A.Laplanche. Adsorption of pesticides onto granular activated carbon: determination of surface diffusivities using simple batch experiments. *Adsorption*, 6:219–228, 2000.
2. M. A. Hossain and D. R. Yonge. Finite Element Modeling of Single Solute Activated Carbon Adsorption. *J. of Environmental Engrg.*, 118(2):238–252, 1992.
3. P. C. Wankat. *Rate–Controlled Separations.* Blackie Academic and Professional, Glasgow, 1994.
4. S. Whitaker. *The method of volume averaging.* Kluwer academic publishers, 1999.

# Injection Vapour Model in a Porous Medium Accounting for a Weak Condensation

J. Pousin[1] and E. Zeltz[1]

[1] MAPLY, UMR CNRS 5585, INSA de Lyon, Centre des Mathématiques, 20 av. A. Einstein, 69621 Villeurbanne cedex, France jerome.pousin@insa-lyon.fr
[2] EricZeltz@aol.com

**Summary.** For studying the impact of a hight pressure vapor on a concrete wall, we propose a stationary 3D homogenized model. We show that the interface evolves as a (shock or rarefaction) wave accordingly with the mobility coefficient values $M$. Moreover, we prove the existence of a finite asymptotical position for the interface when $t$ goes to $+\infty$.

**Key words:** asymptotic expansion, multiphasic flows, Riemann's problem for the interface.

## 1 Motivating Problem and Mathematical Model

The aim of this paper is to provide a simple mathematical model for accounting for what have been reported in experimental studies ([3]): when a concrete wall is subjected to a steady isothermal hight pressure of water vapor, only a finite thickness of the wall is affected by the vapor penetration. Such a situation could arise when accidental context in nuclear power plants is considered. The simplifying physical hypotheses we are assuming to hold true are:
- the concrete wall can be represented with a porous medium made of parallelepiped pores, the thickness of which is negligible compare to the length and to the height. The height is also negligible compare to the length. Thus a Karman-Kozeny model can be used ([1, pp. 164–165]) and the 3D flow in the pore is reduce to a 2D one.
- The vapor viscosity $\eta_2$ is constant and large compare to the dry residual air viscosity $\eta_1$.
- Only a small part of the vapor can condense in the pores.
The previous simplifying physical hypotheses lead us to consider a small mobility coefficient for the experimental data. Nevertheless, the mathematical model derived in this paper also applies for large mobility coefficients. For a general mathematical model we refer to [2] for example.

In a previous paper [7], a mathematical model has been proposed for the injection of resin in thin mold, neglecting the surface tension and assuming that, almost everywhere in a pore the injection front is a regular curve. This last assumption is relevant for the injection of a much viscous fluid than the residual one [5]. Here, since we have gases, no surface tension has to be considered. Starting from this mathematical model, we add some terms due to the vapor condensation, and we show that the vapor injection front reaches an asymptotic limit, which compared to experimental results shows a good qualitative agreement. We consider a constant filter velocity $Q(t) = q$ and thus a constant average injection velocity in the pore domain $\Omega = (0, \varepsilon) \times (0, L)$. Thus the pore is constituted of $\Omega_1(t)$ full of dry air and $\Omega_2(t)$ where the vapor is. The pore is rescaled and for a fixed $z$, we denote by $x_2$ the abscissae of points of the interface $\Gamma(t) = \overline{\Omega_1(t)} \cap \overline{\Omega_2(t)}$ which have $z$ as height. We also

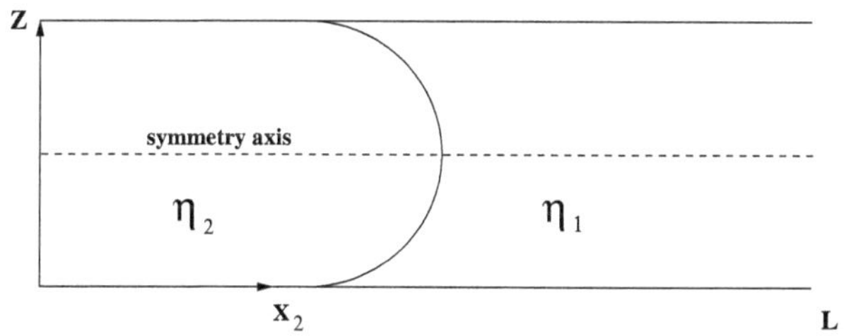


assume the two following hypotheses to be satisfied.

- The interface $\Gamma(t) \subset \Omega$ is symmetric with respect to the axis $z = \frac{1}{2}$;
- For fixed $t$ $\Gamma(t)$ is described with a continuous function $a : (x_2, t) \longmapsto z = a(x_2, t) \in [0, \frac{1}{2}]$.

Let $M = \dfrac{\eta_1}{\eta_2}$ be the constant mobility coefficient, and let the flux function of the conservation law of which the interface is solution, $f_M$ be defined by:
$$f_M(a) = \frac{a^2(2a-3)}{(M-1)(8a^3 - 12a^2 + 6a) - M}.$$
Denoting by $\mathbf{w}$ the flow velocity and by $w_a$ the velocity of the interface $\Gamma$ at point $(x_2, a(x_2, t))$ we have $w_a(a(x_2,t)) = \mathbf{w}(x_2, a(x_2, t))$. In [7] (Theorem 3) it is proved that the function $a$ is the unique entropic solution of a Riemann problem and the following expression for the velocity of the interface is obtained:

$$w_a(a(x_2,t)) = \begin{cases} q\ f'_M(a(x_2,t)) & \text{if } \ M \in [0; \frac{3}{2}] \\ \begin{cases} q\ f'_M(a(x_2,t)) & \text{if } a(x_2,t) \in [0, \alpha_M] \\ q\ f'_M(\alpha_M) & \text{if } a(x_2,t) \in [\alpha_M, \frac{1}{2}] \end{cases} & \text{if } \ M > \frac{3}{2} \end{cases} \tag{1}$$

where $\alpha_M \in ]0, \frac{1}{2}[$ is such that $f_M$ is strictly convex in $[0, \alpha_M]$ and strictly concave in $[\alpha_M, \frac{1}{2}]$. Since the velocity $w_a$ only depends on $z$ (and implicitly of $x_2$), it is straightforward to verify that: the average velocity for fixed $z$

$\frac{1}{x_2}\int_0^{x_2} w_a(a)d\xi = w_a(a)$. In the following, this relation will be generalized to the case with condensation. Let $\delta(t)$ be the fraction of vapor which condenses and denoting by $p_i$ the initial pressure, by $T$ the temperature and by $p_{vs}(T)$ the saturated vapor pressure, then $\delta(t)$ is given by ([6] p. 247): $\delta(t) = \delta = \begin{cases} \frac{p_i - p_{vs}(T)}{p_i} & \text{if } p_i \geq p_{vs}(T) \\ 0 & \text{if } p_i \leq p_{vs}(T) \end{cases}$.
We assume that $\delta$ is small, and that the vapor viscosity $\eta_2$ is not modified, and thus remains constant.

Let $\mathbf{w}$ be the velocity flow without condensation, the velocity flow with condensation $\mathbf{v}$ is defined by:

$$\mathbf{v}(x_2, z) = \mathbf{w}(x_2, z) - k(x_2, z, t)\mathbf{e_2} \tag{2}$$

where $k(x_2, z, t)$ accounts for the change due to condensation and is given by: $k(x_2, z, t) = \begin{cases} \delta^* x_2 & \in \Omega_2(t) \\ \delta^* x_2(z, t) \in \Omega_1(t) \end{cases}$ with $\delta^* = \frac{p_i - p_{vs}(T)}{p_{vs}(T)}$. Consequently we get for the interface velocity: $\mathbf{v}(x_2, a(x_2, t)) = \mathbf{v}(x_2, 1 - a(x_2, t))$ which writes:

$$v(x_2, a(x_2, t)) = w_a(a(x_2, t)) - \delta^* x_2 \tag{3}$$

Along the axis $z = a$ with $a$ fixed in $[0; \frac{1}{2}]$, the interface mean velocity $v_a(x_2, a(x_2, t))$ implicitly depends on $t$ and verifies:

$$v_a(x_2) = \frac{1}{x_2}\left(\int_0^{x_2} (w_a(a) - \delta^* \xi)d\xi\right).$$

It follows that at time $t$, the interface point with abscissae $x_2$ corresponding to $z = a(x_2, t)$ is solution to:

$$x_2 = v_a(x_2)t = \frac{1}{x_2}\left(\int_0^{x_2} (w_a(a) - \delta^* \xi)d\xi\right) t \tag{4}$$

Solving (4) with respect to $x_2$, we get the following result: the interface evolves as a rarefaction wave according to $\mathbf{M} \in [0, \frac{3}{2}]$

$$x_2 = \begin{cases} \left(\frac{q f'_M(z)}{1 + \frac{\delta^*}{2} t}\right) t & \text{if } 0 \leq z \leq \frac{1}{2} \\ \\ \left(\frac{q f'_M(1-z)}{1 + \frac{\delta^*}{2} t}\right) t & \text{if } \frac{1}{2} \leq z \leq 1 \end{cases} \tag{5}$$

For $(M > \frac{3}{2})$ there exists $\alpha_M \in ]0, \frac{1}{2}[$ such that the interface evolves as a shock and rarefaction wave attached at the entrance of the pore according to:

$$x_2 = \begin{cases} \left(\frac{q f'_M(z)}{1 + \frac{\delta^*}{2} t}\right) t & \text{if } 0 \leq z < \alpha_M; \\ \left(\frac{q f'_M(\alpha_M)}{1 + \frac{\delta^*}{2} t}\right) t & \text{if } \alpha_M \leq z \leq 1 - \alpha_M \\ \left(\frac{q f'_M(1-z)}{1 + \frac{\delta^*}{2} t}\right) t & \text{if } 1 - \alpha_M < z \leq 1 \end{cases} \tag{6}$$

Furthermore the following asymptotic property holds.

**Lemma 1.** *When $t$ goes to $+\infty$ the interface $\Gamma$ reaches the following position:*

$$\text{for } M \in [0, \tfrac{3}{2}], \quad x_2 = \begin{cases} \frac{2q}{\delta^*} f'_M(z) & \text{if } 0 \le z < 1/2 \\ \frac{2q}{\delta^*} f'_M(1-z) & \text{if } 1/2 < z \le 1 \end{cases} ;$$
$$\text{for } M \ge \tfrac{3}{2}, \quad x_2 = \begin{cases} \frac{2q}{\delta^*} f'_M(z) & \text{if } 0 \le z < \alpha_M \\ \frac{2q}{\delta^*} f'_M(\alpha_M) & \text{if } \alpha_M \le z \le 1-\alpha_M \\ \frac{2q}{\delta^*} f'_M(1-z) & \text{if } 1-\alpha_M < z \le 1 \end{cases} \tag{7}$$

Denote by $b : (x_2, t) \longmapsto z = b(x_2, t)$ the function describing the interface without condensation (*i.e.*, $\delta^* = 0$). For fixed $z$ by comparing the abscissae $x_2$ given by (4), then we have: $z = b(x_2, t) = a\left(\frac{x_2}{1+\frac{\delta^*}{2}t}, t\right) = a(X_2, t)$ where the change of variable $X_2 = \frac{x_2}{1+\frac{\delta^*}{2}t}$ is used. Since $b$ is solution to a Riemann problem, it follows that $a$ verifies:

$$\begin{cases} \dfrac{\partial a(X_2,t)}{\partial t} + q\dfrac{\partial}{\partial X_2}\left(\dfrac{f_M(a(X_2,t))}{1+\frac{\delta^*}{2}t}\right) = 0 \text{ in } ]0,+\infty[\times]0,T[ \\ a_d = a(X_2,0) = \frac{1}{2} \text{ for } X_2 > 0; \quad a_g = a(X_2,0) = 0 \text{ for } X_2 < 0 \end{cases} \tag{8}$$

Arguing in the same way as in [4], we prove there exists a unique entropic solution to (8). This section is ended with figures depicting the z-curves ($z = a(x_2, t)$) of the interface for $t$ from $0s$ to $1000s$ with a time step of $100s$ and the asymptotic position. Three values of $M$ are presented with $q = 1$ and $\delta^* = 0.01$.

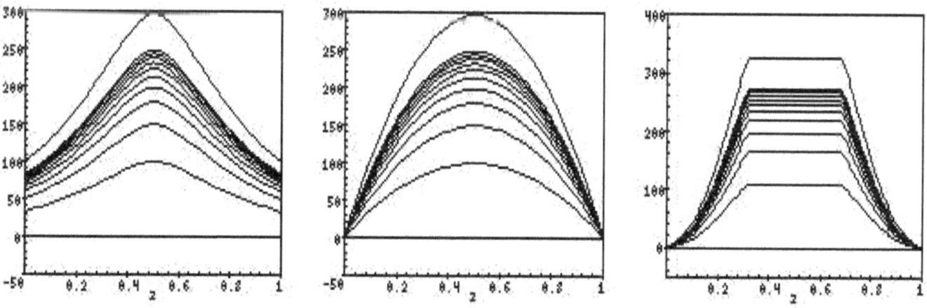

**Fig. 1.** Left: M=0, $\delta^* = 0.01$, $q = 1$; Center: $M = 1$, $\delta^* = 0.01$, $q = 1$; Right: $M = 5$, $\delta^* = 0.01$, $q = 1$.

## 2 Comparisons with Experimental Data

in the same way as in [7], where we have used the proposed model 5) on each pore, when $M \in [0, \frac{3}{2}]$ the mean position $X_{2_{moy}}$ of the interface is defined by

$$X_{2_{moy}}(t) \quad = 2t \int_0^{\frac{1}{2}} \left( \frac{q f'_M(z)}{1 + \frac{\delta^*}{2} t} \right) dz = \frac{qt}{1 + \frac{\delta^*}{2} t}. \tag{9}$$

The experimental case we consider is a cylindrical sample of concrete, the height of which is 1.3 m and the radius of which is 0.25 m. On the top of the cylinder a pressure of water vapor of 15 bars is applied with a temperature equals to $200°C$. The residual gas inside the cylinder is dry air.

The physical values of the parameters used are: a mobility coefficient $M = 0$, $\delta^* = \frac{p_i^* - p_{vs}(473)}{p_i^*} \approx 0.5\%$ and the filter velocity $q$ is evaluated to be : $q \approx 8 \times 10^{-4} m/s$. Then we find the following curve for the average macroscopic interface vapor-air: $t \mapsto \left( 2 \frac{0.08}{2 - 2 \times 0.005 + 0.005t} (1 - 0.005) \right) t$ the asymptotic position of which is : $\lim_{t \to +\infty} x(t) = 2 \frac{0.08(1-0.005)}{0.005} \approx 32$ cm. This result is in good agreement with the experimental data reported in [3]: the thickness of the wall influenced by the vapour penetration is about 25% of the total thickness of wall even for long time period.

## References


1. D. Bear. *Dynamics of Fluids in Porous Media.* Dover Publications, New York, 1988.
2. S. Benzoni-Gavage. Linear stability of propagating Liquid-Vapor interfaces, in capillary fluids. *Physica D*, 155:235–273, 2001.
3. Y. Billard, G. Debicki, L. Granger, and M. Shekarchi. Study of leaktightness integrity of containment wall without liner in high performance concrete under accindental conditions-I. Experimentation. *Nuclear Engineering and Design*, 213:1–9, 2002.
4. P. Concus and W. Proskurowski. Numerical solution of a nonlinear hyperbolic equation by the random choice method. *J. Comp. Physics*, 30:153–166, 1979.
5. Y. Dimakopoulos and J. Tsamopoulos. Trancient displacement of a Newtonian fluid by air in straight or suddenly constricted tubes. *Physics of fluids*, 15:1973–1991, 2003.
6. J.B. Hudson. *Surface Science.* Wiley, New York, 1998.
7. A. Maazouz, A. Mikelic, J. Pousin, and E. Zeltz. Fluid injection model without surface tension for resins in thin molds. *Journal of Comp. and Appl. Math*, 164-165:517–528, 2004.

# Multigrid Solution of Three-Dimensional Radiative Heat Transfer in Glass Manufacturing

M. Seaïd and A. Klar

Fachbereich Mathematik, TU Darmstadt, 64289 Darmstadt, Germany
{seaid,klar}@mathematik.tu-darmstadt.de

**Summary.** We implement a multigrid algorithm to solve the radiative heat transfer equations in glass production. The time, angle and space coordinates are discretized using Crank-Nicolson, discrete-ordinate and Galerkin methods, respectively. Based on the same mesh hierarchy for both heat conduction and radiative transfer, our multigrid algorithm consists on using the Newton-Gmres and Atkinson-Brakhage solvers as smoothers on the coarse meshes.

## 1 Introduction

Developing efficient and accurate techniques to solve Radiative Heat Transfer (RHT) equations attracts many researches from several applications as, radiation hydrodynamics, combustion, or glass manufacturing. In this later field, Rosseland approximation could be the most cheap (as far as the efficiency is concerned) solution for such equations. However, this approximation fails to resolve accurately the boundary layers in the cooling processes. In non diffusive limits (optically thin material) only the solution of the full radiative heat transfer can provide high quality products. In this paper, we present a multigrid algorithm to approximate the full RHT problem in three dimensional enclosure. The algorithm consists on linear and nonlinear multigrid techniques. Thus using same mesh hierarchy for both radiative transfer and heat conduction, the linear system arising from the discretization of radiative transfer is solved by multigrid method using the Atkinson-Brakhage approximate inverse as a preconditioner. On the other hand a multigrid solver, using Newton-Krylov as a smoother, is used for the discretized heat conduction. In both methods linear systems are solved only on the coarse mesh.

The main contribution of the present work is the application of efficient multigrid methods developed in [4, 3] to the three-dimensional RHT problem arising in glass cooling process. Computational results are shown for a cooling glass cube using optical spectrum of 283 frequency bands.

## 2 Radiative Heat Transfer in Glass Manufacturing

In this section we briefly recall the RHT equations used in glass cooling processes. For details on physical aspects and also mathematical studies we refer the reader to [2, 5, 1] and further references are cited therein. Thus the set of equations used in our numerical study is given by

$$\rho c \frac{\partial T}{\partial t} - \nabla \cdot (k \nabla T) = - \int_{\lambda_0}^{\infty} \int_{S^2} \kappa(\lambda) \Big( B_{\text{glass}}(T, \lambda) - I_\lambda \Big) d\Omega d\lambda, \quad (1)$$

$$\forall \lambda > \lambda_0 : \quad \Omega \cdot \nabla I_\lambda + \kappa(\lambda) I_\lambda = \kappa(\lambda) B_{\text{glass}}(T, \lambda), \quad (2)$$

$$k\mathbf{n} \cdot \nabla T + h(T - T_b) = \varepsilon \pi \int_0^{\lambda_0} \Big( B_{\text{air}}(T_b, \lambda) - B_{\text{glass}}(T, \lambda) \Big) d\lambda, \quad (3)$$

$$I_\lambda(\hat{\mathbf{x}}, \Omega) = \varrho(\mathbf{n} \cdot \Omega) I_\lambda(\hat{\mathbf{x}}, \Omega') + (1 - \varrho(\mathbf{n} \cdot \Omega))\, B_{\text{air}}(T_b, \nu), \quad \mathbf{n} \cdot \Omega < 0, \quad (4)$$

$$T(0, \mathbf{x}) = T_0(\mathbf{x}), \quad (5)$$

where $\rho$ is the density, $c$ denotes the specific heat capacity, $\mathbf{x} = (x, y, z)^T$ the position vector, $t$ the time, $T$ the temperature, $I_\lambda$ the spectral intensity, $k$ the thermal conductivity, $\kappa$ the absorption coefficient, $\lambda$ the wavelength, $h$ is the convective heat transfer coefficient, $T_b$ is a given temperature of the surrounding, $\mathbf{n}$ denotes the outward normal in $\hat{\mathbf{x}}$ on the boundary and $\varepsilon$ the mean hemispheric surface emissivity in the opaque spectral region $[0, \lambda_0]$, where radiation is completely absorbed. $B_m(T, \lambda)$ is the spectral intensity of the black-body radiation given by the Planck's function in a medium $m$. In our simulations we used data provided by Schott Glaswerke in Germany and are as follows:

$$\rho = 2200\ kg/m^3, \quad c = 900\ J/kgK$$
$$k = 1, \quad h = 0.001, \quad \varepsilon = 0.92$$
$$T_0 = 1000\ K, \quad T_b = 300\ K$$
$$B_m(T, \nu) = n_m \frac{2\hbar c_0^2}{\lambda^5} \left( e^{\hbar c_0 / \lambda k T} - 1 \right)^{-1}$$
$$n_{\text{air}} = 1, \quad n_{\text{glass}} = 1.46$$

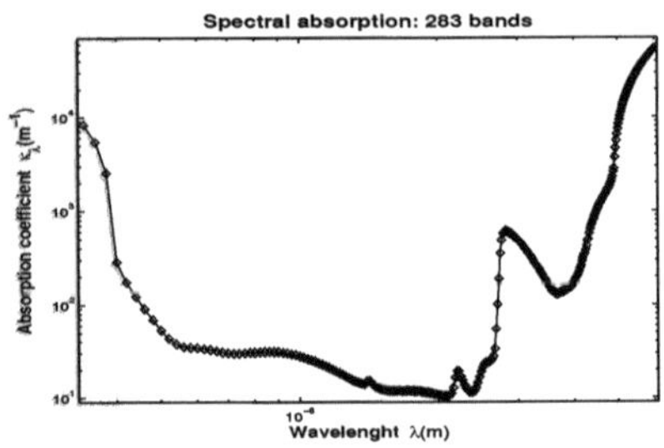


**Fig. 1.** Physical data for glass manufacturing.

In (4), $\Omega' = \Omega - 2(\mathbf{n} \cdot \Omega)\mathbf{n}$ is the specular reflection of the ordinate $\Omega = (\nu, \eta, \xi)^T$, $\varrho \in [0, 1]$ is the reflectivity obtained according to the Fresnel's law. Thus, for an incident angle $\theta_1$ given by $\cos\theta_1 = |\mathbf{n} \cdot \Omega|$ and Snell's law

$$n_{\text{air}} \sin\theta_2 = n_{\text{glass}} \sin\theta_1,$$

the reflectivity $\varrho(\mu)$, $\mu = |\mathbf{n} \cdot \Omega|$, is defined as follows

$$\varrho(\mu) = \begin{cases} \dfrac{1}{2} \left( \dfrac{\tan^2(\theta_1 - \theta_2)}{\tan^2(\theta_1 + \theta_2)} + \dfrac{\sin^2(\theta_1 - \theta_2)}{\sin^2(\theta_1 + \theta_2)} \right), & \text{if} \quad |\sin\theta_1| \le \dfrac{n_{\text{air}}}{n_{\text{glass}}}, \\ 1, & \text{otherwise.} \end{cases}$$

## 3 Multigrid Solution Procedure

The multigrid solver for RHT equation consists on writing the equations (1)-(5) in formal way as a fixed point problem for the temperature $T$ only

$$T = \mathcal{H}\left(\int_{\lambda_0}^{\infty}\int_{S^2}\kappa(\lambda)\Big(B_{\text{glass}}(T,\lambda) - \mathcal{S}\big(I_\lambda, B_{\text{glass}}(T,\lambda)\big)\Big)d\lambda\right), \qquad (6)$$

where $\mathcal{H}(Q)$ is solution operator for the heat conduction problem

$$\rho c\frac{\partial T}{\partial t} - \nabla\cdot(k\nabla T) = -Q, \qquad (7)$$

subject to boundary condition (3) and initial condition (5); and $\mathcal{S}(I_\lambda, q)$ is the solution operator for the radiative transfer problem

$$\forall\lambda > \lambda_0: \quad \Omega\cdot\nabla I_\lambda + \kappa(\lambda)I_\lambda = \kappa(\lambda)q, \qquad (8)$$

subject to boundary condition (4). Note that both problems (7) and (8) can be discretized and solved separately with different discretizations and solvers. In the present work, we have implemented the multigrid solver from [4] to solution operator $\mathcal{S}$, while the solution operator $\mathcal{H}$ has been carried out using mutigrid techniques from [3]. Hence, given a hierarchy of nested meshes, fine-to-coarse (restriction) and coarse-to-fine (prolongation) intergrids transfer operators the Newton's iteration applied to (6) results in

$$T_h^{(k+1)} = T_h^{(k)} - \mathcal{R}_h'\big(T_h^{(k)}\big)^{-1}\mathcal{R}_h\big(T_h^{(k)}\big), \qquad (9)$$

where $\mathcal{R}_h$ is the nonlinear residual associated to fixed point problem (6). The subscripts $H$ and $h$ refer to the coarse and fine mesh, respectively.

An iteration (9) requires, at each time step, both solution of heat conduction (7) and radiative transfer (8). Dividing the spectrum in finite bands with piecewise constant absorption, using discrete-ordinates and Galerkin methods for discretization of the angle and space coordinates the equation (8) can be reformulated as a linear system for the mean intensity $\varphi = \int_{S^2} I_\lambda d\Omega$

$$\big(\mathbf{I} - \mathbf{A}_h\big)\varphi_h = \mathbf{b}_h, \qquad (10)$$

Compare [4] for the construction of the schur matrix $\mathbf{A}$ and the right-hand side $\mathbf{b}$. In (10), $\mathbf{I}$ denotes the identity matrix. The multigrid method we used to solve the linear system (10) can be carried out by the following iterations

$$\varphi_h^{(k+1)} = \varphi_h^{(k)} + \mathbf{B}_H^h\mathbf{r}^{(k)}, \qquad \mathbf{B}_H^h = \mathbf{I} + (\mathbf{I} - \mathbf{A}_H)^{-1}\mathbf{A}_h, \qquad (11)$$

where $\mathbf{r} = \mathbf{b}_h - (\mathbf{I} - \mathbf{A}_h)\varphi_h$ is the residual associated to (10) and $\mathbf{B}_H^h$ is the Atkinson-Brakhage approximate inverse which can be viewed as a smoother.

Using Crank-Nicolson and Galerkin methods for time and space discretizations, equation (7) is transformed to a steady nonlinear heat problem. To solve this problem the Newton's iterations (9) are used along with the Atkinson-Brakhage approximate inverse as a preconditioner in solving the linear system for Newton's directions. For more details on the formulation of these methods and their implementation we refer the reader to [4, 3].

# 4 Results

We present numerical results for a cooling of glass cube. The cube is 1 $m$ length and time duration of the cooling process is 10 sec. The glass parameters used in our simulations are those listed in Fig. 0. These data are kindly suggested by ITWM Kaiserslautern. A wavelength interval of 283 bands, as shown in the right column of Fig. 0, is considered. For the discrete-ordinates we used the $S_n$-approximation sets with $n(n+2)$ directions, the space domain is discretized into $N$ grid points in each dimension and a time step of $\Delta t = 0.01$ sec is used in computations.

In Fig. 2, we plot the temperature distribution on a part of the cube (for better insight) and the temperature profile along a cross section at $y = z = 0.5$ m. We used $S_8$-set with 80 directions, $N = 80$ grid points in the finest mesh and $N = 20$ grid points in the coarsest mesh. For comparison, we have included the results obtained by the Rosseland approach. It is clear that the Rosseland approximation is unable to capture the correct cooling behavior.

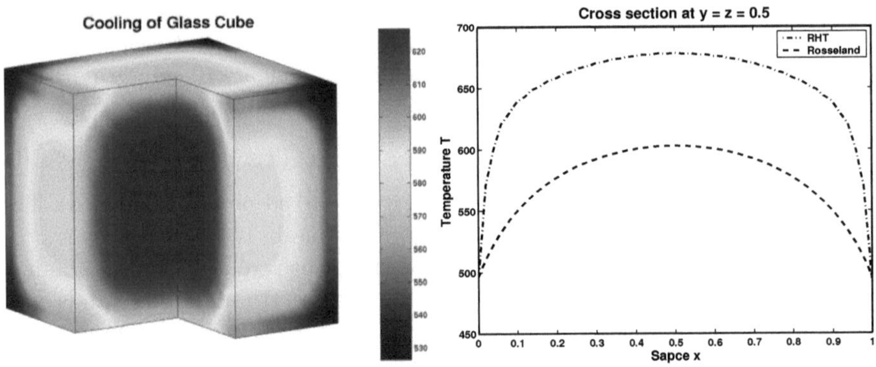


**Fig. 2.** Temperature distribution on the cube (left) and a section at $y = z = 0.5\ m$ for the computed solution by RHT equations and Rosseland approach (right).

Fig. 3 (left) summarizes the computational cost of the multigrid method. The computational cost is distributed in four stages constituting the multigrid algorithm: *RTE* denotes the percent of the total CPU time to solve the radiative transfer equation (8). *Gmres* represents the percent of CPU time involved in the Gmres method for solving linear system in (9). *Grids* denotes the percent of CPU time required for prolongation and restriction by intergrids transfer. *Newt* refers to the percent of CPU time employed in the Newton's algorithm. This includes Jacobian approximation, backtracking linesearch and construction of the right-hand side in (7). The percentage breakdown shown in Fig. 3 (left) is based on a total CPU time of 8 hours assuming no display.

The main features reported in this figure are on one hand, both Newton's operations and the intergrids transfer procedures require very little computa-

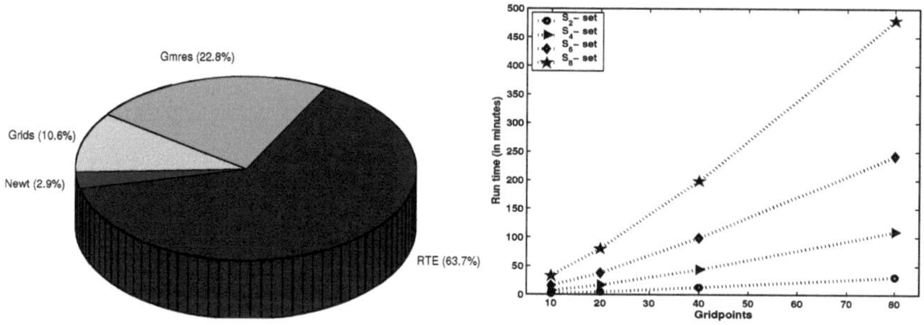


**Fig. 3.** Computational cost of the multigrid algorithm for RHT equations.

tional cost compared to the CPU time needed for the Gmres solver. On the other hand, most of the computational effort goes into solving the radiative transfer problem (8). Therefore, reducing the CPU time in numerical methods for RHT equations (1) – (5) can be hold by constructing more efficient preconditioned iterative solvers for the linear system (10).

Our next concern is to test the influence of spectral discretization on the efficiency of multigrid method. To this end, we have run the algorithm using different $S_n$ sets of discrete-ordinates. In Fig. 3 (right), comparison of run times for $S_2$, $S_4$, $S_6$ and $S_8$ sets are given as a function of gridpoints number $N$ used in computations. As can be seen from Fig. 3 (right), the use of higher order discretizations for the unit sphere leads to considerable increase in the computational work in multigrid algorithm.

## References

1. A. Klar, J. Lang, and M. Seaïd. Adaptive solutions of $\mathrm{SP_n}$-approximations to radiative heat transfer in glass. *Int. J. Thermal Sciences.*
2. E. Larsen, G. Thömmes, A. Klar, M. Seaïd, and Th Götz. Simplified $\mathrm{P_n}$ approximations to the equations of radiative heat transfer and applications. *J. Comp. Phys.*, 183:652–675, 2002.
3. M. Seaïd, M. Frank, A. Klar, R. Pinnau, and G. Thömmes. Efficient numerical methods for radiation in gas turbines. *J. Comp. Applied Math.*, 170:217–239, 2004.
4. M. Seaïd and A. Klar. Efficient preconditioning of linear systems arising from the discretization of radiative transfer equation. *Lecture Notes in Computational Science and Engineering*, 35:211–236, 2003.
5. G. Thömmes, R. Pinnau, M. Seaïd, Th. Götz, and A. Klar. Numerical methods and optimal control for glass cooling processes. *Transp. Theory Stat. Phys.*, 31:513–529, 2002.

# DEM Simulations of the DI Toner Assembly

I.E.M. Severens[1] and A.A.F. van de Ven[2]

[1] Océ Technologies B.V., Venlo, the Netherlands isev@oce.nl
[2] Department of Mathematics, Technische Universiteit Eindhoven, Eindhoven, the Netherlands a.a.f.v.d.ven@tue.nl

**Summary.** This paper describes the modelling of the toner behaviour in the development nip of the Océ Direct Imaging print process. The discrete element method is used as the simulation tool for a quantitative description of the system. The interaction rules and the associated parameters are determined for the toner particles and the surfaces of the development rollers. The model is validated with print quality results. It is shown that it is possible to achieve quantitative agreement between DEM simulations and experimental print quality results.

**Key words:** DEM, electromagnetism, toner.

## 1 Introduction

The Océ Color Technology is called Direct Imaging (DI). The heart of the image development process is formed by a Direct Imaging Unit. The print quality of the DI technology is primarily determined by the toner flow in the region between a DI-drum and an imaging roller; see Fig. 1. The collection of toner between the DI-drum and the imaging roller is called the DI toner assembly. The simulation of toner deposition conducted here is based on the discrete element method, first proposed by Cundall [2] in 1971. In the discrete element method (DEM) all toner particles as well as the rollers, are considered discrete elements. Each element interacts with its neighbouring elements and its surroundings. These interactions are modelled on a microscopic scale: the motion of each particle is tracked numerically. Every time step the forces that act on a particle are summed and from this the speed and the displacement of the particle is calculated by integration of Newton's second law of motion. The macroscopic behaviour of the toner flow and print output is then simulated using DEM.

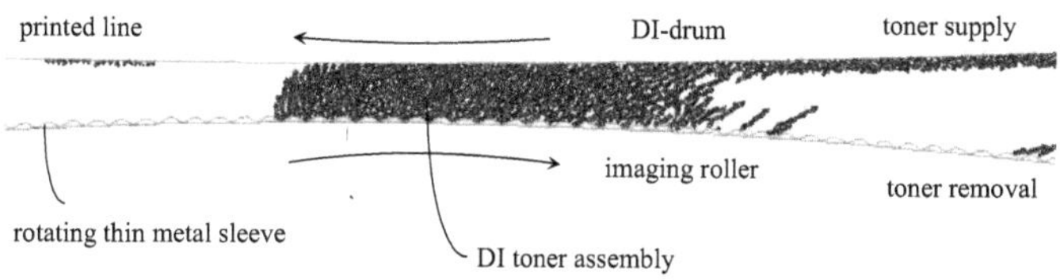


**Fig. 1.** The DI toner assembly.

## 2 Force Models

In this section, we set up models for forces between toner particles. The nature of these forces are due to collisions, friction, adhesion, and electromagnetic actions.

### 2.1 Geometry

In DEM simulations all discrete elements have to be provided with a geometry to indicate the shape of the real object. A toner particle is described by $n$ clustered spheres. More realistic toner geometries can thus be achieved by increasing the number of clustered spheres that form one toner particle.

### 2.2 Collisions

DEM simulations involve modelling each collision between particles and between particles and the boundary objects. During a collision, having a certain contact time, particles deform, energy is dissipated in the form of heat, and particles restore to their original shape. A collision is modelled by penetration of the objects during collision. The penetration is described as a certain overlap $\xi$ between two objects and models the temporary deformation of the objects during collision. A range of contact force models are available which approximate the collision dynamics to various extents, and are of the general form

$$F_n = -k\xi^\alpha - \gamma \frac{d}{dt}(\xi^\beta), \tag{1}$$

where $F_n$ is the normal force between the colliding particles during the collision. The linear spring-dashpot model ($\alpha = \beta = 1$) approximates the collision dynamics to a good extent.

Collisions between particles are in general not head-on, and the particles have angular velocity. Therefore, shear also has to be taken into account.

The shear contact force component $F_s$ is generally modelled with a Coulomb friction model:

$$\begin{aligned} F_s &= -\mu_d F_n \mathrm{sign}(v_s), v_s \neq 0, \\ |F_s| &< \mu_s F_n, v_s = 0, \end{aligned} \tag{2}$$

where $\mu_s$ is the coefficient of static friction, $F_n$ the normal force at the contact ($F_n > 0$ always), $v_s$ the relative tangential velocity of the two particles, and $\mu_d$ the coefficient of dynamic friction ($\mu_d < \mu_s$).

### 2.3 Adhesion Force

When two materials are brought into each others vicinity, they exert an attracting force onto each other. This force is referred to as the adhesion force. Hamaker's theory [3] is used here.

### 2.4 Magnetic Force

The general expression for the magnetic force $\mathbf{F}_{\mathrm{mag}}$ is [4]

$$\mathbf{F}_{\mathrm{mag}} = (\mathbf{m} \cdot \nabla)\mathbf{B}, \tag{3}$$

with $\mathbf{m}$ the magnetic moment of the particle and $\mathbf{B}$ the external magnetic induction. In the DI toner assembly the magnetic field originates from two sources: the field from the magnets within the imaging roller $\mathbf{B}_m$ and the field from the magnetized surrounding toner particles $\mathbf{B}_a$. So, in general, we can write

$$\mathbf{F}_{\mathrm{mag}} = (\mathbf{m} \cdot \nabla)(\mathbf{B}_m + \mathbf{B}_a). \tag{4}$$

### 2.5 Electric Force

To enable electric field toner development, the electric force exerted on the toner particle by the externally applied field strength on the toner particle must be stronger than the magnetic force on that particle. Unfortunately, because of their quadratic nature, electric forces cannot be determined by superposition. We adopt here the approach of [1] and [5] and use bispherical coordinates to solve the problem of calculating the force on a conducting toner particle in the field of an infinitely large electrode.

### 2.6 Charge Model

When a pixel has to be printed, a voltage difference is applied between the imaging roller and a track in the DI-drum. This voltage difference causes toner particles in the DI toner assembly to get charged and to experience an electric force towards the DI-drum, which is stronger than the magnetic force towards the imaging roller. An $SiO_x$ layer, a dielectric layer above the conducting tracks, makes sure that the electric charge on the toner does not leak to the conducting tracks. Due to the dynamics of the DI toner assembly conducting paths are formed and broken. The conducting paths consist of toner-toner contacts and toner-imaging roller contacts. In a first-order approach it is assumed that the electrical contacts can be treated as ideal electric resistances and capacities, and that no charge is transferred in the direction along the ring electrodes through the DI toner assembly.

## 3 Results

We are now able to calculate and visualize the behaviour of the DI toner assembly, which consists of at most 10,000 particles, for a time frame of about 15ms in approximately one night on a PC. As an example we will show here the results for printed dots. Printed dots, also referred to as pixels, do not have perfectly sharp edges. The quality of pixels can be expressed in terms of edge sharpness. Distinction can be made between normal edge sharpness $r_n$ in the direction of the ring electrodes of the DI-drum and axial edge sharpness $r_a$ directed perpendicular to the ring electrodes of the DI-drum.

The ultimate goal is a tool that can predict aspects of print quality correctly. The normal edge sharpness $r_n$ can be split into sharpness of the front edge $r_{n,f}$ and sharpness of the back edge $r_{n,a}$. We will show as an example case the results for the normal edge sharpness when applying DEM for the settings of the black development unit of the Océ CPS700. A total number of fifty simulations were run, where in each simulation a line was printed. From these fifty printed lines an average coverage curve is calculated. From this average curve a sigmoid fit is calculated. The sigmoid fit and the experimentally determined coverage profile for the black unit of the Océ CPS700 are displayed in Fig. 2. Good agreement between the experimental results and the simulation results is observed.

## 4 Conclusion

The dominating forces in the DI toner assembly are due to collisions, friction, adhesion, and electromagnetic actions; all other forces such as gravitational forces and forces due to air flow are neglected. We have derived models for these forces. The force models form the interaction rules between toner particles themselves and between toner particles and rollers.

The parameters of the force models are determined by using experimental data. In cases that no experimental data is available, experiments are set up for the determination of the correct values of the parameters. The model

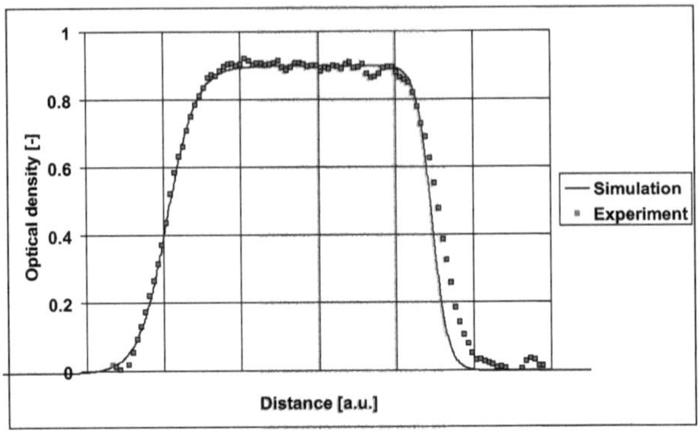


**Fig. 2.** The calculated and measured average coverage curve for the black unit of the Océ CPS700.

parameters form the distinction between the description of the print process of the DI technology and other processes.

The discrete element method has been used for simulating the behaviour of the DI toner assembly in the development nip of the Océ Direct Imaging print process. It is shown that by determining the appropriate interaction rules and the associated parameters, it is possible to gain quantitative agreement between experimental and simulation results, specifically on important aspects as print quality.

The result of the project will be used within Océ as a design tool for fundamentally new ideas in the field of toner development, as a substitute for time-consuming experiments and as an educational tool for new researchers in the field of toner development.

## References

1. J.S. Berry. Electrostatic forces on a conducting sphere due to a charge on a dielectric half-space. *J. Phys. A: Math. Gen.*, 9:1939–1945, 1976.
2. P.A. Cundall. A computer model for simulating progressive large-scale movements in block rock systems. In *Proc. Symp. Int. Soc. Rock Mech.*, 1971.
3. H.C. Hamaker. *Physica*, 4:1058, 1937.
4. K. Hutter and A.A.F van de Ven. *Field Matter Interactions in Thermoelastic Solids: A Unification of Existing Theories Of Electro- Magneto-Mechanical Interactions*, volume 88 of *Lecture Notes in Physics*. Springer, Berlin, 1978.
5. W.E. Warren and R.E. Cuthrell. Electrostatic forces between conducting spheres at constant potential. *Journal of Applied Physics*, 46:4597–4599, 1975.

# Modeling of Drying Processes in Pore Networks

A.G. Yiotis[1,2], A.K. Stubos[1], A.G. Boudouvis[2], I.N. Tsimpanogiannis[3], and Y.C. Yortsos[4]

[1] National Center for Scientific Research "Demokritos", 15310 Athens, Greece
[2] School of Chemical Engineering, National Technical University of Athens, 15780 Athens, Greece
[3] Los Alamos National Laboratory, Earth & Environmental Sciences Division (EES-6), Los Alamos, NM 87545, USA
[4] Department of Chemical Engineering, University of Southern California, Los Angeles, CA 90089-1211, USA

**Summary.** Drying in porous structures is simulated with a 2-D pore network model that accounts for various processes at the pore-scale (mass transfer by advection and diffusion in the gas phase, viscous flow in liquid and gas phases and capillary effects at the gas-liquid interface). We further study the effect of capillarity-driven viscous flow through macroscopic liquid films. It is shown that film flow is a major transport mechanism in drying of porous media, its effect being dominant when capillarity controls the process, which is the case in typical applications.

## 1 Introduction

Drying of porous materials is of interest in many industrial applications such as coatings, food, paper, textile, wood, ceramics, soil remediation and oil recovery. Traditional descriptions of the process rely on phenomenological approaches, in which the porous medium is a continuum, the dependent variables, like moisture content, are volume-averaged quantities and the relation of fluxes to gradients is via empirical coefficients. Such approaches essentially ignore the effect of the pore microstructure, which is of key importance for a quantitative understanding of the two-phase flow process. Many pore-scale mechanisms are involved and should be taken in account: the motion of individual gas-liquid menisci residing in the pore space; diffusion in the gas and the liquid phase; viscous flow in both phases; capillarity and liquid flow through connected films. In earlier experiments using glass-bead packs [3], viscous forces and liquid films were found to be important. Existing pore-network models address mostly slow drying, controlled by capillarity and diffusion, ignoring advection and/or viscous effects [1, 2]. They also neglect the significant role of liquid films. In the first part of this paper we present results from a pore network simulator that accounts for all major mechanisms at the pore scale except liquid films. A detailed description of this first part can be found in [6]. In the second part, we propose a mathematical model that accounts for viscous flow in the liquid films. This part is a short description of the detailed study by [5].

## 2 Pore network modeling of drying without the presence of liquid films

Consider the isothermal drying of a fractured porous medium initially saturated with a volatile liquid that is trapped in the pore space due to capillary forces and may vaporize as a result of an injected purge gas. As the actual overall problem is quite complex, we consider a 2-D square pore network with all but one boundaries impermeable to flow and mass transfer (Fig. 1(a).). At any time during the process, evaporation at the liquid-gas interface leads to the receding of the liquid front (evaporating interface (I) or percolation front (P) in Fig. 1(a)). In general, three different spatial regions can be identified:
(i) a far-field region consisting of the initial liquid (CC);
(ii) a region where the liquid phase is disconnected (DC); and
(iii) a near-field region with continuous gas phase and the liquid in the form of films, the thickness of which is progressively reduced towards a "totally dry" regime.

In Fig. 1(a), the network consists of pore bodies connected via pore throats. A liquid pore is invaded by gas when the pressure difference across its throats exceeds the capillary pressure threshold. Two dimensionless parameters, a diffusion-based capillary number, $Ca_D$, and a Péclet number, $Pe$, in addition to the various geometrical parameters of the pore network, mainly characterize the problem. $Ca_D$ expresses the ratio of viscous to capillary forces, based on a diffusion-driven velocity, while $Pe$ expresses the ratio of inertial to diffusion forces. Liquid films are neglected in this formulation.

We discuss two runs on a $50 \times 50$ pore network that are characteristic of the two limiting regimes that develop in this process. In the first run, the gas flow rate (and the $Pe$ value) through the fracture is very low ($Pe = 0.66$). In this case capillary forces are dominant and mass transfer occurs primarily by diffusion. In the second run the purge gas is injected at a very high flow rate ($Pe = 596$). In this case, viscous forces dominate at the liquid-gas interface while mass transfer occurs primarily by advection. In the low $Pe$ case, every cluster follows the Invasion Percolation (IP) pattern, in which the next throat to be invaded by gas is that with the smallest capillary threshold among all perimeter throats of that cluster. For high $Pe$, the process is controlled by viscous forces and capillarity is negligible at early times. Phase distribution patterns deviate substantially from IP and almost follow a piston-like displacement (PD). However, as the liquid-gas interface recedes in the pore network, viscous forces become weaker and the process gradually becomes of IP type,

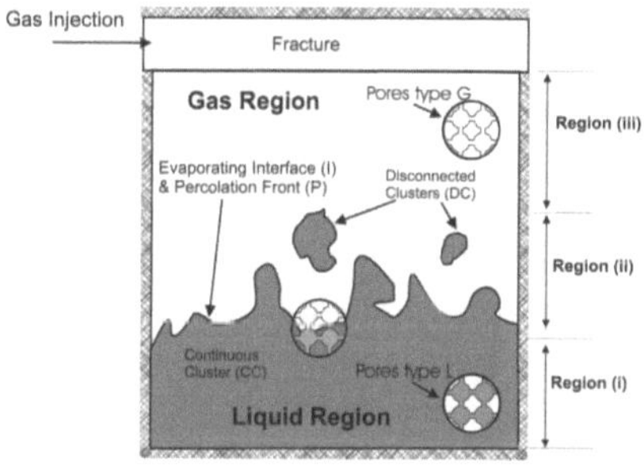


(a) Schematic representation of a drying process in a 2D matrix driven by the injection of a purge gas through a fracture along the upper side of the matrix

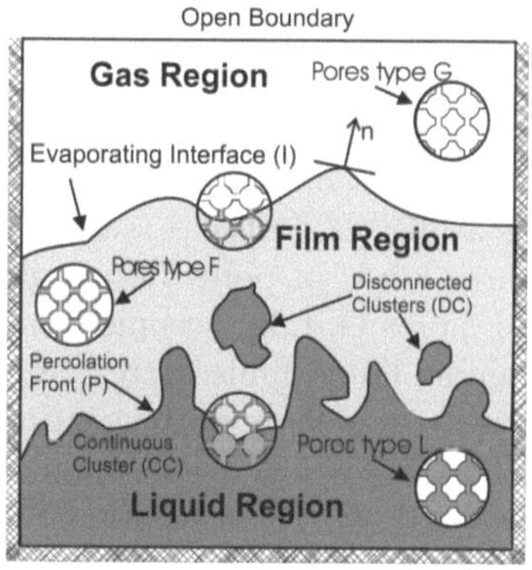


(b) Schematic of liquid and gas phase patterns, indicating the various types of pores in drying used in this study

**Fig. 1.**

namely controlled by capillarity. Drying curves for the process have also been obtained showing (as expected) that the overall drying rate is much smaller in the high $Pe$ case. For details see [6].

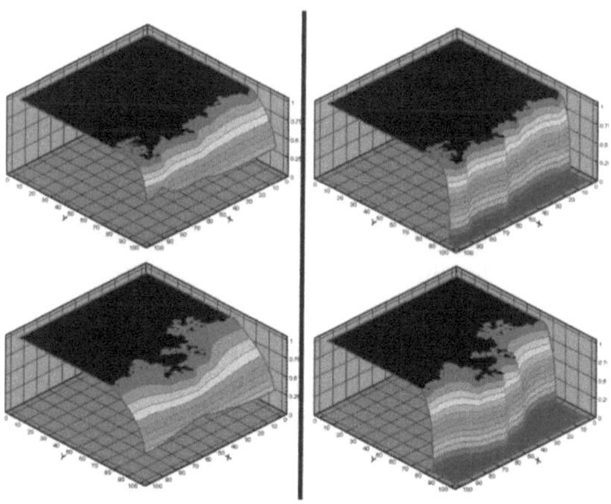

**Fig. 2.** Profiles of the rescaled film radii for $Ca_F = 10^{-4}$ (left) and $Ca_F = 1$ (rigth) at two different stages of the process. Liquid clusters are in black, the fully dry region is in blue.

## 3 The effect of liquid films

In this section we study the role of wetting films in the context of drying. Our focus is on the effect of viscous flow through the liquid films that develop at the corners of the pore network assuming that the velocity of the purge gas is negligible. In the presence of films, the pore space can be characterized by three kinds of pores (close-ups in Fig. 1(b)): pores L, fully occupied by liquid, pores G, fully occupied by gas, and pores F, occupied by gas but also containing liquid films. Here, we account for viscous effects both in the films (F pores), as well as in the continuous liquid phase (region L).

The thickness of each liquid film can be parameterized by its radius of curvature r, which is a function of time and distance. Assuming local capillary equilibrium at the film interface, we can show that the liquid pressure in the film is inversely proportional to its thickness [5]. Any gradient in the film thickness along the capillary results in a pressure gradient along the liquid films. A capillarity-induced flow develops along the film from the cross-section where the film is thicker towards the cross-section where the film is thinner. [5] introduced the capillary number in the form $Ca_F = \frac{\pi D C_e 2\mu_l \beta}{\rho_l C^* r_0 \gamma}$. This capillary number expresses the ratio of the viscous forces due to flow driven by mass transfer to capillary forces.

The gas region of the network contains F pores adjacent to capillaries that contain films (film region) and G pores adjacent to dry capillaries (dry region) (Fig. 1(b)). Assuming that we know the location of the percolation front P at any time, we can solve the full problem using the simple transformation [5]:

$$\Phi \equiv \frac{\rho^3 + \zeta Ca_F}{1 + Ca_F} \tag{1}$$

that satisfies the Laplace equation in regions G and F. We assume that the film thickness $\rho$ is approximately constant at the percolation front P, i.e. $\Phi = 1$ and that drying is driven by imposing $\Phi = 0$ at the open side of the network ($\Phi = 0$ at the open end of the network). Using the above transformation, the solution of the Laplace equation determines the profiles of the film radius and the concentration, the rates of drying through each film, as well as the location of the interface I, where the films terminate and evaporation occurs.

*Phase Distribution Patterns*

For values of $Ca_F$ less than order of 1 (typical case in most physical problems) the effect of the capillary number on the phase distribution patterns is negligible. In that range, the patterns follow IP rules. The left panel in Fig. 2 shows two snapshots of the percolation front for $Ca_F = 10^{-4}$ that correspond to IP patterns. As the capillary number increases, the patterns should eventually depart from IP, particularly at the early times of the process. However, the right panel of Fig. 2 shows that even for $Ca_F = 1$ the pattern is almost identical to IP. It takes larger values, of the order of $Ca_F = 10$ for the effect to be pronounced [5]. Then, the pattern exhibits the expected behavior of viscous "stabilization" [4]. It follows that under typical conditions and for all practical purposes, the drying front can be accurately described as an IP front.

*Extent of liquid films*

For low values of $Ca_F$, the films extend all the way to the open boundary (where practically all evaporation occurs, Fig. 2-left panel). By contrast, when $Ca_F$ is of order 1, the films are short and the film tips (evaporation interface I) reside closer to the liquid cluster interface P (Fig. 2-right panel). This leads to the formation of a completely dry region G, the extent of which increases with time.

*Drying Curves*

The overall drying rate at the open side of the pore network is found to scale as $\frac{Ca_F+1}{Ca_F}$ . The drying rate increases as $Ca_F$ decreases and the film tips are closer to the open boundary. At smaller values of $Ca_F$, capillarity helps to transport liquid over larger distances and to keep the film extent longer.

## 4 Conclusions

We present results from a 2-D pore network model for isothermal drying in porous media that includes mechanisms like mass transfer by advection and diffusion in the gas phase, viscous flow in liquid and gas phases and capillary effects at the gas-liquid menisci in the pore throats. In a further step, we proceed to study the effect of capillarity-driven flow in macroscopic liquid films. Using a novel transformation, it was found that film flow is a major transport mechanism, its effect being dominant when capillarity controls the process, which is the case in typical applications.

## References

1. J.B. Laurindo and M. Prat. Numerical and experimental network study of evaporation in capillary porous media. Drying rates,. *Chem. Eng. Sci.*, 53:2257–2269, 1998.
2. M. Prat and F. Bouleux. Drying of capillary porous media with a stabilized front in two dimensions. *Phys. Rev. E.*, 60:5647–5656, 1999.
3. T.M. Shaw. Drying as an immiscible displacement process with fluid counterflow. *Phys. Rev. Lett.*, 59:1671–1674, 1987.
4. I.N. Tsimpanogiannis, Y.C. Yortsos, S. Poulou, N. Kanellopoulos, and A.K. Stubos. Scaling theory of drying porous media. *Phys. Rev. E.*, 59:4353–4365, 1999.
5. A.G. Yiotis, A.K. Stubos, A.G. Boudouvis, I.N. Tsimpanogiannis, and Y. C. Yortsos. Effect of liquid films on the drying of porous media. *AIChE. J.*, 50:2721–2737, 2004.
6. A.G. Yiotis, A.K. Stubos, A.G. Boudouvis, and Y.C. Yortsos. A 2-D pore-network model of the drying of single-component liquids in porous media. *Adv. Water Resour.*, 24:439–460, 2001.

# Mathematical Modelling of Flow through Pleated Cartridge Filters

V. Nassehi, A.N. Waghode, N.S. Hanspal, and R.J. Wakeman

Department of Chemical Engineering Loughborough University Loughborough, Leicestershire, UK

## Introduction

Mathematical modelling of coupled creeping incompressible Stokes flow and low permeability Darcy flow still presents mathematical and practical challenges. In this paper, we present a finite element model for the prediction and quantitative analyses of the hydrodynamic behaviour of deadend pleated cartridge filters used in aeronautical applications. Elemental discretisation in this scheme is based on the use of unequal order approximation functions for velocity and pressure fields. We show that this discretisation generates unified stabilisation for both Stokes and Darcy equations and prevents 'numerical locking' whilst preserving the geometrical flexibility of the computational grid.

## Mathematical Statement of the Combined Stokes/Darcy Flow Regimes

Consider a flow model consisting of the following equations,

$$\mathbf{A}(\mathbf{u}) + \boldsymbol{\nabla} p = \mathbf{f} \tag{1}$$

$$\boldsymbol{\nabla} \cdot \mathbf{u} = 0 \tag{2}$$

in $\Omega$, where $\Omega \subset \Re^2$ is a bounded domain with a continuous boundary $\Gamma$, $\mathbf{u}$ is the velocity vector, $\mathbf{f}$ is the body force vector and $p$ is the pressure. Depending upon the choice of operator $\mathbf{A}$, we obtain two different flow models,

1. $\mathbf{A}(\mathbf{u}) = \mathbf{I}\eta\mathbf{K}^{-1}\mathbf{u}$, where $\mathbf{I}$ is the identity matrix, $\eta$ is the viscosity of the fluid, $\mathbf{K}$ is the permeability of the medium, which is the Darcy equation
2. $\mathbf{A}(\mathbf{u}) = -2\nabla \cdot \eta\mathbf{D}(\mathbf{u})$, where $\mathbf{D}(\mathbf{u})$ is the symmetric part of the velocity gradients tensor, $\eta$ is the viscosity of the fluid, which is the Stokes equation.

In the modelling of incompressible flow the Ladyzhenskaya-Babuska-Brezzi (LBB) stability condition must be satisfied [4]. This poses a severe restriction on the type of approximating spaces that can be used. In the context of the finite element method, different models based on various strategies have been developed which satisfy the LBB condition. They range from the least squares Galerkin technique to the use of elements generating unequal order interpolation functions for the field unknowns and schemes which depend on a perturbed continuity equation representing slightly compressible fluids [5]. However, some of these techniques fail whilst the others are too complex and pose severe problems for their applicability to modelling of combined flow within cartridge filters [3]. In order to avoid the problems with time stepping schemes, we have used unequal order interpolation functions for velocities and pressure. In this scheme, the incompressibility constraint can be used without any modifications (i.e. eqn (2) remains as $\nabla \cdot \mathbf{u} = 0$). In addition to the requirement of LBB criterion, it is also known that due to the incompatibility of the operators in the Stokes and Darcy equations, the approximating function spaces used for the numerical solution of these equations need to be different. Essentially, as explained by [1], the Darcy equation should be treated as an elliptic Poisson equation where the degrees of freedom should be kept as low as possible. The corresponding approach for the Stokes equation, which would correspond to the use of constant (discontinuous) approximation for the pressure and linear approximation for the velocities, results in generating trivial solutions for the velocity. Our numerical experiments demonstrate that this phenomenon, called numerical locking, can be resolved by using a mixed $P_2/P_1$ approximation for the field variables in the coupled flow regimes. Therefore, the present scheme is developed utilising $C^0$ continuous Taylor-Hood element which is a member of the bubble element family [2] and which corresponds to such approximations for velocities and pressure. The element by element satisfaction of *inf-sup* condition is a necessary and sufficient criterion for the stable and accurate solution of combined free/porous flow problem. This condition is based on the definition of the distance between the exact and finite element approximation defined as,

$$d(u, V_h) = \inf_{v_h \in V_h} \|u - v_h\| = \|u - \tilde{u}_h\|$$

Therefore we need to find conditions on $V_h$ so that,

$$\|u - u_h\| \leq c d(u, V_h)$$

where $c$ is a constant independent of $h$

In the absence of body forces, the creeping, steady state, isothermal flow of a non-Newtonian fluid in coupled free/porous regimes can be described by the Stokes and Darcy equations respectively. To couple the two different flow regimes, the Darcy equation is imposed effectively as the boundary condition for the Stokes equation at the free/porous interface (and vice-versa in the porous/free interface). For a complete mathematical model and the numerical linking procedure, the readers can refer to [3].

## Results and Discussions

The fluid is considered to be a shear-thickening generalised Newtonian fluid with a power law index of 1.15. The consistency coefficient in the power law model is taken to be 80 Pa.s and the density of the fluid is assumed to be 970 kg/m$^3$. The porous matrix is assumed to be homogeneous and isotropic with a permeability value of $10^{-12}$ m$^2$. In the first case, a single pleat with an idealised shape of a filter cartridge is taken as a domain under consideration as shown in Fig. 1. The thickness of the pleat is 0.59 mm and is very small compared to the dimensions of the overall domain, which are 0.005m width and 0.020m overall length. The predicted pressure field for this domain is shown in Fig. 2. The simulated pressure drop in both the free flow regions is nearly zero and the sole contribution to the overall pressure drop value is by the porous matrix where a gradual variation in the pressure drop can be observed in the flow direction. The developed pressure field in a porous region is shown as an enlarged view at the top in Fig. 2. The nature of the pressure variations can be well supported by the velocity vectors over the domain shown in Fig. 3. The flow is observed to be almost parabolic towards the end of the first free flow region. As soon as the fluid approaches the curved boundaries of the porous region, it gets diverted due to the sudden obstruction. Maximum amount of the fluid finds its direction towards the narrow channel formed by the straight part of the porous pleat and the solid impermeable walls of the domain. The fluid then intrudes through the porous walls and plug flow behaviour is observed throughout the porous matrix. In the second free flow domain, the flow again tends towards a parabolic pattern and near the exit it becomes fully developed.

To validate the model, the results were compared against the analytical solutions. Simulations were carried out for the combined Stokes-Darcy flow inside a rectangular duct (Fig. 4). The range of permeability values used in these simulations varies from $10^{-8}$m$^2$ to $10^{-12}$ m$^2$. The analytical solutions were obtained by calculating the pressure drop in the porous regime in the flow direction using the $x$-component of Darcy's relation. Simulated pressure drop figures (as listed in Table 1) were obtained and the analytical pressure drops tally very closely, indicating the validity of the developed model.

Finally, a quarter section of an actual cartridge assembly is considered as a problem domain. Since the pleated cartridge is symmetrical, a quarter section can represent the whole cartridge domain. The cartridge filter assembly consists of three distinct flow regimes. The outer region is the free flow zone between the metal casing and the surface of the porous cartridge. The porous zone is made of the cartridge itself, whereas the inner region is the free flow zone between the porous cartridge and the inner core through which the clean fluid exits. The inlet flow velocity of 0.1 m/s is applied at the inlet i.e. on the periphery of the first free flow zone. Stress free boundary conditions are imposed at the outlet. Again the fluid properties are identical to those used in the previous case. Line of symmetry boundary conditions are imposed on the vertical and horizontal straight sides of the domain. The predicted pressure distribution over the domain is illustrated in Fig. 5. It is evident from Fig. 5 that the nature of developed pressure field exactly corresponds to the pressure variation observed in a single pleat domain shown in Fig. 2. The continuity of mass is preserved with the discrepancy between inflows and outflows is only 0.22%.

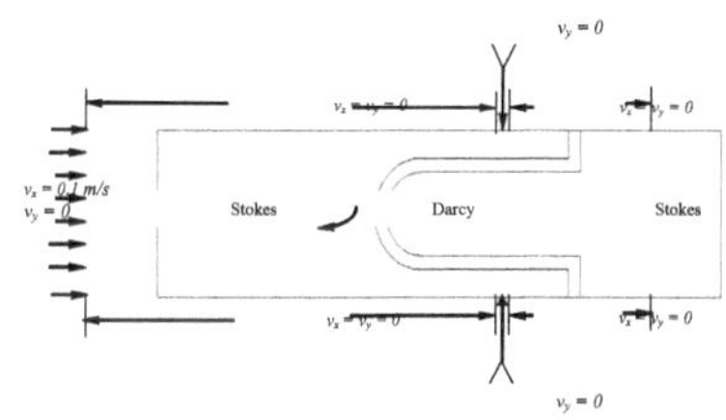


**Fig. 1.** Schematic Representation of an Idealised Pleat of a Pleated Cartridge

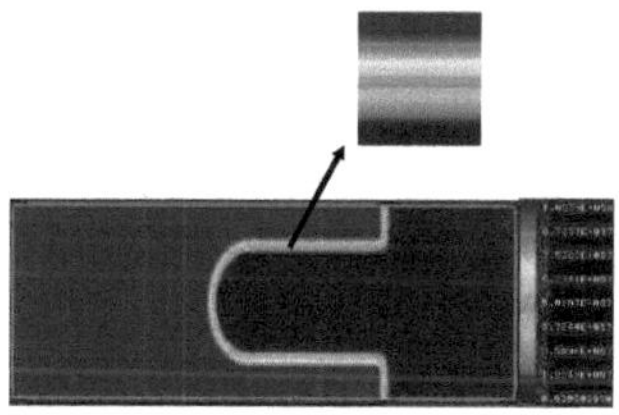

**Fig. 2.** Predicted Pressure Distribution (Pa) in a Single Pleat Domain

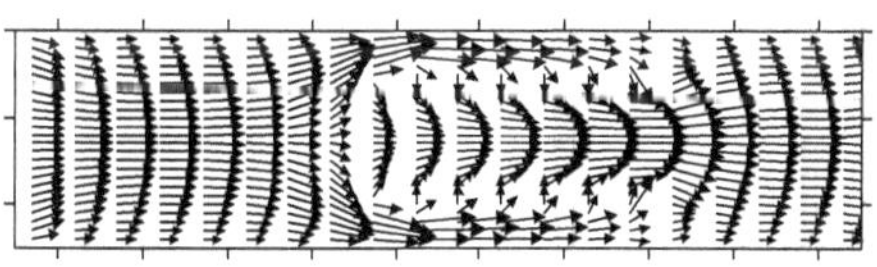

**Fig. 3.** Predicted Velocity Flow Field in the Single Pleat Domain

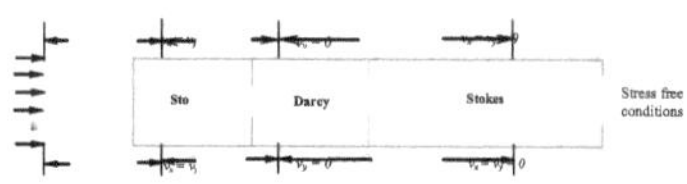


**Fig. 4.** Boundary Conditions on the Stokes/Darcy/Stokes Rectangular Flow Domain

## Conclusions

A finite element model for the solution of the combined free/porous flow problem has been developed without imposition of any artificial boundary conditions at the free/porous interface. It has been concluded that the numerical scheme provides a highly robust and reliable method for the combined flow simulation without any mathematical problems arising from stability requirements, numerical locking or time stepping. The discrepancy between the inlet and outlet flows is found to be less than 0.22% confirming the validity of the

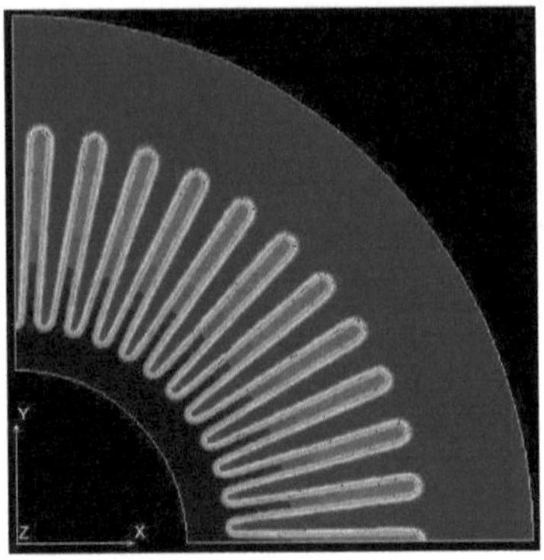


**Fig. 5.** Predicted Pressure Field Distribution (Pa) in the Quarter Cartridge Domain of Pleated Cartridge Assembly

**Table 1.** Comparison of Simulated and Analytical pressure drops at different permeability values

| Permeability ($m^{-2}$) | Pressure Drop according t0 Darcy's Law (Pa) | Simulated Pressure Drop(Pa) |
|---|---|---|
| 10-8 | 6.00E+05 | 6.1307724E+05 |
| 10-9 | 6.00E+06 | 6.0000598E+06 |
| 10-10 | 6.00E+07 | 5.9848202E+07 |
| 10-11 | 6.00E+08 | 5.9832319E+08 |
| 10-12 | 6.00E+09 | 5.9830001E+09 |

developed model. It is expected that through the utilisation of more sophisticated and powerful members of this bubble element family, relatively coarse meshes can be used to maintain computing economy.

## References

1. E. Burman and P. Hansbo. A unified stabilized method for Stokes' and Darcy's equations. Technical report, Chalmers Finite Element Center, Chalmers University of Technology, Göteborg, Sweden,, 2002.
2. M. Fortin, D.N. Arnold, and F. Brezzi. A stable finite element for the Stokes equations. Technical report, Calcolo, 1984.
3. V. Nassehi, N.S. Hanspal, A.N. Waghode, W.R. Ruziwa, and R.J. Wakeman. Finite element modeling of combined free/porous flow regimes: Simulation of flow through pleated cartridge filters. *Chemical Engineering Science Journal*, 2004.
4. J. N. Reddy. *An Introduction to the Finite Element Method.* McGraw-Hill, 2nd edition, 1993.
5. O.C. Zienkiewicz and J. Wu. Incompressibility without tears - how to avoid restrictions on mixed formulation. *International Journal of Numerical Methods in Engineering*, 32:1189–1203, 1991.

# Comparison of Some Mixed Integer Non-linear Solution Approaches Applied to Process Plant Layout Problems

J. Westerlund[1] and L.G. Papageorgiou[2]

[1] Process Design Laboratory, Åbo Akademi University, Turku 20500, Finland
[2] Department of Chemical Engineering, UCL (University College London), London WC1E 7JE, UK

**Summary.** The Process Plant Layout (PPL) problem involves decisions concerning the spatial allocation of equipment items and the required connections among them, [3]. The objective of the PPL problem is to determine the optimal spatial allocation of equipment items and the required connections between them. PPL problems have mostly been solved by heuristic rules but in recent years significant research effort has been put on more rigorous methods, mainly based on mathematical programming techniques. The resulting problem is often subsequently discretised in linear form and solved using linear solvers. In this paper, a non-linear approach to the general PPL problem is investigated. A comparative study between different non-linear solvers is carried out and the performance of the solvers is evaluated.

## 1 Introduction

Optimal plant layout is of great concern both during design of new industrial facilities and for retrofit of existing plants. The geometry of the PPL problem results in large scale combinatorial optimisation problems requiring significant computational effort for their solution. New solution strategies as well as formulation enhancement techniques to cut down the computational complexity is, thus, a research topic of great relevance in order to enable solution of large scale industrial PPL problems. The PPL problem has previously been solved as discretised Mixed Integer Linear Programming (MILP) problems using candidate areas to represent the plant floor area, thereby avoiding non-linear area constraints. Solving the PPL problem as a Mixed Integer Non-Linear Programming (MINLP) problem is one way of approaching the PPL problem from a new angle and at the same time achieving better solution quality. The MINLP approach presented in this paper may also be used as a foundation for different efficient solution approaches such as spatial decomposition methods.

## 2 Problem formulation

The PPL problem formulation used in this study is based on the formulations of [5]. The objective function consists of a number of cost drivers: 1) Fixed connection costs for the connections between the items,[$C^c_{ij}$]. 2) Distance dependent pumping costs for the connected items, [$C^v_{ij}, C^h_{ij}$]. 3) Fixed floor construction cost for each floor, [FC1]. 4) Area dependent floor construction cost for each floor, [FC2]. And 5), area dependent land cost for the plant area [LC]. The objective function formulated mathematically is shown in equation (1) where NF is the total number of floors, $TD_{ij}$ the total rectilinear distance between items $i$ an $j$. Variables $R_{ii}$, $L_{ii}$, $A_{ii}$ and $B_{ii}$ are used to determine the relative distance between items $i$ and $j$, horizontally. A rectangular shape is used as the floor area. In the MINLP approach the dimension of the floor area, FA is determined as a bilinear function (4). While in the MILP approach, candidate areas are used in order to discretise the problem in linear form. For MILP solutions, the floor area is determined as shown in eq. (2) and (3) where $AR_1, \ldots, AR_n$ is a set of proposed, candidate areas. $Q_s$ is a binary variable equal to one if area s is selected; 0 otherwise.

$$\sum_i \sum_{j \neq 1/f_{ij}=1} \left[C^c_{ij} TD_{ij} + C^v_{ij} D_{ij} + C^h_{ij}(R_{ij} + L_{ij} + A_{ij} + B_{ij})\right] + \\ + FC1 \cdot NF + FC2 \cdot NF \cdot FA + LC \cdot FA \quad (1)$$

$$FA = \sum_s AR_s Q_s \quad (2)$$

$$\sum_s Q_s = 1 \quad (3)$$

$$FA = Y^{\max} \cdot X^{\max} \quad (4)$$

To avoid equipment items being allocated outside the plant area and to prevent unit overlap, the area constraints presented in [4] were used, the disjunctive non-overlapping constraints thus being transformed to conjunctive form using big-M constraints.

## 3 Non-Linear Solution Approaches

The following MINLP solvers were used in the comparative work:

1) GAMS/SBB using the Non-Linear Branch and Bound Method.
2) GAMS/Dicopt using the Outer Approximation method.
3) a-ECP using the Generalised Extended Cutting Plane Method.

SBB is based on the Non-Linear Branch and Bound method. One NLP problem is solved in each node. As NLP solver, a NLP code, Conopt, [2] was used. SBB tend to require unreasonable CPU time for solution due to the large number of non-linear function evaluations required to obtain the solution. This is because a large number of nodes must be visited, and a NLP problem must be solved in each node.

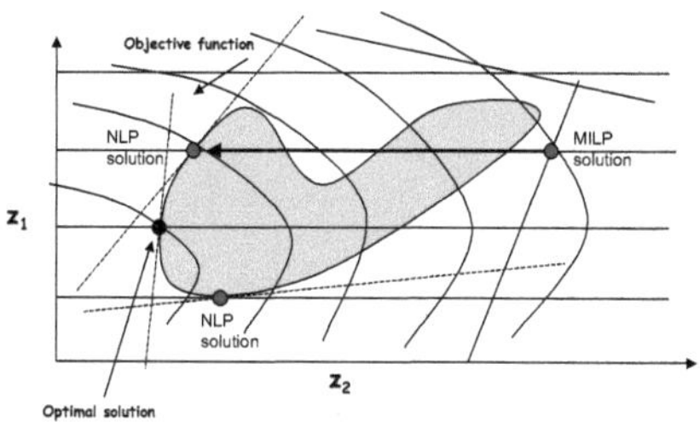


**Fig. 1.**

Dicopt, based on the Outer Approximation method (OA) using an augmented penalty function [6], overestimates the entire feasible region by using supports generated by solving an alternating sequence of MILP and NLP.

The alpha-ECP code uses the Generalised Extended Cutting Plane method [8]. Alpha-ECP encapsulates the optimal solution by cutting planes from non-linear constraints and supports from a reduction constraint when solving a sequence of MILPs, shown in Fig. 1.

When using the alpha-ECP (Fig. 1), the reduction constraints are gained when solving a sequence of MILPs. the MILPs may be solved only to integer feasible MILP solutions (but the final MILP to optimality). Alpha-ECP offers reasonable CPU time as well as good solution quality for the investigated PPL problems. Concerning CPU time, Dicopt appears to be the most efficient of the solvers presented in this paper.

The global optimization code GGPECP (Generalised Geometric Programming Extended Cutting Plane [7]) was used to obtain the global optimum for the illustrative problems. The GGPECP method is able to handle signomial constraints (in this case; bilinear terms) rigorously. The GGPECP code is based on a transformation approach [1] integrated in the Extended Cutting Plane (ECP) method [8]. The solution results obtained using the non-linear formulation is compared to the results, obtained by dicretising the non-linear model, and solving the resulting model using a linear solver. Solution quality and CPU time is evaluated and compared. In the discretised linear case the Mixed Integer Linear Programming (MILP) solver CPLEX, version 8.0 is used.

## 4 Illustrative examples

To enable comparison between the MINLP solvers, two sets of problems were solved. The first set of problems are based on a PPL problem with a 5 unit instant coffee process (ex.1), shown in Fig. 2. The other set is based on a 7 unit ethylene oxide plant (ex.2), shown in Fig. 3.

To highlight the difference in solution performance between the MINLP approach versus an MILP approach, the illustrative problems were also solved using the discretised MILP approach. Two different discretisation grids were used when solving the problems using the MILP approach. In table 1 the solution data for the examples is displayed. Table 1 shows the CPU time and the solution quality for each solver.

The different discretisation grids are: a) coarse grid, fast solution at the expense of the solution quality, 25 candidate areas. b) fine grid, better solution quality at the expense of the CPU time, 100 candidate areas.

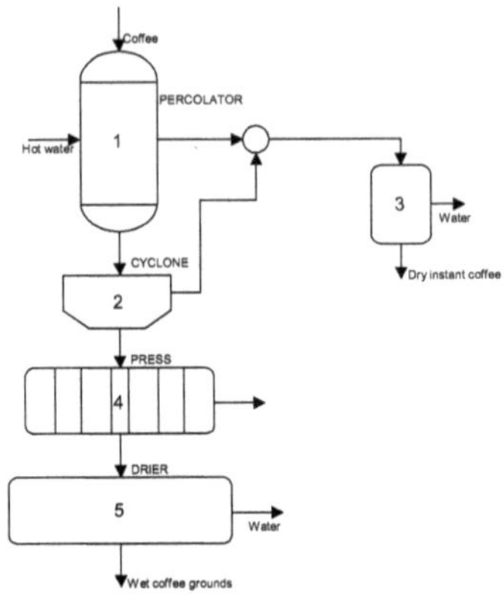


**Fig. 2.**

The solution quality is displayed as a percentage of the global optimum, values over 100% being sub-optimal. Problems $1_i$ are different versions of ex.1 and problems $2_i$ are different versions of ex.2. The major differences being equipment item dimensions and connection costs.

As the MILP approach, using a fine discretisation grid, tend to require exhaustive CPU times, the second set of examples (ex.2) is only solved using the coarse discretisation grid. The coarse grid results in far less complex problems and better solution performance at the expense of the solution quality.

## 5 Conclusions

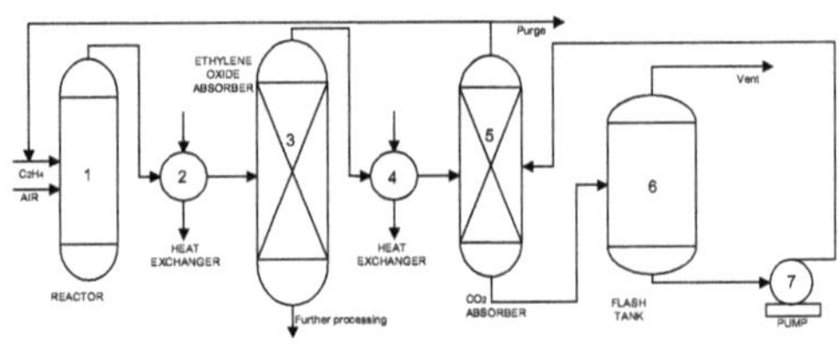


**Fig. 3.**

The PPL problem can be formulated as non-convex MINLP or as discretised MILP problems. In this study some solution approaches for the PPL problem, formulated as MINLP problems were investigated. Solving a PPL problem as an MINLP problem generally result in better solution quality (with the same computational effort). The performance of three different MINLP solvers was investigated using two different sets of PPL problems. The Outer Approximation method using an augmented penalty function, as well as the Extended Cutting Plane method, appears to handle the problems well providing good quality solutions within reasonable CPU times. The non-linear Branch and Bound method on the other hand seem to require excessive CPU time for solution. The solution data of all MINLP solvers was compared to solution data of the same problems solved using a discretised linear approach.

*Acknowledgement.* The first author gratefully acknowledges the financial support from TEKES, the National Technology Agency of Finland.

**Table 1.** CPU time and solution quality for ex.1 and 2

| Problem | MILP(cpoarse) | | MLIP (fine) | | Gams (Dicopt) | | Gams(SBB) | | a-ECP | |
|---|---|---|---|---|---|---|---|---|---|---|
| | CPUs | Quality | CPUs | Quality | CPUs | Quality | CPUs | Quality | CPUs | Quality |
| 1-1 | 11.8 | 113% | 120.4 | 101% | 25.7 | 104% | 130.9 | 194% | 19.5 | 100% |
| 1-2 | 43.4 | 135% | 314.1 | 108% | 8.1 | 100% | 2000* | 100% | 12.6 | 126% |
| 1-3 | 12.4 | 111% | 126.1 | 103% | 4.5 | 100% | 2000* | 107% | 13.9 | 100% |
| 1-4 | 33.6 | 114% | 195.9 | 104% | 13.9 | 103% | 2000* | 100% | 113.4 | 102% |
| 1-5 | 9.1 | 121% | 103.6 | 107% | 7.1 | 102% | 2000* | 120% | 45.9 | 100% |
| 1-6 | 5.2 | 114% | 82.9 | 105% | 3.3 | 106% | 683.3 | 103% | 13.7 | 100% |
| 1-7 | 4.2 | 138% | 41.5 | 105% | 2.4 | 138% | 1691.4 | 128% | 18.1 | 100% |
| 1-8 | 25.3 | 108% | 498.8 | 104% | 4.1 | 100% | 2000* | 100% | 24.7 | 100% |
| 1-9 | 1.7 | 128% | 458.8 | 108% | 4.1 | 106% | 2000* | 106% | 5.2 | 105% |
| 1-10 | 3.7 | 135% | 39.4 | 107% | 2.9 | 109% | 2000* | 119% | 21.7 | 100% |
| Mean 1 | 15.0 | 122% | 198.2 | 105% | 7.61 | 107% | 835.2 | 118% | 28.87 | 103% |
| 2-1 | 773.7 | 109% | – | – | 576.2 | 100% | 42300* | 100% | 125.8 | 100% |
| 2-2 | 3359.94 | 106% | – | – | 850.8 | 100% | 43200* | 104% | 1776.7 | 100% |
| 2-3 | 53262.8 | 109% | – | – | 273.58 | 100% | 43200* | 100% | 875.57 | 100% |
| 2-4 | 1472.69 | 107% | – | – | 150.0 | 101% | 10661.3 | 101% | 819.01 | 100% |
| 2-5 | 14074.5 | 109% | – | – | 8276.4 | 100% | 14200.3 | 106% | 27354 | 105% |
| 2-6 | 12517.1 | 104% | – | – | 6994.7 | 100% | 11237.3 | 114% | 4205.5 | 101% |
| 2-7 | 6379.67 | 106% | – | – | 9817.7 | 101% | 16939.5 | 103% | 5470.7 | 105% |
| 2-8 | 6311.02 | 102% | – | – | 514.06 | 101% | 17990.4 | 111% | 121.11 | 100% |
| 2-9 | 2.6 | 111% | – | – | 38.3 | 100% | 6390.2 | 107% | 16.7 | 100% |
| 2-10 | 8110.45 | 102% | – | – | 43200* | 107% | 43200* | 108% | 10325 | 107% |
| Mean 2 | 10626.4 | 107% | – | – | 3054.7 | 101% | 12903.2 | 105% | 5109 | 102% |

*means that the CPU time limit was reached, 2000s for example 1*i* and 43200 for 2*i*

## References


1. K.M. Björk, P.O. Lindberg, and T. Westerlund. Some convexifications in global optimization of problems containing signomial terms. *Computers and Chemical Engineering*, 27:669–679, 2003.
2. A. Drud. CONOPT: A GRG code for large sparse dynamic nonlinear optimization problems. *Mathematical Programming,*, 31:159–191, 1985.
3. J.C. Mecklenburgh. Process plant layout. Technical report, London: Institution of Chemical Engineers, 1985.
4. L.G. Papageorgiou and G.E Rotstein. Continuous-domain mathematical models for optimal process plant layout. *Ind. Eng. Chem.*, 37:3631–3639, 1998.
5. D.I. Patsiatzis and L.G. Papageorgiou. Optimal multi-floor process plant layout. *Computers and Chemical Engineering*, 26:575–583, 2002.
6. J. Viswanathan and I.E. Grossmann. A combined penalty function and outer approximation method for MINLP optimization. *Computers and Chemical Engineering*, 14:769–782, 1990.
7. T. Westerlund and Westerlund J. GGPECP—a global optimization MINLP algorithm. *Chemical Engineering Transactions*, 3:1045–1050, 2003.
8. T. Westerlund and R. Pörn. Pseudo-convex mixed integer optimization problems by cutting plane techniques. *Optimization & Engng*, 3:253–280, 2002.

# A Mathematical Model of Three-Dimensional Flow in a Scraped-Surface Heat Exchanger

S.K. Wilson[1], B.R. Duffy[1], and M.E.M. Lee[2]

[1] Department of Mathematics, University of Strathclyde, 26 Richmond Street, Glasgow, G1 1XH, UK s.k.wilson@strath.ac.uk, b.r.duffy@strath.ac.uk
[2] School of Mathematics, University of Southampton, Highfield, Southampton, SO17 1BJ, UK m.e.m.lee@soton.ac.uk

**Summary.** We present a simple mathematical model of fluid flow in a Scraped-Surface Heat Exchanger (SSHE). Specifically we consider steady isothermal flow of a Newtonian fluid around a periodic array of pivoted scraper blades in a channel with one stationary and one moving wall, when there is an applied pressure gradient in a direction perpendicular to the wall motion. The flow is fully three-dimensional, but decomposes naturally into a two-dimensional transverse flow driven by the boundary motion and a longitudinal pressure-driven flow.

**Key words:** Mathematical Modelling, Scraped-Surface Heat Exchanger

## 1 Scraped-Surface Heat Exchangers (SSHEs)

Scraped-surface heat exchangers (SSHEs) are widely used in the food industry to cook, chill or sterilize certain foodstuffs quickly and efficiently without causing unwanted changes to the constitution, texture and appearance of the final product. A SSHE is essentially a cylindrical steel annulus whose outer wall is heated or cooled externally; the foodstuff is driven slowly by an axial pressure gradient along the annulus, and a "bank" of blades rotating with the inner wall (the "rotor") is used to scrape it away from the outer wall (the "stator"), preventing fouling, and maintaining mixing and heat transfer. The blades typically are arranged in groups of two (180° apart) or four (90° apart); sometimes pairs of blades are "staggered" axially. The processes that take place inside SSHEs are complex; operating conditions vary with context, and operators are guided largely by experience and empirical correlations. Typically SSHEs are used on highly viscous foodstuffs; examples include purées, sauces, margarines, jams, spreads, soups, baby-foods, chocolate, mayonnaise, caramel, fudge, ice-cream, cream and yoghurt. Such foodstuffs commonly behave as non-Newtonian materials, typically being shear-thinning, viscoplastic and/or viscoelastic, as well as being inhomogeneous, and possibly undergoing

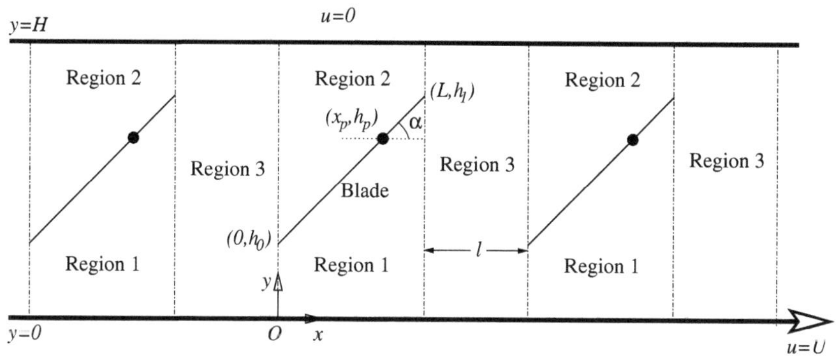


**Fig. 1.** Geometry of the transverse flow problem.

phase changes; also they often have a strongly temperature-dependent viscosity. Moreover, both convection and dissipation of heat can be significant in a SSHE. Extensive literature surveys are given by Härröd [2, 3], and features of the behaviour have recently been analysed by Stranzinger et al. [4], Fitt and Please [1] and Sun et al. [5]. However, despite their widespread use, understanding of the behaviour of the material inside SSHEs is still incomplete.

In the present work we shall concentrate on the fluid flow rather than the heat transfer inside a SSHE, and so we shall restrict our attention to isothermal flow.

## 2 Transverse Flow

First we consider steady two-dimensional flow of an isothermal incompressible Newtonian fluid of viscosity $\mu$ in a long parallel-sided channel of width $H$ in which there is a periodic array of inclined smoothly pivoted thin plane blades, the flow being driven by the motion of one wall of the channel parallel to itself with speed $U$ ($> 0$), the other wall being fixed. Body forces are neglected.

We introduce Cartesian axes $Oxyz$ as shown in Fig. 1, with the wall $y = 0$ moving with velocity $U\mathbf{i}$, and the wall $y = H$ fixed. Suppose a thin plane freely pivoted blade occupies $0 \leq x \leq L$, with its pivot fixed at $(x_{\mathrm{p}}, h_{\mathrm{p}})$, where $0 \leq x_{\mathrm{p}} \leq L$ and $0 < h_{\mathrm{p}} < H \ll L$, and suppose that the separation between the blades is $\ell$ ($\geq 0$), so that the portion $L \leq x \leq L + \ell$ of the channel contains no blades. Let $\alpha$ (which may be positive or negative) denote the angle of inclination of the blade to the $x$ axis, as shown in Fig. 1. In the lubrication-theory approach used here we will assume that $|\alpha| \ll 1$; then the blade is given by $y = h(x)$ for $0 \leq x \leq L$, where $h(x) = h_{\mathrm{p}} + \alpha(x - x_{\mathrm{p}})$.

For steady flow the blade is in equilibrium, subject to forces due to the fluid, the pivot, and (in general) the walls of the channel. Here we consider cases where the ends of the blade are not in contact with the moving wall $y = 0$ of the channel, so that $0 < h_0, h_1 \leq H$, where $h_0 = h(0)$ and $h_1 = h(L)$.

We denote the velocities, pressures and volume fluxes (per unit width in the $z$ direction) by $u_k \mathbf{i} + v_k \mathbf{j}$, $p_k$ and $Q_k$, where $k = 1$ denotes values in $0 \le x \le L$, $0 \le y \le h$, $k = 2$ denotes values in $0 \le x \le L$, $h \le y \le H$, and $k = 3$ denotes values in $L \le x \le L + \ell$, $0 \le y \le H$. A lubrication approximation gives

$$u_1 = \frac{[6Q_1 y + Uh(h - 3y)](h - y)}{h^3}, \tag{1}$$

$$u_2 = \frac{6Q_2(H - y)(y - h)}{(H - h)^3}, \tag{2}$$

$$u_3 = \frac{[6Q_3 y + UH(H - 3y)](H - y)}{H^3}, \tag{3}$$

$$p_1 = \frac{6\mu U}{\alpha}\left(\frac{1}{h_1} - \frac{1}{h}\right) - \frac{6\mu Q_1}{\alpha}\left(\frac{1}{h_1^2} - \frac{1}{h^2}\right) + p_L, \tag{4}$$

$$p_2 = \frac{6\mu Q_2}{\alpha}\left[\frac{1}{(H - h_1)^2} - \frac{1}{(H - h)^2}\right] + p_L, \tag{5}$$

$$p_3 = \frac{6\mu(UH - 2Q_3)}{H^3}(x - L) + p_L, \tag{6}$$

$$Q_1 = \int_0^h u_1 \, \mathrm{d}y = -\frac{h^3 p_{1x}}{12\mu} + \frac{Uh}{2}, \tag{7}$$

$$Q_2 = \int_h^H u_2 \, \mathrm{d}y = -\frac{(H - h)^3 p_{2x}}{12\mu}, \tag{8}$$

$$Q_3 = \int_0^H u_3 \, \mathrm{d}y = -\frac{H^3 p_{3x}}{12\mu} + \frac{UH}{2}. \tag{9}$$

Setting $x = 0$ in (4) and (5) and $x = L + \ell$ in (6), we obtain three representations of $p_0 - p_L$:

$$\begin{aligned} p_0 - p_L &= -\frac{6\mu U}{\alpha}\left(\frac{1}{h_0} - \frac{1}{h_1}\right) + \frac{6\mu Q_1}{\alpha}\left(\frac{1}{h_0^2} - \frac{1}{h_1^2}\right) \\ = -\frac{6\mu Q_2}{\alpha}&\left[\frac{1}{(H - h_0)^2} - \frac{1}{(H - h_1)^2}\right] = \frac{6\mu(UH - 2Q_3)\ell}{H^3}. \end{aligned} \tag{10}$$

Expressions for the $Q_k$ ($k = 1, 2, 3$) and $p_0 - p_L$ are obtained by solving (10) and the global mass conservation condition $Q_1 + Q_2 = Q_3$.

The moment of the forces on the blade about the pivot due to the pressure is of the form $\mathbf{M} = M\mathbf{k}$, where

$$M = \int_0^L (x - x_\mathrm{p})(p_1 - p_2) \, \mathrm{d}x. \tag{11}$$

For equilibrium of the blade we require $M = 0$, which leads to a lengthy algebraic transcendental equation determining $\alpha$ when $L$, $\ell$, $H$, $x_\mathrm{p}$ and $h_\mathrm{p}$

are prescribed. Once $\alpha$ is known, the complete solution is determined. This solution allows us to describe all the qualitative features of the transverse flow. In particular, we can determine when the blades are in contact with the walls of the channel. In addition we can calculate the forces on the blades and on the walls of the channel, and hence make useful estimates of the torque and power required to turn the rotor.

## 3 Longitudinal Flow

In a SSHE the material being processed not only undergoes flow in the transverse direction (caused by the rotation of the rotor), but also is driven by an imposed axial pressure gradient along the annular gap between stator and rotor. To model this fully three-dimensional flow we consider the effect of allowing flow along the channel in the $z$ direction, in addition to the flow in the $(x, y)$ plane, discussed above. We take the blades to be long in the $z$ direction so that the lubrication approximation may again be used. It is found that the motion in the $z$ direction uncouples from that in the $(x, y)$ plane. Thus with velocities and pressures denoted by $u_k\,\mathbf{i} + v_k\,\mathbf{j} + w_k\,\mathbf{k}$ and $P_k$ for $k = 1, 2, 3$ (with $u_k$, $v_k$, $w_k$ and $P_k$ functions of $x$, $y$ and $z$), we find that

$$w_1 = \frac{G}{2\mu} y(h - y), \quad w_2 = \frac{G}{2\mu}(H - y)(y - h), \quad w_3 = \frac{G}{2\mu} y(H - y), \tag{12}$$

and $P_k = -Gz + p_k$, where $G = -\partial P_k/\partial z$ is the (constant) prescribed axial pressure gradient, and the velocity components $u_k = u_k(x, y)$ and $v_k = v_k(x, y)$ and the pressure contributions $p_k = p_k(x, y)$ are exactly as given for the transverse (two-dimensional) flow described above. The volume flux of fluid in the $z$ direction across the section $0 \leq x \leq L + \ell$, $0 \leq y \leq H$ is given by

$$Q_z = \frac{GH}{24\mu}\left[L\left(2(h_0^2 + h_0 h_1 + h_1^2) - 3H(h_0 + h_1)\right) + 2H^2(L + \ell)\right]. \tag{13}$$

The force (per unit axial length) in the $z$ direction on the blade due to the fluid is given by $F_z = GHL/2$, and the forces (per unit axial length) in the $z$ direction on the portions $0 \leq x \leq L + \ell$ of the lower wall $y = 0$ and the upper wall $y = H$ due to the fluid are

$$F_0 = \int_0^L \mu \frac{\partial w_1}{\partial y}\bigg|_{y=0} \mathrm{d}x + \int_L^{L+\ell} \mu \frac{\partial w_3}{\partial y}\bigg|_{y=0} \mathrm{d}x = \frac{G}{4}[2H\ell + (h_0 + h_1)L] \tag{14}$$

and

$$F_H = -\int_0^L \mu \frac{\partial w_2}{\partial y}\bigg|_{y=H} \mathrm{d}x - \int_L^{L+\ell} \mu \frac{\partial w_3}{\partial y}\bigg|_{y=H} \mathrm{d}x = \frac{G}{4}[2H(L+\ell) - (h_0 + h_1)L], \tag{15}$$

respectively.

## 4 Summary

In this short paper we presented a simple mathematical model of fluid flow in a SSHE. Specifically we considered steady isothermal flow of a Newtonian fluid around a periodic array of pivoted scraper blades in a channel with one stationary and one moving wall, when there is an applied pressure gradient in a direction perpendicular to the wall motion. The flow is fully three-dimensional, but decomposes naturally into a two-dimensional transverse flow driven by the boundary motion and a longitudinal pressure-driven flow. In the future we plan to extend our analysis to include other practically important features neglected in this simple model, including blade wear and, of course, non-isothermal effects.

## Acknowledgements

Thanks are due to Profs C. P. Please and A. D. Fitt, School of Mathematics, University of Southampton, Prof. D. L. Pyle and Dr K.-H. Sun, School of Food Biosciences, University of Reading, Dr N. Hall Taylor, Chemtech International Ltd, Reading, Dr J. Mathisson, Tetra Pak, and Dr H. Tewkesbury, Smith Institute for Industrial Mathematics and System Engineering, for many useful discussions. This work forms part of a larger research project supported by the EPSRC (Research Grant GR/R993032), and by Chemtech International and Tetra Pak, under the auspices of the Faraday Partnership for Industrial Mathematics, managed by the Smith Institute.

## References

1. A. D. Fitt and C. P. Please. Asymptotic analysis of the flow of shear-thinning foodstuffs in annular scraped heat exchangers. *J. Eng. Maths*, 39:345–366, 2001.
2. M. Härröd. Scraped surface heat exchangers: a literature survey of flow patterns, mixing effects, residence time distribution, heat transfer and power requirements. *J. Food Proc. Eng.*, 9:1–62, 1986.
3. M. Härröd. Methods to distinguish between laminar and vortical flow in scraped surface heat exchangers. *J. Food Proc. Eng.*, 13:39–57, 1990.
4. M. Stranzinger, K. Feigl, and E. Windhab. Non-Newtonian flow behaviour in narrow annular gap reactors. *Chem. Eng. Sci.*, 56:3347–3363, 2001.
5. K.-H. Sun, D. L. Pyle, A. D. Fitt, C. P. Please, M. Baines, and N. Hall-Taylor. Numerical study of 2D heat transfer in a scraped surface heat exchanger. *Computers and Fluids*, 33:869–880, 2003.

# Part IV

## Theme: Life Sciences

# Transmission Line Matrix Modeling of Sound Wave Propagation in Stationary and Moving Media

M. Bezděk, Hao Zhu, A. Rieder, and W. Drahm

Endress+Hauser GmbH+Co. KG, Am Lohmühlbach 12, 85356 Freising, Germany
bezdek@ehfs.de

**Summary.** The transmission line matrix (TLM) for simulating sound wave propagation in stationary and moving media is presented. TLM is inherently a time-domain approach which does not require solution of a differential equation. TLM and FEM are compared in terms of accuracy and computational complexity. It is concluded that TLM may represent a more efficient alternative to FEM when predicting acoustic fields in stationary media. Furthermore, applicability of TLM to moving media is examined. A TLM model of 2D moving media based on the idea of [3] is introduced.

**Key words:** Transmission line matrix, wave propagation, moving media, finite element method.

## 1 Introduction

Various technical systems (*e.g.* ultrasonic flowmeters) represent large simulation problems which require solution of acoustic fields in stationary and moving media. The existing tools based on the finite element method (FEM) often do not allow to perform a full 3D analysis due to the extensive size of the solution domain with respect to the characteristic acoustic wavelength (see [1]). Furthermore, usage of the boundary element method (BEM) is limited by unavailability of the appropriate integral formulations.

An alternative method, the transmission line matrix (TLM), is presented here. It enables simulation of acoustic wave propagation in fluid media and may reduce the memory demands in comparison with the FEM. It can be coupled with another method (such as FEM) which provides the necessary input data along the boundary of the solution domain typically representing a fluid-structure interface.

TLM was originally developed for problems in the electromagnetic domain. The theoretical background was laid by [4] who introduced discrete circuit models for Maxwell's equations in order to study electromagnetic waveguides. Since then the method has been applied to solution of various problems described by wave and diffusion equations.

An overview of TLM for simulating acoustic wave propagation in stationary media is given by [2], where the author proposed a derivation of TLM which relies on a physical model of the field effects based on the well-known Huygens' principle. One can say that TLM replaces the approximate numerical solution of the governing partial differential equation (when using FEM, for example) by an approximate physical model which can be solved accurately.

In [3] Connor showed that this intuitive approach enables an extension of the standard TLM to moving media. He demonstrated the idea in the case of a 1D moving medium. In this paper, a TLM model of 2D moving media is developed and validated by means of FEM. Furthermore, the standard TLM model of 3D stationary media is verified.

## 2 TLM Model of Stationary Media

According to Huygens, a wavefront consists of a number of secondary point sources which radiate elementary spherical waves. The envelope of these waves forms a new wavefront. Wave propagation can be explained as a repetition of this basic mechanism.

In order to implement this process in a digital computer, the field must be discretized in time and space. One obtains the discrete Huygens' principle which, according to [2], is a synonym for the TLM model. In the TLM model, the continuous acoustic medium is replaced by a regular (equidistant) network of elementary transmission lines (elements). A pressure pulse propagates through the element and reaches the adjacent intersection (node) at the next time step. Due to the impedance discontinuity present at the node, scattering occurs. The energy of the incident pulse is radiated in all four directions in such a way that energy conservation holds. Therefore, TLM is an inherently stable method. Furthermore, conservation of field continuity at the node requires the polarity of the back-scattered pulse to be the opposite. This process is illustrated in Fig. 1.

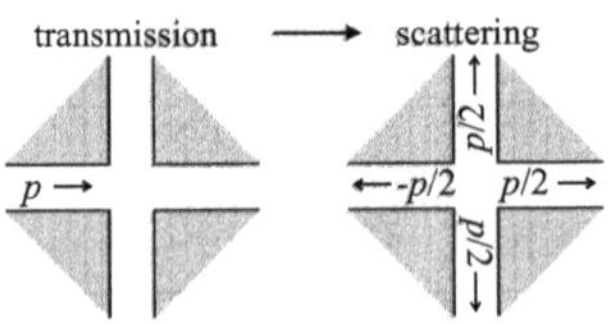


**Fig. 1.** Transmission and scattering of a pressure pulse in a 2D TLM model

The acoustic quantities can be identified from the numerical equivalence of TLM and the finite-difference-time-domain (FDTD) discretization of the wave equation (see [5]).

In order to validate the 3D TLM model, a simulation setup depicted in Fig. 2 is considered. A square piston source of the dimension $10 \times 10$ mm is placed at the end of a water channel entering a large water domain (phase velocity $c = 1500$ m/s). The piston vibrates uniformly in the normal direction and radiates an ultrasonic pulse of the central frequency 100 kHz into the channel. The pulse propagates through the channel and diffraction occurs at the channel opening.

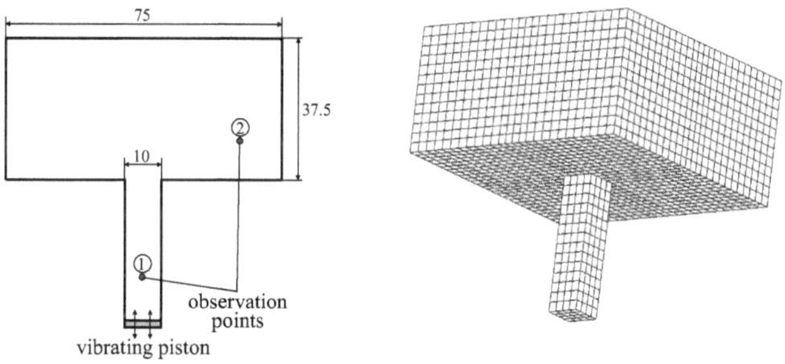


**Fig. 2.** 3D validation setup and its FEM model

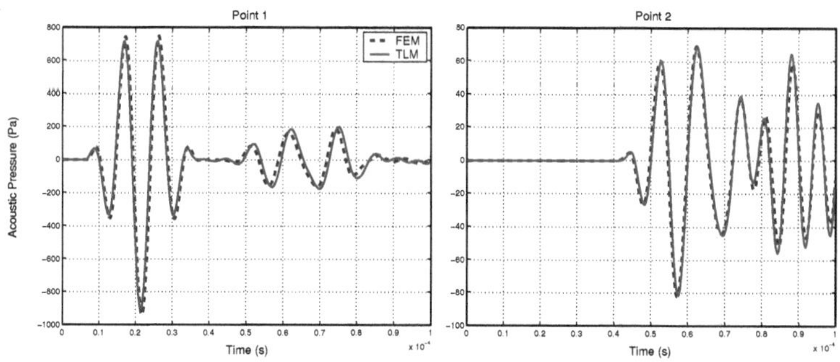


**Fig. 3.** Comparison of the TLM and FEM results in the 3D validation setup

In Fig. 3, the acoustic pressure predicted by TLM is compared to the FEM result at the two observation points. One can see that the results achieved with TLM and FEM correspond very well. In both cases, the element size of $\lambda/30$ was selected, where $\lambda$ denotes the characteristic acoustic wavelength and is equal to 15 mm here. Nevertheless, the FEM simulation using a commercial explicit solver consumed 960 MB of computing memory, whereas the TLM simulation using an interpreted MATLAB® code cost only about 80 MB.

# 3 TLM Model of Moving Media

Following the basic concept of [3], the 2D TLM model of moving media is presented here. The presence of flow in a medium causes the waves to propagate at different velocities in different directions. In order to include this kind of anisotropy into the TLM model, additional unidirectional elements connecting the adjacent nodes are introduced (see Fig. 4).

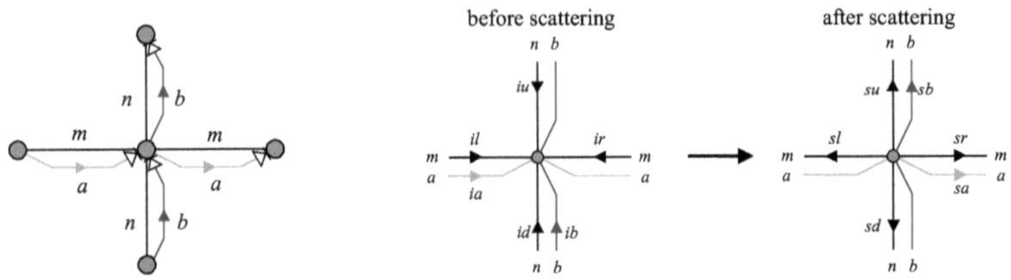


**Fig. 4.** 2D TLM model of moving media and the scattering scheme.

The parameters $m$ and $n$ denote the relative admittances of the original bidirectional elements and the parameters $a$ and $b$ the relative admittances of the new unidirectional elements. By means of $a$ and $b$, the admittance of the medium can be biased for waves propagating in opposite directions. Therefore, the propagation velocities also become different, in this manner simulating the effect of flow. See [6] for details.

In order to validate the 2D TLM model of moving media, the acoustic field generated by a vibrating piston in a uniform shear flow is computed and compared with the results of a FEM simulation (see Fig. 5 left). The piston vibrates uniformly in the normal direction with the amplitude $10^{-8}$ m and the frequency 100 kHz. Water ($c = 1500$ m/s) fills the half-plane above the piston and flows parallel at the velocity $v = 300$ m/s.

The comparison of the FEM and the HIRM results is made in the complex plane for amplitudes and phases of the harmonic signals received at the distance 35 mm from the center of the piston face in the angular range $0° \leq \theta \leq 180°$ (see Fig. 5 right). The displayed TLM results show a certain $\theta$-dependent deviation from the reference FEM results. However, from the qualitative point of view these results are fully satisfactory.

# 4 Conclusion

Application of TLM to simulation of sound wave propagation in stationary and moving media has been described. It has been demonstrated that TLM may represent a more efficient alternative to FEM when calculating acoustic fields in stationary media. However, its major disadvantage—the requirement of an equidistant discretization—must be handled by special techniques.

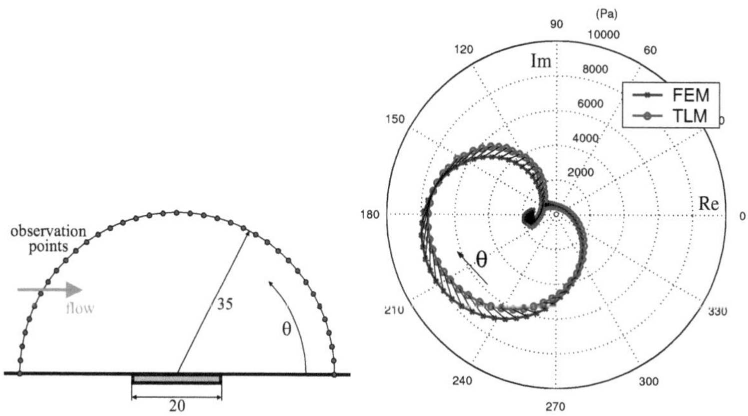


**Fig. 5.** Validation of the 2D TLM model of moving media (left: setup, right: results)

Furthermore, first TLM results for 2D moving media have been presented. Their accuracy is satisfactory, but it seems it can be improved by deriving more precise procedures for setting the TLM parameters $m$, $n$, $a$ and $b$. The used concept can be expanded to the 3D space easily. In the future, the efficiency of TLM for moving media is to be evaluated with respect to FEM.

## References


1. M. Bezděk, A. Rieder, H. Landes, T. Strunz, and R. Lerch. Numerical analysis of wave propagation in an ultrasonic flowmeter. In *Proc. First Congress of Alps Adria Acoustics Association*, pages 573–580, 2003.
2. Y. Kagawa, T. Tsuchiya, B. Fujii, and K. Fujioka. Discrete Huygens' model approach to sound wave propagation. *J. Sound and Vibration*, 218(3):419–444, 1998.
3. W. J. O'Connor. TLM model of waves in moving media. *J. Numerical Modelling*, 15(2):205–214, mar 2002.
4. J. Whinnery and S. Ramo. A new approach to the solution of high-frequency problems. In *Proc. IRE*, volume 32, pages 284–288, 1944.
5. A. Wilde, P. C. Eccardt, and W. J. O'Connor. Modeling acoustic transducer surface waves by transmission line matrix method. In *19th CADFEM-User's Meeting*, Berlin, 2001.
6. H. Zhu. Transmission line matrix modeling of sound wave propagation in stationary and moving media. Master's thesis, Technical University of Munich, Germany, 2003.

# Viscous Drops Spreading With Evaporation And Applications To DNA Biochips

M. Cabrera[1], T. Clopeau[2], A. Mikelić[2], and J. Pousin[3]

[1] LEOM, UMR CNRS 5512, ECL, BP 169, 69131 Ecully Cedex, France,
[2] UFR Mathématiques, Université Claude Bernard Lyon 1, Bât. A, Domaine de Gerland, 50 av. Tony-Garnier, 69366 Lyon Cedex 07, France
clopeau@univ-lyon1.fr , mikelic@univ-lyon1.fr
[3] MAPLY, UMR CNRS 5585, INSA de Lyon, 20 av. A. Einstein, F-69621 Villeurbanne cedex, France, jerome.pousin@insa-lyon.fr

**Summary.** We develop a lubrication model for the viscous drop spreading with evaporation. It is then solved in the quasi-static case and the numerical method is used in the parameter identification in the application to DNA chips.

**Key words:** drop spreading, evaporation, parameter identification, DNA biochips.

## 1 Introduction

The capability to deliver biochip Probe Arrays in a controlled manner across a given 2D substrate (silicium, glass strands, etc) is very important for the industrial production of biochips. In the case of DNA Probe Arrays, one deals with synthetic oligonucleotide sequences typically 12 to 60 base in length, complementary to DNA strands, which presence one wants to detect. One of the possible manufacturing procedures is to pose drops, containing dissolved probes, on the prepared support. After evaporation of the solvent the probes are locally fixed. This procedure depends on many factors, mainly on the drop spreading and solvent evaporation. In this paper we introduce a mathematical model which permits to describe the spreading of a drop on a horizontal surface, in the presence of the evaporation. It is the first step in modeling the fixing of the probes on a 2D predefined region of a substrate. In Section 2 we will use the Navier-Stokes equations and the detailed study of the free surface to write the physical model. It is then reduced to its lubrication approximation. Finally we present numerical simulations of the spreading and determination of the parameters from experimental results.

We note that the regime when the evaporation dominates the spreading is studied in [2] and [5] In our approach both phenomena are present and we get very good agreement with the experimental results from [3] .

## 2 The physical model and the lubrication approximation

We consider the spreading of an axisymmetric liquid drop over a smooth isothermal solid surface. During spreading the evaporation takes place. The spreading is a free boundary incompressible flow described by the Navier-Stokes equations

$$\text{div } \mathbf{v} = 0 \quad \text{and} \quad \rho\left(\frac{\partial \mathbf{v}}{\partial t} + (\mathbf{v}\nabla)\mathbf{v}\right) = -\nabla p + \mu \Delta \mathbf{v} - \rho g \mathbf{k} \tag{1}$$

valid in the drop. In (1) $\rho$ is the mixture density, $g\mathbf{k}$ is the gravitational vector and $\mu$ its viscosity. $\mathbf{v} = (u_r, u_z)$ is the fluid velocity and $p$ is the pressure. The drop is initially axisymmetric with respect to the axis $\{x = 0\} \cap \{y = 0\}$ and posed on the surface $\{z = 0\}$. On the solid surface $\{z = 0\}$ we impose the no-slip condition $\mathbf{v} = 0$. The free surface is defined by $\{z = h(r,t)\}$. The boundary conditions are 2 dynamic conditions and the kinematic condition. We suppose no shear at the free surface and we have

$$\frac{\mu}{1+(\frac{\partial h}{\partial r})^2}\left\{\left[\left(\frac{\partial h}{\partial r}\right)^2 - 1\right]\left(\frac{\partial u_z}{\partial r} + \frac{\partial u_r}{\partial z}\right) + 2\frac{\partial h}{\partial r}\left(\frac{\partial u_r}{\partial r} - \frac{\partial u_z}{\partial z}\right)\right\} = \sigma \mathbf{n}\boldsymbol{\tau} = 0. \tag{2}$$

Next we suppose the free surface surrounded by a tiny layer of the evaporated vapour. Hence following the Kelvin's law ([1, p. 53] ) we have on the free boundary $p_c = p_{eq} \exp\{2V\kappa/(kT)\}$, where $\kappa$ is the mean curvature and $V$ is the molar volume of the volatile substance in the liquid phase. This implies the second dynamic condition

$$p_c = -\sigma \mathbf{n}\mathbf{n} = p - \frac{2\mu}{1+(\frac{\partial h}{\partial r})^2}\left\{\left(\frac{\partial h}{\partial r}\right)^2 \frac{\partial u_r}{\partial r} + \frac{\partial u_z}{\partial z} - \frac{\partial h}{\partial r}\left(\frac{\partial u_z}{\partial r} + \frac{\partial u_r}{\partial z}\right)\right\} \tag{3}$$

It is supposed that in the ambient air there is only a weak concentration of the volatile solvent. Consequently, the mass loss due to the evaporation is given by a surface flux $k_{evap}(p - p_{eq})$ (see [5] and [8, Equation 7] ). Let $\rho_s$ be the solvent density. Then a simple mass conservation argument gives the kinematic condition

$$\frac{\partial h}{\partial t} + u_r \frac{\partial h}{\partial r} - u_z = \frac{k_{evap}}{\rho_s}\left(p|_{z=h(r,t)} - p_{eq}\right)\sqrt{1 + \left(\frac{\partial h}{\partial r}\right)^2} \tag{4}$$

We refer to [5] for a model with a general function $J_s(r,t)$, which is in our situation equal to $k_{evap}(p - p_{eq})$.

The interaction of the drop with the surface $\{z = 0\}$ is characterized by the dynamical contact angle $\theta$, measured from the drop. We make the classical *lubrication hypothesis of the small dynamical angle* $\vartheta$, which is identified with the apparent contact angle. Then $\partial_r h$ is small, $2\kappa \approx -\Delta_r h$ and $p_c \approx p_{eq} - \gamma \Delta_r h$, $\gamma = p_{eq} V/(kT)$. The spreading is axially symmetric for all $t > 0$.

The triple contact line, between the solid, the drop and the ambient air, is located at $r = a(t)$. Initially, $a(0) = a_0$, $h(r, 0) = h_0(r)$ and $h(a_0) = 0$. The contact condition and the value of the apparent contact angle $\vartheta$ are given by

$$h(a(t), t) = 0; \quad \frac{\partial h}{\partial r}(a(t), t) = -\tan \vartheta; \quad \frac{dh_0}{dr}(a_0) = -\tan \theta_0 \approx -\theta_0 \tag{5}$$

The dynamic contact-line boundary condition relating the apparent contact angle $\vartheta$ to the contact line speed $U = \dfrac{da}{dt}$ is given by the modified "Tanner law"

$$U = \kappa_0 (\vartheta - \theta_A)^m, \qquad \vartheta > \vartheta_A \tag{6}$$

where $\vartheta_A$ is the static advancing contact angle, $\vartheta_R$ is the static receding contact angle and $\kappa_0 > 0$ is a constant, estimated in [4] to be $\frac{1}{10}\left(\frac{\pi}{4}\right)^m \frac{\gamma}{\mu}$. The theory predicts $m = 3$ (see [4] or [6]) or values close to 3.

Now, with $\varepsilon = h_0(0)/a_0$, we use the following scales

$$\begin{aligned} &U(\varepsilon) = \kappa_0 \theta_0^m; \ T(\varepsilon) = \tfrac{a_0}{U(\varepsilon)}, r^* = \tfrac{r}{a_0}, \ t^* = \tfrac{t}{T(\varepsilon)}, \ h^*(r^*, t^*) = \tfrac{h(r,t)}{a_0 \varepsilon} \\ &z^* = \frac{z}{a_0 \varepsilon}, \ a^*(t^*) = \frac{a(t)}{a_0}, \Theta(t^*) = \frac{\vartheta(t)}{\theta_0}, \ \Theta_A = \frac{\vartheta_A}{\theta_0}, V^* = \frac{V_0}{2\pi \varepsilon a_0^3} \\ &u_r^*(r^*, t^*) = \tfrac{u_r(r,t)}{U(\varepsilon)}, \ u_z^*(r^*, t^*) = \tfrac{u_z(r,t)}{\varepsilon U(\varepsilon)}, \ p^* = a_0 \varepsilon^2 (p - p_{eq})/(\mu U(\varepsilon)). \end{aligned} \tag{7}$$

When these scalings are introduced into the governing equations, we obtain, at leading order when $\varepsilon \to 0$ the classical lubrication approximation (see [6] and [7]), but with the evaporation effects. As usual, we skip the stars for the lubricated problem. These effects enter into the lubricated kinematic condition and they give the change in the volume of the drop. For the change of the volume we have

$$\begin{cases} V(t) = \int_0^{a(t)} 2\pi r h(r,t)\, dr; \ V(0) = V_0 = \int_0^{a_0} 2\pi r h_0(r)\, dr, \ \dfrac{1}{2\pi}\dfrac{dV}{dt} = \\ -k_{evap} \int_0^{a(t)} (p|_{z=h} - p_{eq}) r \sqrt{1 + (\frac{\partial h}{\partial r})^2}\, dr \approx k_{evap} a(t) \frac{\partial h(a(t),t)}{\partial r} \end{cases} \tag{8}$$

The effective kinematic condition, which includes the evaporation effects, is

$$u_z = \frac{\partial h}{\partial t} - \left\{ \frac{\varepsilon}{U(\varepsilon) a_0} \frac{k_{evap}}{\rho_s \gamma} \Delta_r h - u_r \frac{\partial h}{\partial r} \right\} \tag{9}$$

In our situation, $\theta_0/\varepsilon = O(1)$, the capillary number $\mathbf{Ca} = \frac{\mu U(\varepsilon)}{\gamma \varepsilon^3} \approx C \frac{\mu \kappa_0}{\gamma} \varepsilon^{m-3} << 1$, and we are interested in the quasi-static situation when the triple line moves slowly. This situation, in absence of the evaporation, is studied in [6] . Here we expend once more the solutions to the lubricated approximation with respect to the capillary number **Ca**. For the leading order terms we get the following effective system from [6] , in which we added the evaporation effects. Let **Bo**= $\frac{\rho g a_0^2}{\gamma}$ be the Bond's number, $K_{evap} = \frac{k_{evap}}{\rho_s \gamma}$. Then we search for the functions $h(r,t)$ (the rescaled height), $\lambda(t)$, $V(t)$ (the rescaled volume) and $a(t)$ (the rescaled radius), such that for $r \in (0, a(t))$ and $t \in (0,T)$

$$\begin{cases} \lambda(t) = \ \mathbf{Bo}\ h(r,t) - \frac{1}{r}\frac{\partial}{\partial r}(r\frac{\partial h(r,t)}{\partial r});\ h(a(t),t) = 0 \\ \frac{\partial h}{\partial r}|_{r=0} = 0;\ \Theta(t) = \left[\frac{da}{dt}(t)\right]^{1/m} + \Theta_A = -\frac{1}{2}\frac{\partial h}{\partial r}(a(t),t) \\ a(0) = 1;\ \frac{dV}{dt} = K_{evap}\cdot 2\pi a(t)\frac{\partial h(a(t),t)}{\partial r}, \quad V(0) = V^*. \end{cases} \tag{10}$$

Now we go back to the dimensional variables. The system (10) reduces to the following Cauchy problem for $\{V(t), a(t)\}$ on $(0,T)$:

$$\begin{cases} \frac{da(t)}{dt} = \kappa_0\Big(\arctan\Big(\frac{V(t)\sqrt{Bo}I_1(\frac{a(t)}{a_0}\sqrt{Bo})}{\pi a_0 a(t)^2 I_2(\frac{a(t)}{a_0}\sqrt{Bo})}\Big) - \vartheta_A\Big)^m, \ a(0) = a_0, \\ \frac{dV(t)}{dt} = -4\pi K\frac{V(t)\sqrt{Bo}I_1(\frac{a(t)}{a_0}\sqrt{Bo})}{a(t)I_2(\frac{a(t)}{a_0}\sqrt{Bo})}, \quad V(0) = V_0, \end{cases} \tag{11}$$

where $K = 2\pi a_0 K_{evap}/\sqrt{Bo}$. From $a, \vartheta$ and $V$ it is easy to reconstruct the effective pressure and velocities. The effective drop height $h$ is given by

$$h(r,t) = \frac{V(t)}{\pi}\frac{I_0\Big(\frac{a(t)}{a_0}\sqrt{Bo}\Big) - I_0\Big(\frac{r}{a_0}\sqrt{Bo}\Big)}{a^2(t)I_2\Big(\frac{a(t)}{a_0}\sqrt{Bo}\Big)}, \ 0 \le r \le a(t),\ 0 \le t \le T. \tag{12}$$

## 3 Numerical results and comparison with experimental results

The difficulty with the drop spreading is that a number of parameters are unknown. It is possible to observe experimentally the height $h$ and the foot area $D(t)/\pi = a(t)^2/2$ at various times during spreading. From these data we identify the parameters $Bo$, $K_{evap}$, $m$, $\kappa_0$ and $\vartheta_A$. In the quasistatic situation, the Bond's number $Bo$ determines the shape of the drop. It is estimated from the initial data and for our experiments we have $Bo = 12$. Data from [3] are presented in Fig. 1. Let $\Lambda = (K,\ \kappa_0,\ m,\ \vartheta_A)$ be the vector containing the parameters to identify. We first interpolate the data. Let $\overline{H}$ et $\overline{D}$ be the interpolation polynomials, corresponding respectively to the measurements of the height and of the diameter squared divided by 2. We introduce a subdivision

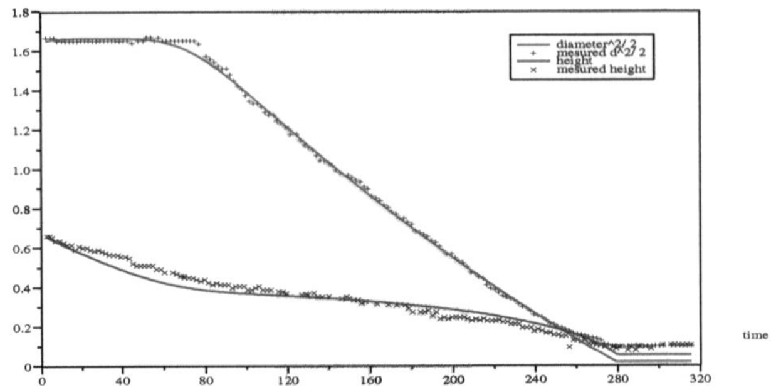


**Fig. 1.** *Experimental data and simulation results.*

of $[0, T]$ in $N$ intervals $[t_i, t_{i+1}]$, with $t_i = i\tau$, $0 \leq i \leq N-1$ and with the discretization step $\tau = T/N$. Then the cost functional to minimize is set to

$$J(\Lambda) = \frac{1}{2}\sum_{i=1}^{N}\left(|\overline{D}_i - D_i(\Lambda)|^2 + |\overline{H}_i - h_i(\Lambda)|^2\right) \tag{13}$$

with $\overline{D}_i = \overline{D}(t_i)$, $D_i(\Lambda) = D(\Lambda, t_i)$, $\overline{H}_i = \overline{H}(t_i)$ et $h_i(\Lambda) = h(\Lambda, t_i)$. Using a non-linear Least Squares method, initiated with $\lambda_0$= (1, 1, 3, 1), giving $J(\Lambda_0) = 3.3512$, we got for the minimum of the cost functional $J(\lambda^*) = 0.1093$. This value corresponds to $\Lambda^*$= (0.9912, 0.8324, 3.0860, 1.1039), and the solution is depicted with a continuous line in Fig. 1.

## References

1. A.W. Adamson and A.P. Gast. *Physical Chemistry of Surfaces.* Wiley New York, 6th edition, 1997.
2. M. Cachile, O. Bénichou, C. Poulard, and M. Cazabat. Evaporating droplets. *Langmuir Letters*, 18:8070– 8078, 2002.
3. S. Cottet Géneau and M. Cabrera. Optimisation d'un procédé de fabrication de puces à ADN. *Rapport de Travail de fin d'étude Laboratoire IFOS UMR CNRS*, 5621, September 2002.
4. P.G. de Gennes. Wetting: statics and dynamics. *Reviews of Modern Physics*, 57:827–863, 1985.
5. R.D. Deegan, O. Bakajin, T.F. Dupont, G. Huber, S.R. Nagel, and T.A. Witten. Contact line deposits in an evaporating drop. *Physical Review E.*, 62(1):756–765, 2000.
6. P. Ehrhard and S.H. Davas. Non-isothermal spreading of liquid drops on horizontal plates. *J. Fluid Mech.*, 229:365, 1991.
7. H.P. Greenspan. On the motion of a small viscous droplet that wets a surface. *J. Fluid. Mech.*, 84:125–143, 1978.
8. M.E.R. Shanahan. Is a Sessile Drop in an Atmosphere Saturated with Its Vapor Really at Equilibrium. *Langmuir Letters*, 18:7763– 7765, 2002.

# Similarity-Based Object Recognition of Airborne Fungi in Digital Images

P. Perner

Institute of Computer Vision and applied Computer Sciences IBaI Körnerstr. 10, 04107 Leipzig ibaiperner@aol.com, www.ibai-institut.de

**Summary.** We propose and evaluate a method for the recognition of airborne fungi spores. We suggest a similarity-based object-recognition method to identify spores in a digital microscopic image. We do not use the gray values of the case, but the object edges instead. The similarity measure measures the average angle between the vectors of the template and the object. Case generation is done semi-automatically by manually tracing the object, automatic shape alignment, similarity calculation, clustering, and prototype calculation.

## 1 Introduction

Airborne microorganisms are ubiquitously present in the various fields of indoor and outdoor environments. The potential implication of fungal contaminants in bioaerosols in occupational health is recognized as a problem in several working environments. Besides the detection of parameters relevant to occupational and public health, in many controlled environments the number of airborne microorganisms has to be kept below the permissible or recommended values (e.g. in clean rooms, operating theaters, domains of the food and pharmaceutical industry). The continuous monitoring of airborne biological agents is consequently a necessity, as well for the detection of risks for human health as for the smooth sequence of technological processes. We describe our first results on the way to develop an automatic image-interpretation system for the detection and interpretation of airborne fungi spores. We describe the developed method for the similarity-based recognition of objects that are probably fungi spores in digital images. Future work will concentrate on classifying the objects. For our study we used six different fungi spores.

## 2 Fungi Images

Six fungal strains representing species with different spore types were used for the study (Fig. 1). The strains were obtained from the fungal stock collection

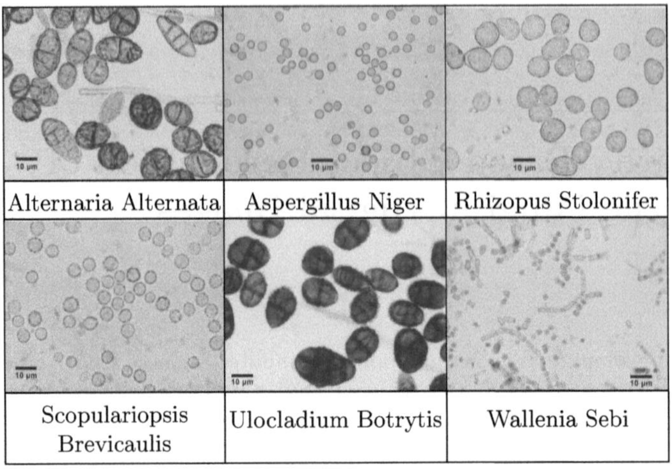


**Fig. 1.** Images of Fungal Strains

of the Institute of Microbiology, University of Jena/ Germany and from the culture collection of JenaBios GmbH. A database of images from the spores of these species was produced.

# 3 Similarity-Based Object Recognition

The objects in the image are highly structured. Our study has shown that these images, represented in Fig. 1, cannot be segmented by thresholding. Biomedical applications have the special characteristic that one object can have a great variation in appearance. Therefore the appearance of this object cannot be generalized by one model as, is well known from model-based object recognition. Instead we decided to apply a similarity-based object recognition procedure for the detection of objects in the image.

The similarity-based object recognition method uses templates that generalize the original contour of the objects and matches these templates against the contour of the objects in the image. During the match a score is calculated that describes the goodness of the fit between the object and the case. We did not use the gray values of the case, but used the object edges instead. For the score of the match between the case and the image we modified the normalized cross correlation in order to measure the average angle between the vectors of the case and the object.

## 3.1 Similarity Measure

We propose a similarity measure based on the cross correlation, using the direction vectors of an image [4]. This approach requires the calculation of the dot product between each direction vector of the case $\mathbf{m}_k = (v_k, w_k)^T$, $k = 1, \ldots, n$ and the corresponding image vector $\mathbf{i}_k = (d_k, e_k)^T$:

$$s_1 = \frac{1}{n}\sum_{k=1}^{n} \mathbf{m}_k \cdot \mathbf{i}_k = \frac{1}{n}\sum_{k=1}^{n} < \mathbf{m}_k, \mathbf{i}_k >= \frac{1}{n}\sum_{k=1}^{n}(\nu_k \cdot d_k + w_k \cdot e_k) \quad (1)$$

The similarity measure of Equation (1) is influenced by the length of the vector. That means that $s_1$ is influenced by the contrast in the image and the case. In order to remove the contrast, the direction vectors are normalized to the length one by dividing them through their gradient. Note that this normalization differs from the normalized cross correlation (NCC): The NCC normalizes each pixel value by the expected mean of all values of the considered pixels. Therefore it is sensitive to nonlinear contrast changes, whereas our method is not. Contrast changes can be ignored if the absolute value of the dot product is calculated:

$$s_2 = \frac{1}{n}\sum_{k=1}^{n} \frac{|\mathbf{m}_k \cdot \mathbf{i}_k|}{\|\mathbf{m}_k\| \cdot \|\mathbf{i}_k\|} \quad (2)$$

### 3.2 Template Generation

A detailed description of the case generation can be found in [2]. The acquisition of the templates is done semi-automatically. Prototypical images are displayed to an expert. The expert manually traces the contour of the object with the help of the cursor of the computer. Afterwards the number of the contour points is reduced for data-reduction purposes by interpolating the marked contour by a $1^{st}$ order polynomial. The marked object shapes are then pair-wise aligned using Procrustes Algorithm [1]. From the set of shapes general groups of shapes are learnt by clustering. Single-linkage technique is used for clustering [3]. The prototype of each cluster is calculated by estimating the mean shape of the set of shapes in the cluster. Each of these prototypes is the representative of a group of similar shapes and will be used as an case template for the recognition process.

#### 3.2.1 Shape Alignment

The aim of the alignment process is to compare the shapes of two objects in order to define a measure of similarity between them. Consider two shape instances $P$ and $O$ defined by the point-sets $p_i \in R^2$, $i = 1, 2, ...N_1$, and $o_j \in R^2$, $j = 1, 2, ...N_2$ respectively. The basic task of aligning two shapes consists of transforming one of them ( say $P$ ), so that it fits in some optimal way the other one ( say $O$ ). Generally the shape instance $P = \{p_i(x, y)\}_{i=1...N}$ is said to be aligned to the shape instance $O = \{o_j(x, y)\}_{j=1...N}$ if a distance $d(P, O)$ between the two shapes can not be decreased by applying a transformation

$\Psi$ to $P$. The alignment of shapes is limited to a similarity transformation in order to eliminate differences in the translation, the rotation and the scale of the two shapes $P$ and $O$. In our application we use the Procrustes distance, a least-squares type distance function:

$$D(P,O) = \sum_{i=1}^{N} \left\| \frac{(p_i - \mu_P)}{\sigma_P} - R(\theta)\frac{(o_i - \mu_O)}{\sigma_O} \right\|^2 \tag{3}$$

where $\theta$ is the rotation matrix, $\mu_p$ and $\mu_o$ are the centroids of the objects $P$ and $O$, respectively and $\sigma_P$ and $\sigma_O$ are the sums of squared distances of each point-set from their centroids.

### 3.2.2 Clustering and Prototype Calculation

The alignment of every possible pair of objects in our database leads us to $N \times N$ pair-wise distances between $N$ shape instances. This matrix is the input for the single-linkage hierarchical cluster analysis [6]. As a result of this process we have divided our set of shape instances $\{P_1,\ P_2,\ \ldots, P_N\}$ into $k$ clusters $C_1,\ C_2, \ldots, C_k$. Each clusters $C_i$, $i = 1,\ 2, \ldots, k$ consists of a subset of $n_i$ shape instances. For each cluster we need to compute a prototype $\overline{\mu}$ that will be the representative of the cluster. This prototype can be calculated by computing the mean over all shapes in a cluster.

## 4 Results

We applied our method to six different airborne fungi spores (see Fig. 1). We labeled a total of 60 objects for each of the six fungal strains. These objects were taken for the case generation according to the procedure as described in Sec. 3. The result was a data base of cases. These cases were applied to images for the particular class which consist of unknown objects.

The threshold for the score was set to 0.8. The recognition rate is defined as the ratio of the number of correct recognized objects to the total number of objects in the image. The results of the matching process are shown in the Tables 1. The highest recognition rate can be achieved for the objects Rhizopus Stolonifer and Scopulariopsis Brevicaulis, since the variation of their shape is expressed well by the number of cases. For those classes where the variation of the shape of the objects is high, the number of the cases is also high. In the other cases the recognition rate shows that we do not have enough cases to recognize the classes with good recognition rate (see Alternaria Alternata and Ulocladium Botrytis). Therefore we need to increase the number of cases. For this task we should like to develop an incremental procedure for the case acquisition in our tool.

**Table 1.** Results of Matching

| Classes | Name of the cases | Recognition Rate |
|---|---|---|
| Alternaria Alternata | 34 | 81.0 |
| Aspergillus Niger | 5 | 89.0 |
| Rhizopus Stolonifer | 22 | 96.2 |
| Brevicaulis Scopulariopsis | 8 | 98.2 |
| Ulocladium Botrytis | 30 | 85 |
| Wallenia Sebi | 10 | 78.8 |

## 5 Conclusions

We have described our method for the recognition of airborne fungi spores in digital microscopic images.

We used a similarity-based recognition method. The case is represented by edges and not by the gray-level itself. The similarity measure is based on the scalar product and is invariant against illumination changes and contrast changes. The case generation was done semi-automatically by manually tracing the contour of the object, automatic shape alignment, shape clustering, and prototype calculation. For future research we intend to develop an incremental case-acquisition procedure that should ensure that we can learn the natural shape variability over time.

*Acknowledgement.* The project "Development of methods and techniques for the image-acquisition and computer-aided analysis of biologically dangerous substances BIOGEFA" is sponsored by the German Ministry of Economy BMWI under the grant number 16IN0147.

## References

1. I.L. Dryden and K.V. Mardia. *Statistical Shape Analysis.* Wiley, New York, 1998.
2. S. Jähnichen and P. Perner. Case acquisition and case mining for case-based object recognition. In P. Funk and P. Gonzales, editors, *Proc. European Conference on Case-Based Reasoning.* Springer, 2004.
3. P. Perner. *Data Mining on Multimedia Data.* Springer Verlag, 2003.
4. P. Perner and A. Bühring. Case-based object recognition. In P. Funk and P. Gonzales, editors, *Proc. European Conference on Case-Based Reasoning.* Springer, 2004.

# Rivalling Optimal Control in Robot-Assisted Surgery

G.F. Schanzer and R. Callies

TU München
Zentrum Mathematik, M2
Boltzmannstraße 3
85748 Garching
schanzer@ma.tum.de and callies@ma.tum.de

**Summary.** Miniaturized robotic manipulators are a key element in future high-precision minimally invasive surgery and telesurgery. This development is supported by the rapidly decreasing size of robotic sensors and actuators. Severe limitations are currently induced by the drives of the micro-joints.

The present paper deals with the optimal control of an advanced six-sectional branched manipulator. Joints are driven by weak, but fast, and strong, but slow, actuators acting in parallel. This results in the new and challenging problem of constrained optimal control of multibody systems subject to rivalling controls. For efficient treatment the differential equations of the state and adjoint variables are recursively defined. Geometric constraints lead to state constraints of second order. The complete problem of optimal control is transferred into a piecewise defined, highly nonlinear multi-point boundary value problem. The numerical solution of the boundary value problem is by the advanced multiple shooting method *JANUS*.

**Key words:** Rivalling optimal control, robot-assisted surgery

## 1 Introduction

For further development of high-precision minimally invasive surgery (e.g. in brain surgery) there is an urgent need for new types of robotic manipulators consisting of miniaturized artificial fingers and hands of increasing complexity. One of the major technical difficulties consists in the development of suitable drives for the micro-joints of the robots. Either actuators are fast and efficient, but produce only small specific forces (e.g. piezo actuators), or they are able to generate high specific forces, but with a slow rate of change or at the cost of relatively high power consumption (e.g. shape memory alloys).

The present paper deals with the optimal control of an advanced branched manipulator consisting of two fingers with three joints each mounted on a joint base. Joints are driven by weak, but fast, and strong, but slow, actuators acting in parallel. For the mathematical treatment of this model, a challenging optimal control problem for cooperating robots has to be solved under control and state constraints and – for the first time – for rivalling controls. From the point of view of optimal control and numerical analysis, this model contains all the relevant difficulties. New control strategies are developed and the structure of the complicated configuration is fully exploited.

# 2 Manipulator Model

Fig. 1 shows a sketch of the geometry of the two-finger hand with three joints at each finger, in total with six degrees of freedom. Joint angles are denoted by $q_i$, angular velocities by $\dot{q}_i$ and actuator torques by $u_i$. As usual, dots indicate time derivatives. The torques $u_i$, $i = 1, \ldots, 6$, are the controls.

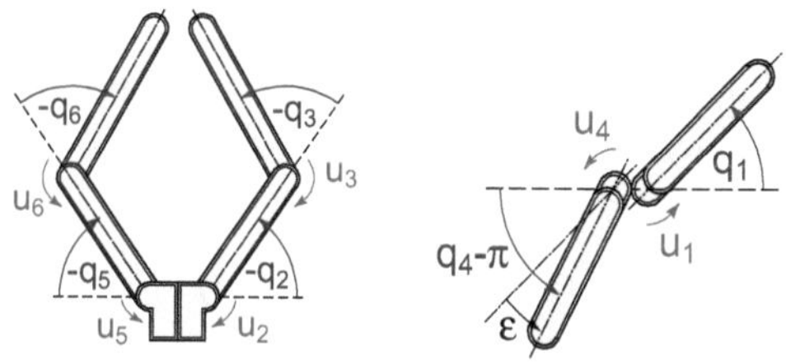


**Fig. 1.** Model of the two-finger hand: front view (left) and top view (right)

The equations of motion are given by

$$M\Big(q(t)\Big) \cdot \ddot{q}(t) + r\Big(q(t), \dot{q}(t)\Big) = u(t)\,, \quad M \in \mathbb{R}^{6\times 6},\ t \in [0, t_F] \qquad (1)$$

with the mass matrix $M$ and the function $r$ containing all the gravity, centrifugal and Coriolis forces.

The evaluation of the equations of motion is efficiently done by the Newton-Euler-algorithm – a recursive formulation which uses geometric informations to evaluate the equations of motion [3].

# 3 Optimal Control

## 3.1 Rivalling Control

In the case of the micro-robot there are two controls acting on the same joint: a weak, but efficient control is rivalling with a strong, but inefficient control. Efficiency in this case means low energy consumption. Therefore, the respective control $u_i$ is split in two new controls

$$u_i = u_i^w + u_i^s$$

with the weak control $u_i^w$ and the strong control $u_i^s$ .

### 3.2 Optimal Control Theory

The (energy) optimal control problem for the model developed in Sect. 2 can be stated as follows:

$$I(u(t)) = \int_0^{t_F} \frac{\sum_i \left((u_i^w)^2 + \beta_i (u_i^s)^2\right)}{||u_{max}||^2} dt \longrightarrow min \quad (\beta_i > 1 \vee \beta_i = 0) \quad (2)$$

The final time $t_F$ has to be fixed. Weight factors $\beta_i > 1$ put controls with higher energy consumption at a disadvantage, the scaling factor $\|u_{max}\|^2$ is a measure of the maximum energy. The objective function $I$ has to be minimized subject to the equations of motion (1), boundary conditions BC for $q$ and $\dot{q}$ and the prescribed control constraints and state constraints. With the Hamiltonian

$$H = \frac{\sum_i \left((u_i^w)^2 + \beta_i (u_i^s)^2\right)}{||u_{max}||^2} + \lambda^T \dot{x} \qquad x := \begin{pmatrix} q \\ \dot{q} \end{pmatrix} \quad (3)$$

the optimal control problem is transformed into the following boundary-value-problem:

$$\left.\begin{aligned} \begin{pmatrix} E^6 & O \\ O & M \end{pmatrix} \cdot \dot{x}(\tau) &= \begin{pmatrix} \dot{q} \\ u(\tau) - r(x(\tau)) \end{pmatrix} \\ \dot{\lambda} = -H_x &= -\left(\frac{\partial}{\partial x}\dot{x}\right)^T \cdot \lambda \end{aligned}\right\} + \text{BC} \quad (4)$$

$\lambda$ denotes the adjoint variables, $E^n := diag(1, \dots, 1) \in \mathbb{R}^{n \times n}$. The controls $u_i^w, u_i^s$ are derived from the *optimality condition* $H_u = 0 \quad \Rightarrow$

$$0 = H_{u_i^w} := \frac{2u_i^w}{||u_{max}||^2} + \lambda^T \left(\frac{\partial}{\partial u_i^w}\dot{x}\right) \quad (5)$$

$$0 = H_{u_i^s} := \frac{2\beta_i u_i^s}{||u_{max}||^2} + \lambda^T \left(\frac{\partial}{\partial u_i^s}\dot{x}\right) \quad (6)$$

and the *Legendre-Clebsch-Condition* ($H_{uu}$ positiv semidefinite).

The matrix $(\frac{\partial}{\partial x}\dot{x})$ is calculated by a modified Newton-Euler-algorithm. It is this decisive modification [4] which increases computational efficiency dramatically over e.g. symbolic differentiation and makes the full application of the calculus of variations possible in practice.

## 4 Optimal Control Constraints

### 4.1 Constraints

The control constraints can be stated as follows

$$|\,u\,| \leq u_{max} \qquad\qquad |\,u^{w,s}\,| \leq u^{w,s}_{max} \tag{7}$$

A state constraint of order two is derived by bionic considerations (cf. Fig. 1)

$$|\,q_1 - q_4 + \pi\,| \leq \varepsilon \tag{8}$$

If geometrical obstacles are present, state constraints of order two of the following type occur

$$h(X,Y,Z) \leq 0 \ \in \mathbb{R}^m \tag{9}$$

After transformation from work space (cartesian coordinates $X, Y, Z$) to joint space *by using manipultor kinematics* [3], these constraints read as

$$h(X,Y,Z) \longrightarrow h(q)$$

By definition state constraints of order $p$ have the following properties

$$h(x) \leq 0\,, \quad \frac{\partial}{\partial u}\left(\frac{\partial^p}{\partial t^p}h\right) \neq 0\,, \quad \frac{\partial}{\partial u}\left(\frac{\partial^i}{\partial t^i}h\right) = 0 \ \text{ for } i = 1,\ldots,p-1 \tag{10}$$

State constraints are treated according to the lemma of Bryson et al. [1] by forming the extended Hamiltonian with the Lagrangian multiplier $\nu$:

$$\tilde{H} = \frac{\sum_i \left((u_i^w)^2 + \beta_i\,(u_i^s)^2\right)}{||u_{max}||^2} + \lambda^T\dot{x} + \nu\frac{\partial^p}{\partial t^p}h$$

The multiplier $\nu$ and the control $u$ are calculated from $\tilde{H}_u = 0$ and $h^p = 0$.

## 4.2 Numerical Realisation

An explicit and possibly instable inversion of the mass matrix $M$ in (1) can be avoided because also $h^p$ is a linear function of $\ddot{q}$. Defining $\ddot{q}$ as a further unknown, $u$ and $\nu$ can be calculated by solving a system of linear equations: Adding (1) to the equations $\tilde{H}_u = 0$ and $h^p = 0$, we obtain the following well-structured system with the equation $h^p = 0$ in the last row

$$\begin{pmatrix} D_1 & M & O \\ D_2 & O & h_u^p \\ & * & \end{pmatrix} \cdot \begin{pmatrix} u \\ \ddot{q} \\ \nu \end{pmatrix} = \begin{pmatrix} -r(q,\dot{q}) \\ -\left(\frac{\partial}{\partial u}\dot{x}\right)^T\lambda \\ * \end{pmatrix}, \quad D_1 = -E^6,\ D_2 = \frac{2}{||u_{max}||^2}\cdot E^6$$

An active state constraint of order two can be interpreted as a DAE of differential index three. To avoid numerical difficulties Baumgarte-stabilisation or extensions are possible. In our calculations no relevant drift-off was observed (rel. deviation $< 10^{-8}$).

Recursive formulations and the solution of a sequence of systems of linear equations provide an elegant way to transform the optimal control problem into a multi-point-boundary-value-problem without ever explicitely formulating the respective, extremely complicated ODE system. Numerical solution of the boundary-value-problem is by the multiple-shooting method *JANUS* [2].

## 5 Example: Constrained Motion and Rivalling Control

In the following example $t_F = 0.137\,[s]$ is fixed and two rivalling controls are considered: $u_2 = u_2^w + u_2^s$, $\beta_2 = 5$, $u_5 = u_5^w + u_5^s$, $\beta_5 = 3$ and $\beta_1 = \beta_3 = \beta_4 = \beta_6 = 0$. Initial and final position of the robotic manipulator are fixed.

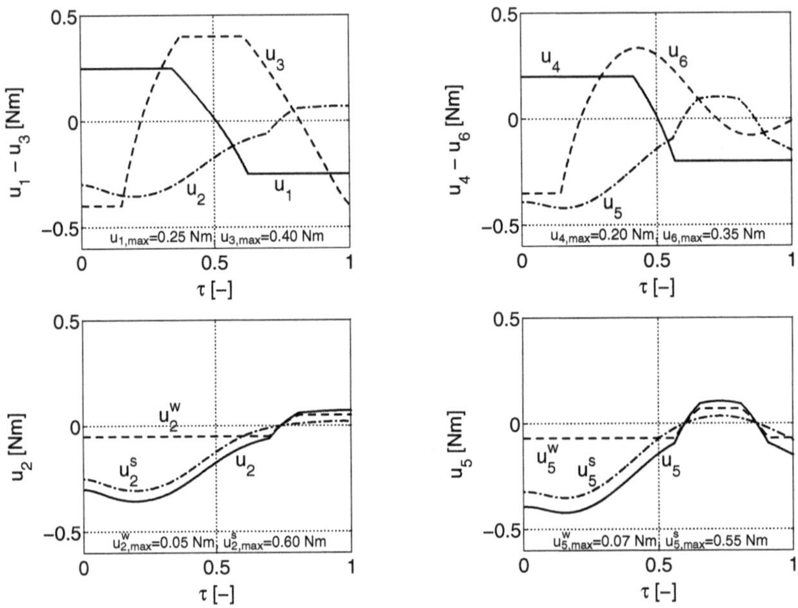


**Fig. 2.** Example problem: Controls as a function of $\tau := t/t_F$

*Acknowledgement.* This work has been supported by the *Bayerische Forschungsstiftung* under contract *ForBild TP II.1.*

## References

1. A. E. Bryson, W. F. Denham, and S. E. Dreyfus. Optimal Programming Problems with Inequality Constraints I: Necessary Conditions for Extremal Solutions. *AIAA 1*, pages 2544–2550, 1963.
2. R. Callies. *Entwurfsoptimierung und optimale Steuerung. Differential-algebraische Systeme, Mehrgitter-Mehrzielansätze und numerische Realisierung.* Habilitationsschrift, Technische Universität München, 2000.
3. J. J. Craig. *Introduction to Robotics.* Addison-Wesley, 1986.
4. T. Schenk. Effiziente Verfahren zur optimalen Steuerung von mehrgelenkigen Industrierobotern. Master's thesis, Technische Universität München, 2002.

# Part V

## Theme: Materials

# A Multiphase Model for Concrete: Numerical Solutions and Industrial Applications

B.A. Schrefler[1], D. Gawin[2], and F. Pesavento[1]

[1] Department of Structural and Transportation Engineering, University of Padova, Padova – Italy
[2] Department of Building Physics and Building Materials, Technical University of Lodz, Lodz – Poland

**Summary.** A mathematical and numerical model to predict the non-linear behaviour of concrete as multiphase porous material is proposed. The model can be usefully applied to several practical cases: evaluation of concrete performance in the high temperature range, e.g. during fire, to early stages of maturing of massive concrete structures, to shotcrete in tunnelling, and to durability. All the important phase changes of water and chemical reactions, i.e. adsorption-desorption, condensation-evaporation, and hydration-dehydration, as well as the related heat and mass sources or sinks are considered. Changes of the material properties caused by temperature and pressure changes, concrete damage or carbonation, fresh concrete hardening, as well as coupling between thermal, hygral and mechanical phenomena are taken into account. This model further allows to incorporate sorption hysteresis. Some relevant applications of the model will be shown in this work.

## Physical and mathematical model

Moist concrete is modelled as a multi-phase material, which is assumed to be in thermo-dynamic equilibrium state locally. The voids of the skeleton are filled partly with liquid water and partly with a gas phase. The liquid phase consists of bound water, which is present in the whole range of moisture content, and capillary water, which appears when water content exceeds the upper limit of the hygroscopic region, $S_{ssp}$. The gas phase is a mixture of dry air and water vapour, and is assumed to be an ideal gas. The chosen primary variables of the model are: gas pressure $p^g$, capillary pressure $p^c = p^g - p^w$ ($p^w$ denotes water pressure), temperature $T$, displacement vector of the solid matrix $\mathbf{u}$, and finally carbon dioxide concentration $\rho_d$ if carbonation process is considered, while the internal variables are: degree of cement hydration $\Gamma_{hydr}$, when hydration or dehydration phenomena are analysed, degree of carbonation $\Gamma_{carb}$, when carbonation is taken into account and mechanical damage d and thermo-chemical damage $V$ when damaging-deterioration processes are considered.

Hence, the general mathematical model of chemo-hygro-thermo-mechanical processes consists of four or five balance equations, depending on the problem analysed. They are completed by an appropriate set of constitutive and state equations, and some thermodynamic relationships. Considering that in this work we do not take into account the case of concrete subjected to carbonation phenomenon (durability mechanics) and that in the case of concrete structures exposed to high temperatures in the range below 600 – 700°C we can neglect the term related to decarbonation process, the final form of the set of governing equations is:

- Mass balance equation of the dry air

$$
\begin{aligned}
&-n\frac{D^s S_w}{Dt}-\beta_s\left(1-n\right)S_g\frac{D^s T}{Dt}+S_g div\,\mathbf{v}^s+\frac{S_g n}{\rho^{ga}}\frac{D^s\rho^{ga}}{Dt}+\frac{1}{\rho^{ga}}div\,\mathbf{J}_g^{ga}\\
&+\frac{1}{\rho^{ga}}div\left(n\,S_g\rho^{ga}\mathbf{v}^{gs}\right)-\frac{\left(1-n\right)S_g}{\rho^s}\frac{\partial\rho^s}{\partial\Gamma_{hydr}}\frac{D^s\Gamma_{hydr}}{Dt}=\frac{\dot{m}_{hydr}}{\rho^s}S_g
\end{aligned}
\tag{1}
$$

- Mass balance equation of the water species

$$
\begin{aligned}
&n\left(\rho^w-\rho^{gw}\right)\frac{D^s S_w}{Dt}+\left(\rho^w S_w+\rho^{gw}S_g\right)\alpha\,div\,\mathbf{v}^s+S_g n\frac{D^s\rho^{gw}}{Dt}\\
&+div\,\mathbf{J}_g^{gw}+div\,\left(n\,S_w\rho^w\mathbf{v}^{ws}\right)+div\,\left(n\,S_g\rho^{gw}\mathbf{v}^{gs}\right)+\\
&-\left(\rho^w S_w+\rho^{gw}S_g\right)\frac{\left(1-n\right)}{\rho^s}\frac{\partial\rho^s}{\partial\Gamma_{hydr}}\frac{D^s\Gamma_{hydr}}{Dt}=\frac{\dot{m}_{hydr}}{\rho^s}\left(\rho^w S_w+\rho^{gw}S_g-\rho^s\right)
\end{aligned}
\tag{2}
$$

- Enthalpy balance equation of the multi-phase medium

$$
\begin{aligned}
&\left(\rho C_p\right)_{eff}\frac{\partial T}{\partial t}+\left(\rho_w C_p^w\mathbf{v}^w+\rho_g C_p^g\mathbf{v}^g\right)\cdot grad\,T-div\left(\lambda_{eff}grad\,T\right)=\\
&-\dot{m}_{vap}\Delta H_{vap}+\dot{m}_{dehydr}\Delta H_{dehydr}
\end{aligned}
\tag{3}
$$

- Linear momentum conservation equation of the multi-phase medium

$$
div\left(\sigma-\alpha p^s\mathbf{I}\right)+\rho\mathbf{g}=0 \tag{4}
$$

where the effective stresses $\sigma''$ is given by:

$$
\sigma=(1-D)\mathbf{\Lambda_0}:\left(\varepsilon_{tot}-\varepsilon_{th}-\varepsilon_0\right) \tag{5}
$$

where the parameter $D$ is the total damage resulting from various material deterioration processes of different nature: mechanical, thermo-chemical, purely chemical. The term $\varepsilon_0$ is formed by two different contributions: chemical strains accounting for thermo-chemical deterioration process in case of elevated temperatures or chemical reactions in all the other cases, and creep strains accounting for mid-long term creep in durability problems and thermal creep in high temperature ranges:

$$
\varepsilon_0=\varepsilon_{chem}+\varepsilon_{creep} \tag{6}
$$

The mass balance equation of carbon dioxide has to be added to the previous ones if carbonation phenomenon is analysed. Furthermore, three (or more) evolution equations, corresponding to the internal variables related to the evolution processes included in the model, can be added to the above described governing equations:

• Hydration/Dehydration process evolution law
When dehydration process is considered (temperature higher than 105°C), taking into account its irreversibility, one may assume that the degree of dehydration depends on the maximum value of temperature reached during heating:

$$\Gamma_{dehydr}(t) = \Gamma_{dehydr}(T_{max}(t)) \tag{7}$$

while, when hydration process is analyzed (below 105°C) the hydration degree is defined in the following way:

$$\Gamma_{hydr} = \frac{\chi}{\chi_\infty} = \frac{m_{hydr}}{m_{hydr\infty}} \tag{8}$$

where $m_{hydr}$ means mass of hydrated water (chemically combined), $\chi$ is the hydration extent and $\chi_\infty$ , $m_{hydr\infty}$ are the final values of hydration extent and mass of hydrated water, respectively.

• Thermo-chemical damage evolution equation (high temperature) The parameter $V$ takes into account both effects of concrete dehydration (chemical component) and material cracking (mechanical component) on material degradation and the Young's modulus decrease with increasing temperature. It is obtained from the experimental results, and is a function of the maximum temperature reached during heating because of the irreversible character of the material structural changes:

$$V(t) = V(T_{max}(t)) \tag{9}$$

• Mechanical damage evolution equation
The mechanical damage parameter $d$ is expressed in terms of the equivalent strain, $\tilde{\varepsilon}$ , and it is given by equations of the classical non-local, isotropic damage theory,

$$d(t) = d(\tilde{\varepsilon}(t)) \tag{10}$$

Similarly to what has been stated for governing equations, if carbonation process is taken into account in the modelling of concrete behaviour, it is necessary to define the corresponding evolution equation. For a full description of the model and its mathematical formulation, see [16, 9, 18].

# 1 Numerical solution

The governing equations of the model (1)-(4) are discretised in space by means of the finite element method, [29, 30, 16]. The unknown variables are expressed in terms of their nodal values as,

$$\begin{aligned} p^g(t) &\cong \mathbf{N}_p \bar{\mathbf{p}}^g(t), & p^c(t) &\cong \mathbf{N}_p \bar{\mathbf{p}}^c(t), \\ T(t) &\cong \mathbf{N}_t \bar{\mathbf{T}}(t), & \mathbf{u}(t) &\cong \mathbf{N}_u \bar{\mathbf{u}}(t). \end{aligned} \tag{11}$$

The variational or weak form of the model equations, was obtained in [16] by means of Galerkin's method (weighted residuals), and can be written in the following concise discretised matrix form,

$$\mathbf{C}_{ij}(\mathbf{x}) \frac{\partial \mathbf{x}}{\partial t} + \mathbf{K}_{ij}(\mathbf{x})\ \mathbf{x} = \mathbf{f}_i(\mathbf{x}), \text{with} \tag{12}$$

$$\mathbf{K}_{ij} = \begin{bmatrix} \mathbf{K}_{gg} & \mathbf{K}_{gc} & \mathbf{K}_{gt} & \mathbf{0} \\ \mathbf{K}_{cg} & \mathbf{K}_{cc} & \mathbf{K}_{ct} & \mathbf{0} \\ \mathbf{K}_{tg} & \mathbf{K}_{tc} & \mathbf{K}_{tt} & \mathbf{0} \\ \mathbf{K}_{ug} & \mathbf{K}_{uc} & \mathbf{K}_{ut} & \mathbf{K}_{uu} \end{bmatrix}, \mathbf{C}_{ij} = \begin{bmatrix} \mathbf{C}_{gg} & \mathbf{C}_{gc} & \mathbf{C}_{gt} & \mathbf{C}_{gu} \\ \mathbf{0} & \mathbf{C}_{cc} & \mathbf{C}_{ct} & \mathbf{C}_{cu} \\ \mathbf{0} & \mathbf{C}_{tc} & \mathbf{C}_{tt} & \mathbf{C}_{tu} \\ \mathbf{0} & \mathbf{0} & \mathbf{0} & \mathbf{0} \end{bmatrix}, \mathbf{f}_i = \begin{Bmatrix} \mathbf{f}_g \\ \mathbf{f}_c \\ \mathbf{f}_t \\ \mathbf{f}_u \end{Bmatrix}, \tag{13}$$

where the vectors $\mathbf{x}^T = \{\bar{\mathbf{p}}^g, \bar{\mathbf{p}}^c, \bar{\mathbf{T}}, \bar{\mathbf{u}}\}$ and $\mathbf{f}_i(\mathbf{x})$ and the non-linear matrices $\mathbf{C}_{ij}(\mathbf{x}), \mathbf{K}_{ij}(\mathbf{x})$ are defined in detail in [16, 9, 18].
The time discretization is accomplished through a fully implicit finite difference scheme (backward Euler),

$$\Psi^i(\mathbf{x}_{n+1}) = \mathbf{C}_{ij}(\mathbf{x}_{n+1}) \frac{\mathbf{x}_{n+1} - \mathbf{x}_n}{\Delta t} + \mathbf{K}_{ij}(\mathbf{x}_{n+1})\ \mathbf{x}_{n+1} - \mathbf{f}_i(\mathbf{x}_{n+1}) = \mathbf{0}, \tag{14}$$

where superscript $i(i = g, c, t, u)$ denotes the state variable, $n$ is the time step number and $\Delta t$ the time step length. The non-linear equation set (14) is linearised and solved by means of a monolithic Newton-Raphson type iterative procedure [16, 9, 18, 30, 25]:

$$\Psi^i(\mathbf{x}^k_{n+1}) = -\left.\frac{\partial \Psi^i}{\partial \mathbf{x}}\right|_{\mathbf{x}^k_{n+1}} \Delta\mathbf{x}^k_{n+1}, \qquad \mathbf{x}^{k+1}_{n+1} = \mathbf{x}^k_{n+1} + \Delta\mathbf{x}^k_{n+1}, \tag{15}$$

where $k$ is the iteration index and $\frac{\partial \Psi^i}{\partial \mathbf{x}}$ is Jacobian matrix.

# Application of the model to young concrete

From the macroscopic point of view, hydration of cement is a complex interactive system of competing chemical reactions of various kinetics and amplitudes. They are associated with complex physical and chemical phenomena at the micro-level of material structure, see e.g. [24, 25], resulting in considerable changes of macroscopic concrete properties. Kinetics of cement hydration (hydration rate) cannot be described properly in terms of equivalent age nor maturity of concrete, if the effect of the reaction rate on temperature (and/or relative humidity) depends upon the hydration degree [25], or chemical affinity of the reaction is affected by temperature variations (and/or relative humidity) [28, 12]. Hence, another thermodynamically based approach has been used instead, similarly as proposed by Ulm and Coussy [27, 28]. In this approach the hydration extent is the advancement of the hydration reaction and its rate is related to the affinity of the chemical reaction through an Arrhenius-type relationship, as usual for thermally activated chemical reactions:

$$\frac{d\chi}{dt} = \tilde{A}_{\chi}(\chi) \exp\left(-\frac{E_a}{RT}\right) \tag{16}$$

where $\widetilde{A}_{\chi}(\chi)$ is normalized affinity (it accounts both for chemical non-equilibrium and for the nonlinear diffusion process), $E_a$ - hydration activation energy, and $R$ - universal gas constant. Equation (16) can be rewritten in terms of hydration degree, defined as in (8), and relative humidity by means of a function $\beta_{\varphi}(\varphi)$, ($\varphi$ is the relative humidity):

$$\frac{d\Gamma_{hydr}}{dt} = \tilde{A}_{\Gamma}(\Gamma_{hydr}) \beta_{\varphi}(\varphi) \exp\left(-\frac{E_a}{RT}\right) \tag{17}$$

An analytical formula for the description of the normalized affinity of the following form,

$$\tilde{A}_{\Gamma}(\Gamma_{hydr}) = A_1 \left(\frac{A_2}{\kappa_{\infty}} + \kappa_{\infty}\Gamma_{hydr}\right)(1 - \Gamma_{hydr}) \exp(-\bar{\eta}\,\Gamma_{hydr}) \tag{18}$$

was proposed by Cervera et al. [6] and is used in our model. The coefficients A1, A2 and can be obtained from the temperature evolution during adiabatic tests.

As far as creep is concerned, fresh concrete is modelled as a visco-elastic material by use of the rate (incremental) formulation of solidification theory [2, 3]. In this theory concrete maturing is attributed to a growth of the volume fraction of load-bearing hydrated cement, which itself is considered as a non-aging visco-elastic material. Usually, the creep on these non-aging constituents is described by a Kelvin chain with a finite number $N$ of Kelvin units. In this case the spectrum is discrete and its identification from test data is an ill-posed problem. To overcome this problem a continuous retardation spectrum technique has been introduced into the model. This means that a continuous Kelvin chain model with infinitely many Kelvin units and retardation times spaced infinitely closely, has been used [3].

It is important to underline that we model hygro-mechanical interactions (*i.e.* capillary shrinkage phenomenon) using effective stresses defined as:

$$\sigma = \sigma + \alpha p^s \mathbf{I}, \tag{19}$$

where $\alpha$ is the Biot's constant and $p^s$ is the solid pressure:

$$p^s = \chi_s^{ws}\left(p^w + s^{ws} J_{ws}^s\right) + \left(1 - \chi_s^{ws}\right)\left(p^g + s^{gs} J_{gs}^s\right) \tag{20}$$

where $\chi_s^{ws}$ is the fraction of skeleton area in contact with water, $J_{ws}^s$, $J_{gs}^s$ the average curvatures of the solid-water and solid-gas phase interfaces, respectively, obtained by integrating the point curvature over the interface within the macro scale volume, and $s^{ws}$, $s^{gs}$are interfacial tension like terms.
Solid pressure accounts for the pressure exerted by pore fluids on solid skeleton. This component of the stress tensor causes an additional deformation of the skeleton (shrinkage strains), hence one can expect that it will contribute to creep strains, as well. Indeed, some experimental studies of autogenous deformations of concrete at early ages, e.g. [17], suggest that a part of the material strains in such a situation (*i.e.* without any external load) can be explained only by creep deformations due to capillary forces.

## Numerical simulation of self-desiccation process

The numerical example deals with self-heating and self-desiccation phenomena in sealed, cylindrical concrete samples made of ordinary and HPC concretes (60 cm long and with diameter of 4 cm), placed in adiabatic conditions. These specimens have been used for testing of autogenous relative humidity change and autogenous deformation, during first 30 days of concrete maturing. The simulation results are compared with available experimental data concerning temperature changes during cement hydration. The deformation is usually measured in a middle part of a sample, which is unaffected by edge effects, hence the element performance is modeled as a 1-D axisymmetric problem. The mesh with 26 (26 × 1) eight-node serendipity finite elements of variable sizes (decreasing towards to the surface and to the axis) was used for space discretisation of the sample. Simulations were performed for two types of concrete: OC tested by Bentz et al. [4] and HPC tested by Laplante [15], which were also used by Cervera et al. [6] for validation of their thermo-chemo-mechanical model of concrete at early ages. The composition of these two types of concrete were similar to those used by Baroghel-Bouny et al. [1] during their measurements of hygral properties, *i.e.* sorption isotherms, intrinsic and relative permeability, thus we assumed these data in our simulations. Main material properties (for dry concrete after 28-days of maturation) used in our calculations are summarized in Table 1. The heat and mass sources related to concrete maturing are expressed as a function of the hydration rate by means of relation (17), where normalized affinity was described by (18) with the parameters determined in [6, 7], see Table 1. Initially, the cylinder has a temperature $T_0 = 293.15$K and relative humidity $0 = 99.9\%$ for OC, while $0 = 99.0\%$ for HPC. It is assumed that the hydration process started about

2-3 hours before, hence the sample had already a certain shape rigidity and initial hydration degree was equal to 0.1. The external surfaces of the sample are sealed and adiabatic. Results of our simulations concerning the changes of temperature, relative humidity and degree of hydration during initial stages of concrete maturing are shown in Figs. 1(a) and 1(b) for OC and HPC. In HP concrete we observe considerably lower values of relative humidity at initial stages of maturing, that shows for HPC concrete an influence of relative humidity on the hydration process and the related autogenous changes of temperature and moisture content (*i.e.* self-heating and self-desiccation phenomena) are of importance. The temperature histories for the analyzed types of concrete are compared to the experimental results from literature [4, 15], showing their good agreement. The temperatures obtained from simulations for HPC for time t 12h are visibly higher than the experimental ones, what is caused by not perfectly adiabatic conditions during the test, as mentioned in [7]. One should underline that the present model takes also into account hygral phenomena, and in particular phase changes and related to them heat effects, which were not considered in [27, 28, 6, 7]. For a deeper investigation of this aspect see [10].

**Table 1.** Characteristic properties of different types of concrete(in dry state, after 28 days of maturing) used in numerical simulations

| Parameter | Symbol | Unit | OC | HPC |
|---|---|---|---|---|
| Water / cement ratio | w/c | [-] | 0.45 | 0.35 |
| Aggregate / cement ratio | c/a | [-] | 4.0 | 4.55 |
| Silica fume / cement ratio | s/a | [-] | 0.00 | 0.09 |
| Porosity | n | [%] | 12.2 | 8.2 |
| Intrinsic permeability | k | $[m^2]$ | $3 \cdot 10 - 18$ | $1 \cdot 10 - 18$ |
| Activation energy | $E_a/R4$ | [K] | 5000 | 4000 |
| Parameter $A_1$ in eq. (13) | $A_1$ | $[1/s]$ | $7.78 \cdot 104$ | $1.11 \cdot 103$ |
| Parameter $A_2$ in eq. (13) | $A_2$ | [-] | $0.5 \cdot 10 - 5$ | $41 \cdot 10 - 4$ |
| Parameter $k_\infty$ in eq. (13) | $k_\infty$ | [-] | 0.72 | 0.58 |
| Parameter $\overline{\eta}$ in eq. (13) | $\overline{\eta}$ | [-] | 5.3 | 6.0 (4.7) |
| Heat of hydration | $Q_{hydr\infty}$ | $[MJ/m^3]$ | 202 | 172 |
| Apparent density | $\rho_{\text{eff}}$ | $[kg/m^3]$ | 2285 | 2373 |
| Thermal conductivity | $\lambda_{\text{eff}}$ | $[W/mK]$ | 1.5 | 1.78 |
| Young's modulus | $E$ | [GPa] | 24.11 | 39.61 |
| Compressive strength | $f_c$ | [MPa] | 26 | 70 |

## 2 Application of the model to concrete structures in high temperature environments

The model has been applied to the analysis of behaviour of concrete structures under severe temperatures and pressures conditions. In these conditions concrete structures experience spalling phenomenon, which results in rapid loss of the surface layers of the concrete at temperature exceeding about 200-300°C. As a result, the core concrete is exposed to these temperatures, thereby increasing the rate of heat transmission to the core part of the element and in particular to the reinforcement, what may pose a risk for the integrity of concrete structure. It is commonly believed that the main reasons of the thermal spalling are: build-up of high pore pressure close to the heated concrete surface as a result of rapid evaporation of moisture, and the release of the stored energy due to the thermal stresses resulting from high values of restrained strains caused by temperature gradients. Nevertheless, relative importance of the two factors is not established yet and still needs further studies, both experimental and theoretical.

The results of the research performed up to now show, that the fire performance of concrete structures is influenced by several factors, like initial moisture content of the concrete, the rate of temperature increase (fire intensity), porosity (density) and permeability of the concrete, its compressive strength, type of aggregate, dimensions and shape of a structure, its lateral reinforcement and loading conditions. The HSC structures are particularly affected by this phenomenon. In fact, HSC provides better structural performance, especially in terms of strength and durability, compared to traditional, normal-strength concrete (NSC).

However, many studies, showed that the fire performance of HSC differs from that of NSC which exhibits rather good behaviour in these conditions. An unloaded sample of plain concrete or cement stone, exposed for the first time to heating, exhibits considerable changes of its chemical composition, inner structure of porosity and changes of sample dimensions (irreversible in part). The concrete strains during first heating, called load-free thermal strains (LFTS) are usually treated as superposition of thermal and shrinkage components, and often are considered as almost inseparable. LFTS are decomposed in three main contributions:

- Thermal dilatation strains,

$$d\varepsilon_{th} = \beta_s(T)\ dT \tag{21}$$

- Capillary shrinkage strains,

$$d\varepsilon_{sh} = \frac{\alpha}{K_T}\left(dx_s^{ws} p^c + x_s^{ws} dp^c\right)\mathbf{I} \tag{22}$$

where $K_T$ is the bulk modulus of the porous medium,

- Thermo-chemical strains,

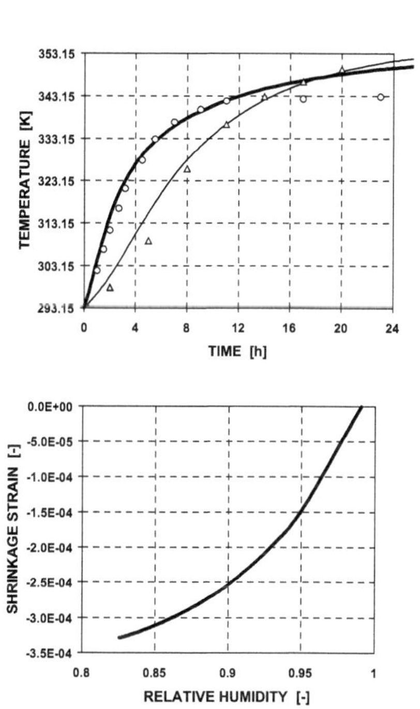


(a) Thin line: Ordinary concrete; solid line: High Performance Concrete: A) Temperature history in the specimen (and comparison with experimental results), B) Shrinkage strains versus Relative Humidity in HPC sample.

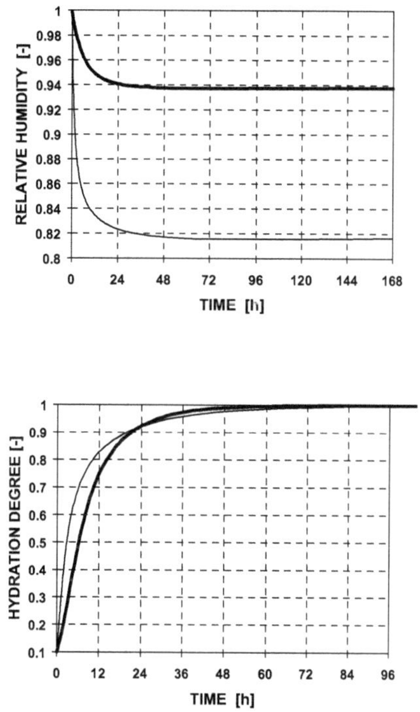


(b) Thin line: Ordinary concrete; solid line: High Performance Concrete: A) Relative Humidity history, B) Hydration degree history

**Fig. 1.**

$$d\varepsilon_{tchem} = \beta_{tchem}\left(V\right)\ dV \tag{23}$$

where $\beta_{tchem}\left(V\right) = \frac{\partial \varepsilon_{tchem}(V)}{\partial V}$ is obtained from experimental tests ($V$ is the thermo-chemical damage parameter).
As far as the first contribution is concerned, the strains are treated in a manner usual in thermo-mechanics, but considering the thermal expansion coefficient $\beta_s$ as a function of temperature. Shrinkage strains are modelled by means of the effective stress principle in the form derived in [9, 18, 16], (19), which for materials with very small pores and well developed internal pore surface, where water is also present as a thin film (like for example in concrete), presents a coefficient $\chi_s^{ws}$ instead of the classical saturation $S$.

In this way, the (capillary) shrinkage represents a load for the skeleton of the material and the related strains are not computed directly in the strain decomposition as it is usual in the classical phenomenological approaches. This coefficient is a function of saturation $S$ and takes into account the disjoining pressure which is important in the range of saturation in which only a thin film of water is adsorbed to the wall of the pores. This treatment of the shrinkage strains is more consistent from thermodynamic point of view. In heated concrete, above the temperature of about 105°C, starts the thermal decomposition of the cement matrix, and at higher temperatures also of aggregate (depending on its type and composition). This is a consequence of several complicated, endothermic chemical reactions, called concrete dehydration. As their result a considerable shrinkage of cement matrix (called chemical shrinkage) and usually expansion of aggregate are observed. Due to these contradictory behaviour of the material components, cracks of various dimensions are developing when temperature increases, causing an additional change of concrete strains (usually expansion). These strains are modelled as function of thermo-chemical damage which takes into account the thermo-chemical deterioration of the material.

Also the so called LITS has been considered in the computation. During first heating, mechanically loaded concrete exhibits greater strains as compared to the load-free material at the same temperature. These additional deformations are referred to as load induced thermal strains (LITS), [14]. A part of them originates just from the elastic deformation due to mechanical load, and it increases during heating because of thermo-chemical and mechanical degradation of the material strength properties. The time dependent part of the strains during transient thermal processes due to temperature changes, is generally called thermal creep.

The formulation employed into the model is due to Thelandersson [26] in its original form, here modified using a coefficient $\overline{\beta}_{tr}(V)$ as a function of thermo-chemical damage $V$ (and not constant) and the effective stresses instead of total stresses, coupling in this way the thermo-chemo-mechanical damage model and capillary shrinkage model with thermal creep model.

$$d\varepsilon_{tr} = \frac{\bar{\beta}_{tr}(V)}{\mathrm{f_c}(T_a)}\mathbf{Q} : \tilde{\sigma}\, dV \tag{24}$$

in eq. (15) $\mathbf{Q}$ is a fourth order tensor, is the effective (in the sense of damage mechanics) stress tensor and finally fc is the compressive strength of the material at 20°C. The model in this form can be successfully applied to several real cases, e.g. the case of fire in tunnels [23]. For further details see [8, 9, 18, 13, 5].

## 3 Numerical simulation of cylindrical specimen exposed to high temperature

This example deals with a comparison between numerical results, obtained using the model described in the previous sections, and experimental results, obtained from compressive tests carried out in United States in the laboratories of NIST (*i.e.* National Institute of Standard and Technology) [19, 21, 22, 20]. The main goal of this comparison is to show the capability of the code to assess spalling phenomena, in particular occurrence of explosive spalling in concrete structures subjected to elevated temperatures.

The specimens were cylinders with diameter of 100mm and height of 200mm, have been tested using three test methods, representing the thermo-mechanical loading conditions: stressed test method (specimens were preloaded, with a load equal to 40% of final compressive strength at room temperature, and then heated), unstressed test method (specimens were directly heated until the time of compressive test), residual property test method (the specimens were heated up to the target temperature and kept at this temperature for a certain period; then they were cooled and tested at room temperature, *i.e.* at residual conditions).

Five target temperatures: 100°C, 200°C, 300°C, 450°C and 600°C were reached during the tests by means of furnace heating rate of 5C/min, in steady state conditions. In this case "steady state" is defined as the temperature state when the temperature at the centre of the specimen is within 10°C of the pre-selected target temperature $T$ and the difference between the surface and centre temperatures of the concrete specimen is less than 10°C.

For further details concerning mix compositions and tests procedures (set-up, instrumentation of the specimens, temperature control), see [19, 21, 22, 20].

Our attention was focused on specimens made of concrete type 1, herein indicated as MIX1 in unstressed conditions with a target temperature equal to 450°C. In fact, for unstressed tests, explosive spalling occurred in all MIX1 specimens heated to 450°C. Initial and boundary conditions used in numerical simulation are listed in [11]. Figure 2(a)A shows the temperature differences between the surface and the centre of the specimens measured during the tests and the corresponding numerical results. The accordance between numerical and experimental results is quite good. The first part of heating shows a strange behaviour with temperature difference between core and surface practically zero for more than one hour.

Figure 2(b)D provides information about damaging of the specimen during heating. Specifically, it shows the history of total damage in three different points (on the surface, in the centre and in the middle of the radius).

Figure 2(a)B shows developments (in three points) of the gas pressure versus temperature compared to the water vapour pressure developments in saturated conditions (red line). The time range between 120 and 150min seems to be the critical range during which the material achieves a state favourable to spalling occurrence; the specimen experienced explosive spalling right in this range of time. Corresponding to the maximum value of $\Delta T$, a sharp increase of mechanical damage parameter $d$ (with a maximum value equal to 80%) may be observed. Similarly to the increase of mechanical damage, the peak of gas pressure corresponds to the maximum value of temperature differences $\Delta T$.

The presented results of numerical simulations, show that both pore pressure and thermally induced strains can be identified as responsible for the spalling occurrence, and that they play a primary or secondary role depending on the particular conditions prevailing. For the analysed HPC concretes, the MIX 1 specimens, having lower value of the $w/c$ ratio, spalled explosively mainly due to the high gas pressure value and relatively high level of thermo-chemical deterioration.

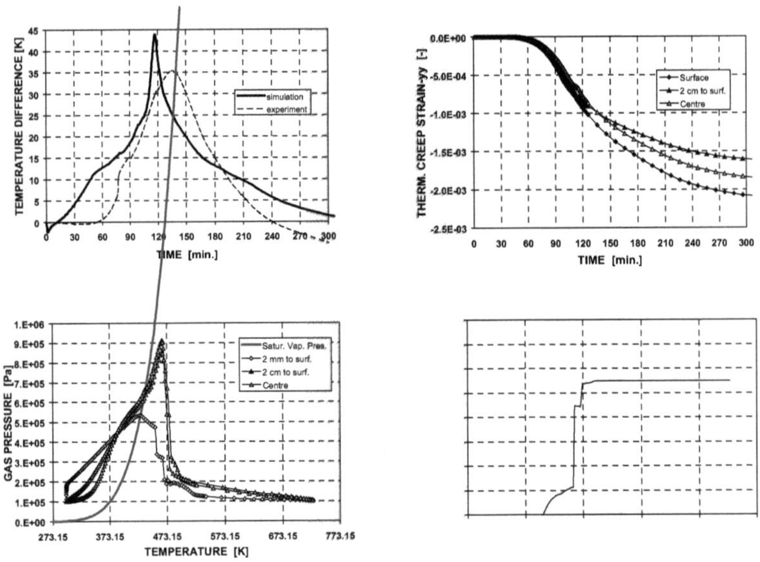


(a) Cylindrical specimen exposed to high temperature: A) Temperature differences history in the specimen (and comparison with experimental results), B) Gas pressure versus temperature in three different points and saturated vapour pressure (red line).

(b) Cylindrical specimen exposed to high temperature: A) Creep strains history according to eq. (19), B) Mechanical damage versus temperature in three different points.

**Fig. 2.**

## Conclusions

A general model for the non-linear modelling of concrete behaviour has been presented in this work. All relevant mass and heat transport phenomena, chemical reactions, phase changes as well as their mechanical effects are taken into account. The richness of the model allows for its application to several practical cases such as the analysis of hydration and aging processes in massive concrete structures and the analysis of the response of concrete structures exposed to high temperatures.

## References


1. V. Baroghel-Bouny, M. Mainguy, T. Lassabatere, and O. Coussy. Characterization and identification of equilibrium and transfer moisture properties for ordinary and high-performance cementitious materials. *Cement Concrete Res.*, 28:1225–1238, 1999.
2. Z.P. Bazant and S. Prasannan. Solidification theory for concrete creep. *J. Eng. Mech.*, ASCE 115:1691–1725, 1989.
3. Z.P. Bazant and Y. Xi. Continuous retardation spectrum for solidification theory of concrete creep. *J. Eng. Mech.*, ASCE 121:281–288, 1995.
4. D.P. Bentz, V. Waller, and F. de Larrard. Prediction of adiabatic temperature rise in conventional and high-performance concretes using a 3-d microstructural model. *Cement Concrete Res.*, 28(2):285–297, 1998.
5. M. Bianco, G. Bilardi, F. Pesavento, G. Pucci, and B.A. Schrefler. A frontal solver tuned for fully-coupled non-linear hygro-thermo-mechanical problems. *International Journal of Numerical Methods in Engineering*, 57(13), 2003.
6. M. Cervera, J. Olivier, and T. Prato. A thermo-chemo-mechanical model for concrete. *J. Eng. Mech.*, ASCE 125(9):1018–1027, 1999.
7. M. Cervera, J. Olivier, and T. Prato. A thermo-chemo-mechanical model for concrete. i: Damage and creep. *J. Eng. Mech.*, ASCE 125(9'):1028–1039, 1999.
8. D. Gawin, F. Pesavento, and B.A. Schrefler. Modelling of hygro-thermal behaviour and damage of concrete at temperature above the critical point of water. *Int.J.Numer*, 26:537–562, 2002.
9. D. Gawin, F. Pesavento, and B.A. Schrefler. Modelling of hygro-thermal behaviour of concrete at high temperature with thermo-chemical and mechanical material degradation. *Comput. Methods Appl. Mech. Engrg.*, 192:1731–1771, 2003.
10. D. Gawin, F. Pesavento, and B.A. Schrefler. Hygro-thermo-chemo-mechanical modelling of concrete at early ages. part i: Hydration and hygro-thermal phenomena. *in parparation*, 2004.
11. D. Gawin, F. Pesavento, and B.A. Schrefler. Modelling of deformations of high strength concrete at elevated temperatures. *Materials and Structures/Concrete Science and Engineering*, 37(268), 2004.
12. O.M. Jensen and P.F. Hansen. Influence of temperature on autogenous deformation and relative humidity change in hardening cement paste. *Cement Concrete Res.*, 29:567–575, 1999.

13. G. Khoury, C.E. Majorana, F. Pesavento, and B.A. Schrefler. Thermo-hydro-mechanical modelling of high performance concrete at high temperatures. *Magazine of Concrete Research*, 54(2), 2002.
14. G.A. Khoury. Strain components of nuclear-reactor-type concretes during first heating cycle. *Nuclear Engineering and Design*, 156:313–321, 1995.
15. P. Laplante. *Mechanical properties of hardening concrete: A comparative analysis of classical and high strength concrete.* PhD thesis, Ecole Nationale des Pontes et Chausses, 1993.
16. R.W. Lewis and B.A. Schrefler. *The Finite Element Method in the Static and Dynamic Deformation and Consolidation of Porous Media.* Wiley & Sons, 1998.
17. P. Lura, O.M. Jensen, and K. van Breugel. Autogenous shrinkage in high-performance cement paste: an evaluation of basic mechanisms. *Cement Concrete Res.*, 32(2):223–232, 2003.
18. F. Pesavento. *Non-linear modelling of concrete as multiphase porous material in high temperature conditions.* PhD thesis, University of Padova, 2000.
19. L.T. Phan. Fire performance of high-strength concrete: a report of the state-of-the-art. *Res. Report NISTIR 5934*, page 105, 1996.
20. L.T. Phan and N.J. Carino. Effects of test conditions and mixture proportions on behavior of high-strength concrete exposed to high temperature. *ACI Materials Journal*, 99(1), 2002.
21. L.T. Phan, D. Duthinh, and E. Garboczi. Proc. int. workshop on fire performance of high-strength concrete. In *NIST Special Publication 919.* Gaitherburg (MD) USA, NIST, feb 1997. NIST Special Publication 919.
22. L.T. Phan, J.R. Lawson, and F.L. Davis. Effects of elevated temperature exposure on heating characteristics, spalling, and residual properties of high performance concrete. *Materials and Structures*, 34, 2001.
23. B.A. Schrefler, P. Brunello, D. Gawin, C.E. Majorana, and F. Pesavento. Concrete at high temperature with application to tunnel fire. *Computation Mechanics*, 29:43–51, 2002.
24. G. de Schutter. Influence of hydration reaction on engineering properties of hardening concrete. *Mater. Struct.*, 35:453–461, 2002.
25. G. de Schutter and L. Taerwe. General hydration model for Portland cement and blast furnace slag cement. *Cement Concrete Res.*, 25(3):593–604, 1995.
26. S. Thelandersson. Modeling of combined thermal and mechanical action on concrete. *J. Eng. Mech.*, ASCE 113(6):893–906, 1987.
27. F.-J. Ulm and O. Coussy. Modeling of thermo-chemo-mechanical couplings of concrete at early ages. *J. Eng. Mech.*, ASCE 121(7):785–794, 1995.
28. F.-J. Ulm and O. Coussy. Strength growth as chemo-plastic hardening in early age concrete. *J. Eng. Mech.*, ASCE 122(12):1123–1132, 1996.
29. O.C. Zienkiewicz and R.L. Taylor. *The Finite Element Method*, volume 1: The Basis, Butterworth-Heinemann. Oxford, 2000.
30. O.C. Zienkiewicz and R.L. Taylor. *The Finite Element Method*, volume 2: Solid Mechanics, Butterworth-Heinemann. Oxford, 2000.

# Modelling the Glass Press-Blow Process

S.M.A. Allaart-Bruin, B.J. van der Linden, and R.M.M. Mattheij

Technische Universiteit Eindhoven, P.O. Box 513, 5600 MB Eindhoven, The Netherlands sbruin@win.tue.nl

**Summary.** For the modelling of the glass press-blow process level set functions are used. Special difficulties arise due to velocity gradients in the domain. A re-initialisation procedure for unstructured triangular meshes is adapted to these difficulties and is applied.

**Key words:** Level Set Method, glass forming process.

## 1 Introduction

A typical stage in the manufacturing of container glass is the blowing stage. At this stage a preform of hot glass is transferred to a blow mould. There it is first given time to sag sufficiently far. Finally pressurised air is used to inflate the preform to form the final bottle or container shape.

This paper briefly describes the equations modeling this stage of the process. In Section 3 a new re-initialisation procedure of the level set function is described. The model is applied to a two dimensional test problem in Section 4. We end with some conclusions.

## 2 Governing equations

The following mathematical model is used to describe the blowing stage of the production process of bottles and jars. Let the domain $\Omega \subset \mathbb{R}^2$ be the interior of the mould. Every point $\mathbf{x} \in \Omega$ is either in air or in glass.

We denote time by $t$, velocity vector by $\mathbf{v}$, pressure by $p$, the dynamic viscosity by $\mu$ and the gravitational force by $\mathbf{g}$. After non-dimensionalising and using the dimensionless Reynolds (Re) and Froude (Fr) numbers, the flow can be described by the Stokes equation together with conservation of mass:

$$\nabla \cdot (\mu(\mathbf{x})\nabla \mathbf{v}) + \frac{\mathrm{Re}(\mathbf{x})}{\mathrm{Fr}(\mathbf{x})}\mathbf{g} = \nabla p, \tag{1}$$

$$\nabla \cdot \mathbf{v} = 0. \tag{2}$$

The viscosity of glass strongly depends on temperature. The temperature dependence of the glass viscosity can be described by the Vogel-Fulcher-Tamman relation [2]. The temperature $T$ can be described by the dimensionless energy balance equation

$$\mathrm{Pé}(\mathbf{x}) \left( \frac{\partial T}{\partial t} + \mathbf{v} \cdot \nabla T \right) = \nabla^2 T, \tag{3}$$

where Pé is the dimensionless Péclet number.

The typical values of glass and air that we use for this problem are: typical length scale $L = 10^{-2}$ m, typical velocity $V = 10^{-2}$ m/s, typical viscosity for glass $\mu_{\text{glass}} = 10^4$ Pa s and air $\mu_{\text{air}} = 10^{-5}$ Pa s, typical densities $\rho_{\text{glass}} = 2.5 \cdot 10^3$ kg/m$^3$ and $\rho_{\text{air}} = 1$ kg/m$^3$, typical temperature jump $\Delta T = 350$°C, typical specific heat for glass $c_{p_{\text{glass}}} = 1.2 \cdot 10^3$ J/(kg K) and air $c_{p_{\text{air}}} = 10^3$ J/(kg K), typical heat conductivities $\kappa_{\text{glass}} = 2.75$ W/(m K) and $\kappa_{\text{air}} = 10^{-2}$ W/(m K). This results in the following values for the Reynolds, Froude and Péclet number for glass (gl) and for air (a)

$$\mathrm{Re}_{\mathrm{gl}} = \mathrm{Re}_{\mathrm{a}} = 2.5 \cdot 10^{-5}, \ \mathrm{Fr}_{\mathrm{gl}} = \mathrm{Fr}_{\mathrm{a}} = 10^{-3}, \ \mathrm{Pé}_{\mathrm{gl}} = 1.1 \cdot 10^2 \text{ and } \mathrm{Pé}_{\mathrm{a}} = 10.$$

Actually, the Reynolds number of air is much bigger, but we replace the air by a fictitious fluid, with viscosity 4 Pa s and the same mass density as air; so $\mathrm{Re} = 2.5 \cdot 10^{-5}$ in the fictitious fluid. Note that the viscosity of the fictitious fluid is much smaller than the viscosity of glass. The inertia terms in the fictitious fluid domain can be neglected, while the pressure drop is still negligible compared to the pressure drop in the glass domain.

The glass position is modelled by two level set functions $\varphi_1(\mathbf{x}, t)$ and $\varphi_2(\mathbf{x}, t)$. These level set functions each capture a glass-air interface and are convected by the flow velocity

$$\frac{\partial \varphi_i}{\partial t} + \mathbf{v} \cdot \nabla \varphi_i = 0 \qquad \text{for } i = 1, 2. \tag{4}$$

At every time $t$ the corresponding interfaces $\Gamma_i(t)$ are given implicitly by $\varphi_i(\mathbf{x}, t) = 0$.

To solve (1), (2), (3) and (4) uniquely we have to prescribe initial and boundary conditions. For glass we assume no-slip when it touches the mould, and for air we prescribe a free-slip boundary condition.

The model is discretized by a finite element method which uses a mesh consisting of triangles.

## 3 Re-initialisation of the level set function

Two level set functions are used to describe the position of the glass. Due to velocity gradients these level set functions become less accurate in describing the interfaces as time evolves. Initially the level set function $\varphi(\mathbf{x}, t)$ is defined as the signed euclidean distance function to the corresponding interface $\Gamma(t)$, i.e.

$$\varphi(\mathbf{x}, 0) := \begin{cases} d(\mathbf{x}, \Gamma(0)), & \text{if } \mathbf{x} \text{ in air,} \\ -d(\mathbf{x}, \Gamma(0)), & \text{if } \mathbf{x} \text{ in glass.} \end{cases} \tag{5}$$

When time evolves the initially nicely shaped level set function can develop steep gradients at one side and can become almost constant on the other side of the domain. This could lead to additional numerical difficulties.

We would like to compute a function $\hat{d}(\mathbf{x})$ which is a distance function and which at the zero level of $d(\mathbf{x})$ coincides with $\Gamma(t)$. Then we replace on every time step $\varphi(\mathbf{x}, t)$ with this new function. This idea of interrupting a level set calculation and rebuilding a new level set function is referred to as re-initialisation. There are several ways to accomplish this re-initialisation [3]. Existing methods use a structured grid consisting of squares or cubes. Our computational mesh consists of unstructured triangles. The re-initialisation procedure used in our computations is based on the Fast Marching Method [3].

Fast Marching Methods rely on building the solution outward, starting with a boundary value. More precisely, knowing one or two value of $\hat{d}$ within an element we would like to compute the value of $\hat{d}$ at the third node. Consider Fig. 1, where a line $l$ and a triangle $ABC$ with angles $\alpha, \beta$ and $\gamma$ are shown. The assumption is made that the values of $d$ are known at points $A$ and $B$, where $d_A$ and $d_B$ are distances from line $l$ to points $A$ and $B$ respectively. Furthermore, we assume that $d_A \geq d_B$.

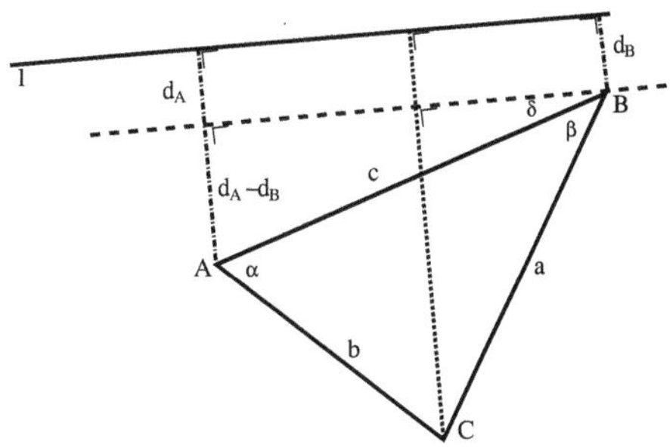


**Fig. 1.** Fast Marching Method as re-initialisation.

First, the angle $\delta$ is computed by $\sin(\delta) = \frac{d_A - d_B}{c}$. Secondly, observe that $\sin(\delta + \beta) = \frac{d_C - d_B}{a}$. This results in the following update equation

$$d_C = a\sin(\delta + \beta) + d_B. \tag{6}$$

We have to make sure that the shortest distance from $C$ to the line $l$ passes trough the interior of the triangle. Hence we require

$$0 \leq \frac{a\cos(\delta + \beta)}{\cos(\delta)} \leq c. \tag{7}$$

If (7) is not satisfied we take $d_C = \min\{d_A + b, d_B + a\}$.

In addition to a triangle with two known values, also triangles with one known $d$ value should be considered. The updating is then done as follows. Assume that the value at $A$ is known, then the values at $B$ and $C$ are assigned as $d_B = d_A + c$ and $d_C = d_A + b$, respectively.

Clearly in (6) we compute the exact distances from point $C$ to line $l$ if the distance $d_A$ and $d_B$ are exact. In practice, the line $l$ is just a line segment of finite length and $d_A$ and $d_B$ are approximations of the distances. So $d_C$ is just an approximation of the real distance.

## 4 Results

The mathematical model is applied to a two dimensional problem. The reason to consider this problem is that in practice the glass thickness of a bottle in circumferential direction varies. We would like to see what is the influence of a temperature gradient on this thickness variation.

The numerical tests are performed with the package Sepran [1]. In this example a quarter of a cross-section of a bottle in the axial direction is considered. At the small arc of our domain (inflow boundary) a pressure is prescribed. The large arc can be considered as the mould, while the other domain boundaries are symmetry axes. We prescribe a temperature gradient in circumferential direction. For this purpose a variable $\theta$ is used, defined as $\theta := \arctan\left(\frac{y}{x}\right)$. The initial temperature field in the domain is defined as

$$T(\theta) := T_{\text{av}} + 2\frac{\theta \Delta T}{\pi} - \frac{\Delta T}{2}, \tag{8}$$

where $T_{\text{av}}$ is the average temperature in the domain. Hence $T_{\text{av}} + \frac{\Delta T}{2}$ is the maximum temperature and $T_{\text{av}} - \frac{\Delta T}{2}$ is the minimum temperature in the domain. In our computations we choose $T_{\text{av}} = 975°\text{C}$ and for $\Delta T$ we take different values, namely 10°C and 20°C.

The computational mesh used consists of 1867 triangular elements. The initial position of the glass is a ring. The results are shown in Fig. 2. The first plot is the final position of the glass, when we have a $\Delta T$ of 10°C. We see that the resulting layer of glass is not uniform. At the top left corner, where the maximum temperature is achieved, the glass is much thinner than at the bottom right corner. This shows that even a relatively small temperature gradient, results in a large differences in thickness, approximately a factor 1.4. For the second and third plot a $\Delta T$ of 20°C is used. The second plot shows the moment just before glass touches the mould. We see that the glass will touch the mould first at the lowest temperature (bottom right corner). The last plot shows that the variations in thickness increase when the initial temperature gradient is increased. For an initial gradient of 20°C the thickness varies almost by a factor 2.

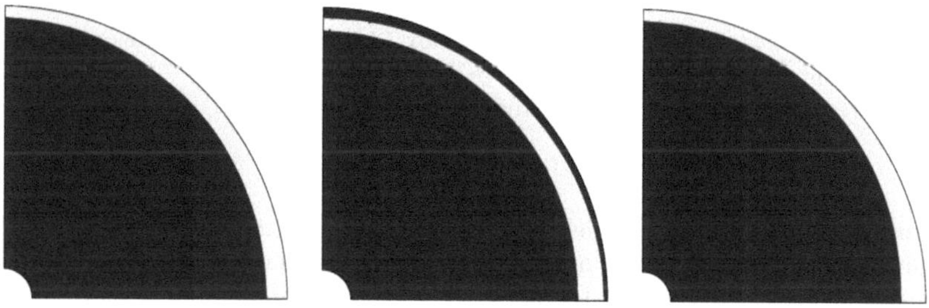

**Fig. 2.** Blowing of the preform.

# 5 Conclusions

The model described here is used to study the blowing phase of the press-blow process of glass. It is implemented and tested to a two-dimensional problem.

The test, a cross-section of a bottle in axial direction, shows that small temperature gradients have a significant impact on the final thickness distribution. This is one of the main reasons we have to consider a full three dimensional model to study the blowing phase.

# References


1. The SEPRAN package. http://ta.twi.tudelft.nl/sepran/sepran.html.
2. R.G.C. Beerkens et al. *Handboek voor de glasfabricage*, 2nd edition, 1997.
3. J.A. Sethian. *Level Set Methods and Fast Marching Methods Evolving Interfaces in Computational Geometry, Fluid Mechanics, Computer Vision, and Material Science.* Cambridge University Press, 1999.

# Real-Time Control of Surface Remelting

M.J.H. Anthonissen[1], D. Hömberg[2], and W. Weiss[2]

[1] Technische Universiteit Eindhoven, P.O. Box 513, 5600 MB Eindhoven, The Netherlands `m.j.h.anthonissen@tue.nl`
[2] Weierstrass Institute for Applied Analysis and Stochastics

**Summary.** We consider a model for laser surface remelting, a process to improve the surface quality of steel components. The mathematical model consists of the two-dimensional heat equation for temperature and an ordinary differential equation for the liquid phase. The equations are coupled via source terms. We study the efficient numerical simulation using adaptive grids, which are especially well-suited for problems with moving heat sources. To account for the local high activity due to the heat source, we introduce local uniform grids and couple the solutions on the global coarse and local fine grids using the local defect correction (LDC) technique.

**Key words:** local defect correction, heat treatment, laser remelting.

## 1 Introduction

Laser surface modification is a widely used technique to increase the strength of the surface of mechanical parts such as cutting tools, gears, machine parts *etc.* There are several ways to alter the properties of the part at hand. The techniques include laser cladding, laser surface alloying, and laser heat treatment. In laser cladding, a high power density is used on a surface to fuse a metal onto another metal. This increases the wear and corrosion resistance of the part. Laser surface alloying is similar to surface melting, but elements are added to the melt pool to change the chemical composition of the surface. Finally, laser heat treatment covers both hardening and surface remelting. These methods improve mechanical properties through microstructural changes. The typical depth of the hardened layer is 0.1–0.2 mm. In this paper we will model and simulate the last technique. Figure 1 sketches the remelting process.

In this paper, we will use the following model for laser surface remelting. We consider a rectangular domain $\Omega = (0, L) \times (0, M)$ and solve the heat equation in the space time domain $\Omega \times (0, t_E)$ with end-time $t_E$ viz.

$$\rho c(T)\frac{\partial T}{\partial t} - \operatorname{div}\left(k(T)\operatorname{grad} T\right) = -\rho L\frac{da}{dt} \qquad \text{in } \Omega \times (0, t_E), \tag{1}$$

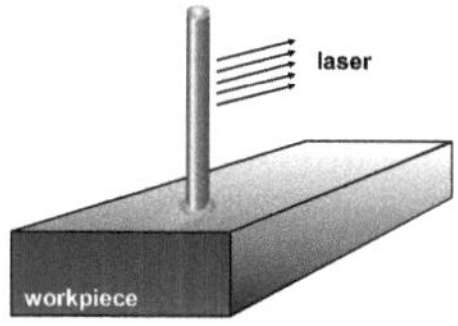


**Fig. 1.** Laser surface remelting

$$\frac{da}{dt} = \frac{1}{\tau(T)}\left(a_{\text{eq}}(T) - a\right). \tag{2}$$

In (1), $\rho$ is the density, $c$ is the specific heat, $T$ is the temperature, $k$ is the heat conductivity, $L$ is the latent heat of the liquid phase, and $a$ is the liquid phase. The function $a_{\text{eq}}$ in (2) is defined by

$$a_{\text{eq}}(T) = \begin{cases} 0, & \text{for } T < T_s \ (T_s\text{: only solid in equilibrium}), \\ \text{linear}, & \text{for } T_s < T < T_l, \\ 1, & \text{for } T < T_l \ (T_l\text{: only liquid in equilibrium}). \end{cases}$$

We model the laser via the boundary condition at the top of the domain

$$-k\frac{\partial T}{\partial n} = \begin{cases} \kappa P F(x - vt), & \text{at the top of the domain}, \\ \mu(T - T_0), & \text{at the bottom}, \\ 0, & \text{at the left and at the right}. \end{cases} \tag{3}$$

Here, $\kappa$ is the absorption coefficient, $P$ is the radiation power, and $F$ is the radiation flux defined by $F(x) = \alpha_1 \exp(-\alpha_2 x^2)$. The velocity $v$ of the laser beam is assumed to be constant during the simulation. The initial condition for the problem is $T = T_0$, $a = 0$. A typical solution of the initial value problem will have large differences in geometric scales: the temperature is very high near the spot at the surface where the laser beam is located and will drop sharply and be almost constant in the larger part of the computational domain. It is evident, that a computational grid for a problem of this type should reflect the solution behavior, *i.e.*, it should have many grid points with fine spacing near the laser, and it may be (much) coarser elsewehere.

## 2 Local grid refinement

The typical depth of the top layer is only 0.1–0.2 mm whereas the computational domain may extend several centimeters. As we have seen, a high spatial resolution is required at the laser position, whereas lower resolution suffices in the rest of the domain. Naturally, the usage of a global uniform fine grid is computationally inefficient. An obvious choice would be to use a truly nonuniform refined grid. However, uniform grids have several advantages over truly nonuniform grids: uniform grids can be represented by simple data structures, simple accurate discretization stencils exist for uniform grids and fast solution techniques are available for solving the system of equations resulting from discretization on uniform grids. For these reasons, so-called *local uniform grid refinement* techniques have been introduced in which a coarse base grid covering the whole computational domain is locally uniformly refined. Some of the better-known techniques are adaptive mesh refinement (AMR) [3]; fast adaptive composite grid (FAC) [7]; local rectangular refinement (LRR) [2]; local uniform grid refinement (LUGR) [8]; and finally local defect correction (LDC) [4]. For our simulations we use an extended version of the LDC method. This technique has a number of advantages. First, the method uses many small structured grids instead of a single unstructured grid (as opposed to LRR). This results in low memory usage and a natural route for parallelization. Our method uses structured grids only. We solve the boundary value problem on a composite grid without explicitly forming the discretization (as opposed to FAC, LRR). Finally, one of the distinctive features of the LDC method is the two-way coupling between grids. It is common in local refinement techniques that the coarse grid solution is used to define boundary conditions for the local fine grid. While this one-way communication suffices for hyperbolic problems [3], it is essential for elliptic problems to transfer information from the fine to the coarse grid, too. As is noted in *e.g.* [6], the naive approach of only coarse-to-fine communication gives the accuracy of the coarse grid alone.

## 3 Local defect correction

In the LDC method the discretization on the *composite grid* is based on a combination of standard discretizations on several uniform grids with different grid sizes that cover different parts of the domain. At least one grid, the coarse grid, should cover the entire domain, and its grid size should be chosen in agreement with the relatively smooth behavior of the solution outside the high activity areas. Apart from this *global coarse grid*, one or several *local fine grids* are used which are also uniform. Each of the local grids covers only a (small) part of the domain and contains a high activity region. The grid sizes of the local grids are chosen in agreement with the behavior of the continuous solution in that part of the domain. The coarse grid solution is used to provide artificial boundary conditions at the interfaces between grids.

The LDC method is an iterative process: the basic global discretization is improved by the local discretizations defined in subdomains. The update of the coarse grid solution is achieved by adding a defect correction term to the right hand side of the coarse grid problem. The defect term is an estimate of the local discretization error of the coarse grid discretization. At each iteration step, the process yields a discrete approximation of the continuous solution on the composite grid. The discrete problem that is actually being solved is an implicit result of the iterative process. Therefore, the LDC method is both a discretization method and an iterative solution method. A detailed description of the LDC algorithm can be found in [1].

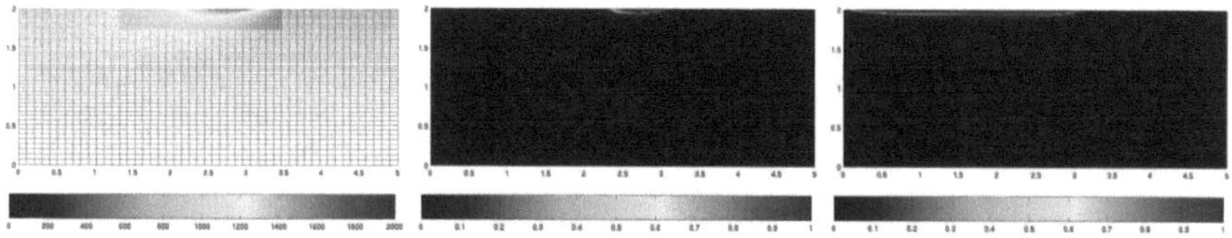


**Fig. 2.** Numerical results for the first simulation

# 4 Simulations

To perform numerical simulation of the surface remelting process, we discretize the differential equations (1)–(2) in time using the $\theta$-method. By using this time discretization, we transform the continuous problem into a series of elliptic boundary value problems that we can solve using the LDC technique. For the space discretization, we use finite differences. The area of refinement is automatically chosen based on (smoothed) temperature values. To solve the resulting nonlinear problems on structured grids, we use a damped Newton solver in which the Jacobi matrix is found via numerical differentiation and the resulting linear systems are solved with BiCGSTAB.

In our first example, we solve the initial value problem on the rectangular domain $\Omega = (0,5) \times (0,2)$. The global coarse grid is a structured grid with 50 grid points chosen equidistant in horizontal, 40 grid points chosen with geometrical refinement in vertical direction. The grid sizes in vertical direction are smallest near the surface; from bottom to top they decrease with a factor of 0.975. We use one level of refinement and the grid sizes for the local grid are half those of the global. We present plots of the temperature field projected on the grid lines of the composite grid, the surface melting, and the formation of the hardening layer in Fig. 2.

Observing the temperature field during the simulation in the first example, one notices that it takes some time for the workpiece to reach maximum surface temperature. Also, since the temperature drops due to self-cooling of the workpiece, the temperature peaks near the right side of the domain, as it is more difficult to loose heat there. This is reflected in the resulting hardening layer in Fig. 2: it is shallow at the left, deeper at the right side of the domain. This occurs even more so if we perform the same simulation for a more challenging problem. In our second example we consider a domain containing a hole. To obtain better results, we employ proportional integral differential (PID) control. Recall that the laser is modeled via the boundary condition at the top, according to $-k\partial T/\partial n = \kappa PF(x - vt)$. We will use the radiation power $P$ as a control variable. We define the error $e$ as $e(t_n) = T^* - T_{\max}(t_n)$, in which $T^*$ is the desired maximum temperature at the surface, $T_{\max}(t_n)$ the real maximum temperature in the simulation. Next we set the power at the next time level to $P(t_{n+1}) = k_P e(t_n) + k_I \int_0^{t_n} e(t)\,dt + k_D \dot{e}(t_n)$, in which the three parameters $k_P$, $k_I$, and $k_D$ are tuned based on a step response of the

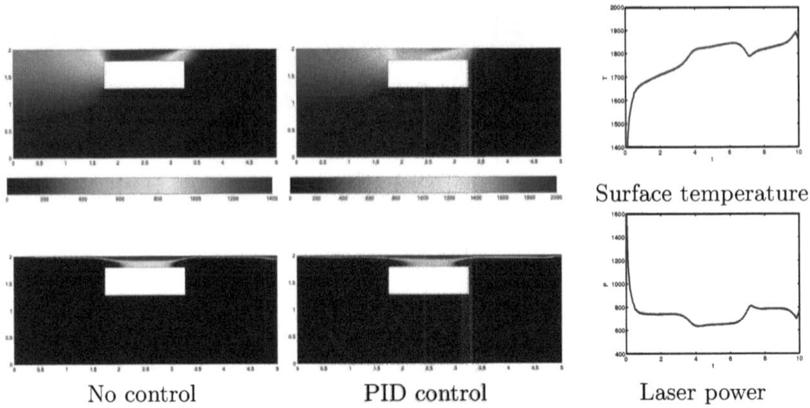


**Fig. 3.** Numerical results for the second simulation: temperature (top), hardening layer (bottom), surface temperature and laser power during PID control

system, see [5] for details as well as a nonlinear variant of PID control for this problem. This results in a more uniform hardening layer, as can be observed from Fig. 3. We also present the laser power and resulting surface temperature found with PID control in the figure.

## References

1. M.J.H. Anthonissen, R.M.M. Mattheij, and J.H.M. ten Thije Boonkkamp. Convergence analysis of the local defect correction method for diffusion equations. *Numerische Mathematik*, 95(3):401–425, 2003.
2. B.A.V. Bennett and M.D. Smooke. Local rectangular refinement with application to axisymmetric laminar flames. *Combust. Theory Modelling*, 2:221–258, 1998.
3. M.J. Berger and J. Oliger. Adaptive mesh refinement for hyperbolic partial differential equations. *J. Comput. Phys.*, 53:484–512, 1984.
4. W. Hackbusch. Local defect correction and domain decomposition techniques. In K. Böhmer and H. J. Stetter, editors, *Defect Correction Methods. Theory and Applications, Computing, Suppl. 5*, pages 89–113, Wien, New York, 1984. Springer.
5. D. Hömberg and W. Weiss. PID-Control of laser surface hardening of steel. Technical Report Preprint No. 876, Weierstraß-Institut für Angewandte Analysis und Stochastik, Berlin, 2003.
6. D.F. Martin and K.L. Cartwright. Solving Poisson's equation using adaptive mesh refinement. Technical Report UCB/ERL M96/66, Univ. Calif. Berkeley, oct 1996.
7. S.F. McCormick and J.W. Thomas. The fast adaptive composite grid (FAC) method for elliptic equations. *Math. Comp.*, 46:439–456, 1986.
8. R.A. Trompert. *Local uniform grid refinement for time-dependent partial differential equations.* PhD thesis, University of Amsterdam, Amsterdam, 1994.

# Fast Shape Design for Industrial Components *

G. Haase[1], E. Lindner[2], and C. Rathberger[3]

[1] Karl-Franzens-Universität Graz, Institute of Mathematics and Computational Sciences, Heinrichstr.36, A-8010 Graz, Austria Gundolf.Haase@uni-graz.at
[2] Johannes Kepler University of Linz, Institute of Computational Mathematics , Altenbergerstr. 69, A-4040 Linz, Austria lindner@numa.uni-linz.ac.at
[3] Pochestr. 1, A-4020 Linz, Austria c.rathberger@inode.at

**Summary.** We consider minimizing the mass of an injection moulding machine, fulfilling certain constraints. The deformation of its frame is described by the plain stress state equations for linear elasticity. The minimization problem is a nonlinear constrained one. When the design parameters change, then also the shape will change. Generating a new finite element mesh for each single shape leads to a non-differentiable objective. Here we deform the mesh elastically.

**Key words:** shape optimization, structural mechanics, finite elements

## 1 Modeling the problem

Various methods for structural optimization using finite elements are given in Haslinger, Mäkinen [5]. For shape optimizaion problems see e. g. Delfour, Zolésio [1] First results of the authors can be found in [2, 3]. The frame of an injection moulding machine is sketched by its 2D-cut $\Omega$ in Fig. 1. The primary goal of the design phase is to minimize the mass of the frame. Let $V_0 = \{v \in H^1(\Omega) \mid v = 0 \text{ on } \Gamma_D, \text{ meas } \Gamma_D > 0\}$ denote the set of admissible displacements where $\partial\Omega = \Gamma_D \cup \Gamma_N$, $\Gamma_D \cap \Gamma_N = \emptyset$. The displacement field $u \in V_0$ fulfills the variational equation

$$a(\rho; u, v) = F(v) \qquad \text{for all } v \in V_0 \ , \qquad \text{with} \tag{1}$$

$$a(\rho; u, v) = \int_\Omega \rho \frac{\partial u_i}{\partial x_j} E_{ijkl} \frac{\partial v_k}{\partial x_l} \,\mathrm{d}x, \quad F(v) = \int_\Omega \langle f, v\rangle \,\mathrm{d}x + \int_{\Gamma_N} g\, v \,\mathrm{d}s$$

where $E_{ijkl}$ denotes the elasticity tensor, $f$ the volume force density and $g$ the surface force density on the part $\Gamma_N$ of the boundary $\partial\Omega$. The design problem reads as follows with $\sigma^{\mathrm{vM}}(u)$ the v. Mises stress, $\sigma^{\mathrm{ten}}(u)$ the tensile stress:

*This work was partially supported by the Austrian Science Fund - 'Fonds zur Förderung der wissenschaftlichen Forschung (FWF)' - SFB F013 'Numerical and Symbolic Scientific Computing', Project F1309

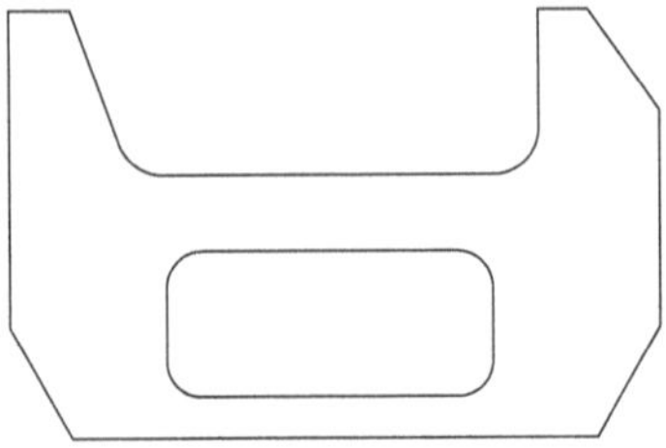

**Fig. 1.** Cross section of the original shape

$$
\begin{aligned}
&\int_\Omega \rho \, dx \longrightarrow \min_{u,\rho} && \\
\text{subject to} \quad & a(\rho; u, v) = F(v) && \text{for all } v \in V_0 \\
& 0 < \underline{\rho} \le \rho \le \overline{\rho}, && \text{a.e. in } \Omega \\
\sigma^{\mathrm{vM}}(u) \le \sigma^{\mathrm{vM}}_{\max}, \quad & \sigma^{\mathrm{ten}}(u) \le \sigma^{\mathrm{ten}}_{\max} && \text{a.e. in } \Omega \\
& \alpha(u) \le \alpha_{\max} &&
\end{aligned}
\tag{2}
$$

The change in the shrinking angle of the clumping unit (vertical edges on top, called *wings*) is denoted by $\alpha(u)$.

We discretize the problem by triangular finite elements with piece-wise constant shape functions for approximating $\rho$ and piece-wise quadratic ones for approximating $u$. We denote the discrete approximation of $\rho$ and $u$ again by $\rho$ and $u$. The upper limits on the angle and the stresses are treated either as constraints or as soft limits. Furthermore, the pointwise constraints on $\sigma^{\mathrm{vM}}$ and $\sigma^{\mathrm{ten}}$ are replaced by using a higher order $\ell^p$ norm. Treating the upper limits as soft constraints leads to:

$$
\begin{aligned}
\mathrm{mass}(\rho) &+ \omega_1 \big( \max (\|\sigma^{\mathrm{vM}}\|_p - \sigma^{\mathrm{vM}}_{\max}, 0) \big)^2 \\
&+ \omega_2 \big( \max (\|\sigma^{\mathrm{ten}}\|_p - \sigma^{\mathrm{ten}}_{\max}, 0) \big)^2 \\
&+ \omega_3 \big( \max (\alpha - \alpha_{\max}, 0) \big)^2 \qquad \longrightarrow \min_{u,\rho} \\
&\text{subject to} \qquad K(\rho)\, u = F \quad \text{and} \quad \underline{\rho} \le \rho \le \overline{\rho} \, .
\end{aligned}
\tag{3}
$$

## 2 A short sketch on the optimization strategy

Problem 3 is a special case of

$$
\begin{aligned}
& J(u, \rho) \longrightarrow \min_{u,\rho} \\
\text{subject to} \quad & K(\rho)\, u = f(\rho) \qquad \text{and} \qquad \underline{\rho} \le \rho \le \overline{\rho} \, ,
\end{aligned}
\tag{4}
$$

where $\rho$ denotes the vector of design parameters and $u$ the solution of the governing finite element (FE) state equation with $K(\rho)$ symmetric and positive definite. $u$ can be formally eliminated which leads to

$$\tilde{J}(\rho) = J(K^{-1}(\rho)\, f(\rho),\, \rho) \longrightarrow \min_{\rho} \qquad \text{subject to} \qquad \underline{\rho} \leq \rho \leq \overline{\rho}\ . \tag{5}$$

The SQP optimizer used in our code, see e.g. Nocedal, Wright [6], is based on a Quasi-Newton approximation of the Hessian using a modified BFGS update formula following Powell [7].

## 3 Calculating the gradient for shape optimization

In [3] a hybrid implementation of the gradient was presented. Here we implement the full gradient calculation in a 2D shape optimization problem, preventing also the problem of remeshing that causes non-differentiability. The shape under investigation is similar to the one in Fig. 1. This shape can be easily described by corner points (x– and y–coordinates), circular parts of the boundary (x– and y–coordinates of the center plus the radius) connected with straight lines, see Fig. 2. Our set of design parameters $P$ contains all

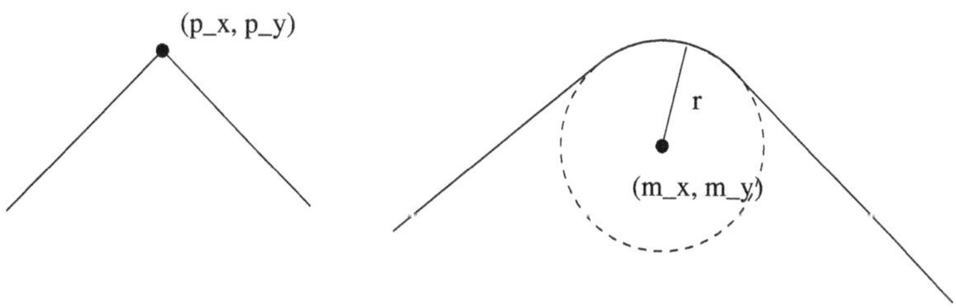


**Fig. 2.** Possible usage of design parameters in Shape Optimization

these parameters $p_x$, $p_y$, $m_x$, $m_y$, $r$ subject to box constraints. More details on topics discussed in the following sections can be found in the master thesis by Rathberger [8].

### 3.1 A second look at the gradient

If we pick an arbitrary design parameter $p \in P$ and assume that our objective $J$ depends only on the mass, the displacement in certain points and the resulting van Mises stress $\sigma^{vM}$ (handling of tensile stresses analogously) we obtain:

$$J = J(p, u(p), \sigma^{vM}(p, u(p))) \tag{6}$$

The total differential with respect to design parameter $p$ reads as

$$\begin{aligned}\frac{dJ}{dp} &= \frac{\partial J}{\partial p} + \frac{\partial J}{\partial u} \cdot \frac{du}{dp} + \frac{\partial J}{\partial \sigma^{vM}} \cdot \left( \frac{\partial \sigma^{vM}}{\partial p} + \frac{\partial \sigma^{vM}}{\partial u} \cdot \frac{du}{dp} \right) \\ &= \frac{\partial J}{\partial p} + \frac{\partial J}{\partial \sigma^{vM}} \cdot \frac{\partial \sigma^{vM}}{\partial p} + \left( \frac{\partial J}{\partial u} + \frac{\partial J}{\partial \sigma^{vM}} \cdot \frac{\partial \sigma^{vM}}{\partial u} \right) \cdot \frac{du}{dp} \end{aligned} \tag{7}$$

Again, we want to eliminate the term $\frac{du}{dp}$ by differentiating the state equation $Ku = f$ with respect to the design variable $p$ and we get

$$\frac{du}{dp} = K^{-1} \left( \frac{df}{dp} - \frac{dK}{dp} \cdot u \right) . \tag{8}$$

Inserting equation (8) into equation (7) results in

$$\begin{aligned}\frac{dJ}{dp} &= \frac{\partial J}{\partial p} + \frac{\partial J}{\partial \sigma^{vM}} \cdot \frac{\partial \sigma^{vM}}{\partial p} + \\ &\quad + \left( \frac{\partial J}{\partial u} + \frac{\partial J}{\partial \sigma^{vM}} \cdot \frac{\partial \sigma^{vM}}{\partial u} \right) \cdot K^{-1} \left( \frac{df}{dp} - \frac{dK}{dp} \cdot u \right) \\ &= \frac{\partial J}{\partial p} + \frac{\partial J}{\partial \sigma^{vM}} \cdot \frac{\partial \sigma^{vM}}{\partial p} + \\ &\quad + \left\langle K^{-1} \cdot \left( \frac{\partial J}{\partial u} + \frac{\partial J}{\partial \sigma^{vM}} \cdot \frac{\partial \sigma^{vM}}{\partial u} \right), \frac{df}{dp} - \frac{dK}{dp} \cdot u \right\rangle \end{aligned} \tag{9}$$

where we have used the fact that $K$ is symmetric in the last transformation. The three principal parts of the derivative in (9) have to be investigated separately, which can be found in [4].

# 4 Numerical results for the shape optimization problem

The formulation for the design problem was already introduced in Section 1 with the only difference that now the geometry $\Omega$ changes but the thickness $\rho$ remains constant. We do not take into account the hole in the middle of the C-frame because we want to simplify the geometry. We use a mesh with 828 triangular finite elements. Considering all the constraints from Section 1, the critical constraints are the angles of the wings. Although $\sigma_{11}$ can reach critical values on some single elements, on most elements the constraints on the stresses are automatically fulfilled if the constraints on the angles are fulfilled. In the optimal design both wings almost reach their maximal allowed deformation, see Fig. 3. The final design fulfills all constraints and the mass has been reduced to 81.83% of the original value. For the original mass of 5.4223t that means a weight reduction of 985.1kg. The optimization process required 79 iterations and 43.2 seconds for a set of 29 design parameters.

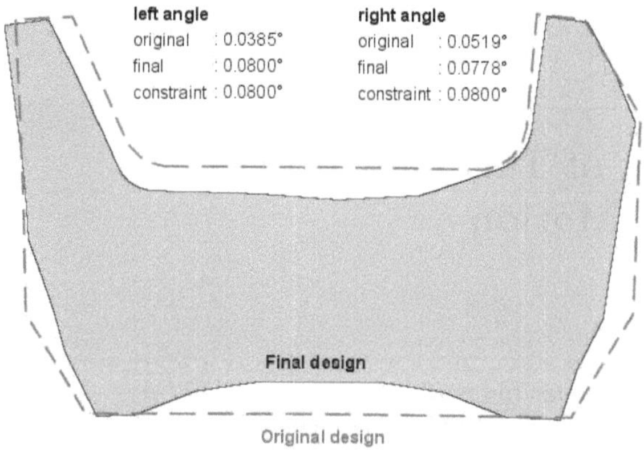


**Fig. 3.** Comparison of the original and of the final deformed geometry

# References

1. M. C. Delfour and J.-P. Zolésio. *Shapes and Geometries: analysis, differential calculus, and optimization.* Advances in Design and Control. SIAM, Philadelphia, 2001.
2. G. Haase and E. H. Lindner. Advanced solving techniques in optimization of machine components. *Computer Assisted Mechanics and Engineering Sciences*, 6(3):337–343, 1999.
3. Gundolf Haase, Ulrich Langer, Ewald Lindner, and Wolfram Mühlhuber. Optimal sizing of industrial structural mechanics problems using automatic differentiation. In G. Corliss, C. Faure, A. Griewank, L. Hascoet, and U. Naumann, editors, *Automatic Differentiation of Algorithms: From Simulation to Optimization*, pages 181–188,, New York, 2002. Springer.
4. Gundolf Haase, Ewald Lindner, Wolfram Mühlhuber, and Christian Rathberger. Optimal sizing ans shape optimization in structural mechanics. Technical Report 2003-6, SFB F013, April 2003.
5. J. Haslinger and R. Mäkinen. *Introduction to Shape Optimization: Theory, Approximation and Computation*, volume 7 of *Advances in Design and Control.* SIAM, Philadelphia, 2003.
6. J. Nocedal and S. Wright. *Numerical Optimization.* Springer Series in Operations Reasearch. Springer, New York, 1999.
7. M. J. D. Powell. A fast algorithm for nonlinear constrained optimization calculations. In G. A. Watson, editor, *Numerical Analysis*, number 630 in Lecture Notes in Mathematics. Springer, Berlin, 1978.
8. C. Rathberger. Fast product design with modern optimization software – shape optimization [4]. Master's thesis, Johannes Kepler University of Linz, Linz, July 2002.

[4]http://www.numa.uni-linz.ac.at/Staff/haase/Thesis/rathberger-diplom.pdf

# Modeling of Turbulence Effects on Fiber Motion

N. Marheineke

Fraunhofer-Institute for Industrial Mathematics (ITWM),
Gottlieb-Daimler-Str.49, D-67663 Kaiserslautern, Germany
marheineke@itwm.fhg.de

**Summary.** This work deals with the motion of a long slender elastic fiber in a turbulent flow. Neglecting the fiber effect on the turbulence, a centered differentiable Gaussian field is derived for the randomly fluctuating component of the flow velocity. The construction of the initial double-velocity correlation tensor is hereby based on the $k$-$\varepsilon$ model and Kolmogorov's universal equilibrium theory. Its dynamic is described by Taylor's hypothesis of frozen turbulence. Using an empirical drag coefficient, the developed fluctuation field leads to a correlated stochastic force that can numerically be treated as white noise with flow dependent amplitude.

**Key words:** Fiber dynamics; Turbulence; $k$-$\varepsilon$ model; Kolmogorov's universal equilibrium theory; Double-velocity correlation tensor; White noise

## 1 Motivation

The understanding of fiber-fluid interactions is of great interest for textiles manufacturing in the melt-spinning process of nonwoven materials. The quality of the nonwoven material depends on the dynamics of hundreds of individual endless polymer fibers that are curled and entangled by turbulent air flows. Fiber-turbulence interaction is governed by many, very complex factors, e.g. nature of flow, turbulent length scales, concentration and size of fibers. For the application, we may assume that the turbulence is not significantly affected by the fibers. Thus, the turbulent flow is determined under neglect of suspended fibers, and its effect is studied on a single long slender fiber.

## 2 Fiber Dynamics

Consider a slender elastic fiber of length $L$ and diameter $d$, $d/L \ll 1$, suspended in a highly turbulent air stream. Its dynamic is described by the Kirchhoff-Love equations for the motion of a Cosserat rod capable of large bending deformations [1]. In terms of these the fiber slenderness allows the formulation of a wavelike system of nonlinear PDEs of 4th order with the algebraic constraint of inexensibility

$$\omega\, \partial_{tt}\mathbf{r}(s,t) = \partial_s[T(s,t)\, \partial_s\mathbf{r}(s,t)] - S_b\, \partial_{ssss}\mathbf{r}(s,t) + \omega\, \mathbf{g} + \mathbf{f}^{air}(\mathbf{r}(s,t)) \quad (1)$$

$$(\partial_s\mathbf{r}(s,t))^2 = 1. \quad (2)$$

Here, $\mathbf{r} : [0,l] \times \mathbb{R}_0^+ \rightarrow \mathbb{R}^3$ might be interpreted as center line with arc length $s$ and time $t$, $\omega$ denotes the line weight. The inner forces stem from bending stiffness $S_b$ and traction $T$ that acts as Lagrangian multiplier in the system. The external forces arise from gravity $\mathbf{g}$ and aerodynamics $\mathbf{f}^{air}$.

The description of the fiber dynamics relies essentially on the model for the drag force $\mathbf{f}^{air}$ imposed on the fiber by the turbulent flow. The high Reynolds number flow (with Re based on $d$) and the presence of small vortices indicated by the relation $\eta < d$ (with Kolmogorov's length of turbulence $\eta$) motivate the use of the empirical Taylor drag [5]. As $\mathbf{f}^{air}$ lies in the plane spanned by fiber's tangent and relative velocity between fluid and fiber, $\mathbf{v}^{rel} = \mathbf{u} - \partial_t\mathbf{r}$, we decompose $\mathbf{f}^{air} = \mathbf{f_n}^{air} + \mathbf{f_t}^{air}$ into a normal $\mathbf{f_n}^{air}$ and a tangential part $\mathbf{f_t}^{air}$ with respect to the fiber's position, $\mathbf{t} = \frac{\partial_s\mathbf{r}}{\|\partial_s\mathbf{r}\|_2}$, $\mathbf{n} = \frac{\mathbf{v}^{rel}-(\mathbf{v}^{rel}\cdot\mathbf{t})\mathbf{t}}{\|\mathbf{v}^{rel}-(\mathbf{v}^{rel}\cdot\mathbf{t})\mathbf{t}\|_2}$

$$\mathbf{f_n}^{air} = 0.5\,\rho\, d\, C_{\mathrm{n}}\, \|\mathbf{v_n}^{rel}\|_2\, \mathbf{v_n}^{rel}, \qquad C_{\mathrm{n}} = 1 + 4\sqrt{\nu/(d\,\|\mathbf{v_n}^{rel}\|_2)}$$

$$\mathbf{f_t}^{air} = 0.5\,\rho\, d\, C_{\mathrm{t}}\, \|\mathbf{v_t}^{rel}\|_2\, \mathbf{v_t}^{rel}, \qquad C_{\mathrm{t}} = 5.4\sqrt{\nu\, \|\mathbf{v_n}^{rel}\|_2/(d\,\|\mathbf{v_t}^{rel}\|_2^2)}$$

with density $\rho$ and kinematic viscosity $\nu$ of the fluid.

## 3 Construction of Fluctuating Flow Velocity

Consider the flow to be subsonic, highly turbulent with small pressure gradients and Mach number Ma $< 1/3$. It can be modeled as an incompressible Newtonian fluid using the Navier-Stokes equations (NSE). Solving NSE with Direct Numerical Simulation (DNS) gives the exact velocity field needed for the determination of the force. However, DNS presupposes the resolution of all vortices ranging from the large $l_{\mathrm{T}}$, energy-bearing ones to the smallest $\eta$, viscously determined Kolmogorov vortices, $l_{\mathrm{T}}/\eta = \mathrm{Re}^{3/4}$. In spite of recent high speed performances, this fact leads still to its impracticality in case of high Re number flow. Stochastic models in contrast represent a reasonable compromise between accuracy and computational efficiency [2]. They are based on the Reynolds-averaged Navier-Stokes equations (RANS) where the instantaneous velocity $\mathbf{u}$ is expressed as sum of mean $\bar{\mathbf{u}}$ and fluctuations $\mathbf{u}'$

$$\mathbf{u}(\mathbf{x},t) = \bar{\mathbf{u}}(\mathbf{x},t) + \mathbf{u}'(\mathbf{x},t). \quad (3)$$

Applying in particular the standard $k$-$\varepsilon$ model yields a deterministic description of mean velocity $\bar{\mathbf{u}}$, turbulent kinetic energy $k$ and dissipation rate $\varepsilon$.

Thereby, the variables $k$ and $\varepsilon$ might be interpreted as parameters of a differentiable random field representing the fluctuations $\mathbf{u}'$

$$k(\mathbf{x},t) = \frac{1}{2}\,\mathbb{E}[\mathbf{u}'(\mathbf{x},t)\cdot\mathbf{u}'(\mathbf{x},t)], \quad \varepsilon(\mathbf{x},t) = \nu\,\mathbb{E}[\nabla\mathbf{u}'(\mathbf{x},t) : \nabla\mathbf{u}'(\mathbf{x},t)]. \tag{4}$$

**Definition 1.** *A turbulent flow is said to be a centered $\mathbb{R}^3$-valued random field $(\mathbf{u}'_{\mathbf{x},t}, (\mathbf{x},t)\in\mathbb{R}^3\times\mathbb{R}_0^+)$ where $\mathbf{u}'_{\mathbf{x},t}\in\mathcal{L}^2(\Omega,\mathcal{A},\mathrm{P})$ represents the velocity fluctuation on the probability space $(\Omega,\mathcal{A},\mathrm{P})$. Its correlation tensor*

$$\Gamma(\mathbf{x},t,\mathbf{y},\tau) = \mathbb{E}[\mathbf{u}'(\mathbf{x},t)\otimes\mathbf{u}'(\mathbf{y},\tau)], \tag{5}$$

*corresponds with the covariance $\mathbb{E}[\mathbf{u}'_{\mathbf{x},t}\ \mathbf{u}'_{\mathbf{y},\tau}]$, where $\mathbb{E}[.]$ denotes the mean.*

In the following we focus on Gaussian flows that are uniquely determined by their correlation tensor. Assuming locally homogeneous and isotropic turbulence the dynamics of $\Gamma$ is modeled by an advection equation whose solution coincides with Taylor's hypothesis of frozen turbulence [4]

$$\Gamma(\mathbf{x},t,\mathbf{y},\tau) = \hat{\Gamma}(\mathbf{x}-\mathbf{y},t-\tau) = \Gamma_0(\mathbf{x}-\mathbf{y}-\bar{\mathbf{u}}(t-\tau)). \tag{6}$$

To construct its initial condition we note that in case of incompressibility the tensor of 2nd order can be expressed by a single 1D function $c\in\mathcal{C}^\infty(\mathbb{R}_0^+)$

$$\Gamma_0(\mathbf{z}) = \left(c(z) + \frac{z}{2}\partial_z c(z)\right)\mathsf{I} - \frac{\partial_z c(z)}{2z}\,\mathbf{z}\otimes\mathbf{z}, \quad z = \|\mathbf{z}\|_2. \tag{7}$$

**Theorem 1.** *Assume turbulent kinetic energy $k$ and dissipation rate $\varepsilon$ to be constant. Let $(\mathbf{u}'_{\mathbf{x},t}, (\mathbf{x},t)\in\mathbb{R}^3\times\mathbb{R}_0^+)$ be an isotropic, homogeneous and incompressible Gaussian flow. Choose its correlation function $c\in\mathcal{C}^\infty(\mathbb{R}_0^+)$ as*

$$c(z) = \frac{2}{z^3}\int_0^\infty G(\kappa)\left[\frac{\sin(\kappa z)-\cos(\kappa z)\kappa z}{\kappa^3}\right]d\kappa \tag{8}$$

*where $G\in\mathcal{C}^2(\mathbb{R}_0^+)$ is given by*

$$G(\kappa) = \begin{cases} K\kappa_1^{-5/3}\sum_{j=4}^{6} a_j\left(\frac{\kappa}{\kappa_1}\right)^j & \kappa<\kappa_1 \\ K\kappa^{-5/3} & \kappa_1\le\kappa\le\kappa_2 \\ K\kappa_2^{-5/3}\sum_{j=7}^{9} b_j\left(\frac{\kappa}{\kappa_2}\right)^{-j} & \kappa<\kappa_2 \end{cases} \tag{9}$$

$$\textit{with} \quad \int_0^\infty G(\kappa)\,d\kappa = k \quad \textit{and} \quad \int_0^\infty \kappa^2 G(\kappa)\,d\kappa = \frac{\varepsilon}{2\nu}.$$

*The parameters are $a_4 = 230/9$, $a_5 = -391/9$, $a_6 = 170/9$, $b_7 = 209/9$, $b_8 = -352/9$, $b_9 = 152/9$ and $K = \varepsilon^{2/3}/2$.*

*Then, $(\mathbf{u}'_{\mathbf{x},t}, (\mathbf{x},t)\in\mathbb{R}^3\times\mathbb{R}_0^+)$ is differentiable and fulfills the requirements of Kolmogorov's energy spectrum $E$ (Fig. 1) as well as of the $k$-$\varepsilon$ model (4).*

The run of the energy spectrum $E$ being a function of wave number $\kappa$ results mainly from Kolmogorov's universal equilibrium theory [3] that states the existence of an inertial subrange between the wave number of energy $\kappa_e$ and dissipation $\kappa_d$. Here, $E$ is described by $E(\kappa) = C_{\mathrm{K}}\,\varepsilon^{2/3}\kappa^{-5/3}$, $\kappa\in(\kappa_e,\kappa_d)$ according to Kolmogorov's 5/3-Law.

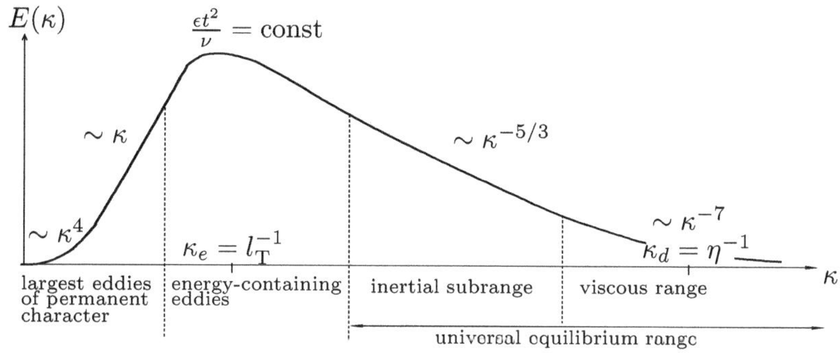


**Fig. 1.** Sketch of energy spectrum for isotropic turbulence

# 4 Stochastic Force Model

The modeling of the turbulence effects on the fiber is geared to the splitting of the velocity field $\mathbf{u}$ into mean $\bar{\mathbf{u}}$ and fluctuating part $\mathbf{u}'$ (3). Introducing the mean relative velocity by $\bar{\mathbf{v}}^{rel} = \bar{\mathbf{u}} - \partial_t \mathbf{r}$, Taylor expansion of $\mathbf{f}^{air}$ yields

$$\mathbf{f}^{air}(\bar{\mathbf{v}}^{rel} + \mathbf{u}') = \underbrace{\mathbf{f}^{air}(\bar{\mathbf{v}}^{rel})}_{\bar{\mathbf{f}},\ \text{deterministic}} + \underbrace{\nabla \mathbf{f}^{air}(\bar{\mathbf{v}}^{rel})\ \mathbf{u}'}_{\mathbf{f}',\ \text{stochastic}} + \mathcal{O}((\mathbf{u}')^2). \tag{10}$$

where $\|\mathbf{u}'\|_\infty^2$ is usually less than 1% of the absolute velocity [4] and thus negligible small. The deterministic force $\bar{\mathbf{f}}$ results from $\bar{\mathbf{u}}$ of RANS and $\partial_t \mathbf{r}$ of (1), (2), whereas the stochastic force $\mathbf{f}'$ inherits its properties from the fluctuation field $\mathbf{u}'$ of Sect. 3 due to linearity. In particular, we gain $\mathbf{f}'$ by restricting $\mathbf{u}'$ on the fiber

$$\mathbf{f}'(s,t) = \nabla \mathbf{f}^{air}(\bar{\mathbf{v}}^{rel}(\mathbf{r}(s,t),t))\ \mathbf{u}'(\mathbf{x},t)|_{\mathbf{x}=\mathbf{r}(s,t)}. \tag{11}$$

Dimensional analysis of fiber and turbulence reveals three characteristic scales. The smallest vortices of size $\eta$ do absolutely not affect the fiber due to its bending stiffness. On the scale of the energy-bearing vortices $l_\mathrm{T}$ in contrast, inner and outer forces are in balance so that the arising entanglement play a decisive role for the fiber dynamic. Over fiber length $L$, the outer forces, in particular $\bar{\mathbf{f}}$ coming from the mean flow, dominate the fiber. As we are only interested in a macroscopic description of the fiber motion, it is sufficient to model the turbulence effects of the meso scale $l_\mathrm{T}$ on the macro scale $L$ instead of resolving them explicitely. Asymptotic analysis of the correlation functions with respect to $l_\mathrm{T}/L \ll 1$ shows that

$$\mathbf{f}'_{approx}(s,t) = \nabla \mathbf{f}^{air}(\bar{\mathbf{v}}^{rel}(\mathbf{r}(s,t),t))\ D(\mathbf{r}(s,t),t)\ \mathbf{w}(s,t) \tag{12}$$

based on white noise, $\mathbf{w}(s,t) \simeq \mathcal{N}(0,\mathsf{I})$, is a good approximation for (11). The fluctuation dependent amplitude $D \sim \int z\, c(z)\, dz$ contains the information of the correlations and carries thus all crucial data of the meso scale.

## 5 Numerical Results with White Noise

Inserting (12) in (1) yields a nonlinear stochastic PDE with additive Gaussian noise. One representative of its solution for a short temporal sequence is visualized in Fig. 2. Apart from usual buckling effects that arise due to gravity and move upwards because of the hyperbolic character of the system (Fig. 2, top), the stochastic force causes additionally fiber entanglement on different scales, random loops and a wide swinging range (Fig. 2, down). The computed statistic quantities, e.g. mean fiber velocity, standard velocity deviation, swinging range, coincide hereby quite well with the experimental measurements.

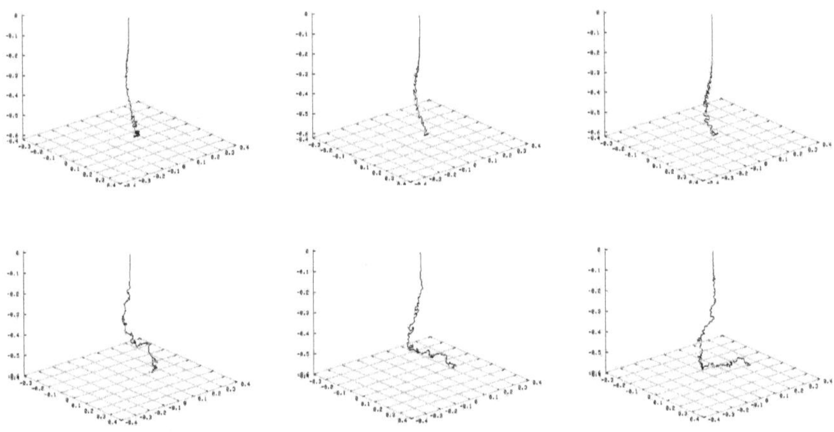

**Fig. 2.** Fiber dynamic without (*top*) and with (*down*) stochastic force $\mathbf{f}'$

## References


1. S. S. Antman. *Nonlinear Problems of Elasticity.* Springer Verlag, New York, 1995.
2. J. H. Ferziger and M. Perić. *Computational Methods for Fluid Dynamics.* Springer, Berlin, 3 edition, 2002.
3. U. Frisch. *Turbulence. The Legacy of A. N. Kolmogorov.* Cambridge University Press, 1995.
4. O. Hinze. *Turbulence.* McGraw-Hill, New York, 2 edition, 1975.
5. G. I. Taylor. Analysis of the swimming of long and narrow animals. *Proceedings of the Royal Society, A*, 214:158–183, 1952.

# Design Optimisation of Wind-Loaded Cylindrical Silos Made from Composite Materials

E.V. Morozov

School of Mechanical Engineering, Howard College, University of KwaZulu-Natal, Durban 4041, South Africa morozov@ukzn.ac.za

**Summary.** The conventional material from which silos are usually constructed is steel, and the existing codes and standards on these structures reflect the design criteria appropriate for an isotropic material. This paper deals with the design optimisation of the silos made from composite materials. The purpose of the present study is to perform the design optimisation of cylindrical composite silos loaded with the unsymmetrical external pressure caused by the action of wind. The design methodology is outlined, and the effectiveness of the optimisation is demonstrated using a particular example. In this case, the resultant optimised design produced a 29% saving in wall thickness, and thus material cost, in comparison with the non-optimised wall thickness.

## 1 Introduction

The standards applicable to the structural design of steel silo structures that have the form of a shell of revolution are presented in reference [2]. This code represents the basis of design, including fundamental requirements, reliability differentiation, limit states, actions and environmental effects, material properties, geometrical data, modelling of the silo for determining action effects, etc. Levels of rigour required for the design of silo structures depend on the reliability of the structural arrangement and the susceptibility to different failure modes. The following actions should be considered in the ultimate limit state design of a silo [2]: filling and storage of particulate solids, discharge of particulate solids (discharge loads), wind when the silo either full or empty, snow, imposed actions or deformations (live loads), thermal loads, imposed deformation (foundation settlement). The optimum design procedures addressing the first two cases (filling, storage, and discharge axisymmetric loading) which include buckling and strength analysis of cylindrical composite silos were set out in reference [3]. The purpose of the present study is to perform the design optimisation of cylindrical composite silos loaded with the unsymmetrical external pressure caused by the action of wind.

## 2 Silo Geometry, Wall Material Structure and Loading Conditions

Typical silo design including a cylindrical main holding section and a conical hopper is shown in Fig. 1, where $H$ and $d_c$ are the height and diameter of the cylindrical part of the structure. The wind actions set down in [1]. The empty silo shell is subjected to wind pressure on the windward side over an arc of and over the rest of the shell to a suction load as shown in Fig. 2 ($\theta \in [-30°, +30°]$). The pressure variation around an isolated silo may be defined in terms of the circumferential coordinate, with its origin at the windward generator (see Fig. 2):

$$p = \rho V_{\infty}^2 C_p$$

where

$$C_p = -0.7 + 0.2d_c/H + 0.4\cos\theta + (1.1 - 0.25d_c/H)\cos 2\theta +$$
$$(0.42 - 0.06d_c/H)\cos 3\theta - (0.14 - 0.04d_c/H)\cos 4\theta - 0.08\cos 5\theta$$

$\rho$ is the density, and $V_{\text{inf}}$ is the wind velocity [1]. The pressure distribution, $p(z, \theta)$ is approximately assumed as constant over the height of the silo (see Fig. 1).

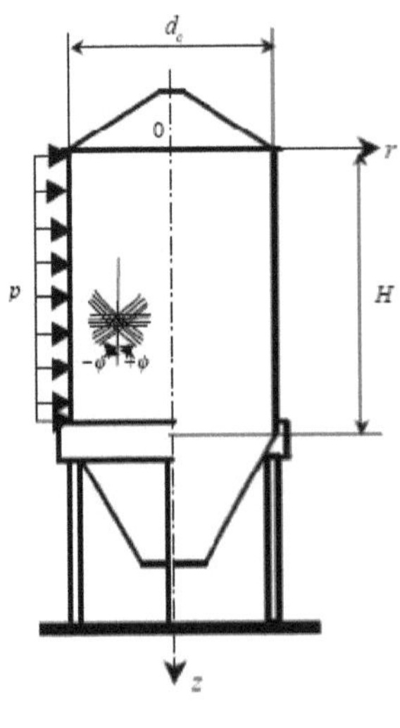


**Fig. 1.** Silo geometry and loading.

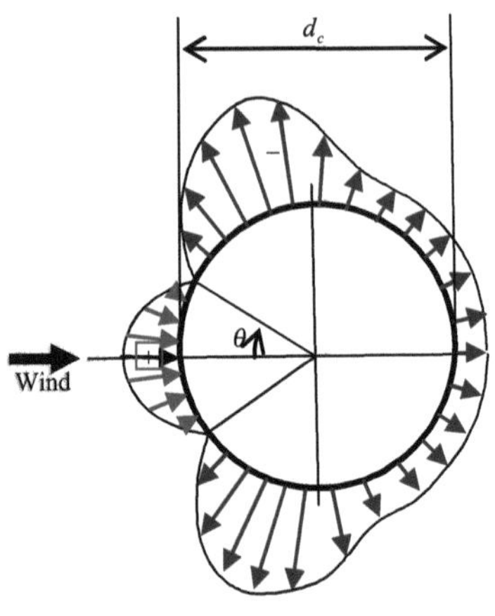


**Fig. 2.** Wind pressure distribution.

The composite material that is considered for the silo wall is a laminate, consisting of the helical layer with $\pm\varphi$ fibre orientation and hoop layer ($\varphi =$

90°) (see Fig. 1). Here the $\pm\varphi$ layer is considered as an angle-ply orthotropic layer consisting of an even number of alternating plies with angles $+\varphi$ and $-\varphi$. The structure of this layer is typical for the process of helical filament winding [4]. The introduction of 90° oriented layers is based on the simplicity and cost efficiency of the implementation of circumferential filament winding for a cylindrical silo, combined with the fact that this type of reinforcement would be used conventionally to withstand the applied internal pressure due to filling, storage, and discharge of particulate solids. Correspondingly, the design variables are the helical angle, $\varphi$ and thicknesses of the layers, *i.e.*, $h_{90}$ and $h_\varphi$.

## 3 Design Optimisation of The Cylindrical Section of The Silo

Minimisation of the material cost as a part of the cost effective silo design was considered as the objective for optimisation. Thus, the minimum mass of the silo structure was selected as the objective function. In terms of the design variables introduced earlier the minimum value of the objective function will be delivered if the following condition is satisfied:

$$h = h_{90} + h_\varphi \quad \rightarrow \quad \textbf{min} \tag{1}$$

where $h$ is the total thickness (which determines the weight and effectively the cost). The buckling constraint due to wind loading is imposed on the design variables. Buckling analysis of the composite cylindrical section of the silo subjected to the lateral pressure caused by the wind load has been performed using MSC Patran finite element software package. Optimum design procedure was based on the construction and analysis of the feasible domains for the design variable $\varphi$, $h_{90}$ and $h_\varphi$. Critical values of the parameter $h_{90}$ were found for given loading conditions (wind velocity), $V_\infty$, helical angle $\varphi$ and thickness $h_\varphi$. from the finite element buckling analysis. The cylinder was modelled with four-node laminate elements. Linear eigenvalue analyses have been performed to determine the buckling (critical) loads and buckled shapes of the cylindrical part of silo. The corresponding critical values of $h_{90}$ were found using the bisection method for given values of helical angle and thickness. These values determined the boundaries of the feasible domains for the design variables, $\varphi$ $h_{90}$ and $h_\varphi$ in the corresponding three-dimensional design space. Owing to the specific analytical form of the objective function (1), it is possible to locate the optimum values for the design parameters $h_\varphi$ and $h_{90}$ for every value of the reinforcement angle $\varphi$. The contour lines, $h = h_{cont}$ of the objective function(1) are determined by the equation

$$h_{90} = h_{cont} - h_\varphi \tag{2}$$

for each value of the angle $\varphi$. Geometrically, this equation represents a family of planes oriented at angles of 45° to the coordinate planes($h_\varphi$, $\varphi$) and ($h_{90}$, $\varphi$)

for different values of $h_{cont}$ , or the corresponding families of lines oriented at angles or to the axes $h_\varphi$ and $h_{90}$ of the two-dimensional coordinate frames for given values of angle $\varphi$. This allows the optimum values of the design parameters within the given feasible domains to be found.

# 4 Example

The relevant geometric and material property data for the particular design example considered in this study are: silo height $H = 7000$ mm, diameter $d_c = 2600$ mm. Composite material (glass-epoxy) properties: $E_1 = 44$ GPa, $E_2 = 9.4$ GPa, $G_{12} = 4$ GPa, $\nu_{21} = 0.26$.

The cylindrical shell was modelled and meshed using laminated shell elements (MSC Patran): 36 elements for the circumference and 20 elements for the length of the cylinder. The bottom section of the shell was considered clamped and the top one reinforced with the rigid ring. Buckling analyses were performed for the critical wind velocity of 100 km/h. Typical buckling mode is shown in Fig. 3. The corresponding critical values of $h_{90}$ were determined using the design procedure described in Section 3. Based on these calculations the limiting critical surface $\bar{h}_{90}(\varphi,\, h_\varphi)$ was built up and optimum values of the design parameters found using conditions specified by the equations (1) and (2). Graphical interpretation of the solution is illustrated in Fig. 4, where all the above-mentioned components (the objective function contour lines and constraints) are plotted for $\varphi = 0°, 20°, 40°, 60°, 90°$ As can be seen, the optimum point has the coordinates $\hat{h}_{90} = 1.28\,\text{mm}$, $\hat{h}_\varphi = 0$. Accordingly, the corresponding minimum thickness of the shell occurs for $h = \hat{h}_{90} = 1.28\,\text{mm}$. The maximum total thickness lying on the design surface is $h = 1.8mm$, which is 29% higher than the minimum value. Comparison

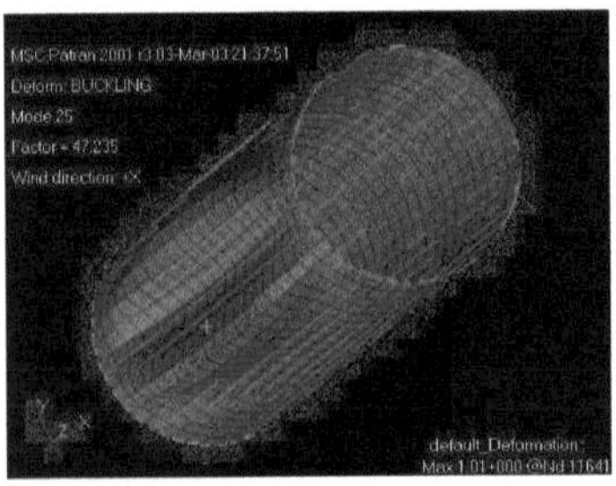

**Fig. 3.** Buckling mode under the wind action.

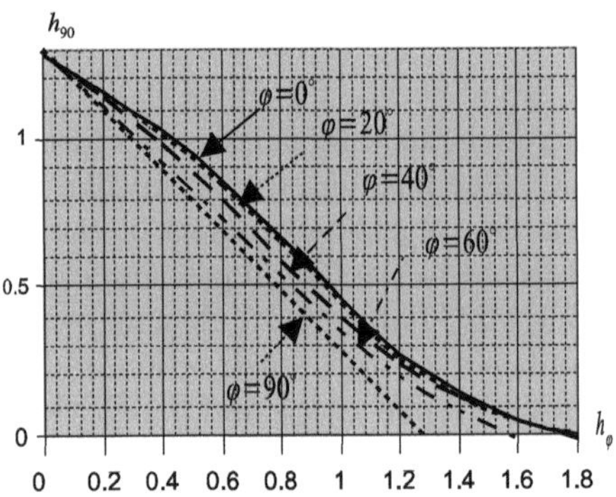


**Fig. 4.** Critical thicknesses $h_{90}$ and $h_\varphi$.

of this result with the optimum design solution obtained in [3] for the same composite cylindrical shell subjected to the axial compressive load caused by filling/storage/discharge (axisymmetrical loading) shows that the thicknesses required to meet the wind buckling limiting constraints are less than those determined by the axial buckling and radial deflection. As was shown in [3], the optimum design parameters were equal to $\hat{\varphi} = \varphi$, $\hat{h}_\varphi = 0$, and $\hat{h}_{90} = 3.44\,\text{mm}$ for angles $\varphi \geqslant 15°$ and provided the same constant value of the objective function $h = h_\varphi + h_{90} = 3.44\,\text{mm}$. The minimum thickness of the shell, $h = 3.32\,\text{mm}$ was delivered by the following parameters: $\hat{\varphi} = 0°$, $\hat{h}_\varphi = 2.9\,\text{mm}$ and $\hat{h}_{90} = 0.4216\,\text{mm}$. As can be seen, the shell thickness values obtained for the axisymmetrical case [3] are well above the critical surface $\bar{h}_{90}(\varphi,\ h_\varphi)$ constructed in the case under consideration (wind-loaded shell) (see Fig. 4).

## 5 Conclusions

Results of the analysis performed in this study indicate that the optimum reinforcement orientation for the composite silo shell operating under wind load is determined by the angle $\varphi = 90°$. The corresponding thickness of the wall can be found from the buckling analysis for the specific shell geometry $(d_c/H)$. In general, the design solution for composite cylindrical silo is controlled mainly by the buckling condition under axial load and the radial deflection constraint.

*Acknowledgement.* The author is grateful to Mr Nicolas Aval (SUPAERO, France) for helping to conduct finite element computations related to this study.

## References

1. *Basis of design and actions on structures*, 1991-2-4 Part 2.4 Wind loads. ENV, 1991 Eurocode 1.
2. *Design of steel structures*, 1993-4-1: Part 4-1: silo. D. prENV, 1997 Eurocode 3.
3. E.V. Morozov, J.L. Henshall, and C.J. Brown. *Design optimisation of silo structures made from composite materials*, chapter Design and Applications - Part L, pages 55–67. Professional Engineering Publishing,London, UK, 2002.
4. V.V. Vasiliev and E.V. Morozov. *Mechanics and Analysis of Composite Materials.* Elsevier, 2001.

# Two-Dimensional Short Wave Stability Analysis of the Floating Process

S. R. Pop

Fraunhofer ITWM, Gottlieb-Daimler-Str. 49, 67663 Kaiserslautern, Germany
spop@itwm.fhg.de

**Summary.** In this paper, we perform a linear stability analysis using normal modes on the two-dimensional system of two superposed fluids confined between two infinite plates in the presence of a large temperature gradient. The movement of the fluids is characterized by a combination of inertial and buoyancy forces, thus we are dealing with a mixed convection problem. The results of the linear stability analysis show that for large wave numbers, the small amplitude waves travel with the interface velocity .

**Key words:** Float glass, mixed convection, stability analysis.

## 1 Mathematical Formulation

The float glass process, since its invention by Allistair Pilkington in 1952, is used to manufacture thin long high quality sheets of glass. However, the increasing demand of thinner glass has dramatic consequences over its optical quality. Our work was motivated by a series of investigations performed over the finite products which showed the existence of many short wave patterns, probably affecting strongly its optical quality.

Convective instabilities for two superposed fluids was treated by many authors in terms of thermocapillarity and buoyancy effects [5]. In the case of horizontal temperature gradient, the basic profile is not trivial, giving rise to a complex flow and a non-linear vertical temperature profile [4].

Short wave limits were first investigated in [3]. Since then, various works analyze the stability of small amplitude waves for different kind of models [1, 2]. In this paper, we emphasize the mixed convection effects over the stability of two superposed fluids that have different characteristics (kinematical viscosity, density).

### 1.1 Governing system of motion

Two immiscible, incompressible, viscous fluids, labelled $j = 1, 2$, are confined between two planes subject to a large horizontal temperature gradient. The upper fluid is denoted with 2 and the lower fluid with 1.

The equations that govern the system of motion are mass, momentum and energy coupled with the Boussinesq approximation:

$$\nabla \cdot \mathbf{u}_j = 0, \quad \bar{\rho}_j \left[ \frac{\partial \mathbf{u}_j}{\partial \bar{t}} + (\mathbf{u}_j \cdot \nabla)\mathbf{u}_j \right] = -\nabla \bar{p}_j + \mu_j \Delta \mathbf{u}_j - \bar{\rho}_j \mathbf{g} \tag{1}$$

$$\frac{\partial T_j}{\partial \bar{t}} + (\mathbf{u}_j \cdot \nabla) T_j = \alpha_j \Delta T_j, \quad j = 1, 2 \tag{2}$$

where $\mathbf{u}_j = (\bar{u}_j, \bar{w}_j)$ is the velocity, $\mathbf{g} = (0, g)$, $g$ is acceleration due to the gravity, $\bar{p}_j$ is the pressure, $\mu_j$ is the dynamic viscosity, $\bar{\rho}_j$ is the density, $\alpha_j$ is the thermal diffusivity and $T_j$ is the temperature of the fluid for each phase.

The Boussinesq approximation is plugged into the gravity term and characterizes the thermal expansion of each fluid, $\bar{\rho}_j = \rho_j \left[1 - \beta_j \left(T_j - T_{ref}\right)\right], \quad j = 1, 2$, where $\beta_j$ is the thermal expansion coefficient and $T^c_{ref}$ is the smallest temperature of the system.

The system is coupled with the interface ($\bar{z} = \bar{h}(\bar{x}, \bar{t})$) and boundary conditions. The interface moves with the velocity of the flow, whereas *normal* and *tangential stresses* are continuous through the interface. We assume *no-slip* and *no-penetration* at the interface. *Heat transfer condition* ensures that fluxes and temperatures are equal at the interface. Moreover, we prescribe the *mass flow rate* conditions and the *kinematical condition* at the interface.

We make dimensionless the velocities, distance, time, pressures for both fluids with respect to $U$ the speed of the upper plate, $d_2$ the height of the upper fluid layer, $d_2/U$, $\mu_1 U/d_2$. Non-dimensional temperatures are given by $\theta = \frac{T - T^c_{ref}}{\Delta T_{ref}}$, where $T^h_{ref}$ and $T^c_{ref}$ are the largest and respectively the smallest temperatures of the system. In the float glass process these temperatures correspond with inlet and outlet temperatures of the float bath. We denote with $\Delta T_{ref}$ the difference between $T^h_{ref}$ and $T^c_{ref}$, (i.e. $\Delta T_{ref} = T^h_{ref} - T^c_{ref}$).

We express all the non-dimensional numbers of the upper fluid with respect to the non-dimensional numbers of the lower fluid. Moreover, we introduce the following notation: $d = \frac{d_1}{d_2}$, $\mu = \frac{\mu_2}{\mu_1}$, $\rho = \frac{\rho_2}{\rho_1}$, $\alpha = \frac{\alpha_2}{\alpha_1}$, $\beta = \frac{\beta_2}{\beta_1}$, $\kappa = \frac{\kappa_2}{\kappa_1}$.

### 1.2 Basic flow

The system of motion for the basic flow has the form:

$$\frac{\partial^3 U_j}{\partial z^3} = N_j \frac{\partial \Theta_j}{\partial x}, \quad U_j = U_j(z), \quad \Theta_j = \Theta_j(x, z) \tag{3}$$

$$U_j \frac{\partial \Theta_j}{\partial x} = M_j \left( \frac{\partial^2 \Theta_j}{\partial x^2} + \frac{\partial^2 \Theta_j}{\partial z^2} \right), \quad j = 1, 2 \tag{4}$$

coupled with the following interface and boundary conditions:

$$U_1(0) = U_2(0), \ \ \mu U_2'(0) = U_1'(0), \ \ \Theta_1(0) = \Theta_2(0), \ \ \kappa\Theta_2'(0) = \Theta_1'(0) \quad (5)$$

$$U_1(-d) = 0, \ \ U_2(1) = 1, \ \ \Theta_1(-d) = \Theta_c^0 + x\Theta_c^1, \ \ \Theta_2(1) = \Theta_h^0 + x\Theta_h^1 \quad (6)$$

where $N_1 = Gr/Re$, $N_2 = \rho Gr\beta/(\mu Re)$, $M_1 = 1/(RePr)$, $M_2 = \alpha/(RePr)$.

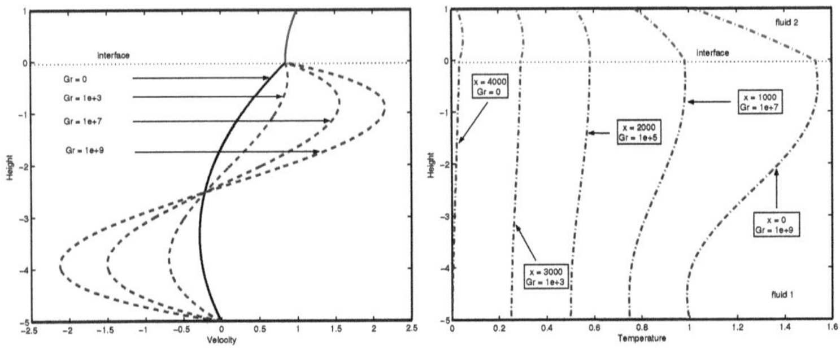


**Fig. 1.** The velocity (left) and temperature (right) profiles for the basic flow.

The solutions of the above system are represented in Fig. 1. In the case when temperature gradient plays no role ($Gr = 0$), the velocity and temperature profiles look similar like in the classic case for low Prandtl numbers (*i.e.*, small kinematical viscosity). Further, the temperature gradient in the horizontal direction is increased and the lower fluid starts to move due to buoyancy effects. For large values of the Grashof number, the profiles describe exactly the recirculation of the fluid in the stable case.

## 2 The Disturbance System of Motion.

We obtain the stability system by perturbing the basic flow by infinitesimal disturbances, linearizing and searching for solutions proportional with $e^{ik(x-ct)}$, where $k$ is the wave number and $c$ is the wave speed. Further, we introduce the stream function defined by $\tilde{u}_j = \frac{\partial \varphi_j}{\partial z}$, $\tilde{w}_j = -ik\tilde{\varphi}_j$, $j = 1,2$ and we obtain the Orr-Sommerfeld and the energy equations which governs the stability of the basic flow.

$$\varphi_j^{(iv)} - 2k^2\varphi_j'' + k^4\varphi_j = ikReD_j(U_j - c)(\varphi_j'' - k^2\varphi_j) - ikReD_j\varphi_j\frac{\partial^2 U_j}{\partial z^2} + ikE_j\frac{Gr}{Re}\tilde{\theta}_j \quad (7)$$

$$ik\tilde{\theta}_j(U_j - c) + \varphi_j'\frac{\partial \Theta_j}{\partial x} - ik\varphi_j\frac{\partial \Theta_j}{\partial z} = \frac{F_j}{RePr}\left(\frac{\partial^2\tilde{\theta}_j}{\partial z^2} - k^2\tilde{\theta}_j\right) \quad (8)$$

where $D_1 = 1$, $D_2 = \rho/\mu$, $E_1 = 1$, $E_2 = \rho\beta/\mu$, $F_1 = 1$, $F_2 = \alpha$ and $j = 1,2$. The system is coupled with the following interface and boundary conditions:

$$\varphi_1 = \tilde{h}(c - U_1), \quad \varphi_2 = \tilde{h}(c - U_2) \to \varphi_1 = \varphi_2 \tag{9}$$

$$\varphi_1' + \tilde{h}\frac{\partial U_1}{\partial z} = \varphi_2' + \tilde{h}\frac{\partial U_2}{\partial z}, \quad \mu\left(\varphi_2'' + k^2\varphi_2\right) = \varphi_1'' + k^2\varphi_1 \tag{10}$$

$$\varphi_1''' - \mu\varphi_2''' - 3k^2(\varphi_1' - \mu\varphi_2') -$$

$$-ikRe_1\left[(U_1 - c)(\varphi_1' - \rho\varphi_2') + \varphi_1 U_1'\left(\frac{\rho}{\mu} - 1\right)\right] = ik^3\frac{Re}{We}\tilde{h} \tag{11}$$

$$\tilde{\theta}_1 = \tilde{\theta}_2, \quad \kappa\left(ik\tilde{h}\frac{\partial\Theta_2}{\partial x} - \frac{\partial\tilde{\theta}_2}{\partial z}\right) = ik\tilde{h}\frac{\partial\Theta_1}{\partial x} - \frac{\partial\tilde{\theta}_1}{\partial z} \tag{12}$$

(BC) $z = -d$: $\varphi_1 = \varphi_1' = \tilde{\theta}_1 = 0$; $z = 1$: $\varphi_2 = \varphi_2' = \tilde{\theta}_2 = 0$

## 3 Short Wave Limit

In the short wave limit, we are looking for waves which has amplitude and wavelength at the same order of magnitude, although much smaller than the characteristic length. In order to keep $x$ and $z$ scales at the same order we change our perspective from the macroscopic to microscopic approach by considering that $d/dz \sim \mathcal{O}(k)$ with $k \to \infty$.

The eigenvectors of the perturbation system of motion should remain bounded, thus we are looking for solutions in the form, $\varphi_1$, $\theta_1 \sim \mathcal{O}(e^z)$; $\varphi_2$, $\theta_2 \sim \mathcal{O}(e^{-z})$.

Using the asymptotic expansions with respect to $k$, the wave number, we look for the solutions of the form $\varphi^j = \varphi_0^j + \frac{1}{k}\varphi_1^j + \frac{1}{k^2}\varphi_2^j + \ldots$, with $1/k \to 0$ and the dimensionless numbers fixed. Analogously we study asymptotic expansions of $\tilde{\theta}^j$, $c$, $\tilde{h}$ with $j = 1, 2$.

Further, we obtain the *leading order system of motion*. The real positive wave speed $c_0$ is the analytical solution of the system. The wave propagates faster when the upper layer is less viscous than the lower fluid layer. Instability occurs whereas the inertia of the lower fluid decreases and buoyancy dominates the flow, hence the disturbance propagates faster when the viscosity effects dominate the lower fluid domain.

The wave velocity grows for higher values of Grashof number, so the temperature destabilizes the system (Fig. 2 (left)). The wave propagates faster when the upper fluid is less viscous than the lower fluid. This results is verified for the no-temperature case, where the wave propagates in the direction of the less viscous flow [2]. At this level, surface tension has no effect over the leading order eigenvalue.

The analytical solution of the *first order system*, (the imaginary eigenvalue $c_1$), is the parameter which influence directly the stability of our problem. From Fig. 2 (right), we can see that the more viscous fluid layer slows down the growth of the disturbance stabilizing the flow, as well as the surface tension. The stability of the system is decreasing with temperature gradient.

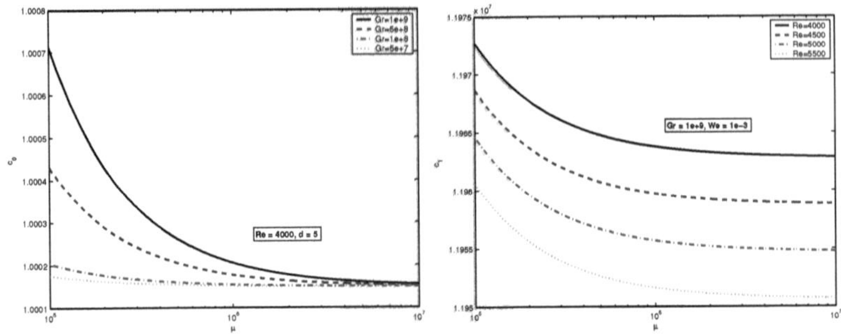


**Fig. 2.** The wave speed $c_0$ with respect to the viscosities ratio for different values of the Grashof number (left) and the eigenvalue $c_1$ with respect to the viscosities ratio for different values of the Reynolds number (right).

Moreover, in our particular case ($d = 5$), the system is unstable for all values of the viscosity ratio. Inertia stabilize the flow when the upper layer is much more viscous than the lower layer. The disturbances decrease in magnitude when the upper layer increases in thickness. The surface tension has a stabilizing effect, which is in perfect agreement with the classical theory. When the influence of the surface tension decreases (*i.e.*, large Weber number) the viscosity apparently plays no role in the stability. Moreover, large temperature gradient and high inertia destroy any effect that may appear due to the viscosity difference.

A prior investigations over the model used in this paper showed that in the long waves limit the system is stable, result which is in perfect agreement with the literature and the real float glass process. Although, the small amplitude waves are found to be unstable, strongly influenced by the governing factors. The short waves travel with the interface velocity forming patterns of standing waves, which later can be seen in the finite products of the process.

## References


1. P.J. Blennerhassett and F. T. Smith. Short-scale waves on wind-driven water (cat's paws). *Proc. R. Soc. London A*, 410:1–17, 1987.
2. F. Charru, P. Luchini, and P. Ern. Instability of a nearly inextensible thin layer in a shear flow. *Eur J. Mech.B/Fluids*, 22:39–50, 2003.
3. A. P. Hooper and W. G. C. Boyd. Shear-flow instability at the interface between two viscous fluids. *J. Fluid Mech.*, 128:507–528, 1983.
4. S. Madruga, C. Perez-Garcia, and G. Lebon. Convective instabilities in two superposed horizontal liquid layers heated laterally. *Phys. Review E*, 68:41607, 2003.
5. R. Narayanan and D. Schwabe, editors. *Interfacial Fluid Dynamics and Transport Processes*, Berlin, 2003. Springer.

# Optimization in high-precision glass forming

M. Sellier

Schott AG, Hattenbergstrasse 10, 55122 Mainz, Germany
mathieu.sellier@itwm.fhg.de

**Summary.** The question of interest in the present study is the inverse problem for high precision glass forming, i.e. 'How to design the mould and the temperature regime so that at the very end of the forming process we will get at room temperature a prescribed glass geometry with a precision in the order of the Micron?' The aim is to eliminate from the manufacturing process the costly and time-consuming post-processing when the final shape does not conform precisely to the desired one.

**Key words:** glass forming, stress/structural relaxation, optimization

## 1 Description of the forward problem

The present study focuses on the cooling stage of the glass forming process and provides a method based on computer-aided simulations to optimize the cooling treatment in order to keep the residual stresses below a given admissible threshold and identify the required initial geometry of the glass piece so that after cooling, it matches precisely the desired one.

The case treated here corresponds to an optical device and the thermo-mechanical analysis is performed using the commercial Finite-Element code Ansys. The geometry and boundary conditions are shown on Figure 1. The glass piece has a symmetry of revolution and occupies the domain $\Omega$ bounded by the surface $\Gamma = \Gamma_1 \cup \Gamma_2 \cup \Gamma_3 \cup \Gamma_4$ at $t = 0$. It is assumed to be initially stress-free and with uniform temperature $T_0 = 873.15$ K. Radiative heat transfer is ignored so that the temperature field within the glass piece is dictated by the heat diffusion equation. Moreover, heat is lost to the surrounding through convective heat transfer characterized by a constant coefficient of heat transfer $h$. In this optimization problem, the time-dependent temperature of the surrounding $T_a(t)$ is the control used to minimize an objective function yet to be defined.

Upon cooling, the glass behavior undergoes drastic changes. At high temperature ($T > T_g + 100$ K) it behaves like a Newtonian liquid while at lower

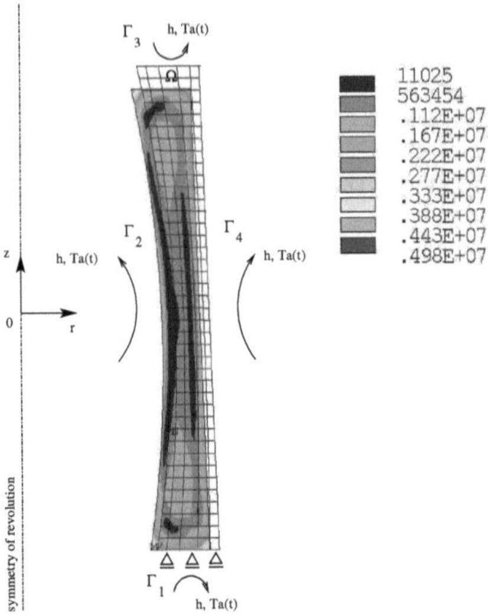


**Fig. 1.** Initial glass piece geometry with corresponding Finite Element mesh and boundary conditions. The deformed glass geometry (after cooling) with the associated map of the residual Von Mises stresses is also shown.

temperatures ($T < T_g - 50$ K), classical linear elasticity applies. In the intermediate temperature range, the glass is best described as a viscoelastic solid where stress and structure relaxation occur. The state of the structure of the glass is characterized by the fictive temperature $T_f(\mathbf{x}, t)$ a concept well established after the work of [2]. Accordingly, constitutive laws may be expressed in the following integral form,

$$s_{ij}(\mathbf{x}, t) = 2 \int_0^t G\left(\xi(\mathbf{x}, t) - \xi(\mathbf{x}, t')\right) \frac{\partial e_{ij}(\mathbf{x}, t')}{\partial t'} \mathrm{d}t' \ , \tag{1}$$

$$\sigma(\mathbf{x}, t) = 3K\left(\varepsilon(\mathbf{x}, t) - \varepsilon_{th}(\mathbf{x}, t)\right) \ , \tag{2}$$

where $s_{ij}$, $\sigma$ and $e_{ij}$, $\varepsilon$ are the deviatoric and volumetric parts of the stress and strain tensor respectively. The bulk modulus $K$ is chosen to be constant while the shear modulus $G$ is a function of the elapsed *reduced time*, $\xi(\mathbf{x}, t) - \xi(\mathbf{x}, t')$. A classical Arrhenius model is used to represent the influence of the temperature on the relaxation behavior so that the relaxation time is expressed as:

$$\xi(\mathbf{x}, t) = \int_0^t \frac{\tau_{ref}}{\tau(T, T_f, t')} \mathrm{d}t' = \int_0^t e^{\frac{\Delta H}{R}\left(\frac{1}{T_0} - \frac{\beta}{T(\mathbf{x}, t')} - \frac{1-\beta}{T_f(\mathbf{x}, t')}\right)} \mathrm{d}t' \ , \tag{3}$$

where $\tau_{ref}$ and $\tau(T, T_f, t')$ are the relaxation times at the initial temperature and the temperature $T$, respectively. $\beta$ is a constant ($0 < \beta < 1$), $\Delta H$ the activation energy and $R$ the ideal gas constant. The shear modulus and fictive temperature are expressed in the form of Prony series, viz

$$G(\xi) = G_\infty + \sum_{i=1}^{n} G_i e^{-\xi/\lambda_i} \text{ with } G_i = \nu_i(G_0 - G_\infty) \text{ and } \sum_{i=1}^{n} \nu_i = 1\,, \quad (4)$$

$$T_f(\mathbf{x}, t) = \sum_{i=1}^{m} \omega_i T_{f\,i}(\mathbf{x}, t) \text{ with } \sum_{i=1}^{m} \omega_i = 1\,. \quad (5)$$

In eqs. (4) and (5), $G_0$ and $G_\infty$ are the initial and final shear moduli, respectively, while $G_i$ and $\omega_i$ are weights and $\lambda_i$ are constants associated with a discrete relaxation spectrum in shear. $T_{f\,i}$ are the partial fictive temperatures and these must satisfy the following ODE, [1]:

$$\frac{\mathrm{d}T_{f\,i}}{\mathrm{d}t} = -\frac{T_{f\,i} - T}{\mu_i}\frac{\mathrm{d}\xi}{\mathrm{d}t}\,, \quad (6)$$

where $\mu_i$ are constants associated with a discrete structural relaxation spectrum. Finally, the thermal strain in eq. (2) is given by:

$$\varepsilon_{th} = \alpha_g\,(T - T_0) + (\alpha_l - \alpha_g)\,(T_f - T_0)\;, \quad (7)$$

where $\alpha_g$ and $\alpha_l$ are the coefficients of thermal expansion of solid and liquid glass respectively. The glass transition occurs around $T_g = 773.15$ K.

The glass piece is assumed to be traction free and slides without friction on its base. Numerical simulations proceed by first computing the temperature field until the temperature in the glass is uniform and equal to the room temperature (293.15 K) and then impose it as a load to the structural analysis. A typical map of the residual Von Mises stresses can be seen on Figure 1.

## 2 Optimization of the cooling curve

A first step in the identification of the required initial shape consists in optimizing the cooling curve in order to reduce the permanent stresses produced by temperature gradients. To this end, the algorithm proposed by [3] was employed which attempts to reduce the total cooling period while keeping residual stresses below a prescribed threshold, $\sigma_{adm}$.

As seen on Figure 2, three regions define the cooling curve. The stresses only have the ability to relax in the first region characterized by $T_g - 50 < T_a < T_g + 100$. The optimization is therefore restricted to this part of the cooling curve. In the region to be optimized, the cooling curve is defined by $N = 7$ key-points with locations $(i\Delta t, T_i), i = 1, \ldots, N$, where $\Delta t$ is a time interval. The initial and final temperatures are fixed to $T_a(t = 0) = T_0$ and $T_a(t_f = (N+1)\Delta t) = 723.15$K and the temperature is interpolated linearly between the key-points. The optimization problem has therefore $N+1$ degrees of freedom, namely the $N$ values of the temperature $T_i$ and $\Delta t$.

The aim of the optimization is to reduce the total cooling time $t_f$ subject to the constraint that the maximum value of the Von-Mises stresses should not

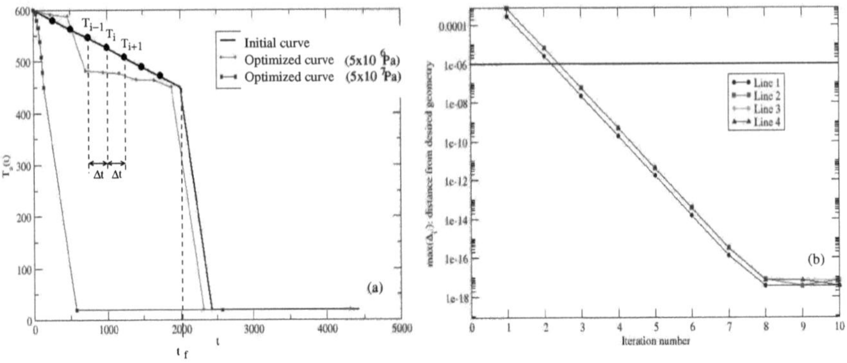


**Fig. 2.** **(a)**: Initial and optimized cooling curves for $\sigma_{adm} = 5 \times 10^6$ Pa, $\sigma_{adm} = 5 \times 10^7$ Pa; **(b)**: Convergence history of the proposed algorithm for each line $\Gamma_i$.

exceed $\sigma_{adm}$ at room temperature. Moreover, realistic cooling curves should be monotonically decreasing and the slope bounded by a constant $\kappa = 1\text{K/s}$ in order to avoid exceedingly large temporary stresses. Accordingly, the optimization problem is formulated as follows:

$$\min_{T_i, \Delta t} \quad t_f \left(1 + P_1 + P_2 + P_3\right) , \tag{8}$$

where the $P_i$ correspond to the penalty functions associated with each constraint. These are given by:

$$P_1 = \begin{cases} C_1 \frac{\max\limits_{\mathbf{x}\in\Omega}(\sigma_{VM}(\mathbf{x},t=t_e)) - \sigma_{adm}}{\sigma_{adm}} & \text{if } \max\limits_{\mathbf{x}\in\Omega}\left(\sigma_{VM}(\mathbf{x}, t = t_e)\right) > \sigma_{adm} \\ 0 & \text{if } \max\limits_{\mathbf{x}\in\Omega}\left(\sigma_{VM}(\mathbf{x}, t = t_e)\right) \leq \sigma_{adm} \end{cases}$$

$$P_2 = \begin{cases} C_1 \frac{T_{i+1} - T_i}{T_0} & \text{if } T_{i+1} > T_i \\ 0 & \text{if } T_{i+1} \leq T_i \end{cases} , \quad P_3 = \begin{cases} C_1 \frac{T_{i+1} - T_i}{\kappa \Delta t} & \text{if } \|T_{i+1} - T_i\| > \kappa \Delta t \\ 0 & \text{if } \|T_{i+1} - T_i\| \leq \kappa \Delta t \end{cases}$$

The constant $C_1$ is chosen to be equal to $10^6$. The Nelder-Mead simplex direct search method from Matlab was adopted to minimize eq. (8).

The initial and optimized cooling curves for $\sigma_{adm} = 5 \times 10^6$ Pa and $\sigma_{adm} = 5 \times 10^7$ Pa are shown on Figure 2 **(a)**. As expected when the constraint on the maximum admissible Von Mises stress is least severe, a much quicker cooling is possible. The shape of the cooling curve for $\sigma_{adm} = 5 \times 10^6$ Pa is also as expected: after an initial rapid cooling, the temperature remains approximately constant. This feature allows the stresses to relax to the admissible level. Finally, the map of the residual Von Mises stresses when $\sigma_{adm} = 5 \times 10^6$ Pa shown on Figure 1 confirms that, the level of stress is kept below $\sigma_{adm}$.

# 3 Identification of the required initial geometry

In order to describe the algorithm which tackles the inverse problem of identifying the required initial geometry, notations are introduced. Let $M_1^d, \ldots, M_L^d$ denotes the $L$ boundary nodes of the desired glass geometry at room temperature. The required initial geometry is found by updating the location of the boundary nodes $M_i^j$ iteratively (the superscript indicates the iteration number). At each iteration $N_i^j$ corresponds to new location of the node $M_i^j$ in the deformed geometry and $\mathbf{U}_\mathrm{i}^\mathrm{j}$ is the associated displacement. The algorithm is defined in pseudo-code notation as follows:

1. `for i=1 to L` {
   $\mathbf{OM}_\mathrm{i}^1 = \mathbf{OM}_\mathrm{i}^\mathrm{d}$; $\mathbf{ON}_\mathrm{i}^1 = \mathbf{OM}_\mathrm{i}^1 + \mathbf{U}_\mathrm{i}^1$; $\mathbf{\Delta}_\mathrm{i}^1 = \mathbf{OM}_\mathrm{i}^\mathrm{d} - \mathbf{ON}_\mathrm{i}^1$; } `j=2`;
2. `Do` {
   `for i=1 to L` {
   $\mathbf{OM}_\mathrm{i}^\mathrm{j} = \mathbf{OM}_\mathrm{i}^{\mathrm{j}-1} + \mathbf{\Delta}_\mathrm{i}^{\mathrm{j}-1}$; $\mathbf{ON}_\mathrm{i}^\mathrm{j} = \mathbf{OM}_\mathrm{i}^\mathrm{j} + \mathbf{U}_\mathrm{i}^\mathrm{j}$; $\mathbf{\Delta}_\mathrm{i}^\mathrm{j} = \mathbf{OM}_\mathrm{i}^\mathrm{d} - \mathbf{ON}_\mathrm{i}^\mathrm{j}$; }
   `j=j+1;`}
   `While` $\max(||\mathbf{\Delta}_\mathrm{i}^\mathrm{j}||) > \varepsilon$

Stated in simpler terms, the initial guess for the required initial boundary node locations is taken to be the location of the nodes of the desired geometry at room temperature. At each iteration the *residual vector* ($\mathbf{\Delta}_\mathrm{i}^\mathrm{j}$) which measures how far the deformed geometry is from the desired one is evaluated and added to the previous guess of the required initial boundary node location.

This algorithm was tested for the case when the desired geometry at room temperature corresponds to the initial geometry on Figure 1 and the cooling treatment is as shown on Figure 2 **(a)** with $\sigma_{adm} = 5 \times 10^6$ Pa. The convergence history is displayed on Figure 2 **(b)**. For each of the $\Gamma_i$ of the contour $\Gamma$, the maximum of the Euclidean norm of $\mathbf{\Delta}_\mathrm{i}^\mathrm{j}$ is plotted against the number of iteration. The convergence rate is around two decades per iteration which is very satisfactory and the Micron threshold is achieved after three iterations.

*Acknowledgement.* The author gratefully acknowledges the funding of the European Union through the MAGICAL project.

## References

1. A. Markovsky and T.F. Soules. An efficient and stable algorithm for calculating fictive temperatures. *C. Am. Ceram. Soc.*, 67:C–56–C–57, 1984.
2. O.S. Narayanaswamy. A model of structural relaxation in tempering glass. *J. Am. Ceram. Soc.*, 54(10):491–498, 1971.
3. F.O. Sonmez and E. Eyol. Optimal post-manufacturing cooling paths for thermoplastic composites. *Composites: Part A*, 33:301–314, 2002.

# A Mathematical Model for the Mechanical Etching of Glass

J.H.M. ten Thije Boonkkamp

Technische Universiteit Eindhoven, Department of Mathematics and Computer Science tenthije@win.tue.nl

**Summary.** A nonlinear first-order PDE describing the displacement of a glass surface subject to solid particle erosion is presented. The analytical solution is derived by means of the method of characteristics. Alternatively, the Engquist-Osher scheme is applied to compute a numerical solution.

**Key words:** solid particle erosion, kinematic condition, single PDE of first order, characteristic-strip equations, Engquist-Osher scheme

## 1 Introduction

Some modern television displays have a vacuum enclosure, that is internally supported by a glass plate. This plate may not hinder the display function. For that reason it has to be accurately patterned with small trenches or holes so that electrons can move freely from the cathode to the screen. One method to manufacture such glass plates is to cover it with an erosion-resistant mask and blast it with an abrasive powder. In Section 2 we present a nonlinear first-order PDE modelling this so-called *solid particle erosion* process. Next, in Section 3, we present the analytical solution using the method of characteristics. Alternatively, in Section 4, we briefly describe a numerical solution procedure.

## 2 Mathematical Model for Powder Erosion

In this section we outline a mathematical model for solid particle erosion, to produce thin trenches in a glass plate; for more details see [4].

Consider an initially flat substrate of brittle material, covered with a line-shaped mask. We introduce an $(x, y, z)$-coordinate system, where the $(x, y)$-plane coincides with the initial substrate and the positive $z$-axis is directed into the material. A continuous flux of alumina ($Al_2O_3$) particles, directed in the positive $z$-direction, hits the substrate at high velocity and removes material. The position $z = \zeta(x,t)$ of the trench surface at time $t$ is governed by the kinematic condition

$$\zeta_t + \Phi(x) f(\zeta_x) = 0, \quad 0 < x < 1, \quad t > 0, \tag{1}$$

where $x$ is the transverse coordinate in the trench, and where $\Phi(x)$ is the particle mass flux, which will be specified later. The spatial variables $\zeta$ and $x$ are scaled with the trench width and the time $t$ with a characteristic time needed to propagate a surface at normal impact over this width. The function $f = f(p)$ in (1) is defined by

$$f(p) := -\left(1 + p^2\right)^{-k/2}, \tag{2}$$

with $k$ a constant ($2 \leq k \leq 4$). A theoretical model predicts the value $k = 7/3$, [3]. Equation (1) is supplemented with the following initial and boundary conditions:

$$\zeta(x,0) = 0, \qquad 0 < x < 1, \tag{3a}$$

$$\zeta(0,t) = \zeta(1,t) = 0, \quad t > 0. \tag{3b}$$

The boundary conditions in (3b) mean that the trench cannot grow at the ends $x = 0$ and $x = 1$.

## 3 Analytical Solution Method

We can write equation (1) in the canonical form

$$F(x,t,\zeta,p,q) := q - \Phi(x)\left(1 + p^2\right)^{-k/2} = 0, \tag{4}$$

with $p := \zeta_x$ and $q := \zeta_t$. The solution of (4) can be constructed from the following IVP for the characteristic-strip equations [1]

$$\frac{\mathrm{d}x}{\mathrm{d}s} = F_p \qquad = \Phi(x)\,\frac{kp}{(1+p^2)^{k/2+1}}, \qquad x(0;\sigma) = \sigma, \tag{5a}$$

$$\frac{\mathrm{d}t}{\mathrm{d}s} = F_q \qquad = 1, \qquad t(0;\sigma) = 0, \tag{5b}$$

$$\frac{\mathrm{d}\zeta}{\mathrm{d}s} = pF_p + qF_q \quad = \Phi(x)\,\frac{1+(k+1)p^2}{(1+p^2)^{k/2+1}}, \qquad \zeta(0;\sigma) = 0, \tag{5c}$$

$$\frac{\mathrm{d}p}{\mathrm{d}s} = -(F_x + pF_\zeta) = \Phi'(x)\,\frac{1}{(1+p^2)^{k/2}}, \qquad p(0;\sigma) = 0, \tag{5d}$$

$$\frac{\mathrm{d}q}{\mathrm{d}s} = -(F_t + qF_\zeta) = 0, \qquad q(0;\sigma) = \Phi(\sigma), \tag{5e}$$

where $s$ and $\sigma$ are the parameters along the characteristics and the initial curve, respectively. Note that the solution of (5b) and (5e) is trivial, and we find $t(s;\sigma) = s$ and $q(s;\sigma) = \Phi(\sigma)$.

In order to model the finite particle size, which makes that particles close to the mask are less effective in the erosion process, we introduce transition regions of thickness $\delta$. We assume that $\Phi(x)$ increases continuously and monotonically from 0 at the boundaries of the trench to 1 at $x = \delta, 1-\delta$. The parameter $\delta$ is characteristic of the (dimensionless) particle size and a typical value is $\delta = 0.1$. We adopt the simplest possible choice for $\Phi(x)$, *i.e.*,

$$\Phi(x) = \begin{cases} x/\delta & \text{if } \ 0 \leq x < \delta, \\ 1 & \text{if } \ \delta \leq x \leq 1-\delta, \\ (1-x)/\delta & \text{if } \ 1-\delta < x \leq 1. \end{cases} \tag{6}$$

As a result of (6), the growth rate of the surface position close to the mask is smaller than in the middle of the hole. Since $\Phi(0) = \Phi(1) = 0$, we obtain from (5) the solutions $x(t;0) = \zeta(t;0) = 0$ and $x(t;1) = 1$, $\zeta(t;1) = 0$, implying that the boundary conditions (3b) for $\zeta$ are automatically satisfied.

By introducing transition regions, we create intersecting characteristics. Therefore, the solution of (4) can only be a weak solution and it is anticipated that shocks will emerge from the edges $x = \delta$ and $x = 1-\delta$. Let $x = \xi_{\mathrm{s},1}(t)$ and $x = \xi_{\mathrm{s},2}(t)$ denote the location of the shocks at time $t$ originating at $x = \delta$ and $x = 1-\delta$, respectively. Each point $(\xi_{\mathrm{s,i}}(t), t)$ $(i = 1,2)$ on these shocks is connected to two different characteristics that exist on both sides of the shocks. The speed of these shocks is given by the jump condition

$$\frac{\mathrm{d}\xi_{\mathrm{s,i}}}{\mathrm{d}t}[\![p]\!] = -[\![\Phi(x)(1+p^2)^{-k/2}]\!], \quad (i = 1,2), \tag{7}$$

where $[\![p]\!]$ denotes the jump of $p$ across the shock. Thus, we can distinguish the following five regions in the $(x,t)$-plane: the left transition region $0 \leq x \leq \delta$

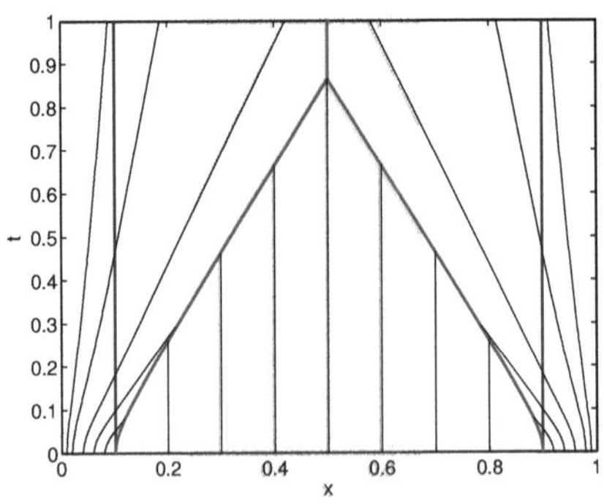


**Fig. 1.** Characteristics and shocks of (5), for $\delta = 0.1$ and $k = 2.33$.

(region 1), the right transition region $1-\delta \leq x \leq 1$ (region 2), the interior domain left of the first shock (region 3), the interior domain right of the second shock (region 4) and the region between the two shocks (region 5); see Fig. 1. Note, that the location of the shocks depends on the solution through (7).

We can derive the analytical solution of (5) in the regions 1, 3 and 5, coupled with a numerical solution of (7). The solution in the other two region follows by symmetry; for more details see [4]. The results are collected in Fig. 2, which gives the solution for $\zeta$ and $p$ at time levels $t = 0.0, 0.1, \ldots, 1.0$ for $\delta = 0.1$ and $k = 2.33$. This figure nicely displays the features of the solution: a slanted surface in the transition regions, a flat bottom in the interior domain and a curved surface in between. Also, inwardly propagating shocks are clearly visible.

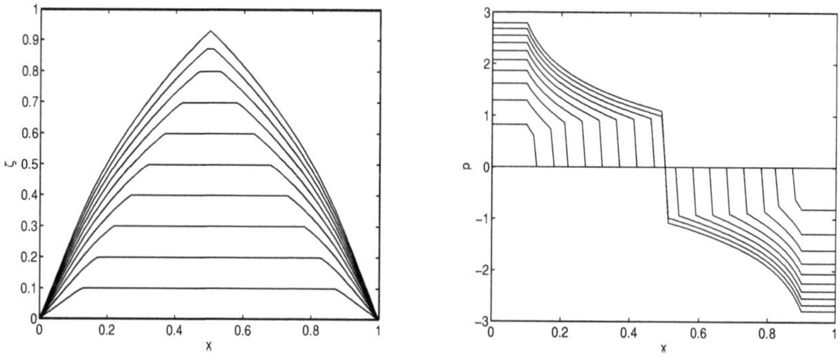


**Fig. 2.** Analytical solution for the surface position (left) and its slope (right). Parameter values are $\delta = 0.1$ and $k = 2.33$.

# 4 Numerical Solution Method

Alternatively, we will compute a numerical solution of (1). To that purpose, we cover the domain $[0, 1]$ with control volumes $V_j = [x_{j-1/2}, x_{j+1/2})$ of equal size $\Delta x = x_{j+1/2} - x_{j-1/2}$. Let $x_j$ be the grid point in the centre of $V_j$. Furthermore, we introduce time levels $t^n = n\Delta t$, with $\Delta t$ being the time step. Let $\zeta_j^n$ denote the numerical approximation of $\zeta(x_j, t^n)$. A finite volume numerical scheme for (1) can be written in the generic form

$$\zeta_j^{n+1} = \zeta_j^n - \Delta t\, \Phi(x_j) F\big(p_{j-1/2}^n, p_{j+1/2}^n\big), \tag{8}$$

with $p_{j\pm1/2}^n$ a numerical approximation of $p(x_{j\pm1/2}, t^n)$ and $F = F(p_\ell, p_\mathrm{r})$ the numerical flux function, that we assume to depend on two values of $p$. The numerical flux $F\big(p_{j-1/2}^n, p_{j+1/2}^n\big)$ is an approximation of $f(p(x_j, t^n))$. We approximate $p_{j\pm1/2}^n$ by central differences and take the Engquist-Osher numerical flux function [2]. For our particular function $f(p)$, which is not convex, it reduces to

$$F(p_\ell, p_\mathrm{r}) = \begin{cases} f(p_\ell) & \text{if} \quad p_\ell, p_\mathrm{r} \geq 0, \\ f(p_\mathrm{r}) & \text{if} \quad p_\ell, p_\mathrm{r} < 0, \\ f(0) & \text{if} \quad p_\ell < 0 \leq p_\mathrm{r}, \\ f(p_\ell) + f(p_\mathrm{r}) - f(0) & \text{if} \quad p_\mathrm{r} < 0 \leq p_\mathrm{r}. \end{cases} \tag{9}$$

The numerical solution computed with (8) and (9) is presented in Fig. 3. It is computed for $k = 2.33$ and $\delta = 0.1$ with grid size $\Delta x = 5 \times 10^{-3}$ and time step $\Delta t = 5 \times 10^{-3}$. Clearly, it is an accurate approximation of the exact solution in Fig. 2. Finally, both the analytical and numerical solution are in good qualitative agreement with experimental results; see also [4].

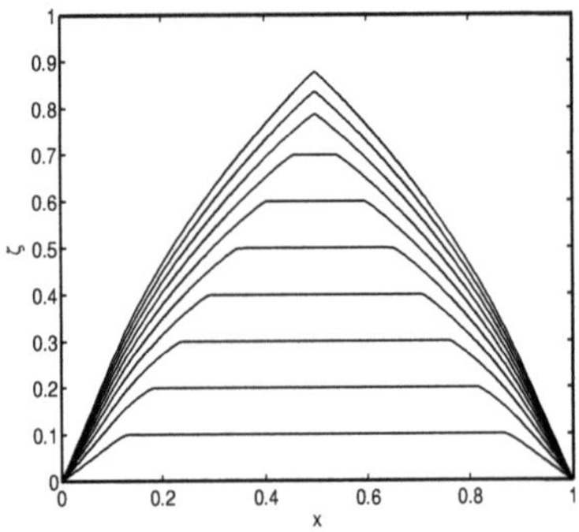


**Fig. 3.** Numerical solution for the surface position. Parameter values are $\delta = 0.1$ and $k = 2.33$.

## References

1. J. Kevorkian. *Partial Differential Equations: Analytical Solution Techniques.* Brooks/Cole, Pacific Grove, 1990.
2. J.A. Sethian. *Level Set Methods and Fast Marching Methods.* Cambridge University Press, Cambridge, 1999.
3. P.J. Slikkerveer, P.C.P. Bouten, F.H. in 't Veld, and H. Scholten. Erosion and damage by sharp particles. *Wear*, 217:237–250, 1998.
4. J.H.M. Ten Thije Boonkkamp. An analytical solution for mechanical etching of glass by powder blasting. *Journal of Engineering Mathematics*, 43:385–399, 2002.

# FPM + Radiation = Mesh-Free Approach in Radiation Problems

A. Wawreńczuk

Fraunhofer Institut Technound Wirtschaftsmathematik Gottlieb-Daimler Str. 49 Kaiserslautern, Germany wawrencz@itwm.fhg.de

**Summary.** This article discusses mathematical outlines of the numerical project combining particle method with radiation models in order to simulate glass cooling process. Its initial part gives a sketch of the particle Finite Pointset Method (FPM) [1], the next one debriefs the radiation models considered to implement in the method framework and the final one presents some preliminary, qualitative results of current research.

**Key words:** mesh-free methods, particle methods, FPM, radiation, glass cooling

## 1 Project

Current research concentrates on numerical simulations of glass cooling process. Glass has many applications and some of them raise high demands in respect to its expected quality and properties. Hence, there is a strong interest of glass-industry in tools which could help to master controlling and time-predicament of this process. The ultimate goal of this project is possible creation of industrially applicable tool addressing this subject. Technically we try to achieve this by combining the power of FPM, a method extensively developed at Fraunhofer Institut [1], with radiation models presented briefly in the next section [2].

During cooling glass undergoes a transition from viscous fluid to amorphous solid body. Therefore at the initial stages of the process we have to deal with moving boundaries. That is the place where mesh-free methods can be helpful. Their main advantage is that they allow to avoid the mesh and all mesh-related operations, like reconstructions and/or surface tracking, which significantly simplifies code structure. Simply, the changes in surface geometry may be followed by straightforward observation of moving particles.

The cooling itself obviously involves heat transfer by conduction or/and radiation. While the first mechanism does not create any problems from mathematical standpoint then the situation is different for radiation. The latter one is dominant for high temperatures, so is required for full description, and its 3D and global character creates additional difficulties which must be solved in order to deal with it successfully.

## 2 FPM

FPM is a particle method, where particles used as a representation of a system carry physical properties (energy, momentum, etc.) and possess their own characteristics (density, temperature, velocity etc.). The prescribed quantities change according to the governing equations:

- mass conservation equation

$$\frac{d\rho}{dt} + \rho \cdot \nabla \mathbf{u} = 0 \tag{1}$$

- momentum equation

$$\frac{d(\rho \mathbf{u})}{dt} + (\rho \mathbf{u}) \cdot \nabla \mathbf{u} = \nabla(\mathbf{S}) - \nabla(p) + \rho \mathbf{g} \tag{2}$$

- energy equation

$$\frac{d(\rho E)}{dt} + (\rho E) \cdot \nabla \mathbf{u} = \nabla(\mathbf{S} \cdot \mathbf{u}) - \nabla(p \cdot \mathbf{u}) + \rho \mathbf{g} \cdot \mathbf{u} \tag{3}$$

- eventual additional conditions (e.g. incompressibility constraint).

  Other most important features of the method may be itemized as follows:

- Particles are understood only as a **mathematical representation** of macroscopic fluid elements (they are NOT fluid elements in a physical sense).
- Approximation by Weighted Least Squares (WLS) – allows to compute values of physical fields and **discretise differential operators** at **arbitrary** point of the computational domain using only particles (no mesh necessary here).
- Boundary conditions introduced by boundary particles (possessing proper characteristic: density, velocity, etc.) and boundary elements (segments composing poly-line in 2D, patches in 3D).

## 3 Radiation models

FPM in our project is a general dynamics engine for solving set of equations describing time behavior of the considered system. This section presents shortly the models of radiation that are supposed to extend the capabilities of the basic program by delivering all radiation-related data required for time integration of the problem.

The main task is to solve following Energy Transfer Equation (ETE):

$$c_m \rho_m \frac{\partial T}{\partial t} = \nabla \cdot (k_h \nabla T - \mathbf{q}(\mathbf{r}, T)) \tag{4}$$

with boundary condition:

$$k_h \frac{\partial T}{\partial n} = h(T_a - T) + \varepsilon \pi \left( \frac{n_a}{n_g} \right)^2 \cdot \int_{opaquc} [B(T_a, \lambda) - B(T, \lambda)] \, \mathrm{d}\lambda \tag{5}$$

where $c_m$ is a specific heat at constant pressure, $\rho_m$ density, $k_h$ thermal conductivity, $T$ temperature, $n$ refractive index (a–surrounding, g–glass), $B$ Planck's function and $\lambda$ wavelength. We are particularly interested in function: $\mathbf{q}$ which represents radiative heat flux. All models mentioned in following sections have only one purpose: determine $\mathbf{q}$ or more specifically $\nabla \cdot \mathbf{q}$ (divergence of radiative heat flux).

### 3.1 Rosseland approximation

This rather straightforward approximation describes the radiative term in a diffusion-like manner:

$$\mathbf{q} = -k_{rad} \nabla T, \tag{6}$$

where

$$k_{rad} = \frac{16 n^2 \sigma}{3 \kappa_{ros}} T^3 \text{ and } \frac{1}{\kappa_{ros}(T)} = \frac{\int_0^\infty \frac{1}{\kappa(\lambda)} \frac{d\,B(T,\lambda) d\lambda}{d\,T}}{\int_0^\infty \frac{d\,B(T,\lambda) d\lambda}{d\,T}}$$

Its main disadvantage is that its usage, though easy, should be restricted only to optically thick media which not always takes place in industrial applications.

### 3.2 Radiative Transfer Equation (RTE) approximations

The next approximations may be classified into a single category as having a common point: all try to find directly $\nabla \cdot \mathbf{q}(\mathbf{r}, T)$ using additional Radiative Transfer equation (RTE):

$$\Omega \cdot \nabla I(\mathbf{r}, \Omega) = \kappa B(T(\mathbf{r})) - \kappa I(\mathbf{r}, \Omega) \tag{7}$$

with boundary condition (b.c.):

$$I(\mathbf{r}, \Omega) = \varrho I(\mathbf{r}, \Omega') + (1 - \varrho) B(T_a)) \tag{8}$$

Here $I$ stands for intensity, $\mathbf{r}$ for position, $\Omega$ a direction unit vector (versor), $T_a$ surrounding temperature, $\varrho$ reflectivity and $B(T)$ is a Planck's function integrated over wavelengths. Found intensity function may be used directly for calculation of $div\ (\mathbf{q})$ according to formula:

$$\nabla \cdot \mathbf{q}(\mathbf{r}, T) = \kappa \left[ 4\pi\, B(T(\mathbf{r})) - \int_{S^2} I(\mathbf{r}, \Omega) \mathrm{d}\Omega \right] \tag{9}$$

($S^2$ means integration over unit sphere).

## Formal Solution (FS)

The approximation uses the fact that formally the solution of equation (7) may be written as:

$$I(\mathbf{r}, \Omega) = B(T_a) \exp\left(-\kappa d(\mathbf{r}, \Omega)\right) + \kappa \int_0^{d(\mathbf{r},\Omega)} B(T(\mathbf{r} - s\Omega))\ \exp\left(-\kappa s\right)\ \mathrm{d}s \tag{10}$$

(with $d(\mathbf{r}, \Omega)$ as a distance between point r and boundary in given direction) and further:

$$\begin{aligned} I(\mathbf{r}, \Omega) &= B(T_a) \exp\left(-\kappa d(\mathbf{r}, \Omega)\right) + B(T(\mathbf{r}))\ \left(1 - \exp\left(-\kappa d(\mathbf{r}, \Omega)\right)\right) \\ &- \frac{1}{\kappa} \frac{\mathrm{d}B(T(\mathbf{r}))}{\mathrm{d}t} \nabla T(\mathbf{r}) \cdot \Omega \left[1 - (1 + \kappa d(\mathbf{r}, \Omega)) \exp\left(-\kappa d(\mathbf{r}, \Omega)\right)\right]. \end{aligned} \tag{11}$$

## Improved approximation

In this method the divergence of the radiative heat flux may be computed explicitly from (see [2]):

$$\begin{aligned} \nabla \cdot \mathbf{q}(\mathbf{r}) = \kappa \left[B(T(\mathbf{r})) - B(T_a)\right] \cdot \int_{S^2} (1 - \rho(\Omega)) \exp(-\kappa d(\mathbf{r}, \Omega)) \mathrm{d}\Omega \\ - \frac{\mathrm{d}B(T)}{\mathrm{d}T} \cdot \int_{S^2} (1 - \varrho(\Omega)) \exp(-\kappa d(\mathbf{r}, \Omega)) \nabla T \cdot \Omega \mathrm{d}\Omega \\ - \nabla \cdot \left( \frac{1}{\kappa} \frac{\mathrm{d}B(T)}{\mathrm{d}T} \mathbf{A} \cdot \nabla T \right), \end{aligned} \tag{12}$$

where

$$\mathbf{A} = \begin{pmatrix} a_1 & 0 & 0 \\ 0 & a_2 & 0 \\ 0 & 0 & a_3 \end{pmatrix} \tag{13}$$

$$a_i(\mathbf{r}) = \int_{S^2} \Omega_i^2 \left[1 - (1 + \kappa_k d(\mathbf{r}, \Omega)) \exp(-\kappa_k d(\mathbf{r}, \Omega))\right] \mathrm{d}\Omega. \tag{14}$$

## Ray tracing

This methods seeks for approximate solution of RTE in a step-wise manner following the rays along the points $\mathbf{r}_l$, $l = 0 \ldots n$ ($\mathbf{r}_0$ – boundary, $\mathbf{r}_n = \mathbf{r}$). In this case we obtain formula:

$$\begin{aligned} I(\mathbf{r}_{l+1}, \Omega_i) = \left[I(\mathbf{r}_l, \Omega_i) - B(T(\mathbf{r}_l))\right] \exp(-\kappa h_l) + B(T(\mathbf{r}_{l+1})) - \\ - \frac{B(T(\mathbf{r}_{l+1})) - B(T(\mathbf{r}_l))}{\kappa h_l} (1 - \exp(-\kappa h_l)), \\ h_l = |\mathbf{r}_{l+1} - \mathbf{r}_l| \end{aligned} \tag{15}$$

# 4 Results

Numerical schemes presented in previous sections were used in preliminary simulations of glass cooling. The result temperature fields (initial and intermediate one) are presented on Fig. 1. The observed behavior of temperature

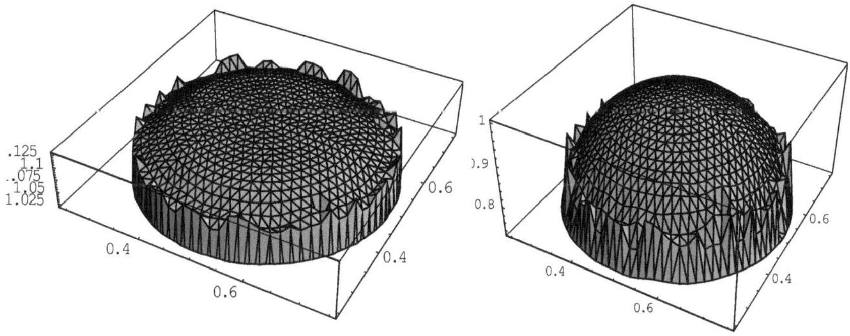


**Fig. 1.** Cooling from initial constant temperature of $T_0 = 873\text{K}$. Temperature fields for two time instants (left – initial, right – intermediate). Both pictures are normalised in respect to the latter time.

field was satisfactory. It allows to hope that full simulations, actually in preparation, will yield also good quantitative results.

*Acknowledgement.* The research is financed by EU in the frame of "**MAGICAL**" program. Author wants to thank J. Kuhnert and N. Siedow for their assistance during the research.

## References


1. Tiwari S. and Kuhnert J. *Grid free method for solving Poisson equation.* preprint berichte des Fraunhofer ITWM Nr.25. Kaiserslautern, Germany, 2001.
2. Lentes T. and Siedow N. Three–dimensional radiative heat transfer in glass cooling processes. *Glastech. Ber. Sci. technol.*, 72(6), 1999.

# Part VI

## Theme: Geophysics

# Multiscale Methods and Streamline Simulation for Rapid Reservoir Performance Prediction

J.E. Aarnes, V. Kippe, and K.-A. Lie

SINTEF ICT, Applied Mathematics, P.O. Box 124 Blindern, N-0314 Oslo, Norway.
Email: `Jorg.Aarnes@sintef.no`, `Vegard.Kippe@sintef.no`,
`Knut-Andreas.Lie@sintef.no`

**Summary.** We introduce a novel multiscale approach for reservoir simulation as an alternative to industry-standard upscaling methods. In our approach, reservoir pressure and total velocity is computed separately from the fluid transport. Pressure is computed on a coarse grid using a multiscale mixed-finite element method that gives a mass-conserving velocities on a fine subgrid. The fluid transport is computed using streamlines on the underlying fine geogrid.

**Key words:** multiscale methods, porous media, upscaling, streamlines.

## 1 Introduction

The size of geomodels used for reservoir description typically exceeds by several orders of magnitude the capabilities of conventional reservoir simulators based upon finite differences. These simulators therefore employ upscaling techniques that construct coarsened reservoir models with a reduced set of geophysical parameters. In this way the size of the simulation model is reduced so that simulations can run within an acceptable time-frame.

Streamline methods are gaining in popularity and have a potential of simulating much larger reservoir models than what is possible using traditional finite difference simulators. Streamline methods are based upon a fractional flow formulation, where the model is split into an elliptic/parabolic pressure equation and hyperbolic fluid transport equations. For immiscible, incompressible fluids and negligible gravity and capillary forces, the equations read

$$\nabla \cdot v = q, \qquad v = -K\lambda_t \nabla p, \tag{1}$$

$$\varphi \frac{\partial S}{\partial t} + \nabla \cdot f_w(S) v = q_w. \tag{2}$$

Here $p$ denotes pressure, $v$ the total velocity, $S$ water saturation, $K$ rock permeability, $\lambda_t(S)$ total mobility, and $\varphi$ rock porosity. The two equations are solved sequentially: first the pressure equation (1) is solved to give a velocity field, by which the saturations can be transported according to (2), and so on.

A major obstacle in applying streamline methods to large geomodels is the need for accurate and efficient solution of the pressure equation (1). In particular, the pressure solver must be locally (and globally) mass-conservative and should handle: (i) irregular grids that conform to geological structures; (ii) strongly heterogeneous and anisotropic formations; and (iii) flows with large dynamic aspect ratios. Mixed finite element methods (MFEM) and multi-point flux-approximation finite-volume methods (MPFA) are examples of methods that handle these properties, and cover the most widely used methods for elliptic problems where mass preservation is an issue.

Here we present a new simulation method for incompressible, immiscible two-phase flow on Cartesian grids. Pressure and velocities are computed using a multiscale, mixed finite-element method (MsFEM) [1, 3], where the pressure is computed on a coarse grid and a mass-conservative velocity field is computed on the underlying fine grid, using numerically constructed base functions with subgrid resolution on the coarse grid. Together with streamline computation of fluid transport, this gives an efficient and robust method that resolves detailed flow patterns on the underlying fine grid. A more detailed study of this multiscale method is presented in [2]. Our main point here is to indicate that the combination of multiscale pressure solvers and streamline methods has a great potential for bridging the gap between high-resolution geomodels and the capabilities of current reservoir simulators.

## 2 Streamline Method

Streamlines are flow-paths traced out by a particle being passively advected by an external flow field such that the streamline is tangential to the flow velocity at every point. The streamlines can be parametrised by the *time-of-flight* $\tau$, which measures the travel time along each streamline. In our case,

$$v \cdot \nabla \tau = \varphi \qquad \text{or equivalently} \qquad \partial \tau = \varphi / |v| \, ds. \tag{3}$$

Together with the bistream functions $\psi$ and $\chi$, which satisfy $u = \nabla\psi \times \nabla\chi$, the streamlines define a formal spatial coordinate transform. Applied to the saturation equation (2), for which $u = v/\varphi$, this transformation gives

$$S_t + f(S)_\tau = 0. \tag{4}$$

Streamline simulators compute the fluid transport by solving one-dimensional equations like (4) along streamlines in 3D. Here we use a very efficient *front-tracking method* [6] to solve (4). The method starts from piecewise initial data, approximates the flux function by a piecewise linear function, and solves the corresponding Cauchy problem exactly.

## 3 Multiscale Mixed Finite-Elements

The mixed formulation of (1) over a domain $\Omega \in \mathbb{R}^3$ reads: find $(p, v) \in L^2(\Omega) \times H_0^{1,\mathrm{div}}(\Omega)$ such that

$$\begin{aligned} \iiint_\Omega (K\lambda_t)^{-1} v \cdot u \, dx - \iiint_\Omega p \, \nabla \cdot u \, dx = 0, \\ \iiint_\Omega l \, \nabla \cdot v \, dx = \iiint_\Omega ql \, dx, \end{aligned} \tag{5}$$

for all $u \in H_0^{1,\mathrm{div}}(\Omega)$ and $l \in L^2(\Omega)$. In a mixed-finite element method, the approximation space for $v$ is spanned by a finite set of base functions $\{\psi\} \subset H_0^{1,\mathrm{div}}(\Omega)$; for instance, a set of piecewise linear functions as in the Raviart–Thomas elements of lowest order. In the multiscale method, the base functions are computed numerically by solving a subgrid problem for each interface $\Gamma_{ij}$ between two coarse grid blocks $T_i$ and $T_j$

$$\begin{aligned} (\nabla \cdot \psi_{ij})|_{T_i} &= \begin{cases} 1/|T_i|, & \text{if } \int_{T_i} q \, dx = 0, \\ q / \int_{T_i} q \, dx, & \text{otherwise,} \end{cases} \\ (\nabla \cdot \psi_{ij})|_{T_j} &= \begin{cases} -1/|T_j|, & \text{if } \int_{T_j} q \, dx = 0, \\ -q / \int_{T_j} q \, dx, & \text{otherwise} \end{cases} \end{aligned} \tag{6}$$

with no-flow boundary conditions $\psi_{ij} \cdot n = 0$ on $\partial(T_i \cup T_j)$. These numerically generated base functions guarantee a velocity approximation with subgrid resolution. The approximation is mass conservative on the subgrid if the subgrid problems (6) are solved with a mass-conservative method. The base functions $\psi_{ij}$ will generally be time dependent since they depend on $\lambda_t$, which is time dependent through $S(x,t)$. For incompressible two-phase flow it is sufficient to regenerate only a small portion of the base functions in each pressure step since the mobility only varies significantly near strong saturation fronts.

## 4 Numerical Results

To demonstrate that our multiscale method is a viable and robust approach, we present numerical results for Model 2 in the tenth SPE comparative solution project [4]. The model was designed as a benchmark for various upscaling techniques and contains a stack of two heterogeneous formations, see Fig. 1. Both formations have large permeability variations, 8–12 orders of magnitude, but are qualitatively different. The Tarbert formation is smooth, and therefore not too hard to upscale. The Upper Ness formation is fluvial with narrow and intertwined flow channels of high permeability.

We compare our simulation results with a reference solution obtained by direct simulation on the fine grid using a standard two-point finite-volume

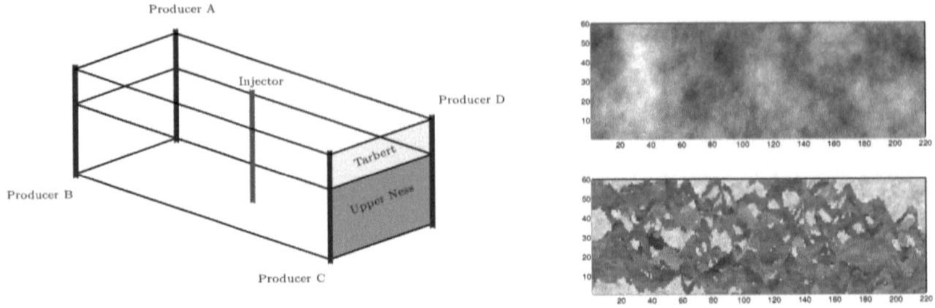


**Fig. 1.** Schematic of the reservoir model used in [4]. The reservoir dimensions are $1200 \times 2200 \times 170$ ft., and the model consists of $60 \times 220 \times 85$ grid cells. The top and bottom plots to the right depicts the logarithm of the horizontal permeability in the top layer of the Tarbert formation and the bottom layer of the Upper Ness formation.

method. We also compare with the nested gridding method of Gautier et al. [5], which can be considered as the upscaling-based analogue of our method. In the nested-gridding method the absolute mobility $(K\lambda_t)$ is upscaled by solving local flow problems. Secondly, the pressure equation is solved on the coarse grid using the upscaled absolute mobilities. Finally, the coarse-grid fluxes are used to determine boundary conditions for local subgrid problems that are solved to obtain a mass-conservative velocity on the subgrid scale. The fluid transport is solved using streamlines for all three methods.

Figure 2 shows a plot of the fraction of water in the produced fluid (water cut) as a function of time for 2000 days of production. The time steps are 25 days up to day 250, 50 days up to day 500, 100 days up to day 1000, and then 200 days. The performance of our multiscale method is remarkably good; the match is almost exact for all four producers and the fine-scale flow channels are reproduced to a large extent as can be seen in Fig. 3. Although the nested-gridding method has subgrid resolution, it does not account for the coupling between fine-grid and coarse-grid effects and therefore fails to reproduce the individual water cuts correctly.

*Acknowledgement.* The research was funded by the Research Council of Norway under grant number 158908/I30.

## References

1. J. E. Aarnes. On the use of a mixed multiscale finite element method for greater flexibility and increased speed or improved accuracy in reservoir simulation. *Multiscale Model. Simul.*, 2(3):421–439, 2004.
2. J.E. Aarnes, V. Kippe, and K.-A. Lie. Mixed multiscale finite elements and streamline methods for reservoir simulation of large geomodels. *Adv. Wat. Resour., submitted.*

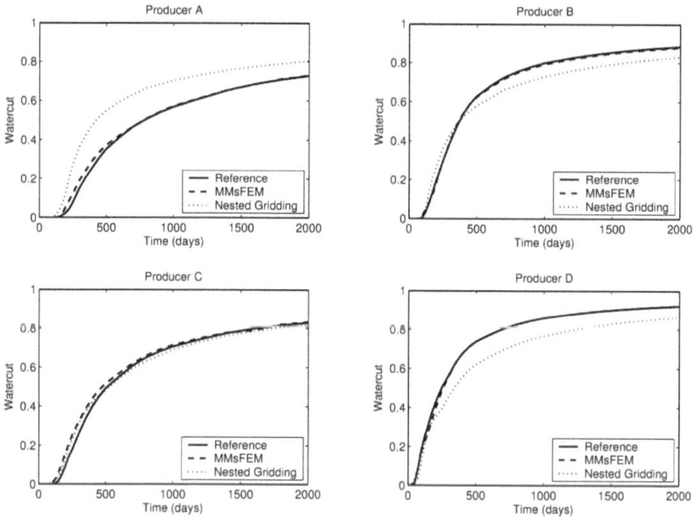


**Fig. 2.** Water cut curves after 2000 days of simulation.

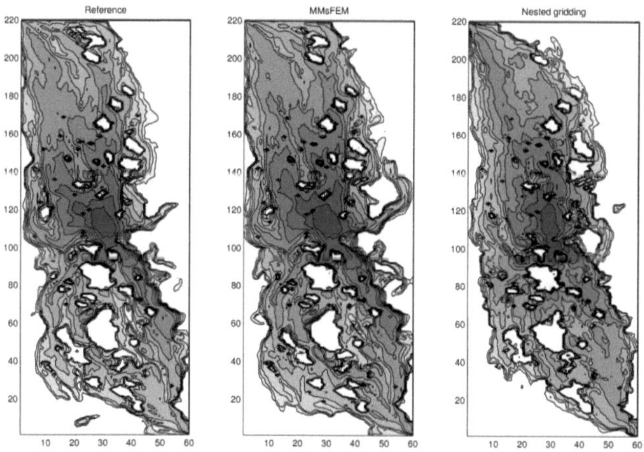


**Fig. 3.** Water saturation in the bottom layer after 800 days.


3. Z. Chen and T.Y. Hou. A mixed multiscale finite element method for elliptic problems with oscillating coefficients. *Math. Comp.*, 72:541–576, 2003.
4. M.A. Christie and M.J. Blunt. Tenth SPE comparative solution project: A comparison of upscaling techniques. SPE 72469, url: www.spe.org/csp, 2001.
5. Y. Gautier, M. J. Blunt, and M.A. Christie. Nested gridding and streamline-based simulation for fast reservoir performance prediction. *Comp. Geosci.*, 3:295–320, 1999.
6. H. Holden and N.H. Risebro. *Front Tracking for Hyperbolic Conservation Laws.* Springer, New York, 2002.

# Part VII

## Theme: Financial Mathematics

# ONE FOR ALL
# The Potential Approach to Pricing and Hedging

L.C.G. Rogers

University of Cambridge `lcgr1@cam.ac.uk`

## 1 Introduction

Any academic working in mathematics, physics or engineering will know examples of bright PhD students who have turned their backs on the profession that nurtured them, and sought out instead the more uncertain but more lucrative world of financial services. What do they find themselves doing when they get there? This largely depends on the job they get (for a large investment bank is a diverse employer, with many different roles and required skills), but most often banks are employing people with outstanding quantitative skills because those are the skills they need.

In particular, a big part of the business of an investment bank is in making and selling various *derivatives*[1]. A very simple example is a (zero-coupon) *bond*, where the seller undertakes to pay 1 at a fixed time $T$ in the future, in return for a payment (typically less than 1!) made by the buyer at time 0. In more complicated examples, the amounts to be paid can be random, as in a European put option, which gives the buyer the right to sell one unit of a named stock at a specified time (the *expiry* of the contract) for a specified price (the *strike* price). However, the holder is not compelled to exercise this right, so he will clearly do so at expiry if and only if the stock is trading for less than the strike, because he is then able to buy the stock, and pocket the difference between the current price and the strike price he gets by selling it to the option writer. The timings of payments can also be random, as in an American put, where the holder of the option is allowed to sell the stock for the strike price at *any* time before the expiry of the option.

What should be the price the bank charges for a derivative? At one level, the answer is 'As much as the market will bear!' but competition precludes arbitrary profit, and the bank needs to have some idea of how cheaply they can sell the derivative and still make a profit. Quite viable models and methods

[1] A derivative is a well-defined financial deal between two parties where the timings and amounts of payments to be made are specified in the contract.

exist to help banks answer such questions, and the job of a *quant* (as your former PhD student is now known) is to use such models to come up with prices, and to extend them to deal with the novel derivatives that are often requested by clients. Such models will be used to find lower bounds for the prices the banks will charge. In very liquid (active) markets, the actual price charged may be very close to that computed in the model, but in less liquid markets it will typically be a lot higher. This is because the bank does not simply sell a derivative and then wait for events to determine what their part of the deal is going to cost them; they engage in *hedging*, which is to say they take some offsetting position in other financial assets so as to cancel out (as best they can) any gains and losses to be made on the derivative. If the market is highly liquid, they are better able to do this than in an illiquid market, and the price charged reflects this. Indeed, it is largely true that the price a bank will charge for a derivative is the cost they face in hedging it, rather than anything that a model tells them.

So pricing and hedging forms a large part of the work your former PhD student will be doing. In this introductory paper, we will see in Section 2 some of the basic notions of pricing developed from a perfectly plausible (though slightly unconventional) axiomatic standpoint; and we will see in Section 3 how the resulting expressions suggest an approach to modelling asset prices. This approach (known as the *potential approach*) is again rather unconventional, but has overwhelming advantages in the modelling of complex cross-currency derivatives (for example), that more conventional approaches struggle with; Section 5 explains why. The potential modelling approach is actually extremely general, and specific choices have to be made to apply it in practice - these are discussed in Sections 4 and 6. We discuss how such models can be calibrated in Section 7, and present the results of such a calibration in Section 8. As befits an unconventional approach, the form that hedging takes (explained in Section 9) is also unconventional, but perfectly tractable. Section 10 concludes and presents further directions for research.

Very little that is in this paper is new; indeed, most of it is in one or more of [6, 5, 7, 9, 10]. I have said that the potential approach is unconventional, and it is; for fine expositions of the conventional approach to derivative pricing and hedging you can consult [1, 4, 2] (for an account with more emphasis on the economic origins), or many others - there are plenty. These accounts will tell you a lot about how pricing and hedging is done today; this account will tell you a little about how it may well be done tomorrow.

## 2 Generalities about pricing

We put ourselves in a filtered probability space $(\Omega, (\mathcal{F}_t)_{t\geq 0}, \mathbb{P}))$ and consider *market pricing* operators $(\pi_{tT})_{0\leq t\leq T}$ for contingent claims[2]:

[2] A contingent claim is local jargon for a random variable.

$$\pi_{st} : L^\infty(\mathcal{F}_t) \to L^\infty(\mathcal{F}_s) \qquad (0 \le s \le t).$$

The idea is that if $Y$ is some bounded random variable which is $\mathcal{F}_t$-measurable, then the time-$s$ market price of $Y$ is $\pi_{st}(Y)$, again a random variable (because what we had observed up to time $s$ would affect what we thought this contingent claim was worth), and again bounded - obviously.

We shall assume that the pricing operators $(\pi_{st})_{0\le s\le t}$ satisfy certain axioms:

(A1) Each $\pi_{st}$ is a bounded positive linear operator from $L^\infty(\mathcal{F}_t)$ to $L^\infty(\mathcal{F}_s)$;
(A2) If $Y \in L^\infty(\mathcal{F}_t)$, $Y \ge 0$, then

$$\pi_{0t}(Y) = 0 \iff P(Y > 0) = 0.$$

(*no arbitrage*)
(A3) For $0 \le s \le t \le u$, $Y \in L^\infty(\mathcal{F}_u)$, $X \in L^\infty(\mathcal{F}_t)$,

$$\pi_{su}(XY) = \pi_{st}(X\pi_{tu}(Y))$$

(*intertemporal consistency*)
(A4) If $(Y_n) \in L^\infty(\mathcal{F}_t)$, $|Y_n| \le 1$, $Y_n \uparrow Y$ then $\pi_{st}(Y_n) \uparrow \pi_{st}(Y)$ (*continuity*)

REMARKS. Axiom (A1) says that the price of a non-negative contingent claim will be non-negative, and the price of a linear combination of contingent claims will be the linear combination of their prices - which are reasonable properties for a market price. Axiom (A2) says that a contingent claim that is almost surely worthless when paid, will be almost surely worthless at all earlier times (and conversely) - again reasonable. The third axiom, (A3), is a 'consistency' statement; the market prices at time $s$ for $XY$ at time $u$, or for $X$ times the time-$t$ market price for $Y$ at time $t$, should be the same, for any $X$ which is known at time $t$. The final axiom is a natural 'continuity' condition which is needed for technical reasons.

**Theorem 1.** *Assuming Axioms (A1)–(A4), there exists a strictly positive process $(\zeta_t)_{t\ge 0}$ such that the pricing operators $\pi_{st}$ can be expressed as*

$$\pi_{st}(Y) = \frac{E_s[\zeta_t Y]}{\zeta_s} \qquad (0 \le s \le t). \tag{1}$$

*If we also assume (A5) For all $0 \le s \le t$, $\pi_{st}(1) \le 1$ (where 1 denotes the constant function identically equal to 1) then $\zeta$ is a positive supermartingale:*

$$\zeta_s \ge E_s\zeta_t \qquad (0 \le s \le t)$$

REMARKS. This result is the famous 'Fundamental Theorem of Asset Pricing' (FTAP) - or at least its *conclusion* is the same as that of the FTAP, though its hypotheses are quite different. The FTAP is proved from the hypothesis that the market does not admit any arbitrage[3]. This is a perfectly sensible axiomatic staring point, but not the only one possible, and the approach taken here shows that if we adopt the equally-sensible axioms (A1)–(A4) (which need no subtle modification for a continuous-time setting), then we harvest the conclusion of the FTAP using little more than basic measure theory. Notice how short the proof is!

*Proof.* Firstly, for any $T > 0$, the map

$$A \mapsto \pi_{0T}(I_A)$$

defines a non-negative measure on the $\sigma$-field $\mathcal{F}_T$, from the linearity and positivity (A1) and the continuity property (A4). Moreover, this measure is absolutely continuous with respect to $\mathbb{P}$, in view of (A2). Hence there is a non-negative $\mathcal{F}_T$-measurable random variable $\zeta_T$ such that

$$\pi_{0T}(Y) = E[\zeta_T Y]$$

for all $Y \in L^\infty(\mathcal{F}_T)$. Moreover, $\mathbb{P}[\zeta_T > 0] > 0$, because of (A2) again. Now we exploit the consistency condition (A3); we have

$$\pi_{0t}(X\pi_{tT}(Y)) = E[X\zeta_t\pi_{tT}(Y)] = \pi_{0T}(XY) = E[XY\zeta_T].$$

Since $X \in L^\infty(\mathcal{F}_t)$ is arbitrary, we deduce that

$$\pi_{tT}(Y) = E_t[Y\zeta_T]/\zeta_t,$$

as claimed. The final statement that $\zeta$ is a positive supermartingale under (A5) is now immediate.

REMARKS. (i) The form (1) shows that if we write $Y_t \equiv \pi_{tT}(Y)$ for some fixed $Y \in L^\infty(\mathcal{F}_T)$ then

$$\zeta_t Y_t = E_t[\zeta_T Y] \text{ is a martingale.}$$

Conventionally, the process $\zeta$ (known as the *state-price dnsity process*) is represented as

$$\zeta_t = \exp(-\int_0^t r_s ds) Z_t,$$

where $r_t$ is the instantaneous rate of interest at time $t$, and $Z_t$ is a positive martingale, which is interpreted as a change of measure, from the reference probability $\mathbb{P}$ to some new 'pricing' probability, also referred to as an *equivalent martingale measure*, because it is equivalent to the original measure (both have

[3]The intuitive idea of an arbitrage as 'something for nothing' is easy to formalise mathematically in discrete time, but the exact definition for a continuous-time setting was elusive and subtle, and was indeed a key part of the difficulty experienced in proving this celebrated result. See Delbaen-Schachermayer [3] for the definitive form.

the same null sets), and because in the new measure the discounted prices of all traded assets become martingales.

(ii) Though we have looked at pricing systems which are linear in the contingent claim, there is good reason not to restrict exclusively to this property, because individual agent's prices for contingent claims are generally concave - you might be prepared to pay \$2 for 1l of icecream, but does this mean you would be prepared to pay \$200 for 100l of icecream?! Taking this into account leads us into ideas of economic equilibrium; often, the anaylsis of an equilibrium can be enormously complicated, and the equilibrium prices arrived at will depend on the nature of all the agents in the market. However, the equilibrium prices, being *marginal* prices, will be linear in the contingent claim.

## 3 The potential approach

Theorem 1 and the form (1) of the price of a contingent claim suggests a simple and natural approach to modelling (and pricing) in a financial market: *model* $\zeta$*!*

The process $\zeta$ is a positive supermartingale (if we make the natural further assumption A5). The expression (1) allows us to write down the price $B(t,T)$ at time $t$ of a zero-coupon bond maturing at later time $T$; in this case, $Y \equiv 1$, so we have

$$B(t,T) = E_t[\zeta_T]/\zeta_t. \tag{2}$$

If we make the further assumption (financially very natural) that as the maturity $T$ of the bond tends to infinity the current value of it tends to zero, we see that the positive supermartingales $\zeta$ that we are considering have to satisfy the further condition

$$\lim_{T\to\infty} E\zeta_T = 0; \tag{3}$$

a positive supermartingale satisfying this condition is called a *potential*, whence the name of this approach. Under a mild further condition[4] a potential can be represented as

$$\zeta_t = E_t[A_\infty - A_t] = E_t(A_\infty) - A_t \tag{4}$$

for some previsible increasing integrable process $A$. The potential approach therefore requires us to find tractable forms of previsible increasing process to build models. We do not need to look very far; by (1), prices are to be expressed in terms of conditional expectations of random variables whose values are yet to be revealed, and for tractability we will want such conditional expectations to be expressible simply in terms of a few variables. Thus we are inevitably drawn towards modelling in the context of Markov processes.

[4]Explicitly, that the process $\zeta$ should be of class (D) - see, for example, [10].

## 4 Markov processes and potentials

If $(X_t)$ is a Markov process on a state space $\mathcal{X}$, and $f : \mathcal{X} \to [0, \infty)$, then for any $\alpha > 0$ we may consider the increasing process

$$A_t = \int_0^t e^{-\alpha s} f(X_s)\, ds. \tag{5}$$

This is adapted and continuous (therefore previsible), and under mild conditions on $f$ (uniform boundedness will be sufficient but far from necessary) it will also be integrable. From the discussion of Section 3 we can use this to build a pricing model; we find that

$$\zeta_t = E_t\Big[ \int_t^\infty e^{-\alpha s} f(X_s) ds \Big] = e^{-\alpha t} R_\alpha f(X_t), \tag{6}$$

where $(R_\alpha)_{\alpha>0}$ is the so-called *resolvent*[5] of the Markov process.

Though this is not by any means the only way that we could use a general Markov process to build a potential pricing model (see [6] for other ideas), it is sufficiently explicit for us to appreciate immediately how flexible and simple this modelling methodology will be:

(i) We can choose *any* non-negative function $f$ on the state space, and *any* positive $\alpha$.

(ii) The decomposition (4) of $\zeta$ into a martingale less an increasing process takes a very simple form. If we make the usual interpretion of the supermartingale $\zeta$ as the product of a positive change-of-measure martingale $Z$ times the discount factor $\exp(-\int_0^t r_s ds)$, then we have two decompositions of $\zeta$ using Itô's formula:

$$\begin{aligned} d\zeta_t &= \zeta_t(dM_t - r_t dt) \\ &= dN_t - e^{-\alpha t} f(X_t) dt, \end{aligned}$$

where $M$ and $N$ are two local martingales, so equating the finite-variation parts gives us

$$r_t = \frac{f(X_t)}{R_\alpha f(X_t)}, \tag{7}$$

an explicit expression for the spot-rate process $r$ as a function of the underlying Markov process $X$.

(iii) There are few examples where the resolvent of a Markov process can be written in closed form (though see Section 6). Nevertheless, using the relation

$$R_\alpha = (\alpha - \mathcal{G})^{-1}$$

[5]Equation (6) is the definition of the resolvent. This is an important and familiar concept from the theory of Markov processes; see, for example, [8].

between the resolvent and the infinitesimal generator $\mathcal{G}$ of the Markov process, we may build examples by firstly choosing $g \equiv R_\alpha f$ and then recovering $f$ by the recipe $f = (\alpha - \mathcal{G})g$. There is no guarantee that the $f$ so constructed will be non-negative, but choice of $\alpha$ allows considerable leeway here. See [6] for this approach in use in a number of examples.

Before we take up the theme of explicit construction of models based on Markov chains, we digress to point out how simply foreign exchange (and more general asset classes) can be incorporated in the potential approach.

## 5 Foreign exchange in the potential approach

Suppose now that we wish to consider the pricing of assets in many countries at once, each asset's price being expressed in the currency of its home country. This kind of problem arises quite frequently in practice; we may be asked to price a swap which swaps floating USD interest payments for fixed EUR interest payments. In a conventional approach to such a problem, one would firstly build a model for the interest rates in the US, then a a model for the interest rates in Euroland, and then try to model the USD/EUR exchange rate. Even using extremely simple models, a conventional approach would need one driving Brownian motion for the USD yield curve, one for the EUR yield curve and one for the exchange rate - three Brownian motions in total. Bearing in mind that a pricing calculation is in effect an integration, we are beginning to hit problems of dimensionality; a pricing calculation is an integration over three dimensions, and a pricing calculation for an American-style option is an optimal stopping problem in three dimensions. Add to this the facts that no-one would use a one-factor model to model the USD yield curve (unless forced to by tractability considerations); and that there might well be some knockout feature based on some other exchange rate, and the complexity of pricing such an asset becomes very real.

Or at least it does if you want to use a conventional approach. But let's see how easy it becomes using the potential approach. To introduce some notation, suppose that

$$1 \text{ unit of currency } j \quad = \quad Y_t^{ij} \text{ units of currency } i$$

Now if $(S_t^j)$ is a traded asset in country $j$, then

$$\zeta_t^j S_t^j \text{ is a martingale;}$$

also, by converting its currency-$j$ price into currency $i$, it becomes a traded asset in country $i$, and so

$$\zeta_t^i Y_t^{ij} S_t^j \text{ is a martingale.}$$

Therefore

$$N_t^{ij} \equiv \frac{\zeta_t^i Y_t^{ij}}{\zeta_t^j}$$

is a martingale orthogonal to the space of martingales of the form $\zeta^j S^j$. Thus we can express the exchange rate $Yij$ as

$$Y_t^{ij} = \frac{N_t^{ij} \zeta_t^j}{\zeta_t^i}.$$

In a complete market, $N^{ij}$ must be constant, so we have the simple and appealing result that *in a complete market, the exchange rate between two countries is the ratio of the state-price densities in the two countries.* More generally, there is the possibility of some exchange-rate risk not hedgeable through other assets, represented by the martingale $N^{ij}$.

The beauty of the potential approach based on Markov processes is that *adding another country does* not mean adding extra sources of randomness; we simply need to build another state-price density over the *same* Markov process, which requires us only to choose a new $f$ and $\alpha$. Thus adding extra countries to the model deos not need to cause problems of dimensionality (though one may well find that the treatment of the martingales $N^{ij}$ needs to be handled cleverly so as not to lose the simplicity of the methodology.)

## 6 Markov chain potential models

What makes a good model for an academic is not the same as what makes a good model for a practitioner. The academic is looking for something with simple features and closed-form expressions for basic derivatives, whereas the practitioner recognises that most derivative prices and hedges will have to be computed numerically, so demands quick and accurate numerical algorithms for doing these calculations, and a decent fit to market data. It seems in general that the better a model is for one purpose the worse it is for the other!

When it comes to using the potential approach and a Markovian model, we see from (1) that any pricing calculation is an *integration*, and that if we are to do this numerically then we have somehow to compute a finite weighted sum over the state space of the Markov process. Since this is so, it seems natural to work from the start with a Markov process with a finite state space, that is, a Markov chain!

Making the assumption that the state space $\mathcal{X}$ is a finite set of size $N$ has several very clear advantages:

(i) the generator of the chain is a $N \times N$ matrix Q, in terms of which the transition semigroup can be expressed as

$$(p_t(x, y))_{x,y,\in\mathcal{X}} = \exp(tQ);$$

(ii) all calculations reduce to calculations with finite matrices, and are therefore *fast*;

(iii) no splining of functions onto some finite subset of $\mathcal{X}$ will ever be needed;

(iv) pricing of American-style options becomes an optimal-stopping problem for a finite Markov chain, which is easy to handle;

(v) we are not restricted to possibly irrelevant propoerties of the underlying process (such as path continuity in the case of a diffusion).

Opposed to this are two disadvantages:

(i) the size of the parameter space is $O(N^2)$, so gets quite large quite quickly;

(ii) a given model will only admit $N$ possible values for the price of any given asset.

This latter is apparently quite restrictive, because if we were working with a 9-state chain, then the model says that only 9 possible yield curves could ever be observed, which is simply incompatible with a casual daily observation of the interest rates reported in any decent newspaper. Our resolution of this is to interpret those prices as being in some sense a 'market average' of the 'pure' prices that would apply if we knew with certainty what state we were in. We shall explain in more detail how this may be handled in the next Section on calibration, where we address the key question, '*Does this work?*'

## 7 Calibration

The methodology outlined here is very similar to that of [9], with a couple of important variations that substantially improve the performance of the fitting. The first is to drop the restriction to symmetrizable Markov chains, used in [9] to ensure that the diagonal matrices to be computed remain real, and the second is to allow the constant $\alpha$ of Section 4 to become a function of the state. Thus inatead of the additive functional $A$ defined at (5), we shall be using

$$A_t = \int_0^t \exp\Big(-\int_0^s \alpha(X_u)du\Big) f(X_s)\, ds.$$

This change was introduced as a result of experience with the calibration presented in [9]; the goodness of fit seemed to depend quite sensitively on the (previously assumed constant) value of $\alpha$, and thus allowing $\alpha$ to depend on state seemed a natural (and as it turned out helpful) variation to consider.

The model is parametrised by a vector[6] $\theta$. On day $n$ we have a vector $y_n$ of observations[7]. If the model were correct, the value of this observation vector $y_n$ would be exactly equal to the model values $Y(X_n, \theta)$, but we suppose

[6] For us, $\theta$ is the stacked vector of the off-diagonal entries of $Q$, the vector $g \equiv (\alpha - Q)^{-1} f$ and the vector $\alpha$ for each country involved.

[7] These observations will be market prices of certain assets.

that the $\log(y_n)$ are $\log Y(X_n, \theta)$, plus some independent Gaussian noise. We adopt a Bayesian standpoint, and suppose that the initial law of $X$ is given by $\pi = (\pi_i)_{i=1}^N$, and the initial law of $\theta$ is given by density $f_0(\theta)$; conceptually, $\theta$ is unchanging with time, even though our knowledge of it varies[8]. The notation $\mathbf{z}_n \equiv (z_0, \ldots, z_n)$ serves to make formulae more compact.

Based on the assumptions above, and ignoring irrelevant constants, the likelihood $\Lambda_n$ of $(\mathbf{X}_n, \mathbf{y}_n, \theta)$ is

$$\begin{aligned} \Lambda_n &\equiv \Lambda_n(\mathbf{X}_n, \mathbf{y}_n, \theta) \\ &= f_0(\theta)\, \pi_{X_0} \prod_{j=1}^{n} p_{X_{j-1}X_j}(s_j;\theta) \exp[-b(y_j, Y(X_j;\theta))] \end{aligned} \tag{8}$$

where $p_{ij}(s;\theta) = P_\theta(X_s = j | X_0 = i)$, and $b(y,y') \equiv \frac{1}{2}\log(y/y')\cdot V^{-1}\log(y/y')$, where $V$ is the covariance matrix of the Gaussian errors. We have also used the notation $s_j = t_j - t_{j-1}$ for the time between the $(j-1)$th and $j$th observations. We shall be more interested in the posterior distribution of $(X_n, \theta)$ given $\mathbf{y}_n$, so we introduce the notation

$$L_n(x, \mathbf{y}_n, \theta) = \sum_{\mathbf{X}_n : X_n = x} \Lambda_n(\mathbf{X}_n, \mathbf{y}_n, \theta), \tag{9}$$

and notice that

$$L_n(x, \mathbf{y}_n, \theta) = \sum_{\xi} L_{n-1}(\xi, \mathbf{y}_{n-1}, \theta) p_{\xi x}(s_n;\theta) \exp[-b(y_n, Y(x;\theta))]. \tag{10}$$

It is clear that for the Markov chain model in mind this expression will be far too complicated to allow exact analysis, so we make some simplifying assumptions, *specifically we assume that the likelihood $L_n$ has the product form*

$$L_n(x, \mathbf{y}_n, \theta) = \pi_n(x, \mathbf{y}_n)\, l_n(\theta, \mathbf{y}_n). \tag{11}$$

The justification is that if we have seen so much data that we have a pretty good idea what the values of the parameters must be, then the values of $\theta$ will largely be determined by the long-run historical average behaviour of the system. On the other hand, the posterior distribution of $X_n$ will be more influenced by recent history, because of the ergodicity of the Markov chain, and so some approximate conditional independence is reasonable; recent history tells us all we can know of $X_n$, distant history tells us all we can know of $\theta$. We shall further assume that

$$l_n(\theta, \mathbf{y}_n) \propto \exp(-\frac{1}{2}(\theta - \hat{\theta}_n)\cdot S_n(\theta - \hat{\theta}_n)) \tag{12}$$

for some positive-definite symmetric matrix $S_n$. If we think that we have nearly identified the true value of $\theta$, then such a quadratic approximation to the likelihood is quite natural.

[8] We shall soon consider what happens if we modify this assumption.

The values $\hat{\theta}_n$, $S_n$, and $\pi_n(\cdot, \mathbf{y}_n)$ are computed recursively, using (11). Supposing that we know already $\hat{\theta}_{n-1}$, $S_{n-1}$, and $\pi_{n-1}(\cdot, \mathbf{y}_{n-1})$, returning to (10) and using (11) gives

$$\begin{aligned} L_n(x, \mathbf{y}_n, \theta) &= \sum_\xi \pi_{n-1}(\xi, \mathbf{y}_{n-1})\, l_{n-1}(\theta, \mathbf{y}_{n-1}) p_{\xi x}(s_n; \theta) \exp[-b(y_n, Y(x;\theta))] \\ &\propto \sum_\xi \pi_{n-1}(\xi, \mathbf{y}_{n-1})\, p_{\xi x}(s_n; \theta) \exp[-b(y_n, Y(x;\theta))] \\ &\qquad\qquad . \exp\left[-\frac{1}{2}(\theta - \hat{\theta}_{n-1}) \cdot S_{n-1}(\theta - \hat{\theta}_{n-1}))\right] \end{aligned} \tag{13}$$

Sum this expression over $x$, and numerically pick $\theta$ to maximise; the maximising value is our new estimate $\hat{\theta}_n$ of $\theta$. By computing the second derivative matrix with respect to $\theta$ at $\hat{\theta}_n$ we find the value of $S_n$, and finally we get $\pi_n$ from

$$\pi_n(x, \mathbf{y}_n) \propto \sum_\xi \pi_{n-1}(\xi, \mathbf{y}_{n-1}) p_{\xi x}(s_n; \hat{\theta}_n) \exp[-b(y_n, Y(x; \hat{\theta}_n))].$$

Strictly speaking, the posterior $\pi_n$ for $X_n$ should be obtained by integrating the likelihood (13) with respect to $\theta$, but we approximate this by assuming that the posterior distribution for $\theta$ can be replaced by the point mass at $\hat{\theta}_n$, to avoid the need to integrate over a large number of dimensions.

As we indicated in the previous Section, this approach was modified in one vital respect; the model values $Y(x;\theta)$ were replaced by averaged values, averaging with respect to the weights

$$\sum_\xi \pi_{n-1}(\xi, \mathbf{y}_{n-1})\, p_{\xi x}(s_n; \theta) . \exp\left[-\frac{1}{2}(\theta - \hat{\theta}_{n-1}) \cdot S_{n-1}(\theta - \hat{\theta}_{n-1}))\right]$$

# 8 Evidence from bond data

The data used here is daily yield curve data covering the period from 2nd January 1992 to 1st March 1996[9].

For each day we have values of the yield of bonds with maturity 1 month, 3 months, 6 months, 1 year, 2 years, 5 years, 7 years and 10 years. We shall use daily yield curve data for three currencies; these are sterling (GBP), the US dollar (USD) and the German Mark (DEM).

This introductory account is not the place to present an exhaustive analysis of the results of the fit. We simply report that on average the fitted values were within 2-5 basis points[10] of the actual values, with extreme bad fits of

---

[9] We are grateful to Dr Simon Babbs for supplying the GBP and DEM data. The USD data was taken from the website http://www.stls.frb.org/fred/index.html

[10] 1 basis point $= 10^{-4}$.

the order of 25 basis points for some bonds on some days. The number of states of the Markov chain used was not large, varying between 5 and 11. What we did find is that the calibration was good up to a point, and then the methodology we were using would 'lose the plot' - this would typically be after 15-20 days of running the algorithm. The graph of the mean error per fitted bond price (using $N = 9$) indexed by day is given in Fig. 1.

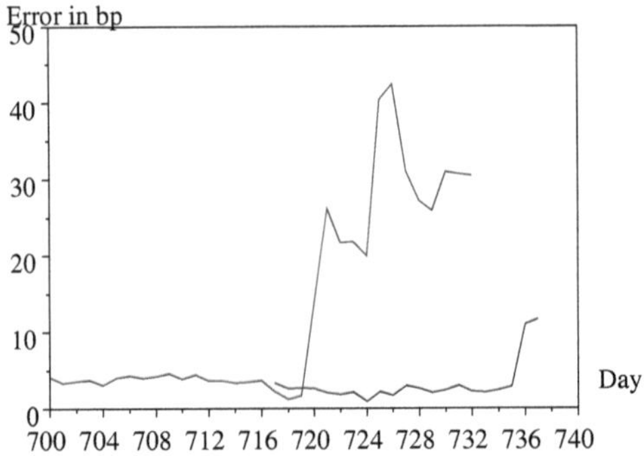


**Fig. 1.** Mean absolute deviation of prices of 8 bonds in 3 countries.

We see here two plots, one starting on day 700, which seems to stay close to the data until about day 718, after which it goes astray. Also plotted is the result of a fit begun on day 717, which tracks the data well until about day 736, when it in its turn loses the data. This instability of the fitting process is of course very undesirable; recent work on improvements of the methodology allow us to track the data with a mean error per fitted bond of 2-5 basis points over prolonged periods; we manage to hold this quality of fit over all the subsets of the data that we have tried (currently up to 150 trading days).

This is a remarkably good fit based on an extremely simple model, using only 9 possible states. At one level, it is surprising that such a simple system can do such a good job (if we were able to fit to within 1 basis point, then we would have something that could be traded off). On the other hand, Fig. 2 show why this may not be so surprising; a plot of 1-month LIBOR along with the Bank of England's base rate shows close agreement; if 1m LIBOR is really very close to the base rate, then we should be able to do a good job modelling interest rates if we had done a good job modelling the base rate, and it is not unreasonable to consider a model for the base rate that takes only a few possible values - indeed, we expect that whatever today's base rate is, the base rate in three months from now will differ by 25, 50, 75 or 100 basis points!

# 9 Hedging

In conventional models, the standard way to hedge a derivative is to *delta-hedge* it. The idea here is to compute the differential of the price of the derivative with respect to the prices of the underlying instruments (so in the case of a put option, we differentiate with respect to the stock price). The differential tells us how many units of the underlying to hold to protect (to leading order) against the moves in the underlying. In the case of a complete market, this hedging methodology is exact, in the sense that if we follow it perfectly, then we will perfectly replicate the contingent claim we were trying to hedge.

If we are using a Markov chain potential model, the notion of differentiating has no meaning, nevertheless the philosophy of immunising our portfolio against possible changes will work just as well. Suppose that we have a derivative $Z$, and hedging instruments $z^{(1)}, z^{(2)}, \ldots$. Suppose that if the state of the chain at time $t$ is $i$ and it jumps to $j$ then the value of $Z$ changes by $\Delta Z_{ij}(t)$. Then what we will do is to hold $w_r(t)$ units of asset $r$ so that

$$\Delta Z_{ij}(t) + \sum_{r=1}^{m} w_r(t) \Delta z_{ij}^{(r)}(t) = 0 \quad \forall j \quad (X_t = i).$$

Thus whatever jumps of the chain occur, our hedging portfolio will be immune to them. Of course, we do not in practice claim to be able to know $X_t$, but thie does not alter the hedging methodology; we would now make a portfolio of more hedging assets so as to ensure that

$$\Delta Z_{ij}(t) + \sum_{r=1}^{M} w_r(t) \Delta z_{ij}^{(r)}(t) = 0 \quad \forall i, j.$$

Following this recipe in the case of (say) a 9-state chain would entail taking a position in 72 different hedging instruments (if that many were available!) So

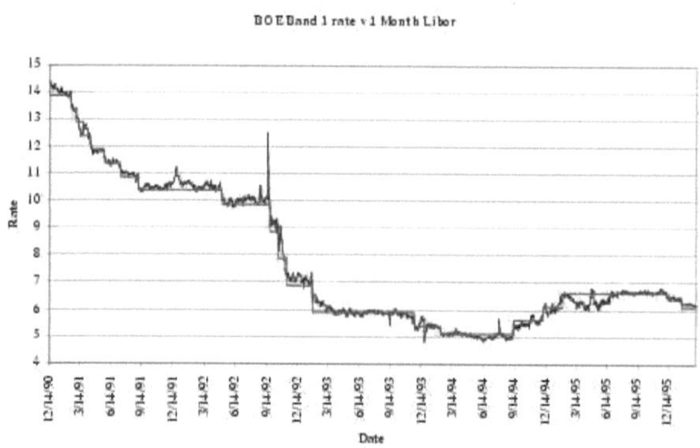


**Fig. 2.** 1m LIBOR and Bank of England base rate

we see that the practice of this methodology may not be quite so simple as the theory, but we can expect that the general approach will be as effective as the delta-hedging methodology is for diffusion-based models.

## 10 Conclusions and future directions

This brief introduction to the potential approach has shown that this modelling methodology has clear advantages:
(i) pricing is easy;
(ii) hedging is easy;
(iii) handling many currencies is easy;
(iv) calibration is perfectly feasible.

It is my belief that if there is ever to be a universal modelling methodology, it will have to look something like this. By 'universal' I mean a model that will account for all the different asset classes that an investment bank deals with - equity, foreign exchange, fixed income, commodities, credit risks - and the reason that a bank would like such a model is principally for what is known as *risk management* (though more properly called risk measurement). This refers to the regulatory requirements placed on the bank to assess the riskiness of their positions, and this needs to be understood on a firm-wide basis, as well as by business unit. The potential approach to modelling really can embrace such a wide swathe of the bank's business, and even if the calculations may have to be approximate to deal with such a wide sweep, at least the approach is consistent over the whole, rather than being some patched-together pastiche of wholly different models.

So far, the potential approach has been tested only on some bond data, and the next stage of the checking has to be to try to fit other fixed-income derivatives, then extend to other asset classes, notably equities. The only obstacle here is in obtaining decent data to work with, as many of the products that are important for calibration are not traded on exchanges, so price data is hard to come by. But given such data, we can more further to the final verdict on the potential approach: is it simply the right way to model prices, or is it a nice idea that cannot cope with the full complexity of the real data?

## References

1. M. Baxter and A. Rennie. *Financial Calculus.* Cambridge University Press,Combridge, 1996.
2. R. Dana and M. Jeanblanc. *Financial Markets in Continuous Time.* Springer, Berlin, 2003.
3. F. Delbaen and W. Schachermayer. A general version of the fundamental theorem of asset pricing. *Mathematische Annalen*, 300:463–520, 1994.
4. I. Karatzas and S. E. Shreve. *Methods of Mathematical Finance.* Springer, New York, 1998.

5. L. C. G. Rogers. One for all. *RISK*, 10:57–59, 1997.
6. L. C. G. Rogers. The potential approach to the term structure of interest rates and foreign exchange rates. *Mathematical Finance*, 7:157–176, 1997.
7. L. C. G. Rogers. *Handbook of Risk Management and Analysis*, chapter The origins of risk-neutral pricing and the Black-Scholes formula. Wiley, Chichester, 1998.
8. L. C. G. Rogers and D. Williams. *Diffusions, Markov Processes, and Martingales.* Cambridge University Press, Cambridge,, 2000.
9. L. C. G. Rogers and F. A. Yousaf. *Mathematical Finance - Bachelier Congress 2000*, chapter Markov chains and the potential approach to modelling interest rates and exchange rates. Springer, Berlin, 2002.
10. L. C. G. Rogers and O. Zane. *Vasicek and Beyond*, chapter Fitting potential models to interest rates and foreign exchange rates. RISK Publications, London, 1997.

# The Largest Claims Treaty ECOMOR

S.A. Ladoucette[1] and J.L. Teugels[2]

[1] EURANDOM, P.O. Box 513, NL–5600 MB Eindhoven, The Netherlands
ladoucette@eurandom.tue.nl

[2] EURANDOM and Katholieke Universiteit Leuven, University Center for Statistics, W. De Croylaan 54, B–3001 Heverlee, Belgium
jef.teugels@wis.kuleuven.ac.be

**Summary.** In the seventies of the previous century, the reinsurance treaty ECOMOR used to enjoy some limited popularity. However, since then the treaty has been largely neglected by most reinsurers, partly because of its technical complexity. In this paper, we give results pertaining to asymptotic properties of this reinsurance form. In particular, we formulate asymptotic estimates for the tail of the distribution of the ECOMOR-quantity. Furthermore, we give its weak laws.

**Key words:** asymptotic behaviour, Extreme Value Theory, heavy-tailed distribution, reinsurance treaty, ECOMOR.

## 1 Introduction

Assume that $\{X_1, X_2, \ldots\}$ forms the sequence of successive claim sizes in an homogeneous portfolio. This means that the random variables are independent and identically distributed (i.i.d.) as generated by the distribution $F$ of a generic random variable $X$. The consecutive claims occur according to a counting process $\{N(t);\ t \geq 0\}$ which is, for convenience, assumed to be independent of the claim size process $\{X_i;\ i \geq 1\}$. These claims determine the aggregate claim amount in the random sum $S_{N(t)} := \sum_{i=1}^{N(t)} X_i$. We denote by $(X_1^*, X_2^*, \ldots, X_{N(t)}^*)$ the order statistics of the random vector $(X_1, X_2, \ldots, X_{N(t)})$ of successive claim sizes up to time $t$.

One of the main goals of a reinsurance treaty is the coverage against large claims. It is somewhat surprising that classical reinsurance contracts (proportional, surplus, excess-of-loss, stop-loss) are not expressed in terms of the largest claims. This may be due to the mathematical intractability of the latter quantities. We will indicate how extreme value theory is capable to overcome part of this problem. We restrict our attention to ECOMOR. For an extended version of the present work as well as for the largest claims reinsurance treaty, we refer to [5]. For an overview of most of the currently employed reinsurance forms with their properties, see [7].

The reinsurance treaty ECOMOR (excédent du coût moyen relatif) has been introduced in 1950 by the French actuary Thépaut [8]. The treaty, defined by the upper order statistics of the random sample, uses as reinsured amount the quantity:

$$R_r(t) := \sum_{i=1}^{r} X^*_{N(t)-i+1} - rX^*_{N(t)-r} = \sum_{i=1}^{N(t)} \left\{X_i - X^*_{N(t)-r}\right\}^+, \quad r \geq 1,\ t \geq 0 \tag{1}$$

if $N(t) > r$ and $R_r(t) := 0$ otherwise. The quantity $R_r(t)$ is thus a function of the $r+1$ upper order statistics $X^*_{N(t)-r} \leq \cdots \leq X^*_{N(t)}$. The second form in (1) indicates that ECOMOR can also be considered as an excess-of-loss treaty with the $(r+1)$th largest claim as a (random) retention. The number of reinsured claims is equal to the deterministic value $r$ while their values are diminished by the random retention $X^*_{N(t)-r}$. Note that the first line insurer carries the responsibility for the random retention.

For a survey of the literature on ECOMOR, we refer to [5]. Most of the relevant papers are given within a Poisson-Pareto setting. The first publication that deals with an analytic treatment of the weak convergence of $R_r(t)$ is [6]. For a probabilistic approach to the same problem, see [4].

We first state results dealing with the *asymptotic equivalence* between the tail of the claim size distribution $F$ and the tail of the distribution of $R_r(t)$ for a fixed $t \geq 0$. We then deal with the distribution of the quantity $R_r(t)$ when $t$ tends to infinity, touching on the question of *convergence in distribution.* The latter results are particularly important when one wants to replace the complicated distribution of $R_r(t)$ by a simpler expression.

## 2 Results

We formulate four types of results: bounds, asymptotic equivalence, weak convergence and moment convergence. All these results are proved in Ladoucette et al. [5].

### 2.1 Bounds

We start by giving an asymptotic upper bound for the ratio of the tail of the distribution of $R_r(t)$ and that of the generic variable $X$. For most of the concepts below, see [3] or [4]. Recall that a distribution $F$ with support $(0, \infty)$ is in the class $\mathcal{S}$ of *subexponential distributions* if

$$\lim_{x\to\infty} \frac{1 - F^{*2}(x)}{1 - F(x)} = 2$$

where $F^{*2}$ denotes the 2-fold convolution of $F$ with itself. We then have the following result.

**Theorem 1.** *Assume that $F \in \mathcal{S}$ and that $\mathbb{E}\left\{e^{\varepsilon N(t)}\right\} < \infty$ for some $\varepsilon > 0$. Then for $t \geq 0$ we have*

$$\limsup_{s\to\infty} \frac{\mathbb{P}[R_r(t) > s]}{\mathbb{P}[X > s]} \leq \mathbb{E}N(t).$$

The condition on $N(t)$ means in particular that the probability of a large number of claims in $[0, t]$ is exponentially small. In the other direction, we only need weaker conditions on $F$ and on $N(t)$. A distribution $F$ on $\mathbb{R}$ belongs to the class $\mathcal{L}$ of *long-tailed distributions* if for all $y \in \mathbb{R}$:

$$\lim_{x\to\infty} \frac{1 - F(x + y)}{1 - F(x)} = 1.$$

It is well known that $\mathcal{S}$ is a proper subset of $\mathcal{L}$ on the positive half-line, see for example Embrechts et al. [4].

**Theorem 2.** *Assume that $F \in \mathcal{L}$ and that $\mathbb{E}N^r(t) < \infty$. Then for $t \geq 0$ we have*

$$\liminf_{s\to\infty} \frac{\mathbb{P}[R_r(t) > s]}{\mathbb{P}[X > s]} \geq \mathbb{E}N(t).$$

### 2.2 Asymptotic Equivalence

Here we give a result that states the asymptotic equivalence between the tail of the distribution of the quantity $R_1(t)$ and that of the generic claim size $X$ under the long-tailed assumption.

**Theorem 3.** *Assume that $F \in \mathcal{L}$ and that $\mathbb{E}N(t) < \infty$. Then for $t \geq 0$:*

$$\mathbb{P}[R_1(t) > s] \sim \mathbb{E}N(t)\,\mathbb{P}[X > s] \quad as \quad s \to \infty.$$

With a stronger assumption, we also get a further extension. The last part involves the maximum $X^*_{N(t)}$ of the random sample as well as the random sum $S_{N(t)}$.

**Theorem 4.** *Assume that $F \in \mathcal{S}$ and let $t \geq 0$ be fixed.*

*(i) If $\mathbb{E}N(t) < \infty$, then:*

$$\mathbb{P}[R_1(t) > s] \sim \mathbb{E}N(t)\,\mathbb{P}[X > s] \quad as \quad s \to \infty.$$

*(ii) If $\mathbb{E}\left\{e^{\varepsilon N(t)}\right\} < \infty$ for some $\varepsilon > 0$, then for every fixed $r \geq 2$:*

$$\mathbb{P}[R_r(t) > s] \sim \mathbb{E}N(t)\,\mathbb{P}[X > s] \quad as \quad s \to \infty.$$

*(iii) If $\mathbb{E}\left\{e^{\varepsilon N(t)}\right\} < \infty$ for some $\varepsilon > 0$, then for every fixed $r \geq 1$ we have*

$$\mathbb{P}[R_r(t) > s] \sim \mathbb{P}[X^*_{N(t)} > s] \sim \mathbb{P}[S_{N(t)} > s] \sim \mathbb{E}N(t)\,\mathbb{P}[X > s] \quad as \quad s \to \infty.$$

### 2.3 Weak Convergence of $R_r(t)$

In this subsection, we deal with the asymptotic behaviour of $\mathbb{P}[R_r(t) > a(t)s]$ for an appropriate norming function $a(t)$ when $t \to \infty$. It is not surprising that we need the weak convergence of the largest claim but for a deterministic sample size. As shown for example in Beirlant et al. [2], this is guaranteed under the *extremal domain of attraction condition* expressed in terms of the *tail quantile function* $U$ of the claim size distribution $F$ which is defined by $U(y) := \inf\{x : F(x) \geq 1 - 1/y\}$. The distribution $F$ on $\mathbb{R}$ with tail quantile function $U$ belongs to the *extremal class* $\mathcal{C}_\gamma(a)$ if there exists a constant $\gamma \in \mathbb{R}$ and an ultimately positive auxiliary function $a(.)$ such that

$$\lim_{x\to\infty} \frac{U(ux) - U(x)}{a(x)} = \int_1^u v^{\gamma-1}\, dv =: h_\gamma(u)$$

for all $u > 0$. In addition, we need to assume that the claim number process is *mixed Poisson, i.e.*, $\{N(t);\ t \geq 0\} = \{\tilde{N}(\Lambda t);\ t \geq 0\}$ where the mixing variable $\Lambda$ is nonnegative and $\{\tilde{N}(t);\ t \geq 0\}$ is a homogeneous Poisson process with intensity 1, independent of $\Lambda$. It is convenient to formulate the weak convergence in terms of Laplace transforms.

**Theorem 5.** *Let $r \geq 1$ be fixed. Assume that $F \in \mathcal{C}_\gamma(a)$ with $\gamma \in \mathbb{R}$ and that $\{N(t);\ t \geq 0\}$ is a mixed Poisson process with mixing random variable $\Lambda$. Then for $\theta \geq 0$ we have*

$$\lim_{t\to\infty} \mathbb{E}\left\{\exp\left(-\theta \frac{R_r(t)}{a(t)}\right)\right\} = \frac{1}{r!}\int_0^\infty w^r\, q_{r+1}(w) \left(\int_0^1 e^{-\theta w^{-\gamma} h_\gamma(1/z)}\, dz\right)^r dw,$$

*where $q_{r+1}(w) := \mathbb{E}\left\{e^{-w\Lambda}\Lambda^{r+1}\right\}$.*

Due to the explicit expression for the function $q_{r+1}$, it is easy to check that the right-hand side equals 1 when $\theta = 0$. The limiting distribution can be given explicitly in the case $r = 1$. However, no inversion of the Laplace transform seems possible for general $r \geq 2$ except for the case $\gamma = 0$ for which we can readily show that we end up with a limiting gamma distribution.

### 2.4 Moment Convergence

By using a similar approach as in Theorem 5, we can derive the expressions for the limiting value of the first few moments. For instance, we get for the mean:

$$\lim_{t\to\infty} \frac{\mathbb{E}R_r(t)}{a(t)} = \frac{\Gamma(r-\gamma+1)}{(r-1)!(1-\gamma)}\,\mathbb{E}\left\{\Lambda^\gamma\right\}$$

under the condition $\gamma < 1$, where $\Gamma$ denotes the gamma function. Similarly for the moment of second order, we get:

$$\lim_{t\to\infty} \frac{\mathbb{E}\left\{R_r^2(t)\right\}}{a^2(t)} = \frac{\{1+r(1-2\gamma)\}\Gamma(r-2\gamma+1)}{(r-1)!(1-\gamma)^2(1-2\gamma)} \mathbb{E}\left\{\Lambda^{2\gamma}\right\},$$

where we need to assume $\gamma < 1/2$. Writing the limiting results in this form permits us to illustrate the role played by the mixing random variable $\Lambda$.

## 3 Conclusion and Remarks

We finish with a few observations.

(*i*) There is some need to get further information on the accuracy of the approximations given above. For example, it would be interesting to get remainder results for the case where the claim size distribution $F$ belongs to the extremal class $\mathcal{C}_\gamma(a)$, $\gamma \in \mathbb{R}$, with remainder as treated in Beirlant et al. [2].

(*ii*) We point out that most of our results can be extended to the case where the counting process $\{N(t);\ t \geq 0\}$ satisfies the condition $\frac{N(t)}{t} \stackrel{\mathcal{D}}{\to} \Lambda$ as $t \to \infty$. For example, this condition is fulfilled with $\Lambda$ degenerate at a positive value if $\{N(t);\ t \geq 0\}$ is a renewal process. The same is true when the process is infinitely divisible. See for example in Bingham et al. [3].

(*iii*) As indicated by Beirlant [1], it is worth noting that the condition $t \to \infty$ can be replaced by a condition of the form $\mathbb{E}N(t) \to \infty$.

## References

1. J. Beirlant. Largest claims and ECOMOR reinsurance. In J.L. Teugels and B. Sundt, editors, *Encyclopedia of Actuarial Science.* John Wiley & Sons, Chichester, 2004.
2. J. Beirlant, Y. Goegebeur, J. Segers, and J.L. Teugels. *Statistics of Extremes: Theory and Applications.* John Wiley & Sons, Chichester, 2004.
3. N.H. Bingham, C.M. Goldie, and J.L. Teugels. *Regular Variation.* Cambridge University Press, Cambridge, 1987.
4. P. Embrechts, C. Klüppelberg, and T. Mikosch. *Modelling Extremal Events for Insurance and Finance.* Springer-Verlag, Berlin Heidelberg, 1997.
5. S.A. Ladoucette and J.L. Teugels. Reinsurance of large claims. *J. Comput. Appl. Math.*, 2004. To appear.
6. J.L. Teugels. Selected Topics in Insurance Mathematics. Technical report, Katholieke Universiteit Leuven, Belgium, 1985.
7. J.L. Teugels. Reinsurance Actuarial Aspects. Technical Report 2003-006, EURANDOM, Technical University of Eindhoven, The Netherlands, 2003.
8. A. Thépaut. Une nouvelle forme de réassurance: le traité d'excédent du coût moyen relatif (ECOMOR). *Bull. Trim. Inst. Actu. Français*, 49:273–343, 1950.

# American Options With Discrete Dividends Solved by Highly Accurate Discretizations

C.C.W. Leentvaar and C.W. Oosterlee

Delft Institute for Applied Mathematics, Delft University of Technology, the Netherlands.e-mail: c.c.w.leentvaar, c.w.oosterlee@ewi.tudelft.nl

**Summary.** We present an accurate numerical solution for the discrete Black-Scholes equation with only a few grid points. European and American option problems with deterministic discrete dividend modelled by a jump condition at the ex-dividend date are solved. Fourth order finite differences are employed, as well as a grid stretching in space and a Lagrange interpolation at the ex-dividend date.

**Key words:** American options, Black-Scholes, 4th order discretization, stretched grid, discrete dividend.

## 1 Black-Scholes Equation, Discretization

Research in option pricing theory concerns, among other issues, the computation of the fair price of an option. We aim for accurate American option values by 4th order finite differences and grid stretching in space. The Black-Scholes equation represents a simple model for pricing a put or a call option. Option value $u$ depends on asset price $s$ and is influenced by exercise price $K$, expiration time $T$ $(0 \leq t \leq T)$, interest rate $r$, and volatility $\sigma$. The equation reads

$$\frac{\partial u}{\partial t} + \frac{1}{2}\sigma^2 s^2 \frac{\partial^2 u}{\partial s^2} + rs\frac{\partial u}{\partial s} - ru = 0, \quad 0 \leq s < \infty, \quad 0 \leq t < T \ . \tag{1}$$

It is valid under the assumption of a geometric Brownian motion for the asset price process $\{S_t\}$ and comes with boundary and final conditions distinguishing a put from a call and a European from an American option. We adopt the technique of modelling discrete dividends $D$ by a jump condition at the ex-dividend date $t_d$ [9]:

$$u(s, t_d^-) = u(s - D, t_d^+), \tag{2}$$

where $t_d^-$, $t_d^+$ represent the times just before and after the ex-dividend date.

Boundary conditions arise naturally from financial arguments [9].

$$\text{For a call: } u(0,t) = 0, \text{ and } u(s_{max}, t) = s_{max} - Ke^{-r(T-t)}.$$

It is shown in [5] that the boundary condition at $s = s_{max}$ for a call after dividend payment changes into $u(s_{max}, t) = s_{max} - Ke^{-r(T-t)} - De^{-r(t_d - t)}$. The final condition at $t = T$ for the European call reads $u(s, T) = \max(s - K, 0)$.

The exercise of American-style options is permitted at any time during the lifetime of an option, $0 < t \leq T$. When early exercise is permitted, a constraint "$u(s,t) \geq$payoff" plus a smooth pasting condition at the payoff must be imposed to the solution of (1), giving rise to a free boundary problem. A special $s$-value exists, the optimal exercise price $s_f(t)$, which is not known in advance and needs to be computed. For the American call, early exercise may be optimal just before the dividend payment *only if* the dividend payment is large enough, *i.e.*, $D > K\left(1 - e^{-r(T-t_d)}\right)$. The boundary condition at $s = s_{max}$ for the American call before the dividend payment, $t < t_d$ reads

$$u(s_{max}, t) = \max\{s_{max} - Ke^{-r(T-t)} - De^{-r(t_d-t)},\ s_{max} - Ke^{-r(t_d-t)}\}, \quad (3)$$

and the final condition before the dividend payment reads:

$$u(s, t_d) = \max\{s - Ke^{-r(T-t_d)} - D,\ s - K\}, \quad (4)$$

which is also the condition whether the option should be exercised at $t_d^-$.

For the American put option the boundary condition at $s = 0$ changes after the dividend payment from $u(0,t) = Ke^{-r(T-t)}$ into

$$u(0,t) = \max\{Ke^{-r(T-t)} + De^{-r(t_d-t)},\ K\},$$

and we have $u(s_{max}, t) = 0$. The payoff for a put after a dividend has been paid remains $u(s,T) = \max(K - s,\ 0)$. Early exercise of a put option may be optimal at any time within the option's lifetime.

### 1.1 Grid Transformation and Discretization

The implicit 4th order accurate backward differentiation formula, BDF4 [2], with time discretization is employed on an equidistant grid with time step $k$. It is preceded by two Crank-Nicholson and one BDF3 step. Crank-Nicholson is unconditionally stable, whereas BDF3 and BDF4 have a stability region. For our applications these stability constraints are not problematic.

In space, we use an analytic coordinate transformation, which results in an a-priori stretching of the grid around $s = K$, *i.e.*, at the non-differentiability in the final condition. An equidistant grid discretization can be used after the analytic transformation, as only the coefficients in front of the derivatives in (1) change. The spatial transformation used for Black-Scholes originates from [1, 8]:

$$y = \psi(s) = \sinh^{-1}(\mu(s - K)) + \sinh^{-1}(\mu K). \tag{5}$$

The parameter $\mu$ determines the rate of stretching; $\mu K = 15$ is satisfactory in many cases.

A fourth order "long stencil" finite difference discretization in space on the equidistant $y$-grid based on Taylor's expansion is employed. First-order derivatives are discretized by central differences. For the fourth order approximation, near-boundary points $y_1$, $y_{N-1}$ need special treatment by means of one-sided differences. Important for our future applications is a small discretization error with only a few grid points. This has been achieved in [7] with the techniques proposed: for some reference Black-Scholes option pricing problems with known exact solutions only 20 – 40 space- and time-steps were necessary to get an accuracy of less than one cent (€ 0.01).

The strategy to solve the Black-Scholes equation with discrete deterministic dividends numerically is as follows (see also [8]):

- Start solving from $t = T$ to $t = t_d$ with the usual pay-off.
- Apply an interpolation to calculate the new asset and option price on the $s$-grid discounted with $D$.
- Restart the numerical process with the PDE from the interpolated price as final condition from $t_d$ tot $t = 0$.

In our computations we place $t_d$ exactly on a time line, $t_d^-$ and $t_d^+$ are assumed to lie on the same line.

## 2 Numerical Results with Discrete Dividend

### 2.1 European Call

We present European call results for multiple discrete dividends, as in [3], with problem parameters $s_0 = K = 100$, $r = 0.06, \sigma = 0.25$ and multiple dividends of 4 (ex-dividend date is each half year). Table 1 presents numerical results for $T = 1$, $T = 2$ and $T = 3$, with one, two and three dividend payments, respectively. It compares the numerical approximation to the exact solution from [3] (HHL in the table). The value $s_{max} = 3K$ is chosen according to a well-known formula [4] and the stretching parameter is set to $\mu = 0.15$. For larger values of $T$ the number of points in time increases proportionally. The numerical results obtained with only a few grid points agree very well with the reference.

Other discrete dividend results from [3] can also be confirmed with our discretization techniques, whereas binomial tree approaches may need special techniques based on financial arguments to get the correct price.

### 2.2 American Put

Next, we consider an American put with parameters from [6]: $K = 100$, $\sigma = 0.4$, $r = 0.08$, $D = 2$, $t_d = 0.3$, $T = 0.5$ and $\mu = 0.15$, $s_{max} = 3K$. Results

**Table 1.** Multiple discrete dividends payments, $K = 100,\ d = 4,\ \mu = 0.15$.

| | $T = 1$ | | $T = 2$ | | $T = 3$ |
|---|---|---|---|---|---|
| Grid | $u\ (s_0, t = 0)$ | Grid | $u\ (s_0, t = 0)$ | Grid | $u\ (s_0, t = 0)$ |
| $10 \times 10$ | 10.612 | $10 \times 20$ | 15.177 | $10 \times 30$ | 18.698 |
| $20 \times 20$ | 10.660 | $20 \times 40$ | 15.202 | $20 \times 60$ | 16.607 |
| $40 \times 40$ | 10.661 | $40 \times 80$ | 15.201 | $40 \times 120$ | 18.600 |
| HHL | 10.661 | | 15.199 | | 18.598 |

for $s = 80,\ 100,\ 120$ at $t = 0$ on a $20^2$ – and $40^2$– grid are compared to those in [6] in Table 2. With only 20 points in space and time the results from [6] are obtained.

**Table 2.** American put reference problem from [6], $K = 100,\ D = 2,\ \mu = 0.15$.

| Grid | $u_h\ (80, t = 0)$ | $u_h\ (100, t = 0)$ | $u_h\ (120, t = 0)$ |
|---|---|---|---|
| $20 \times 20$ | 0.223 | 0.105 | 0.043 |
| $40 \times 40$ | 0.223 | 0.105 | 0.043 |
| Meyer (J. C. Fin. 2001) [6]: | 0.223 | 0.105 | 0.043 |

Finally, we consider an American put with two ex-dividend dates, $t_{d_1} = 0.3,\ t_{d_2} = 0.8\ (T = 1)$ and visualize the free boundary as a function of time for different dividend payment strategies (as in [6]). Figure 1 shows the free boundary for an American put with problem parameters: $K = 100,\ \sigma = 0.4,\ r = 0.08,\ T = 1$. In the figure the free boundary functions presented are: one without any dividend payment ($D = 0$, solid line), with a fixed dividend payment $D = 2$ at $t_{d_1},\ t_{d_2}$ (dashed line) and with a payment proportional to the asset price $D = 0.98s$ (dotted line) at the ex-dividend dates. It can be seen that after the discrete dividend payment the free boundary may disappear, and reappear, indicating that early exercise is not always favorable after an ex-dividend date.

## 3 Conclusion

In this paper we have solved the Black-Scholes equation for a European and American option with discrete dividend with only a few grid points. Fourth order accurate space and time discretizations have been employed, using spatial grid stretching by means of an analytical coordinate transformation. The discrete dividend payment is handled very satisfactorily by the stretched grid discretization and a 4th order Lagrange interpolation at the ex-dividend date. Reference results for a European call with multiple dividends and American puts from the literature are retained with only 20 – 40 grid points in space and time.

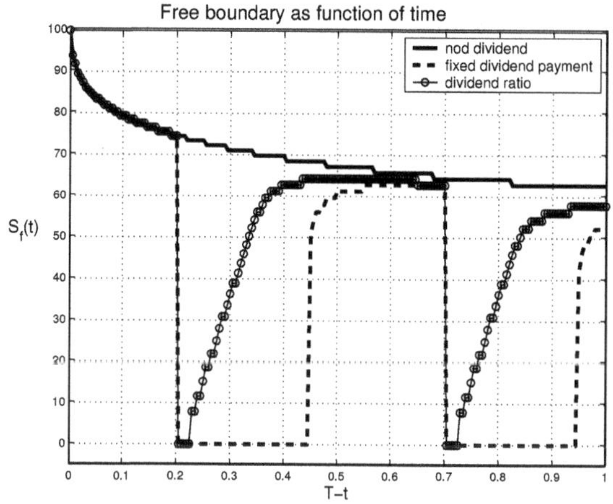


**Fig. 1.** Free boundary as function of time with two ex-dividend dates and different forms of dividend payment: $D = 0$ (solid), $D = 2$ (dashed) vs. $D = 0.98s$ (dotted).

*Acknowledgement.* Coenraad C.W. Leentvaar wishes to thank the Dutch Technology Foundation (STW) for financial support.

## References

1. N. Clarke and K. Parrot. Multigrid for American option pricing with stochastic volatility. *Appl. Math. Finance*, 6:1999, 177–179.
2. E. Hairer and K. Wanner. *Solving Ordinary Differential Equations. Vol. 2. Stiff and Differential-Algebraic Problems.* Springer-Verlag, Heidelberg, 1996.
3. E.G. Haug, J. Haug, and A. Lewis. Back to basics: a new approach to the discrete dividend problem. *Wilmott Magazine*, pages 37–47, September 2003.
4. R. Kangro and R. Nicolaides. Far field boundary conditions for Black-Scholes equations. *SIAM J. Numerical Analysis*, 38(4):1357–1368, 2000.
5. Y.K. Kwok. *Mathematical Models of Financial Derivatives.* Springer Verlag, 1998, Heidelberg.
6. G.H. Meyer. Numerical investigation of early exercise in American puts with discrete dividends. *J. Comp. Finance*, 5(2), 2001.
7. C.W. Oosterlee, C.C.W. Leentvaar, and A. Almendral Vázquez. Pricing options with discrete dividends by high order finite differences and grid stretching. In P. Neittaanmäkki et al., editor, *ECCOMAS 2004*, Jyväskylä, Finland, 2004.
8. D. Tavella and C. Randall. *Pricing Financial Instruments, The Finite Difference Method.* Wiley, New York, 2000.
9. P. Wilmott, S. S. Howison, and J. Dewynne. *The Mathematics of Financial Derivatives.* Cambridge University Press, Cambridge, 1997.

# Semi-Lagrange Time Integration for PDE Models of Asian Options

A.K. Parrott[1] and S. Rout[2]

[1] University of Greenwich, 30 Park Row, London SE10 9LS, U.K
a.k.parrott@gre.ac.uk
[2] University of Greenwich, 30 Park Row, London SE10 9LS, U.K
s.rout@gre.ac.uk

**Summary.** Semi-Lagrange time integration is used with the finite difference method to provide accurate stable prices for Asian options, with or without early exercise. These are combined with coordinate transformations for computational efficiency and compared with published results.

**Key words:** Semi-Lagrange time integration, Asian American Options, finite difference, coordinate transformation

## 1 Asian Options

Conventional Call and Put options have payoffs that depend on the instantaneous price of the underlying security, either at maturity (European case) or throughout the contract (American case); this creates an incentive for traders to attempt price manipulation on the underlying security. Asian options avoid this by using payoffs that depend on the average price of the underlying security, rather than the instantaneous price. The PDE for the price $V(A, S, t)$ of an Asian option is an "ultra-parabolic equation" given by

$$-\frac{\partial V}{\partial t} = \frac{1}{2}\sigma^2 S^2 \frac{\partial^2 V}{\partial S^2} + rS\frac{\partial V}{\partial S} + \frac{1}{t}(S-A)\frac{\partial V}{\partial A} - rV \; ; \quad S \geq 0, \; A \geq 0, \; t \in (0, T]$$

with final condition (payoff) $V(S, A, T) = g(S, A, T)$, assuming a geometric Brownian motion model for the asset price, $S_t$:

$$dS_t = \mu S_t dt + \sigma S_t dW_t \tag{1}$$

where $W_t$ is a Brownian motion. $A_t$, the continuously sampled arithmetic average of $S_t$ over $[0, t]$, is defined by

$$A_t = \frac{1}{t}\int_0^t S_\tau \, d\tau, \;\; \text{where } A_0 = S_0, \;\; \text{and } dA_t = \frac{1}{t}\Big(S_t - A_t\Big)dt \tag{2}$$

### 1.1 Semi-Lagrangian Time Integration

The Semi-Lagrangian scheme [6] uses a different set of trajectories at each time-step, choosing them such that they arrive at the points of the regular grid at the end of the time-step. It was first applied to option pricing problems in [4] and has been recently applied to jump process models in [3].
The Lagrangian derivative along a path in the $A-t$ plane is given by

$$\frac{dV}{dt} = \frac{\partial V}{\partial t} + \frac{\partial V}{\partial A}\frac{dA}{dt}$$

It follows that the Asian pricing equation simplifies to an $A$-parameterised pricing PDE with identical spatial derivatives to Black-Scholes, namely

$$-\frac{dV}{dt} = \frac{1}{2}\sigma^2 S^2 \frac{\partial^2 V}{\partial S^2} + rS\frac{\partial V}{\partial S} - rV = \mathcal{L}_S V \tag{3}$$

for paths $\mathcal{P}(A,t;S)$ such that

$$\frac{dA}{dt} = \frac{1}{t}(S-A) \tag{4}$$

In the case of early exercise

$$-\frac{dV}{dt} \geq \mathcal{L}_S V, \text{ and } V(S,A,t) \geq g(S,A,t) \tag{5}$$

leading to the linear complementarity problem

$$(V-g)(-\frac{dV}{dt} - \mathcal{L}_S V) = 0$$

### 1.2 Discretisation

Take the integral of (3) along the path $\mathcal{P}_k^m(A,t;S_j) \equiv (\tilde{A}_k, t^m) \curvearrowright (A_k, t^{m-1})$,

$$V(S_j, A_k, t^{m-1}) - V(S_j, \tilde{A}_k, t^m) = \int_{\mathcal{P}_k^m(A,t;S_j)} \mathcal{L}_S V(S,A,t)\, dt$$

If $\mathcal{L}_S^h$ is an $O(h^2)$ approximation to $\mathcal{L}_S$, where $h = \min(\Delta S)$, and $\{V_{j,k}^m\}$ are finite difference mesh prices on $\Omega = \{S_0, \ldots, S_N\} \otimes \{A_0, \ldots, A_N\}$ such that

$$V_{j,k}^{m-1} = \tilde{V}_{j,k}^m + \Delta t\left(\theta \mathcal{L}_S^h(V_{j,k}^{m-1}) + (1-\theta)\tilde{\mathcal{L}}_S^h(V_{j,k}^m)\right)$$

then $V_{j,k}^m \approx V(S_j, A_k, t^m)$ to $O(\Delta t^2) + O(h^2)$ for $\theta = 0.5$ and is unconditionally stable for $\theta \geq 0.5$. $\tilde{V}_{j,k}^m$ is the mesh price interpolated to $\tilde{A}_k$.

**Determining $\tilde{A}_k$**

Integrating (4) with respect to time along the path $\mathcal{P}_k^m(A, t; S)$ gives

$$\int_{A_k}^{\tilde{A}_k} \frac{1}{S_j - A}\, dA = \int_{t^{m-1}}^{t^m} \frac{1}{t}\, dt$$

thus

$$\tilde{A}_k = S_j - \left(\frac{t^{m-1}}{t^m}\right)(S_j - A_k) = \alpha A_k + (1-\alpha) S_j, \quad \text{where } \alpha = t^{m-1}/t^m.$$

### 1.3 Boundary Conditions for the Fixed-Strike Call

An average-rate fixed-strike call has payoff $g(S, A, T) = \max\,(A - K,\ 0)$. If the asset price reaches zero at some time $t^\star$, then it stays at zero for all $t \in [t^\star, T]$. The final average price is then

$$A_T = \frac{1}{T}\int_0^T S_t dt = \frac{1}{T}\int_0^{t^\star} S_t dt = \frac{t^\star}{T} A_{t^\star}$$

and hence the payoff, $\forall A \geq 0$ is also known. For an average rate call, the boundary condition at $S = 0,\ \forall A \geq 0$, is

$$V(0, A, t) = e^{-r(T-t)} \max\,(\frac{t}{T}A - K, 0) \tag{6}$$

Alternatively one could use the exact solution of the PDE for $S = 0$ (it becomes linear hyperbolic). The original domain is the infinite quarter plain and must be truncated before it can be meshed: eg. at $S = S_{max}$ and $A = A_{max}$. A second derivative truncation condition is applied at some $S = S_{max}$, i.e.

$$\frac{\partial^2 V(S_{max}, A, t)}{\partial S^2} = 0 \qquad \forall\ A \in [0, A_{max}]; \quad t \in (0, T] \tag{7}$$

### 1.4 Co-ordinate Stretching

A stretched coordinate $x$ is used (see [2]), defined by

$$S = \frac{K}{\lambda}\ \sinh(x - L) + K\ ; \qquad x = \sinh^{-1}\left(\frac{\lambda}{K}(S - K)\right) + L$$

where $K$ is the strike and $\lambda = \sinh L$ is a parameter controlling the degree of stretch. Substituting these expressions into the pricing equation gives:

$$-\frac{dV}{dt} = \frac{1}{2}\sigma^2\left(\mathcal{T}_\lambda(x)\right)^2 \frac{\partial^2 V}{\partial x^2} V - \frac{1}{2}\sigma^2 \tanh(x + L)\mathcal{T}_\lambda(x)\frac{\partial V}{\partial x} + r\mathcal{T}_\lambda(x)\ \frac{\partial V}{\partial x} - rV$$

## 2 Results

Table 1 displays convergence results corresponding to the fixed strike European call Asian when $r = 0.1$, and prices are quoted at $S_0 = 100$. $\sigma$, $K$ and expiry, $T$ vary as shown in the table. Richardson extrapolations are also shown and give basis point accuracy using meshes of 40 and 80 points with a stretching parameter of ten, in 10 time steps. Table 2 shows a comparison with [5], [7] and also [1], where available. Table 3 displays convergence results corresponding to the fixed strike Asian put with early exercise. The comparison results are from [1].

**Table 1.** Calculated Semi-Lagrangian (S-L) prices at $S = 100$ showing convergence and Richardson extrapolation for Fixed Strike Call Asians where $r = 0.1$, $\sigma = 0.2$. $nt = 10$ for all maturities and $\lambda = 10$.

| $T$ | $K$ | $N = 20$ | $N = 40$ | $N = 80$ | $N = 160$ | (20,40) | (40,80) | (80,160) |
|---|---|---|---|---|---|---|---|---|
| 0.25 | 95 | 6.462 | 6.474 | 6.477 | 6.478 | 6.478 | 6.478 | 6.478 |
| 0.25 | 100 | 2.884 | 2.922 | 2.931 | 2.933 | 2.935 | 2.934 | 2.934 |
| 0.25 | 105 | 0.920 | 0.942 | 0.948 | 0.950 | 0.949 | 0.951 | 0.951 |
| 0.50 | 95 | 7.898 | 7.894 | 7.893 | 7.893 | 7.893 | 7.893 | 7.893 |
| 0.50 | 100 | 4.502 | 4.505 | 4.506 | 4.506 | 4.506 | 4.506 | 4.506 |
| 0.50 | 105 | 2.197 | 2.206 | 2.208 | 2.209 | 2.209 | 2.209 | 2.209 |
| 1.00 | 95 | 10.338 | 10.305 | 10.297 | 10.295 | 10.295 | 10.294 | 10.294 |
| 1.00 | 100 | 7.086 | 7.054 | 7.046 | 7.044 | 7.043 | 7.043 | 7.043 |
| 1.00 | 105 | 4.534 | 4.518 | 4.513 | 4.512 | 4.512 | 4.512 | 4.512 |

**Table 2.** Calculated Semi-Lagrangian (S-L) results (Richardson extrapolation (40,80)) for Fixed Strike Call Asians where $r = 0.1$, at a spot price of $S_0 = 100$. $nt = 10$ for all maturities and $\lambda = 10$.

| $\sigma$ | $T$ | $K$ | S-L | (RS, 1995) | (ZFV, 1997) | (BP, 1996) |
|---|---|---|---|---|---|---|
| 0.20 | 0.25 | 95 | 6.478 | 6.476 | 6.501 | 6.5 |
| | | 100 | 2.934 | 2.932 | 2.928 | 2.96 |
| | | 105 | 0.951 | 0.947 | 0.971 | 0.966 |
| | 0.50 | 95 | 7.893 | 7.891 | 7.921 | 7.793 |
| | | 100 | 4.506 | 4.505 | 4.511 | 4.548 |
| | | 105 | 2.209 | 2.211 | 2.229 | 2.241 |
| | 1.00 | 95 | 10.294 | 10.295 | 10.309 | 10.336 |
| | | 100 | 7.043 | 7.042 | 7.042 | 7.079 |
| | | 105 | 4.512 | 4.509 | 4.519 | 4.539 |

**Table 3.** Semi-Lagrangian (S-L) convergence for Fixed Strike Put Asians with early exercise, where $r = 0.1$, quoted at strike $K = 100$. $nt = 40$, for all maturities. $\lambda = 5$.

| $\sigma$ | $T$ | Calculated by S-L | | | | (BP, 1996) |
|---|---|---|---|---|---|---|
| | | 20x20 | 40x40 | 80x80 | 160x160 | |
| 0.20 | 0.25 | 1.721 | 1.959 | 2.084 | 2.090 | 2.066 |
| | 0.50 | 2.286 | 2.621 | 2.667 | 2.667 | 2.629 |
| | 1.00 | 2.981 | 3.243 | 3.256 | 3.255 | 3.181 |
| 0.40 | 0.25 | 4.360 | 4.619 | 4.614 | 4.614 | 4.581 |
| | 0.50 | 6.084 | 6.128 | 6.124 | 6.122 | 6.078 |
| | 1.00 | 7.850 | 7.862 | 7.856 | 7.855 | 7.761 |

## 3 Conclusions

Semi-Lagrange time integration simplifies the Asian pricing PDE into a $A$-parameterised set of one-factor problems. The asset price process can be easily extended e.g. for volatility surfaces. The method is unconditionally stable when combined with implicit finite differences. The resulting algebraic problems are linear and block tri-diagonal and stretched meshes are easily incorporated and give very accurate prices. Early exercise leads to A-parameterised one-factor LCP's that can be solved by PSOR.

## References


1. J. Barraquand and T. Pudet. Pricing of American path-dependent contingent claims. *Mathematical Finance*, 6(1):17–51, 1996.
2. N.C. Clarke and A.K. Parrott. Multigrid for American option pricing with stochastic volatility. *Applied Mathematical Finance*, 6:177–195, 1999.
3. Y. d'Halluin, P.A. Forsyth, and G. Labahn. A semi-Lagrangian approach for American Asian options under jump diffusion. *to appear in SIAM J. Sci. Comp.*
4. A.K. Parrott and Clarke N.C. Parallel solution of American Asian options. In *Proceedings of the 11th Domain Decomposition Conference*, Bergen, 1999. Domain Decomposition Press.
5. L.C.G. Rogers and Z. Shi. The value of an Asian option. *Journal of Applied Probability*, 32:1077–1088, 1995.
6. A. Staniforth and J. Cote. Semi-Lagrangian integration schemes for atmospheric models - a review. *Monthly Weather Review*, 119:2206–2223, 1991.
7. R. Zvan, P. Forsyth, and K.R. Vetzal. Robust numerical methods for PDE models of Asian options. *Journal of Computational Finance*, 1(2):39–78, 1997/98.

# Fuzzy Binary Tree Model for European Options

S. Muzzioli[1] and H. Reynaerts[2]

[1] Department of Economics, University of Modena and Reggio Emilia, Italy muzzioli.silvia@unimore.it
[2] Department of Applied Mathematics and Computer Science, University of Gent, Belgium huguette.reynaerts@UGent.be

**Summary.** The derivation of the risk neutral probabilities in a binary tree, in the presence of uncertainty on the underlying asset moves, boils down to the solution of dual fuzzy linear systems. The issue has previously been addressed and different solutions to the dual systems have been found. The aim of this paper is to apply a methodology which leads to a unique solution for the dual systems.

**Key words:** Dual fuzzy linear systems, vanilla option, binary tree.

## 1 Introduction

In option pricing theory, the Cox-Ross-Rubinstein [2] formula is a fundamental result for option pricing in a binomial model. The formula is derived under the assumption that the stock price follows a binomial process characterised by the two jump factors: "up", $u$, and "down", $d$. In the literature, various generalisation of the Cox-Ross-Rubinstein formula characterised by more rough assumptions on the stock price moves have been proposed. Kolokoltsov [3] examines the case in which only minimal information on the future evolution of the stock is available, namely the case in which only the bounds of the possible stock moves are known. He supposes that the possible stock moves belong to the closed interval $[d, u]$ and derives, by resorting to the non-expansive maps theory, an interval of possible prices for the option contract. Muzzioli and Torricelli [5] analyse the case in which more information is available on the two jump factors. They suppose that $u$ and $d$ are represented by two triangular fuzzy numbers $(u_1, u_2, u_3)$ and $(d_1, d_2, d_3)$, *i.e.*, that $u$ and $d$ can take values on the closed intervals $[u_1, u_3]$ and $[d_1, d_3]$ and have a most possible value $u_2$, and $d_2$, respectively. In this paper we follow the approach by Muzzioli and Torricelli [5] and analyse the problem of the derivation of the risk neutral probabilities in a binomial tree. The problem boils down to the solution of a fuzzy linear system. The issue has previously been addressed and different solutions for the same system have been found. The aim of this paper is twofold. First we highlight that the different solutions proposed arise from different versions of the original system. Second we apply a methodology which leads to a unique solution for all the different versions of the fuzzy linear system. The plan of the paper is the following. In section 2 we recall the financial problem. In section 3 we highlight that the system has no solution if one applies standard fuzzy arithmetic and we derive the vector solution. The last section concludes.

## 2 European-style Plain Vanilla Options in the Presence of Uncertainty

A call option is a financial security that provides its holder, in exchange for the payment of a premium, the right but not the obligation to buy a certain underlying asset at a certain date in the future for a specified price $K$. In the binary tree model of Cox-Ross-Rubinstein [2] the following assumptions are made: (A1) the markets have no transaction costs, no taxes, no restrictions on short sales, and assets are infinitely divisible; (A2) the lifetime $T$ of the option is divided into $N$ time steps of length $T/N$; (A3) the market is complete; (A4) no arbitrage opportunities are allowed, which implies for the risk-free interest factor, $1+r$, over one step of length $T/N$, that $d < 1+r < u$, where $u$ is the up and $d$ the down factor. The European call option price at time zero, has a well-known formula in this model, $EC(K,T) = \frac{1}{(1+r)^N}\sum_{j=0}^{N}\binom{N}{j}p_u^j p_d^{N-j}\left(S(0)u^j d^{N-j} - K\right)_+$, where $K$ is the exercise price, $S(0)$ is the price of the underlying asset at time the contract begins. $p_u$ and $p_d$ are the resp. up and down risk-neutral transition probabilities which are solutions to the system:

$$p_u + p_d = 1 \qquad up_u + dp_d = 1 + r. \tag{1}$$

The solution is given by

$$p_u = ((1+r) - d)(u-d)^{-1}, \qquad p_d = (u - (1+r))(u-d)^{-1} \tag{2}$$

The standard methodology (see Cox et al. [2]) leads to set $u = d^{-1} = e^{\sigma\sqrt{T/N}}$, where $\sigma$ is the volatility of the underlying asset. If there is some uncertainty about the value of the volatility, then it is also impossible to precisely estimate the up and down factors. Muzzioli et al. [5] analyse the case in which $u$ and $d$ are represented by triangular fuzzy numbers. A triangular fuzzy number is uniquely defined by three numbers $(f_1, f_2, f_3)$ or can be written in terms of its $\alpha$-cuts, $f(\alpha) = [\underline{f}(\alpha), \overline{f}(\alpha)] = [\underline{f}, \overline{f}] = [f_1 + \alpha(f_2 - f_1), f_3 - \alpha(f_3 - f_2)]$, $\alpha$ in $[0,1]$. Since the $\alpha$-cuts of a triangular fuzzy number are compact intervals of the set of real numbers, interval calculus can be applied on them. In this setting the up (resp. down) factor is represented by the triangular fuzzy numbers

$u = (u_1, u_2, u_3)$ (resp. $d = (d_1, d_2, d_3)$). Assumptions (A1), (A2), (A3) still hold, while assumption (A4) becomes: $d_1 \le d_2 \le d_3 < 1 + r < u_1 \le u_2 \le u_3$. Note that this condition guarantees that the fuzzy matrix $F$ has full rank $\forall d \in [d_1, d_3]$ and $\forall u \in [u_1, u_3]$ :

$$F = \begin{bmatrix} 1 & 1 \\ d & u \end{bmatrix}.$$

## 3 Solving Fuzzy Linear Systems

A fuzzy version of the two equations of the system (1) should now be introduced. This can be done (for each equation) in two different ways, since for an arbitrary fuzzy number $f$ there exists no fuzzy number $g$ such that $f + g = 0$ and for all non-crisp fuzzy numbers $f + (-f) \neq 0$. Therefore the linear system (1) can be rewritten in four different ways.

$$\begin{cases} p_u + p_d = 1 \\ up_u + dp_d = 1 + r \end{cases} \qquad \begin{cases} p_u = 1 - p_d \\ up_u + dp_d = 1 + r \end{cases} \tag{3}$$

$$\begin{cases} p_u = 1 - p_d \\ dp_d = (1 + r) - up_u \end{cases} \qquad \begin{cases} p_u + p_d = 1 \\ dp_d = (1 + r) - up_u \end{cases} \tag{4}$$

Buckley et al. [1] propose the following procedure to solve the fuzzy matrix equation $Ax = b$, where the elements, $a_{ij}$, of the $n \times n$-matrix $A$ and the elements, $b_i$, of the $n \times 1$-vector $b$ are triangular fuzzy numbers: (1) solve the linear system by using fuzzy number arithmetic, (2) if no such solution exists use the vector solution $X_J$, with $X_J(\alpha) = \{x \mid A_\alpha x = b_\alpha, (A_\alpha)_{ij} \in a_{ij}(\alpha), (b_\alpha)_i \in b_i(\alpha)\}$. Moreover, Muzzioli and Reynaerts [4] generalize the vector solution of Buckley et al. [1] to the fuzzy linear systems $A_1 x + b_1 = A_2 x + b_2$, where the elements of the $n \times n$-matrices $A_1$ and $A_2$ and the elements of the $n \times 1$-matrices $b_1$ and $b_2$ are fuzzy numbers. They prove that the fuzzy system $A_1 x + b_1 = A_2 x + b_2$ has a vector solution $X_J^*$, with

$$\begin{aligned} X_J^*(\alpha) = \{x \mid \; & A_{1\alpha} x + b_{1\alpha} = A_{2\alpha} x + b_{2\alpha}, (A_{1\alpha})_{ij} \in a_{1,ij}(\alpha), \\ & (A_{2\alpha})_{ij} \in a_{2,ij}(\alpha), (b_{1\alpha})_i \in b_{1,i}(\alpha), (b_{2\alpha})_i \in b_{2,i}(\alpha)\}, \end{aligned}$$

if all matrices $A_{1,0} - A_{2,0} = [a^0_{1,ij} - a^0_{2,ij}]$, with $a_{1,ij} \in a_{1,ij}(0)$ and $a_{2,ij} \in a_{2,ij}(0)$, are nonsingular. They also prove that the linear systems $Ax = b$ and $A_1 x + b_1 = A_2 x + b_2$, with $A = A_1 - A_2$ and $b = b_2 - b_1$ have the same vector solution. Let us apply the above mentioned procedure in order to find the solution to the fuzzy linear system. We have first to solve the systems by using fuzzy number arithmetic. One can easily prove that the four fuzzy linear systems have no solution if one applies fuzzy arithmetic. Moreover, by solving system (3) and system (4) respectively, Muzzioli and Torricelli [5] and

Reynaerts and Vanmaele [6], found different solutions for the original fuzzy linear system. We now investigate the vector solution. This solution leads to one and the same result for all four linear systems. It is obtained by solving the first system in (3):

$$\begin{aligned}(A_\alpha)_{1,1} &= (A_\alpha)_{1,2} = (b_\alpha)_1 = 1 \qquad (b_\alpha)_2 = 1 + r\\ (A_\alpha)_{2,1} &= u_1 + \alpha(u_2 - u_1) + \lambda_1(u_3 - u_1 - \alpha(u_3 - u_1)), \quad \lambda_1 \in [0,1]\\ (A_\alpha)_{2,2} &= d_1 + \alpha(d_2 - d_1) + \lambda_2(d_3 - d_1 - \alpha(d_3 - d_1)), \quad \lambda_2 \in [0,1].\end{aligned}$$

The vector solution is:

$$\begin{aligned}p_u(\alpha) &= \frac{(1+r) - (d_1 + \alpha(d_2 - d_1) + \lambda_2(d_3 - d_1)(1-\alpha))}{f}\\ p_d(\alpha) &= \frac{u_1 + \alpha(u_2 - u_1) + \lambda_1(u_3 - u_1)(1-\alpha) - (1+r))}{f}, \lambda_1, \lambda_2 \in [0,1],\\ f &= u_1 - d_1 + \alpha((u_2 - u_1) - (d_2 - d_1)) + (1-\alpha)(\lambda_1(u_3 - u_1) - \lambda_2(d_3 - d_1)).\end{aligned}$$

By minimising and maximising those functions over $\lambda_1, \lambda_2$ we get the marginals:

$$\begin{aligned}p_u(\alpha) &= [\frac{1 + r - d_3 + \alpha(d_3 - d_2)}{u_3 - \alpha(u_3 - u_2) - d_3 + \alpha(d_3 - d_2)},\\ &\qquad \frac{1 + r - d_1 - \alpha(d_2 - d_1)}{u_1 + \alpha(u_2 - u_1) - d_1 - \alpha(d_2 - d_1)}]\\ p_d(\alpha) &= [\frac{u_1 + \alpha(u_2 - u_1) - (1+r)}{u_1 + \alpha(u_2 - u_1) - d_1 - \alpha(d_2 - d_1)},\\ &\qquad \frac{u_3 - \alpha(u_3 - u_2) - (1+r)}{u_3 - \alpha(u_3 - u_2) - d_3 + \alpha(d_3 - d_2)}]\end{aligned}$$

Thus for $\alpha = 1$ one gets:

$$p_u(1) = ((1+r) - d_2)(u_2 - d_2)^{-1} \qquad p_d(1) = (u_2 - (1+r))(u_2 - d_2)^{-1}$$

which means that the most possible value for the fuzzy transition probabilities is equal to the transition probabilities (2) in the crisp case. For $\alpha = 0$ one gets:

$$\begin{aligned}p_u(0) &= \frac{(1+r) - d_1 - \lambda_2(d_3 - d_1)}{u_1 - d_1 + \lambda_1(u_3 - u_1) - \lambda_2(d_3 - d_1)}, \quad \lambda_1, \lambda_2 \in [0,1]\}\\ &= [\frac{(1+r) - d_3}{u_3 - d_3}, \frac{(1+r) - d_1}{u_1 - d_1}]\\ p_d(0) &= \frac{u_1 + \lambda_1(u_3 - u_1) - (1+r)}{u_1 - d_1 + \lambda_1(u_3 - u_1) - \lambda_2(d_3 - d_1)}, \quad \lambda_1, \lambda_2 \in [0,1]\}\\ &= [\frac{u_1 - (1+r)}{u_1 - d_1}, \frac{u_3 - (1+r)}{u_3 - d_3}]\end{aligned}$$

Note that this is the solution found in Muzzioli and Torricelli [5].
Muzzioli and Reynaerts [4] propose a practical algorithm to find directly the marginals for each unknown that involves the solution of $2^k$ systems, where $k$ is the number of fuzzy parameters in the original fuzzy system. Each element of the extended coefficient matrix of those systems is either the lower or the upper bound of the $\alpha$-cut of the corresponding element of the original fuzzy extended coefficient matrix. If one applies this algorithm to the financial example, one should solve four linear systems, with respective solutions:

$$\begin{cases} p_u = \frac{(1+r)-\underline{d}}{\underline{u}-\underline{d}} \\ p_d = \frac{\underline{u}-(1+r)}{\underline{u}-\underline{d}} \end{cases} \quad \begin{cases} p_u = \frac{(1+r)-\underline{d}}{\overline{u}-\underline{d}} \\ p_d = \frac{\overline{u}-(1+r)}{\overline{u}-\underline{d}} \end{cases} \quad \begin{cases} p_u = \frac{(1+r)-\overline{d}}{\underline{u}-\overline{d}} \\ p_d = \frac{\underline{u}-(1+r)}{\underline{u}-\overline{d}} \end{cases} \quad \begin{cases} p_u = \frac{(1+r)-\overline{d}}{\overline{u}-\overline{d}} \\ p_d = \frac{\overline{u}-(1+r)}{\overline{u}-\overline{d}}. \end{cases}$$

The final solution is obtained by taking the minimum and maximum for each unknown: $\left( [\frac{(1+r)-\overline{d}}{\overline{u}-\overline{d}}, \frac{(1+r)-\underline{d}}{\underline{u}-\underline{d}}]\, [\frac{\underline{u}-(1+r)}{\underline{u}-\underline{d}}, \frac{\overline{u}-(1+r)}{\overline{u}-\overline{d}}] \right)$.

## 4 Conclusions

The derivation of the risk neutral probabilities in a lattice framework, in the presence of uncertainty on the underlying asset moves, boils down to the solution of a fuzzy linear system. In this paper we have investigated the solution of such a system by using the methodology proposed by Muzzioli and Reynaerts [4]. The solution, that is the same solution as found in Muzzioli and Torricelli [5], is here given in terms of vector solution.

*Acknowledgement.* S. Muzzioli acknowledges support from COFIN2001. H. Reynaerts acknowledges support from the BOF-project 001104599, UGent.

## References


1. J.J. Buckley and Y. Qu. Solving systems of linear fuzzy equations. *Fuzzy sets and systems*, 43:33–43, 1991.
2. J. Cox, S. Ross, and S. Rubinstein. Option pricing, a simplified approach. *Journal of Financial Economics*, 7:229–263, 1979.
3. V. Kolokoltsov. Nonexpansive maps and option pricing theory. *Kybernetica*, 34:713–724, 1998.
4. S. Muzzioli and H. Reynaerts. Fuzzy linear systems of the form $a_1x+b_1 = a_2x+b_2$. Submitted to Fuzzy sets and systems, 2004.
5. S. Muzzioli and C. Torricelli. A multiperiod binomial model for pricing options in a vague world. *Journal of Economic Dynamics and Control*, 28:861–887, 2004. Special Issue on Financial Modelling.
6. H. Reynaerts and M. Vanmaele. A sensitivity analysis for the pricing of European call options in a binary tree model. In *Proceedings of the third International Symposium on Imprecise Probabilities and Their Applications*, pages 467–481, 2003.

# Effective Estimation of Banking Liquidity Risk

P. Tobin and A. Brown

Swinburne University of Technology, Hawthorn, Australia

**Summary.** We present an effective way to estimate liquidity risk.

**Key words:** liquidity risk, estimation, extreme value theory.

## 1 Introduction

Banks commonly identify four specific forms of financial risk—credit risk, operational risk, market risk and liquidity risk [1]. The first three of these are often incorporated into existing capital allocation frameworks (see *e.g.*, [5]). Liquidity risk is risk that much of depositors' funds may be withdrawn in a short period of time. This is partly confounded with market risk. Sometimes (*e.g.*, [5]), liquidity risk is treated as a part of market risk associated with the Financial Market line of business. Others (*e.g.*, [4]) treat liquidity risk separately. Allen ([2]) divided liquidity risk into Funding and Asset components identifying the latter with market risk. No technique for modelling liquidity risk currently has wide acceptance, but use of modified Value at Risk models (popular in modelling market risk (see *e.g.*, [5]) has been suggested ([6],[3]). Liquidity problems can occur in normal times as no market is perfectly liquid or in crisis times where the most severe outcomes can be expected. Management of liquidity in normal times can include use of derivative instruments [3] as well as sensible control on cash flows. The cost of holding reserve funds must be built into banking product prices. Liquidity risk is assessed independently for each product to avoid unintended subsidy across products.

Here we examine a specific banking case and review performance on four products on offer - personal transactions, savings accounts, term deposits both carded (those with standard interest rates offered on fixed periods) and non-carded (where the deposit is negotiated on an ad hoc basis). The specific task was to assess the level at which the reserves would suffice, with a probability of 99.97%, to cover the withdrawals in a week for these four retail banking products.

## 2 Data Handling

Transaction data by amount and numbers for product and type were supplied for the period October 29, 2000 to June 30, 2001, a total of 245 days (35 weeks). Working on weekly data smoothed out some of irregularities associated with holidays. Data were rescaled using average values taken over the 35 week period to ensure confidentiality. These average values are named the *mob*, $M_t$, the *clip* $C_t$ and the *bag* $B_t$ . The mob measures numbers of withdrawals and the clip the size of those withdrawals. The bag is a total from these. Weekly mean values were calculated giving rise to weekly factors with a mean of 1, despite the incidence of public holidays. Dispersion of these weekly factors was the subject of further analysis.

From extreme value theory we realise that the worst case outcome can be far worse than anything "normal" distributions of data may predict. For Liquidity risk this has two implications. In extreme circumstances like the Wall street crash of 1929 market forces cause massive liquidity problems. Then a run on all products in a bank will coincide and usually the outcome will be handled politically. Such a liquidity crisis will affect all banks at once. There is literally no way to trade in normal times if we must be able to guard against such a crisis. Some banks have faced their own major crises alone however through mismanagement or very bad luck. Usually liquidity shortages arise in more normal times and these can be handled by conventional means. For this reason we will not be concerned about the problems in using univariate extreme value theory.

The sample moments of each distribution can be readily calculated from the data. The standard deviation is the same as the coefficient of variation when the mean is 1. The skewness and kurtosis are based on the third and fourth moments. The Normal Power approximation can be used to estimate the 99.97% point of the amount distribution. This approximation to the normal distribution, $y$, for the number of bags required, $x$ is given by

$$\frac{x-\mu}{\sigma} = y + \frac{\gamma}{6}(y^2-1) + \frac{\kappa}{24}(y^3-3y) + \frac{\gamma}{6}\left(2y^3-5y\right) + O(n^{-1.5}),$$

where $\Phi(y) = 1-\varepsilon$ gives the normal variate in terms of its tail, $\varepsilon$. Here $\Phi(y) = 99.97\%$. The inverse normal distribution then gives $y = 3.432$. The values of the mean $\mu$, standard deviation $\sigma$, skewness $\gamma$ and kurtosis $\kappa$ from the distribution of amount data are then used to find $x$. The number of reserve bags is given by $x-1$.

We see that any error can be made smaller by increasing the size of the sample data, $n$. The use of just 35 weeks of data to estimate probabilities at the 3 in 10,000 level involves a high degree of extrapolation as we are attempting to estimate the reserves required for the worst week in 64 years!

The following table gives the reserves at 99.97% confidence level for the four products as well as data on the distribution moments.

| Product | P Trans. | Sav A/C's | Carded | Noncarded | All |
|---|---|---|---|---|---|
| Mean bag, $\mu$ | 1.000 | 1.000 | 1.000 | 1.000 | 1.000 |
| Std dev'n, $\sigma$ | 0.096 | 0.146 | 0.259 | 0.355 | 0.150 |
| Skewness, $\gamma$ | 0.333 | -0.488 | 1.161 | 1.468 | 0.689 |
| Kurtosis, $\kappa$ | 6.372 | 2.086 | 1.454 | 3.573 | 1.976 |
| Reserve Bags | 1.138 | 0.692 | 1.284 | 2.397 | 0.945 |

The high kurtosis in the amount of personal transactional withdrawal is due to an Easter effect. The term deposit withdrawals have strong positive skewness in amount due to positive skewness in their size. The overall reserve indicated for all products is less than the sum of the reserves for each individual product. This is due to less than perfect correlation for movements by all the products in the data.

## 3 Correlations

In extreme circumstances we cannot assume products all move similarly. We can examine the correlation effect in detail. The correlation of the withdrawal amounts (in the same week) between products is shown. This is fairly high within a week, but falls away rapidly at weekly lags 1 and 2. This indicates a tendency for high withdrawals to occur for all products at the same time.

| Product | P Trans | Sav A/C's | Carded | Noncarded | All |
|---|---|---|---|---|---|
| **Pers Trans** | 1.00 | 0.85 | 0.38 | 0.63 | 0.81 |
| **Savings A/C's** | 0.85 | 1.00 | 0.28 | 0.49 | 0.69 |
| **Carded T.D.** | 0.38 | 0.28 | 1.00 | 0.51 | 0.79 |
| **Noncarded T.D.** | 0.63 | 0.49 | 0.51 | 1.00 | 0.85 |
| **All Retail** | 0.81 | 0.69 | 0.79 | 0.85 | 1.00 |

The correlation of the number of withdrawals and their size between products is shown following. The personal transactional and non-carded term deposit products show strong positive correlations between the number and size of the withdrawal. The other products do not conform to this pattern and the overall result for all products is a modest 32% but correlations of this magnitude cannot be ignored. Assuming independence between number and size of withdrawal is clearly not tenable. A real difficulty is determining the strength of this dependence in a crisis, although it may be expected to follow the worst case scenario.

| Product | P Trans | Sav A/C's | Carded | Noncarded | All |
|---|---|---|---|---|---|
| **Pers Trans** | 0.65 | 0.50 | -0.21 | 0.41 | 0.34 |
| **Savings A/C's** | 0.58 | 0.13 | -0.20 | 0.14 | 0.21 |
| **Carded T.D.** | 0.46 | 0.68 | -0.37 | 0.31 | 0.51 |
| **Noncarded T.D.** | 0.78 | 0.60 | 0.10 | 0.51 | 0.83 |
| **All Retail** | 0.66 | 0.44 | -0.22 | 0.37 | 0.32 |

The expected values of the mob, clip and bag are fixed at 1 by definition. It follows immediately that $E[B_t] = E[M_t] \times E[C_t]$. We cannot assume that $E[B_t] = E[M_t \times C_t]$ unless we can show that mob and clip are independent.

We have good reason to suspect the contrary case. In fact we may assume that some association exists. Using logarithms we obtain the new equation

$$\ln B_t = \ln M_t + \ln C_t - k$$

The value of $k$ can be positive, zero or negative. This measure is related to the coefficient of correlation between number and size of withdrawals incorporating any skewness effect existing in either variable. This linear structural equation for the data gives equal weight to variations in number and size. The parameter $k$ provides a measure of how much association between size and number influences the total amount. Using our expected values we have $k \approx \ln E[N_t \times Y_t] - \ln(E[N_t] \times E[Y_t])$.

If size and number are independent variables, then $k$ would be close to zero. The sample weekly data retail products provides us with a range of examples. No association: Savings accounts ($k = 0$) (This may be spurious as there is distinct evidence of a fortnightly cycle.) Negative association: Carded term deposits ($k = -.019$). Positive association: Personal transactional ($k = .002$) and non carded term deposits ($k = .019$).

Does variation in size or number of withdrawals provide the bigger risk? Recall that the log clip and log mob are simple measures of size and number of withdrawals, and a scatterplot of these gives a sense of their relationship. A wider scatter in data points in either axis direction implies a greater degree of risk associated with the measure along that axis. For the personal transactional product both number and size are influential. The data for the Savings account product has strong fortnightly cycle. For this product the number of transactions is more variable than the size. There is a distinct lack of values close to the average which can be readily explained by the fortnightly cycle.

The time series for the carded term deposit data exhibits an intertwined pattern due to the negative association between number and size. The non-carded term deposits exhibit features that increase liquidity risk. The size and number of withdrawals have positive association. Surprisingly, for this product, it seems the number of withdrawals is more influential on the liquidity risk than their size.

## 4 Conclusion

The bank management should be pleased with the results for carded term deposits. The carded rates seem to smooth fluctuations in the withdrawal amounts. However the result for non-carded term deposits, often larger deposits, is disappointing. The investors who operate in this market may be increasing the liquidity risk of the bank.

The structural equation developed provides a simple inexpensive tool for supervisory control of the on-going liquidity risk of the bank. Liquidity risk can be reduced by controlling the manner in which the book is built. It is desirable that number and size of withdrawals have a negative association at each future date. A methodology to calculate reserves to cover the liquidity risk has been demonstrated. More data is required before we rely on the calculated values as parameter estimates obtained from small samples can contain significant biases.

## References

1. J.N. Allan, P.M. Booth, R.J. Verrall, and D.E.P. Walsh. The management of risks in banking. *British Actuarial Journal*, 4:707–802, 1998.
2. S. Allen. *Financial Risk Management.* Wiley, 2003.
3. K. Dowd. *Beyond Value at Risk: The New Science of Risk Management.* Wiley, 1998.
4. P. Embrechts. Extremes in economics and economics of extremes. In B. Finkenstadt and H. Rootzén, editors, *Extreme Values in Finance, Telecommunications and the Environment*, pages 169–183. Chapman & Hall, 2003.
5. E.F. Kupper. Risk management in banking. In C. Cassidy, editor, *Risk and Capital Management, Proceedings of the Conference on Risk and Capital Management in Financial Institutions.* Australian Prudential Regulation Authority, 1998.
6. H. Monkkonen. Managing the spread. *Risk*, (October):109–112, 2000.

# Part VIII

## Theme: Water Flow

# Multiphase Flow and Transport Modeling in Heterogeneous Porous Media

R. Helmig[1], C.T. Miller[2], H. Jakobs[1], H. Class[1], M. Hilpert[2], C. E. Kees[2], and J. Niessner[1]

[1] Institut für Wasserbau, Lehrstuhl für Hydromechanik und Hydrosystemmodellierung, Universität Stuttgart, Pfaffenwaldring 61, 70569 Stuttgart, Germany rainer@iws.uni-stuttgart.de

[2] Department of Environmental Sciences and Engineering, CB 7431, 104 Rosenau Hall, University of North Carolina at Chapel Hill Chapel Hill, NC 27599-7431, U.S.A. casey_miller@unc.edu

**Summary.** We focus on the inter-related roles of scale and heterogeneity of porous medium properties for fluid flow and contaminant transport in isothermal and non-isothermal multiphase systems across a range of scales. Multiscale network and macro-scale continuum models, and detailed laboratory experiments are used to support the investigation. We demonstrate the critical role of scale in determining the dominant forces in a porous medium system, the importance of heterogeneity across a range of scales, and the dominant role of block heterogeneities on macro-scale fluid flow and non-isothermal contaminant remediation. We give special attention to the numerical approximations of pressure-saturation-conductivity relations in heterogeneous systems, and we show the effects of interface approximation schemes on both the ability to resolve phenomena of concern and on the efficiency of the numerical simulator.

**Key words:** multiphase flow, heterogeneous systems, multi–scale problems, numerical solution

## 1 Motivation

Environmental remediation and protection has provided an especially important motivation for multiphase research in the course of the last 15 years [23, 45]. The release of non-aqueous phase liquids, both lighter and denser than water (LNAPLs and DNAPLs), into the environment is a problem of particular importance to researchers and practitioners alike [44]. Of late, such work has focused on the construction of mathematical models which can be used to test and advance our understanding of complex multiphase systems, evaluate risks to human and ecological health, and aid in the design of control and remediation methods.

One of the foremost problems facing the reliable modeling of multiphase porous medium systems is the problem of scale. Roughly speaking, a model is assembled from a set of conservation equations and constitutive, or closure, relations. One must identify constitutive relations and system-specific parameters that are appropriate for the spatial and temporal scales of interest. Often, however, a disparity exists between the measurement scale in the field or laboratory and the scale of the model application in the field. Furthermore, neither the measurement nor the field application scales are commensurate with the scale of theoretical or empirical process descriptions. Both closure relation forms and parameters are subject to change when the system of concern is heterogeneous in some relevant respect.

Figure 1 graphically depicts the range of spatial scales of concern in a typical porous medium system. It illustrates two important aspects of these natural systems: several orders of magnitude in potentially relevant length scales exist, and heterogeneity occurs across the entire range of relevant scales. A similar range of temporal scales exists as well, from the pico-seconds over which a chemical reaction can occur on a molecular length scale to the decades of concern in restoring sites contaminated with DNAPLs.

A careful definition of relevant length scales can clarify any investigation of scale considerations, although such definitions are a matter of choice and modeling approach [26]. We define the following length scales of concern: the molecular length scale, which is of the order of the size of a molecule; the microscale, or the minimum continuum length scale on which individual molecular interactions can be neglected in favor of an ensemble average of molecular collisions; the local scale, which is the minimum continuum length scale at which the microscale description of fluid movement through individual pores can be neglected in favor of averaging the fluid movement over a representative elementary volume (REV) – therefore this scale is also called the REV-scale; the mesoscale, which is a scale on which local scale properties vary distinctly and markedly; and the megascale or field-scale. Measurements or observations can yield representative information across this entire range of scales, depending on the aspect of the system observed and the nature of the instrument used to make the observation. For this reason, we do not specifically define a measurement scale.

For the minimum continuum length scale, we take the boundaries of the different grains directly into account. For the microscale, we look at a variety of pore throats and pore volumes. Note that, for both scales, we average over the properties of the fluids only (achieving for example density, viscosity).

When looking at the REV-scale, we average over both fluid–phase properties and solid–phase properties. In Fig. 2, we show schematically the averaged properties (e.g. the porosity). While averaging over a representative elementary volume (REV), we assume that the averaged property $P$ does not oscillate significantly. In Fig. 2 this is the case in the range of $V_0$ to $V_1$ with $V_0 < V_1$, so any volume $V$ with $V_0 \leq V \leq V_1$ can be chosen as REV. Accordingly, we do not assume any heterogeneities on the REV-scale. For our model, we

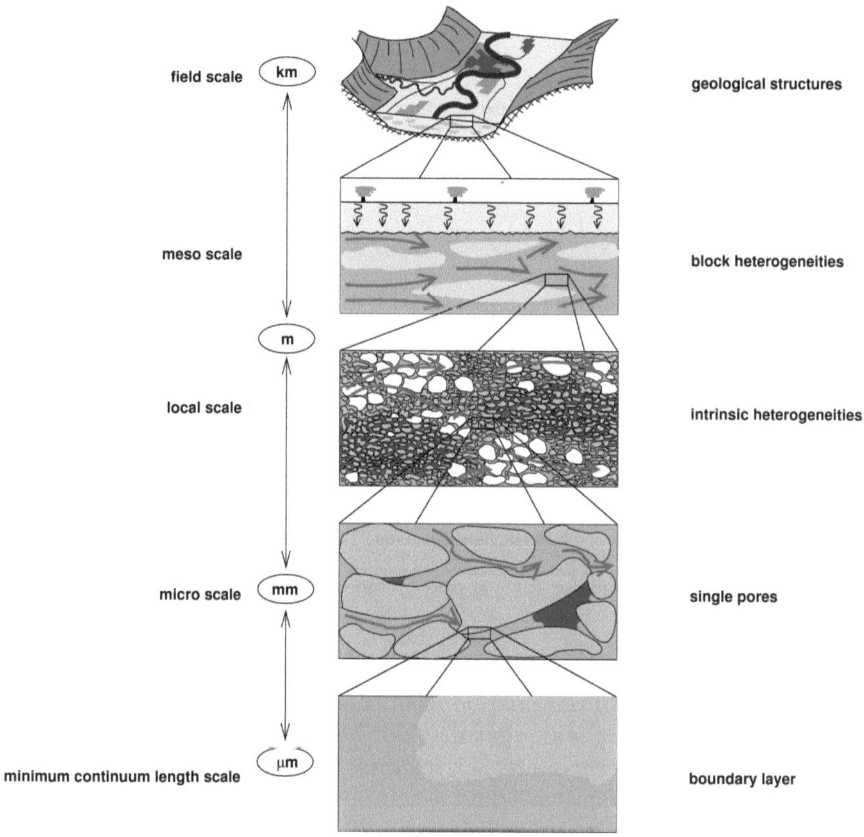


**Fig. 1.** Different scales for flow in porous media.

assume that the effects of the sub–REV–scale heterogeneities are taken into account by effective parameters. The super–REV–scale heterogeneities have to be taken into account by applying different parameters to the domain of interest. Both steady transitions as well as jumps have to be considered for the parameters. We denominate those heterogeneities with jumps within the spatial parameters as block heterogeneities. Within the context of this work, we assume that block heterogeneities can be described by subdomains with well–defined interfaces. In this paper, we do not consider heterogeneities on the field scale.

Because the scale of interest in this paper is ultimately the meso–scale, one can usually ignore molecular-scale phenomena, although these effects are embodied in continuum–conservation equations and associated closure relations. However, we must consider all other important and relevant scales in the current study of multiphase porous–medium systems.

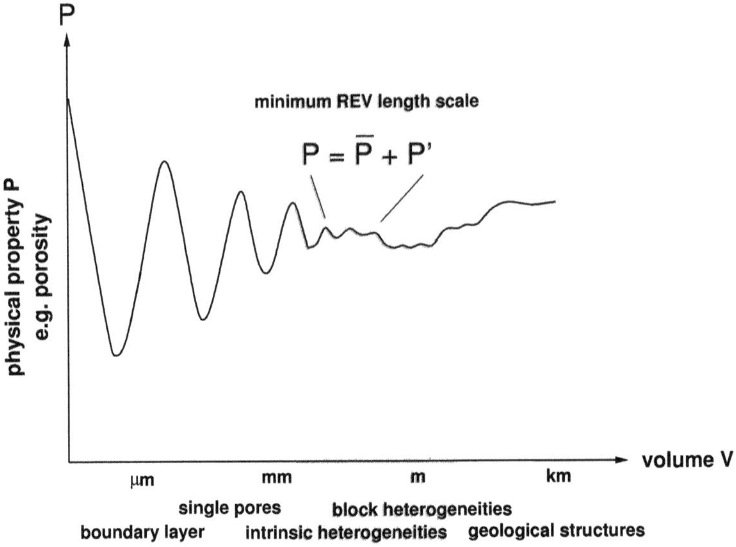


**Fig. 2.** Different scales for flow in porous media (schematically for Fig. 1).

Conceptually, one wishes to describe phenomena at a given scale using the minimum amount of information from smaller scales. This process gives rise to quantities at each scale that may not be meaningful at smaller scales. For example, fluid pressures are not relevant to individual collisions at the molecular scale, and point-wise fluid saturations or volume fractions do not necessarily reflect the microscale fluid composition at that point. A conceptually satisfying theoretical approach – one that could fundamentally increase the field's maturity – must provide a method for incorporating models on a given scale sparingly into models on the next larger scale using rigorous mathematics and sound physical reasoning.

For example, microscale models can be developed to describe fluid flow in individual pores by solving the Navier-Stokes equations [1] or Boltzmann equation [13] over an appropriate domain. These methods can in turn be used to model systems consisting of many pores, even of a size equivalent to an REV for a REV-scale porous–medium system. Such approaches have been used to develop REV-scale closure relations based upon microscale processes [25].

As yet, this kind of connection does not exist across relevant length scales for all the phenomena considered in multiphase porous–medium systems. Valid questions remain about the importance of heterogeneities for specific processes, the appropriate form and parameterization of closure relations for heterogeneous multiphase porous–medium systems, and effective ways of simulatint such systems economically.

In spite of the problems of scale, we need reliable efficient multiphase flow and transport simulators that represent the dominant flow and transport mechanisms in heterogeneous multiphase porous–medium systems. The REV-scale modeling problem has been operationally separated from the more general problem of cascading scales, although the two problems are formally entwined. The two have been split apart because of the urgent need to respond practically to such problems, even before we understand them fully. The operational separation of local scale modeling from a more comprehensive theoretical modeling methodology has resulted in many practical models and experimental studies of complex multiphase phenomena [48, 34, 30, 35, 31, 32]. Engineering has played an important role in implementing this practical response.

From the mesoscopic perspective, two basic classes of multiphase applications have received attention in the literature and deserve further consideration: the imbibition of DNAPL into a heterogeneous porous–medium system [31, 32, 22] and the removal of a DNAPL originally in a state of residual saturation [39, 40]. The former class determines the morphology of the DNAPL distribution at residual saturation, which, therefore, determines the initial condition of the latter problem. While the public is greatly concerned with remediating DNAPL–contaminated soils, many questions concerning DNAPL imbibition and removal still hinder our remediation efforts.

The overall goal of this work is to advance our understanding of models for heterogeneous multiphase porous–medium systems across a range of scales. Our specific objectives are (1) to evaluate the role of the spatial scale in determining the dominant process for multiphase flow; (2) to investigate the influence of pore-scale heterogeneity on microscale and REV-scale flow processes; (3) to summarize conventional continuum-scale mathematical models; (4) to evaluate the accuracy and efficiency of a set of spatial and temporal discretization approaches for solving multiphase flow and transport; (5) to compare numerical simulations with experimental observations for heterogeneous mesoscopic systems; and (6) to point the way toward important future areas of research in the field. The following sections address these goals by combining modeling at the pore and porous-medium continuum scales with observations of heterogeneous systems.

## 2 Scales and forces

Porous medium properties are commonly heterogeneous across a wide range of scales in subsurface systems, and such systems consequently exhibit complex fluid flow and species transport behavior. Porous medium heterogeneity can induce various physical phenomena: stable immiscible displacement; fluid entrapment; capillary-induced by-passing and pooling; and viscous, gravity, and dissolution fingering [41]. Moreover, multiple physical phenomena may be exhibited at a single spatial scale. In this section, we present experimental evidence that shows the importance of heterogeneity on fluid flow and demonstrates the operative phenomena. We then interpret the results in a larger context to demonstrate the scale-dependent nature of dominant forces.

We present results from a heterogeneous mesoscale experiment conducted at the VEGAS research facility in Stuttgart, Germany, described in more detail by [9]. Fig. 3 shows the experimental apparatus, which is a 7-m (long) $\times$ 3-m (high) $\times$ 1-m (deep) steel cell equipped with a glass front panel to facilitate visualization. On both the inflow and the outflow side, the flume was equipped with wells that extended over the whole width of the flume, leaving 6.35 [m] between the screens for the actual aquifer. Water reservoirs were maintained at a constant level on each end of the cell, resulting in a 1% gradient in the water potential, which decreased from left to right. The dimensions of the heterogeneous system within the container are 6.35-m (long) $\times$ 2.4-m (high) $\times$ 0.4-m. The cell was packed with coarse, medium, and fine sands arranged in the precise pattern shown in Fig. 3. The sands were installed in layer thicknesses of approximately $0.05[m]$, moisturized and compressed. The chief features of this heterogeneous packing included four fine sand lenses, three of which rose at gradients of 1%, 5%, and 10%, respectively, from left to right; and a horizontal medium sand layer bounded by vertical fine sand blocks. After the saturation of the aquifer, a base flow with partially deaired water was established for some time before the infiltration of TCE to dissolve the entrapped air.

According to our definitions of scale, this is a mesoscale experiment: the experimental system is large enough relative to microscale phenomena to be treated as a porous medium continuum, yet not large enough with respect to the physical dimensions of the fine sand lenses to be represented as a homogeneous porous medium continuum while still adequately resolving system processes. The physical properties of the sands used for the experiment are given in Table 1. The coarse sand's multiphase properties were measured for a water-trichloroethylene (TCE) system, and the medium and fine sand's parameters for the constitutive relationships were measured for a water-air system. The BC parameters $p_d$ for the medium and the fine sand were scaled according to [49]. The BC parameter $\lambda$ was obtained by a hand fit from measurements for one probe using the controlled outflow cell technique described in [38]. Recent comparisons between measurements taken by the controlled outflow cell technique and measurements for the same sand taken by gamma radiation system technique [18] suggest that there is a systematic error for the measurements taken by controlled outflow cell technique. Further research work on this topic is carried out within the VEGAS facility. The confidence intervals for one measurement for one probe, respectively, are given in Table 2. The results for the hydraulic conductivities measurements and consequently the values for intrinsic permeabilities vary from $-50\%$ to $+100\%$ of the mean values for different probes of the respective sands. No confidence intervals can be given for the residual saturation of water. The residual saturation for the non wetting phase measured for drainage only was zero, for imbibition

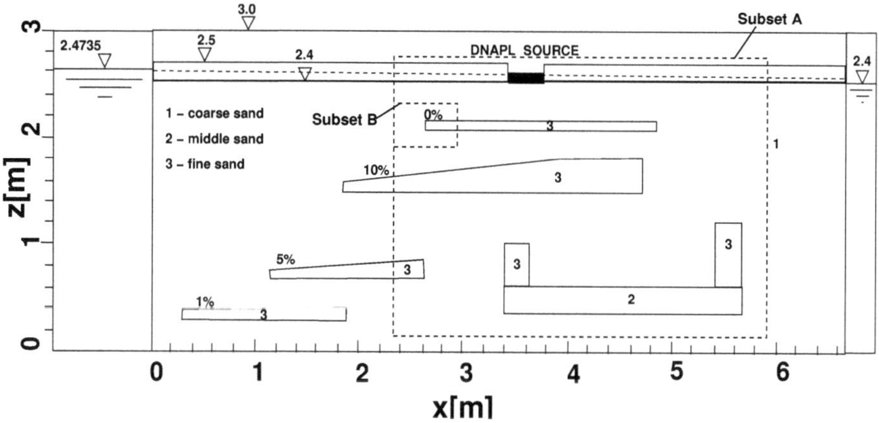


**Fig. 3.** Experimental setup for mesoscale VEGAS experiment.

hysteretic effects must be taken into account ([49, 36, 37]). Table 3 shows the fluid properties. The fluid parameters were taken from literature for water [27] and provided by the manufacturer for the TCE [21] for 20 $^oC$. The confidence intervals for these measurements are quite small ($< 1\%$) and the temperature of the system was well controlled (20 $^oC$). TCE was dyed with a hydrophobic and fluorescent compound to allow for visualization of flow patterns and of the residual saturation of TCE. The definition of all variables is summarized in the notation section.

We investigated the distribution of TCE resulting from a release of TCE into a water-saturated system under steady flow. TCE was released at a rate of approximately $175[kg/h] \pm 15\%$ from the top of the cell over an approximately $0.3[m]$ long region centered in the domain as seen in Fig. 3 for a period of 1 hour 30 minutes and allowed to redistribute for a further period of two hours

**Table 1.** Physical Properties of the Sands.

| | $p_d$ [Pa] | $\lambda$ [-] | $S_{wr}$ [-] | $K[m^2]$ | $\varphi$ [-] |
|---|---|---|---|---|---|
| coarse sand (1) | 200.0 | 2.0 | 0.05 | 4.6 E-10 | 0.39 |
| medium sand (2) | 700.0 | 2.3 | 0.15 | 3.1 E-11 | 0.35 |
| fine sand (3) | 1800.0 | 3.5 | 0.18 | 9.0E-12 | 0.43 |

**Table 2.** Confidence Intervals for one probe respectively.

| | $p_d$ [Pa] | $\varphi$ [-] |
|---|---|---|
| coarse sand (1) | $\pm 10\%$ | $\pm 7\%$ |
| medium sand (2) | $\pm 10\%$ | $\pm 10\%$ |
| fine sand (3) | $\pm 10\%$ | $\pm 2\%$ |

**Table 3.** Physical Properties of the Fluids.

| | | water | TCE |
|---|---|---|---|
| density, $\rho$ | [kg/m$^3$] | 1000.0 | 1460.0 |
| dynamic viscosity, $\mu$ | [Pa s] | 1E-3 | 5.8E-4 |

and 45 minutes. The water flow from the left to the right decreased linearly during the infiltration from 210 [l/h] to 160 [l/h]. After the infiltration was ended the water flow increased with one half hour to 180 [l/h] and remained constant. Results for saturation from the experiment were observed visually and recorded using the time domain reflectometry technique based on the techniques described in [14] and [28].

The final TCE distribution is depicted for Subset A of the cell in Fig. 4 and for Subset B of the cell in Fig. 5. The fractions of the cell represented by domain Subsets A and B are shown in Fig. 3. Several phenomena can be clearly observed: (1) gravity-motivated flow, shown by the pronounced vertical distribution of TCE (Fig. 4 overall); (2) capillary residual trapping, shown by regions of relatively evenly distributed TCE entrapped at small saturations (Fig. 4(A)); (3) capillary by-passing, shown by horizontal transport around the fine sand layers, with relatively high TCE saturations pooled on top of the fine layers (Fig. 5); and (4) entering of TCE into the fine sand lenses after the threshold saturation has been reached (Fig. 4(B)). These features result from the following physicochemical factors: (1) the balance of capillary, gravity, and viscous forces; (2) the variability in capillary properties according to porous medium type; (3) pore morphology variability that exists even within homogeneous layers; and (4) the unstable nature of vertical DNAPL migration into a water-saturated medium. The net result of these factors is a morphologically complex distribution of entrapped TCE.

These experimental results show that DNAPL migration is not only governed by the block heterogeneities but also by the variability of media properties within the presumed homogeneous regions. This indicates the important effect of heterogeneity at several scales on DNAPL movement and residual establishment. Another important feature is the formation of DNAPL pools, which are morphologically stable regions with locally large DNAPL saturations that can be extremely difficult to remediate because of mass transfer limitations.

One can evaluate multiphase systems using the principles of scaling, whereby the operative forces in a system are considered in a nondimensional form using dimensional analysis. While dimensional analysis does not provide the same level of information as solving a sound mathematical model of the system, it can provide significant insights regarding the dominant forces as functions of system scale. We use dimensional analysis to this end. The relevant forces in a multiphase system are capillary, gravitational, and viscous forces. We follow the analysis in [24] consistently. The relevant non-

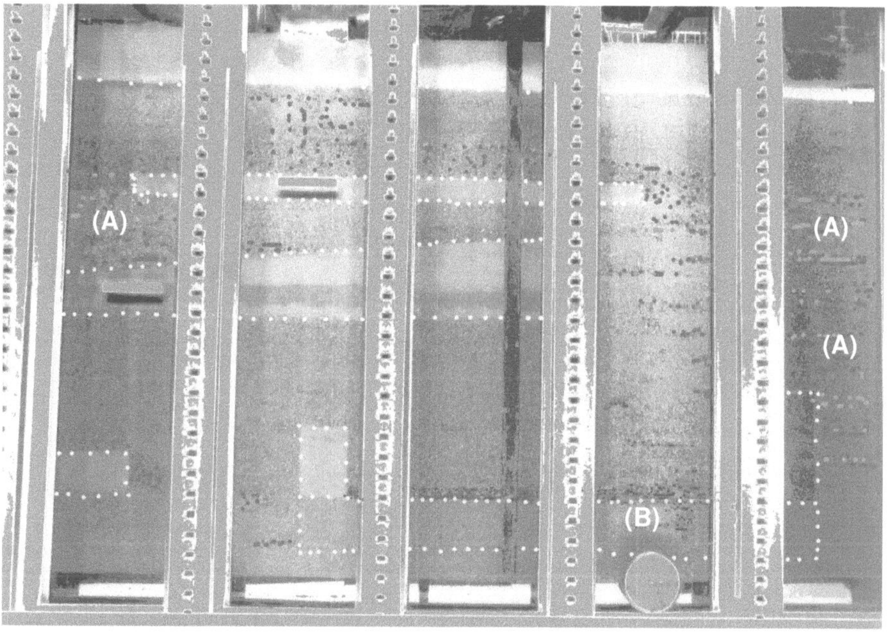


**Fig. 4.** TCE distribution for Subset A (TCE regions highlighted by red spots).

dimensional variables can be summarized as:

$$\text{capillary number} \quad Ca := \frac{\text{viscous forces}}{\text{capillary forces}} = \frac{\mu \cdot v \cdot L}{K \cdot p_c^*} \tag{1}$$

$$\text{gravitational number} \quad Gr := \frac{\text{viscous forces}}{\text{gravity forces}} = \frac{\mu \cdot v}{\Delta\varrho \cdot g \cdot K} \tag{2}$$

$$\text{bond number} \quad Bo := \frac{\text{buoyancy forces}}{\text{capillary forces}} = \frac{\Delta\varrho \cdot g \cdot L}{p_c^*}. \tag{3}$$

These numbers show that the predominance of a force depends on three kinds of variables:

1. parameters describing fluid properties, such as the density $\rho$ and the dynamic viscosity $\mu$;
2. solid phase parameters, such as the intrinsic permeability $K$;
3. parameters describing the fluid matrix interaction, as the characteristic capillary pressures $p_c^*$, here we choose the entry pressure, and a typical flow velocity $v$, which is correlated to the permeability; and
4. the characteristic length $L$.

For the viscosity and density difference, we consider a water-TCE system, i.e. we use $\mu = 0.001[Pas]$ and $\Delta\rho = 460$ $[kg\ /\ m^3]$. We use the measured material properties for the coarse sand reported in Table 1 for the

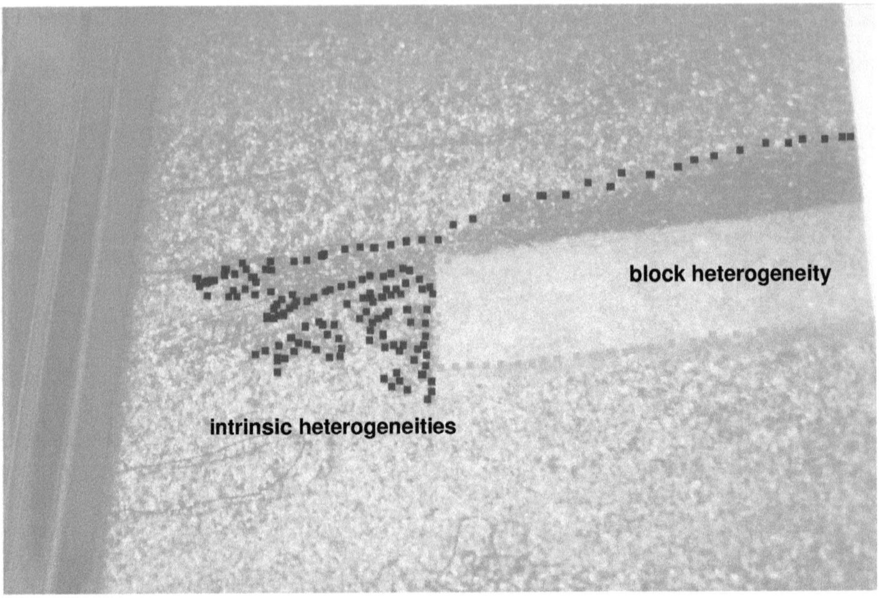


**Fig. 5.** TCE distribution for Subset B (TCE regions highlighted by red spots, and the fine sand lenses by yellow spots).

entry pressure and the intrinsic permeability. We approximate the water velocity using the inflow $q_{begin} = 210[l/h]$, $q_{min} = 160[l/h]$ and $q_{end} = 180[l/h]$ as $v_{begin} = 6.1E - 5[m/s]$, $v_{min} = 4.6E - 5[m/s]$ and $v_{end} = 5.2E - 5[m/s]$ (subscripts *begin*, *min* and *end* indicating the begin of the infiltration, the end of the infiltration – at which time the water flow is minimal – and the end of the experiment). When using the hydraulic gradient from the left to the right and the intrinsic permeability of the coarse sand, we can approximate the water flow velocity as $v_w = 4.6E-5[m/s]$ ($-50\%, +100\%$, considering the confidence interval for the permeability, indeed the good agreement between the velocity derived from the inflow rate and the velocity derived from the the hydraulic gradient and the intrinsic permeability shows us that the confidence interval is much smaller than that derived by the Darcy experiments for several probes). As a rough approximation of the fluid velocity we take $v = 5E - 5[m/s]$ for the horizontal direction. For the dimensional analysis, we assume the vertical component of the water velocity to be zero, neglecting the influence of the heterogeneities. At the time the pictures shown in Fig. 4 und 5 were taken – i.e. two hours and 45 minutes after the infiltration of DNAPL into the system was ended – the velocity of the DNAPL was essentially zero $v_n = 0$. As the capillary forces are in equilibrium with the gravitational forces the bond number must be equal one. From this results a characteristic length for the coarse sand of $L_{vertical} = 0.044[m]$ ($-10\%, +12\%$, considering the confidence interval for $p_d$ and assuming a confidence interval of 1% for the density difference). For this characteristic length we get a value for the capillary number of $Ca \approx 0.024$, i.e. the capillary forces dominate the viscous forces by factor of 42. Indeed we see in Fig. 4 that the migration of TCE is hardly influenced by the water flow from the left to the right. We don't consider the gravitational number $Gr$ for the analysis of the experiment since the viscous forces take effect in horizontal direction while the gravitational forces take effect in vertical direction.

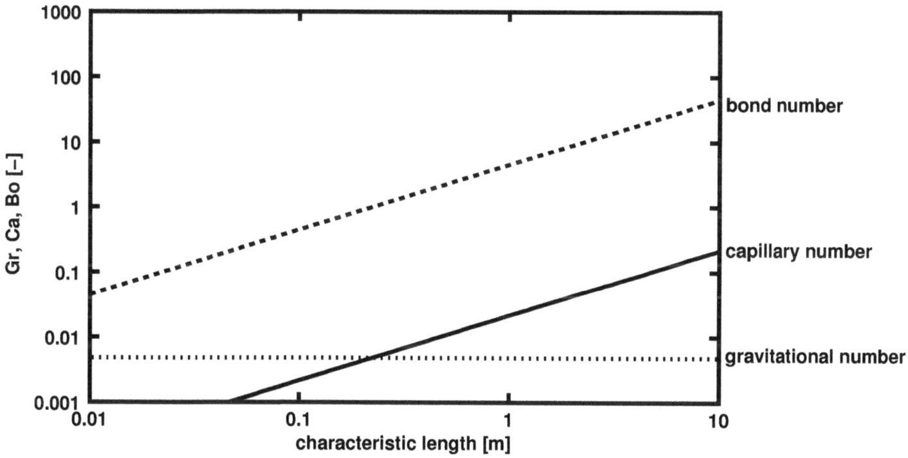


**Fig. 6.** Dimensionless numbers as a function of the characteristic length for the coarse sand according to Table 1, logarithmic scales for both axes.

Figure 6 is a plot of $Ca, Gr, Bo$ for the coarse sand used in the VEGAS experiment as a function of the characteristic length used to define these quantities in equation (2). An equilibrium between the forces is attained when the numbers equal one. For $Ca < 1$, capillary dominate viscous forces, while for $Ca > 1$, viscous dominate capillary forces. Similar reasoning can be applied to the $Gr$ and $Bo$. At small length scales, capillary forces will dominate, whereas at large length scales gravity forces will dominate, for velocities and media of the type considered here. The importance of both gravity and capillary forces can clearly be observed in the experimental results. However, the dimension analysis as used in this context does not cover the effects of heterogeneities. An extension to this end would be desirable.

Figure 4 shows NAPL flowing mainly downward due to gravity. At this scale, however, we also observe a distinct lateral spreading of the NAPL. Figure 5 shows that after reducing the scale of observation only a few decimeters, the lateral spreading of the NAPL due to capillary pressure differences is locally stronger than the density-driven downward migration of the NAPL. We attribute the lateral spreading of DNAPL in the coarse sand to anisotropic entry pressures at the pore scale.

A continuum-scale model used to describe multiphase flow and transport, for example the one presented in Sec. 4, should account for all types of heterogeneity at the various scales. While such models readily account for heterogeneous and anisotropic phase permeabilities, it is not clear whether they account appropriately for anisotropic entry pressures at the pore scale.

## 3 Anisotropy at the pore scale

As discussed in the previous section, a variety of physical phenomena are important in heterogeneous multiphase systems across a range of scales. While the VEGAS experiment showed several of these phenomena, alternative approaches can be used to investigate certain phenomena of importance. In this section, we use quasi-static pore-scale modeling to show the importance of pore-scale heterogeneities on macroscale properties of concern, namely relations among capillary fluid pressure, fluid saturations, and relative permeability. Such heterogeneities are always present in nature and can lead to macro-scale behavior that the closure relations described in Sec. 4 do not describe well.

Very few studies investigate how anisotropy of a pore space's geometry and topology at the pore scale affects macro-scale closure relations [50]. [7] showed that an anisotropic network of capillary tubes implies an anisotropic effective permeability tensor. [19] showed that both an anisotropic coordination number and pore-size distribution in a pore network lead to anisotropic permeability and diffusivities. [52] showed that size-, connectivity-, and spatial correlation-induced anisotropy at the pore scale demands a tensorial form of the Forchheimer equation. [50] derived anisotropic relative permeabilities by upscaling Miller-similar [42] porous media. The impact of pore-scale anisotropy on other closure relations for two-phase flow, such as the capillary pressure-saturation relation, has not yet received sufficient attention.

To investigate the impact of a pore space's pore-scale anisotropy on macro-scale closure relations, we performed pore-network model simulations. We used a network model with cubic pore bodies and square pore throats as described in [47] except for one extension: we introduced spatial correlations among the pore bodies. We restricted ourselves to geometric anisotropy, i.e. the semivariogram has the functional form

$$\gamma(\mathbf{h}) = \gamma_0(\mathbf{h}^T\mathbf{Q}\mathbf{h}), \tag{4}$$

where $\gamma_0$ is an isotropic semivariogram model, $\hat{Q}$ a positive definite matrix, and $\mathbf{h}$ the distance vector. We assumed that the principal directions of anisotropy are aligned with the coordinate axis and that there is only one horizontal range. Further, we assumed an exponential model. Thus,

$$\gamma(\mathbf{h}) = c\left[1 - \exp\left(-\frac{3}{a_h}\sqrt{h_x^2 + h_y^2 + \left(\frac{a_h h_z}{a_v}\right)^2}\right)\right] \tag{5}$$

where $a_h$ is the effective horizontal and $a_v$ the effective vertical range. Such a model resembles, for example, a pore space obtained by sedimentation as typical for fluvial aquifers. For $a_h/a_v \to \infty$, one obtains zonal or stratified anisotropy. The pore-body and -throat sizes were normally distributed. The pore-body radii were $r_b = (0.3 \pm 0.04)\lambda$, and the throat radii were $r_t = (0.15 \pm 0.02)\lambda$, where $\lambda$ is the lattice constant of the network. We prescribed pore-body correlations according to Eqn. (5). We used the GSLIB software package [17] to generate the pore-body radius field given the mean and variance of the pore-body radius as well as $a_h$ and $a_v$. Unlike prior investigations, we did not use periodic boundary conditions in the horizontal directions, because GSLIB does not support them; instead we removed the pore throats at the vertical boundaries of the network. We simulated two-phase flow in both the horizontal and vertical direction by attaching pressure reservoirs on opposite sides of the network.

We performed simulations in small networks with $10^3$ nodes, where $a_h$ exceeded the domain size, to understand the impact of anisotropy. Figure 7 shows that vertical flow in a horizontally stratified network is a piston-type displacement: one horizontal layer is invaded after another. More residual nonwetting phase is left behind for vertical than for horizontal flow for the following reason: suppose wetting phase is stuck in front of a layer with large mean pore-body size that is followed by a layer with smaller pore-body size. As we gradually decrease the external capillary pressure, wetting phase invades exactly one pore body. Then, the wetting phase invades the pore throat leading to the next layer and almost the entire next layer, because its pore-body sizes are, on average, smaller than the pore body originally invaded. Thus, nearly the entire nonwetting phase in the layer with large pore-body size becomes residual. Figure 8 shows that horizontal flow in a horizontally stratified network is finger type (the fingers are planes in a three-dimensional network). Less entrapment occurs for horizontal than vertical flow, because the wetting phase invades one layer after another without significant vertical flow, which could disconnect the nonwetting phase. Anisotropy at the pore scale obviously encourages layers of residual NAPL, i.e. large trapped NAPL blobs, which have been observed experimentally [39, 41]. Thus, we expect pore-scale anisotropy to lengthen the time required for NAPL dissolution compared to more evenly distributed NAPL residual in isotropic media that have a greater specific interfacial area.

To investigate the impact of anisotropy on hysteretic capillary pressure-saturation relations and wetting phase permeability $k^{rw}$, we generated pore networks with $50^3$ nodes, where $a_v = 2\lambda$ and $a_h = 2\lambda, 5\lambda$ and $50\lambda$. Assuming that the principal axis of permeability are aligned with the lattice generating vectors, we may compute a relative permeability for horizontal and verti-

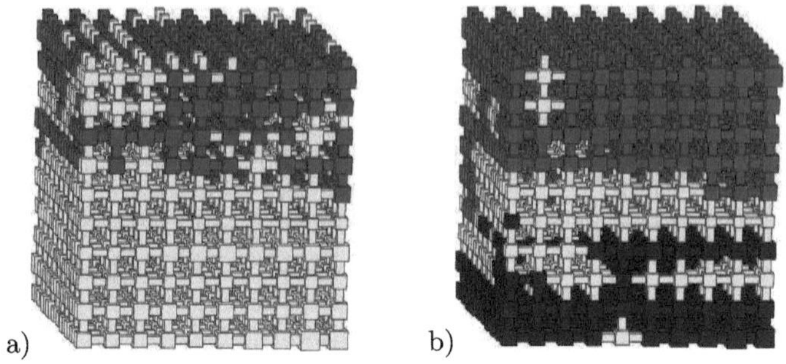


**Fig. 7.** Vertical flow through stratified pore network. The nonwetting and wetting phase reservoirs are on top and bottom, respectively. Medium and dark gray indicate connected and residual nonwetting phase, respectively; light gray indicates wetting phase. The displacement is piston type. a) During drainage. b) After main imbibition.

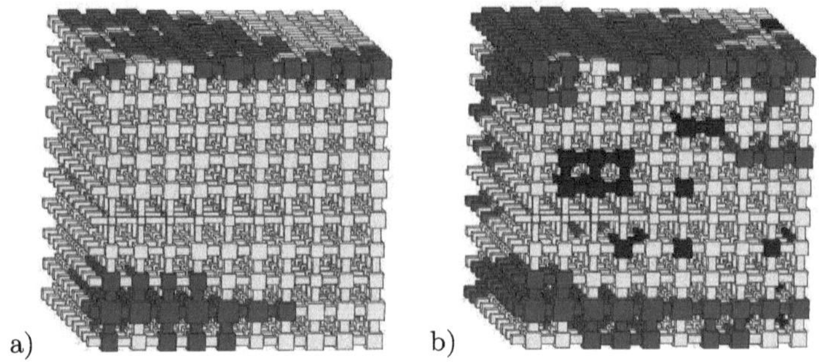


**Fig. 8.** Horizontal flow through stratified pore network. The nonwetting and wetting phase reservoirs are on the left and right, respectively. Colors as in Fig. 7. The displacement is finger type. a) During drainage. b) After main imbibition.

cal flow, $k_h^{rw}$ and $k_v^{rw}$ [7]. Likewise, we may compute the capillary pressure-saturation relation for horizontal and vertical flow, $p_h^c$ and $p_v^c$. Figure 9 shows that the amount of residual nonwetting phase obtained by main imbibition is larger for vertical than horizontal flow, consistent with the observations presented in Figs. 7 and 8 for the smaller networks. The residual increases as the effective horizontal range increases (for constant network size). Because entire horizontal layers tend to be left behind during vertical flow, $k_v^{rw} < k_h^{rw}$. The primary drainage curve for vertical flow tends to be more discontinuous than for both horizontal flow and the isotropic case, because nonwetting phase remains trapped in front of the layers with small pore-throat size until the capillary pressure exceeds a certain threshold value. We present simulation results for only one random realization of a pore network specified by its statistical parameters. With more simulations, the results usually scatter, particularly in networks with a range close to or larger than the system size. The exact shape of the primary drainage curve for vertical flow, for example, depends on how the horizontal layers of different pore-body sizes are arranged. Our comments on residual entrapment and wetting phase permeability, however, would not differ for other random realizations.

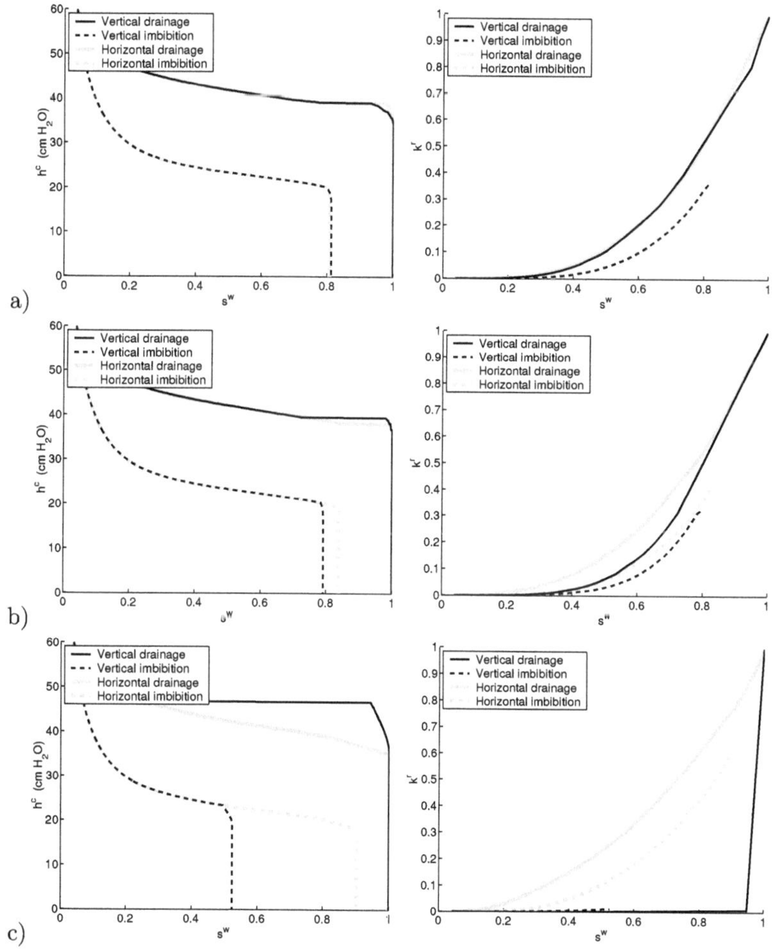


**Fig. 9.** Simulation of primary drainage and main imbibition in pore networks with $50^3$ pore bodies. a) $a_h = 2\lambda$. b) $a_h = 5\lambda$. c) $a_h = 50\lambda$. $a_v = 2\lambda$ for all cases. Both the capillary pressure $p^c$ and the wetting phase permeability $k^{rw}$ become more directionally dependent as $a_h/a_v$ deviates from 1.

The directional dependence of the capillary pressure-saturation relation $p^c(S^w)$ suggests that it is in fact a second rank tensor $\mathbf{\Pi}^c(S^w)$ as is true for the phase permeabilities [52, 50]. The tensor $\mathbf{\Pi}^c(S^w)$ depends on the wetting properties and morphology of the pore space. For our stratified pore-network model, with the $z$ axis in the vertical direction, this tensor is diagonal,

$$\mathbf{\Pi}^c = \begin{pmatrix} p_h^c & 0 & 0 \\ 0 & p_h^c & 0 \\ 0 & 0 & p_v^c \end{pmatrix} \tag{6}$$

and the degree of anisotropy is determined by the anisotropic spatial correlations of the pore space, i.e. by the difference between $a_h$ and $a_v$. The question is whether the continuum modeling approach of Sec. 4 can be modified to account for this generalization. If we want to keep a scalar capillary pressure-saturation relation, the following form will do:

$$p^n - p^w = \mathbf{n}^T \mathbf{\Pi}^c(S^w)\mathbf{n} \tag{7}$$

In our pore-network model, which resembles a retention cell, the unit vector $\mathbf{n}$ specifies the direction pointing from the nonwetting to the wetting phase reservoir. In a real porous medium, however, we do not have reservoirs. Instead, we must express $\mathbf{n}$ in terms of the continuum modeling approach's independent variables. We can use vector variables to form $\mathbf{n}$. The only vector variable, the velocity $\mathbf{v}$, however, is not appropriate, because one cannot distinguish the anisotropic cases for the static case $\mathbf{v} = 0$. But we can form gradients of scalar variables. A suitable choice is

$$\mathbf{n} = \frac{\nabla S^w}{|\nabla S^w|} \tag{8}$$

Equation (7) is of course purely hypothetical, although supported by numerical simulations. To validate our hypothesis, media with anisotropic capillary pressure-saturation relations would need to be created and specified by a non-standard retention cell experiment. Then, a two-phase flow experiment would need to be performed in this anisotropic porous medium, and the macroscale equations could be used to model two-phase flow. Better agreement between experiment and simulation by using the modified Equation (7) would then support our hypothesis.

# 4 Dynamic Macroscale Model Formulation

We have examined multiphase flow in heterogeneous porous media from the perspective of a dimensional analysis of forces at a mesoscopic scale, and we have also considered certain aspects of macroscale capillary pressure-saturation-relative permeability relations for multiphase flow in heterogeneous systems by considering microscale models. Both of these approaches demonstrated the importance of heterogeneity for fundamental multiphase fluid flow processes, but both approaches were limited to quasi-static systems. Furthermore our microscale modeling approach only yielded information about the effect of microscale variability on closure relations and did not address macroscale variability in macroscale quantities such as permeability and closure relations. In this section, we consider dynamical model formulation approaches valid for the macroscale, or porous-medium-continuum scale, and larger scales. These approaches are based partly on conservation principles. Conservation principles alone, however, yield a set of fundamental balance equations with more unknowns than equations. Constitutive relations, which should model average or integrated microscale behavior, are used to formally close the system of balance equations. To apply the model equations to laboratory or field simulations we then specify appropriate auxiliary conditions and material properties.

Since the focus of this work is on heterogeneous systems, we are concerned with REV-scale to mesoscale systems for which closure relation parameters are distributed in space and must be resolved and accurately accounted for in the formulated model. We do not consider multiple–length scales of heterogeneity in this section, although such systems are certainly important and may comprise the majority of natural systems. We also do not consider closure relations, such as those postulated in the previous section, which incorporate some effects of microscale heterogeneity such as hysteresis and anisotropic relative permeability. Instead we focus on the complexities of representing mesoscale heterogeneity using widely studied approaches to REV-scale balance formulations and closure relations.

We will consider a standard isothermal two-phase flow model coupled with two widely used sets of closure relations, and include a detailed discussion of the role of heterogeneity in two-phase flow.

## 4.1 Multiphase Mass Balance Equations

Several quantities are conserved in nature, including mass, momentum, and energy. Conservation of these quantities forms the basis for modeling transport phenomena in multiphase systems. While elegant formulations based upon a comprehensive set of conservation equations for volumes, interfaces, curves, and points are evolving, we consider the traditional approach of volume–based conservation equations.

$$\frac{\partial(\varphi_\alpha \varrho_\alpha)}{\partial t} = -\nabla \cdot (\varphi_\alpha \varrho_\alpha \mathbf{v}_\alpha) + \mathcal{I}_\alpha + \mathcal{S}_\alpha \tag{9}$$

where $\varphi_\alpha = \varphi S_\alpha$ is a volume fraction of the $\alpha$ phase, $\varrho_\alpha$ is a density, $t$ is time, $\mathbf{v}_\alpha$ is the mean macroscopic pore velocity vector, $\mathcal{I}_\alpha$ represents all interphase mass transfer that can occur between the $\alpha$ phase and all other $\beta$ phases for $\beta \neq \alpha$, and $\mathcal{S}_\alpha$ represents a source of mass. Mass balance, geometric considerations, and our definition of sources and mass transfer also imply the following constraints:

$$\sum_\alpha \varphi_\alpha = 1, \qquad \sum_\alpha \mathcal{I}_\alpha = 0 \tag{10}$$

### 4.2 Multiphase Momentum Balance Equations

While it is straightforward to derive a conservation of momentum equation based on a control volume for a macroscopic porous medium system, it is common practice to use an approximate momentum equation based upon a posited extension to Darcy's law for multiphase systems of the form

$$\mathbf{v}_\alpha = -\frac{\mathbf{k}_{r\alpha}}{\varphi_\alpha \mu_\alpha} \mathbf{K}_i \cdot (\nabla p_\alpha - \varrho_\alpha \mathbf{g}) \tag{11}$$

where $\mathbf{k}_{r\alpha}$ is the relative permeability tensor, $\mu_\alpha$ is the dynamic viscosity, $p_\alpha$ is the fluid pressure (each for phase $\alpha$), $\mathbf{K}_i$ is the intrinsic permeability tensor, and $\mathbf{g}$ is the gravity vector, which is assumed to be oriented in the opposite direction to the vertical coordinate direction, $z$. While Darcy's law is considered a momentum balance, it is, in fact, a closure relation that we assume represents the microscale dynamics averaged over the REV under a variety of assumptions about microscale dynamics. Microscale heterogeneities must, therefore, be captured through the functional relations in Darcy's law. Macroscale heterogeneity is captured by the spatial variability of the parameters. We will return to the issue of heterogeneity shortly.

### 4.3 Multiphase Flow Equations

Substitution of Eq. (11) into Eq. (9) yields the standard multiphase flow equations

$$\frac{\partial(\varphi_\alpha \varrho_\alpha)}{\partial t} = \nabla \cdot \left( \varrho_\alpha \frac{\mathbf{k}_{r\alpha}}{\mu_\alpha} \mathbf{K}_i \left(\nabla p_\alpha - \varrho_\alpha \mathbf{g}\right) \right) + \mathcal{I}_\alpha + \mathcal{S}_\alpha \tag{12}$$

For two-phase flow this formulation yields two equations so that $\alpha = n, w$. The set of dependent variables in the equation is $\{\varphi_\alpha, \varrho_\alpha, \mathbf{k}_{r\alpha}, \mu_\alpha, p_\alpha\}$, hence further assumptions and relations are necessary to achieve formal closure of the system of equations. We next consider our choice of primary dependent variables.

First, we define saturation as

$$S_\alpha = \varphi_\alpha / \varphi \tag{13}$$

where $\varphi$ is the porosity of the medium and $\varphi_\alpha$ the part of the volume occupied by phase $\alpha$. For this work, we assume $\varphi$ is a fixed spatially variable property of the soil matrix. Furthermore we define the capillary pressure as

$$p_c = p_n - p_w \tag{14}$$

While alternative sets of primary variables are valid, we choose the wetting phase pressure, $p_w$, and the nonwetting–phase saturation, $S_n$, as primary variables. Combining our definitions of saturation and capillary pressure with Eqns. (10) and (12) yields:

$$-\varphi \varrho_w \frac{\partial S_n}{\partial t} - \nabla \cdot \left( \frac{\varrho_w \mathbf{k}_{rw}}{\mu_w} \mathbf{K}_i (\nabla p_w - \varrho_w \mathbf{g}) \right) - q_w = 0 \tag{15}$$

$$\varphi \frac{\partial (\varrho_n S_n)}{\partial t} - \nabla \cdot \left( \frac{\varrho_n \mathbf{k}_{rn}}{\mu_n} \mathbf{K}_i (\nabla p_w + \nabla p_c - \varrho_n \mathbf{g}) \right) - q_n = 0 \,. \tag{16}$$

Note that we take $\varphi$ out of the time derivative term since it is constant. We do the same with $\varrho_w$ assuming incompressibility of water. We did not assume incompressibility of the nonwetting phase, which allows this formulation to be applied with the nonwetting phase being a compressible gas or liquid. Some alternatives to this pressure–saturation formulation exist; for an overview see, e.g., [23].

### 4.4 Constitutive Relationships

To complete the multiphase flow equations, a set of constitutive relationships is needed to describe how the secondary variables depend on the primary variables $\{S_n, p_w\}$. We distinguish between constitutive relations that describe the fluid properties and those that quantify the interaction between the phases and the porous medium. First, we consider relations of the former type.

Viscosity is a property of the fluids that can be treated as a constant for isothermal conditions. For the phase densities, we assume that water ($w$) and NAPL ($n$) are incompressible.

We now move to the task of describing closure relations that capture fluid–porous medium interactions, namely relations among capillary pressure, saturation, and relative permeability. The development of closure relations of this type is an open area of research. Hence, a growing variety of functional forms and corresponding parameterizations is found in the literature. While these relationships are interdependent, we first focus on the relationship between capillary pressure and saturation that will be used in our two-phase model.

Among the most well-known approaches for describing capillary pressure-saturation are those of [11] and [51]. The Brooks–Corey (BC) approach is formulated as

$$p_c = p_d \, S_e^{-1/\lambda} \tag{17}$$

and the van Genuchten (VG) approach as

$$p_c = \frac{1}{\alpha} \left( S_e^{-1/m} - 1 \right)^{1/n} \tag{18}$$

with

$$S_e = \frac{S_w - S_{w,r}}{1 - S_{w,r}} \tag{19}$$

and

$$m = 1 - \frac{1}{n} \,. \tag{20}$$

$S_{w,r}$ is the residual wetting phase saturation and $S_e$ the effective saturation of the wetting phase, which are local macroscale constants for the porous medium/fluid system determined from equilibrium experimental data. The parameters $p_d$, $\lambda$, $\alpha$, and $n$ are determined by fitting the functionals to experimental data.

The parameterized functionals above differ most significantly as $S_e$ approaches 1, that is as the medium becomes water saturated. We see from Equation. (17) that the BC model yields a non–zero capillary pressure, $p_d$, in this case. The parameter, $p_d$, called the displacement or entry pressure, describes fluid/porous media systems that exhibit a non–zero pressure that the nonwetting phase must exceed before it can penetrate a water-saturated porous medium. In contrast, the VG-curve for $S_e \rightarrow 1$ approaches the value $p_c = 0$. When both functionals are fitted to experimental systems, it can be seen that the BC functional models the rapidly changing relationship between saturation and capillary pressure near water saturation with a discontinuity, whereas the VG-curve approximates this relationship with a continuous curve in such a way that $1/\alpha \approx p_d$. The parameters $\lambda$ and $m$ are used to account for the geometric variability of the microscale pore morphology.

Permeability-saturation behavior can be described by coupling the BC relations with the approach of [12] or the van Genuchten relations with the approach of [43]. The BC functions for the wetting and the nonwetting phases yield

$$k_{rw} = S_e^{\frac{2+3\lambda}{\lambda}} \tag{21}$$

$$k_{rn} = (1 - S_e)^2 \left( 1 - S_e^{\frac{2+\lambda}{\lambda}} \right) \tag{22}$$

and the VG functions

$$k_{rw} = \sqrt{S_e} \left[ 1 - \left( 1 - S_e^{1/m} \right)^m \right]^2 \tag{23}$$

$$k_{rn} = (1 - S_e)^{\frac{1}{3}} \left[ 1 - S_e^{1/m} \right]^{2m} . \tag{24}$$

To summarize, we have closed the system of mass balance equations for two fluid phases by employing a form of Darcy's law for the description of fluid velocities, standard physico-chemical models of microscale fluid density, constant values of fluid viscosity and media porosity, and empirically derived relationships between macroscale pressure–saturation–permeability. The effects of microscale heterogeneity must then be captured solely by the intrinsic permeability tensor and the form of the pressure–saturation–permeability. As the results in Section 3 suggest, the above commonly used models may in fact be incapable of representing well-known effects of microscale heterogeneity such as tensorial and hysteretic relationships. It is partly for this reason that recent research in pressure-saturation-permeability relationships has yielded a large variety of alternative functional forms; however, we ignore this significant open issue in what follows in order to investigate the complex problem of incorporating macroscale heterogeneity, that is spatial variability in the standard parameterizations over mesoscale and field scale simulations.

### 4.5 Inclusion of Microscale Heterogeneity

The constitutive relationships shown in this section so far assume that the parameters obtained by averaging over a REV are homogeneous and isotropic and therefore are best described by scalars rather than tensors. However, the experiment shown in Section 2 and the constitutive relationships derived in Section 3 make it clear that those assumptions are not acceptable for natural systems in most cases. The relationships upscaled from the microscale for $\mathbf{k}_{rw}$ and $p_n - p_w = \mathbf{n}^T(S^w)\mathbf{\Pi_c}(S_w)\mathbf{n}$ are more adequate, as they take into account the anisotropy of the medium. For anistropic relative permeability function, see for example [2, 8].

In both papers, the anisotropic relative permeabilities have been upscaled from a smaller scale where an REV has already been applied. However, anisotropic capillary–pressure relationships are the result of an upscaling process from the pore size scale, where we may assume an anisotropic distribution of pore throats. For a steady change of the upscaled parameters, which may be the result of non–stationary distributions for pore bodies and pore throats rather than stationary distributions like those in Section 3, it makes sense to use the effective permeability $\mathbf{k}_\alpha = \mathbf{k}_{r\alpha} \cdot \mathbf{K}_i$ and to weight it harmonically in the numerical model. Therefore, we reformulate the Equations (15) and (16) to

$$-\varphi \varrho_w \frac{\partial S_n}{\partial t} - \nabla \cdot \left( \frac{\varrho_w \mathbf{k}_w}{\mu_w} (\nabla p_w - \varrho_w \mathbf{g}) \right) - q_w = 0 \qquad (25)$$

$$\varphi \frac{\partial (\varrho_n S_n)}{\partial t} - \nabla \cdot \left( \frac{\varrho_n \mathbf{k}_n}{\mu_n} (\nabla p_w + \nabla p_c - \varrho_n \mathbf{g}) \right) - q_n = 0 \, . \qquad (26)$$

### 4.6 Inclusion of Macroscale Heterogeneity

Smoothly varying material properties naturally preserve continuity in both saturation and pressure; however, in realistic situations, such as the DNAPL experiment presented in Section 2, discontinuous materials are a good approximation of the variation in material properties for macroscale models.

As shown in [22], it is advantegeous to upwind the mobility $\lambda_\alpha = k_{r\alpha}/\mu_\alpha$. Therefore, to deal with flow across the interfaces of block heterogeneities, we reformulate Equations (15) and (16):

$$-\varphi\varrho_w \frac{\partial S_n}{\partial t} - \nabla \cdot (\varrho_w \lambda_w \mathbf{K}_i(\nabla p_w - \varrho_w \mathbf{g})) - q_w = 0 \tag{27}$$

$$\varphi \frac{\partial(\varrho_n S_n)}{\partial t} - \nabla \cdot (\varrho_n \lambda_n \mathbf{K}_i(\nabla p_w + \nabla p_c - \varrho_n \mathbf{g})) - q_n = 0 \,. \tag{28}$$

Since $\lambda_\alpha$ is a scalar, we have to take into account the anisotropy of the relative permeability. With **n** as the normal vector of the interface of the block heterogeneity we use $\lambda_\alpha = \frac{\mathbf{n}^T \mathbf{k}_{r\alpha} \mathbf{n}}{\mu_\alpha}$. If we use capillary pressure relationships as introduced in Equation 7, we also take the normal vector of the interface rather than the saturation gradient.

Let us now abstract Subset C shown in Fig. 3 as an example of variability for the mesoscale in soil properties: Figure 10 shows DNAPL entering a water-saturated column containing a fine–sand lens embedded into ambient coarse sand. We divide the domain into three subdomains $G_1 = [y_0, y_1], G_2 = (y_1, y_2), G_3 = [y_2, y_3]$ with interfaces at $y_1$ and $y_2$. $G_1$ and $G_3$ are coarse sand, $G_2$ is fine sand, that is there is a discontinuity in material properties at $y_1$ and $y_2$ such that $p_d|G_2 > p_d|G_1$ and $p_d|G_1 = p_d|G_2$. We first define several limits at the interface:

$$S_w|G_1^{y_1} := \lim_{y \to y_1^+} S_w, \quad S_w|G_2^{y_1} := \lim_{y \to y_1^-} S_w, \tag{29}$$

$$p_c|G_1^{y_1} := \lim_{y \to y_1^-} p_c, \quad p_c|G_2^{y_1} := \lim_{y \to y_1^+} p_c \tag{30}$$

For $v_{alpha}$ defined as in Equation (11) to remain finite given our definition of capillary pressure, we must require that $p_c$ be continuous at discontinuities in material properties, and we adopt the term capillary equilibrium condition for this continuity requirement. Hence, $p_c|G_1^{y_1} = p_c|G_2^{y_1}$. This requirement, however, implies that spatially variable BC and VG functions, and therefore $S_w$, can be discontinuous at media discontinuities. For the system in Fig. 10, we have $S_w|G_1^{y_1} = p_c^{-1}|G_1(p_c^{y_1}) \neq p_c^{-1}|G_2(p_c^{y_1}) = S_w|G_2^{y_1}$. For VG functionals, the saturation is continuous for $p_c = 0$, regardless of the material properties and hence a fully water-saturated domain obeys the capillary equilibrium condition. On the other hand, the BC functional, because of its explicit inclusion

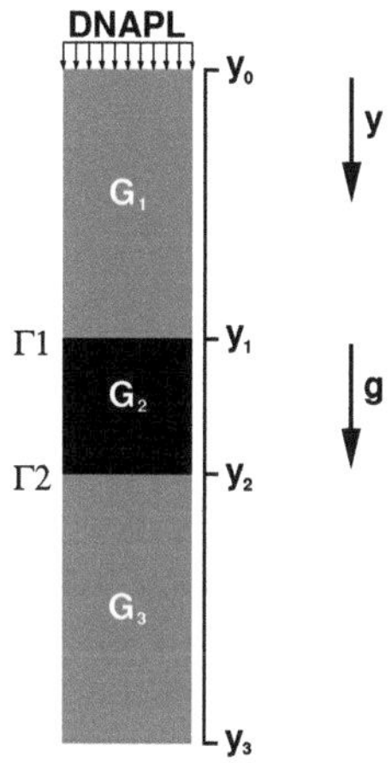


**Fig. 10.** DNAPL spreading at the interface of two domains: setup of the experiment

of finite entry pressure, cannot permit a fully saturated system while preserving the capillary equilibrium condition. As a result, in our example domain, we then have $p_c|G_1(1) = p_d|G_1 < p_d|G_2 = p_c|G_2(1)$ at full water saturation.

In such cases, we can develop numerical techniques that yield discrete solutions that both preserve the discontinuity in saturation and maintain a discrete capillary equilibrium condition. We will, therefore, revisit this issue when we present numerical methods for solving the flow equations.

## 5 Numerical Model

The mathematical description of the physical processes yields a system of coupled partial differential equations with a hyperbolic/parabolic character. They exhibit a high degree of nonlinearity. We handle this with a Newton–Raphson method. We can write the system of two equations given by Equations (15) and 16 in the following simplified functional form:

$$\mathbf{F}(\mathbf{x}) = \mathbf{0} , \tag{31}$$

where the vector $\mathbf{x}$ holds the primary variables.

The damped inexact Newton–Raphson algorithm for solving the resulting nonlinear system of equations is given by:

Choose $\mathbf{x}^{k+1,0}$; set $m = 0$;
**while** $(||\mathbf{F}(\mathbf{x}^{k+1,m})||_2 \;/\; ||\mathbf{F}(\mathbf{x}^{k+1,0}||_2 > \varepsilon_{nl}))$
{
Solve $\mathbf{K}(\mathbf{x}^{k+1,m})\mathbf{u} = -\mathbf{F}(\mathbf{x}^{k+1,m})$
with accuracy $\varepsilon_{lin}$;
$\mathbf{x}^{k+1,m+1} = \mathbf{x}^{k+1,m} + \eta\mathbf{u}$;
$m = m + 1$;
}

$\mathbf{F}(\mathbf{x}^{k+1,m})$ represents the defect term obtained at time level $k+1$ and iteration $m$ depending on the nonlinear functions $\mathbf{F}$ and the vector of primary variables $\mathbf{x}$. $\varepsilon_{nl}$ and $\varepsilon_{lin}$ are the accuracy criteria of the nonlinear and the linear solution respectively. $||\cdot||$ is the Euclidean vector norm. The damping factor $\eta = (1/2)^q$ is chosen such that

$$||\mathbf{F}(\mathbf{x}^{k+1,m+1})||_2 \leq \left[1 - \frac{1}{4}\left(\frac{1}{2}\right)^q\right] ||\mathbf{F}(\mathbf{x}^{k+1,m})||_2 \tag{32}$$

is valid for the smallest possible $q \in \{0, 1, ..., n_{ls}\}$ with the maximum number of line search steps $n_{ls}$ being between 4 and 6.

For time discretization, we use a fully implicit Eulerian approach. The time discretization is applied to the storage term, e.g.

$$\frac{\partial S}{\partial t} \approx \frac{S^{k+1} - S^k}{\Delta t^{k+1}}. \tag{33}$$

A time-step reduction with a given factor *dtscale* is applied if no $q$ as described in (32) can be found within 6 line searches. A time-step extension with factor *dtscale* is applied if such a $q$ can be found within the first line search. For the algorithm, we set a starting value $\Delta_{start}$, a minimum value $\Delta_{min}$ which acts as a stopping criterium if it is undershot, and a maximum value $\Delta_{max}$ which restricts time steps from getting too large. Within these bounds the size of the actual time step $\Delta t^{k+1}$ can be interpreted as a rough measure of the convergence behavior of the nonlinear algorithm.

As seen in the inexact Newton–Raphson algorithm,

$$\mathbf{K}\mathbf{u} = \mathbf{f} \tag{34}$$

is the Jacobian system to be solved by a linear solver. As the linear solver we use the Bi-Conjugate Gradient Stabilized solver (Bi–CGSTAB) (e.g. [3]) with a preconditioner based on a multigrid technique [4]. The multigrid mesh hierarchy yields a Jacobian system on each grid level $l$. We need linear mappings, restriction $\mathbf{R}_l$ and prolongation $\mathbf{P}_l$, for interpolation between the grid levels. A multigrid algorithm (V-cycle) for an iterative improvement of a given vector $\mathbf{u}_l$ can be written as follows:

```
mgc (l, u_l, f_l)
{
  if (l == 0) u_0 = K_0^{-1} f_0;
  else {
    Apply ν_1 smoothing iterations to K_l u_l = f_l;
    d_{l-1} = R_l (f_l - K_l u_l);
    e_{l-1} = 0;
    mgc (l - 1, e_{l-1}, d_{l-1});
    u_l = u_l + P_l e_{l-1};
    Apply ν_2 smoothing iterations to K_l u_l = f_l;
  }
}
```

For smoothing iterations, for example, $\nu_1 = \nu_2 = 2$ ILU steps (incomplete decomposition, e.g. [20]) can be chosen.

For a more detailed explanation of the discretization and solution methods implemented in MUFTE_UG, we recommend, for example, [5] or [6].

### 5.1 Adaptive Time Discretization

In addition to the backward Euler time discretization in Equation (33), we examine the behavior of numerical models based on higher order backward difference (BDF) discretization formulas, which are a generalization of Equation (33) given by

$$\frac{\partial S}{\partial t} \approx \sum_{i=k+1-b}^{k+1} \beta^i S^i \tag{35}$$

where $b$ is the order of the BDF discretization and the $\beta^i$ are coefficients of the BDF method. Just as with backward Euler, higher order BDF discretizations lead to a nonlinear system which is solved iteratively with an inexact Newton-Raphson algorithm [10]. However, the time step is selected according to heuristics that use an estimate of the local truncation error of the semi-discrete system, which has the form

$$\varepsilon^{k+1} \approx \kappa^b \|\mathbf{x}^{k+1} - \mathbf{x}_p^{k+1}\| \tag{36}$$

where $\kappa$ is an error coefficient for the BDF method and $x_p^{k+1}$ is the initial guess of the Newton algorithm, the so-called predictor. The predictor is obtained by extrapolating to level $k+1$ with a $b$-th order Lagrange polynomial; thus, $\varepsilon^{k+1}$ is essentially an error estimate based on comparing two methods of order $b$, a common approach to error estimation for automatic time step control. With this approach, the user–specified truncation error, $\tau$, is maintained by requiring $\varepsilon^{k+1} < \tau$. Furthermore, as the error also has the form $C^b h^{b+1}$, we can further exploit the truncation error estimate to choose the order and maximize the stepsize chosen for the next step. Further details on multistep time discretizations can be found in [10] and for the code in particular see [29].

### 5.2 Subdomain collocation finite volume method (box method)

We derive a finite volume formulation (box method) based on equations (15) and (16). We assume the model domain $G$ to be discretized by a set of vertices $V = \{v_1, \dots, v_n\}$ with $n$ the number of vertices and a set of adjoined elements $E = \{e_1, \dots, e_m\}$ with $m$ the number of elements. For the sake of simplicity we do not distinguish here between the vertices and their indices or elements and their indices.

The boundary of the model domain is divided into parts holding a Dirichlet boundary condition $\Gamma_{\alpha,D}$ and parts holding a Neumann boundary condition $\Gamma_{\alpha,N}$, so that $\partial B = \Gamma_{\alpha,D} \cup \Gamma_{\alpha,N}$ with $\alpha \in \{w, n\}$.

The finite volume method requires the construction of a secondary mesh. In the present case of vertex centered finite volumes, the finite volume mesh is constructed by connecting element barycenters with edge midpoints as shown in Fig. 11 in two dimensions. In three dimensions, first the element barycenters are connected to element face barycenters and then these are connected with edge midpoints. Each control volume $B_i$ belongs to a grid vertex $v_i$, the intersection of a control volume $B_i$ with element $k$ is denoted by $b_i^k$ (sub control volume).

In Fig. 11 (a) we see the solid line representing the finite elements and the dashed lines representing the control volumes. Note, that the boundary $\Gamma$ between two subdomains of $G$ with different properties runs along the control volume, i.e. each control volume may have its own property and there may be different properties within each element.

We define weighting functions as

$$W_i(x) = \begin{cases} 1 & \text{if } x \in B_i \\ 0 & \text{if } x \notin B_i \end{cases} \tag{37}$$

and the shape functions $N_i$ as linear interpolation of $\delta_{ij}$ with $i, j \in V$. This can be seen in Fig. 12 for one dimension. With this definitions we approximate equations (15) and (16) using the Euler scheme described above for the new time step $k+1$ and according to the subdomain collocation finite volume method (box method) with $|B_i|$ area of $B_i$ in 2D or volume of $B_i$ in 3D as follows:

**Wetting phase:**

$$\begin{aligned} &((S_n \varrho_w)_i^{k+1} - (S_n \varrho_w)_i^k) \tfrac{\varphi}{\Delta t} |B_i| = \\ &- \oint_{\partial B_i} \varrho_{wij}^{k+1} \lambda_{wij}^{k+1} \mathbf{K} (\nabla p_w^{k+1} - \varrho_w^{k+1} \mathbf{g})_i \mathbf{n} \, d\Gamma_{B_i} - (q_w)_i^{k+1} |B_i| \end{aligned} \tag{38}$$

**Non–wetting phase:**

$$- ((S_n\varrho_n)_i^{k+1} - (S_n\varrho_n)_i^k)\frac{\varphi}{\Delta t}|B_i| =$$
$$- \oint_{\partial B_i} \varrho_{nij}^{k+1}\lambda_{nij}^{k+1}\mathbf{K}(\nabla p_w^{k+1} + \nabla p_c^{k+1} - \varrho_n^{k+1}\mathbf{g})_i\mathbf{n}d\Gamma_{B_i}$$
$$- (q_n)_i^{k+1}|B_i| . \tag{39}$$

Here $|B_i|$ denotes the area (2D) or the volume (3D) of $B_i$, $\mathbf{n}$ denotes the outer normal vector of $\partial B_i$ and $\Gamma_{B_i}$ the integration path around $B_i$.

In practice, the global stiffness matrix is constructed according to the finite element formulation, i.e. all line integrals within an element are computed by a loop over all elements. The integral over the segment of a straight line is approximated by the midpoint rule, i.e. the value at the midpoint of the subcontrol volume face between the vertices $v_i$ and $v_j$ – the integration point $x_{ij}^{FUe_l}$ – is multiplied by the length of the corresponding subcontrol volume face. In Fig. 11 (b) we see the local element of arbitrarily chosen element $e_3$ from Fig. 11 (a) with the integration point between vertix $v_2$ and vertix $v_3$ according to the midpoint rule (number according to the local numbering of the vertices; $v_3$ in local notation is $v_i$ in global notation). We approximate the flux from sub control volume $b_2^3$ to sub control volume $b_3^3$ along the sub control volume face indicated by the bracket.

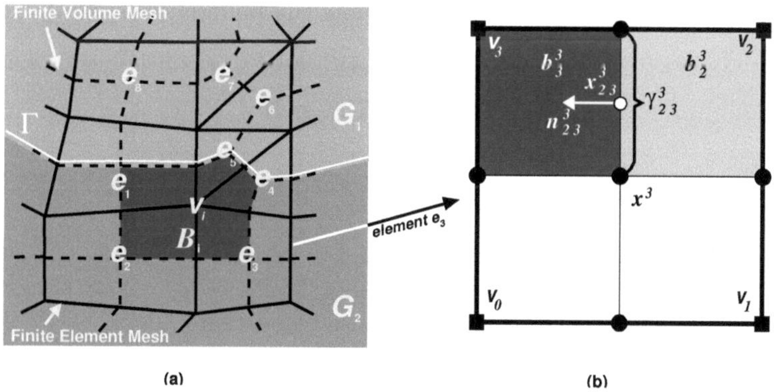


**Fig. 11.** (a) Overlap of FE- and FV-mesh; (b) one FE with corresponding sub control volumes

With

$$
\begin{aligned}
p_w &\approx \sum_{i\in V} p_{wi}\, N_i \\
p_c &\approx \sum_{i\in V} p_{ci}\, N_i \\
S_n &\approx \sum_{i\in V} S_{ni}\, N_i
\end{aligned}
\tag{40}
$$

the integration of equations (38) and (39) over a box $B_i$ yields:

$$
\begin{aligned}
g_{\alpha i}(S_{ni}^{k+1}; S_{n,i}^{k}; p_{wi}^{k+1}; p_{wj}^{k+1}) :=& \\
&- \underbrace{(-1)^{\delta_{\alpha w}}\{(S_n\,\varrho_\alpha)_i^{k+1} - (S_n\,\varrho_\alpha)_i^{k}\}\frac{\varphi}{\Delta t}|B_i|}_{\text{term 1}} \\
&- \underbrace{\sum_{l\in E_i}\sum_{j\in\eta_i} \lambda_{\alpha ij}^{FUe_l}\,\varrho_{\alpha ij}^{k+1}\,\gamma_{ij}^{FUe_l}(\psi_{\alpha j}^{k+1} - \psi_{\alpha i}^{k+1})}_{\text{term 2}} \\
&- \underbrace{q_{\alpha i}^{k+1}|B_i|}_{\text{term 3}} - \underbrace{m_{\alpha i}}_{\text{term 4}} = 0\,, \\
&\alpha \in \{w, n\}
\end{aligned}
\tag{41}
$$

with term 1 as accumulation term, term 2 as internal flux term, term 3 as sink and source term, and term 4 describing the boundary flux.

We set

$$
\gamma_{\alpha ij}^{FUe_l} := \oint_{\partial B_i\setminus\Gamma_{\alpha,F}} \mathbf{K}\nabla N_j\,\mathbf{n}\,d\Gamma_{B_i} \tag{42}
$$

$$
\psi_{\alpha\, i}^{k+1} := p_{wi}^{k+1} + \delta_{\alpha n}p_{ci}^{k+1} - \varrho_{\alpha i}^{k+1}\, g\, z_i\,. \tag{43}
$$

$E_i$ is the set of elements which are adjoined to vertex $v_i$, e.g. $\{e_1, e_2, e_3, e_4, e_5\}$ for $v_i$ in Fig. 11. $\eta_i$ is the set of neighbor nodes of $v_i$ whose boxes share subcontrol volume faces with $B_i$, i.e. whose respective integral $\gamma_{\alpha ij}^{FUe_l}$ is non–zero. $(\psi_{\alpha j} - \psi_{\alpha i})$ is the direction of the discrete flow of phase $\alpha$. $z_i$ is the geodetic hight of vertix $v_i$.

$m_{\alpha i}$ is the flow over $\partial B_i \cap \Gamma_{\alpha,N}$. The product must be considered in analogy to the finite element formulation.

The Fully Upwind finite volume method (FU–box) is given by:

$$
\lambda_{\alpha ij}^{FUe} = \begin{cases} \lambda_{\alpha j} & \text{if} \quad (\psi_{\alpha j} - \psi_{\alpha i}) \geq 0 \\ \lambda_{\alpha i} & \text{if} \quad (\psi_{\alpha j} - \psi_{\alpha i}) \leq 0 \end{cases} \tag{44}
$$

Note that the integral over the interface $\partial B_i \cap \partial B_j$ of box $B_i$ is of the same magnitude as the respective boundary integral for box $B_j$, but has the opposite sign. Hence, the method is locally mass conservative at each control volume (box).

For a more detailed description of the box method we refer to [5] or [6].

As we see in Fig. 12 we have a fully upwinding of the mobility term. However, as we will show in Section 6.1, even the Fully Upwind box method makes an error depending on the grid width.

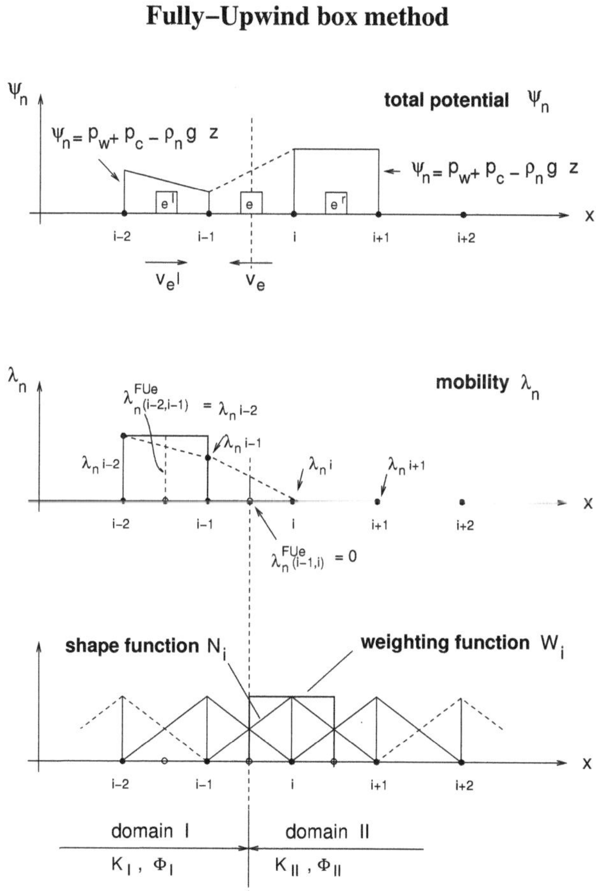


**Fig. 12.** Fully Upwind Box discretization for the non–wetting phase at the interface between two different geological structures.

## Numerical Implementation of the Interface Conditions

By using the upwinding of mobility it is possible to establish a correct reproduction of the entering of the non-wetting phase into the low permeable area. However, by using this method the convergence behaviour of the numerical solver worsens significantly, when the non-wetting phase reaches the low permeable area. The reason is that with the entering of the non–wetting phase into the low permeable domain the upwind vertex oscillates between the vertexs $i-1$ and $i$. For the state, we see in Fig. 12 the upwind vertex is still set to vertex i.

Furthermore, even when using the upwinding scheme it is not possible to reproduce the discontinuity of saturation as described in subsection 4.6 at the interface while using a $p_w - S_n$–formulation. It is possible if we use a $p_w - p_n$–formulation or a $p_w - p_c$– formulation, but these formulations have the disadvantages mentioned in section 4.

A further possibility to reproduce the discontinuity of saturation at the interface is to use the capillary pressure $p_c$ as kind of quasi primary variable at the interface between the two subdomains.

We use the definitions of subsection 4.6, i.e. $G_1$ being the subdomain with the higher permeability and the lower entry pressure in comparisonto the subdomain $G_2$. The threshold saturation $S_w^*$ is defined via $p_c^{G_1}(S_w^*) = p_e^{G_2}$, with $S_w|G_1^\Gamma = 1 - S_n|G_1^\Gamma$ and $S_w|G_2^\Gamma = 1 - S_n|G_2^\Gamma$ and we recalculate $S_w|G_2^\Gamma$ at the interface using the extended capillary pressure condition:

$$S_w|G_2^\Gamma = \begin{cases} 1 & \text{if } S_w|G_1^\Gamma \geq S_w^* \\ Inv_{pc}^{G_2} & \text{if } S_w|G_1^\Gamma < S_w^* \end{cases} \tag{45}$$

with $Inv_{p_c}^{G_2}$ as the inverse of the capillary pressure saturation function of subdomain $G_2$.

Thus, it is made sure that the non wetting phase does not enter into subdomain $G_2$ until the threshold saturation has been reached. After the threshold saturation has been reached the continuity of capillary pressure at the interface is guaranteed.

We refer to [16] und [15] for a detailed discussion of this method and for validating this method on the basis of a theoretical solution for a self similar problem with a discontinuity.

We use the subdomain collocation finite volume method described above, but establish the interface condition. For the incorporation of the interface condition into the discretization the parameters have to be evaluated elementwise. The difference between a patch oriented approach and an element oriented approach in the evaluation of parameters can be seen when comparing the boundary $\Gamma$ in Fig. 11 and Fig. 13. Assuming that capillary pressure is higher in domain $G_2$, the elements in domain $G_2$ are evaluated with a recalculated saturation for the vertices located on the interface $\Gamma$ as seen in Fig. 13. This saturation can (but need not) be different from the saturation for the same vertices when evaluating the elements located in subdomain $G_1$.

In Fig. 13 we see the two subdomains $G_1$ and $G_2$ detached at the interface $\Gamma$. The vertex $v_i$ still exists only once, but the associated saturation values $S_n|v_i^{G_1}$ and $S_n|v_i^{G_2}$ can vary between $v_i^{G_1}$ and $v_i^{G_2}$, as $S_n|v_i^{G_2}$ is computed by

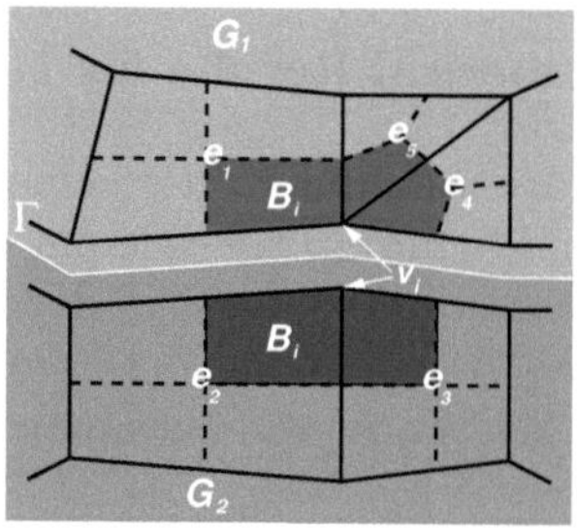


**Fig. 13.** For the PPSIC method we evaluate the capillary pressure over the adjoined elements (here $e_1 ... e_5$) of vertex $v_i$ with the saturation at $v_i$.

the inverse capillary pressure saturation function of $G_2$ using $1 - S_n|v^{G_1}$. Since the pressure $p_w$ is continuous—we assume a mobile wetting phase here—this value is identical for both virtual vertices, i.e. we have

$$p_w|v_i^{G_1} = p_w|v_i^{G_2}. \tag{46}$$

For the actual computation of $S_n|v_i^{G_1}$ from $S_n|v_i^{G_2}$ we first define the minimal capillary pressure $p^i_{c,min}$ for vertex $v_i$ with regard to all elements which have $v_i$ as a corner. Therefore, we denote $E(i)$ as the set of indices of those elements which have $v_i$ as a corner (e.g. $E(i) = \{1, ..5\}$ in Fig. 13).

$$p^i_{c,min} = \min_{k \in E(i)} p_c(\mathbf{x}^k, 1 - \mathbf{S}_{n,i}). \tag{47}$$

$\mathbf{x}^k$ is the barycenter of element $e_k$. For the determination of $p^i_{c,min}$ at vertex $v_i$, we evaluate the capillary pressure function in all elements that vertex $v_i$ is part of for the saturation that is associated with $v_i$ in element $e_k$ and compute the minimum value.

$p^i_{c,min}$ is used to compute the saturation $S_n$ at vertex $v_i$ with respect to element $e_k$. The extended capillary pressure condition interface condition can be used as follows:

$$S_{n,i,k} = \begin{cases} S_{n,i} & \text{if } p_c(\mathbf{x}^k, 1 - S_{n,i}) = p^i_{c,min} \\ 0 & p^i_{c,min} < p_c(\mathbf{x}^k, 1) \\ 1 - S_w & \text{where } S_w \text{ solves } p_c(\mathbf{x}^k, S_w) = \mathbf{p}^i_{c,min} \end{cases} \tag{48}$$

With this definition it is possible for more than two subdomains to meet at vertex $v_i$.

We now evaluate the secondary variables (besides the capillary pressure) as shown in section 4. However, for the evaluation of the domain dependent parameters we use the barycenter of the element, not the position of the nodes.

For a more detailed description of the implementation of the PPSIC method we refer to [5].

# 6 Examples

## 6.1 Examination of Numerical Results for 1D

In this section we investigate the numerical schemes described in Section 4.6 applied to a simple system. We have three different domains in a column with a length of $2.5[m]$ and a width of $0.5[m]$. The domain is averaged over $1[m]$ depth. The upper and the lower domains consist of coarse sand, the middle of fine sand (see Fig. 14). We choose the parameters for the sands according

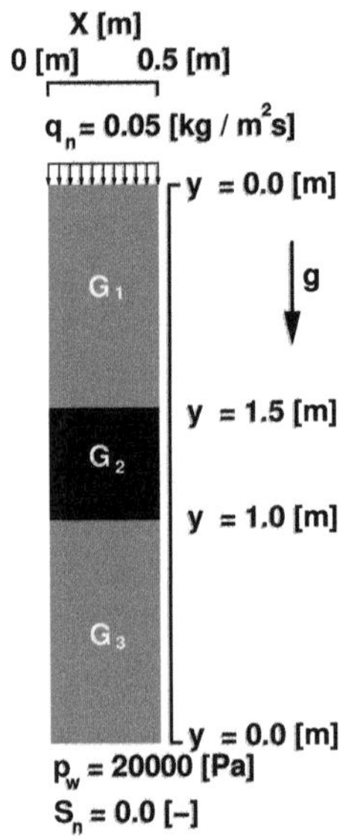


**Fig. 14.** Setup of 1D example

to those of the coarse and the medium sand described in [33]. DNAPL spills from the top into the fully water saturated system with $q_n = 0.05\ [kg/(m^2 s)]$. The correspondent parameters for the fluids and the porous media, initial conditions and boundary conditions can be seen in Tables 4 to 7. In Fig. 15 on the left hand side, we see the capillary pressure saturation relationships after Brooks and Corey according to the parameters given in Table 5. The threshold saturation as defined in Section 4.6 is $S_w^* \approx 0.085$. In Fig. 15 on the right hand side, we see the relative permeability saturation functions for the two different sands according to Brooks and Corey. As spatial discretization methods we use the subdomain collocation finite volume method described in Section 5.2 with and without interface condition described in Section 5.2. The discretization scheme without interface condition will be referred to as Phase Pressure Saturation formulation (PPS), the discretization scheme with

**Table 4.** Fluid Parameters

| | Density | Viscosity |
|---|---|---|
| Water | $\rho_w = 998\ [kg/m^3]$ | $\mu_w = 1.0E-03\ [kg/(ms)]$ |
| DNPL | $\rho_n = 1621\ [kg/m^3]$ | $\mu_n = 0.9E-03\ [kg/(ms)]$ |

**Table 5.** Parameters of sands

| Sand | | $p_d[Pa]$ | $\lambda$ | $S_{wr}$ | $K[m^2]$ | Porosity |
|---|---|---|---|---|---|---|
| $G1, G3$ | coarse sand | 370 | 3.86 | 0.078 | $5.04*10^{-10}$ | 0.40 |
| $G2$ | fine sand | 1324 | 2.49 | 0.098 | $5.26*10^{-11}$ | 0.39 |

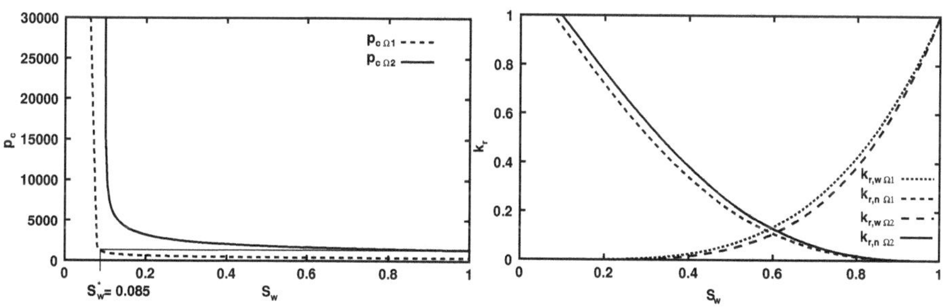


**Fig. 15.** Left hand side: Capillary pressure of 1D example; right hand side: Relative permeability function after Brooks and Corey for $G_1$ and $G_2$

interface condition will be referred to as Phase Pressure Saturation formulation with Interface Condition (PPSIC). We use two different sets of element lengths, one set with $\Delta h = 0.25/2^3[m]$ and one set with $\Delta h = 0.25/2^6[m]$.

For time discretization we use the implicit Euler scheme given in Equation (33) with a start (and maximum) value of $\Delta t_{start} = 80.0\ [s]$ for $\Delta h = 0.25/2^3[m]$ and $\Delta t_{start} = 10.0\ [s]$ for $\Delta h = 0.25/2^6[m]$. The minimum value is chosen small enough so it is never reached $\Delta t_{start} = 10.0E-8\ [s]$. The computation was carried until $6800[s]$ of simulated time were exceeded. After $6800[s]$ the DNAPL has almost reached the bottom of the column.

We show the DNAPL front $S_n(y)$ at five time steps in the Figs. 16 and 17. These have been chosen according to location of the DNAPL front:

1. The front is in $G_1$, arbitrarily chosen at simulated time $t \approx 1500[s]$.
2. The front reaches $G_2$ and would enter it, if $G_2$ had the same parameters as $G_1$. This happens at the simulated time of $t \approx 2200[s]$.
3. The front actually enters $G_2$. This happens at $t \approx 2750[s]$.
4. The front leaves $G_2$. This happens at $t \approx 4600[s]$.
5. The front is in $G_3$, arbitrarily chosen at simulated time $t \approx 6350[s]$.

We can see that the DNAPL saturation at $y = 1.5[m]$ according to the PPS formulation (Figs. 16 and 17 on the left) is lower than according to the PP-

**Table 6.** Initial Conditions

| | |
|---|---|
| Water | $p_w = 20000$ $[Pa]$ |
| DNPL | $S_n = 0.0$ $[-]$ |

**Table 7.** Boundary Conditions

| | Water | DNAPL |
|---|---|---|
| left, right | $q_w = 0.0[kg/m^2s]$ | $q_n = 0.0[kg/m^2s]$ |
| top | $q_w = 0.0[kg/m^2s]$ | $q_n = 0.05[kg/m^2s]$ |
| bottom | $p_w = 20.000[Pa]$ | $S_n = 0.0[-]$ |

SIC formulation (Figs. 16 and 17 on the right). The threshold saturation $S_w^* \approx 0.085$ is equivalent to a DNAPL saturation of $S_n = 0.915$. This DNAPL saturationis not reproduced if we use the PPS formulation, but it is reproduced if we use the PPSIC formulation both for the coarse and the fine grid. The PPS method approximates the threshold saturation somewhat better for the finer discretization (comparing Fig. 16 PPS with Fig. 17 PPS). This corresponds with the theory as explained in Section 4.6 and Section 5.2. According to

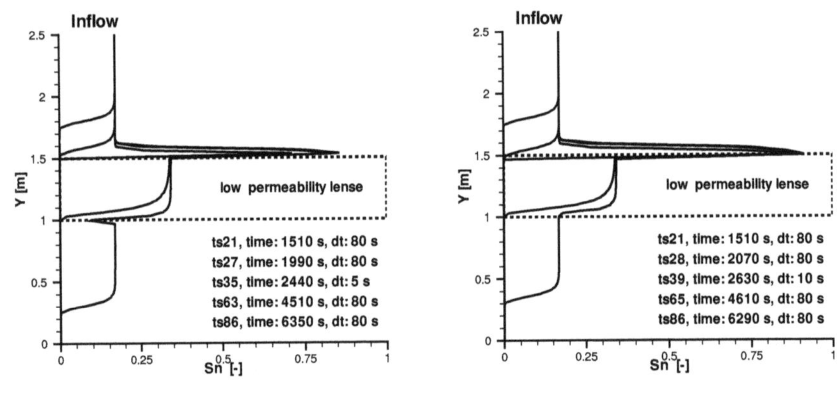


**Fig. 16.** DNAPL saturation for different time steps; $\Delta h = 0.25[m]/2^3$; left: without Interface Condition (PPS); right: with Interface Condition (PPSIC).

the results, we can conclude that the PPSIC method is able to reproduce the threshold saturation at interfaces correctly even for coarse spatial discretizations, while for the PPS method the reproduction of the threshold saturation at interfaces is dependent on $\Delta h$.

For comparisons of the PPS and the PPSIC scheme regarding the time step as a rough measure of convergence as well as investigations on the number of

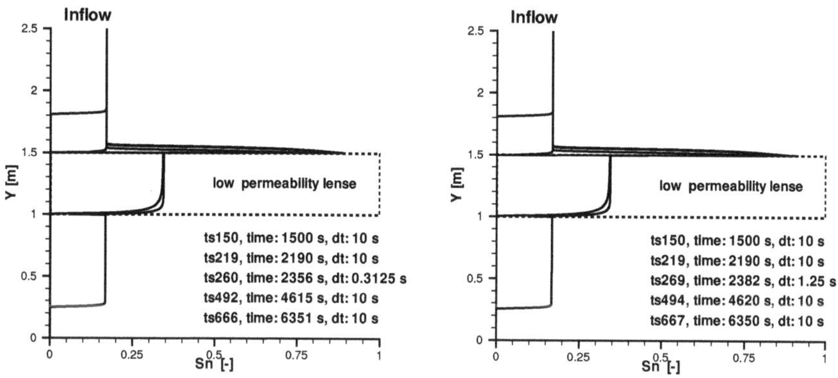


**Fig. 17.** DNAPL saturation for different time steps; $\Delta h = 0.25[m]/2^6$; left: without Interface Condition (PPS); right: with Interface Condition (PPSIC).

nonlinear and linear iterations within the inexact Newton-Raphson algorithm and a two-dimensional example, we refer to [46].

# 7 Conclusions

In this paper we investigated the influence of heterogeneities on multiphase flow in porous media at the macro scale with emphasis on subsurface DNAPL contamination. To motivate the work we began by considering a mesoscale laboratory experiment that exhibited complex DNAPL flow processes. The laboratory experiment was engineered to contain both macroscale and microscale heterogeneities such as those found in natural and engineered mesoscale systems. We then characterized heterogeneous multiphase porous media systems using dimensional analysis, porescale modeling, and a macroscale continuum approach. The goal of our characterization was to develop models appropriate at the meso- and field-scales that can answer questions such as

- How far does the DNAPL contamination spread?
- Does it enter into lower permeability zones?
- If so, what kinds of remediation techniques can we apply to remove it from the lower permeable zones?

Our characterization demonstrates many of the challenges involved in addressing such questions through modeling.

The laboratory experiment with which we began our investigation contained a regular packing of four widely differing porous media types. Hence, in addition to the variation of the microscale pore geometry within each porous medium (microscale or intrinsic heterogeneity), our experimental system contained strong variations at the macroscale, which influenced flow and transport behavior. The dominant DNAPL flow phenomena observed during the experiment included 1) residual entrapment 2) lateral spreading 3) vertical fingering and 4) ponding at low permeability layers.

A dimensional analysis of the relevant forces in multiphase porous media yielded a characterization of these systems in terms of non-dimensional capillary, gravity, and gravillary numbers. For length scales from the microscale to the mesoscale our dimensional analysis showed that capillary forces and gravity are active in determining behavior. Hence, gravity and capillary forces must be accounted for in the model representation of fluid momentum if modeling is to be used to faithfully reproduce flow processes. While gravity is straightforward to incorporate into macroscale models, capillary forces in particular are generally included via empirical closure relations relating capillary pressure and saturation.

Instead of obtaining capillary pressure-saturation relations from macroscale observations of multiphase systems, we can also study such relationships using models of the porous media at the microscale. To this end a pore scale model of multiphase flow was used to investigate the form of capillary pressure–saturation relations in porous media with heterogeneity and anisotropy in the microscale pore morphology. We represented the pore scale geometry as a network of cubic pore bodies and square pore throats. The pore bodies and throats were normally distributed with an exponential semivariogram. Even given this fairly simple geometric representation the microscale pore morphology we found that corresponding macroscale soil characteristics would need to account for 1) anisotropic permeability tensor 2) pressure-saturation hysteresis and 3) non-wetting phase residual entrapment.

While standard macroscale continuum models rarely include all of the effects above via the constitutive equations, heterogeneous macroscale permeability and pressure-saturation relationships should at least be capable of reproducing some flow phenomena produced by macroscale heterogeneities. We formulated macroscale continuum models of multiphase, multicomponent transport and investigated numerical methods for incorporating macroscale heterogeneity. In particular we showed that the correct reproduction of capillary equilibrium conditions is crucial for reproducing flow processes, such as lateral spreading and ponding, which occur at macroscale heterogeneities. The capillary equilibrium condition can be reproduced using either an upwinding scheme or by directly implementing the condition. Details for implementing both methods as well as numerical experiments validating their effectiveness were included.

We concluded with numerical simulations and laboratory examples of thermally enhanced soil vapor extraction for remediating DNAPL-contaminated, heterogeneous soils. Both our previous laboratory experiments and our macroscale continuum models suggested that if the DNAPL has indeed entered into low permeability zones with high entry pressures, then extracting entrapped DNAPL from these zones may be difficult using fluid flow alone. In this case, a more sophisticated remediation technique making use of component transport and interphase mass transfer may be the only effective means of removing

DNAPL contamination. Specifically, while DNAPL flow may be difficult to induce in low permeability/high entry pressure zones, air flow and mass transfer from DNAPL to air may be effective at reducing trapped DNAPL saturations in low permeability zones. We showed that in both laboratory and theoretical models the remediation technique was effective.

*Acknowledgement.* The German efforts were supported by the Federal Ministry of Economics and Technology (BMWi) under the identification number 02E9370 and the Deutsche Forschungsgemeinschaft (German Research Foundation SFB 404). The US efforts were supported by National Institute of Environmental Health Sciences Grant 2 P42 ES05948 and computational resources by an allocation from the North Carolina Supercomputing Center. CEK was supported by a fellowship from the University Corporation for Atmospheric Research Visiting Scientist Program. We are grateful for valuabe editorial contributions made by L.F. Kees.

## References

1. P. M. Adler. *Porous Media: Geometry and Transports*. Butterworth-Heinemann, Boston, 1992.
2. M.A. Celia B. Ataie-Ashtiani, S.M. Hassanizadeh. Effects of heterogeneities on capillary pressure–saturation–relative permeability relationships. *Journal of Contaminant Hydrology*, 56:175 – 192, 2002.
3. R. Barrett, M. Berry, T. F. Chan, J. Demmel, J. Donato, J. Dongarra, V. Eijkhout, R. Pozo, C. Rominc, and H. Van der Vorst. *Templates for the Solution of Linear Systems: Building Blocks for Iterative Methods, 2nd Edition.* SIAM, Philadelphia, PA, 1994.
4. P. Bastian. *Parallele adaptive Mehrgitterverfahren.* Teubner–Verlag, 1996.
5. P. Bastian. Numerical computation of multiphase flows in porous media, 1999. Habilitation thesis, Christian-Albrechts-Universität Kiel.
6. P. Bastian and R. Helmig. Efficient Fully-Coupled Solution Techniques for Two Phase Flow +in Porous Media. Parallel Multigrid Solution and Large Scale Computations. *Advances in Water Resources*, 23:199–216, 1999.
7. J. Bear, C. Braester, and P. C. Menier. Effective and relative permeabilities of anisotropic porous media. *Transport in Porous Media*, 2:301–316, 1987.
8. C. Braun, R. Helmig, and S. Manthey. Determination of constitutive relationships for two–phase flow processes in heterogeneous porous media with emphasis on the relative permeability–saturation–relationship. *submitted to Journal of Contaminant Hydrology*, 2002.
9. J. Braun. Ausbreitung von NAPL in gesättigten und ungesättigten Böden. In *VEGAS-Workshop und BMBF/PWAB-Seminar In-Situ Technologien zur Grundwasser- und Altlastensanierung*, University of Stuttgart, 1996.
10. K. E. Brenan, S. L. Campbell, and L. R. Petzold. *The Numerical Solution of Initial Value Problems in Differential-Algebraic Equations.* Society for Industrial and Applied Mathematics, Philadelphia, PA, 1996.
11. R. T. Brooks and A. T. Corey. Hydraulic properties of porous media. Technical Report paper number 3, Colorado State University, Fort Collins, CO, 1964.

12. N.T. Burdine. Relative permeability calculations from pore–size distribution data. Technical report, Petroleum Transactions, AIME, 1953.
13. S. Chen and G. D. Doolen. Lattice Boltzmann method for fluid flows. *Annual Review of Fluid Mechanics*, 30:329–364, 1998.
14. S. Dasberg and F. N. Dalton. Time domain reflectometry field measurements of soil water content and electrical conductivity. *Soil Sci. Soc. Am. J.*, 49:293–297, 1985.
15. M. J. de Neef and J. Molenaar. Analysis of dnapl infiltration in a medium with a low–permeable lens. *Computational Geosciences*, 1:191–214, 1997.
16. Michel de Neef. *Modelling capillary effects in heterogeneous porous media.* PhD thesis, Technische Universiteit Delft, 2000.
17. C. V. Deutsch and A. G. Journel. *GSLIB: Geostatistical Software Library and User's Guide.* Oxford University Press, New York, 1992.
18. A. Färber and C. Betz. Thermisch unterstützte Bodenluftabsaugung: Planung, Aufbau und Messtechnik der Versuchsstände, 1995. Scientific report, University of Stuttgart.
19. S. P. Friedman and N. A. Seaton. On the transport properties of anisotropic networks of capillaries. *Water Resources Research*, 32(2):339–347, 1996.
20. W. Hackbusch. *Multi–Grid Methods and Applications.* Springer–Verlag, 1985.
21. 2002. WWW: http://www.hugohaeffner.com/haeffnerdatenbank/ (04/02/02).
22. R. Helmig and R. Huber. Comparison of Galerkin-type discretization techniques for two-phase flow in heterogeneous porous media. *Advances in Water Resources*, 21(8):697–711, 1998.
23. Rainer Helmig. *Multiphase flow and transport processes in the subsurface - a contribution to the modeling of hydrosystems.* Springer Verlag, 1997.
24. R. Hilfer and P. E. Oren. Dimensional analysis of pore scale and field scale immiscible displacement. *Transport in Porous Media*, 22(1):53–72, 1996.
25. M. Hilpert, J. F. McBride, and C. T. Miller. Investigation of the residual-funicular nonwetting-phase-saturation relation. *Advances in Water Resources*, 24(2):157–177, 2001.
26. D. T. Hristopulos and G. Christakos. An analysis of hydraulic conductivity upscaling. *Nonlinear Analysis*, 30(8):4979–4984, 1997.
27. International Formulation Committee. A formulation of the thermodynamic properties of ordinary water substance. Technical report, IFC Sekretariat, Düsseldorf, Germany, 1967.
28. H. Fl"uhler K. Roth, R. Schulin and W. Attinger. Calibration of time domain reflectometry for water content measurement using a composite dielectric approach. *Water Resources Research*, 26:2267 – 2273, 1990.
29. C. E. Kees and C. T. Miller. C++ implementations of numerical methods for solving differential-algebraic equations: Design and optimization considerations. *Association for Computing Machinery, Transactions on Mathematical Software*, 25(4):377–403, 1999.
30. B. H. Kueper, W. Abbott, and G. Farquhar. Experimental observations of multiphase flow in heterogeneous porous media. *Journal of Contaminant Hydrology*, 5:83–95, 1989.
31. B. H. Kueper and E. O. Frind. Two-phase flow in heterogeneous porous media, 1. Model development. *Water Resources Research*, 27(6):1049–1057, 1991.
32. B. H. Kueper and E. O. Frind. Two-phase flow in heterogeneous porous media, 2. Model application. *Water Resources Research*, 27(6):1059–1070, 1991.

33. B.H. Kueper and E.O. Frind. Two-Phase Flow in Heterogeneous Porous Media: 1. Model Development. *Water Resources Research*, 6:1049–1057, 1991.
34. R. J. Lenhard and J. C. Parker. Measurement and prediction of saturation-pressure relationships in three-phase porous media systems. *Journal of Contaminant Hydrology*, 1:407–424, 1987.
35. R. J. Lenhard, J. C. Parker, and J. J. Kaluarachchi. A model for hysteretic constitutive relations governing multiphase flow 3. Refinements and numerical simulations. *Water Resources Research*, 25(7):1727–1736, 1989.
36. R.J. Lenhard, J.H. Dane, J.C. Parker, and J.J. Kaluarachchi. Measurement and Simulation of One-Dimensional Transient Three-Phase Flow for Monotonic Liquid Drainage. *Water Resources Research*, 24:853 – 863, 1988.
37. R.J. Lenhard and J.C. Parker. A Model for Hysteric Constitutive Relations Governing Multiphase Flow 3. Refinements and Numerical Simulation. *Water Resources Research*, 25(7):1727 – 1736, 1989.
38. S.A. Lorentz, D.S. Durnford, and A.T. Corey. *Liqud retention measurement on porous media using a controlled outflow cell.* Dept. of Chemical and Biorecource Engineering, Colorado state University, Fort Collins, Colorado, 1992. Copy of manuscript submitted to Soil Sci. Soc. Am. J.
39. A. S. Mayer and C. T. Miller. An experimental investigation of pore-scale distributions of nonaqueous phase liquids at residual saturation. *Transport in Porous Media*, 10(1):57–80, 1993.
40. A. S. Mayer and C. T. Miller. The influence of mass transfer characteristics and porous media heterogeneity on nonaqueous phase dissolution. *Water Resources Research*, 32(6):1551–1567, 1996.
41. C. T. Miller, G. Christakos, P. T. Imhoff, J. F. McBride, J. A. Pedit, and J. A. Trangenstein. Multiphase flow and transport modeling in heterogeneous porous media: Challenges and approaches. *Advances in Water Resources*, 21(2):77–120, 1998.
42. E. E. Miller and R. D. Miller. Physical theory of capillary flow phenomena. *Journal of Applied Physics*, 27(4):324–332, 1956.
43. Y. Mualem. A new model for predicting the hydraulic conductivity of unsaturated porous media. *Water Resources Research*, 12:513–522, 1976.
44. National Research Council. Groundwater contamination: Overview and recommendations. In National Research Council, editor, *Groundwater Contamination*, pages 3–20. National Academy Press, Washington, DC, 1984.
45. National Research Council. *Natural Attenuation for Groundwater Remediation.* National Academy Press, Washington, DC, 2000.
46. J. Niessner, R. Helmig, H. Jakobs, and J. E. Roberts. Interface Condition and Linearization Schemes in the Newton Iterations for Two-Phase Flow in Heterogeneous Porous Media. *submitted to: Advances in Water Resources*, 2004.
47. C. Pan, M. Hilpert, and C. T. Miller. Pore-scale modeling of saturated permeabilities in simulated porous media. *in press, Physical Review E*, 2001.
48. J. C. Parker and R. J. Lenhard. A model for hysteretic constitutive relations governing multiphase flow 1. Saturation-pressure relations. *Water Resources Research*, 23(12):2187–2196, 1987.
49. J.C. Parker and R.J. Lenhard. A Model for Hysteric Constitutive Relations Governing Multiphase Flow 1. Saturation-Pressure Relations. *Water Resources Research*, 23(12):2187 – 2196, 1987.

50. N. Ursino, K. Roth, T. Gimmi, and H. Fluhler. Upscaling of anisotropy in unsaturated Miller-similar porous media. *Water Resources Research*, 36(2):421–430, 2000.
51. M.Th. van Genuchten. A closed-form equation for predicting the hydraulic conductivity of unsaturated soils. *Soil Sci. Soc. Am. J.*, 44:892–898, 1980.
52. X. Wang, F. Thauvin, and K. K. Mohanty. Non-Darcy flow through anisotropic porous media. *Chemical Engineering Science*, 54(12):1859–1869, 1999.

# The Unsteady Expansion and Contraction of a Two-Dimensional Vapour Bubble Confined Between Superheated or Subcooled Plates

K.S. Das and S.K. Wilson

Department of Mathematics, University of Strathclyde, 26 Richmond Street, Glasgow, G1 1XH, UK ra.kdas@maths.strath.ac.uk, s.k.wilson@strath.ac.uk

**Summary.** In this paper we analyse the unsteady expansion and contraction of a long, two-dimensional bubble confined between superheated or subcooled parallel plates, whose motion is driven by mass transfer between the liquid and the vapour.

**Key words:** Unsteady Expansion and Contraction, Vapour Bubble, Boiling

## 1 Introduction

Microscale boiling occurs in a number of industrial situations, including aerospace science, micro-electro-mechanical systems (MEMS), compact heat exchangers and chemical microreactors, and as a result there is considerable interest in the dynamics of confined vapour bubbles. Wilson, Davis and Bankoff [5] studied the dynamics of a long, two-dimensional vapour bubble confined between two parallel plates held at, in general, different temperatures. Unlike Bretherton's [4] classical model, in which the steady translation of the bubble is driven by an externally imposed pressure gradient, they studied the unsteady expansion and contraction of a vapour bubble whose motion is driven by mass transfer between the liquid and the vapour. As in Bretherton's isothermal model, the velocity of the bubble determines the initial thickness of the thin films of liquid laid down on both plates as the bubble expands, but unlike in Bretherton's model the evaporation from and/or condensation onto those films (which may break up into disconnected patches of liquid as they evaporate) determine the velocity of expansion and/or contraction of the bubble, and so there is a nonlinear coupling with a delay character between the profiles of the thin films and the overall dynamics of the bubble. Ajaev and Homsy [1, 2] studied a steady vapour bubble in a rectangular channel with a prescribed temperature distribution on its walls in which there is a balance between evaporation from the hotter parts of the bubble interface and condensation onto the colder parts. Subsequently Ajaev, Homsy and Morris

[3] considered a steady two-dimensional vapour bubble between two parallel plates held at different temperatures, and then investigated its dynamic response to temporally varying plate temperatures.

## 2 Problem Formulation

Following the approach of Wilson *et al.* [5], we consider a long, two-dimensional vapour bubble of inviscid and incompressible vapour of density $\rho^{(V)}$ surrounded by its condensate and confined between two parallel plates a distance $2d$ apart. The condensate is assumed to be a Newtonian liquid with constant viscosity $\mu$, kinematic viscosity $\nu$, density $\rho$, surface tension $\sigma$, thermal diffusivity $\kappa$ and thermal conductivity $k$, and the latent heat of vaporization is denoted by $\mathcal{L}$. The two plates are held at (in general different) uniform temperatures $T_{\text{u}} = T_{\text{s}} + \Delta T_{\text{u}}$ and $T_{\text{l}} = T_{\text{s}} + \Delta T_{\text{l}}$, respectively, which may be either above or below the saturation temperature $T_{\text{s}}$.

Global conservation of mass of liquid and vapour means that the rate of change of the mass of the bubble is equal to the total mass flux into the bubble and so

$$2U = DE \int J \, \mathrm{d}s, \tag{1}$$

where $U = U(t)$ is the velocity of the bubble, $D = \rho/\rho^{(V)}$ is the ratio of the liquid density to the vapour density, $E = k|\Delta T_{\text{l}}|/\rho\nu\mathcal{L}$ is the non-dimensional evaporation number, and the integral is over the entire liquid-vapour interface in $x > 0$, parameterized by its arclength $s$.

We consider the limit of strong surface tension (i.e., the limit of small capillary number $C = \mu\nu/\sigma d \to 0$) in which the solution in $x > 0$ and $y > 0$ is composed of three different regions, namely a "capillary-statics" region in $R(t) < x < R(t) + 1$, a "transition" region near $x = R(t)$ and a "thin-film" region in $0 < x < R(t)$. In the capillary-statics region the liquid-vapour interface is a semi-circular cap of radius unity that fits between the two plates. If $P = \nu/\kappa = o(1)$ then the contribution to the integral in (1) from the capillary-statics region is given by $(\Delta T_{\text{u}} + \Delta T_{\text{l}})S$, where $S = S(K)$ is a numerically determined monotonically decreasing function of $K$, a non-dimensional kinetic parameter which measures the degree of nonequilibrium at the interface. Viscous effects become significant in the narrow transition region, and a straightforward extension of Bretherton's [4] isothermal analysis shows that the initial thickness of the film as it is laid down on the plate as the bubble expands is proportional to $U^{2/3}$. For a retreating bubble (i.e., when $U < 0$) the liquid on the plate is swept up by the transition region and so the details of the solution in this region are unimportant. Once a thin liquid film has been laid down on the plate by an expanding bubble it begins either to evaporate (if the plate is superheated) or to be condensed onto (if the plate is subcooled), and the thickness of the film is given by

$$h(x,t) = h_0(x) - \frac{\hat{E}\Delta T_l}{K}(t - t_0(x)), \tag{2}$$

in which $\hat{E} = C^{-2/3}E$ is assumed to be $O(1)$ in the limit $C \to 0$, and where if $x < R(0)$ then $h_0 = h_0(x)$ denotes the initial profile of the thin film at $t = t_0 = 0$, while if $x > R(0)$ then $h_0 = h_0(x) = U^{2/3}$ denotes the thickness of the film laid down at position $x$ at time $t = t_0(x) = R^{-1}(x)$. Equation (2) shows that if the plate is superheated (i.e., if $\Delta T_l > 0$) then liquid evaporates from the thin film which dries out locally at position $x$ at time $t = t_0 + Kh_0(x)/\hat{E}\Delta T_l$, while if the plate is subcooled (i.e., if $\Delta T_l < 0$) then vapour condenses onto the thin film and local dry-out never occurs. The dominant contributions to the integral in (1) are those from the capillary-statics and thin-film regions. Note that this situation is different from that studied by Wilson *et al.* [5] for whom the contributions from the thin-film regions dominated those from the other two regions.

## 3 Both Plates Superheated

If both plates are superheated (i.e., if $\Delta T_u > 0$ and $\Delta T_l > 0$) then the bubble always expands and the dynamics of the expansion are governed by (1) which takes the form

$$U = \frac{dR}{dt} = \frac{DE}{2K}\left[(\Delta T_u + \Delta T_l)KS + \Delta T_u L_u + \Delta T_l L_l\right] > 0, \tag{3}$$

where $L_u = L_u(t)$ and $L_l = L_l(t)$ $(0 \le L_u, L_l \le R)$ denote the total lengths of film on the upper and lower plates in $x > 0$, respectively.

### 3.1 Delay-Equation Formulation for Continuous Films

The liquid films laid down on the plates as the bubble expands will not, in general, remain as continuous films as they dry. However, it is still very informative to investigate the solution in this special case analytically before analysing the more general situation numerically. For continuous films we denote the position of the front of the film (where $h = h_0 = U^{2/3}$) by $x = R(\tau_u + \mathcal{T}_u(\tau_u)) = R(\tau_l + \mathcal{T}_l(\tau_l))$, the position of the back of the film on the upper plate (where $h = 0$) by $x = R(\tau_u)$, and the position of the back of the film on the lower plate (where again $h = 0$) by $x = R(\tau_l)$, where $\mathcal{T}_u = \mathcal{T}_u(\tau_u)$ and $\mathcal{T}_l = \mathcal{T}_l(\tau_l)$, given by

$$\mathcal{T}_u(\tau_u) = \frac{Kh_0}{E\Delta T_u} = \frac{KU(\tau_u)^{2/3}}{E\Delta T_u} \quad \text{and} \quad \mathcal{T}_l(\tau_l) = \frac{Kh_0}{E\Delta T_l} = \frac{KU(\tau_l)^{2/3}}{E\Delta T_l}, \tag{4}$$

are the lengths of time it takes for the liquid deposited on the upper and lower plates at times $t = \tau_u$ and $t = \tau_l$, respectively, to dry out. Adopting this new notation (3) can be written as

$$U(\tau_u + T_u(\tau_u)) = \frac{DE}{2K}\left[(\Delta T_u + \Delta T_l)KS + \Delta T_u \int_{\tau_u}^{\tau_u + T_u(\tau_u)} U(\hat{\tau})\,d\hat{\tau} \right.$$
$$\left. + \Delta T_l \int_{\tau_l}^{\tau_l + T_l(\tau_l)} U(\hat{\tau})\,d\hat{\tau}\right]. \quad (5)$$

Equation (5) is an integro-delay equation for $U$ with non-constant delays $T_u$ and $T_l$ which depend on the solution for $U$ at times $t = \tau_u$ and $t = \tau_l$, respectively, according to (4).

### 3.2 Constant-Velocity Solutions and their Stability

Equation (5) permits an exact travelling-wave solution with constant delays and constant velocity $U_0(> 0)$, where $U_0$ satisfies

$$U_0 = \frac{DE}{2}(\Delta T_u + \Delta T_l)S + DU_0^{5/3}. \quad (6)$$

Figure 1 shows $U_0$ plotted as a function of $D$ for a range of values of $\Delta T_u$ when $\Delta T_l = 1$, and shows that for $0 < D < D_c$ there are two branches of positive solutions, namely a "fast" mode satisfying $U_0 > U_{0c}$ and a "slow" mode satisfying $0 < U_0 < U_{0c}$, but that there are no positive solutions for $D > D_c$, where

$$D_c = \left(\frac{3}{5}\right)^{3/5}\left[\frac{4}{5E(\Delta T_u + \Delta T_l)S}\right]^{2/5} \quad \text{and} \quad U_{0c} = \left(\frac{3}{5D_c}\right)^{3/2}. \quad (7)$$

In particular, the slow mode satisfies $U_0 = O(D) \to 0^+$ and the fast mode satisfies $U_0 \sim D^{-3/2} \to \infty$ in the limit $D \to 0^+$, while both modes satisfy $U_0 - U_{0c} = O(D_c - D)^{1/2}$ in the limit $D \to D_c^-$. For both modes the profiles of the films on both plates are linear in $x$. On the lower plate the profile increases from the value $h = 0$ at the back $x = R(t - T_{l0}) = U_0 t - KU_0^{5/3}/E\Delta T_l$ to the value $h = U_0^{2/3}$ at the front $x = R(t) = U_0 t$ according to

$$h = U_0^{2/3} + \frac{E\Delta T_l}{KU_0}(x - U_0 t). \quad (8)$$

The corresponding results for the profile of the film on the upper plate can be readily deduced. A linear stability analysis shows that the fast mode is always *unstable* while the slow mode is always *stable*. These results are qualitatively different from those obtained by Wilson *et al.* [5] who found a single unstable mode with velocity $U_0 = D^{-3/2}$ for *all* values of $D$.

## 4 Summary

In this paper we analysed the unsteady expansion and contraction of a long, two-dimensional bubble confined between superheated or subcooled parallel

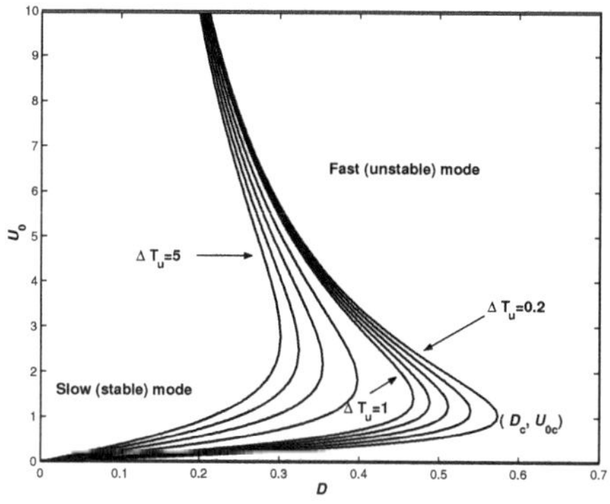


**Fig. 1.** The velocity of the constant-velocity solutions $U_0$ plotted as a function of $D$ for $\Delta T_u = 0.2, 0.4, 0.6, 0.8, 1, 2, 3, 4$ and $5$ when $\Delta T_l = 1$.

plates, whose motion is driven by mass transfer between the liquid and the vapour. Specifically, we extended the analysis of Wilson *et al.* [5] to include significant mass transfer from and/or to the semi-circular cap regions at the nose of the bubble as well as from and/or to the thin liquid films attached to the plates. When both plates are superheated the bubble always expands. In this case there are two possible constant-velocity travelling-wave solutions for the expansion of the bubble when $0 < D < D_c$, namely an unstable fast mode with velocity $U_0$ satisfying $U_0 > U_{0c}$ and a stable slow mode with velocity $U_0$ satisfying $0 < U_0 < U_{0c}$, but none for $D > D_c$.

## Acknowledgement

The present work is supported by United Kingdom Engineering and Physical Sciences Research Council (EPSRC) Research Grant GR/R74468.

## References

1. V. S. Ajaev and G. M. Homsy. Steady vapor bubbles in rectangular microchannels. *J. Colloid Interface Sci.*, 240:259–271, 2001.
2. V. S. Ajaev and G. M. Homsy. Three-dimensional steady vapor bubbles in rectangular microchannels. *J. Colloid Interface Sci.*, 244:180–189, 2001.
3. V. S. Ajaev, G. M. Homsy, and S. J. S. Morris. Dynamic response of geometrically constrained vapor bubbles. *J. Colloid Interface Sci.*, 254:346–354, 2002.
4. F. P. Bretherton. The motion of long bubbles in tubes. *J. Fluid Mech.*, 10:166–188, 1961.
5. S. K. Wilson, S. H. Davis, and S. G. Bankoff. The unsteady expansion and contraction of a long two-dimensional vapour bubble between superheated or subcooled parallel plates. *J. Fluid Mech.*, 391:1–27, 1999.

# Animating Water Waves Using Semi-Lagrangian Techniques

M. El Amrani[1] and M. Seaïd[2]

[1] Dpto. Matemáticas, Univ. Rey Juan Carlos, c/Tulipán s/n, 28933 Mostoles-Madrid, Spain mofdi@escet.urjc.es
[2] Fachbereich Mathematik, TU Darmstadt, 64289 Darmstadt, Germany seaid@mathematik.tu-darmstadt.de

**Summary.** Semi-Lagrangian techniques are proposed for animating water waves in realistic events. The two-dimensional shallow water equations are considered to model the motion of water flow and a second order time marching procedure which combines the characteristic method with a finite differencing discretization is used to integrate the model. Numerical results are carried out on a squared pool without and with obstacles. The obtained results show that our algorithm is robust, stable and highly accurate.

## 1 Introduction

We present a comprehensive methodology for realistically animating water waves. Our approach is based on the shallow water equations which result from the depth averaged incompressible Navier-Stokes equations and consequently describe water motion. The method we propose in this paper consists of an Eulerian-Lagrangian splitting of the equations along the characteristic curves. The Lagrangian stage of the splitting is treated by the modified method of characteristics, while the Eulerian stage is approximated by an implicit time integration scheme using finite differencing for spatial discretization. The combined two stages lead to a semi-Lagrangian method which is robust, second order accurate, and simple to implement for water flows over fixed solid obstacles. Computational results are shown for two test problems on animating water waves in squared pools without and with obstacles.

The model we consider in this paper for animating waves is the two-dimensional shallow water equations with no Coriolis effect and zero viscosity, compare [1] for details. In Lagrangian form these equations are given by

$$\begin{aligned} \frac{DU}{Dt} + g\nabla h &= 0, \\ \frac{Dh}{Dt} - U \cdot \nabla b + d\nabla \cdot U &= 0, \end{aligned} \qquad (1)$$

where $h = h(t, x, y)$ is the water height, $U = (u, v)^T$ is the velocity field with $u = u(t, x, y)$ and $v = v(t, x, y)$ are the velocities in $x-$ and $y$–direction, respectively. $b = b(x, y)$ is height of the bottom ground assumed to be time independent, $d(t, x, y) = h(t, x, y) - b(x, y)$ is the depth of the water above the bottom, and $g$ is the gravitational acceleration.

In (1), $\nabla = (\frac{\partial}{\partial x}, \frac{\partial}{\partial y})^T$ is the gradient operator, and $\frac{D}{Dt} = \frac{\partial}{\partial t} + U \cdot \nabla$ is the material derivative. In addition, the equations (1) have to be solved subject to appropriate boundary and initial conditions for both $h$ and $U$.

## 2 Semi-Lagrangian Techniques

To construct our semi-Lagrangian algorithm we cover the spatial domain with gridpoints $(x_i, y_j)$ using uniform space sizes $\Delta x$ and $\Delta y$, and divide the time interval into subintervals $[t_n, t_{n+1}]$ of equal length $\Delta t$. We use the notation $w_{ij}^n = w(t_n, x_i, y_j)$. Following [2], the characteristics curves associated to equations (1) are the solution of initial-value problem

$$\begin{aligned} \frac{d\mathbf{X}_{ij}}{d\tau} &= U\left(\tau, \mathbf{X}_{ij}(\tau; t_{n+1}, \mathbf{x}_{ij})\right), \quad \tau \in [t_n, t_{n+1}], \\ \mathbf{X}_{ij}(t_{n+1}; t_{n+1}, \mathbf{x}_{ij}) &= \mathbf{x}_{ij}. \end{aligned} \tag{2}$$

Note that $\mathbf{X}_{ij}(\tau; t_{n+1}, \mathbf{x}_{ij}) = (X_i(\tau; t_{n+1}, \mathbf{x}_{ij}), Y_j(\tau; t_{n+1}, \mathbf{x}_{ij}))^T$ is the departure point at time $\tau$ of a water particle that will reach the gridpoint $\mathbf{x}_{ij} = (x_i, y_j)^T$ at time $\tau = t_{n+1}$. To approximate solutions to (2) we used a method first proposed in the context of semi-Lagrangian schemes to integrate the weather prediction equations [4]. Details on the implementation of this step in viscous incompressible flows can be found in [3].

Once the characteristic curves are computed, the value of a solution function $w$ at the characteristic feet, $\tilde{w}_{ij}^n = w(t_n, \mathbf{X}_{ij}(t_n; t_{n+1}, \mathbf{x}_{ij}))$, is approximated by interpolation from known values at gridpoints of the host element where the feet are localized. To perform this step in our algorithm we used the bicubic spline interpolation. The complete discretization we propose in the present work for equations (1) reads

$$\frac{u_{ij}^{n+1} - \tilde{u}_{ij}^n}{\Delta t} + \frac{1}{2} g \mathcal{D}_x \tilde{h}_{ij}^n + \frac{1}{2} g \mathcal{D}_x h_{ij}^{n+1} = 0, \tag{3}$$

$$\frac{v_{ij}^{n+1} - \tilde{v}_{ij}^n}{\Delta t} + \frac{1}{2} g \mathcal{D}_y \tilde{h}_{ij}^n + \frac{1}{2} g \mathcal{D}_y h_{ij}^{n+1} = 0, \tag{4}$$

$$\begin{aligned} &\frac{h_{ij}^{n+1} - \tilde{h}_{ij}^n}{\Delta t} - \frac{1}{2} \tilde{u}_{ij}^n \mathcal{D}_x b_{ij} - \frac{1}{2} \tilde{v}_{ij}^n \mathcal{D}_y b_{ij} - \frac{1}{2} u_{ij}^{n+1} \mathcal{D}_x b_{ij} - \frac{1}{2} v_{ij}^{n+1} \mathcal{D}_y b_{ij} + \\ &\qquad d_{ij}^* \left( \frac{1}{2} \mathcal{D}_x \tilde{u}_{ij}^n + \frac{1}{2} \mathcal{D}_y \tilde{v}_{ij}^n + \frac{1}{2} \mathcal{D}_x u_{ij}^{n+1} + \frac{1}{2} \mathcal{D}_y v_{ij}^{n+1} \right) = 0, \end{aligned} \tag{5}$$

where $d_{ij}^* = \frac{3}{2} \tilde{d}_{ij}^n - \frac{1}{2} \tilde{d}_{ij}^{n-1}$, $\mathcal{D}_x$ and $\mathcal{D}_y$ denote the centered difference operators

$$\mathcal{D}_x w_{ij} = \frac{w_{i+1j} - w_{i-1j}}{2\Delta x}, \qquad \mathcal{D}_y w_{ij} = \frac{w_{ij+1} - w_{ij-1}}{2\Delta y}. \tag{6}$$

Note that the fully disctization (3) – (5) is second order accurate in both space and time. A simple way to solve the equations (3) – (5) is to use the first and second equations to eliminate the flow velocity and its divergence from the third equation. Hence, (3) and (4) give

$$u_{ij}^{n+1} = \tilde{u}_{ij}^n - \frac{\Delta t}{2} g \mathcal{D}_x \tilde{h}_{ij}^n - \frac{\Delta t}{2} g \mathcal{D}_x h_{ij}^{n+1}, \tag{7}$$

$$v_{ij}^{n+1} = \tilde{v}_{ij}^n - \frac{\Delta t}{2} g \mathcal{D}_y \tilde{h}_{ij}^n - \frac{\Delta t}{2} g \mathcal{D}_y h_{ij}^{n+1}. \tag{8}$$

Inserting (7) and (8) in (5) leads to the following Helmholtz equation for $h^{n+1}$

$$h_{ij}^{n+1} + \frac{(\Delta t)^2}{4} g \mathcal{D} b_{ij} \cdot \mathcal{D} h_{ij}^{n+1} - \frac{(\Delta t)^2}{4} g d_{ij}^* \mathcal{D}^2 h_{ij}^{n+1} = \tilde{h}_{ij}^n + \Delta t \tilde{U}_{ij}^n \cdot \mathcal{D} b_{ij} -$$
$$\Delta t d_{ij}^* \mathcal{D} \cdot \tilde{U}_{ij}^n - \frac{(\Delta t)^2}{2} g \mathcal{D} b_{ij} \cdot \mathcal{D} \tilde{h}_{ij}^n + \frac{(\Delta t)^2}{2} g d_{ij}^* \mathcal{D}^2 \tilde{h}_{ij}^n, \tag{9}$$

where $\mathcal{D} = (\mathcal{D}_x, \mathcal{D}_y)^T$ with $\mathcal{D}_x$ and $\mathcal{D}_y$ are given in (6), and $\mathcal{D}^2$ is the central discretization of the Laplace operator

$$\mathcal{D}^2 w_{ij} = \frac{w_{i+1j} - 2w_{ij} + w_{i-1j}}{(\Delta x)^2} + \frac{w_{ij+1} - 2w_{ij} - w_{ij-1}}{(\Delta y)^2}.$$

The implementation of semi-Lagrangian algorithm to solve the shallow water equations (1) is carried out in the following steps:

*(a) For all gridpoints $\mathbf{x}_{ij}$ compute the departure points $\mathbf{X}_{ij}(t_n; t_{n+1}, \mathbf{x}_{ij})$ in (2) and identify the grid elements where such points are located.*
*(b) Compute the approximation $\tilde{h}_{ij}^n$, $\tilde{u}_{ij}^n$ and $\tilde{v}_{ij}^n$ employing the bicubic spline interpolation and formulate the right-hand side term in (9).*
*(c) Solve for $\tilde{h}_{ij}^{n+1}$ the linear system (9) using for instance the Bicgstab method.*
*(d) Update the velocity field $u_{ij}^{n+1}$ and $v_{ij}^{n+1}$ using (7) and (8), respectively.*

Note that the steps (*a*)-(*d*) can straightforwardly be implemented in parallel.

## 3 Numerical Results

The semi-Lagrangian algorithm has been implemented to animate water waves in a squared pool of 200 m length. The problem starts with a circular wave created in the upper-left corner of the pool. Then, the wave is propagated freely along the main diagonal according to the shallow water equations (1). The wave is centered at (10 m, 190 m) and the bottom ground $b$ is set to 5 m. Initially the water is at rest and $g = 9.8$ m/s$^2$. The computational domain is discretized in $50 \times 50$ grid points. On the boundaries we impose solid wall boundary conditions and a time step of $\Delta t = 0.1$ sec is used in computations.

The numerical results are presented in Fig. 1 at times $t = 25$, 50 and 100 sec. Next we introduce a $\Gamma$-shaped obstacle in the pool and numerical results are shown in Fig. 2. In these figures the left and right columns represent the water height and the velocity vectors, respectively.

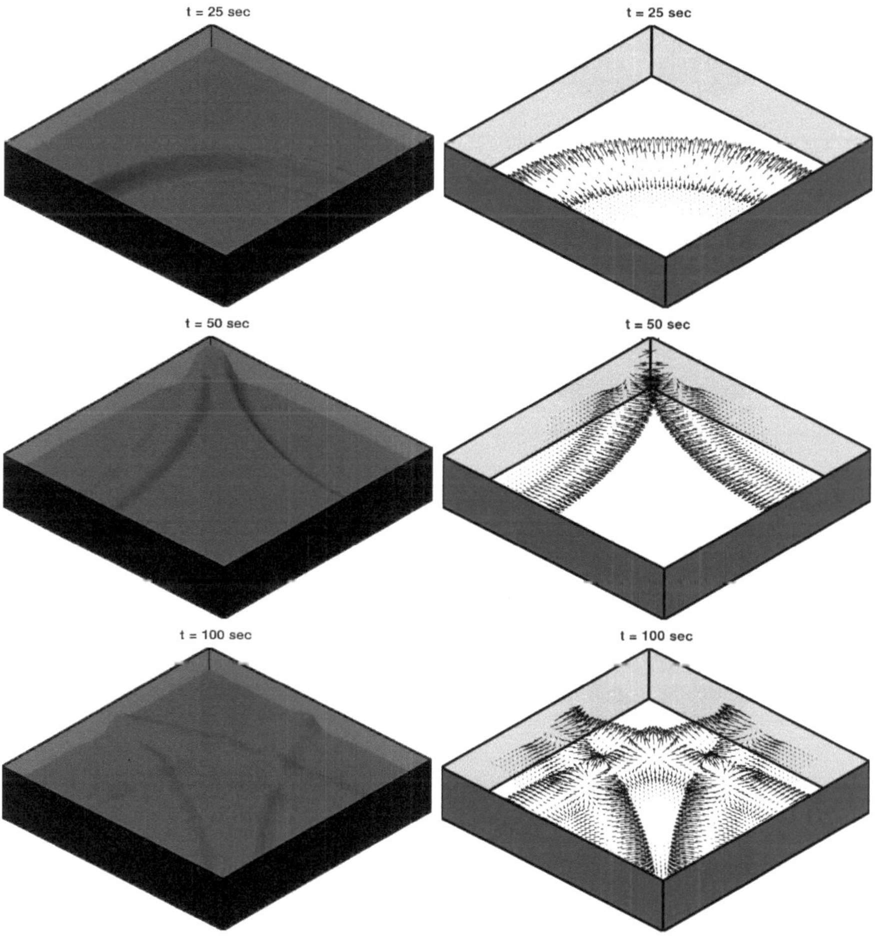


**Fig. 1.** Animating water waves in a squared pool.

For both examples, the semi-Lagrangian method captures the correct water-flow structures with no oscillations or extra numerical dissipation. Furthermore, the water-flow symmetry in the first example and water-wave reflections from the solid walls in the second example are well resolved by the algorithm using coarse mesh and large time step as those used in our computations. These facts make the semi-Lagrangian very attractive as numerical tools for animating water waves in interactive computer simulations.

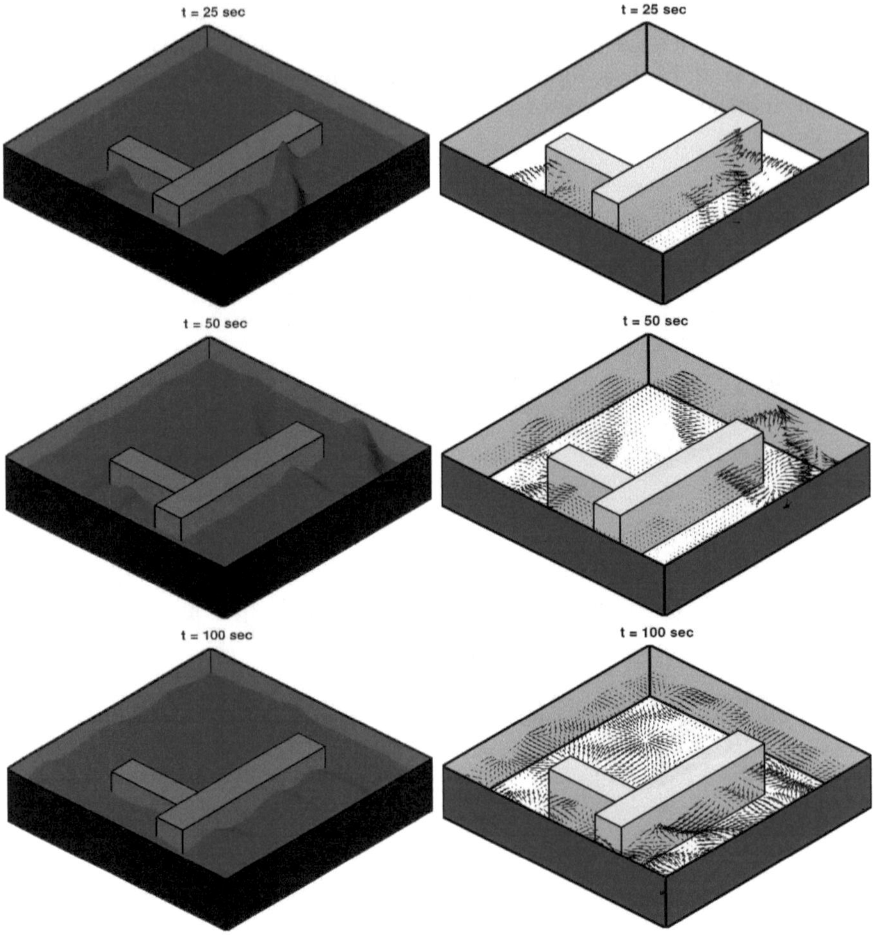


**Fig. 2.** Animating water waves in a squared pool with fixed obstacles.

## References

1. D.R. Durran. *Numerical Methods for Wave Equations for Geophysical Fluid Mechanics*. Springer, 1998.
2. A. Robert. A stable numerical integration scheme for the primitive meteorological equations. *Atmos. Ocean*, 19:35–46, 1981.
3. M. Seaïd. Semi-lagrangian integration schemes for viscous incompressible flows. *J. Comp. Methods in App. Math.*, 4:392–409, 2002.
4. C. Temperton and A. Staniforth. An efficient two-time-level semi-Lagrangian semi-implicit integration scheme. *Quart. J. Roy. Meteor. Soc.*, 113:1025–1039, 1987.

# A Filtered Renewal Process as a Model for a River Flow

M. Lefebvre

École Polytechnique de Montréal, Canada mlefebvre@polymtl.ca

**Summary.** Models based on a filtered Poisson process are used for the flow of a river. The aim is to forecast the next peak value of the flow, given that another peak was observed not too long ago. The most realistic model is the one when the time between the successive peaks does *not* have an exponential distribution, as it is often assumed. An application to the Delaware River, in the USA, is presented.

**Key words:** filtered Poisson process, Rayleigh distribution, forecast, peaks.

## 1 Introduction

In [1] (see also [4]), a filtered Poisson process was used to forecast the various peaks of rivers. Let $\{N(t),\ t \geq 0\}$ be a homogeneous Poisson process and let $X(t)$ be the river flow at time $t$. It was assumed, in the previous references, that

$$X(t) = \sum_{n=1}^{N(t)} Y_n e^{-(t-\tau_n)/c}, \qquad (1)$$

where the random variables $\tau_n$ are the arrival times of the Poisson events, $Y_n$ is the magnitude of the signal that occurred at time $\tau_n$ and $c$ is a constant which characterizes the river system. The authors also assumed that the random variables $Y_n$ have an exponential distribution. The stochastic process $\{X(t),\ t \geq 0\}$ defined by (1) is indeed a particular case of what is known as filtered Poisson processes. This type of stochastic process has been used to model various phenomena; see [2]. In civil engineering, filtered Poisson processes have served as models for stochastic rainfall ([5]) and seismic hazard ([3]), in particular.

The model set up in [1] worked only relatively well, mainly because the correlation coefficient between the successive peaks is rather weak, in general. It is well known that trying to predict the next peak value of a river flow is a very difficult task. However, we believe that we can at least improve the results obtained so far by rendering the model more realistic. Indeed, many mathematical assumptions made in the formulation of the model are often not realistic at all or are only used to make the model tractable, or for lack of better alternatives.

There are two main criticisms that one can state with regard to the $\{X(t),\ t \geq 0\}$ process above. First, it assumes that an event that occurs at time $\tau_n$ has an immediate maximum effect and that this effect decreases with time. In practice, a more or less steep increase of the river flow is almost always observed before it begins to decrease. Therefore, the choice of an exponential function as a "response function" can surely be criticized. Next, the main assumption in the model above is that the time between two consecutive peaks has an exponential distribution, so that events occur according to a Poisson process. Again, in practice this will almost surely be false. Remember that the density function of an exponential random variable is *strictly decreasing* from zero. In practice, the density function of the times between the flow peaks increases toward a maximum value and then is strictly decreasing until infinity. Therefore, a Poisson process is *not* appropriate.

In the next section, the notion of filtered renewal process will be introduced. We will see how the next peak flow value could be forecasted, based on the most recent peak observed. An application to the Delaware River will then be presented in Section 3 and a few conclusions will be drawn in Section 4.

## 2 Filtered Renewal Process

A renewal process is such that the times $T_1$, $T_2$, ... between the consecutive events are independent and have the same distribution. We propose the model

$$X(t) = \sum_{n=1}^{N(t)} w(Y_n, t - \tau_n)\,,$$

where $w(\cdot,\cdot)$ is the response function, $\{N(t), t \geq 0\}$ is a renewal process and $\tau_n = \sum_{k=1}^{n} T_k$. The random variables $T_k$ $(\geq 0)$ are general.

Next, because there is almost always a period during which the flow increases before decreasing again, we consider the response function

$$w(Y_n, t - \tau_n) = Y_n(t - \tau_n)^k e^{-(t-\tau_n)/c}\,.$$

To estimate the unknown parameters $k$ and $c$, let $g(t) = t^k e^{-t/c}$. This function attains its maximum at $t_{max} = kc$. Hence, we can estimate $kc$ by computing the mean time taken by the flow to reach a peak from the preceding minimum.

If the time between the consecutive peaks is large enough, we can neglect the effect of the signals $Y_1, \ldots, Y_{N(t)-1}$ and write that

$$X(t+\delta) \simeq Y_{N(t)}(t + \delta - \tau_{N(t)})^k e^{-(t+\delta-\tau_{N(t)})/c}\,.$$

We then deduce that

$$\frac{X(t+\delta)}{X(t)} \simeq e^{-\delta/c}\left\{1+\frac{\delta}{t-\tau_{N(t)}}\right\}^k,$$

which is valid for values of $t$ and $t+\delta$ between two consecutive peaks. If $t$ is the time at which the most recent peak was observed, we may write that

$$\frac{X(t+\delta)}{X(t)} \simeq e^{-\delta/c}\left\{1+\frac{\delta}{kc}\right\}^k. \tag{2}$$

Since $kc$ can be estimated, we solve for $k$ in (2) and obtain that

$$k \simeq \ln\left(\frac{X(t+\delta)}{X(t)}\right) \Big/ \left\{\ln\left(1+\frac{\delta}{kc}\right)-\frac{\delta}{kc}\right\}.$$

To estimate $k$ (and $c$), we will compute the mean value of $k$ if $t+\delta$ is the time at which the minimum following the last recorded peak was observed.

Our aim will be to forecast the next peak flow value, based on the preceding peak, once we observe that the river flow has started to increase again. As a predictor we will use

$$\widehat{Peak}_1 = Max(\bar{N}_I + N_D)^{\hat{k}} e^{-(\bar{N}_I+N_D)/\hat{c}} + \bar{I},$$

where $Max$ is the value of the most recent peak flow, $\bar{N}_I$ is the average number of days taken by the river to go from a minimum to a maximum flow, $N_D$ is the number of days between $Max$ and the following minimum flow, and $\bar{I}$ is the average difference between the various peaks and the preceding minima. We will compare the results obtained with $\widehat{Peak}_1$ to the ones when $k = 0$. Based on this (original) model, a simple estimator of the next peak flow is

$$\widehat{Peak}_2 = Min + \bar{I},$$

where $Min$ is the minimum flow that has just been observed. As a criterion to assess the quality of the estimators considered in the paper, we will use the *correlation coefficient* $r$, defined, for the pairs $(x_1, y_1), \ldots, (x_n, y_n)$, by

$$r = \frac{\sum_{i=1}^n (x_i-\bar{x})(y_i-\bar{y})}{\sqrt{\sum_{i=1}^n (x_i-\bar{x})^2 \sum_{i=1}^n (y_i-\bar{y})^2}}$$

## 3 An Application

To test our model on real data, we have chosen the Delaware River. During the years 1993-2002, there have been 91 peak flow values $\geq$ 10000 ft$^3$/s at the Montague, NJ, station, of which 61 were followed by another peak in a short enough interval. Our objective will be to first find a model for the flow of the Delaware River. Next, we will use the data from the years 1993-1997 to estimate the various parameters and quantities in the model, and then we will forecast the 33 peak flows that were preceded by another peak a few days beforehand during the 1998-2002 time period.

### 3.1 Model fitting

The first step is to find a distribution for the $T_n's$. If we denote by $T$ the random variable representing the time between two events, we find, using the 61 data points, that

$$\bar{t} \simeq 11.689 \quad \text{and} \quad s_T \simeq 6.125\,. \tag{3}$$

We notice that an exponential distribution is *not* an appropriate model, since we should have $\bar{t} \simeq s_T$. Therefore, we should not consider a filtered Poisson process as a model for the flow of the Delaware River. We fit a Rayleigh distribution to the data. That is,

$$f_T(t) = (t/\alpha^2)e^{-t^2/2\alpha^2} \quad \text{for } t \geq 0\,.$$

We have: $E[T] \overset{(*)}{=} (\pi/2)^{1/2}\alpha$ and $\text{STD}[T] \overset{(**)}{=} [2 - (\pi/2)^{1/2}]\alpha$. From (3), (*) implies that $\alpha \simeq 9.3265$, while (**) yields $\alpha \simeq 9.3492$. Hence, the model seems very good. A chi-square goodness-of-fit test was performed with $\alpha = 9.33$. We obtained a test statistic $D^2 = 1.026$, which corresponds to a $p$-value of 0.60.

Because $T^{1/2}$ has an exponential distribution, we should take the square root of all the time variables before estimating the parameters $k$ and $c$ in the model. The transformed process is then a filtered Poisson process, for which many exact and explicit results are known. We find that

$$\widehat{X}(t) = \sum_{n=1}^{N(t)} Y_n(t - \tau_n)^{1.983} e^{-(t-\tau_n)/0.973}\,, \tag{4}$$

in which $t$ is measured in square roots of days.

### 3.2 Forecasting

Based on the model fitted above, the value of the forecasted peak flow following the current minimum flow is given by

$$\widehat{Peak_1} = Max(1,93 + N_D)^{1.983} e^{-(1.93+N_D)/0.973} + 15468\,,$$

where 15468 is the mean difference between the peaks and the preceding minima during the years 1993-2002. Using this predictor, we find that the correlation coefficient between the observed and forecasted peaks is $r = 0.489$. The value of $r$ for the consecutive pairs of flows is actually 0.416. Therefore, the model (4) enabled us to improve the forecasts of peak flow values. However, the usefulness of the model is more apparent when we *forecast* the peak flows. We first estimate the parameters $k$ and $c$ in the model by using the data from the years 1993-1997, obtaining

$$\widehat{X}(t) = \sum_{n=1}^{N(t)} Y_n(t - \tau_n)^{2.178} e^{-(t-\tau_n)/0.926}\,,$$

from which we deduce that

$$\widehat{Peak}_1 = Max(2.02 + N_D)^{2.178} e^{-(2.178+N_D)/0.926} + 18163\,.$$

The value of $r$ for the forecasted and observed peak flows obtained for the years 1998-2002 is 0.347. This result is more impressive when we compute $r = -0.206$ for the observed peak flows during this time period (whereas $r = 0.516$ for the years 93-97). Thus, the filtered renewal process model has been able to transform a negative $r$ into a relatively high (and positive) $r$. We obtain $r = 0.084$ when we use the estimator

$$\widehat{Peak}_2 = Min + 18163\,.$$

## 4 Conclusion

We developed a filtered renewal process for the flow of a river which is intended to be used to forecast the oncoming peak flow when we notice that the flow has begun to increase from a minimum value. The data set used to compare the predictors is special: the correlation coefficient between the consecutive peaks during the first five years is relatively high and positive (0.516), while it is small and negative (−0.206) for the last five years. We feel that it is in such a challenging situation that the quality of an estimator can be established. One way of rendering the filtered renewal process even more realistic would be to choose a response function that is not deterministic. Finally, another subject on which more work is needed is a method to estimate the parameters in the filtered renewal process when we cannot neglect all the signals that occurred before the most recent one.

*Acknowledgement.* Work supported by the Natural Sciences and Engineering Research Council of Canada. The author also wishes to thank the referee.

## References

1. M. Lefebvre, J. Ribeiro, J. Rousselle, O. Seidou, and N. Lauzon. Probabilistic prediction of peak flood discharges. In A. Der Kiureghian, S. Madanat, and J.M. Pestana, editors, *Proceedings of the ICASP9 Conference, San Francisco, USA, July 6-9, 2003*, volume 1, pages 867–871, Rotterdam, The Netherlands, 2003. Millpress.
2. E. Parzen. *Stochastic Processes.* Holden-Day, San Francisco, 1962.
3. S. Rahman and M. Grigoriu. Markov model for seismic reliability analysis of degrading structures. *J. Struct. Eng.*, 119:1844–1865, 1993.
4. J. Ribeiro-Corréa. *Étude de quelques problèmes reliés à l'estimation des débits de crue.* PhD thesis, École Polytechnique de Montréal, Canada, 1994.
5. J. Yoon and M.L. Kavvas. Probabilistic solution to stochastic overland flow equation. *J. Hydraul. Eng.*, 8:54–63, 2003.

# A Parallel Finite Element Method for Convection-Diffusion Problems

J.M.L. Maubach[1]

Technische Universiteit Eindhoven (TU/e), P.O. Box 513, 5600 MB Eindhoven, The Netherlands. J.M.L.Maubach@tue.nl **

**Summary.** The robust Parallel Finite Element Method examined in [5] and [4]. It is an element-wise parallel iterative solution method based on a Red-Black domain decomposition. Convection-diffusion problems are solved in an optimal order for a method which makes use of not more than local communication. For the parallellism, the recent paper [8] shows that a near perfect load-balance can be obtained for two-dimensional problems. This paper proves that one of the conditions which is sufficient in the two-dimensional case, unexpectedly is not so for the three-dimensional case.

**Key words:** Finite Elements, Parallel Iterative Method, constrained problems.

## 1 The computational mesh

Assume that the domain of interest can be covered with a coarse tensor product mesh, each cell covered with $n$-simplices, and that it is refined with the use of local bisection as introduced in [7] (or alternatively as in [2] or [10]). This is the case for a multitude of challenging applications such as marker-optics and on-chip-interconnects.

## 2 The parallel finite element method

Because the size of the discretized problem, it is desirable to solve difficult problems, such as convection-diffusion problems, in parallel. The parallel finite element method applied to a convection-diffusion equation colours each element either with Red or Black as in figure which shows a mesh and its 2-coloured variant. Because integration over the domain is a sum of integration over the Red and integration over the Black domain, the coefficient matrix $A$

**In part supported by NWO-RFBR Grant 047.016.008.

splits: $A = A_R + A_B$. Assume that we use non-conforming piece-wise linear (or higher order) finite element basis functions for the discretization ([5]).

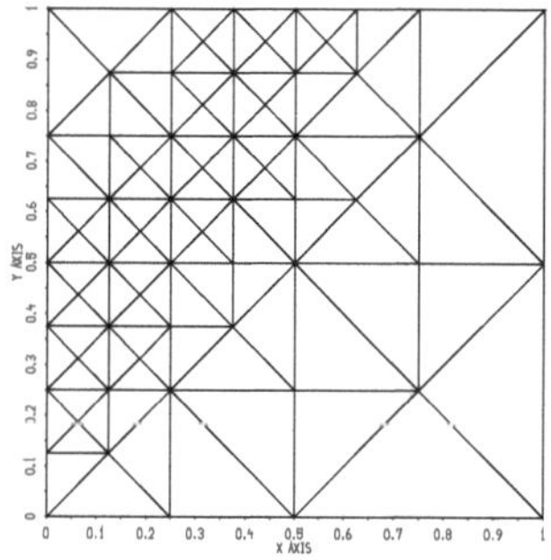


The mesh.

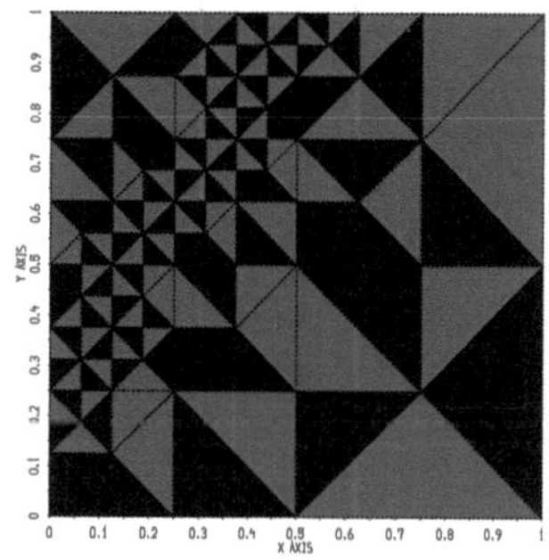


The 2-coloured mesh.

Then, a discrete solution can be computed with iterative method such as [6] or [9] In all such cases, this leads to iterations of the form ($\rho \in (0, \infty)$):

$$\begin{aligned}(\rho I_N + A_R)\mathbf{v} &= (\rho I_N - A_B)\mathbf{u}^{(k)} + \mathbf{b} \\ (\rho I_N + A_B)\mathbf{u}^{(k+1)} &= (\rho I_N - A_R)\mathbf{v} + \mathbf{b},\end{aligned}$$

where $A_R$ and $A_B$ are block diagonal (under a permutation). Each diagonal block corresponds to a (Red respectively Black) cluster of elements, in the best possible case each cluster contains just 1 element (see [3]).

## 3 Load balance

Because of the block-diagonal structure of $A_R$ and $A_B$, it is possible to use one processor per cluster in the iterative algorithm. However, in practice there are fewer processors than clusters and a kind of load-balance is required.

With this in mind, we proved in [8] that for local refined meshes described in [7] there exists a connected space-filling curve in two dimensions, (see the figure with the space-filling curve and the partition over four processors).

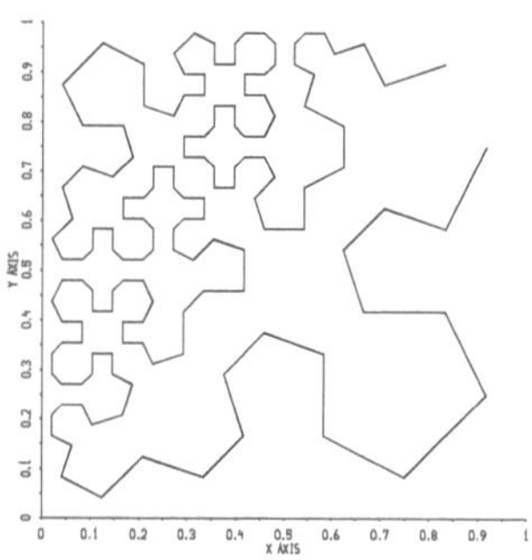


The space-filling curve.

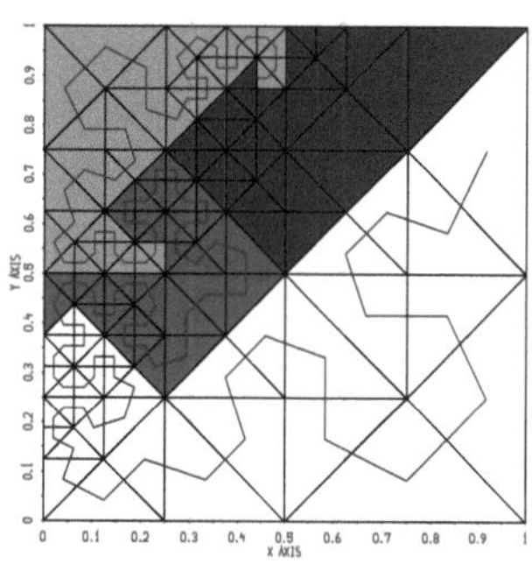


Partition over 4 processors.

To this end, it is shown that it is sufficient if a curve which passes an element through 2 of its facets, also passes through the inside elements created with bisection through the same facets.

In three dimensions, we now prove that this sufficient condition does not hold. The proof does not require the concept of the level of a facet in [8].

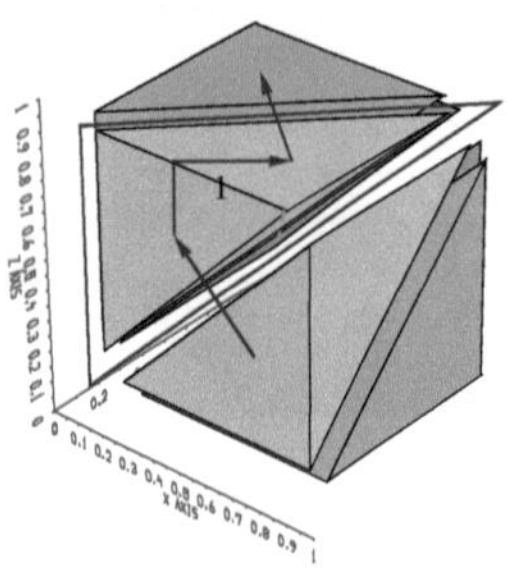


The curve on mesh 1.

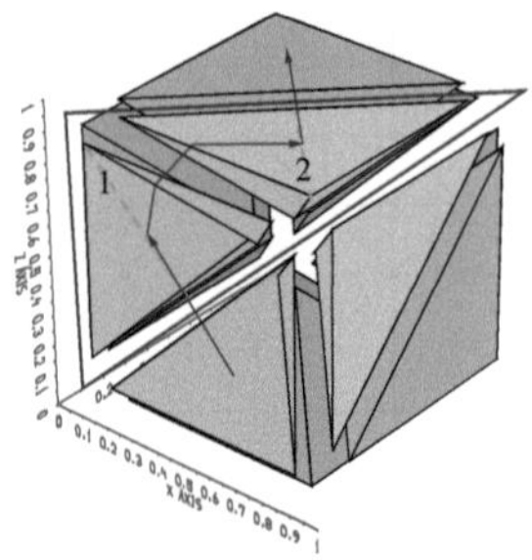


The curve on mesh 2.

The first mesh on the unit-cube consists of the standard 6 tetrahedral elements ([7]). These have one common axis (edge) (0,0,0)(1,1,1). In addition, each element has precisely 2 neighbours inside the unit cube ([1]). Hence, the sole possible curve through these 6 elements is a 'circle' around the common axis, which enters each elements through one of its neighbours, and leaves through the unique other neighbour. Part of this curve – which would be invisible because it remains in the inside of the unit-cube – is shown in the first figure. It shows how the curve enters and exists the element called 1.

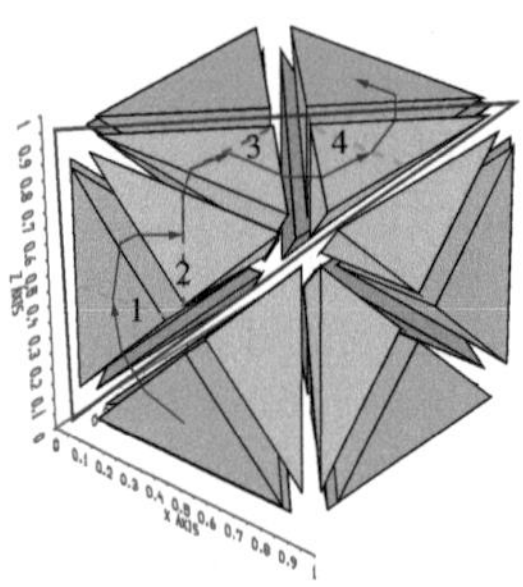


The curve on mesh 3.

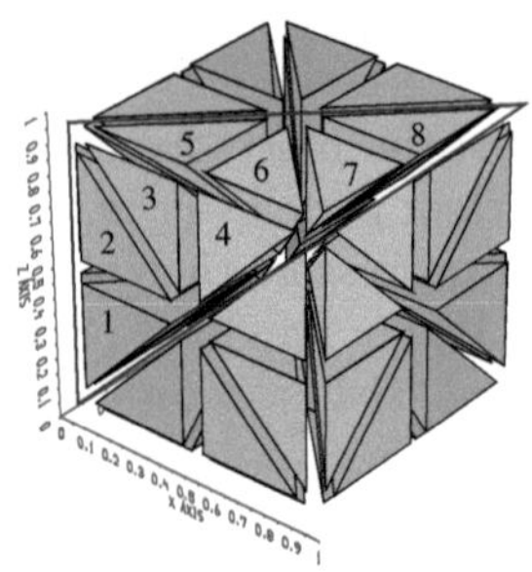


Mesh 4.

Now, for the first bisection step in the next figure, the two-dimensional property holds: All elements are bisected into two elements. The curve through parent 1 of mesh 1 now passes its two descendants, called 1 and 2.

Also, after the next bisection step to obtain mesh 3, there is a unique manner to let the curve pass the descendants 1 – 4 of 1 and 2. Still the two-dimensional property holds. In the figure with mesh 3, the green and red lines are the edges which will be created with the bisection step ([7]) for the creation of mesh 4.

Mesh 4 shows the result of the third bisection step. Here it turns out to be impossible that the new children 1 – 8 are connected with one curve through their facets. The curve which entered element 1 of mesh 1 would have to enter element 1 or element 4 of mesh 4. If the curve would enter (element) 1 of (mesh) 4, it would have to pass in order through facets between elements 1, 2, 3, 5, 6, ... of mesh 4 or elements 1, 2, 3, 4, 6 ... of mesh 4. In the first case, it would then have to skip element 4, or terminate in element 4 (which contradicts a space-filling/connected curve). The other case ends up similarly.

## References


1. E. Allgower and K. Georg. Generation of triangulations by reflections. *Utilitas Mathematica*, 16:123–129, 1979.
2. E. Bänsch. Local mesh refinement in 2 and 3 dimensions. *Impact of Computing in Science and Engineering*, 3:181–191, 1991.
3. V. Ervin, W. Layton, and J. Maubach. Some graph coloring problems. In Levelt A.H.M., editor, *Algoritmen in de Algebra – A Seminar on Algebraic Algorithms*, pages 39–48. University of Nijmegen, the Netherlands, 1993.
4. W. Layton, J.M. Maubach, and P. Rabier. Parallel algorithms for maximal monotone operators of local type. *Numerische Mathematik*, 71:29–58, 1995.
5. W. Layton, J.M. Maubach, and P. Rabier. Robustness of an elementwise parallel finite element method for convection diffusion problems. *SIAM Journal of Scientific Computing*, 19:1870–1891, 1998.
6. P.-L. Lions and B. Mercier. Splitting algorithms for the sum of two non-linear operators. *SIAM Journal on Numerical Analysis*, 16:964–979, 1979.
7. J. Maubach. Local bisection refinement for n-simplicial grids generated by reflections. *SIAM Journal on Scientific Computing*, 16:210–227, 1995
8. J.M.L. Maubach. Space-filling curves for 2-simplicial meshes created with bisections and reflections. submitted, preprint or electronic version available upon request.
9. D.W. Peaceman and H.H. Rachford. The numerical solution of parablokic and elliptic differential equations. *Journal of the Society for Industrial and Applied Mathematics SIAM*, 1955.
10. M.C. Rivara and C. Levin. A 3-D refinement algorithm suitable for adaptive and multi-grid techniques. *Journal of Computational and Applied Mathematics*, 8:281–290, 1992.

# Modelling The Flow And Solidification of a Thin Liquid Film on a Three-Dimensional Surface

T.G. Myers[1], J.P.F. Charpin[1], and S.J. Chapman[2]

[1] Department of Mathematics and Applied Mathematics, University of Cape Town, Rondebosch 7701, South Africa myers@maths.uct.ac.za, jcharpin@maths.uct.ac.za

[2] OCIAM, Mathematical Institute, 24-29 St. Giles', Oxford, OX1 3LB, UK chapman@maths.ox.ac.uk

**Summary.** A mathematical model for the flow and solidification of a thin liquid film is briefly described. Typical results for ice accretion due to incoming rain droplets on a flat surface and aerofoil are shown.

**Key words:** thin film, phase change, ice accretion, lubrication theory.

## 1 Introduction

The flow and solidification of a thin liquid film has important applications in a number of physical processes, ranging from ice accretion to lava flow and the industrial process of spray forming. In this paper a mathematical model for this process is briefly described and sample results presented. This work is a summary of the models developed in a series of papers by Myers *et al.*

## 2 Mathematical model

When supercooled fluid droplets impact on a substrate which is below the solidification temperature, the droplets will initially solidify and, depending on the energy in the system, may subsequently form a thin liquid layer on top of the solid. In the following we denote the solid layer thickness by $b(x, y, t)$ and the fluid layer thickness by $h(x, y, t)$. The fluid flow is driven by gravity, surface shear, pressure gradient and surface tension.

In general free surface thin film flows subject to surface tension forces are described by a fourth-order nonlinear degenerate partial differential equation. These are notoriously difficult to solve. For this reason we focus first on the flow problem. The solidification is then a relatively simple extension.

## 2.1 Thin film flow

We first consider the problem of flow on a flat surface and subsequently modify the solution to deal with arbitrary shapes. Employing the standard lubrication approximation, the Navier-Stokes equations reduce to

$$\mu_f \frac{\partial^2 u}{\partial z^2} = \frac{\partial p}{\partial x} - \rho_f g \hat{\mathbf{g}} \cdot \hat{\mathbf{x}} \ , \ \mu_f \frac{\partial^2 v}{\partial z^2} = \frac{\partial p}{\partial y} - \rho_f g \hat{\mathbf{g}} \cdot \hat{\mathbf{y}} \ , \ 0 = \frac{\partial p}{\partial z} + \rho_f g \hat{\mathbf{g}} \cdot \hat{\mathbf{z}} \ , \quad (1)$$

where the fluid velocity in the $(x, y, z)$ directions is $\mathbf{u} = (u, v, w)$ and $p$ is the fluid pressure. The fluid viscosity is denoted $\mu_f$, the density $\rho_f$ and $\hat{\mathbf{g}}$ is the unit gravity vector.

This system of equations requires solving subject to the following boundary conditions. At the solid-liquid interface, $z = b$, there is no slip, hence $u = v = 0$. A mass balance leads to

$$w|_{z=b} = \left(1 - \frac{\rho_s}{\rho_f}\right) \frac{\partial b}{\partial t} \ , \quad (2)$$

where $\rho_s$ is the density of the solid phase (so a normal fluid velocity only occurs at the interface if the solid and fluid densities are different).

At the liquid-air interface, $z = b + h$ there is continuity of shear stress, $\mu_f u_z = A_1$, $\mu_f v_z = A_2$. The pressure jump across the interface depends on the surface tension, $\sigma$, and is proportional to the free surface curvature, $p = p_a - \sigma \nabla^2 (b + h)$. Note, the ambient pressure $p_a$ may vary with space and time. Again a mass balance determines the velocity $w$:

$$w|_{z=b+h} = \left(1 - \frac{\rho_A}{\rho_f}\right) \left(\frac{\partial b}{\partial t} + \frac{\partial h}{\partial t}\right) + u \left(\frac{\partial b}{\partial x} + \frac{\partial h}{\partial x}\right) + v \left(\frac{\partial b}{\partial y} + \frac{\partial h}{\partial y}\right) - J, \quad (3)$$

where the rate at which fluid enters the system is represented by $J$ and $\rho_A$ is the density of the air-droplet mixture. In general $\rho_A / \rho_f \ll 1$.

Expressions for $u, v, p$ are readily obtained by integrating (1) subject to the boundary conditions. These may then be used when integrating the continuity equation (for an incompressible fluid) across the film. Imposing the boundary conditions on $w$ leads to a mass balance for the film:

$$\frac{\partial h}{\partial t} + \nabla \cdot \mathbf{Q} = \frac{J}{\rho_f} - \frac{\rho_s}{\rho_f} \frac{\partial b}{\partial t} \ , \quad (4)$$

where the fluid flux, $Q$, is given by

$$\mathbf{Q} = \left( -\frac{h^3}{3\mu_f} \left( \frac{\partial p}{\partial x} + \rho_f g \hat{\mathbf{g}} \cdot \hat{\mathbf{x}} \right) + \frac{h^2}{2\mu_f} A_1, -\frac{h^3}{3\mu_f} \left( \frac{\partial p}{\partial y} + \rho_f g \hat{\mathbf{g}} \cdot \hat{\mathbf{y}} \right) + \frac{h^2}{2\mu_f} A_2 \right).$$

Equation (4) shows that the film height $h$ varies due to the fluid flow, the rate at which fluid enters the system, and the rate at which it solidifies. The fluid flow is driven surface tension (through the pressure gradient), gravity and surface shear. The rate at which fluid enters the system is usually determined by an outer flow routine, see the work described in [3, 2]. The solidification rate comes from considering the thermal problem.

## 2.2 Thermal problem

When an incoming supercooled fluid impacts on a surface which is below the solidification temperature there will always be an initial period where all the fluid solidifies (this may be very short, depending on the energy in the system). For brevity we consider only the stage where both phases are present. The single phase solution is then a special case of this.

We now assume the solidified region is also thin (for ice layers thin may be of the order 2 cm, see [1]). In this case the leading order heat equations in the ice and water layers are

$$\frac{\partial^2 T}{\partial z^2} = 0 \qquad\qquad \frac{\partial^2 \theta}{\partial z^2} = 0 \ , \tag{5}$$

where terms of order $\mathcal{O}(\varepsilon^2, Pe)$ ($Pe$ is the Péclet number) have been neglected. If the substrate temperature is fixed at $T_s$ the temperature in the two phases is

$$T = T_s + (T_f - T_s)\frac{z}{b} \qquad\qquad \theta = T_f + \frac{q_0 + q_1 T_f}{1 - q_1 h}(z - b) \ , \tag{6}$$

where $T_f$ is the solidification temperature and $q_i$ are constants which incorporate the various energy terms, such as convective heat transfer, evaporation and kinetic energy of the droplets, see [1]. To $\mathcal{O}(\varepsilon^2)$ the solid thickness is determined by a standard Stefan condition

$$\rho_s L_f \frac{\partial b}{\partial t} = k_s \frac{T_f - T_s}{b} - k_f \frac{q_0 + q_1 T_f}{1 - q_1 h} \ . \tag{7}$$

The problem therefore reduces to solving equations (4), (7) for the two unknowns $b$ and $h$. Once these are known the temperatures follow via (6).

## 2.3 Extension to an arbitrary substrate

The extension of the flat surface model to an arbitrary substrate model is described in [2]. In this case the Stefan condition (7) remains unchanged however the flow model becomes:

$$\frac{\partial h}{\partial t} + \nabla_s \cdot \mathbf{Q} = \frac{J}{\rho_f} - \frac{\rho_s}{\rho_f}\frac{\partial b}{\partial t} \ , \tag{8}$$

where the surface operator is

$$\nabla_s \cdot \mathbf{Q} = \frac{1}{(EG)^{1/2}} \left( G^{1/2} \frac{\partial}{\partial s_1} Q_1 + E^{1/2} \frac{\partial}{\partial s_2} Q_2 \right) \ , \tag{9}$$

where $Eds_1^2 + Gds_2^2$ is the first fundamental form and $s_i$ are the principal directions on the surface. The flux is now given by

$$Q_1 = \int_b^{b+h} u \, d\eta = -\left( \frac{1}{E^{1/2}} \frac{\partial p}{\partial s_1} - \mathbf{g} \cdot \mathbf{e}_1 \right) \frac{h^3}{3\mu_f} + A_1 \frac{h^2}{2\mu_f} \tag{10}$$

with a similar expression for $Q_2$. The fluid pressure by

$$p = p_0 - \sigma \left( \kappa_1 + \kappa_2 + \varepsilon(b+h)(\kappa_1^2 + \kappa_2^2) + \varepsilon \left[ \frac{1}{E} \frac{\partial^2}{\partial {s_1}^2}(b+h) + \frac{1}{G} \frac{\partial^2}{\partial {s_2}^2}(b+h) \right] \right) , \tag{11}$$

where $\kappa_i$ are the principal curvatures. This set of equations reduces to the flat surface result by setting $s_1 = x$, $s_2 = y$, $E = G = 1$ and the curvatures $\kappa_i = 0$. The main difference between this form and that on a flat surface is the presence of the substrate curvature terms $\kappa_i$. In particular, if the curvature is non-constant then it is the substrate curvature that dominates the surface tension driven flow. A good example of this is when painting corners; paint is pulled into internal corners, leaving a thicker layer, and pulled away from external ones, leaving a thinner layer.

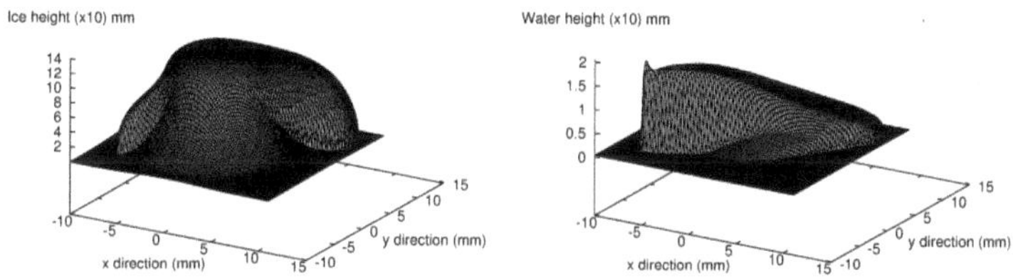


**Fig. 1.** a) Ice and b) water layers on a flat surface: flow driven by gravity and surface shear.

## 3 Results

The numerical solution of equations (7), (8) is described in [3, 2]. A typical solution is shown on Fig. 1. This is for a case where the incoming fluid is represented by a Gaussian profile. Since the incoming fluid initially freezes an approximation to the Gaussian can be seen in the central part of the ice accretion (Fig. 1a). The humps on either side are caused by the water flow (shown on Fig. 1b) on top of the ice. The flow is driven by gravity and air shear which act in opposite directions along the diagonal, gravity acts to the left. The distinct ridge on the right hand side is a standard consequence of surface tension, an interesting feature is the lack of a similar ridge on the right hand side.

Figures 2 a–d show the results of a simulation on an aerofoil. In this case the external air flow and droplet trajectories were calculated using FLUENT V. The droplet impact region can be inferred from the early time solution, Fig. 2 a) since there is no significant water flow at this time. As time progresses the accretion builds up at the front but the water flow also allows it to extend backwards. After 15 minutes it has reached beyond $x = -0.11$. The water flow calculation is important since it shows that ice will build up past the impact region and so aids engineers in determining how far back heating systems must be installed.

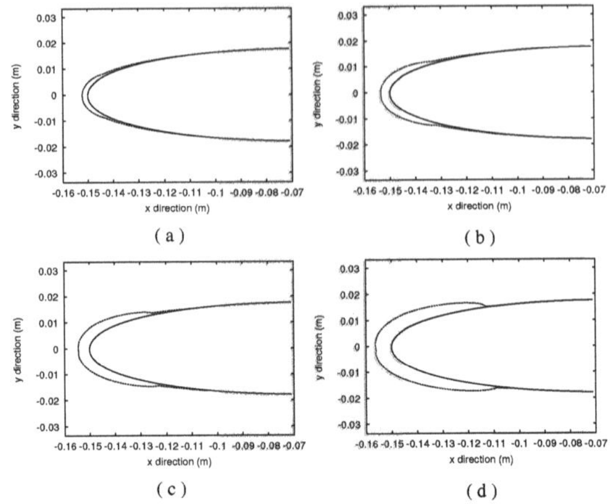


**Fig. 2.** Ice layers on a NACA0012 at (a) $t = 3$ (b) 6 (c) 9 (d) 15 minutes.

# 4 Conclusions

The series of papers by Myers *et al* build up to describing a novel model for the flow and solidification of a thin liquid film. The model is valid within the limits of lubrication theory and small Péclet number. It may be used to describe solidification on an arbitrary 3D surface and hence is appropriate for most practical configurations. A testimony to the model is that it is currently employed in a commercial aircraft icing code, ICECREMO.

Future extensions include modelling flow over a rough surface and the inclusion of heating from below the substrate which will permit modelling of anti-icing systems.

*Acknowledgement.* T.G. Myers acknowledges support of this work under the National Research Foundation of South Africa grant number 2053289 and also the support of the University of Cape Town Research Council. J.P.F. Charpin acknowledges the support of the Claude Leon Harris.

# References

1. T.G. Myers. An extension to the Messinger model for aircraft icing. *AIAA J.*, 39(4):211–218, 2001.
2. T.G. Myers, J.P.F. Charpin, and S.J. Chapman. The flow and solidification of a thin fluid film on an arbitrary three-dimensional surface. *Phys. Fluids*, 14(8):2788–2803, 2002.
3. T.G. Myers, J.P.F. Charpin, and C.P. Thompson. Slowly accreting glaze ice due to supercooled droplets impacting on a cold substrate. *Phys. Fluids*, 14(1):240–256, 2002.

# Numerical Schemes for Degenerate Parabolic Problems

I.S. Pop[1]

Technische Universiteit Eindhoven, CASA, P.O. Box 513, 5600 MB Eindhoven, The Netherlands I.Pop@tue.nl

**Summary.**

**Key words:** porous medium, Stefan problems, numerical schemes, regularization, error estimates, convergence.

## 1 Introduction

Degenerate parabolic equations are encountered as mathematical models for several phenomena in physics, chemistry, biology or economy (see, *e.g.*, [2], or [3]). In this sense, the simplest example is the porous medium equation, describing the flow of an ideal gas in a homogeneous porous medium. More complex situation are encountered in petroleum reservoir and groundwater aquifer simulations, or in the design of industrial filters and battery management. Phase change problems corresponding to processes of heat transfer involving melting or solidification lead to equations of the same type.

Compared to regular parabolic problems – like the heat equation, in the degenerate case the diffusive term may vanish or blow up, depending on the solution. This leads to a possible change of the parabolic character of the equation into an elliptic or even hyperbolic one. The interfaces separating the domains of regularity – also called free boundaries – have finite speed of propagation. Generally these are not known in advance and have to be determined together with the solution.

Therefore the solutions of such problems are lacking regularity. The singularities do not smooth out as time evolves and, in fact, they may even develop, giving the problem a strongly nonlinear character. With respect to the numerical approximation of such solutions, this fact requires adequate algorithms being able to deal both with the free boundary and the singularities of the solution.

In this paper we discuss two simple time discretization algorithms to solve the following degenerate parabolic problem:

**Problem P:**

$$\begin{aligned} \partial_t u - \nabla \cdot (\nabla \beta(u) + F(u)) = r(u), & \quad \text{in } Q_T \equiv (0,T) \times \Omega, \\ u(0,\cdot) = u^0(\cdot), & \quad \text{in } \Omega, \\ u = 0, & \quad \text{on } \partial\Omega. \end{aligned} \tag{1}$$

Here $0 < T < \infty$ is fixed, $\Omega$ is a bounded domain in $\mathbb{R}^d (d \geq 1)$ with a Lipschitz continuous boundary and $Q_T \equiv (0,\ T) \times \Omega$. The function $\beta : \mathbb{R} \to \mathbb{R}$ is non-decreasing and differentiable. By degeneracy we mean a vanishing diffusion, namely $\beta'(u) = 0$ for some $u$. Growth conditions are also imposed on $F$ and $r$. Specifically, we work under the following assumptions:

**(A1)** $\beta$ is Lipschitz and differentiable, $\beta(0) = 0$, $0 \leq \beta'(u) \leq L_\beta$.

**(A2)** $u^0 \in L^2(\Omega)$.

**(A3)** $r : \mathbb{R} \to \mathbb{R}$ and $F : \mathbb{R} \to \mathbb{R}^d$ are continuous in $u$ and it holds

$$|r(u) - r(v)|^2 + |F(u) - F(v)|^2 \leq C_F(u - v)(\beta(u) - \beta(v))$$

for any $u, v \in \mathbb{R}$, where $C_F > 0$ does not depend on $x, t, u$ and $v$. Moreover, $r(0) = 0$ and $F(0) = \bar{0} = (0, \ldots, 0)$.

*Remark 1.* Non-homogeneous Dirichlet or natural boundary conditions may be considered without any problem here. The assumption (A3) is slightly less restrictive than the commonly used Lipschitz continuity w.r.t. $\beta(u)$. In this setting, existence and uniqueness of a weak solution is proved in [1] and [6].

We use standard notations for the spaces of functions, norms and scalar products: $L^2(\Omega)$, $H_0^1(\Omega)$, or its dual $H^{-1}(\Omega)$, or $L^2(0,\ T;\ X)$ with $X$ being one of the spaces before. We let $(\cdot, \cdot)$ stand for the inner product on $L^2(\Omega)$, or the duality pairing between $H_0^1(\Omega)$ and $H^{-1}(\Omega)$. $\|\cdot\|$ denotes the norm in $L^2(\Omega)$, while by $\|\cdot\|_X$ we mean the norm in $X$. We often write $u$ or $u(t)$ instead of $u(t,\ x)$ and use $C$ to denote a generic positive constant.

## 2 The Numerical Approaches

Both schemes discussed here build on regularization, which means that the originally degenerate problem is perturbed to a regular parabolic one. Specifically, the nonlinearity $\beta$ is approximated by a function $\beta_\varepsilon$ satisfying $\beta'_\varepsilon \geq \varepsilon > 0$ for all $u$. For example, taking

$$\beta'_\varepsilon(u) \equiv \max\{\beta'(u), \varepsilon\} \quad \text{and} \quad \beta_\varepsilon(u) = \int_0^u \beta'_\varepsilon(v) dv \tag{2}$$

we end up with a perturbation satisfying

$$\varepsilon \leq \beta_\varepsilon'(u) \leq L_\beta \quad \text{and} \quad 0 \leq \beta_\varepsilon'(u) - \beta'(u) \leq \varepsilon \tag{3}$$

for any real $u$. Clearly, $\beta_\varepsilon$ is a strictly increasing approximation of $\beta$ and admits an inverse that is differentiable.

*Remark 2.* Generally an explicit formula for $\beta_\varepsilon$ or its inverse may not be available, or it can be extremely complicated. Moreover, function calls are increasing the computing time significantly. Therefore when implementing the numerical schemes proposed below we first construct a look-up table of values of $\beta_\varepsilon$ for a range of points, at the expense of an additional computer memory requirement. Together with a simple (linear) interpolation step for values not present in the table, this reduces significantly the time of computation, while the errors are controlled by an appropriate choice of the interpolation knots. Because of the monotonicity of $\beta_\varepsilon$, searching in this table is fast.

The next step is to consider the problem P in terms of the more regular unknown, $\beta(u)$. Since $\beta'$ is not bounded away from 0, the numerical schemes are constructed in terms of $\beta_\varepsilon$. Due to the lack in regularity of solutions we restrict ourselves to first order time discretization methods. Given a natural number $n$, in what follows $\tau = T/n$ will denote the time step, which is assumed fixed for the ease of presentation. Variable or adaptive time stepping can also be considered.

We acknowledge here the works [5] and [8], where regularization based algorithms are discussed for the fully implicit discretization approach. A relaxation scheme is proposed in [4]. A linear approach is discussed *e.g.*, in [10]. Here we work in a more general framework, and also present improved convergence estimates.

With $t_k = k\tau$ $(k = 1, \dots, n)$, by $\theta^k$ we denote the approximation of the average $\frac{1}{t_k - t_{k-1}} \int_{t_{k-1}}^{t_k} \beta(u(t))dt$. Approximating first $\beta(u)$ is motivated by its better regularity when compared to the original unknown $u$, leading to better convergence results than computing directly $u$. After determining $\theta^k$, $u$ is approximated by $\beta_\varepsilon^{-1}(\theta^k)$.

The first scheme discussed here is fully implicit. At each time step $t_k$, $k = \overline{1, n}$, one has to solve the following problem:

**Problem PI$_k$:**

$$\begin{aligned} \beta_\varepsilon^{-1}(\theta^k) - \beta_\varepsilon^{-1}(\theta^{k-1}) &= \tau\nabla \cdot (\nabla\theta^k + F(\beta_\varepsilon^{-1}(\theta^k))) + \tau r(\beta_\varepsilon^{-1}(\theta^k)), \\ \theta^k|_{\partial\Omega} &= 0. \end{aligned} \tag{4}$$

Here $\theta^0 = \beta_\varepsilon(u^0)$.

The above approach is nonlinear. A simpler approach is to solve

**Problem PL$_k$:**

$$\begin{aligned} \sigma_{k-1}(\theta^k - \theta^{k-1}) &= \tau\Delta\theta^k + \tau(\beta_\varepsilon^{-1})'(\theta^{k-1})F'(\beta_\varepsilon^{-1}(\theta^{k-1})) \cdot \nabla\theta^k \\ &\quad + \tau r(\beta_\varepsilon^{-1}(\theta^{k-1})), \\ \theta^k|_{\partial\Omega} &= 0, \\ \sigma_k &= (\beta_\varepsilon^{-1})'(\theta^k), \end{aligned} \tag{5}$$

for $k = \overline{1,n}$, where $\sigma_0 = (\beta_\varepsilon^{-1})'(\theta^0)$.

We should mention that in the second scheme the initial data $\theta^0$ must be chosen more carefully. Specifically, the analysis requires $\theta^0 \in H^1$. If this regularity does not hold for $u^0$, then this can be replaced by a $H^1$ approximation. A practical choice is given by the solution of the heat equation after a (small) time step, with initial data $u^0$.

*Remark 3.* In the linearized scheme we approximate the convection as

$$\nabla \cdot F(\beta_\varepsilon^{-1}(\theta^k)) \approx (\beta_\varepsilon^{-1})'(\theta^{k-1})F'(\beta_\varepsilon^{-1}(\theta^{k-1}))\nabla\theta^k.$$

The assumption (A3) together with (2) prevents the 'speed' on the right hand side above from becoming unbounded:

$$|(\beta_\varepsilon^{-1})'(\theta)F'(\beta_\varepsilon^{-1}(\theta))| \leq \sqrt{C_F/\varepsilon}.$$

At each time step, both schemes imply solving an elliptic problem. By the (nonlinear) Lax-Milgram Lemma, these problems have a unique solution. Moreover, the following estimates can be given:

**Theorem 1.** *Assume (A1), (A2) and (A3). If $\theta^k$ is the weak solution of Problem $PI_k$ we have*

$$\sum_{k=1}^{n} \{(\beta^{-1}(\theta^k) - \beta^{-1}(\theta^{k-1}), \theta^k - \theta^{k-1}) + \|\theta^k - \theta^{k-1}\|^2\} + \tau \sum_{k=1}^{n} \|\nabla\theta^k\|^2 \leq C.$$

*Further, if $u$ is the weak solution of Problem P, then*

$$\int_0^T (\beta_\varepsilon(u(t)) - \theta_\Delta(t), u(t) - \beta_\varepsilon^{-1}(\theta_\Delta(t)))dt + \|\beta(u) - \theta_\Delta\|^2_{L^2(Q_T)} \leq C\{\tau + \varepsilon\},$$

*where $\theta_\Delta(t) = \theta^k$ for $t \in (t_{k-1}, t_k]$ and $k = \overline{1,n}$.*

The proof makes use of the ideas in [5] and is given in [7].

*Remark 4.* The above theorem shows that the implicit approach is convergent. The error estimates are of order $\tau^{1/2}$, at least as good as those for the algorithms in [5] or [4]. Based on the results in [9], in some certain cases better estimates can be obtained. For example, if Problem P is considered without convection or reaction, and if $\beta$ is a maximal monotone graph having $\mathbb{R}$ as its range, then the estimates become optimal:

$$\|\beta(u) - \theta_\Delta\|^2_{L^2(Q_T)} + \tau\|\beta(u) - \theta_\Delta\|^2_{L^2(0,T;H_0^1(\Omega))} \leq C\{\tau^2 + \varepsilon^2\}.$$

For the linear scheme we assume additionally that both $\beta'$ and $F$ are Lipschitz continuous. If positive solutions are sought, the last assumption can be removed if $\beta$ is convex (see for details [7, Chapter 3]). This gives:

**Theorem 2.** *Assume (A1), (A2), (A3) and one of the alternatives mentioned above. If $\theta^k$ is the weak solution of Problem $PL_k$, for any $0 \leq p \leq n$ we have*

$$\sum_{k=1}^{n} \|\sqrt{\sigma_{k-1}}(\theta^k - \theta^{k-1})\|^2 + \tau\|\nabla\theta^p\|^2 + \tau\sum_{k=1}^{n} \|\nabla(\theta^k - \theta^{k-1})\|^2 \leq C\tau.$$

*Moreover, if $u$ is the weak solution of Problem P, then*

$$\int_0^T (\beta_\varepsilon(u(t)) - \theta_\Delta(t), u(t) - \beta_\varepsilon^{-1}(\theta_\Delta(t)))dt + \|\beta(u) - \theta_\Delta\|^2_{L^2(Q_T)}$$
$$\leq C\left(\tau/\varepsilon^2 + \varepsilon\right),$$

*with $\theta_\Delta$ being defined above.*

The proof can be found again in [7].

*Remark 5.* The linear scheme is similar to the one considered for example in [10], where no convection or reaction terms are included. In the simplified setting there, using again Rulla's result [9], the estimates can be improved to the order $O(\tau^2/\varepsilon^2 + \varepsilon)$.

*Remark 6.* The results discussed here can be extended straightforwardly to fully discrete scheme by assuming, for example, a finite element discretization in space (see [7]).

## References


1. H. W. Alt and S. Luckhaus. Quasilinear elliptic-parabolic differential equations. *Math. Z.*, 183:311–341, 1983.
2. J. Bear. *Dynamics of Fluids in Porous Media.* American Elsevier, New York, 1972.
3. A. Friedman. *Variational Principles and Free-Boundary Problems.* Wiley, New York, 1982.
4. W. Jäger and J. Kačur. Solution of doubly nonlinear and degenerate parabolic problems by relaxation schemes. *Math. Model. Numer. Anal.*, 29:605–627, 1995.
5. R.H. Nochetto and C. Verdi. Approximation of degenerate parabolic problems using numerical integration. *SIAM J. Numer. Anal.*, 25:784–814, 1988.
6. F. Otto. $l^1$-contraction and uniqueness for quasilinear elliptic-parabolic equations. *J. Differ. Equations*, 131:20–38, 1996.
7. I.S. Pop. *Regularization Methods in the Numerical Analysis of Some Degenerate Parabolic Equations.* PhD thesis, Universitatea "Babeş-Bolyai" Cluj-Napoca, Romania, 1998.
8. M.E. Rose. Numerical methods for flows through porous media I. *Math. Comp.*, 40:435–467, 1983.
9. J. Rulla. Error analysis for implicit approximations to solutions to Cauchy problems. *SIAM J. Numer. Anal.*, 33:68–87, 1996.
10. M. Slodička. Solution of nonlinear parabolic problems by linearization. Technical Report Preprint M3-92, Comenius University, Bratislava, 1992.

# Finite Element Modified Method of Characteristics for Shallow Water Flows: Application to the Strait of Gibraltar

M. González[1] and M. Seaïd[2]

[1] Dept. de Oceanografía y Medio Ambiente Marino, 20110 Gipuzkoa Spain
mgonzalez@pas.azti.es
[2] Fachbereich Mathematik, TU Darmstadt, 64289 Darmstadt, Germany
seaid@mathematik.tu-darmstadt.de

**Summary.** A Finite Element Modified Method of Characteristics (FEMMOC) is proposed for numerical solution of the two-dimensional shallow water equations. The method is formulated and implemented for mean flow and hydraulics in the strait of Gibraltar. Preliminary results presented in this work show that the FEMMOC is able to provide stable, accurate and efficient solutions.

**Key words:** FEMMOC, shallow water equations, strait of Gibraltar.

## 1 Introduction

The strait of Gibraltar connects the Atlantic ocean with the Mediterranean sea. The differences on density, salinity and temperature of the two water bodies lead to a flow exchange through the strait. This flow exchange consists of two counter-flowing layers: an upper layer of Atlantic water flowing into the Mediterranean sea and a lower layer of Mediterranean water flowing into Atlantic ocean. For comprehensive contributions on oceanography of the strait of Gibraltar we refer the reader to the proceedings [1]. Here, we are concerned with numerical study of inflow contributed by the Atlantic ocean into the Mediterranean sea which takes place on the free water surface.

Our goal in the present work is to develop a robust numerical method to approximate solutions to the equations governing mean flow in the strait of Gibraltar. The key idea is to combine the modified method of characteristics and finite element discretization. This technique make use of the transport nature of the governing equations and greatly reduces the time truncation errors in the Eulerian methods. In addition, the method can be implemented in complex geometry and alleviates the restrictions on the Courant number, thus allowing for large time steps in the simulations and reduces artificial numerical dispersion, see [5, 4] and further references are cited therein.

The mathematical equations, widely used in literature to model mean flow in the strait of Gibraltar, are given by [1]: continuity equation

$$\partial_t \eta + \partial_x\big((\eta + h)U\big) + \partial_y\big((\eta + h)V\big) = 0, \tag{1}$$

and momentum equations

$$\partial_t U + U\partial_x U + V\partial_y U - fV = -g\partial_x \eta - \frac{r}{\eta + h} U\sqrt{U^2 + V^2} + K_H \nabla^2 U, \tag{2}$$

$$\partial_t V + U\partial_x V + V\partial_y V + fU = -g\partial_y \eta - \frac{r}{\eta + h} V\sqrt{U^2 + V^2} + K_H \nabla^2 U, \tag{3}$$

where $\eta$ is the free surface height, $\mathbf{U} = (U, V)^T$ is the vertically integrated velocity, $h$ is the water depth measured from the mean sea level, $g$ is the gravity acceleration, $K_H$ is the horizontal eddy viscosity, $r$ denotes the drag coefficient on the bottom, $f$ is the Coriolis parameter defined by $f = 2\omega \sin\varphi$, with $\omega$ is the angular velocity of the earth and $\varphi$ is the geographic latitude, and $\nabla^2$ denotes the two-dimensional Laplace operator.

Equations (1)-(3) are defined in a spatial domain $\Omega$ bounded by the Tangier-Barbate line at the west and Ceuta-Gibraltar at the east as shown in Fig. 1. This domain contains the Camarinal Sill (the interface that separates the Mediterranean sea and Atlantic ocean) where exchange of the water body takes place.

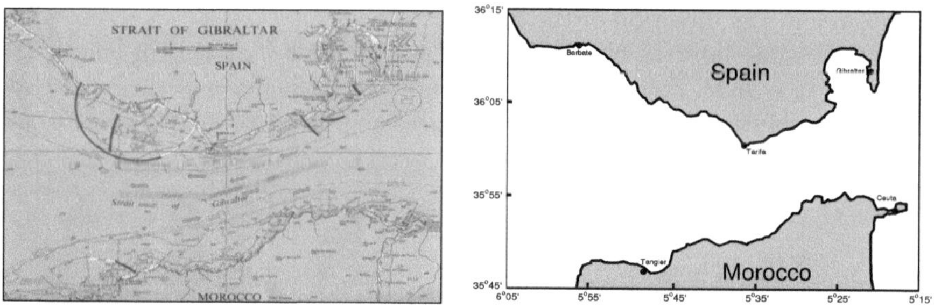


**Fig. 1.** Definition of the strait of Gibraltar.

Initially, the water flow is at rest and boundary conditions are given by

$$\eta = \eta_D, \quad \mathbf{x} \in \Gamma_D, \qquad \mathbf{U} \cdot \mathbf{n} = 0, \quad \mathbf{x} \in \Gamma_N, \tag{4}$$

where $\Gamma_D$ and $\Gamma_N$ are boundaries of $\Omega$ as shown in Fig. 2. In (4), $\mathbf{n}$ is the outward normal on the boundary and $\eta_D$ are prescribed data obtained by interpolation from measurements on the strait, see [2].

## 2 Formulation of FEMMOC

In order to formulate the FEMMOC we rewrite the equations (1)-(3) in vector-valued form, as

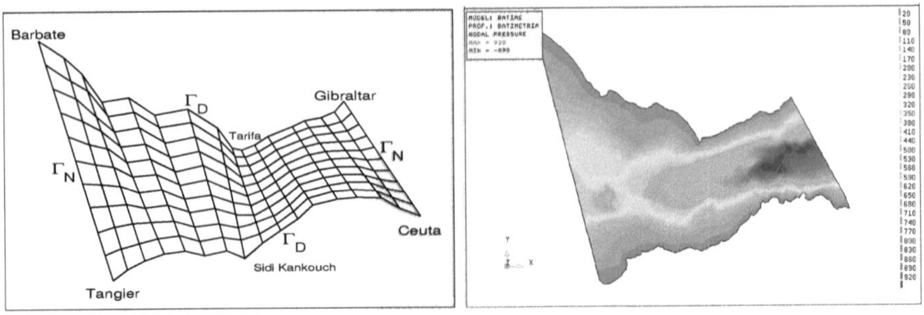


**Fig. 2.** Computational mesh and bathymetry of the strait.

$$
\begin{aligned}
&D_t\eta + \mathbf{U}\cdot\nabla h + d\nabla\cdot\mathbf{U} = 0,\\
&D_t\mathbf{U} + g\nabla\cdot\eta - K_H\nabla^2\mathbf{U} = \mathbf{S}(\eta,\mathbf{U}),
\end{aligned}
\tag{5}
$$

where $d = \eta + h$, the source term is defined as

$$
\mathbf{S}(\eta,\mathbf{U}) = \left(fV - \frac{r}{\eta+h}U\sqrt{U^2+V^2}, -fU - \frac{r}{\eta+h}V\sqrt{U^2+V^2}\right)^T, \tag{6}
$$

$\nabla = (\partial_x, \partial_y)^T$, and $D_t w = \partial_t w + \mathbf{U}\cdot\nabla w$ is the material derivative of the function $w$ in the direction of the flow $\mathbf{U}$.

Let the time interval be divided into subintervals $[t_n, t_{n+1}]$ of length $\Delta t$ such that $t_n = n\Delta t$. Following [3], the characteristics curves of the equations (5) are the solution of initial-value problem

$$
\begin{aligned}
&\frac{\partial\mathbf{X}(\tau; t_{n+1}, \mathbf{x})}{\partial\tau} = \mathbf{U}\left(\tau, \mathbf{X}(\tau; t_{n+1}, \mathbf{x})\right), \quad \tau\in[t_n, t_{n+1}],\\
&\mathbf{X}(t_{n+1}; t_{n+1}, \mathbf{x}) = \mathbf{x}.
\end{aligned}
\tag{7}
$$

Note that $\mathbf{X}(\tau; t_{n+1}, \mathbf{x})$ is the departure point at time $\tau$ of a fluid particle that will arrive at $\mathbf{x}$ at time $t_{n+1}$. Once the characteristics feet $\mathbf{X}(t_n; t_{n+1}, \mathbf{x})$ are known, the semi-discretization of (5), we consider in the present work, reads

$$
\frac{\eta^{n+1} - \hat{\eta}^n}{\Delta t} + \mathbf{U}^{n+1}\cdot\nabla h + d^n\nabla\cdot\mathbf{U}^{n+1} = 0, \tag{8}
$$

$$
\frac{\mathbf{U}^{n+1} - \hat{\mathbf{U}}^n}{\Delta t} + g\nabla\cdot\eta^{n+1} - K_H\nabla^2\mathbf{U}^{n+1} = \mathbf{S}(\eta^{n+1}, \mathbf{U}^{n+1}), \tag{9}
$$

A simple way to solve the equations (8)-(9), is to use the first equation (8) to eliminate the $\eta^{n+1}$ and $\nabla\cdot\eta^{n+1}$ terms from the second equation (9). These procedure yields to a fixed point problem for $\mathbf{U}$ only

$$
\mathbf{U} = \mathcal{H}(\mathbf{U}), \tag{10}
$$

We have dropped the $n+1$ superscript for ease of notation. Newton's method applied to (10) results in the following iteration

$$\mathbf{U}^{(k+1)} = \mathbf{U}^{(k)} - \mathcal{R}'\big(\mathbf{U}^{(k)}\big)^{-1}\mathcal{R}\big(\mathbf{U}^{(k)}\big), \tag{11}$$

where $\mathcal{R}(\mathbf{U}) = \mathbf{U} - \mathcal{H}(\mathbf{U})$ is the nonlinear residual and $\mathcal{R}'$ is the system Jacobian approximated by a forward difference quotient. We used the GMRES method to compute the Newton's direction. The free surface height $\eta^{n+1}$ can be updated by backsubstituting $\mathbf{U}^{n+1}$ in the first equation from (5).

The variational formulation for solution of (8)-(9) is based on the spaces

$$H^1_{\Gamma_D}(\Omega) = \{\xi \in H^1(\Omega) : \quad \xi = 0 \quad \text{on} \quad \Gamma_D\},$$
$$H_{\Gamma_N}(\text{div}, \Omega) = \{\mathbf{V} \in H(\text{div}, \Omega) : \quad \mathbf{V}\cdot\mathbf{n} = 0 \quad \text{on} \quad \Gamma_N\}.$$

Thus, finite element subspaces $Q_h \subset H^1_{\Gamma_D}(\Omega)$ and $\mathbf{V}_h \subset H_{\Gamma_N}(\text{div}, \Omega)$ are selected such as standard $H^1$-conforming piecewise polynomial functions for $Q_h$ and the $H(\text{div})$-conforming Raviart-Thomas elements for $\mathbf{V}_h$.

## 3 Preliminary Results

The FEMMOC has been implemented for a test case kindly provided by the University of Malaga (Spain). Further results and comparisons will be considered in a future work. As part of an ongoing project, this method will be implemented for the full model and obtained results will be compared to measurements done in the strait of Gibraltar. In Fig. 3, we show the flow field for the main diurnal and semidiurnal K2, M2, N2 and S2, compare [2] for details. The computational mesh and the bathymetry are shown in Fig. 2. The gravity acceleration $g = 9.81\ m/s^2$, the drag coefficient on the bottom $r = 10^{-3}$, the Coriolis parameter $f = 8.55 \times 10^{-5}\ s^{-1}$, and the horizontal eddy viscosity coefficient $K_H = 10^2\ m^2/s$. The model is integrated for a time period of three months using a time step size $\Delta t = 0.5$ hour.

The FEMMOC resolves the water flow accurately without introducing extra numerical dissipation. We can see that the small complex structures of the water flow being captured by the FEMMOC.

## References

1. J.I. Almazán, H. Bryden, T. Kinder, and G. Parrilla, editors. *Seminario Sobre la Oceanografía Física del Estrecho de Gibraltar*, Madrid, 1988. SECEG.
2. M. González. Un modelo numérico en elementos finitos para la coriente inducida por la marea. Aplicaciones al Estrecho de Gibraltar. Tesina de especialidad, University of Barcelona, 1994.
3. A. Robert. A stable numerical integration scheme for the primitive meteorological equations. *Atmos. Ocean*, 19:35–46, 1981.

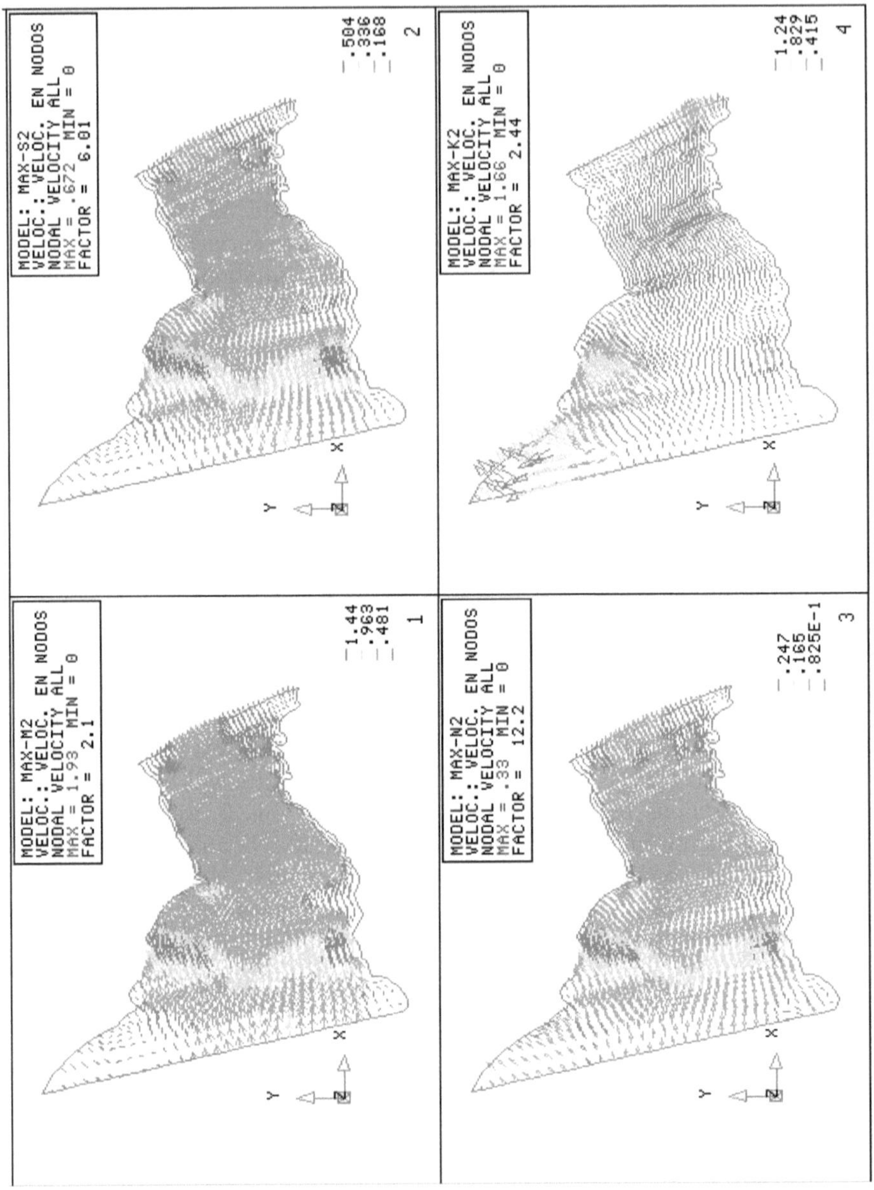


**Fig. 3.** Flow field for the main diurnal and semidiurnals K2, M2, N2 and S2.

4. M. Seaïd. Semi-Lagrangian integration schemes for viscous incompressible flows. *J. Comp. Methods in App. Math.*, 4:392–409, 2002.
5. A. Staniforth and J. Côté. Semi-Lagrangian integration schemes for the atmospheric models: A review. *Wea. Rev.*, 119:2206–2223, 1991.

# LDC with compact FD schemes for convection-diffusion equations

M. Sizov[1], M.J.H. Anthonissen, and R.M.M. Mattheij

Technische Universiteit Eindhoven M.Sizov@tue.nl

**Summary.** We discuss an algorithm for convection-diffusion equations with high activity areas which combines the Local Defect Correction technique with high order compact finite difference schemes.

**Key words:** Local defect correction, compact finite difference schemes

## 1 Introduction

Many boundary value problems produce solutions that have highly localized properties. In this paper we consider boundary value problems with solutions that have one or a few small regions with high activity.

We study a method based on a combination of high order compact finite difference discretizations on several uniform grids with different grid sizes that cover different parts of the domain. At least one grid should cover the entire domain; the mesh size of this global coarse grid is chosen in agreement with the relatively smooth behaviour of the solution outside the high activity regions. Apart from this global coarse grid, one or several local grids are used which are also uniform. Each of these local grids covers only a (small) part of the domain and contains a high activity region. This refinement strategy is known as local uniform grid refinement. The solution is approximated on the composite grid, which consists of the uniform coarse grid and subgrid(s). Note that such composite grids are highly structured and hence very simple data structures can be used.

The boundary value problem is solved on the composite grid by the local defect correction method (LDC) (see [2]). In this method, which is an iterative process, a basic global discretization is improved by local discretizations defined in the subdomains. This update of the coarse grid solution is achieved by performing a defect correction on the right hand side of the coarse grid problem. The discrete problem that is actually being solved is an implicit result of the iterative process. Therefore the LDC method is both a discretization and a solution method.

It should be noted, however, that LDC was previously used with standard finite difference schemes like second order central differences. Nowadays, especially for Direct Numerical Simulation (DNS) of turbulent flows, high order compact finite difference (HOCFD) schemes are becoming more and more popular. We would like to apply LDC technique to these schemes.

The paper is built up as follows. In Section 2, we formulate a stationary convection diffusion problem and describe the LDC algorithm. In Section 3 we briefly introduce HOCFD schemes. In Section 4 we present an algorithm which combines LDC with HOCFD schemes. We conclude with some examples, in which we compare theoretical results from Section 3 with those obtained from numerical simulations and we also discuss the efficiency and accuracy of the LDC in combination with HOCFD schemes.

## 2 Problem description and formulation of the LDC algorithm

The problem we study in this paper is given by

$$\begin{cases} Lu = -\varepsilon \nabla^2 u(\mathbf{x}) + \mathbf{c} \cdot \nabla u(\mathbf{x}) = f(\mathbf{x}) \text{ in } \Omega, \qquad c, \varepsilon > 0, \\ u = g \text{ on } \Gamma. \end{cases} \tag{1}$$

In (1), $L$ is a linear elliptic differential operator, and $f$ and $g$ are the source term and Dirichlet boundary condition, respectively, $\varepsilon$ is the diffusion coefficient, $\mathbf{c}$ - convection coefficient, $u$ is the unknown function of $\mathbf{x}$, $\Omega$ is the domain of interest and $\Gamma$ is the boundary of this domain. Most of details on the Local Defect Correction technique one can find in [1], below we just outline main features.

In order to discretize (1), we first choose a global coarse grid with grid size $H$, which we denote by $\Omega^H$ . The next step is to find an initial approximation $u_0^H$ on $\Omega^H$ by solving the system

$$L^H u_0^H = f^H, \tag{2}$$

which is a discretization of boundary value problem (1). In (2), the right hand side $f^H$ incorporates the source term $f$ as well as the Dirichlet boundary condition $g$.

Now assume that the continuous solution $u$ of (1) has a high activity region in some (small) part of the domain. This high activity can either be caused by the boundary conditions or by the source term. We would like to capture this high activity of $u$ by discretizing (1) on a composite grid. So we choose $\Omega_l \subset \Omega$ such that the high activity region of $u$ is contained in $\Omega_l$. If we have more than one high activity region, one may take more regions of refinement.

In $\Omega_l$, we choose a local fine grid with grid size $h$, denoted by $\Omega_l^h$. The fine grid is chosen such that $\Omega^H \cap \Omega_l \subset \Omega_l^h$, i.e., grid points of the global coarse grid that lie in the area of refinement belong to the local fine grid too.

Now we have to define a local discrete problem on $\Omega_l$. So we define artificial boundary conditions on $\Gamma$, the interface between $\Omega_l$ and $\Omega \setminus \Omega_l$. On $\Gamma$ we have more fine grid points than coarse grid ones, so we prescribe artificial Dirichlet boundary conditions by applying an interpolation operator $P^{h,H}$. In practice we take $P^{h,H}$ to be the linear interpolation operator on the interface for simplicity. In this way we find the following approximation $\mathbf{u}_{l,i}^h$ , $i = 0$, on $\Omega_l^h$. After boundary value problem (1) has been discretized and solved on a coarse grid and an area of the coarse grid has been refined and a local solution has been calculated on the finer grid, we can define a composite grid approximation $\mathbf{w}_0^{H,h}$ as

$$\mathbf{w}_0^{H,h} := \begin{cases} \mathbf{u}_{l,0}^h(x,y), & (x,y) \in \Omega_l^h, \\ \mathbf{u}_0^H(x,y), & (x,y) \in \Omega^H \setminus \Omega_l^h. \end{cases} \tag{3}$$

So for the coarse grid points within the region of the refinement we have two solutions, one coming from the coarse grid and another from the fine grid. We will now use the local fine grid solution to update the coarse grid approximation. In order to do so we introduce an discrete approximation of the local defect $\mathbf{d}^H := \mathbf{L}^H(u|_{\Omega^H}) - \mathbf{f}^H \approx \mathbf{L}^H\mathbf{w}_0^H - \mathbf{f}^H =: \mathbf{d}_0^H$. We can update the coarse grid approximation by placing the estimate at the right hand side of the coarse grid equation (1). This leads to the coarse grid correction step to find $\mathbf{u}_i^H$ , $i = 1$, on the coarse grid

$$\mathbf{L}^H\mathbf{u}_i^H = \mathbf{f}_{i-1}^H, \tag{4}$$

where

$$\mathbf{f}_i^H(x,y) := \begin{cases} \mathbf{f}^H(x,y) + \mathbf{d}_i^H(x,y), & (x,y) \in \Omega_l^H := \Omega^H \cap \Omega_l, \\ \mathbf{f}^H(x,y), & (x,y) \in \Omega^H\{\}\Omega_l^H. \end{cases}$$

The correction step (4) produces a new solution $\mathbf{u}_1^H$ on the coarse grid.

## 3 High order compact schemes

The basic idea of high order compact finite difference schemes is to employ not only the function values but also the values of the derivatives as unknowns. This gives a possibility to obtain higher accuracy or better spectral resolution while keeping the stencil relatively small [3]. Restricting ourselves to three-point expressions and one dimensional approximations, the general form of an implicit finite difference relation between a function and its first two derivatives would read

$$a_+ u_{i+1} + a_0 u_i + a_- u_{i-1} + b_+ (u_x)_{i+1} + b_0 (u_x)_i + b_- (u_x)_{i-1} \\ + c_+ (u_{xx})_{i+1} + c_0 (u_{xx})_i + c_- (u_{xx})_{i-1} = 0. \quad (5)$$

By imposing different constraints on the coefficients $a$, $b$ and $c$, we can tune the numerical scheme.

## 4 Combination of LDC with HOCFD

We would like to discretize our two dimensional variant of the equation (1) using HOCFD schemes. We introduce the vector of unknowns $\mathbf{x}^T = (\mathbf{u}_{xx}^T, \mathbf{u}_{yy}^T, \mathbf{u}_x^T, \mathbf{u}_y^T, \mathbf{u}^T)$, with the size $5N$ where $N$ is the number of grid points. As a result we get an algebraic system of the form

$$\mathbf{A}_i j \mathbf{x} = \mathbf{f} \quad (6)$$

The matrix $\mathbf{A}$ is a $5 \times 5$ block matrix. Entries $\mathbf{A}_{1,i}$ represent the discretization of $u_{xx}$ by one of the possible discretization schemes, entries $\mathbf{A}_{2,i}$ for $u_{yy}$, entries $\mathbf{A}_{3,i}$ for $u_x$, entries $\mathbf{A}_{4,i}$ for $u_y$, entries $\mathbf{A}_{5,i}$ for $u$. The entries $\mathbf{A}_{5,i}$ represent the equation (1) as well as the boundary conditions. Depending on the type of discretization used, some of the off-diagonal submatrices $\mathbf{A}_{i,j}$ could be zero or singular blocks. The matrix $\mathbf{A}$ has quite a large condition number, so we use equilibration of rows in order to reduce it. The basic idea is to multiply rows of the matrix such that we get $O(1)$ values on the main diagonal.

In its original form (see Section 2) the LDC technique is not directly applicable to the system (6). So we need a reformulation, like the following algorithm

1. Solve the coarse grid problem
   a) Construct a matrix $\mathbf{A}^H$ and a right hand side $\mathbf{f}^H$.
   b) Perform block LU-decomposition of the matrix $\mathbf{A}^H$. As a result we have $\mathbf{A}^H = \mathbf{L}^H \mathbf{U}^H$ and we get $\mathbf{L}^H$ and $\mathbf{U}^H$. Define the vector $\mathbf{y}^H := \mathbf{U}^H \mathbf{x}$. Solve $\mathbf{L}^H \mathbf{y}^H = \mathbf{f}^H$. Get $\mathbf{y}_5^H$. Solve$\mathbf{U}_{55} \mathbf{u}^H = \mathbf{y}_5^H$ and get $\mathbf{u}^H$, the coarse grid solution.
2. Solve the fine grid problem
   a) Get the boundary conditions for the fine grid boundary value problem (from the coarse grid solution) and construct $\mathbf{A}_l^h$ and $\mathbf{f}_l^h$.
   b) Solve the local grid problem $\mathbf{A}_l^h \mathbf{x}_l^h = \mathbf{f}_l^h$ and get $\mathbf{x}_l^h$.Extract $\mathbf{u}_l^h$.
3. Calculate the defect
   a) Construct the vector $\mathbf{w}^H$
   b) Construct the defect $\mathbf{d}_0^H = \mathbf{U}_{55} \mathbf{w}^H - \mathbf{y}_5^H$. Restrict thr defect by setting it to zero outside the area of refinement.
4. Solve the updated coarse grid problem $\mathbf{U}_{55} \mathbf{u}_1^H = \mathbf{y}_5^H + \mathbf{d}_0^H$ and get the new coarse grid solution $\mathbf{u}_1^H$.

## 5 Numerical results

We apply the LDC algorithm to the boundary value problem

$$\begin{cases} -\varepsilon_1 \frac{\partial^2 u}{\partial x^2} - \varepsilon_2 \frac{\partial^2 u}{\partial y^2} + a_1 \frac{\partial u}{\partial x} + a_2 \frac{\partial u}{\partial y} + cu = f \qquad (x,y) \in \Omega = (0,1) \times (0,1), \\ u(x,0) = g_1(x),\ u(x,1) = g_2(x),\ u(0,y) = g_3(y),\ u(1,y) = g_4(y) \end{cases} \tag{7}$$

In (7) boundary conditions and the source term have been chosen such that $u(x,y) = \tanh[25(x+y-1/3)] + 1$. We choose a uniform coarse grid $\Omega^H$ in $\Omega$ with grid sizes $\Delta x = \Delta y = 1/(N-1)$, with $N = 11, 16, 21$. We choose a uniform fine grid $\Omega_l^h$ with grid sizes $\Delta x = \Delta y = h$ with $h = H/2, H/4$. For the coarse grid discretization we tested different combinations of convective and diffusive terms discretizations. For the fine grid we used the same schemes as for the coarse grid, although it is not required. The results of computations can be summarized in Table 1 (accuracy of the algorithm compared to uniform fine grid) and Table 2 (efficiency in comparison with uniform fine grid). First we should mention that the accuracy of LDC is of the same order as for the uniform fine grid. As for the performance of the LDC technique compared to uniform fine grid method, LDC is much faster, and the difference in speed increases with the size of the problem.

**Table 1.** $\|\mathbf{u}^{exact} - \mathbf{u}^H\|_\infty$ for LDC algorithm and equivalent uniform grid

| Grid size | init | 1 iteration | uniform grid |
|---|---|---|---|
| Coarse: $11 \times 11$, fine: $11 \times 11$, uniform: $21 \times 21$ | $2.56 * 10^{-2}$ | $1.30 * 10^{-3}$ | $1.26 * 10^{-3}$ |
| Coarse: $11 \times 11$, fine: $21 \times 21$, uniform: - | $2.56 * 10^{-2}$ | $1.28 * 10^{-3}$ | |
| Coarse: $21 \times 21$, fine: $21 \times 21$, uniform: $41 \times 41$ | $1.26 * 10^{-3}$ | $3.30 * 10^{-5}$ | $3.46 * 10^{-5}$ |

**Table 2.** Calculation time for LDC algorithm and equivalent uniform grid

| Grid size | 1 iteration | uniform grid |
|---|---|---|
| Coarse: $11 \times 11$, fine: $11 \times 11$, uniform: $21 \times 21$ | 2.60 | 7.71 |
| Coarse: $11 \times 11$, fine: $21 \times 21$, uniform: - | 1.84 | - |
| Coarse: $21 \times 21$, fine: $21 \times 21$, uniform: $41 \times 41$ | 9.988 | 530 |

## References


1. M.J.H. Anthonissen, R.M.M. Mattheij, and J.H.M. ten Thije Boonkkamp. Convergence analysis of the local defect correction method for diffusion equations. *Numerische Mathematik*, 95(3):401–425, 2003.
2. W. Hackbush. *Elliptic Differential Equations. Theory and Numerical Treatment.* Springer, Berlin, 1992.
3. S.K. Lele. Compact finite difference schemes with spectral-like resolution. *J. Computational Physics*, 103:16–42, 1991.

# A Finite-Dimensional Modal Modelling of Nonlinear Fluid Sloshing

A. Timokha and M. Hermann[1]

Institut für Angewandte Mathematik, Friedrich-Schiller-Universität Jena, Ernst-Abbe-Platz 1-2, Jena, 07745, Germany
email: tim@imath.kiev.ua; hermann@mathematik.uni-jena.de

**Summary.** Since steady-state nonlinear fluid sloshing in moving tanks is caused by a finite set of natural modes, approximate solutions of the original free boundary value problem can be found from a system of nonlinear ordinary differential equations (modal system) coupling time dependent amplitudes of these leading modes. We focus on two-dimensional flows in a rectangular tank. We present an extensive literature survey and examine bifurcations of periodic (steady-state) solutions of a single-dominant modal system derived by [1].

**Key words:** fluid sloshing, modal systems, bifurcations.

## 1 Single-dominant Modal System

As shown in [1], solutions of the nonlinear sloshing problem in a rectangular basin with two-dimensional flows can be presented as the Fourier expansion

$$z = \sum_{i=1}^{\infty} \beta_i(t) \cos(\pi i(x + 0.5)), \tag{1}$$

which governs the surface wave motions, *i.e.*, the free boundary $\Sigma(t)$, in the $Oxz$-system, such that the $Ox$-axis is rigidly fixed with the horizontal equilibrium plane of the fluid. Each generalised coordinate $\beta_i(t)$ determines an amplitude-evolution of the $i$th natural standing mode and, following to asymptotic schemes, which are widely accepted in the surface wave theory, it should be ruled out by an inter-ordering in scale of the highest asymptotic order $\tau$. The dominating character of the lowest natural mode $\beta_1(t)$ to realistic free-standing and resonant waves has been used by many researchers for postulating $\beta_1(t)$ to have the lowest asymptotic order and deriving various asymptotic periodic solutions of the original free boundary problem. Adopting the asymptotic analysis by [1], one can show that both Stokes' and Moiseyev's third-order asymptotic solutions require the relationships

$$\beta_1 = O(\tau^{1/3});\quad \beta_2 = O(\tau^{2/3});\quad \beta_i \le O(\tau),\ i \ge 3 \tag{2}$$

and, as a consequence, the nonlinear free-standing Stokes' waves are up to $O(\tau)$ described by the modal system

$$\ddot{\beta}_1 + D\dot{\beta}_1 + (1-\delta_1(\lambda))\beta_1 + d_1(\ddot{\beta}_1\beta_2 + \dot{\beta}_1\dot{\beta}_2) + \\ +d_2(\ddot{\beta}_1\beta_1^2 + \dot{\beta}_1^2\beta_1) + d_3\ddot{\beta}_2\beta_1 = 0; \tag{3}$$

$$\ddot{\beta}_2 + (4-\delta_2(\lambda))\beta_2 + d_4\ddot{\beta}_1\beta_1 + d_5\dot{\beta}_1^2 = 0; \tag{4}$$

$$\ddot{\beta}_3 + (9-\delta_3(\lambda))\beta_3 + q_1\ddot{\beta}_1\beta_2 + q_2\ddot{\beta}_1\beta_1^2 + q_3\ddot{\beta}_2\beta_1 + \\ +q_4\dot{\beta}_1^2\beta_1 + q_5\dot{\beta}_1\dot{\beta}_2 = 0 \tag{5}$$

(higher-order equations for $\beta_i(t)$, $i \ge 4$, are linear). In this system, the coefficients $d_i, q_i, i \ge 1$, are known functions of the mean fluid depth $h$,

$$\delta_i = \delta_i(\lambda) = i^2 - \mu_i^2(1-\lambda),\quad i \ge 1;\quad (\delta_1 \equiv \lambda), \tag{6}$$

where

$$\mu_i = \frac{\bar{\sigma}_i}{\bar{\sigma}_1};\quad \sigma_i^2 = g\pi i \tanh(\pi i h) \tag{7}$$

is the dispersion and the constant $D =$ represents the damping. Furthermore, we study periodic solutions of (5) which satisfy the condition

$$\beta_i(0) = \beta_i(2\pi);\quad \dot{\beta}_i(0) = \dot{\beta}_i(2\pi),\quad i \ge 1. \tag{8}$$

Since the modal system is linear in terms of the modal function $\beta_3$ as well as (3)-(4) do not contain $\beta_3$, the nonlinear bifurcation analysis of periodic (steady-state) solutions will be focused on (3)-(4), (8). The key difficulty in handling the two-point boundary problem consists of either the non-uniqueness of its periodic solutions occurring due to the phase shift invariance as $D = 0$ or the absence of non-trivial solutions for $D \ne 0$. This is a typical situation in modelling conservative ($D = 0$) or dissipative ($D \ne 0$) mechanical models. Pursuing non-trivial solutions, to avoid the non-uniqueness, we introduce the Poincaré boundary condition, which reads

$$\dot{\beta}_1(0) = 0, \tag{9}$$

and adequately enlarge the dimension of the modal system (3), (4) by adding the formal, artificial equation

$$\dot{D} = 0. \tag{10}$$

Setting $B = (\beta_1, \beta_2, \dot{\beta}_1, \dot{\beta}_2, D)^T$ makes it possible to transform (3), (4), (8), (9) and (10) to a parametrised nonlinear two-point boundary value problem that permits the following operator formulation

$$T(B,\lambda) = 0,\ B \in X,\quad -\infty < \lambda < 1, \tag{11}$$

in suitable Banach spaces $X$ and $Y$ (see [3]). Apparently, the operator problem (11) has a trivial solution which does not depend on the parameter $\lambda$. Bifurcating points of the trivial solution were found by [3] at $\lambda_0(k,i) = 1 - k^2/\mu_i^2$, $k = 1, 2;\ i = 1, 2, \ldots$.

## 2 Local and Non-Local Bifurcation Analysis

Since $\mu_1^2 = 1$ and $2 < \mu_2^2 < 4$, there do not exist two integers $k_1$ and $k_2$ such that $\lambda_0(1, k_1) = \lambda_0(2, k_2)$. By using the Lyapunov-Schmidt reduction, [3] obtained two independent families of the bifurcating points, for $i = 1$ and 2, respectively. These are easily interpreted in terms of the "asymptotic" norm $||B|| = O(s)$. The locally bifurcating branches in the $(\lambda, s)$-plane (two types of backbones, solid and dashed lines) are shown in Fig. 1 (a). The type of branching depends on $h$ so that a passage of the curves from "hard-spring" (solid lines) to "soft-spring" (dashed lines) occurs at $h_R = 0.3368\ldots$. The relationship between the two families at $\lambda_0(1, k)$ and $\lambda_0(2, k)$ also changes with $h$ and, therefore, their inter-ordering along the $\lambda$-axis is hardly predictable, in general. However, it can be shown, the three lowest values from the resulting set $\{|\lambda_0(i, k)|, i = 1, 2;\ k \in \mathbb{N}\}$ are $\lambda(1, 1) = 0$, $\lambda(2, 2) \in (-1, 0)$ and $\lambda(2, 1) \in (1/2, 3/4)$ and they are linked as $\lambda_0(2, 2) < \lambda < \lambda_0(2, 1)$. Moreover, $\lambda(2, 2) \to 0$ as $h \to 0$, but it is bounded away from zero for finite depths $h$. The local branching related to these three bifurcation points is shown in Fig. 1 (b).

We found it useful to give a three-dimensional presentation of the local bifurcating curves by operating independently with norms of $||\beta_1||$ and $||\beta_2||$. Corresponding local branching in the $(\lambda, ||\beta_1||, ||\beta_2||)$-space is shown in Figs. 1 (c,d), where (c) implies $h > h_R$ and (d) corresponds to $h < h_R$, respectively. One important conclusion, based on this three-dimensional bifurcating diagram, is that equivalence of $\lambda$ to $\lambda_0(1, k)$ or $\lambda_0(2, k)$ leads to "orthogonal" bifurcations in appropriate spaces.

Another important point is that, while the branches at $\lambda_0(2, k)$ are of "linear nature" with vertical strain lines in the $(\lambda, ||\beta_1||, ||\beta_2||)$-space, the curves bifurcating at $\lambda_0(1, k)$ become of three-dimensional nature with increasing $s$. This constitutes a very interesting mathematical problem on description of the strictly three-dimensional curves, which has not been considered by [3], for the neighbourhood of the primary bifurcating point $\lambda_0(1, 1) = 0$. The numerical analysis of these non-local curves utilises a path-following procedure using the RWPM-package developed by [4]. This package can be used to study parametrised two-point boundary value problems. It is based on two numerical shooting techniques (multiple shooting and stabilised march, see *e.g.*, [2]) and enables the computation of isolated solutions of two-point boundary value problems as well as path-following and detection and determination of turning and bifurcation points. The calculations used the asymptotic solutions by [3] as an approximation of a non-trivial periodic solution on the approximating branch.

Four typical results for different $h$ are presented in Figs. 2 (a-d). The numerical analysis establishes that the periodic third-order Stokes' waves obtained from the single-dominant modal system may be qualitatively different from the asymptotic prediction, even if the wave amplitude (norm of the periodic solutions) is relatively small. If $h > h_R = 0.3368\ldots$, the numerically determined periodic solutions characterise not only the primary bifurcation

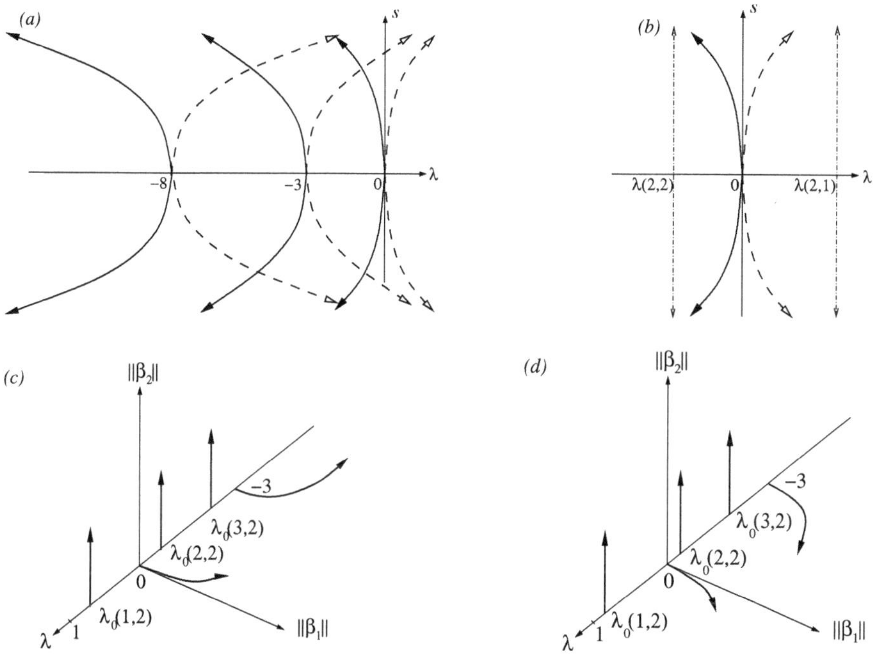


**Fig. 1.** (a) represents the local branches at $\lambda_0(1, k) = 1 - k^2,\ k \in \mathbb{N}$ in the $(\lambda, s)$-plane. Solid lines correspond to $h > h_R$, dashed lines represent the case $h < h_R$, where $h_R = 0.3368\ldots$. (b) sketches the local branches in the neighbourhood of the origin $(0, 0)$, imposed always by three bifurcation points $\lambda_0(1, 1) = 0$ (nonlinear standing waves dominated by the mode $f_1(x)$ with the frequency close to the lowest natural tone) and $\lambda_0(2, 1),\ \lambda_0(2, 2)$ (standing waves associated with the second mode $f_2(x)$); (c) and (d) give 3D treatment of the local branches in the $(\lambda, ||\beta_1||, ||\beta_2||)$-space for $h > h_R$ and $h < h_R$, respectively.

of the trivial solution, but also the secondary bifurcations arising as a secondary turning point $S$. It appears when $||\beta_2|| \sim ||\beta_1||$. The presence of $S$ makes our numerical results similar to the fifth-order theory by [5] capturing the case of the critical depth $h \approx h_R$. However, the secondary bifurcation point $S$ does not disappear for large $h$. The numerical failure of the Stokes third-order ordering (2) at the second bifurcation point $S$ makes it possible to quantify mathematically the applicability of the single-dominant model. The forthcoming studies should expand this numerical analysis to alternative, multi-dominant modal theories in which some higher modes can be of the same order as the primary $\beta_1$.

*Acknowledgement.* The present studies are in part supported by DFG. The speaker's (A.T.) attendance at the conference was sponsored by the Alexander von Humboldt Foundation.

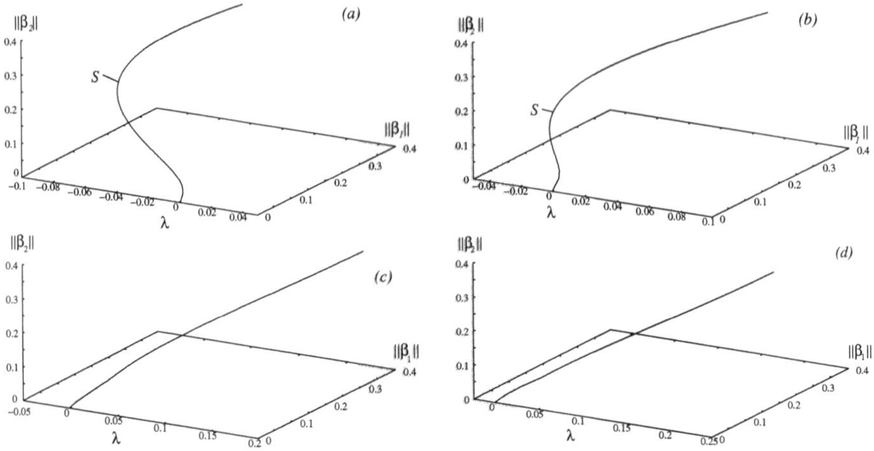


**Fig. 2.** Typical branching of the free-standing periodic waves at $\lambda_0(1,1) = 0$ for some typical $h$. (a) – $h = 1.0$, (b) – $h = 0.5$, (c) – $h = h_R$, (d) – $h = 0.3$.

## References


1. O.M. Faltinsen, O.F. Rognebakke, I.A. Lukovsky, and A.N. Timokha. Multidimensional modal analysis of nonlinear sloshing in a rectangular tank with finite water depth. *J. Fluid Mech.*, 407:201–234, 2000.
2. M. Hermann. *Numerik Gewöhnlicher Differentialgleichungen.* Oldenburg Verlag, 2004.
3. M. Hermann and A. Timokha. Modal modelling of the nonlinear resonant sloshing in a rectangular tank I: A single dominant model. Mathematical Models and Methods in Applied Sciences, to appear.
4. M. Hermann and K. Ullrich. RWPKV: a software package for continuation and bifurcation problems in two-point boundary value problems. *Appl. Math. Lett.*, 5:57–61, 1992.
5. D.D. Waterhouse. Resonant sloshing near a critical depth. *J. Fluid Mech.*, 281:313–318, 1994.

# Part IX

# Other Contributions

# On the Reliability of Repairable Systems: Methods and Applications

F. Ruggeri

CNR – IMATI, Via Bassini 15, 20133 Milano, Italy, fabrizio@mi.imati.cnr.it

**Summary.** Failures in repairable systems are often described by means of renewal or non-homogeneous Poisson processes, depending upon the repair policy. In the former case repairs bring the system reliability back to its initial value, whereas in the latter they restore the same reliability the system had just before the failure. We focus on the latter process, illustrating some properties and applications, mainly in a Bayesian framework.

**Key words:** Bayesian analysis, non-homogeneous Poisson processes, model selection, sensitivity analysis.

## 1 Introduction

Repairable systems are those systems (machines, industrial plants, software, etc.) which, in the event of a failure, can be repaired, e.g., by replacing a component, and be returned to regular operation. In some cases, the reliability of a system, after a repair, returns to the same state as before the failure. Conversely, "perfect" repairs bring the reliability back to the state the system had at the start of the operation. Failures of the former repairable systems are often described by means of non-homogeneous Poisson processes (NHPP), whereas renewal processes usually describe the latter systems. In this paper we will focus on NHPP's and we will illustrate some of their properties and applications. Here we provide an overview of problems and ideas related to the use of NHPP's in reliability, with a quick illustration of the mathematical techniques involved. A thorough presentation of the mathematical methods and the statistical analyses is not possible within the limited space of a review paper; we refer the interested reader to the works mentioned in the paper. A general review of NHPP's and their applications to the reliability of repairable systems can be found in [28].

In Section 2 we present some examples of repairable systems, whereas properties of NHPP's are described in Section 3 along with some issues related to their applications in reliability. In particular, estimation of parameters and reliability measures will be discussed, along with classes of NHPP's and more sophisticated models based on the introduction of change points, on superposition of NHPP's and Bayesian nonparametric methods. In Section 4 NHPP's will be used to analyse data about gas escapes in a city network of steel pipelines, applying some models described in Section 3. We will mainly focus on comparisons between parametric and nonparametric models, on model selection and sensitivity analysis. The paper ends with some remarks in Section 5.

## 2 Repairable systems

Systems subject to failures can be divided into two main groups: the non-repairable and the repairable ones. When a hydraulic pump in a power plant fails, it is replaced; the *hydraulic pump* is a non-repairable system, whereas the *power plant* is a repairable one.

Once a system experiences a failure, different repair strategies have different influences on the system reliability, usually defined as the probability of no failures in time intervals. Under "perfect" repair, the system is brought back to its initial reliability, whereas an instantaneous, minimal repair of a small component restores the same reliability the system had just before the failure. The former strategy corresponds to a condition commonly called "good-as-new" or "same-as-new", whereas the latter corresponds to the "bad-as-old" or "same-as-old" condition. System failures under the two strategies are usually modelled with renewal processes and NHPP's, respectively (see, e.g., [28]).

The reliability could be constant over time, as it happens in the homogeneous Poisson process (HPP), the only NHPP which is a renewal process as well. Systems are subject to reliability decay or growth (with steadiness as a special case of both); in their lifetime they can either experience only one of them or alternate them at some change points. Sequential detection of bugs in software, without introduction of new ones, implies reliability growth all over the testing phase. Conversely, there are systems subject to many early failures, then a decrease in them is followed by a long period of rare failures and by a final period with an increasing number of failures. After the initial phase of "burn in" followed by reliability growth, those systems experience a constant reliability ("useful life") followed by a final phase of reliability decay; the term "bath-tub" is used to describe this behaviour because of the shape of the intensity function (described later) of the corresponding NHPP's.

The two repair strategies illustrated before are two extreme, opposite ones. In practice, repairs usually increase the reliability of the system with respect to (w.r.t.) the one it had just before the failure but they do not bring it back to its initial value. Furthermore, repairs are not, in general, instantaneous and unavailability needs being modelled as well.

Finally, it is worth mentioning that complex systems could be split into components and their reliability could be the result of the superposition of the reliability of their components.

We illustrate these notions with examples considered in past works; some of them will be used in the next sections.

*Example 1. (Non-repairable system)* [24] considered failure times of hydraulic pumps, replaced after each failure, in different power plants.

*Example 2.* (*HPP*) [4], [5] and [6] considered escapes ("failures") in a city network of *old cast* gas pipes. 150 escapes were observed in 6 years in a 320 Km long network. Since repairs involved a very small component of the network and they were performed in a very short period w.r.t. the lifetime of the network (some pipes were more than 100 years old), the assumption of instantaneous, minimal repair was made and an NHPP was deemed suitable for modelling the escapes. Since *old cast* pipes were not subject to ageing (corrosion), the reliability of the system was considered constant over time, so that the HPP was the natural choice in this case.

*Example 3.* (*NHPP*) [26] considered failures in underground trains. 40 trains were observed over a 8-years period and failures of doors were recorded. Since repairs were almost instantaneous and minimal, then an NHPP was chosen to model the failures.

*Example 4.* (*NHPP with change points*) [32] considered a well known dataset about the dates of serious coal-mining disasters ("failures"), between 1851 and 1962. Conditions to model "failures" with an NHPP applied, but there were strong feelings about the change in the reliability of the "system" at some points. A NHPP allowing for change points was the chosen model in this case.

*Example 5.* (*Superposition of NHPP's*) [25] considered escapes in a city network of steel gas pipes. 53 escapes were observed in 30 years in an expanding network, currently 380 Km long. As in Example 2, an NHPP was considered a suitable model for describing escapes but, in this case, reliability was not constant since steel pipes were subject to corrosion and ageing. Furthermore, the expansion of the network over time, combined with the ageing property of the pipes, compelled to distinguish among pipes with different ages. Therefore, the complex system was split into subsystems (determined by installation year of the pipes) and its reliability followed from the reliability of its components.

## 3 Non-homogeneous Poisson processes

In this section, we illustrate properties of NHPP's and models based on them.

### 3.1 Main properties

We recall the well known definition of NHPP, along with few properties. More detailed illustration of NHPP's applied to reliability can be found in [28], whereas general Poisson processes are described in, e.g., [18].

Let $N(s,t)$ denote the number of failures in the interval $(s,t]$, whereas the notation $N(t)$ is used instead of $N(0,t)$.

**Definition 1.** *A counting process $N(t)$ is said to be an NHPP with intensity $\lambda(t)$ if*

1. $N(0) = 0$;
2. *it has independent increments;*
3. $\mathcal{P}\{N(t,t+h) \geq 2\} = o(h)$;
4. $\mathcal{P}\{N(t,t+h) = 1\} = \lambda(t)h + o(h)$.

When the intensity $\lambda(t)$ is constant over time, then a HPP is obtained.

The mean value function (m.v.f.) of the NHPP is defined as the nondecreasing, nonnegative, function $M(s,t) = E\{N(s,t)\}$, $0 \leq s < t$, with $M(t) = E\{N(t)\}$, $t \geq 0$. Assuming that $M(t)$ is differentiable, then $\mu(t) = \dfrac{dM(t)}{dt}$ is the rate of occurrence of failures (ROCOF) for the NHPP. The property 3) in Definition 1 implies that $\mu(t) = \lambda(t)$ a.e. so that $M(y,s) = \int_y^s \lambda(t)dt$.

The NHPP owes its name to the fact that

$$\mathcal{P}\{N(y,s) = k\} = \frac{M(y,s)^k}{k!} e^{-M(y,s)}$$

for any integer $k$.

We illustrate two examples of NHPP's. The first NHPP, widely used in reliability, is the Power Law process (PLP), sometimes called Weibull process and described in, *e.g.*, [13]. Its intensity is given by $\lambda(t) = M\beta t^{\beta-1}$, $M, \beta > 0$, with $M(t) = Mt^{\beta}$. As shown in Fig. 1, the intensity function is decreasing for $\beta < 1$, constant for $\beta = 1$ and increasing for $\beta > 1$. The three behaviours correspond, respectively, to growth, steadiness and decay in reliability. Inference on $\beta$ is crucial to understand the nature of the data at hand. Such richness of behaviours, the simple mathematical form and the Weibull distribution of the first failure time are, probably, the main reasons for the relevant role played by the PLP in reliability.

The second example of NHPP is due to [26]. They considered the NHPP with intensity

$$\lambda(t) = \beta_0 \frac{\log(1+\beta_1 t)}{(1+\beta_1 t)}, \beta_0 > 0, \beta_1 > 0,$$

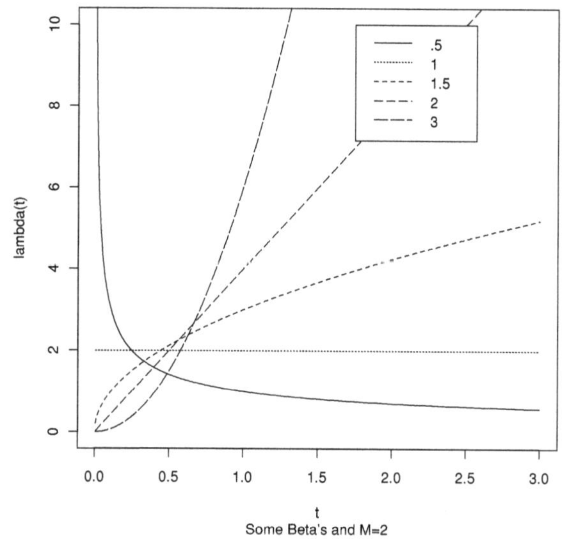


**Fig. 1.** Intensity function of the Power Law process.

*i.e.* a function which is increasing up to its maximum point at $(e-1)/\beta_1$ and then goes to zero as $t$ goes to infinity. This behaviour represents the case of reliability decay followed by reliability growth, and this NHPP could be used to model the system lifetime when excluding its final portion.

### 3.2 Statistical analysis of simple NHPP's

Consider an NHPP with intensity function $\lambda(t;\theta)$. Suppose we observe the system up to time $y$ and let $n$ be the number of failures, occurred at times $t_1 < t_2 \ldots < t_n$; then the likelihood function is given by

$$L(\theta;\boldsymbol{t}) = \prod_{i=1}^{n} \lambda(t_i) \exp\{-\int_0^y \lambda(t)dt\},$$

where $\boldsymbol{t} = (t_1, \ldots, t_n)$.

At least two experiments are possible: observation of the system up to a given time or until the $n$-th failure occurs. The former case is called *time truncation*, whereas the latter is called *failure truncation*, with $y = t_n$. Conceptually different, the two experiments lead to the same estimates in a Bayesian framework, whereas some differences are possible following a frequentist approach as in the maximum likelihood estimation of the parameters of the PLP.

Illustration of frequentist and Bayesian results about parameter estimation, (Bayesian) confidence intervals, hypothesis testing, etc. are well beyond the scope of the current paper. We refer the interested reader to [28]. The PLP has been considered, among the others, by [1, 12, 13, 14, 27] in a frequentist framework, and by [2, 7, 17] and [19] in a Bayesian framework.

Another experiment was considered in [8] and [22], in a frequentist and a Bayesian framework, respectively. Only count data were available from $k$ identical systems, i.e. the number of failures but not the failure times. The systems were modelled with PLP's with equal parameters and their parameters were estimated.

### 3.3 Reliability measures

Parameter estimation is not, probably, the most important goal of a statistical analysis in reliability. The major interest in a statistical reliability study relies on the possibility of forecasting future failures, either on the system under observation or a new, similar one. An example of quantity of interest is the system reliability, which is defined as $R(y,s) = \mathcal{P}\{N(y,s) = 0\}$ for the system observed up to time $y$ or $R(s) = \mathcal{P}\{N(s) = 0\}$ for a new one. For a PLP, the system reliability becomes $R(y,s) = \exp\{-Ms^\beta + My^\beta\}$ and $R(s) = \exp\{-Ms^\beta\}$, respectively. A related quantity is the prediction of the number of failures in some time interval.

Another quantity of interest is the expected number of failures in future intervals, i.e. $\mathcal{E}[N(y,s)]$ for the same system or $\mathcal{E}[N(s)]$ for a new one. For a PLP, the expected number of failures in future intervals is given by $\mathcal{E}[N(y,s)] = M[s^\beta - y^\beta]$ and $\mathcal{E}[N(s)] = Ms^\beta$, respectively.

Finally, estimation of the intensity function at $y$ is important. NHPP's are sometimes used for describing failures in prototype testing. A new product is tested before being marketed. Achievement of satisfactory reliability, excessive cost of further testing, risk of obsolescence of the product are some of the causes which imply interruption of testing at time $y$. Once the product is marketed, no further testing is allowed to improve it and its future reliability is the same it had at time $y$.

### 3.4 Covariates in NHPP's

The NHPP's described earlier refer to a unique system with homogeneous characteristics. Sometimes systems can differ for some features, like the location of the power plants and the operating conditions of the hydraulic pumps analysed in Example 1. In [24] failure times were described by an exponential density $f(t;\lambda)$ whose parameter $\lambda$, following the Bayesian paradigm, was given a prior distribution $\pi(\lambda)$, namely a Gamma one. Failures were recorded for each set of homogeneous pumps (i.e. with same location and operating conditions) and for each of them the posterior distribution of $\lambda$ was computed. Posterior distributions for parameters of different sets of pumps were compared using the Kullback–Leibler (K–L) divergence w.r.t. a baseline distribution and pumps were clustered according to their K–L values.

A different approach was taken in Example 2. As discussed in [4, 5] and [6], the network of *old cast* gas pipes could be divided according to some features identified by a preliminary exploratory analysis. In particular, diameter, laying depth and laying location were identified as the three most influential factors (covariates) on the gas escapes. Two levels were assigned to each covariate and their combination led to 8 classes. The behaviour of the pipes in each class was considered as not influent on the other pipes, so that 8 independent HPP's were considered to describe failures in the classes. Each HPP was identified by its constant intensity $\lambda$ and Gamma (or Lognormal) priors were considered for the 8 $\lambda$'s, with their own hyperparameters determined through interviews with 26 company experts and combination of their opinions via Analytic Hierarchy Process (AHP), described in [33]. The posterior mean of the parameter $\lambda$ was computed for each class and their values were compared and ranked, determining the class with the lowest reliability (or highest posterior mean) and, therefore, the first to be replaced.

Independence among classes could be an unrealistic assumption. A more complex structure was proposed in [21] where a Cox model - type approach was taken. The distribution of the parameter $\lambda$ depended upon the covariates $\underline{X} = (X_1, \ldots, X_n)$, $X_i = 0, 1$, for all $i$, i.e. it was a Gamma distribution with parameters $\alpha \exp\{\underline{X}_i^T \underline{\beta}\}$ and $\alpha$, so that their prior mean was $\exp\{\underline{X}_i^T \underline{\beta}\}$. The link among the classes was therefore via the hyperparameters $\alpha$ and $\underline{\beta}$, and their joint prior distribution. An alternative, presented in [21], considered $\alpha$ fixed and estimated $\underline{\beta}$ via an empirical Bayes approach.

Finally, [34] considered NHPP's whose parameters were functions of the covariates $\underline{X}$. As an example, the parameter $M$ in the PLP, with intensity $\lambda(t) = M\beta t^{\beta-1}$, was considered as $M = M(\underline{X}) = M_0 \prod_{i=1}^n \delta_i^{X_i}$, $\delta_i > 0$, for all $i$.

## 3.5 Classes of NHPP's

As shown in Fig. 1, the same NHPP, a PLP namely, describes completely different behaviours (from reliability decay to reliability growth) for different values of its parameters. A plethora of NHPP's, differing in their intensities, has been defined in literature (as listed by Chris McCollin, private communication) and then applied to different problems. Unifying properties could be sought and classes of NHPP's have been actually identified and their common properties have been investigated; two examples of such classes are illustrated below.

### A general class

A general class of NHPP's can be described (see [32]) by the intensity function $\lambda(t; \alpha, \beta) = \alpha g(t, \beta)$, with $\alpha, \beta > 0$, such that their m.v.f. is

$M(t;\alpha,\beta) = \alpha G(t,\beta)$, with $G(t,\beta) = \int_0^t g(u,\beta)du$. This class contains well-known processes, such as the Musa-Okumoto, the Cox-Lewis and the Power Law processes.

The first process, described in [23], has been widely used in modelling software reliability; it has intensity function $\lambda(t;\alpha,\beta) = \alpha/(t+\beta)$ and m.v.f. $M(t;\alpha,\beta) = \alpha\log(t+\beta)$. The second process, described in [11], is such that $\lambda(t;\alpha,\beta) = \alpha\exp\{\beta t\}$ and $M(t;\alpha,\beta) = (\alpha/\beta)\,[\exp\{\beta t\} - 1]$. As shown before, the intensity and mean value functions of the PLP are given, respectively, by $\lambda(t;\alpha,\beta) = \alpha\beta t^{\beta-1}$ and $M(t;\alpha,\beta) = \alpha t^{\beta}$, $\alpha, \beta > 0$.

A unified treatment of this class from a Bayesian viewpoint is possible. Consider the failures $\boldsymbol{t} = (t_1,\ldots,t_n)$ in $(0,y]$, then the likelihood becomes $M^n \prod_{i=1}^{n} g(t_i,\beta)e^{-MG(y,\beta)}$. Consider any prior distribution $\pi(\beta)$ for $\beta$ and a Gamma prior $\mathcal{G}(\mu,\rho)$ for $M$ or a Gamma distribution $\mathcal{G}(\mu,\rho\beta)$ for $M \mid \beta$. Under these assumptions, we obtain a conditional Gamma posterior $\mathcal{G}(n+\mu, G(y,\beta)+\rho[\beta])$ for $M \mid \beta, \boldsymbol{t}$, whereas the posterior density of $\beta \mid \boldsymbol{t}$ is proportional to $\dfrac{\prod_{i=1}^{n} g(t_i,\beta)\pi(\beta)}{(G(y,\beta)+\rho[\beta])^{n+\mu}}$. Note that the term $[\beta]$ is included when the conditional prior of $M \mid \beta$ is chosen.

General results for this class were found in [32] when considering change points in NHPP's.

### A class based on differential equations

Another class has been proposed in [31], stemming from the consideration that [23] found the process named after them by postulating the following relation between $\lambda(t)$ and $M(t)$:

$$\lambda(t) = \lambda e^{-\theta M(t)}.$$

The relation can be expressed as the first order differential equation $[M(t)]' = \lambda e^{-\theta M(t)}$, whose solution is $M(t) = \log(\lambda\theta t + 1)/\theta$ when $M(0) = 0$.

Similar relations can be found for other NHPP's. The PLP has intensity function $\lambda(t) = M\beta t^{\beta-1}$ with $M(t) = Mt^{\beta}$; their relation is given by $[M(t)]' = \beta M(t)/t$, with $M(0) = 0$. Similarly, the relation $[M(t)]' = b[M(t) + at]$ holds for the process with $\lambda(t) = a(e^{bt} - 1)$ and $M(t) = \frac{a}{b}(e^{bt} - bt - 1)$, whereas $[M(t)]' = b\,(M(t) + at)\,/\,(1 + bt)$ holds for the NHPP with $\lambda(t) = a\log(1+bt)$ and $M(t) = (a/b)\,(1+bt)\log(1+bt) - at$.

All the above relations belong to the class of first order differential equations:

$$y' = \frac{\alpha y + \varepsilon + \beta x}{\rho y + \gamma + \delta x}.$$

(Note that we express the differential equations using $x$ and $y(x)$.)

[31] focussed on all the NHPP's whose mean value functions were solutions of the differential equation

$$y' = \frac{\alpha y + \beta x}{\gamma + \delta x}, \tag{1}$$

where all the parameters were nonnegative, and both parameters, either at the numerator or the denominator, could not be zero at the same time.

The solution of (1) is

$$y = e^{\int \alpha/(\gamma+\delta x)dx} \left\{ \int \frac{\beta x}{\gamma + \delta x} e^{-\int \alpha/(\gamma+\delta x)dx} dx + c \right\}.$$

Among all possible combinations of parameters, we present only those leading to actual mean value functions. We present the corresponding mean value functions and intensities in Table 1.

**Table 1.** NHPP solution of the differential equation (1).

| $M(t)$ | $\lambda(t)$ |
|---|---|
| $\frac{t}{\delta}$ | $\frac{1}{\delta}$ |
| $\frac{t^2}{2\gamma}$ | $\frac{t}{\gamma}$ |
| $\frac{t}{\delta} - \frac{\gamma}{\delta^2} \log\left(1 + \frac{\delta}{\gamma} t\right)$ | $\frac{t}{\gamma + \delta t}$ |
| $\lvert c \rvert t^{\alpha/\delta}$ | $\lvert c \rvert \frac{\alpha}{\delta} t^{\alpha/\delta\ 1}$ |
| $\beta\gamma\left(e^{t/\gamma} - \frac{t}{\gamma} - 1\right)$ | $\beta\left(e^{t/\gamma} - 1\right)$ |
| $\frac{\beta}{\delta - 1}\left\{t + \gamma\left[1 - \left(1 + \frac{\delta}{\gamma}t\right)^{1/\delta}\right]\right\}$ | $\frac{\beta}{\delta - 1}\left\{1 - \left(1 + \frac{\delta}{\gamma}t\right)^{1/\delta - 1}\right\}$ |
| $\beta\gamma\left(1 + \frac{t}{\gamma}\right)\log\left(1 + \frac{t}{\gamma}\right) - \beta t$ | $\beta \log\left(1 + \frac{t}{\gamma}\right)$ |

## 3.6 Change points in NHPP's

Example 4 describes a system whose behaviour changes at some time point. [32] considered two different types of change point models. In the first, they considered models allowing changes in reliability level after each failure, as the system is repaired and put to operation, e.g., in software reliability. In the second, they considered a model allowing changes at random points in time, due to break down of a component without causing the failure of system or due to interventions by maintenance squad at unknown (for the statistician) time points.

[32] considered change points in PLP's right after each failure (at times $t_i^+$'s), modifying the value of $\beta$. Changes in $M$ could be considered in a similar, but cumbersome, manner. We denote the parameter value at time $t_i^+$, $i = 1, \ldots, n$, right after a failure, by $\beta_i$, identifying the process over $(t_i, t_{i+1}]$. We denote by $\beta_0$ the parameter value over $(t_0, t_1]$. Here we take $t_0 = 0$ and $t_{n+1} = y$, *i.e.* the endpoints of the observation interval.

[32] considered both a hierarchical model and a dynamic one. In the first model it was assumed that, given $(\varphi, \sigma^2)$, the $\beta_i$'s were i.i.d. with a lognormal distribution $\mathcal{LN}(\varphi, \sigma^2), i = 0, \ldots, n$. At the second stage of the hierarchical model $\varphi$ and $\sigma^2$ had, respectively, a normal prior $\mathcal{N}(\mu, \tau^2)$ and an inverse Gamma prior $\mathcal{IG}(\rho, \gamma)$.

The likelihood became

$$M^n \prod_{i=1}^{n} \beta_{i-1} t_i^{\beta_{i-1}-1} \exp\{-M \sum_{i=1}^{n+1} (t_i^{\beta_{i-1}} - t_{i-1}^{\beta_{i-1}})\}.$$

Assuming a gamma prior for $M$, then the posterior conditionals of $M$, $\sigma^2$ and $\varphi$ were, respectively, a Gamma, an inverse Gamma and a normal distribution. The conditional distributions of the $\beta_i$'s were known apart from a normalising constant. This is a typical situation in which the full conditionals can be used to simulate from the joint posterior via Metropolis-Hastings and Gibbs sampling.

In the second model considered by [32], *i.e.* the dynamic model, the parameter $\beta_{i-1}$ was modified at time $t_i^+$, $i = 1, \ldots, n$, according to

$$\log \beta_i = \log a + \log \beta_{i-1} + \varepsilon_i,$$

where $a$ was a positive constant and $\varepsilon_i$ was a normally distributed random variable with mean 0 and variance $\sigma^2$. Suitable choices of priors lead to simulation from the joint posterior via Gibbs sampling with Metropolis steps within.

The proposed method to describe failures at random points relied on the Reversible Jump MCMC technique developed in [16]. We refer to [32] for more details on the algorithm developed for changes in the parameters of the general class described in Section 3.5.

### 3.7 Superposition of NHPP's

Example 5 is about a complex system with identifiable components which function independently and are repairable systems themselves. Components are modelled by independent NHPP's with intensity $\lambda_s(t; \theta_s)$, $s \in \mathcal{S}$ (in example 5, $\mathcal{S}$ coincides with the set of all the years in which portions of the gas network were installed). The superposition of these independent NHPP's is again an NHPP with intensity $\lambda(t; \underline{\theta}) = \sum_{s \in \mathcal{S}} \lambda_s(t; \theta_s)$, as a consequence of the Superposition Theorem (see, e.g. [18]).

Many situations are possible depending on the characteristics of the system under examination: the components can perform different operations, and thus be subject to different types of failures, or they can be identical units with minor differences only. In the first case it is appropriate to have different intensity functions with different parameters, whereas in the second one the intensity functions would have the same functional form and would differ only for a subset of the $\theta_s$'s or for some known constant (for example, one that is related to the size of the component). [25] considered both cases assuming PLP's with intensity

$$\lambda_s(t;\theta_s) = l_s M_s \beta_s (t-s)^{\beta_s} \ {}^1\mathrm{I}_{[s,+\infty)}(t),$$

with $M_s, \beta_s, l_s > 0$, where $s$ is the installation date (year) and $l_s$ is the known length of the pipes installed at $s$. In the first case each PLP had its own parameters $M_s$ and $\beta_s$, whereas two situations were considered in the second one: the parameters were the same for all PLP's or they were drawn from the same distribution (*exchangeability*).

In this case, we observe the system up to time $y$ and data are given by both $n$ failure times $t_k$ and installation dates $\delta_k$ of failed parts, with $r$ being the number of different installation dates $s_i$. Considering PLP's with same parameters $M$ and $\beta$, the likelihood is given by $L(M,\beta;\underline{t},\underline{\delta})$:

$$M^n \beta^n \prod_{k=1}^{n} l_{\delta_k}(t_k-\delta_k)^{\beta-1} e^{-M\sum_{i=1}^{r} l_{s_i}(y-s_i)^{\beta}}.$$

In a Bayesian framework Gamma priors can be chosen for both parameters $M$ and $\beta$, but MCMC methods are needed to compute their posterior distributions and estimate the reliability measures illustrated in Section 3.3.

When each PLP has its own parameters, then the likelihood $L(\underline{M},\underline{\beta};\underline{t},\underline{\delta})$ becomes, for $\underline{M} = (M_1,\ldots,M_r)$ and $\underline{\beta} = (\beta_1,\ldots,\beta_r)$,

$$\prod_{k=1}^{n} \beta_{\delta_k} l_{\delta_k} M_{\delta_k}(t_k-\delta_k)^{\beta_{\delta_k}-1} e^{-\sum_{i=1}^{r} l_{s_i} M_{s_i}(y-s_i)^{\beta_{s_i}}}.$$

When exchangeability of parameters is assumed, then each $M_s$ (and $\beta_s$) is drawn from the same prior distribution $\pi(M;\theta_M)$ ($\pi(\beta;\theta_\beta)$). Prior distributions on $\theta_M$ and $\theta_\beta$ are chosen as well. In [25], exponential distributions were chosen for $M_s$ and $\beta_s$ and the hyperparameters $\theta_M$ and $\theta_\beta$.

### 3.8 Nonparametric models

Sometimes a parametric assumption, like the intensity function $\lambda(t;\theta)$ in an NHPP, is a burden since the parametric model could be unable to describe data at hand very well. Failure data from Example 5 were considered all together in [9], where an NHPP with logarithmic intensity was considered and the estimated model gave a very poor fit to the data. The nonparametric approach based on weighted Gamma processes, and proposed in [9], gave a very good fit to the data.

To understand the nonparametric approach, consider the distribution of the number of events in a given interval $(s,t]$ under an NHPP with m.v.f. $M(s,t)$: it is a Poisson distribution with parameter $M(s,t)$. In a parametric model, *i.e.* an NHPP with intensity $\lambda(t;\theta)$, we saw in Section 3 that $M(s,t) = \int_s^t \lambda(u;\theta)du$. In a nonparametric approach we consider $M(s,t)$ as a Gamma distributed random variable for each choice of $0 \leq s < t$.

Stemming from results in [20], then [9] and [10] considered the m.v.f. $M$ as the distribution function of a random measure $\Lambda$, the intensity measure of the process.

Given a measure $\mu$ and a measurable subset $B$, we define $\mu B := \mu(B)$.

**Definition 2.** *Let $\alpha$ be a finite, $\sigma$-additive measure on $(\mathbf{S}, \mathcal{S})$. The random measure $\mu$ follows a* **Standard Gamma** *distribution with* shape $\alpha$ *(denoted by $\mu \sim \mathcal{GG}(\alpha, 1)$) if, for any family $\{S_j, j = 1, \ldots, k\}$ of disjoint, measurable subsets of $\mathbf{S}$, the random variables $\mu S_j$ are independent and such that $\mu S_j \sim \mathcal{G}(\alpha S_j, 1)$, for $j = 1, \ldots, k$.*

**Definition 3.** *Let $\beta$ be an $\alpha$-integrable function and $\mu \sim \mathcal{GG}(\alpha, 1)$. The random measure $\nu = \beta\mu$ follows a* **Generalised Gamma** *distribution, with* shape $\alpha$ *and* scale $\beta$ *(denoted by $\nu \sim \mathcal{GG}(\alpha, \beta)$).*

The Generalised Gamma distributions are conjugate for the Poisson processes.

**Theorem 1 ([20]).** *Let $\underline{\xi} = (\xi_1, \ldots, \xi_n)$ be $n$ Poisson processes with intensity measure $\Lambda$. If $\Lambda \sim \mathcal{GG}(\alpha, \beta)$ a priori, then $\Lambda \sim \mathcal{GG}(\alpha + \sum_{i=1}^n \xi_i, \beta/(1 + n\beta))$ a posteriori.*

In the next we consider the observations $\{y_{ij}, i = 1 \ldots k_j\}_{j=1}^n$ from $n$ Poisson processes $\underline{\xi} = (\xi_1, \ldots, \xi_n)$ on $\mathbf{S}$. Note that $y_{ij}$ denotes the failure time when $k_j$ failures are recorded in the $j$-th system, $j = 1, \ldots, n$, observed over $\mathbf{S} = [0, T]$.

**Theorem 2 ([20]).** *Under the above assumptions and squared loss function, the Bayesian estimator of $\Lambda \sim \mathcal{GG}(\alpha, \beta)$ is given by the measure $\widetilde{\Lambda}$ such that*

$$\widetilde{\Lambda}S = \int_S \frac{\beta(x)}{1 + n\beta(x)} \alpha(dx) + \sum_{j=1}^n \sum_{i=1}^{k_j} \frac{\beta(y_{ij})}{1 + n\beta(y_{ij})} \mathbf{I}_S(y_{ij}), \quad \forall S \in \mathcal{S}.$$

**Theorem 3 ([10]).** *Given $n$ Poisson processes $\underline{\xi}$ with common intensity function $\Lambda \sim \mathcal{GG}(\alpha, \beta)$, the Bayesian estimator of $R$, under squared loss function, is given by $\widetilde{R}$ such that, $\forall S \in \mathcal{S}$,*

$$\widetilde{R}S = \exp\left\{ -\int_S \ln(1 + \frac{\beta(x)}{1 + n\beta(x)}) \alpha(dx) - \sum_{j=1}^n \sum_{i=1}^{k_j} \ln(1 + \frac{\beta(y_{ij})\mathbf{I}_S(y_{ij})}{1 + n\beta(y_{ij})}) \right\}.$$

# 4 Examples

In this section we illustrate the models presented earlier with two examples. Other examples are thoroughly discussed in [25, 26, 4, 5] and [6]. The first example compares a parametric model with a nonparametric one, whereas the latter considers model selection and sensitivity analysis. Both examples consider subsets of the data presented in Example 5, *i.e.* gas escapes in a city network of steel pipes. More details on the data can be found in [15].

It is worth noticing that steel pipelines have very strong mechanical properties (failures are very rare), but they are easy prey for corrosive agents unless they are correctly protected. Available data and experts' opinions led to identify three principal elements that may be related to the failures: the age of the pipe, the type of corrosion that leads to the rupture of pipe, and the lay location of the pipe. In the first example only the first element is considered, whereas the second element is considered in the second example. As mentioned before, the choice of an NHPP is justified since corrosion develops progressively, reducing the thickness of the pipe walls until the pipe breaks and the gas escapes.

## 4.1 Parametric vs. nonparametric models

We consider 33 failures, due to corrosion, in the period 1978–1997 over a network of 275 kilometres.

### Parametric model

[9] and [10] compared three parametric NHPP's to model all the failures: the HPP, the PLP and, mainly, the logarithmic NHPP with $\lambda(t) = a\ln(1+bt)+c$.

The m.v.f. of the logarithmic NHPP is given by

$$M(t) = \int_0^t \lambda(s)\,ds = \frac{a}{b}(1+bt)\ln(1+bt) + (c-a)t.$$

MLE and Bayesian (under independent Gamma priors) estimates are presented in Table 2.

**Table 2.** MLE and Bayesian estimates.

| Parameter | MLE | Mean | Median | Mode | Variance |
|---|---|---|---|---|---|
| a | 0.000 | 1.243 | 1.141 | 0.962 | 0.493 |
| b | 0.724 | 1.662 | 1.374 | 0.799 | 1.454 |
| c | 4.755 | 2.797 | 2.780 | 2.862 | 1.021 |

The MLE lead actually to a HPP, whereas the Bayesian approach favoured the nonhomogeneous component. In this case MLE performed better than the posterior mean and median, *i.e.* the Bayesian estimators under two different loss functions. In fact, [10] showed that the estimated m.v.f. of the HPP was closer to the sample m.v.f. than the one from the NHPP and the Bayes factor favoured the HPP w.r.t. the NHPP's, specially the PLP. This is an example of model selection: when different models are entertained, tools are used to choose among them, like the posterior probability of each model or the Bayes factor, defined as the ratio between the marginal distributions of the data under two models (see, e.g., [3] for more details).

### Nonparametric model

The findings of Section 4.1 lead to relax the parametric assumption, considering the nonparametric model described in 3.8 and choosing its parameters so that it was "centred" at the best estimate of the logarithmic NHPP, whose m.v.f. was the Bayesian estimator $\hat{M}_\theta$, with intensity $\hat{\lambda}_\theta(s)$. Therefore, [10] considered a weighted Gamma process $\mathcal{GG}(\hat{M}_\theta/\sigma, \sigma)$, with $\sigma = 0.509$. The choice of the parameters lead to $\hat{M}_\theta$ as the expected value of the process (this explains the notion of "centering" we have been using). Furthermore $\sigma$ determines how much the process is spread around its expected value. More details are available in [10], including the expressions for the Bayesian estimators of the m.v.f. and the reliability, depicted in Figs. 2 and 3.

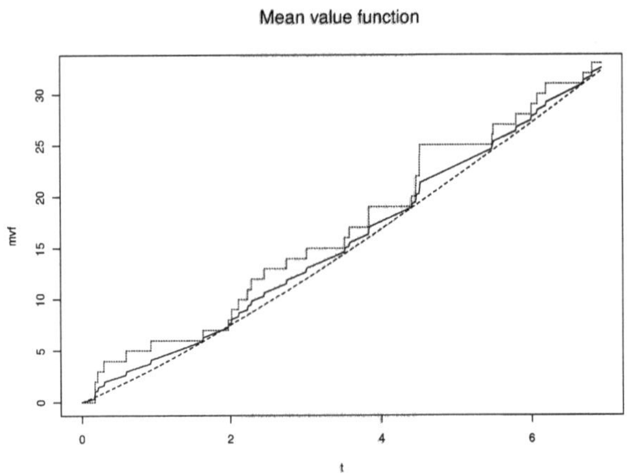


**Fig. 2.** Nonparametric (solid) and parametric (dashed) estimators of $N[0,t]$ and cumulative $N[0,t]$ (dotted).

It should be observed that the estimation of the m.v.f. improves significantly upon the parametric case. The finding is confirmed when computing the Bayes factor between the parametric and the nonparametric models, as shown in [10].

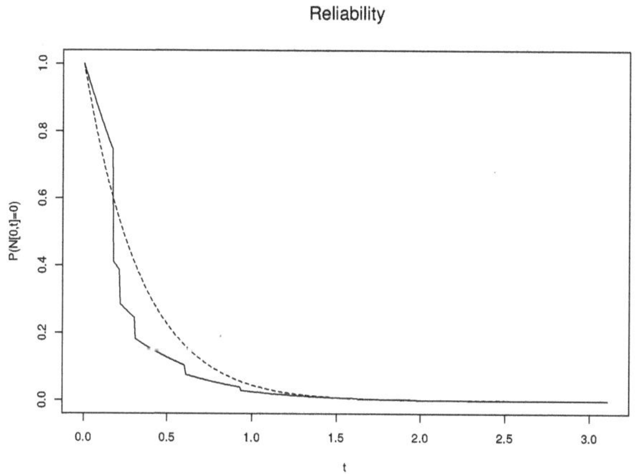


**Fig. 3.** *Nonparametric (solid) and parametric (dashed) estimators of the reliability (dotted).*

## 4.2 Model selection and sensitivity analysis

We now combine the notion of model selection introduced in Section 4.1 with the issue of sensitivity w.r.t. the choice of the prior distribution. The choice of prior distributions is a crucial (and, somehow, controversial) issue in Bayesian analysis. Usually, it is impossible to specify a distribution which represents exactly the prior beliefs over all the parameter space. Approximations are therefore necessary: the prior is often chosen in a tractable class (e.g. normal distributions) to match some known features (e.g. some quantiles). Bayesian robustness stems from this practical impossibility of specifying the "true" prior. Classes of distributions, compatible with the prior knowledge, are considered and some measure of robustness (e.g. range of quantities of interest, like the posterior mean) are computed to assess the influence of the choice of the prior in the class. More details are well beyond the scope of this paper; we refer to [29] for a thorough illustration of the robust Bayesian approach.

Failure data from 1978 to 1998 in steel pipelines are now split into three sets according to the different types of corrosion which caused them: natural, galvanic and by stray currents. See [15] for more details on the problem. In Table 3, we present the data; times are computed as elapsed time since the first failure (for each type) in the interval.

We consider the models corresponding to the following choice of parameters in the class of NHPP's defined in Section 3.5: $\alpha = 0$, $\beta > 0$, $\gamma \geq 0$ and (for identifiability) $\delta = 1$. For $\gamma = 0$, we get the HHPP with rate $\beta$, whereas $\gamma > 0$ implies the NHPP with $\lambda(t) = \beta t/(\gamma + t)$. We consider a gamma distribution $\mathcal{G}(a, b)$ on $\beta$ and a prior measure $\Pi$ on $\gamma$ in the quantile class $\Gamma = \{\Pi : \Pi \text{ has median } \theta_M = 1\}$ (see [30]). We compare the two models, using the Bayes factor.

**Table 3.** Failures (1978 - 1998)

| Corrosion | | | | | | | |
|---|---|---|---|---|---|---|---|
| Galvanic | 2.1233 | 3.5205 | 4.3945 | 8.9041 | | | |
| Natural | 2.8438 | 4.1534 | 7.2383 | 9.5232 | 9.8082 | 9.8191 | 9.8219 |
| | 12.4931 | 13.8904 | 14.4136 | 15.7890 | 16.1013 | | |
| Stray Currents | 0.0027 | 0.1041 | 0.3507 | 1.1753 | 3.9726 | 5.0329 | 5.2932 |
| | 5.7616 | 7.0219 | 11.7425 | 11.7616 | 15.3918 | 16.1644 | |

Given $n$ failures at time $t_i, i = 1, \ldots, n$ observed in an interval $[0, y]$, the likelihood functions are

$$L(\beta, 0) = \beta^n e^{-\beta y}$$

and

$$L(\beta, \gamma) = \beta^n \prod_{i=1}^{n} \frac{t_i}{\gamma + t_i} \left(1 + \frac{y}{\gamma}\right)^{\beta\gamma} e^{-\beta y},$$

for the HPP and the NHPP, respectively.

The Bayes factor of the HPP vs the NHPP is given by

$$\begin{aligned} BF &= \frac{\int L(\beta, 0) \Pi(d\beta)}{\int L(\beta, \gamma) \Pi(d\beta) \Pi(d\gamma)} \\ &= \left\{ \int \prod_{i=1}^{n} \frac{1}{1 + \gamma/t_i} \frac{1}{\left[1 - \gamma \log\left(1 + y/\gamma\right)/(b + y)\right]^{a+n}} \Pi(d\gamma) \right\}^{-1} \end{aligned} \quad (2)$$

after integrating with respect to the distribution on $\beta$.

We now compute upper and lower bounds on the Bayes factor (2); they are achieved (see [30]) when considering two-point distributions $(1/2)\delta_{\theta_1} + (1/2)\delta_{\theta_2}$, with $\theta_1 \leq \theta_m \leq \theta_2$. Bounds are considered in a sensitivity analysis to check if the preference for a model, as measured by the Bayes factor, is not affected by the choice of the prior in $\Gamma$.

As a further illustration of sensitivity analysis, consider the estimation of the parameters in the NHPP. Under a squared loss function, the (optimal) Bayesian estimators are given by the posterior means, which, in our case, correspond to

$$E\beta|d = \frac{(a+n) \int \prod_{i=1}^{n} \{1 + \gamma/t_i\}^{-1} \{b + y - \gamma \log\left(1 + y/\gamma\right)\}^{-a-n-1} \Pi(d\gamma)}{\int \prod_{i=1}^{n} \{1 + \gamma/t_i\}^{-1} \{b + y - \gamma \log\left(1 + y/\gamma\right)\}^{-a-n} \Pi(d\gamma)}$$

and

$$E\gamma|d = \frac{\int \gamma \prod_{i=1}^{n} \{1 + \gamma/t_i\}^{-1} \{b + y - \gamma \log\left(1 + y/\gamma\right)\}^{-a-n} \Pi(d\gamma)}{\int \prod_{i=1}^{n} \{1 + \gamma/t_i\}^{-1} \{b + y - \gamma \log\left(1 + y/\gamma\right)\}^{-a-n} \Pi(d\gamma)}.$$

Upper and lower bounds for the three sets are given in Table 4.

**Table 4.** Lower and Upper Bounds; $a = 1$, $b = 1$

| Corrosion | Bayes Factor | $E\beta\|d$ | $E\gamma\|d$ |
|---|---|---|---|
| Galvanic | $(0.6788, 0.8174)$ | $(0.5948, 1.0958)$ | $(0.5913, 8.0836)$ |
| Natural | $(0.2495, 0.5428)$ | $(0.8704, 2.4025)$ | $(0.7096, 22.6397)$ |
| Stray Currents | $(1.9997, 13968.0225)$ | $(0.8156, 0.9970)$ | $(0.000003, 0.1612)$ |

Observe that there is a slight preference for the NHPP for galvanic and natural corrosion data, whereas the preference for the HPP in the third case goes from slight to very convincing. In estimating posterior means, we observe that the ranges for the first two types of corrosion are larger than in the third case, the less sensitive to prior changes.

Upper and lower bounds could be computed for the measure of reliability illustrated in 3.3, as well.

# 5 Discussion

We have motivated and illustrated various uses of NHPP's in the analysis of the reliability of repairable systems. Others are available, like the exploratory data analysis and the intensity depending on a double scale (time and kilometers) used for failures in doors of underground trains, discussed in [26]. Further work is needed to explore properties of the various NHPP's proposed in literature and listed by Chris McCollin (personal communication). An important field, but still quite unexplored, is represented by the use of nonparametric models. There is a need for tools allowing practitioners to choose among these models, express their beliefs on parameters of interest and perform sensitivity analysis. That is actually one of the major challenges for the future.

*Acknowledgement.* The author wishes to acknowledge the contribution by his colleague Antonio Pievatolo, Enrico Cagno, Franco Caron, Mauro Mancini, Siva Sivaganesan, and the former students Raffaele Argiento, Davide Cavalry, Massier Fumagillin, Christina Mazzard, Lorentz Maine and Emmanuel Saccular for the joint work on the reliability of repairable systems.

# References

1. H.E. Ascher and H. Feingold. *Repairable Systems Reliability.* Marcel Dekker, New York, 1984.
2. S. Bar-Lev, I. Lavi, and B. Reiser. Bayesian inference for the Power Law process. *Ann. Inst. Statist. Math.*, 44:623–639, 1992.
3. J.O. Berger. *Statistical Decision Theory and Bayesian Analysis, 2nd ed.* Springer Verlag, New York, 1985.

4. E. Cagno, F. Caron, M. Mancini, and Ruggeri F. On the use of a robust methodology for the assessment of the probability of failure in an urban gas pipe network. In S. Lydersen, G.K. Hansen, and Sandtorv H.A., editors, *Safety and Reliability, vol. 2*, pages 913–919. Balkema, Rotterdam, 1998.
5. E. Cagno, F. Caron, M. Mancini, and Ruggeri F. *Robust Bayesian Analysis*, chapter Sensitivity of replacement priorities for gas pipeline maintenance, pages 335–350. Springer Verlag, New York, 2000.
6. E. Cagno, F. Caron, M. Mancini, and F. Ruggeri. Using AHP in determining the prior distributions on gas pipeline failure in a robust Bayesian approach. *Reliability Engineering and System Safety*, 67(3):275–284, 2000.
7. R. Calabria, M. Guida, and G. Pulcini. Bayesian estimation of prediction intervals for a Power Law process. *Comm. Statist. Theory Methods*, 19:3023–3035, 1990.
8. R. Calabria, M. Guida, and G. Pulcini. Reliability analysis of repairable systems from in-service failure count data. *Applied Stochastic Models and Data Analysis*, 10:141–151, 1994.
9. D. Cavallo. Nuovi modelli bayesiani nell'analisi dell'affidabilità dei sistemi riparabili. B.Sc. Thesis, University of Milano, Milano, 1999.
10. D. Cavallo and F. Ruggeri. Bayesian models for failures in a gas network. In E. Zio, M. Demichela, and N. Piccinini, editors, *Safety and Reliability - ESREL 2001*, pages 1963–1970, 2001.
11. D.R. Cox and P.A. Lewis. *Statistical Analysis of Series of Events.* Methuen, London, 1966.
12. L.H. Crow. *Reliability and Biometry*, chapter Reliability analysis for complex repairable systems, pages 379–410. SIAM, Philadelphia, 1974.
13. L.H. Crow. Confidence interval procedures for the Weibull process with applications to reliability growth. *Technometrics*, 24:67–72, 1982.
14. M. Engelhardt and L.J. Bain. Prediction intervals for the Weibull process. *Technometrics*, 240:167–169, 1978.
15. M. Fumagalli. Valutazione bayesiana della probabilità di corrosione di tubazioni interrate di acciaio di una rete per la distribuzione di gas in ambito metropolitano. B.Sc. Thesis, Politecnico di Milano, Milano, 1999.
16. P. Green. Reversible jump Markov Chain Monte Carlo computation and Bayesian model determination. *Biometrika*, 82:711–732, 1995.
17. M. Guida, R. Calabria, and G. Pulcini. Bayes inference for a nonhomogeneous Poisson process with power intensity law. *IEEE Transactions on Reliability*, 38:603–609, 1989.
18. J.F.C. Kingman. *Poisson Processes.* Oxford University Press, Oxford, 1993.
19. J. Kyparisis and N. Singpurwalla. Bayesian inference for the Weibull process with applications to assessing software reliability growth and predicting software failures. In L. Billard, editor, *Proceedings of the 16th Computer Science and Statistics Symposium on the Interface*, pages 57–64. North-Holland, Amsterdam, 1985.
20. A.Y. Lo. Bayesian nonparametric statistical inference for Poisson point processes. *Z. Wahrsch. Verw. Gebiete*, 59:55–66, 1982.
21. L. Masini. Un approccio bayesiano gerarchico nell'analisi dell'affidabilità dei sistemi riparabili. B.Sc. Thesis, University of Milano, Milano, 1998.
22. C. Mazzali and F. Ruggeri. Bayesian analysis of failure count data from repairable systems. Technical Report 98.19, CNR - IAMI, Milano, 1998.

23. J.D. Musa and K. Okumoto. A logarithmic Poisson execution time model for software reliability measurements. In *Proceedings of the Seventh International Conference on Software Engineering*, pages 230–238, 1984.
24. L. Olivi, R. Rotondi, and F. Ruggeri. Measures of dissimilarities for contrasting information sources in data fusion. In V. Colombari, editor, *Reliability Data Collection and Use in Risk and Availability Assessment*, pages 671–681. Springer Verlag, Berlin, 1989.
25. A. Pievatolo and F. Ruggeri. Bayesian reliability analysis of complex repairable systems. *Appl. Stoch. Models Bus. Ind.*, 20:253–264, 2004.
26. A. Pievatolo, F. Ruggeri, and R. Argiento. Bayesian analysis and prediction of failures in underground trains. *Quality Reliability Engineering International*, 19:327–336, 2003.
27. S.E. Rigdon and Basu A.P. The Power Law process: a model for the reliability of repairable systems. *Journal of Quality Technology*, 10:251–260, 1989.
28. S.E. Rigdon and Basu A.P. *Statistical Methods for the Reliability of Repairable Systems*. Wiley, New York, 2000.
29. D. Rios Insua and F. Ruggeri, editors. *Robust Bayesian Analysis*. Springer Verlag, New York, 2000.
30. F. Ruggeri. Posterior ranges of functions of parameters under priors with specified quantiles. *Comm. Statist. Theory Methods*, 19:127–144, 1990.
31. F. Ruggeri. Sensitivity issues in the Bayesian analysis of the reliability of repairable systems. In *Second International Conference on Mathematical Methods in Reliability*, pages 916–919. Université Victor Segalen, Bordeaux, 2000.
32. F. Ruggeri and S. Sivaganesan. On modeling change points in non-homogeneous Poisson processes. Tentatively accepted by Stat. Inference Stoch. Process., 2005.
33. T.L. Saaty. *The Analytic Hierarchy Process*. McGraw-Hill, New York, 1980.
34. E. Saccuman. Modelli che incorporano covariate nell'analisi bayesiana dell'affidabilità di sistemi riparabili. B.Sc. Thesis, University of Milano, Milano, 1998.

# New Schemes for Differential-Algebraic Stiff Systems.

E. Alshina[1], N. Kalitkin[1], and A. Koryagina[2]

[1] Institute for Mathematical Modelling, Russian Academy of Science, Miusskay pl. 4A, Moscow, Russia, 125047 alshina@gmx.co.uk
[2] Moscow Institute of Electronic Technique koralina@mail.ru

**Summary.** We present new efficient schemes of Rosenbrock's type for numerical solution of differential-algebraic stiff systems. For these schemes, we develop an algorithm for accuracy control.

**Key words:** Stiff systems, numerical methods, difference schemes, accuracy control.

## 1 Introduction

In practice it is often necessary to solve differential algebraic systems or singular differential problems with a small parameter $\varepsilon$, which become differential-algebraic when $\varepsilon$ tends to zero:

- electric current in circuits is governed by differential equations together with an algebraic Kirchhoff law
- hydrodynamic flows are described by Euler's equation supplemented by algebraic equations of state
- Van der Pol's equation for non-linear oscillations in a limit case becomes a differential algebraic system;
- the Navier-Stokes equations lose their higher derivative, when viscosity tends to 0 and so become differential algebraic system.

In practice differential algebraic systems often are formulated in an implicit form so that the algebraic equations cannot be explicitly separated from differential. In the most general case, such problems can be written in the following form:

$$M\frac{du}{dt} = f\left(u, t\right), \tag{1}$$

where the vector-function $u = \left\{u_j\right\}, 1 \le j \le J$ is unknown, $M$ is an unknown singular matrix is unknown . If the function $f$ in the right-hand side of this system does not depend on the time $t$, then the system is called autonomic.

All problems mentioned above are stiff, which leads to additional difficulties for numerical solution. Explicit schemes are unsuitable for stiff systems. The presence of fast varying and slowly damping components of a solution is typical for stiff systems. The time characteristics of different physical processes in a stiff system are varying in a wide range.

## 2 Accuracy control

For testing numerical methods and for practical application of numerical results, it is necessary to control accuracy. The unique known method of accuracy control is the Richardson's method from 1927. The detail review of practical aspects of applying the Richardson's method can be found in the monograph [4]. In practice, this method is applied not so often and researchers tend not to appreciate its potential possibilities. The reason is that Richardson offered his method only for equidistant grids. During the last 4 years our research group [2] offered so-called quasi-equidistant grids for calculation of initial-boundary value problems.

Use of quasi-equidistant grids in calculations allows estimation of the accuracy by Richardson method. Let us carry out two calculations on neighbour grids with number of nodes $N$ and $2N$, respectively . All nodes of the sparse quasi-equidistant grid are identical to the even nodes of the dense grid. One can compare two solutions in those points and estimate the accuracy by applying the following formula:

$$\Delta_{2n}^{(2N)} \approx \frac{u_{2n}^{(2N)} - u_n^{(N)}}{2^p - 1}, \tag{2}$$

where $u_{2n}^{(2N)}$ and $u_n^{(N)}$ denote numerical solutions in conterminous knots of neighbour grids, $p$ is the effective accuracy order of the numerical method, and $\Delta_{2n}^{(2N)}$ its numerical solution error.

The formula (2) is asymptotically exact when $N \to \infty$. We interpret it as a single-error correction and increase the efficient order of accuracy. The error of the corrected numerical solution is already $O\left(N^{-p-\sigma}\right)$, where $\sigma = 1$ for non-symmetrical difference schemes and $\sigma = 1$ - for symmetrical difference schemes. This increase of accuracy requires only few arithmetical operations and so it is very cheap. The corrected numerical solution can be interpreted as a calculation with scheme of accuracy order $p + \sigma$. For smooth enough solutions, the accuracy can be increased one more time. The main idea of our algorithms is to sequentially double the number of grid points until we reach of an acceptable error level.

## 3 Rosenbrock Schemes

According to our experience for numerical solution of pure differential stiff systems the family of Rosenbrock schemes is very effective. Transformation to new temporal level in Rosenbrock's schemes requires the solution of a linear algebraic system. Formally those schemes are implicit. However, the matrix of this linear system is very well conditioned and can be inverted by a direct method. So the solution of this linear system requires a finite number of beforehand known operations. The so-called method of $\varepsilon$-enclosure [1] allows us to modify those schemes for numerical solution of differential-algebraic stiff systems. The method of $\varepsilon$ -enclosure is correct for autonomous systems only. Any non-autonomous system can be easily transformed to the autonomous form by introducing additional unknown function $u_{J+1} \equiv t$. Differential equations for this new function are easy to obtain: $du_{J+1}/dt = 1$. The dimension increase of the system (1) leads to non-crucial increase of calculation volumes. However, the effective accuracy order of our method is higher for systems written in autonomic form.

The best results for differential-algebraic stiff systems in tests following schemes are shown below.

CROS: 1-stage $L2$-stable scheme: this complex parameter has accuracy $O\left(N^{-2}\right)$:

$$\hat{u} = u + \tau \mathrm{Re} k;\ (M - \alpha\tau f_u)\, k = f\,(u)\,.$$

Here $f_u \equiv \partial f / \partial u$ is the Jacobi matrix.

ROS2.3: 2-stage $A$-stable scheme with accuracy $O\left(N^{-3}\right)$:

$$\left[M - \tau a_{ll} f_u \left(u + \tau \sum_{m=1}^{l-1} c_{lm} k_m\right)\right] k_l = f\left(u + \tau \sum_{m=1}^{l-1} a_{lm} k_m\right), l = 1, 2.$$

$$a_{11} = a_{22} = \left(3 + \sqrt{3}\right)\Big/6, b_1 = b_2 = 1/2, c_{21} = \left(3 - \sqrt{3}\right)\Big/6, a_{21} = -1\Big/\sqrt{3}.$$

CROS was constructed for first time by Rosenbrock [5]. Optimal parameters for ROS2.3 for pure differential systems were obtained by Kalitkin and Panchenko [3].

For testing our new schemes we applied them to simulations of a transistor amplifier. The differential-algebraic system in this case in the autonomous form is as follows:

$$\begin{gathered}
\tfrac{U_e(u_6)}{R_0} - \tfrac{u_1}{R_0} + C_1\left(u_2' - u_1'\right) = 0, \\
\tfrac{U_b}{R_2} - u_2\left(\tfrac{1}{R_1} + \tfrac{1}{R_2}\right) + C_1\left(u_1' - u_2'\right) - 0.01 f\left(u_2 - u_3\right) = 0, \\
f\left(u_2 - u_3\right) - \tfrac{u_3}{R_3} - C_2 u_3' = 0, \\
\tfrac{U_b}{R_4} - \tfrac{u_4}{R_4} + C_3\left(u_5' - u_4'\right) - 0.99 f\left(u_2 - u_3\right) = 0, \\
-\tfrac{u_5}{R_5} + C_3\left(u_4' - u_5'\right) = 0, u_6' = 1.
\end{gathered}$$

The dimension of the singular matrix $M$ is 6 and the rank of $M$ equals 4.

Such simulation were carried out in the classical monograph [1] using the program RADAU5. To find out the effective accuracy order of our Rosenbrock schemes we carried out tests on embedded grids. The decrease of errors when doubling the number of grid nodes is shown in Fig. 1 in double logarithmic scale. Only for differential-algebraic system transformed to autonomous form all methods realize their theoretical order of accuracy.

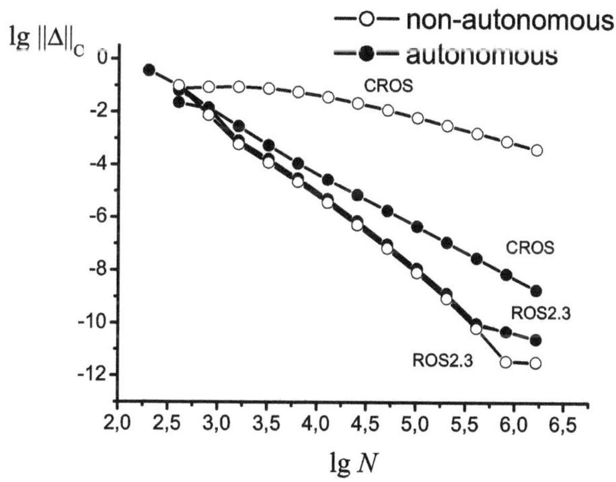


**Fig. 1.** Test on embedded grids confirm that only for differential-algebraic systems written in autonomous form Rosenbrock's schemes realize theoretical order of accuracy.

*Acknowledgement.* This work is supported by RFBR (projects 02-01-00066, 03-01-00439), Russian Science Support Foundation, presidential supporting program for scientific schools (project 1918.2003.1) and young Ph.D. (project 1907.2004.9).

## References

1. E. Hairer and G. Wanner. *Solving Ordinary Differential Equations II, Stiff and Differential-Algebraic Problems.* Springer Verlag, Berlin, 2nd edition, 1996.
2. N.N. Kalitkin, A.B. Alshin, E.A. Alshina, and B.V. Rogov. *Computation on Quasi-Equidistant Grids.* PhysMathLit, Moscow, 2004. In Russian.
3. N.N. Kalitkin and S.L. Panchenko. Optimal 2-stage schemes for stiff systems. *Doklady Mathematics*, 45(3):307–310, 1998. Translation from Doklady Academii Nauk, 1998, v. 361, N 3, p. 307-310.
4. G.I. Marchuk and V.V. Shaydurov. *Increasing the Accuracy of Difference Scheme's Solutions.* Nauka, Moscow, 1979. In Russian.

5. H.H. Rosenbrock. Some general implicit processes for the numerical solution of differential equations. *Comput. J.*, 5:329–330, 1963.

# Wavelet and Cepstrum Analyses of Leaks in Pipe Networks

S.B.M. Beck[1], J. Foong, and W.J. Staszewski[2]

[1] Department of Mechanical Engineering, University of Sheffield, Mappin Street, Sheffield, S1 3JD, UK. s.beck@shef.ac.uk
[2] w.j.staszewski@shef.ac.uk

**Summary.** It is well known that discontinuities in pipe networks give reflections to pressure waves that can be analysed to find the time delay between the original signal and the reflected one. A leak in a pipe will also give a reflection point, though possibly a more diffuse one. It is a reasonably straightforward task (using, say, a cross correlation) to measure the time delay of the first reflection, but more complicated methods are required to extract data about further reflections from, for example, the end of the pipe which has a leak in it.

Cepstrum techniques were used to find the common pipe lengths in the network. Latterly, this has been used in conjunction with wavelet analysis to filter the data. Finally, continuous wavelets are being used. These help to explain many of the results that have previously been produced. These were conducted on both real (experimental) and modelled networks.

**Key words:** Leak detection, water hammer, wavelets, cepstrum, waves in pipes, signal analysis.

## 1 Introduction

The aim of this work is to use to two modern forms of signal analysis (cepstrum and wavelet analysis) to find the position of reflection points in fluid pipeline networks. This is of great interest to the chemical, gas and water distribution industries to work out the state of the system and also for monitoring for leaks and blockages. The basic premise is to use water hammer to create waves, which propagate round the system. It is possible to capture the wave, and reflections of the wave, using a single sampling point. By using a cepstrum analysis, the characteristic lengths between junctions can be extracted. Comparing these lengths with those found in the original system will show the position of the new reflection point, and hence allow corrective action to be initiated.

As waves bounce around a pipe network system, they encounter features such as local changes in section, resistances, and junctions where three or more pipes join. At each of these features, three effects occur to the incident wave. These are: reflection where some of the wave is reflected back down the pipe, absorption where the amplitude of the transmitted wave is decreased due to the feature, and transmission, where the wave continues down the other pipes in the junction.

A fuller description of waves in pipes and the modelling techniques employed by the authors can be found in [1].

The work described in the present paper represents a continuation of the investigations and modelling due to Beck and various colleagues [3] and [2] who described a technique whereby the features of a pipe network were detected by use of an artificially generated pressure wave and a single pressure transducer. Two other recent pieces of work on the same general subject can be found in [4] and [5]

## 2 Theory

A wavelet transform decomposes a signal into a variety of different waves, centred around different frequencies. A full description of these can be found in [6]. The continuous wavelet transform form is described as:

$$W_{\Psi}\left[x\left(t\right)\right] = W_{\Psi}\left(a,b\right) = \frac{1}{\sqrt{a}}\int_{-\infty}^{+\infty} x_m\left(t\right)\Psi^{\star}\left(\frac{t-b}{a}\right), \tag{1}$$

where $b$ is the translation indicating locality, $a$ is the dilation or scale parameter and $W_{\Psi}$ is the analyzing or mother wavelet, with $\Psi^{\star}$ being its complex conjugate.

The sum of all the wavelet levels is equal to the original signal, so:

$$x\left(t\right) = \sum_{m} x_m\left(t\right). \tag{2}$$

Cepstrum analysis is the name given to a range of techniques all involving functions which can be considered as a s“”spectrum of a logarithmic spectrum”. It was proposed as a better alternative to the autocorrelation function for the detection of echoes in seismic signals. The definition of the complex cepstrum is [7]:

$$C_A = F^{-1}\left(\log A\{f\}\right), \tag{3}$$

where $A(f)$ is the complex spectrum of $a(t)$ and can be represented by

$$A\{f\} = F\{a(t)\} = A_R + JA_I\{f\}. \tag{4}$$

The cepstrum analysis has the ability to detect periodic structures in the logarithmic spectrum, for example families of harmonics and/or sidebands with uniform spacing. Another of the cepstrum effects is that it is capable of identifying the presence of echoes. The power of this technique even extends to reflections that are not perfect copies of the original signal. It is thus ideal for extracting information on the time delay between the creation of a wave and the receiving of an echo in pipe network systems.

# 3 Experiment

A simple T-shaped network was set out in the laboratory as shown in Fig. 1. The solenoid valve was opened and shut with a signal generator, and the

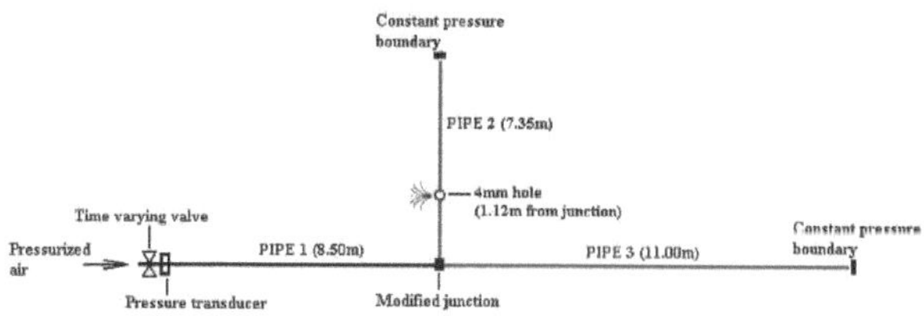


**Fig. 1.** Diagram of network

signals were acquired using a pressure transducer attached to a signal analyser.

# 4 Comparison between theory and experiment

First we shall look at the continuous wavelet transform of the theoretical (ideal) pressure history shown in Fig. 2.

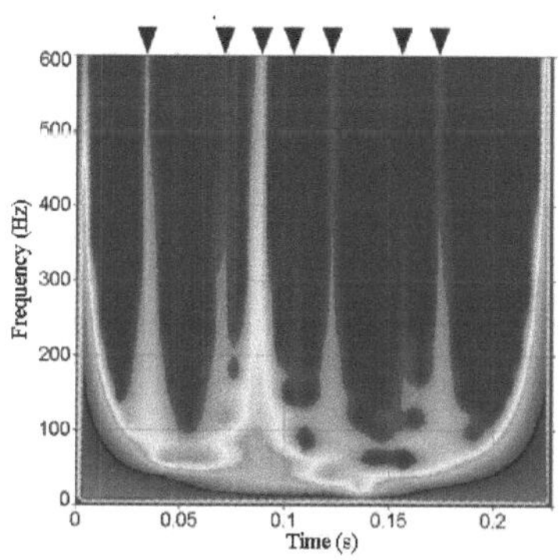


**Fig. 2.** Continuous wavelet of modelled network.

The light vertical lines on this show the discontinuities in the original signal. The expected time delays are identified with triangles. It will be seen that the discontinuities and the time delays tie up very well.

However, when the same analysis is applied to the experimental signal (Fig. 3), it will be seen that the expected results do not agree with the lengths in the network. This is due to the the wave spreading out due to dispersion.

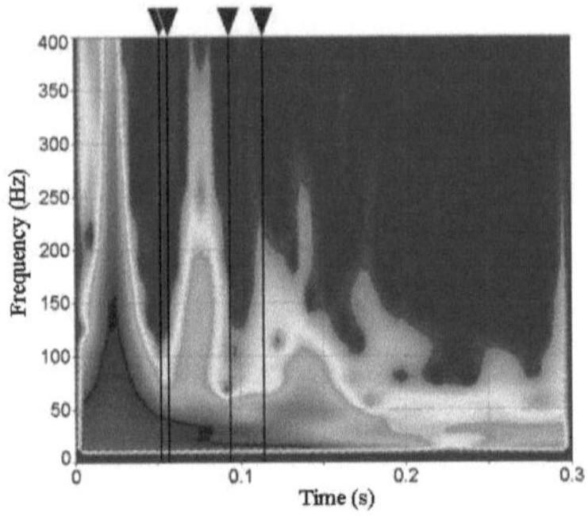


**Fig. 3.** Continuous wavelet of experimental network.

The reflected wave is different to the created one which means that the characteristic frequency of the wavelet is constantly decreasing, making it difficult to track.

When the cepstrum analysis was applied to the experimental signal, the reflection points were extracted as shown in Fig. 4. When the main reflection

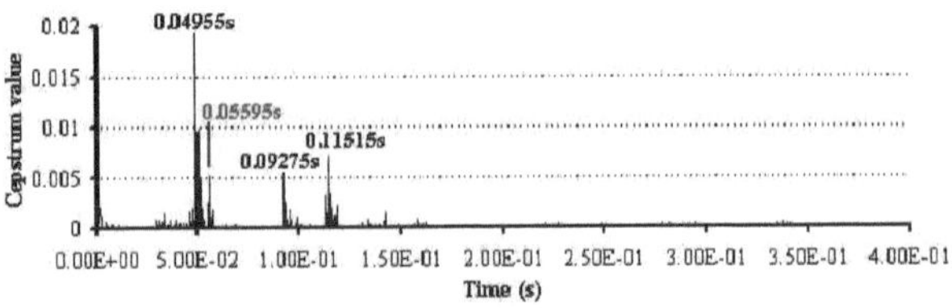


**Fig. 4.** Cepstrum of experimental network.

points are identified and compared with the measured lengths (table 1), it will be seen that the predicted and analysed lengths agree very well.

**Table 1.** Experimental and predicted pipe lengths from cepstrum analysis.

| Analysed time (s) | Analysed length (m) | Twice measured distance (m) | Accuracy % |
|---|---|---|---|
| 0.04995 | 16.996 | 17.00 | 99.98 |
| 0.05595 | 19.023 | 19.24 | 98.87 |
| 0.09275 | 31.535 | 31.40 | 99.57 |
| 0.11515 | 39.151 | 39.00 | 99.61 |

## 5 Conclusions

The results from the show that wavelet analysis, without extensive modification, is unsuitable for finding features in pipeline networks.

It will be seen that the cepstrum analysis method is capable of extracting useful information from real experimental networks with an extremely high degree of accuracy.

Further work is ongoing to improve the method and test it on more complicated and longer circuits using a variety of fluids.

*Acknowledgement.* The Authors would like to thank Mr Simon Wiles for his help with the experimental work and Dr A Tijsseling for his encouragement.

## References

1. S.B.M. Beck, H. Haider, and Boucher. R.F. Transmission line modelling of simulated drill strings undergoing water-hammer. *J Mechanical Engineering Science*, 102(Part C):419–427, 1995.
2. S.B.M. Beck and W.J. Staszewsky. Wavelet analysis of features in fluid pipeline networks. In *Proceedings of the 9th International Conference Pressure Surges*, pages 199–209, Chester, March 2004.
3. S.B.M. Beck, N.J. Williamson, N.D. Sims, and R. Stanway. Pipeline system identification through cross-correlation analysis. *J Process Mechanical Engineering*, 216(Part E):113–142, 2001.
4. D. Covas, H. Ramos, B. Brunone, and A. Young. Leak detection in water trunk mains using transient pressure signals: field tests in scottish water. In *Proceedings of the 9th International Conference Pressure Surges*, pages 185–198, Chester, March 2004.
5. M. Ferrante and B. Brunone. Pipe system diagnosis and leak detection by unsteady-state tests. 2. wavelet analysis. *Journal of advances in water resources*, 26:107–116, 2003.
6. S. Mallat. *Wavelet Tour of Signal Processing*. San Diego: Academic Press, 1998.
7. R.B. Randall. *Frequency analysis*. Bruel and Kjær, Denmark, 1987.

# Robust Design Using Computer Experiments

R.A. Bates[1], R.S. Kenett[2], D.M. Steinberg[3], and H.P. Wynn[1]

[1] London School of Economics, London,UK
[2] KPA Ltd., Raanana, ISRAEL
[3] Tel Aviv University, Tel Aviv, ISRAEL and KPA Ltd.

**Summary.** In this paper we compare several different strategies for robust design when the experiment is carried out via a computer simulator.

**Key words:** robust design, computer experiments.

## 1 Introduction

Competitive products must meet strict standards for quality and reliability. Robust design experiments are an important quality engineering tool for developing low-cost, yet high quality products and processes. The name "robust design" derives from the idea of making products insensitive, or "robust", to the effects of natural variations, thereby reducing variation. Robust design was pioneered by [11] in Japan and has been embraced by many engineers around the world in the last 20 years. See [9] for detail on robust design and relevant chapters in [5] for background on design of experiments, robust design and reliability analysis. In this paper we compare several different strategies for robust design when the experiment is carried out via a computer simulator. In these experiments, many different levels can be assigned to each factor and (ignoring numerical issues) running the simulator at the same set of inputs gives the same output. See [1, 6, 7] for background and insight on the statistical aspects of computer experiments. The next section provides details of the simulation model that will be used throughout the paper. The following sections include a description of the methods and details of the results achieved with each method. We then compare the methods for a particular robust design task.

## 2 The Piston Simulator

We analyze a simulator of a piston developed by [5]. The response variable is cycle time of a complete revolution of the piston's shaft. The piston's performance can be regulated by changing seven control factors: A) Piston weight, B) Piston surface area, C) Initial gas volume, D) Spring coefficient, E) Atmospheric pressure, F) Ambient temperature and G) Filling gas temperature. Each factor is constrained to lie within a stated range. For given factor settings, the cycle time is computed deterministically. However, the ability to set each factor to a nominal level is limited and normally distributed variations of the seven control factors about their nominal values are used to induce randomness in cycle times (see [5]). This mechanism for randomness is quite different from that used in simulations of stochastic phenomena such as queueing networks.

## 3 Robustness Strategies

In this section we briefly describe several alternative methods that have been proposed for achieving robust design. We will denote the $k$ inputs to the simulator by $X_1, \ldots, X_k$ and will focus on a single output $Y$. Some of the inputs will be controllable factors (*e.g.*, the thickness of an auto bumper or the material from which it is made) whereas other factors may be uncontrollable (*e.g.*, the angle of impact with a crash barrier). The goal of the robustness strategies is to find nominal settings of the controllable input factors for which a desired output distribution is obtained. The output goal is often to achieve a given mean value with minimal variance. Taguchi's strategy for *robust design experimentation* is to include both design (controllable) factors and noise factors, which represent either uncontrollable factors or tolerances of controllable factors about their nominal levels. Separate orthogonal arrays for the design factors and noise factors are combined to form a crossed-array design. The data are analyzed by computing signal-to-noise (SN) ratios that summarize the results at each control factor setting. Data from a crossed-array experiment can also be analyzed by the *response model* approach, which uses main effects for the noise factors and interactions between design factors and noise factors to study variation ([8, 10]). The *dual response surface approach* [12] provides a direct way to assess jointly how the mean value and dispersion of $Y$ depend on the design factors. An experimental plan is prepared for the design factors only. Replicate measurements are generated from the simulator at each design point and are summarized by their average and a measure of spread. [12] recommended a crossed-array design to obtain the replicates, but noted that they could also be sampled at random from distributions that are specified for the noise factors. Finally statistical models are built to relate the average and the measure of spread to the design factors. The *stochastic emulator approach* is based on the idea of replacing the original simulator, which may be expensive to evaluate, by a fast empirical surrogate, known as an emulator. The first step in this approach is to model the dependence of the output $Y$ on the full set of input values, whether or not they are controllable, to produce an emulator $\hat{Y}(X_1, \ldots, X_k)$. Subsequent investigation uses the emulator to generate output rather than the simulator. The output distribution for any set of factor values can be approximated by sampling values for the noise factors and evaluating the emulator rather than the actual simulator. The final step involves choosing a feature of the output distribution, called the *stochastic response*, that represents the required robustness criteria. The rationale for the stochastic emulator approach is that complex simulators may take considerable time and computing resources to generate even a single output value. Thus the project sample size is determined by computing resources. The Stochastic Emulator approach has the potential advantage of devoting all of the computing resources to studying how the simulator output depends on the input factors. The first step is to run an experiment using all the input factors. Research to date has recommended "space filling" designs like Latin Hypercube Sampling (LHS) designs or lattice designs. See [7, 1, 4] for more detail on designs. The emulator can be of any model type such as kriging estimators [7], radial basis functions [3] or polynomial functions [2].

## 4 Comparison Of Robustness Strategies on the Piston

We applied the various robust design methods outlined above to the piston simulator. We purposely used designs with the same number of function calls for each method, in line with our comment that computing resources should dictate experiment size. The objective in all cases was to achieve a mean cycle time of 0.20 seconds while minimizing SD. Table 1 presents recommended factor settings derived from each of the analysis methods along with the results of 1000 actual simulator runs at those conditions. We begin with a brief explanation of how the recommended settings were obtained. The Taguchi analysis indicated that setting the surface area (B) to a high value is beneficial for both the SD and mean responses and setting the ambient temperature (F) to its highest value reduces variation and has almost no effect on the mean. Increasing the initial gas volume (C) to its highest value also reduces variation but moves the mean well above 0.2. A small amount of trial and error shows that setting C to 0.0044 achieves an estimated mean of about 0.2. The recommendations from the response model analysis for reducing variation were to set factors B, C and D, which have the strongest effects on cycle time, to high values. However, to achieve the desired average cycle time of 0.2 seconds, it was necessary to adopt a lower nominal value for C. The response model analysis did not find any important effect for F, which was set arbitrarily to its mid-range. The dual response surface designs, with replicates sampled at random from the noise distributions, were less successful than the cross-product design of the first two analyses in identifying factor settings that provide a mean value of 0.20 with a small SD. The analysis of the $2^{7-3}$ design favoured high settings for B, which reduce both the mean value and the SD. As no other effects were found on the SD, the remaining factors could be set to adjust the mean to the target value. Table 1 shows three possible solutions, each based on using factors C and D, to adjust the mean. All of the proposed solutions have means that are more than one SD above the target value. In addition, the SD's are about 25% larger than those achieved in the other analyses. For the 32-run LHS design, we used kriging models for the mean and SD as inputs to an optimizer, with the goal of minimizing the SD subject to maintaining the mean at 0.20. The resulting solution is shown in Table 1. Although these factor settings had excellent predicted performance (based on the kriging models), their actual performance is not good, with the mean more than 2.5 SDs off target. The stochastic emulator used a 64-point LHS design to build the emulator from simulator results. Monte Carlo analysis of the emulator was then used to estimate the response distribution at input factor values, using a 128-point LHS design to cover the factor space with 200 point samples from the noise distributions at each LHS point. New stochastic emulators were then built to predict the mean and SD. A constrained optimization was then performed, minimizing the stochastic emulator of the SD, while requiring that the emulator of the mean satisfy the constraint of equality to 0.2. The recommended factor settings and the results from actually running the simulator are shown in Table 1.

**Table 1.** Results for all methods for the problem of minimizing cycle time standard deviation about a target mean value of 0.2 seconds.

| | A | B | C | D | E | F | G | Mean | Std. Dev. |
|---|---|---|---|---|---|---|---|---|---|
| Taguchi | 30.3 | 0.017 | 0.00440 | 4850 | 100,000 | 295.6 | 350.0 | 0.204 | 0.0106 |
| Response Model | 30.3 | 0.017 | 0.00440 | 4850 | 100,000 | 293.0 | 350.0 | 0.204 | 0.0110 |
| Dual Response $2^{7-3}$ | 45.0 | 0.017 | 0.00275 | 3426 | 100,000 | 293.0 | 350.0 | 0.218 | 0.0137 |
| | 45.0 | 0.017 | 0.00488 | 4850 | 100,000 | 293.0 | 350.0 | 0.263 | 0.0137 |
| | 45.0 | 0.017 | 0.00382 | 4138 | 100,000 | 293.0 | 350.0 | 0.242 | 0.0132 |
| Dual Response LHS | 34.5 | 0.014 | 0.00346 | 4245 | 109,700 | 294.6 | 355.9 | 0.230 | 0.0154 |
| Stochastic Emulator I | 30.3 | 0.017 | 0.00359 | 3924 | 101,610 | 294.5 | 340.4 | 0.199 | 0.0107 |

For the piston example, the Taguchi method, the response model analysis and the stochastic emulator all provided better solutions than the dual response method. In particular, they did a much better job of keeping the mean cycle time on target. The dual response surface methods estimate the mean value at a given design point by taking a small sample of results from the noise distributions for the input factors. Our analysis suggests that small random samples do not provide sufficiently precise estimates of the mean cycle time. The stochastic emulator has the advantage of modelling the simulator directly from function evaluations and produced an excellent robust design solution for our trial problem. The results highlight the need to consider how best to allocate resources when conducting computer experiments. Our results with the stochastic emulator approach indicate that this may be an efficient method for reducing the number of simulations in a robust design study.

*Acknowledgement.* This research was supported by the European Union grant TITOSIM (Time to Market via Statistical Information Management, project number: GRD1-2000-25724).

## References

1. R.A. Bates, R.J. Buck, E. Riccomagno, and H.P. Wynn. Experimental design and observation for large systems. *Journal of the Royal Statistical Society*, 58:77–94, 1996.
2. R.A. Bates, B. Giglio, and H.P. Wynn. A global selection procedure for polynomial interpolators. *Technometrics*, 45:246–255, 2003.
3. R.A. Bates and H.P. Wynn. Adaptive radial basis function emulators for robust design. In I.C. Parmee, editor, *Evolutionary Design and Manufacture*, pages 343–350. Springer–Verlag, 2000.
4. V.C.P. Chen, K.-L. Tsui, R.R. Barton, and J.K. Allen. A review of design and modeling in computer experiments. In R. Khattree and C.R. Rao, editors, *Handbook of Statistics*, volume 22, pages 231–261. Elsevier Science, 2003.
5. R.S. Kenett and S. Zacks. *Modern Industrial Statistics: Design and Control of Quality and Reliability*. Duxbury Press, San Francisco, 1998.
6. D. Romano and G. Vicario. Reliable estimation in computer experiments on finite element codes. *Quality Engineering*, 14(2), 2002.
7. J. Sacks, W.J. Welch, T.J. Mitchell, and H.P. Wynn. Design and analysis of computer experiments. *Statistical Science*, 4:409–435, 1989.
8. A.C. Shoemaker, K.-L. Tsui, and C.F.J. Wu. Economical experimentation methods for robust parameter design. *Technometrics*, 33:415–428, 1991.
9. D.M. Steinberg. Robust design: Experiments for improving quality. In S. Ghosh and C.R. Rao, editors, *Handbook of Statistics*, volume 13, pages 199–240. Elsevier Science, 1996.
10. D.M. Steinberg and D. Bursztyn. Dispersion effects in robust design experiments with noise factors. *J. Qual. Technology*, 26:12–20, 1994.
11. G. Taguchi. *Introduction to Quality Engineering*. Kraus International Publications, White Plains, NY, 1986.
12. G.G. Vining and R.H. Myers. Combining Taguchi and response-surface philosophies – a dual response approach. *J. Qual. Technology*, 22(1):38–45, 1990.

# Non-Classical Shocks for Buckley-Leverett: Degenerate Pseudo-Parabolic Regularisation

C. M. Cuesta[1], C. J. van Duijn[2], and I. S. Pop[2]

[1] Vienna University of Technology carlota@aurora.anum.tuwien.ac.at
[2] Technische Universiteit Eindhoven c.j.v.duijn@tue.nl, ipop@win.tue.nl

**Summary.** We consider oil-water flow in porous media, with a dynamic capillary pressure relation. This leads to a pseudo-parabolic degenerate regularisation of the Buckley-Leverett (BL) equation. It is known that linear pseudo-parabolic regularisations of BL lead to shock solutions that do not satisfy the Oleinik condition. In this note we analyse the existence of travelling wave solutions that violate the Oleinik condition, taking special care of the degeneracy of the problem.

**Key words:** Buckley-Leverett equation, dynamic capillary pressure, non-classical shocks, degenerate diffusion

## 1 Introduction

In this paper we study travelling wave solutions of equation

$$0 = \frac{\partial u}{\partial t} + \frac{\partial}{\partial x}\left\{ f(u) + N_c H(u) \frac{\partial}{\partial x}\left( p_c(u) - \frac{L_c}{N_c}\frac{\partial u}{\partial t}\right)\right\}. \qquad (1)$$

Equation (1) is written in dimensionless form, and arises in two-phase flow in porous media as a model of oil recovery by water-drive in a one-dimensional horizontal flow, with a dynamic capillary pressure relation. In this context, the unknown $u$ stands for water saturation, and is expected to take values in $[0, 1]$. $f$ is the water fractional flow, and $H$ is the capillary induced diffusion, they are given by

$$f(u) = \frac{\lambda_w(u)}{\lambda_w(u) + M\lambda_o(u)}, \quad \text{and} \quad H(u) = \lambda_o(u) f(u). \qquad (2)$$

$M$ is the viscosity ratio, and $\lambda_o$ and $\lambda_w$ are the dimensionless relative permeabilities, which we take

$$\lambda_w(u) = u^{p+1}, \quad \lambda_o(u) = (1-u)^{q+1}, \quad \text{with } p,\ q > 0. \qquad (3)$$

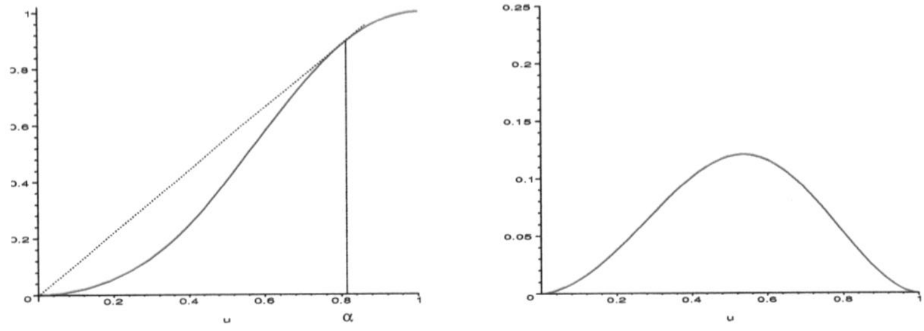

(a) The flux $f(u)$ (b) The diffusivity $H(u)$

**Fig. 1.** The non-linear functions $f$ and $H$

$N_c$ is the capillary number and accounts for capillary forces.

Equation (1) results from coupling conservation mass equations, and generalised Darcy's laws for the water and the oil phases, and the pressure relation

$$P_c = P_o - P_w, \tag{4}$$

where $P_w$ and $P_o$ are the water pressure and the oil pressure, respectively, and $P_c$ is the capillary pressure function. $P_c$ is typically determined experimentally as a function of the water saturation $u$. In equation (1), however, we have considered the dynamical capillary pressure approach suggested by Hassanizadeh and Gray in [2], where the *empirical* capillary pressure is extended by a relaxation term, this reads

$$P_c = p_c(u) - L_c \frac{\partial u}{\partial t}. \tag{5}$$

We observe that the damping term ($L_c > 0$) gives the mixed derivatives term in equation (1). $p_c$ denotes the *static* capillary pressure, this is a positive function that vanishes at $u = 1$ and that becomes unbounded at $u = 0$. In this note, however we take, in dimensionless form, $p_c(u) = 1 - u$.

Equation (1) for $N_c = 0$ is often considered as the limit of vanishing capillary forces. It is called *Buckley-Leverett* equation (BL), and is an example of a scalar conservation law with a convex-concave flux, see Fig. 1(a). Equation (1) for $L_c = 0$, is a non-linear degenerate diffusion equation, with double degeneracy; the *diffusivity* $H$ vanishes at $u = 0$ and at $u = 1$, see Fig. 1(b). This property implies the existence of fronts, i.e. lines on the $(x, t)$-plane separating the regions where $u = 0$ (oil region), where $1 < u < 0$ (both fluid present) and where $u = 1$ (water region).

Coming back to BL, it is well-known that solutions to the Riemann problem are constructed according to the Oleinik condition (cf. [5]); for admissible shock solutions this reads

$$\frac{f(u) - f(u_l)}{u - u_l} \geq s \tag{6}$$

where $s$ is the shock speed, given by the Rankine-Hugoniot condition

$$s = \frac{f(u_r) - f(u_l)}{u_r - u_l}, \tag{7}$$

and $u_l$ and $u_r$ denote the left and right values aside the shock. When a conservation law with a non convex flux, such as the BL equation, is regularised by a diffusion term, with diffusion coefficient $\varepsilon$, say, the condition (7) ensures the existence of travelling wave solutions connecting the level $u_l$ as $\eta := (x - st)/\varepsilon \to -\infty$ to the level $u_r$ as $\eta \to \infty$. Other regularisations, however, are possible. For instance, diffusive-dispersive regularisations, where the dispersive term consists of a third-order in space term. These give rise to shock solutions violating the condition (6), see for instance [3]. Also for the following linear pseudo-parabolic regularisation of BL

$$\frac{\partial u}{\partial t} + \frac{\partial f(u)}{\partial x} = \varepsilon \frac{\partial^2 u}{\partial x^2} + \delta \frac{\partial^3 u}{\partial x^2 \partial t}, \tag{8}$$

travelling wave solutions giving rise to non-classical shocks exist in the regime $\delta = O(\varepsilon^2)$, see van Duijn *et al.* [6].

In this work we analyse existence of travelling wave solutions of (1) violating (6), in the regime $L_c = O(N_c^2)$, where now the regularisation is non-linear and degenerate; special care of fronts is therefore taken.

## 2 Travelling waves

We introduce the travelling coordinate $\eta = (x - st)/N_c$, and the parameter $\tau = L_c/N_c^2$. The travelling wave equation reads

$$-su' + \left\{ f(u) + H(u)\left((1-u) + s\tau u'\right)' \right\}' = 0, \tag{9}$$

we look for solutions such that

$$u(+\infty) = u_r = 0, \ u(-\infty) = u_l, \quad \text{with } 0 < u_l \leq 1. \tag{10}$$

We are interested in the cases where $u_l > \alpha$, with $\alpha$ such that $f'(\alpha) = f(\alpha)/\alpha$, see Fig. 1(a), i.e. a travelling wave connecting 0 to such $u_l$'s violates condition (6).

We integrate (9), use the boundary conditions (10), divide by $H(u)$, and introduce the new unknown $w(u) = -u'(\eta(u))$. After these manipulations equation (9) becomes

$$s\tau w \frac{dw}{du} + w = g(u) \quad \text{in } (0, \ u_l). \tag{11}$$

where $s$ is given by (7), and $g(u) := \frac{F(u)}{H(u)}$ with $F(u) := su - f(u)$.

In terms of $w$, we seek solutions of (11) such that

$$w(u) > 0 \quad \text{for } u \in (0, u_l), \tag{12}$$

$$w \to 0 \quad \text{as } u \to 0^+ \text{ and } u \to u_l^- . \tag{13}$$

We can see equation (11) as a dynamical system in the $(u, w)$-plane. The zeros of $g$ give the critical points in the $(u, w)$-plane. For $u_l > \alpha$, $F$ has three zeros, $u = 0$, $u = u_m$ and $u = u_l$, with $0 < u_m < u_l$. $u_l$ is a zero of $g$ only if $u_l < 1$, in this case the corresponding critical point $(0, u_l)$ is a saddle point whenever $u_l \neq \alpha$. $u_m$ is always a zero of $g$, and $(0, u_m)$ is a sink, except if $u_l = u_m = \alpha$, in that case it is a saddle-node. Due to the degeneracy of the equation, the

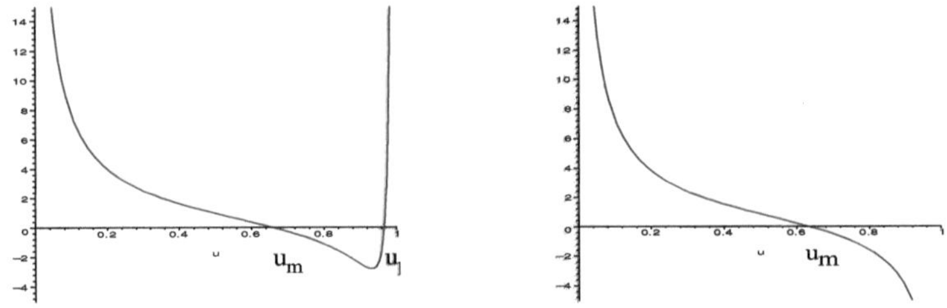


(a) $g(u)$ for $u_l < 1$.

(b) $g(u)$ for $u_l = 1$.

**Fig. 2.** The isocline $g$ in the $(u, w)$-plane.

points $(0, 0)$ and $(0, 1)$ (when $u_l = 1$) are not critical points of (11) ($g$ blows up at $u = 0$ and $u = u_l = 1$, see Fig. 2). Thefore the condition (10) will be fulfilled at finite values of $\eta$, i.e. solutions have fronts.

The power laws (3) play a crucial role in the construction of travelling waves; we namely have the following non-existence result.

**Theorem 1 (non-existence).** *If $p \geq 1$ or $q \geq 1$ solutions of (11) satisfying (12) and (13) cannot be constructed. Moreover, if $p$, $q < 1$,*

$$0 < \int_0^{u_l} g(y)\mathrm{d}y < \infty \tag{14}$$

*is a necessary condition for existence.*

The proof of the first statement is based on the integrability of the function $g$, by noticing that $g(u) \sim u^p$ as $u \to 0^+$, and $g(u) \sim -(1-u)^q$ as $u \to 1^-$ (for $u_l = 1$). The second statement holds by integration of equation (11) using the conditions (12) and (13), since this implies $\int_0^{u_l} g(y)\mathrm{d}y = \int_0^{u_l} w(y)\mathrm{d}y$. Finally, condition (12) implies the statement.

We are ready to state our existence result

**Theorem 2 (existence).** *If $p$, $q < 1$ and (14) holds, then*

*(i) For $u_l = \alpha$ there exists a $\tau_*$ such that for all $\tau \leq \tau_*$ there exists a solution $w_\tau$ of (11) satisfying (12) and (13).*

*(ii) For $u_l = 1$ there exists a $\tau^*$ for which (11) has a solution satisfying (12) and (13).*

*(iii) For each $u_l \in (\alpha, 1)$ there exist a unique $\tau \in (\tau_*, \tau^*)$ such that (11) has a solution satisfying (12) and (13).*

To prove this theorem, it is first necessary to prove local existence of solutions of (11) that satisfy $w \to 0$ as $u \to 0$. This is achieved by a local transformation of equation (11) that brings the point $(0, 0)$ of the $(u, w)$-plane to a saddle-point in the transformed equation; its unstable manifold corresponds to the desired orbit $w$. The rest of the proof uses continuity on the parameter $\tau$, and the fact that orbits are ordered with respect to $\tau$ in different regions of the phase-plane. The nature of the point $(0, u_l)$ determines whether the connection is possible for a unique value of $\tau$ ($u_l \neq \alpha$) or not ($u_l = u_m = \alpha$, saddle-node case).

*Remark 1.* The conditions (13) are restrictive. In terms of $u$, these give front solutions with $u' = 0$ at the front. However, front solutions with $u' \neq 0$ are also possible. These *generic* fronts were encountered by Hulshof and King in [4], and by Cuesta *et al.* in [1], in the unsaturated flow case. In the latter a rigorous lifting regularisation argument serves to select the smoothest fronts ($u' = 0$) as the relevant ones. Based on these results, we have restricted our attention to the smoothest fronts only.

## References


1. C. Cuesta, C. J. van Duijn, and J. Hulshof. Infiltration in porous media with dynamic capillary pressure: travelling waves. *Euro. J. Appl. Math.*, 11:381–397, 2000.
2. S. M. Hassanizadeh and W. G. Gray. Thermodynamic basis of capillary pressure in porous media. *Water Resour. Res.*, 29:3389–3405, 1993.
3. Brian T. Hayes and Philippe G. LeFloch. Non-classical shocks and kinetic relations: scalar conservation laws. *Arch. Rational Mech. Anal.*, 139:1–56, 1997.
4. Josephus Hulshof and John R. King. Analysis of a Darcy flow model with a dynamic pressure saturation relation. *SIAM J. Appl. Math.*, 59:318–346, 1999.
5. O. A. Oleĭnik. Uniqueness and stability of the generalized solution of the Cauchy problem for a quasi-linear equation. *Uspehi Mat. Nauk*, 14:165–170, 1959.
6. C. J. van Duijn, L. A. Peletier, and I. S. Pop. A new class of entropy solutions of the Buckley-Leverett equation. In preparation.

# A Multi-scale Approach to Functional Signature Analysis for Product End-of-Life Management

T. Figarella[1] and A. Di Bucchianico[2]

[1] EURANDOM, P.O. Box 513, 5600 MB Eindhoven, The Netherlands
figarella@eurandom.tue.nl
[2] EURANDOM and Technische Universiteit Eindhoven, Department of Mathematics, P.O. Box 513, 5600 MB Eindhoven, The Netherlands
a.d.bucchianico@tue.nl

**Summary.** Electronic products tend to be economically outdated before their technical end-of-life has been reached. The ability to analyze and predict the (remaining) technical life of a product would make it possible either to re-use sub-assemblies in the manufacture process of new products, or to design products for which the technical and economical life match. This requires models to predict and monitor performance degradation profiles. In this paper we report on designed experiments to obtain such models. We show how wavelet analysis can be used to extract features from electrical signals. These features are analyzed using the Analysis of Variance in order to establish relations between these features and performance degradation.

**Key words:** Signature analysis, wavelet analysis, Analysis of Variance.

## 1 Introduction

Nowadays, the economical life of electronic products tends to be shorter than their technical end-of-life as a result of the fast evolution of electronics and related software. As a consequence a high amount of waste ends up in landfill sites. New EU legislation addresses the implication of producers for the recycling and disposal of their products. A reliable monitoring system would allow prediction of components performance and re-use. Therefore, Flextronics Netherlands, EURANDOM and the Eindhoven University of Technology decided to joint a project to develop methods to predict performance rather than failures of electrical appliances. These methods should be used to monitor products and assess the re-use of components. Signature analysis is the general name for condition monitoring using electrical signals (cf. [1]).

In this paper we describe an experiments performed on the submodule Main Tray to obtain degradation profiles. We cannot directly use Analysis of Variance (ANOVA) to analyse the experiments, because the response variable is a time series of approximately one million observations. We show how wavelet analysis can be used to extract features from these signals. ANOVA is then applied to suitable maxima of wavelet coefficients over different scales, in order to determine which factors influence dominant features of the signal.

This paper is organized as follows. In Section 2 we describe the experimental setup. Section 3 contains the details of our multiscale approaches. We finish the report in Section 4 with some conclusions and recommendations for future work.

## 2 Experimental Setup

In this paper we perform an analysis of one of the parts of the finisher module: the *stapler motor*. The stapler motor stitches three staples in each piece of paper (for full details we refer to [3]).

### 2.1 Main Tray Experiment

The experiment on the main tray is a replicated $2^{7-3}$ fractional screening experiment with "centre points" (Table 1 contains the settings of the factors and the choice we made for the middle settings of the qualitative factors). The factors are based on an Failure Modes and Effects Analysis (FMEA). The goal of the experiment is to identify the influence of these factors on the features extracted from the current signals. The results of this experiment will be input for further tests to obtain precise functional relationships. The final result is a monitoring scheme for the dominant parameters.

**Table 1.** Factors and their settings in fractional factorial screening experiment.

| Factor | −1 Level | 0 Level | 1 Level |
|---|---|---|---|
| Supply voltage 24 Vdc | 22 | 24 | 26 |
| Stapler set size | 5 | 27 | 50 |
| Feed rolls load | low | low | high |
| PWBA modification | mod I | mod I | mod II |
| PWBA temperature (°C) | 20 | 40 | 60 |
| Supply voltage 5 Vdc | 4.5 | 5.0 | 5.5 |
| Belt tension | low | low | high |

### 2.2 Measurements and Feature Extraction

During the experiment the current consumption of the stapler motor was measured per run in the experiment. Classical multivariate analysis cannot be performed with signals as a response variables because the signal consists of approximately one million observations. Instead we are looking for features in the signals. The first peak of the current signal corresponds to the action *spring load*, at this point the stapler anvil goes down against the paper before stitching itself. A high current consumption indicates a deterioration of the sub-module. Therefore, we used the *maximum amplitude* of the first peak of the current signal as a feature. Direct manual extraction of the features is time consuming and inaccurate. Since the nature of the information contained in the signals is local (*e.g.*, a peak) we prefer wavelet analysis over time series to get reliable features. Figure 1 shows the current signal of the stapler motor.

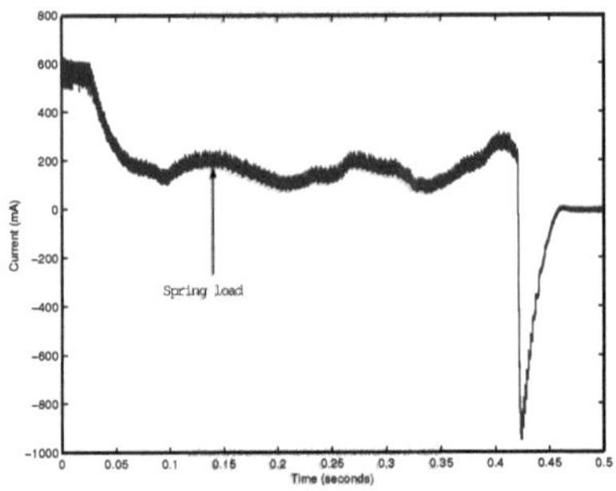


**Fig. 1.** Current signal of the stapler motor and spring load peak

## 3 Wavelet Approach for Analysis of Stapler Motor Data

We have chosen wavelet analysis because it enables the analysis of localized areas of a larger signal [2]. By means of wavelet analysis we first simplify the description of a signal in terms of a small number of wavelet coefficients, and afterwards use them as features to perform the Analysis of Variance. Since we get too many significant coefficients, we extract the maximum amplitude of the first peak current with a more accurate methodology than the manual extraction in order to improve the results. In the following we present the results of two wavelet approaches.

### 3.1 Approach 1: Rough Denoising - Extracting the Features Using $A_6$

Rough denoising consists of decomposing the signal at several levels, removing all high-frequency components at each level, and then reconstruct the signal. Afterwards, we obtain a smooth signal and we extract the maximum of the first peak. At the 6th approximation level, $A_6$, almost no noise is present and it still keeps the main features of the signal visualizing the strength of the wavelet analysis, see Fig. 2. Therefore, the maxima are extracted from approximation level $A_6$.

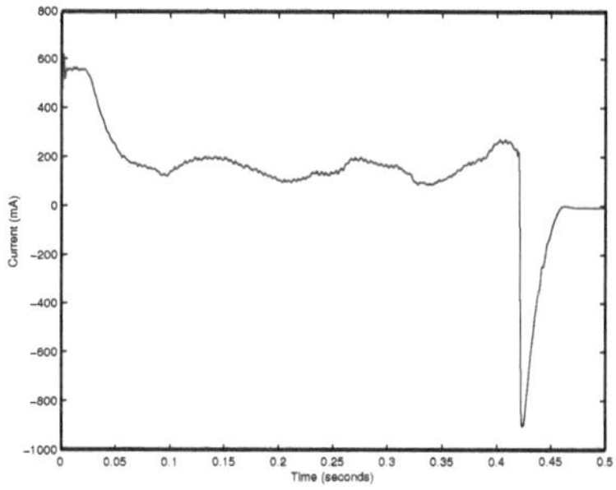


**Fig. 2.** Reconstructed approximation at level 6

### 3.2 Approach 2: Extracting the Features Using the Average of Approximation Coefficients

In this approach we calculate the wavelet coefficients in each level and we calculate the maximum of the first peak of the coefficients at levels 4 up to 8. Then the weighted average of the maximum of the wavelet coefficients of each level is computed. The weights are given by $2^{-j/2}$ for levels $j = 4, \dots, 8$, so the maximum of the different levels are at the same scale. Table 2 shows the results of the ANOVA for the first peak using the features extracted manually, the first and the second wavelet approach. We see that few factors affect the maximum amplitude of the first peak. This is favourable for translating this peak back to internal degradation parameters of the machine, which is the subject of future research. Taking the average of the maximum of the wavelet coefficients of the 5 levels we obtain the same significant factors and interactions as with the first approach.

## 4 Conclusions

The experiment on the main tray was a screening experiment in which seven factors vary systematically. Since classical multivariate analysis cannot be performed with signals as a response variables, it was decided to extract the maximum amplitude of of the first peak of the current signal as a feature to perform ANOVA. We presented two wavelet-based approaches to analyse the experiment. The first approach is based on the reconstructed approximation at level 6 because it contains much less noise than the original signal, and it still keeps the main characteristics of the signal. In the second approach

**Table 2.** Summary of the ANOVA results for the spring load peak

| Factors and interactions | Manual extraction | Approach 1 | Approach 2 |
|---|---|---|---|
| Supply voltage 24 Vdc | 0.00 | 0.00 | 0.00 |
| Number of sheets | 0.28 | 0.00 | 0.00 |
| Feed roll load | 0.40 | 0.82 | 0.79 |
| PWBA modification | 0.00 | 0.00 | 0.00 |
| PWBA temperature | 0.08 | 0.69 | 0.76 |
| Belt tension | 0.01 | 0.65 | 0.60 |
| Supply voltage 5 Vdc | 0.85 | 0.26 | 0.20 |
| Supply 24 Vdc:number of sheets | 0.08 | 0.02 | 0.00 |
| Supply 24 Vdc:feed roll load | 0.47 | 0.73 | 0.41 |
| Supply 24 Vdc:PWBA modification | 0.19 | 0.31 | 0.43 |
| Supply 24 Vdc:PWBA temperature | 0.41 | 0.38 | 0.38 |
| Number of sheets:feed roll load | 0.15 | 0.07 | 0.08 |
| Number of sheets:PWBA modification | 0.79 | 0.10 | 0.96 |
| Feed roll load:PWBA modification | 0.02 | 0.32 | 0.32 |
| Residual standard error | 9.86 | 6.10 | 5.69 |

we use directly the wavelet coefficients at 5 levels and we average them. For the first peak of the stapler motor, averaging the maxima of the wavelet coefficients appears to be the best approach since the residual standard error is the smallest, and because it considers the information from several levels of decomposition assuring stability of the feature. Besides the reduction of the residual standard deviation and the number of outliers, the computation time during the wavelet analysis is negligible. Therefore, our method can be used for on-line extraction of signal features.

*Acknowledgements*

We would like to thank Stef van Eijndhoven and Maarten Jansen for inspiring discussions on wavelet analysis. The research reported in this paper has been carried out as part of the Signature Analysis project EETK20015 of the *Economy, Ecology and Technology* programme.

## References

1. M.E. Benbouzid. A review of induction motors signature analysis as a medium for faults detection. *IEEE Trans. Ind. Electron.*, 47(5):984–993, 2000.
2. C.S. Burrus, R.A. Gopinath, and H. Guo. *Introduction to Wavelets and Wavelet Transforms : A Primer.* Prentice Hall, Upper Saddle River, 1998.
3. T. Figarella. Signature analysis: Modelling the characteristics of the main tray of the Lake finishers. Master of Technological Design Thesis, Eindhoven University of Technology, Eindhoven, 2003.

# Aspects of Multirate Time Integration Methods in Circuit Simulation Problems

A. El Guennouni[1,3], A. Verhoeven[2], E.J.W. ter Maten[2,3], and T.G.J. Beelen[3]

[1] Yacht, Eindhoven, The Netherlands
[2] Technische Universiteit Eindhoven, Eindhoven, The Netherlands
[3] Philips Research, Eindhoven, The Netherlands. Jan.ter.Maten@philips.com

## 1 Introduction

We present a new robust compound step strategy for multirate time integration methods. The strategy applies to DAEs and to time-dependent PDEs as well. We will compare this to other known strategies like Slowest First and Fastest First and modern ones that involve some form of a compound step integration during the process.

The circuit equations can be written as a DAE system of equations

$$\frac{d}{dt}[\mathbf{q}(t,\mathbf{x})] + \mathbf{j}(t,\mathbf{x}) = 0. \tag{1}$$

We assume that some partition is given that distinguishes between "slow" and "fast" varying components $\mathbf{x}_S$ and $\mathbf{x}_F$. In this case (1) can be written as

$$\frac{d}{dt}[\mathbf{q}_F(t,\mathbf{x}_F,\mathbf{x}_S)] + \mathbf{j}_F(t,\mathbf{x}_F,\mathbf{x}_S) = 0 \tag{2}$$

$$\frac{d}{dt}[\mathbf{q}_S(t,\mathbf{x}_F,\mathbf{x}_S)] + \mathbf{j}_S(t,\mathbf{x}_F,\mathbf{x}_S) = 0 \tag{3}$$

Ordinary multirate time integration aims to integrate (2)–(3) to a same accuracy using different time-steps $H$ and $h$, in which $H \gg h$. We intend to apply the approach to mixed signal simulation in which digital and analog circuitry are combined. The digital part often shows latent time behaviour, while the analog part often shows time varying activity. In addition, on the digital part less accuracy is needed than on the analog part. This gives way to combine multirate time integration with distributed tolerances.

Because circuit simulators usually apply Backward Differentiation Formula (BDF) methods as time integrator we will consider multirate time integration for the most simple one, the Euler Backward method.

The Slowest First strategy integrates first (2) for $\mathbf{x}_S$ using extrapolated values for $\mathbf{x}_F$ and step-size $H$. Next (3) is integrated repeatedly for $\mathbf{x}_F$ using interpolated values of $\mathbf{x}_S$ and step-size $h$.

The Fastest First strategy simply starts with (3). The first approach benefits from being better suited when dealing with automatic step-size control mechanism. However both methods have weak stability properties, due to the extrapolation involved [2]. For this reason we were led to study implicit methods. Interesting ones can be cast in the following General Compound Strategy, in which $q = \frac{H}{h}$ and $0 < \alpha \leq 1$ is just a parameter: Here (4)-(7)

---

**ALGORITHM 1** *A General Compound (G.C.) Strategy*

**Compound phase**: *Solve for* $\mathbf{x}_S^{n+q}$ *and* $\mathbf{x}_F^{n+\alpha q}$:

$$\mathbf{q}_F(\mathbf{x}_F^{n+\alpha q}, \hat{\mathbf{x}}_S^{\alpha}) - \mathbf{q}_F(\mathbf{x}_F^{n}, \mathbf{x}_S^{n}) + \alpha H \mathbf{j}_F(\mathbf{x}_F^{n+\alpha q}, \hat{\mathbf{x}}_S^{\alpha}) = 0 \tag{4}$$

$$\hat{\mathbf{x}}_S^{\alpha} - \mathbf{x}_S^{n} - \alpha(\mathbf{x}_S^{n+q} - \mathbf{x}_S^{n}) = 0 \tag{5}$$

$$\mathbf{q}_S(\hat{\mathbf{x}}_F^{\alpha}, \mathbf{x}_S^{n+q}) - \mathbf{q}_S(\mathbf{x}_F^{n}, \mathbf{x}_S^{n}) + H \mathbf{j}_S(\hat{\mathbf{x}}_F^{\alpha}, \mathbf{x}_S^{n+q}) = 0 \tag{6}$$

$$\hat{\mathbf{x}}_F^{\alpha} - \mathbf{x}_F^{n} - \frac{1}{\alpha}(\mathbf{x}_F^{n+\alpha q} - \mathbf{x}_F^{n}) = 0 \tag{7}$$

**Refinement phase**: *Solve for* $\mathbf{x}_F^{n+j+1}$ $(j = 0, \ldots, q-1)$:

$$\mathbf{q}_F(\mathbf{x}_F^{n+j+1}, \hat{\mathbf{x}}_S^{n+j+1}) - \mathbf{q}_F(\mathbf{x}_F^{n+j}, \hat{\mathbf{x}}_S^{n+j}) + h \mathbf{j}_F(\mathbf{x}_S^{n+j+1}, \hat{\mathbf{x}}_S^{n+j+1}) = 0 \tag{8}$$

$$\hat{\mathbf{x}}_S^{n+j+1} - \mathbf{x}_S^{n} + \frac{j+1}{q}(\mathbf{x}_S^{n+q} - \mathbf{x}_S^{n}) = 0 \tag{9}$$

---

form a "Compound Step" in which $\mathbf{x}_S$ is determined at $t = t_n + qh = t_n + H$, together with implicitly determined $\mathbf{x}_F$. If $\alpha = 1$, this "Compound Step" is just the result of Euler Backward with a large step $H$, which is easy to implement. If $\alpha = \frac{1}{q}$, the solutions $\mathbf{x}_S^{n+q}$ and $\mathbf{x}_F^{n+1}$ are simultaneously calculated. This option corresponds to the multirate method described in [1, 3] (but for Runge-Kutta and Rosenbrock-Wanner methods).
Integration of (8) is the"Refinement Step" for the fast part. It uses interpolated values $\hat{\mathbf{x}}_S^{n+j+1}$ as expressed in (9).

Clearly the GC step methods solve a larger system during the Compound Step phase than in the Slowest First or the Fastest First strategies. However, in most mixed signal applications the size of the digital parts exceeds that of the analog part several times. The GC step methods have much better stability properties than the Slowest First or the Fastest First strategies as conjectured by [4] and which is proved in [5]. For instance, considering the next two-dimensional test equation

$$\begin{pmatrix} \dot{x}_F \\ \dot{x}_S \end{pmatrix} = \underbrace{\begin{pmatrix} a_{11} & a_{12} \\ a_{21} & a_{22} \end{pmatrix}}_{\mathbf{A}} \begin{pmatrix} x_F \\ x_S \end{pmatrix} \tag{10}$$

the following stability conditions for $\mathbf{A}$ are derived in [5]:

| SF | GC | GC ($\alpha = 1$) |
|---|---|---|
| $a_{11} < 0$ | $a_{11} < 0$ | $a_{11} < 0$ |
| $a_{22} < 0$ | $\alpha a_{11} + a_{22} < 0$ | $a_{11} + a_{22} < 0$ |
| $\lvert a_{12}a_{21}\rvert < \lvert a_{11}a_{22}\rvert$ | $-a_{11}a_{22} - 2\alpha a_{11}^2 < a_{12}a_{21}$ | $-a_{11}a_{22} - 2a_{11}^2 < a_{12}a_{21}$ |
| | $a_{12}a_{21} < a_{11}a_{22}$ | $a_{12}a_{21} < a_{11}a_{22}$ |

In the sequel we assume $\alpha = 1$, which is the most robust choice. We will demonstrate that this strategy elegantly fits hierarchical circuit definition. Furthermore, the impact of the partiton will be considered.

## 2 Model Problem

Using Modified Nodal Analysis, in circuit simulation applications $\mathbf{q}(t, \mathbf{x}) = \sum_e \mathbf{B}_e q_e(t, \mathbf{B}_e^T \mathbf{x})$, in which $q_e$ is a local branch function. For instance, for a (linear) capacitor $C(a, b)$ between nodes $a$ and $b$, $q_e = C$ and $\mathbf{B}_e = \mathbf{e}_a - \mathbf{e}_b$, in which $\mathbf{e}_a$ (resp $\mathbf{e}_b$) is a canonical unit vector with a 1 at place $a$ (resp. $b$), and zeros elsewhere. The operators $\mathbf{B}_e$ are defined by the topology of the network. They do not depend on $t$ or $\mathbf{x}$. Similar results hold for the function $\mathbf{j}$. Assembly can be grouped to sub-circuits to fit hierarchical circuit definitions. In this way

$$\mathbf{q}(t, \mathbf{x}) = \sum_s \mathbf{B}_s \mathbf{q}_s(t, \mathbf{B}_s^T \mathbf{x})$$

where $\mathbf{B}_s^T \mathbf{x}$ selects the unknowns at the sub-circuit level and

$$\mathbf{q}_s(t, \mathbf{y}) = \sum_e \mathbf{B}_e q_e(t, \mathbf{B}_e^T \mathbf{y})$$

defines the assembly of $\mathbf{q}_s$ inside the sub-circuit.

A simple model problem is shown in Fig. 1. With $\mathbf{x} = (V_1, V_2, i_E, V_3, V_4)^T$, the functions $\mathbf{q}$ and $\mathbf{j}$ are given by

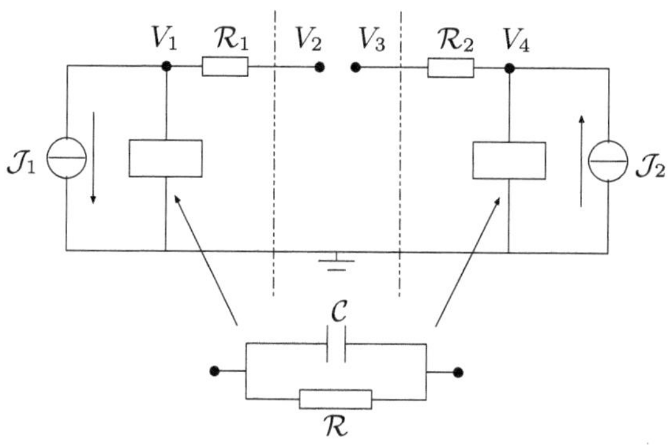


**Fig. 1.** Simple model circuit; Fast at the left, Slow at the right

$$\mathbf{q}(t,x) = \begin{bmatrix} \mathcal{C}\mathbf{x}_1 \\ 0 \\ 0 \\ 0 \\ \mathcal{C}\mathbf{x}_5 \end{bmatrix}, \; J(t,x) = \begin{bmatrix} \mathcal{J}_1(t,\mathbf{x}_1) - \frac{(\mathbf{x}_1-\mathbf{x}_2)}{\mathcal{R}_1} \\ \frac{(\mathbf{x}_1-\mathbf{x}_2)}{\mathcal{R}_1} - \mathbf{x}_3 \\ \mathbf{x}_2 - \mathbf{x}_4 \\ \frac{(\mathbf{x}_4-\mathbf{x}_5)}{\mathcal{R}_2} - \mathbf{x}_3 \\ \mathcal{J}_2(t,\mathbf{x}_5) - \frac{(\mathbf{x}_5-\mathbf{x}_4)}{\mathcal{R}_2} \end{bmatrix} \tag{11}$$

The source functions $\mathcal{J}_k(t,x)$ may be defined by $\mathcal{J}_k(t,x) = \sin(\omega_k t)$ with $\omega_1 \gg \omega_2$. The role of $i_E$ of the short $E$ serves to explicitly obtain the terminal current between the fast and the slow part (the short $E$ can be included automatically as a "virtual" glue-elements by the simulator, see also section 3). In the refinement phase a current source can be used to define the outgoing current of the slow part.
Several partitions $\mathcal{P}_k$ may be considered. Here the following are considered:

- $\mathcal{P}_1$: $x_f = [V_1, V_2, i_E, V_3]$ and $x_s = [V_4]$.
- $\mathcal{P}_2$: $x_f = [V_1, V_2, i_E]$ and $x_s = [V_3, V_4]$.

If a partition $\mathcal{P}$ is chosen, topological matrices $\mathbf{B}_f$ and $\mathbf{B}_s$ can be defined in the same style as before to define for instance $q(t, \mathbf{B}_s\mathbf{x})$.
The Euler Backward Compound method ($\alpha = 1$) proved to be very stable

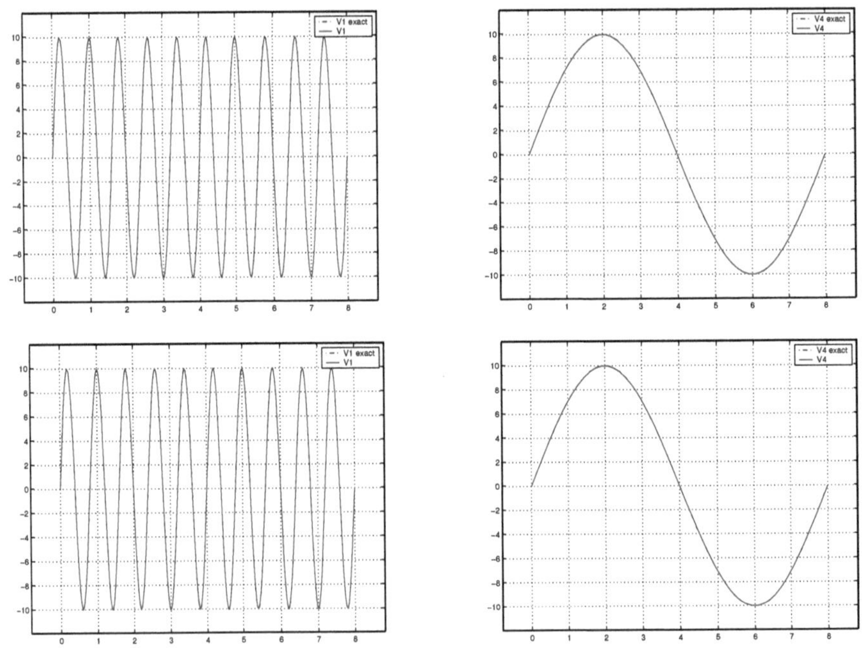


**Fig. 2.** Results with Partition $\mathcal{P}_1$ (at the top) and Partition $\mathcal{P}_2$ (at the bottom). $\mathcal{R}_i = 10^5$, $\mathcal{R} = 10$, $\mathcal{C} = 2.10^{-10}$, $H = 0.16$ and $h = 0.032$. $Error(\mathcal{P}_1(V_1)) \approx Error(\mathcal{P}_2)(V_1) \approx 10^{-10}$. $Error(\mathcal{P}_1(V_2)) = 9.63\ 10^{-7} < Error(\mathcal{P}_2(V4)) = 8.92\ 10^{-5}$

due to the implicit extrapolation. Considering the last method more closely for different partitions, we observe that $\mathcal{P}_1$ performs best. Results are shown in Fig. 2. However, despite the less accurate results, due to the interpolation error, partition $\mathcal{P}_2$ is very attractive, because it elegantly fits an existing hierarchical evaluator: the main part of the partition is along the boundary of the known sub-circuits $S_1$ and $S_2$ and thus this partition may even be given by the user. In the algorithm $S_1$ and $S_2$ are treated as in ordinary transient simulation (with their own step-size).

## 3 Interface treatment fitting hierarchical sub-circuits

At the interface, the partition concentrates in the "glue elements" . A more general glue-elements is shown in Fig. 3, from which it is clear how it can be generated. For instance, it applies to sub-circuits of which two terminals of one are connected to the same terminal of another. Note also that the connections between the sub-circuits are treated more symmetrically. Now each short can measure a particular terminal current. In the glue-elements this facilitates particular implementations for multirate integration, also when dealing with partitions with more different multirate behaviour.

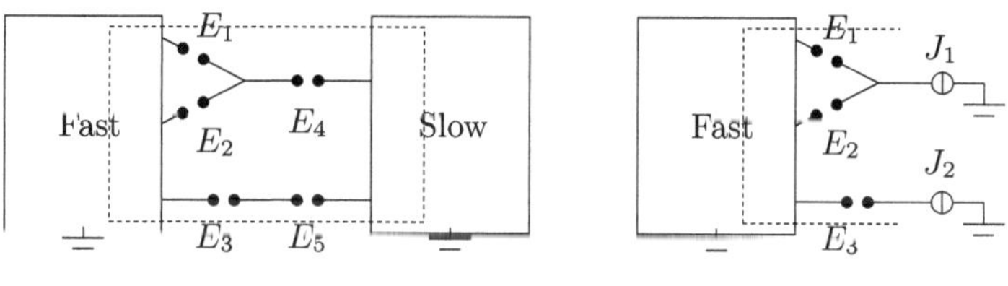


**Fig. 3.** Generation and different role of glue-elements. In the compound phase (at the left) shorts $E_4$ and $E_5$ allow to measure currents at the slow boundary. In the refinement phase (at the right) these current values are used to define the current sources $J_1$ and $J_2$. The boxes "fast" and "slow" can be treated as black boxes.

## References

1. A. Bartel and M. Günther. A multirate W-method for electrical networks in state-space formulation. *J. of Comput. and Applied Maths.*, 147:411–425, 2002.
2. C.W. Gear and D.R. Wells. Multirate linear multistep methods. *BIT*, 24:484–502, 1984.
3. M. Günther, A. Kværnø, and P. Rentrop. Multirate partitioned Runge-Kutta methods. *BIT*, 41:504–515, 2001.
4. S. Skelboe. Accuracy of decoupled implicit integration. *SIAM J. Sc. Comput.*, 21(6):2206–2204, 2000.

5. A. Verhoeven, A. El Guennouni, , E.J.W. ter Maten, and R.M.M. Mattheij. A general compound multirate method for circuit simulation problems. Presented at *Scientific Computing in Electrical Engineering*, Capo D'Orlando, Sicily, Italy, 2004.

# Exploiting Features for Finite Element Model Generation

O. Hamri[1,2], J.-C. Léon[1], F. Giannini[2], and B. Falcidieno[2]

[1] Laboratory 3S – Integrated Design Project, UMR CNRS 5521 – INPG – UJF, Domaine Universitaire, 38041 Grenoble Cedex 9 (France).
[okba.hamri,jean-claude.leon]@hmg.inpg.fr

[2] Istituto per la Matematica Applicata-CNR, Via de Marini, Genova (Italy).
[giannini,Falcidieno]@ge.imati.cnr.it

**Summary.** The preparation of simulation models from CAD models is still a difficult task since shape changes are often required to adapt a component or a mechanical system to the hypotheses and specifications of the simulation task. Detail removal or idealization operations are among the current treatments performed during the preparation of simulation models. In this paper we introduce the concept of simplification features, which allows a user to improve the efficiency of the analysis model generation process. As a result, form feature semantics and simulation data are attached to a polyhedral model during the preparation phase to ease the Finite Element(FE) details identification and removal as well as to maintain the consistency between a CAD model and its associated F.E models.

**Key words:** Simplification features, details feature, CAD/FEA link, consistency, F.E. model preparation.

## 1 Introduction

Currently, a CAD system contains only a part of the information required for structural analysis, namely, geometricl data. To generate a FE model, the CAD geometry needs to be adapted to fit the hypotheses of the mechanical models needed. Additional information about boundary conditions is also required. This task cannot be performed on the basis of a geometric model only [1, 2] but thus require also engineering knowledge [3]. Therefore, a direct automatic transition from a CAD model to a finite element model is not feasible[6]. To increase the efficiency of the preparation phase of analysis models, it is desirable to create robust links among the various models generated for a given simulation. This paper proposes a model preparation process for structural analysis using feature information. This process is the basis of our proposal and directly contributes to improve the details' identification and removal steps on a polyhedral model. The paper is organized as follows. Section

2 addresses the process of simulation model generation without form feature information. Sections 3 introduces the advantages gained by using feature information for the FE model generation and consistency preservation between CAD and FE models.

## 2 Analysis model preparation

CAD and Finite Element Analysis (FEA) are two significantly different disciplines, therefore they require considerably different object model representations, that is to say a series of models (Fig. 1). The first model of the chain is the case of study, which depicts the design results available at a time (see Fig. 1a). At this point, mechanical hypotheses, simulation objectives, are inserted to generate the domain of study compatible with the simulation requirements [3, 5]. The second model of the chain is the polyhedral model

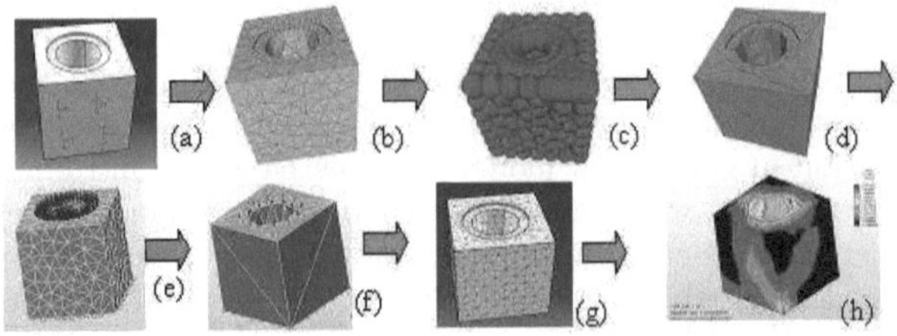


**Fig. 1.** The workflow of the analysis model generation: (a) case of study, (b) tesselated model, (c) envelope around the polyhedron, *i.e.* FE map of sizes, (d) adapted model (simulation model), (e) polyhedral model with boundary conditions, (f) transfert of boundary conditions on the adapted model, (g) mesh of the adapted model, (h) analysis result.

generated by the tessellation process [6], this model represents the first step towards CAD/FEM integration (Fig. 1b). The third model of the preparation chain is the adapted model (Fig. 1d) generated by the simplification process according to the simulation objectives. The mechanical concept used to identify the shape details is the map of FE sizes specified by the user for the planned FE mesh since FE size states the distribution of strain energy (for example) in the structure for a given load configuration (Fig. 1c). According to the type of modifications that must be carried out on the polyhedral model to remove details, three classes categorised the details and their associated operators [3]: skin details, topological details, and abstraction details. Skin details designate those details that can be removed by performing only continuous transformations like deforming a clay model. Topological details designate those ones affecting the genus of the object, like holes, that cannot be removed by continuously deforming the object surface. Abstraction details refer to those areas of the object that can be idealized by using 2D or 1D geometry through a reduction of the manifold dimension.

## 3 Exploiting feature attributes for FE model preparation

Having the objective of ensuring a robust link between CAD and simulation models, our approach intends to exploit all the information available in the CAD environment. Considering the fact that design by features is now largely adopted in the modeling phase, such information includes not only the surface topological description but also form feature data (blind holes, through holes,...). This has the advantage of:

- Reducing the complexity of the detail identification by allowing a reasoning directly on a set of geometric elements belonging to a specific feature instead of the low level elements only, *i.e.* vertices of a polyhedral model [3]. A feature brings high level information either to complement or supersede polyhedral model data structures,
- Tracking shape topology changes during the simplification process, using the concept of HLT (High Level Topology) [4], which makes it possible to take into account physical attributes attached to the case of study (BC's, materials,...) as well as form features or B-Rep CAD topology parameters to describe how the component is initially modelled. All the attributes attached to the CAD model are propagated to the polyhedral model. This allows a better monitoring of the model preparation and also the identification and characterization of the shape changes during the simplification process.

### 3.1 Simplification features

In our context, we define a simplification feature as a form feature whose removal does not affect the analysis results. The criteria specifying which form features are also simplification features are based on the FE map of sizes expressed a priori by the user. Therefore, being a simplification feature is governed by the mechanical configuration under evaluation, represented by this map of sizes. Then, a feature defined on the polyhedron model represents a simplification feature if it can be considered a detail if and only if the map of sizes associated to the feature fully contains it (Fig. 2d). As such, a simplification feature aggregates geometric as well as mechanical data through:

- at least, a connected set $F$ of geometric elements forming one or more form features, and possibly parts of features which are adjacent to them in the feature graph,
- a set of mechanical data characterised at least by a map of FE sizes to reflect, a priori, the user's view of the discretization of the structure in accordance to the objectives of the simulation.

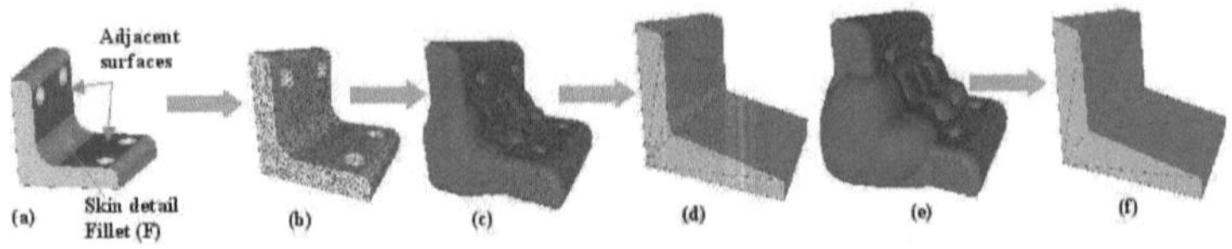


**Fig. 2.** The concept of simplification features: (a) initial CAD model(B-Rep), (b) polyhedral model, (c) small map of FE sizes attached to (b), (d) simplification result of (c), (e) large map of FE sizes attached to (b), (f) simplification result of (e)

### 3.2 Detail feature categories

The categories of details listed at the end of Section 2 are based mainly on the effects that their elimination has on the object topology. Their corresponding removal operators are only exploiting the geometry of the polyhedral model thus requiring several simplification loops for the removal of entities corresponding to a single simplification feature. Taking advantage of the CAD model information, such operations could be done in a single step. The same categorization adopted for details can be used for the simplification features. In our current development, two main categories can be considered: skin, topological simplification features. The first category includes features such as fillets, blends and chamfers that can be compared to skin details. For such kind of features, the high level information concerning the primitive surfaces of the adjacent B-Rep faces is clearly the critical information to characterize the removal operation. For example in Fig. 2, two possible removal operations associated to the surface simplification feature corresponding to a fillet are shown. The first is obtained by the extension of the adjacent faces forming a 90° angle (Fig. 2f), the other by the substitution of the fillet with one planar face (Fig. 2d). In both cases the new configuration to be applicable must be inside the given map of FE sizes. These two configurations are useful complements to the skin detail operator since they can provide a means to better guide the shape adaptation process according to the user preferences. The second category includes those features changing the topology of the object or of the face to which they are applied (*e.g.* blind and through holes). In this case a form feature is considered a detail feature if and only if its virtual volume is fully contained in the map of sizes, this guarantees that there is no difference with the analysis in considering the corresponding region as containing material. Knowing that, its removal can occur in a single operation by removing the set of polyhedral faces belonging the form features and filling the opening on the remaining polyhedron.

## 4 Conclusion

In this paper an FE model preparation process has been described that exploits the various information accessible in the CAD object description. It represents a first step towards semantic driven integrated product development systems. The concept of simplification feature has been proposed as semantic entities guiding the simplification process. They allow taking advantage of the form feature information, which are commonly available in most advanced CAD systems or obtainable through recognition processes. A classification of the simplification features has been provided to form the basis for the development of the related simplification operators. To ensure a robust link between CAD and simulation models in our approach we combine mechanical and CAD data in an attributed representation for improving the efficiency of FE model preparation. Our future work will focus on the realization of the mechanism on the attribute structure for tracing the CAD and mechanical semantics through all the simplification process. It will also include the specification and development of the removal operators for the identified simplification features.

*Acknowledgement.* The work is carried out within a bi-lateral collaboration agreement between L3S of the INPG in Grenoble (France) and CNR IMATI-GE in Genova (ITALY). It is partially supported by the EU through the Network of Excellence AIM@SHAPE Contract IST 506766.

## References

1. M. Belaziz, A. Bouras, and J.M. Brun. Morphological analysis for product design. *Computer-Aided Design*, 32(5-6):377–388, 2000.
2. P. Dabke, V. Prabhakar, and S. Sheppard. Using features to support Finite Element Idealisation. *Int. Conference ASME, Minneapolis*, 1:183 193, 1994.
3. L. Fine. *Processus et méthodes d'adaptation et d'idéalisation de modèles dédiés à l'analyse de structures mécaniques.* PhD. thesis, INPG, Grenoble, France, July 6th 2001.
4. O. Hamri, J.-C. Léon, F. Giannini, and B. Falcidieno. From CAD models to F.E. simulations through a feature-based approach. volume Sept. 28th – Oct. 3rd, Salt Lake City, 2004.
5. J-C. Léon and L. Fine. A new approach to the preparation of models for F.E. analyses. *Int. J. of Comp. Appl.*, 2004.
6. A. Sheffer, T. Blacker, J. Clements, and M. Bercovier. Virtual topology operators for meshing. *Int. J. of Computational Geometry and App.*, 10(3):309–331, 2001.

# Implicit Subgrid-Scale Models in Space-Time VMS Discretisations

S. J. Hulshoff

Faculty of Aerospace Engineering, Delft University of Technology, Kluyverweg 1, 2629 HS Delft, The Netherlands S.J.Hulshoff@LR.TUDelft.NL

**Summary.** The effects of discretisation parameters on the performance of a space-time VMS FEM are investigated. A moving-wave solution of the one-dimensional viscous Burgers equation is used to limit the influence of SGS modelling errors. Factors influencing the magnitude of the implicit SGS model are discussed.

**Key words:** Variational multiscale, space-time finite elements, large-eddy simulation.

## 1 Introduction

Variational-multiscale (VMS) discretisations [4], show promise for application to large-eddy simulations (LES) due to their consistency with the governing equations at large scales, and their consistent treatment of scale separation near boundaries. When combined with a finite-element method (FEM) [1], VMS discretisations can also be applied to complex domains. Time-discontinuous space-time FEM are particularly advantageous in this regard, as they naturally incorporate mesh movement, and allow arbitrary re-meshing from one time step to the next [3].

To account for the effects of unresolved scales, VMS methods normally employ a physical subgrid-scale (SGS) model. Like all discretisation techniques, however, VMS discretisations introduce errors which are greatest in magnitude for the smallest resolved scales. These result in an implicit SGS model, which competes with the physical SGS model in the description of the effects of the unresolved scales.

This paper examines the performance of a space-time VMS FEM discretisation using a test case for the Burgers equation with minimal model calibration errors. The effects of discretisation parameters on the behaviour of the implicit SGS model is investigated.

## 2 Discretisation

For a space-continuous, time-discontinuous Galerkin discretisation on a space-periodic domain, the viscous Burgers equation may be represented as:

$$B(w,u) = (w, u_t + uu_x - \nu u_{xx} - f)_Q = 0 \quad (1)$$

$$= -(w_t, u)_Q - (w_x, \frac{u^2}{2} - \nu u_x)_Q - (w, f)_Q$$
$$+(w, u^-)_{\Omega^{n+1}} - (w, u^+)_{\Omega^n} + (w, (u^+ - u^-))_{\Omega^n} \quad (2)$$

Here $(\ ,\ )_Q$ represents the $L_2$ inner product on the space-time domain, while $(\ ,\ )_{\Omega^n}$ and $(\ ,\ )_{\Omega^{n+1}}$ are the $L_2$ inner products on the boundary of the domain at times $t = t^n$ and $t = t^{n+1}$. The solution is advanced in time by solving a sequence of space-time domains with thickness $\Delta t = t^{n+1} - t^n$ (slabs), with the $\Omega^{n+1}$ boundary of each slab providing the initial condition for the following slabs. The initial condition is imposed weakly, using the last term in (2), with $u^+ - u^-$ being the jump in solution across the $\Omega^n$ boundary (see [3] for details).

The discretisation is implemented as a VMS method by employing a hierarchical basis, and interpreting its components in terms of large and small scales [4, 1]. Here the hierarchical basis consists of the standard bilinear functions supplemented with $r$ Legendre polynomials in space. From the argument of scale separation, it is assumed that the unresolved scales have negligible influence on the large resolved scales. In contrast, the effects of the unresolved scales on the small resolved scales are represented by an additional physical SGS model, $M$. The large- and small-scale equations are then:

$$B(\bar{w}, \bar{u}) = -B(\bar{w}, u') + (\bar{w}_x, (\bar{u}u'))_Q \quad (3)$$
$$B(w', u') = -B(w', \bar{u}) + (w'_x, (\bar{u}u'))_Q + M \quad (4)$$

where $\bar{w}, \bar{u}$ are associated with the linear and low-order Legendre components of the interpolation and $w', u'$ are associated with the high-order Legendre components. For the current study a Smagorinsky-like physical SGS model is employed, $M = -(w'_x, \nu_T u'_x)_Q$, with the turbulent viscosity coefficient $\nu_T$ defined to reproduce the dissipation of the unresolved scales.

## 3 Burgers Test Case

The test case is defined by (1) with $\nu = \frac{2\pi}{1000}$ and $f(x,t) = A\ \sin(x - Ut)$, where $U = 1$, $A = 0.1$. This produces a moving wave with a fixed profile. DNS computations were performed on a sequence of meshes to ensure grid independence. The finest of these used 8192 bilinear space-continuous elements on a domain of length $L = 2\pi$, with a constant time step of $\frac{\pi}{8192}$. The computations were started with a uniform flow of $U = 1$, and advanced to

$t = 8\pi$. The constant profile of the final solution allows the parameters of the physical SGS model model to be accurately estimated. In particular, given the wave-number range to which the physical SGS model will be applied, the finest DNS solution is used to estimate $\nu_T$ with

$$\nu_T = \nu \, \frac{\int_{k_c}^{\infty} k^2 \, E(k) \, \mathrm{d}k}{\int_{k_m}^{k_c} k^2 \, E(k) \, \mathrm{d}k}, \tag{5}$$

where $E = \frac{u^2}{2}$ is the kinetic energy, $k_m$ is the starting wavenumber of the physical SGS model range, $k_c$ is the cutoff wavenumber, and the spectral definition of the energy dissipation rate has been used [2]. For VMS-FEM, the cut-off wave number is defined by $k_c = \frac{qN}{2}$, where $N$ is the number of elements, and $q$ is the number of unique unknowns per element along a constant-time boundary. Here, the first spatial interpolation function to which the physical SGS model is applied is denoted by $m$. The gradients for the model are computed using all scales above and including $m$, and the model is applied to all scales above and including $m$. $k_m$ is estimated by multiplying the ratio of scales without modelling to the total number of scales by $k_c$. The procedure described above minimises the calibration error of the physical SGS model, so that differences in results can be related to the choice of discretisation parameters.

# 4 Computed Results

## 4.1 Spatial discretisation effects at small time steps

The behaviour of the discretisation at a time step equal to that of the finest DNS is considered first. Figure 1 shows the error in the time average of the total kinetic energy relative to the DNS between $t = 6\pi$ and $t = 8\pi$. This interval is equal to the period necessary for the wave to completely traverse the mesh. The errors are plotted versus the inverse of the number of degrees of freedom used to represent the solution in space, which varies from 16 to 128. The solid line labelled "Basic LES" indicates the results obtained for $q = 1,\ m = 1$.

The errors of the $q = 2$ and $q = 3$ solutions were found to be almost identical to those for $q = 1,\ m = 1$ when $M$ is applied to all scales. Releasing the first scale from the physical SGS model ($q = 2,\ m = 2$), however, results in a large increase in accuracy. The latter implies that the errors of the Basic LES cases are dominated by the application of the model to the largest resolved scales, rather than by specifics of the discretisation. The advantage of the VMS approach is that it can limit the distortion of the large resolved scales by shifting the application of the model to the smaller resolved scales. Increasing the order of the method while keeping $\frac{k_m}{k_c}$ constant ($q = 4,\ m = 3$ and $q = 8,\ m = 5$) results ultimately in a faster convergence rate, but larger

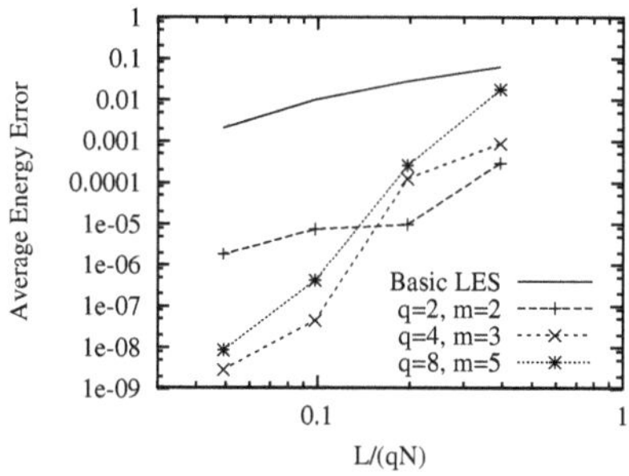


**Fig. 1.** Average energy error for different numbers of scales

errors at low numbers of degrees of freedom. The corresponding errors in the solution profile have high spatial frequencies, and arise from the non-smooth solution of the LES problem.

## 4.2 Implicit SGS model

One method for estimating the implicit SGS dissipation is to subtract the resolved change in energy due to viscosity per time step from the work done by the body force per time step, $\Delta E_f$:

$$\Delta E_{ISGS} = \Delta E_f - \sum_e^N \int_{Q_e} (\nu (u_x)^2 + \nu_T (u'_x)^2)\, dQ_e \tag{6}$$

where $Q_e$ is the space-time domain of element $e$. An alternative approach is to estimate the change in the energy of the linearised system with $\nu = \nu_T = 0$. In Fig. 2(a) the normalised implicit dissipation, $\frac{\Delta E_{ISGS}}{\Delta E_f}$ computed with (6) for $q = 2$, $m = 2$ is compared with the estimate from the $\nu = \nu_T = 0$ linearised system for varying values of the Courant number, $\frac{U \Delta t}{\Delta x}$. In spite of the relatively large perturbation to the mean convection speed, the estimates agree reasonably well.

The influence of spatial discretisation parameters on the implicit SGS model is shown in Fig. 2(b). Here the normalised implicit dissipation based on (6) is plotted for different values of $q$ and $m$ for $qN = 32$ and a Courant number of 1. Independent of $q$, there is an initial drop in the implicit dissipation as $m$ is increased from 1. When larger numbers of scales are released from the model, however, the implicit dissipation begins to increase. In general, the primary action of the implicit SGS model is to damp high-frequency components of the solution [3]. As $m$ is increased, $M$ is applied to a higher range of frequencies with larger values of $\nu_T$. This biases the energy towards low frequencies, where the implicit model is less active. For the largest values of $m$, however, the increase in spatial order results in strong high-frequency errors, and correspondingly increased implicit dissipation. The action of the implicit SGS model is thus minimised at intermediate values of $m$.

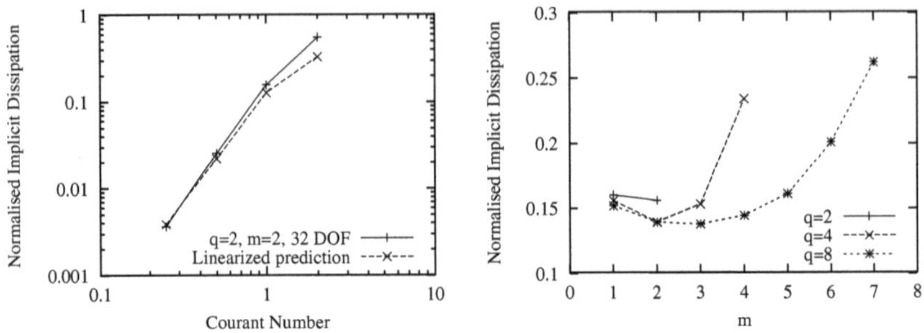


(a) (b)

**Fig. 2.** Normalised implicit SGS energy dissipation

# 5 Conclusions

Using a one-dimensional test case, this paper has demonstrated effects of discretisation parameters on the performance of a space-time VMS FEM, and on its associated implicit SGS model. The VMS method was shown to be advantageous in that it can improve the accuracy of large-scale data by limiting the application of the physical SGS model to the smaller resolved scales. There is a limit to the number of scales which can be released from the model, however, due to the lack of smoothness of the underlying solution. The magnitude of the implicit SGS dissipation was primarily influenced by the timestep, but was also shown to be influenced by the number of model-free scales. Again low numbers of free scales per element proved best, as larger numbers resulted in increased implicit dissipation associated with the high-frequency errors of higher-order discretisations.

# References

1. S. S. Collis. The DG/VMS method for unified turbulence simulation. AIAA Paper No. 2002-3124, 2002.
2. P. G. Saffman. *Lectures on Homogeneous Turbulence.* Topics in Nonlinear Physics. Springer, Berlin, 1968.
3. F. Shakib and T. J. R. Hughes. A new finite element formulation for computational fluid dynamics: IX: Fourier analysis of space-time Galerkin/least-squares algorithms. *Comput. Methods Appl. Mech. Engrg.*, 87:35–58, 1991.
4. L. Mazzei T. J. R. Hughes and K.E. Jansen. Large eddy simulation and the variational multiscale method. *Comput. Visual Sci.*, 3:47–59, 2000.

# Multiscale Change-Point Analysis of Inhomogeneous Poisson Processes Using Unbalanced Wavelet Decompositions

M. Jansen

Technische Universiteit Eindhoven, Department of Mathematics, PO Box 513, 5600 MB Eindhoven, the Netherlands mjansen@win.tue.nl
and KU Leuven, Department of Computer Science, Celestijnenlaan 200A, 3001 Leuven, Belgium www.cs.kuleuven.ac.be/~maarten

**Summary.** We present a continuous wavelet analysis of count data with time-varying intensities. The objective is to extract intervals with significant intensities from background intervals. This includes the precise starting point of the significant interval, its exact duration and the (average) level of intensity. We allow multiple change points in the intensity curve, without specifying the number of change points in advance. We extend the classical (discretised) continuous Haar wavelet analysis towards an unbalanced (*i.e.*, asymmetric) version. This additional degree of freedom allows more powerful detection. Locations of intensity change points are identified as persistent local maxima in the wavelet analysis at the successive scales. We illustrate the approach with simulations on low intensity data. Although the method is presented here in the context of Poisson (count) data, most ideas (apart from the specific Poisson normalization) apply for the detection of multiple change points in other circumstances (such as additive Gaussian noise) as well.

**Key words:** Change point, wavelet, Poisson process, unbalanced, maxima.

## 1 Introduction

In many applications [3, 6], data are counts with time (or space) varying intensities. Typical examples are photons emitted by a device or by a natural source and captured by scanners, or microscopes. The goal is to detect the presence of a significant signal against background noise and to characterize the signal by its intensity, location and duration. Figure 1 shows a test example.

A statistical description of this problem is known as change point detection. Change point analysis using wavelets has been investigated in several papers including [5]. The work summarized in this paper concentrates on finding change points in Poisson data. Similar problem descriptions appear in earlier papers such as [7] but none of these applied a multiscale analysis to the problem.

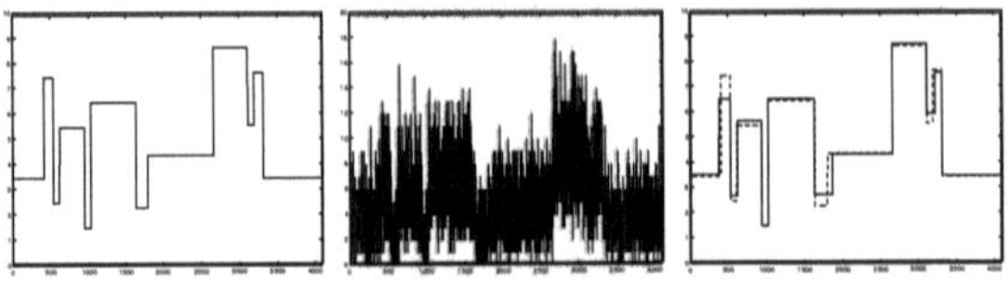

**Fig. 1.** A simulated example of Poisson data with time varying intensities. On the left the plot of the intensity curve. This is a scaled and vertically translated version of the well-known "Blocks" test example [1]. In the middle a random realization. On the right the estimation from that realization, using the procedure proposed in this paper.

The contribution of this paper consists of the combination of broad assumptions and an original approach applying new techniques in wavelet analysis. More specifically, on the assumption side, we allow more than one change point and we do not specify how many of them are present in the data. The background noise level is assumed unknown. The method is not restricted to high intensity signals. On the other hand, we scan the local maxima in the (discretised) continuous wavelet analysis of the data, and we extend the analysis to unbalanced transforms. As explained later, this increases the (statistical) power of the detection procedure.

Although the techniques are presented in the framework of Poisson count data, the method can easily be extended towards other types of random data.

Due to space limitations, this text can only summarize the ideas of this new approach. For a full description, we refer to a paper still to be written about this subject.

## 2 Multiscale binning

Suppose we are given observations $x_i, i = 0, \ldots, n-1$ of random variables $X_i$ that are Poisson distributed, *i.e.*,

$$P(X_i = k) = \frac{e^{-\mu_i} \mu_i^k}{k!},$$

where $\mu_i = EX_i$ is the expected value of the $i$th observation, also called the *intensity* of the counting process at location $i$.

The intensity $\mu_i$ is not constant, but depends on the location (or time point) $i$. More specifically, we assume that the intensity is a piecewise constant, *i.e.*, consecutive observations have the same intensity, except at some transition points. Obviously, the exact values of $\mu_i$ are unknown. We want to estimate those values from the observations. A central question in this estimation is to find good estimates for the locations of jumps. Not only are these locations crucial as such for several applications, they also allow good estimates of the intermediate intensities.

In the first instance, we apply a recursive binning of these observations, *i.e.*, we consider the following decomposition:

1. Call $S_{J,k} = X_k$ the finest level scaling coefficients. We let $J$ denote the finest scale. Subsequent scales (or levels) get a lower number.
2. Let $S_{j,k} = S_{j+1,2k+1} + S_{j+1,2k}$ be the summed values at level $j$ of binned pairs at finer level $j+1$.
3. Let $W_{j,k} = S_{j+1,2k+1} - S_{j+1,2k}$ be the differences between scaling coefficients at level $j$.

If such a difference is significantly different from zero, we know that somewhere in the interval covered by the associated two bins, a change point must have occurred. Given the information that there must be a change point somewhere, it is easier to locate it at finer scales than without this additional information, even if the differences at those finer scales are no longer significant.

The decomposition is the Haar wavelet transform. To check whether a wavelet coefficient $W_{j,k}$ is significant, it is interesting to normalize it:

$$Z_{j,k} = W_{j,k}/\sqrt{S_{j,k}}.$$

The values $Z_{j,k}$ have an asymptotically normal distribution [2] and their variance is constant if one leaves out the cases where $S_{j,k} = 0$:

$$V(Z_{j,k}|S_{j,k} \neq 0) = 1.$$

## 3 Wavelet maxima

The Haar decomposition presented in the previous section is a *dyadic* transform. This means that the computation of $W_{j,k}$ involves a dyadic number of observations, *i.e.*, the number of observations involved is an integer power of two. Also, the subsequent coefficients at a given level are based on mutually disjoint sets of observations:

$$W_{j,k} = \sum_{i=0}^{2^{j^*}-1} X_{k2^{j^*+1}+2^{j^*}+i} - \sum_{i=0}^{2^{j^*}-1} X_{k2^{j^*+1}+i},$$

where $j^* = J - j - 1$. These two properties obviously limit the power of the statistical test: if the coefficients would not be limited to dyadic locations and dyadic bins, we would certainly find a coefficient where at least one of the adjacent bins would perfectly coincide with a complete interval of constant intensity. Such a full analysis is known as a maximal discretization of the continuous Haar wavelet transform. It leads to the following wavelet coefficients:

$$W_{j,k} = \sum_{i=0}^{j^*-1} X_{k+j^*+i} - \sum_{i=0}^{j^*-1} X_{k+i},$$

where $j^* = N - j$ and $j = N-1, \ldots, 1$.

Such a complete analysis allows to find at each scale $j$ the locations $k$ where the normalized wavelet coefficient $Z_{j,k}$ reaches a local maximum. If this maximum is sufficiently large (say, if its absolute value is larger than 3), the corresponding location is considered as a candidate change point. Before the final selection of significant change points takes place, we first want to select the optimal scale for each candidate. To this end, the locations of local maxima are linked into lines of local maxima across scales [4], as in Figure 2. On all maxima lines, we select the scale and location with the highest absolute normalized coefficient value.

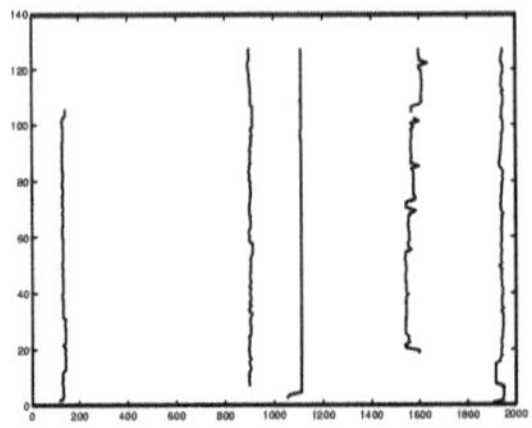

**Fig. 2.** Line of local maxima across scales. The horizontal axis contains the locations, the vertical axis depicts the successive scales.

## 4 Unbalanced wavelet analysis

Even the continuous wavelet transform may not be able to detect all change points. Figure 3(a) shows an example of a wavelet analysis where one interval of constant intensity is fully covered but the other (left) interval is not completely observed by this wavelet coefficient. If the difference between the intensities in the two intervals is small, it may be crucial to estimate both intensities with the highest possible accuracy, *i.e.*, the lowest possible variance. Searching for significant values through all possible unbalanced coefficients

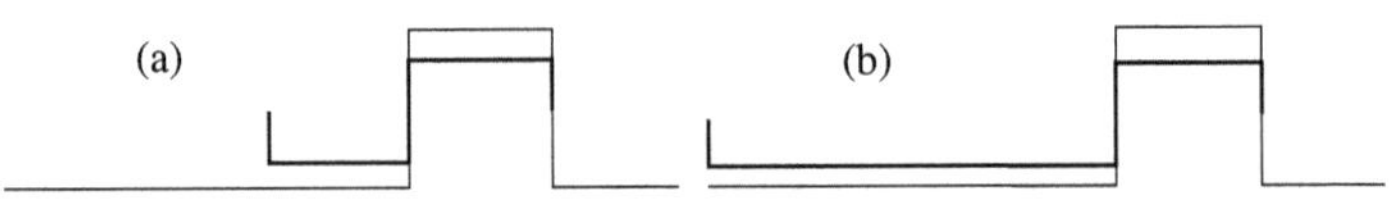


**Fig. 3.** A symmetric wavelet analysis (bold line) is not optimal in detecting change points. If one allows unbalanced analyzes, the resulting wavelet coefficients may be more significant if the analysis covers two complete, adjacent intervals of constant intensity. This may be crucial in change points with small jumps.

would be computationally impossible. However, if we start our search from the previously selected wavelet maxima, this becomes an easy optimization.

## 5 Elimination of false maxima and results

The last phase of the algorithm is the selection of change points. Candidate change points are given by locations of local wavelet maxima and the unbalanced extension indicates the range of a change point. As Figure 4 indicates, two successive change points may reinforce each other's significance, by sharing observations. This occurs if two successive change points are both jumps up or both jumps down, thereby forming a staircase. In that case, the most significant one is selected as primary covariate, and its location is now considered as an inpenetrable boundary: the significance of the adjacent change point is recomputed within this new situation. As soon as the ranges of the selected change points cover all observations, we stop the selection of change points. As illustrated in Figure 1, the procedure finds good estimates of all change points. Moreover, it removes all spurious points.

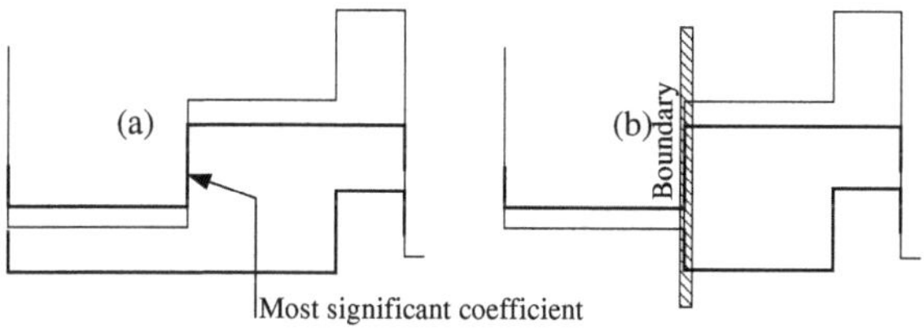


**Fig. 4.** Selection of most significant change point and update of the range and significance of the remaining candidates

## References

1. D.L. Donho and I. M. Johnstone. Ideal spatial adaptation via wavelet shrinkage. *Biometrika*, 81:425–455, 1994.
2. P. Fryźlewicz and G. Nason. A wavelet-Fisz algorithm for Poisson intensity estimation. *Journal of Computational and Graphical Statistics*, 13:621–638, 2003.
3. T. Herberts and U. Jensen. Optimal detection of a change point in a Poisson process for different observation schemes. *Scand. J. Stat.*, 31(3):347–366, September 2004.
4. S. Mallat and W. L. Hwang. Singularity detection and processing with wavelets. *IEEE Transactions on Information Theory*, 38(2):617–643, 1992.
5. T. Ogden and E. Parzen. Change-point approach to data analytic wavelet thresholding. *Statistics and Computing*, 6:93–99, 1996.
6. M. Raimondo and N. Tajvidi. A peaks over threshold model for change-point detection by wavelets. *Statistica Sinica*, 14:395–412, 2004.
7. J. Scargle. Studies in astronomical time series analysis. Bayesian blocks, a new method to analyze structure in photon counting data. *Astrophys. J.*, 504:405–418, 1997.

# Robust Soft Sensors Based on Ensemble of Symbolic Regression-Based Predictors

E. Jordaan[1], A. Kordon[2], and L. Chiang[2]

[1] Dow Benelux B.V., Terneuzen, The Netherlands EMJordaan@dow.com
[2] Dow Chemical, Freeport(TX), USA {AKKordon,HChiang}@dow.com

**Summary.** One way to increase the robustness of soft sensors is to use ensembles of symbolic regression-based predictors. Ensembles can increase the robustness because it gives a more consistent estimate of the output, it allows the derivation a measure of confidence and it can be used for problem detection. In this paper we will demonstrate the robust soft sensor by using ensembles of symbolic regression-based predictors in an industrial application.

**Key words:** Soft sensor, ensembles, symbolic regression, Pareto front.

## 1 Introduction

Inferential or soft sensors are mathematical models that are used to predict the outcome of processes. These sensors are often needed because online measurements of the outcome of the processes can not be made with a low cost or ease. One factor limiting the widespread use of soft sensors in the process industry is their inability to cope with noisy data and process variability.

There are a number of steps that can be taken to improve the robustness of soft sensors. One way is to use explicit nonlinear functions that are derived by Genetic Programming (GP) [4]. A major advantage of this approach is that there is a potential physical interpretation of the model. Other advantages are the ability to examine the extrapolation behavior of the model and to impose external constraints on the modeling process. Furthermore, process engineers are more open to take the risk of implementing such type of models. In a second approach the idea of balancing the modeling performance and complexity is used to increase the robustness. This approach uses the Pareto front to find the best trade-off between the model's performance and its complexity [7]. The third approach that could improve robustness is the use of an ensemble of predictors. The key idea is to use combined predictors and their statistics as a confidence indicator of the performance.

In this the paper we will demonstrate how to design ensembles of GP-based predictors in order to improve the robustness of the soft sensors. An successful industrial application of the proposed approach is also given.

## 2 Ensemble of GP-generated Predictors in Soft Sensors

The section describes the nature of GP-based models and focuses on the novel model procedure for selecting the models on the Pareto-front of the performance-complexity plane. Thus, the ensemble is based on predictors with the best generalization capabilities and potential for robust performance in industrial conditions.

### 2.1 Genetic Programming

One modeling approach that is increasingly being used in the industry is Genetic Programming (GP). GP is of special interest to soft sensor development due to its capability for symbolic regression [4]. GP-generated symbolic regression is a result of simulation of the natural evolution of numerous potential mathematical expressions. The final results is a list of "the best and the brightest" analytical forms according to the selected objective function. Of special importance to industry are the following of unique features of GP [3]: no *a priori* modeling assumptions derivative-free optimization; few design parameters; natural selection of the most important process inputs; and parsimonious analytical functions as a final result.

### 2.2 Ensembles of GP Generated Predictors

Ensembles consist of several models that are used to predict future measurements. For a given input the final prediction is the average of the predictions of all the models in the ensemble. The advantages for using ensembles of predictors instead of a single model are all related to the robustness requirements.

Firstly, since the prediction is a combination of a number of predictions, one obtains an estimate of the output that is reduced in variance. The soft sensor is more robust as the predicted outcome does not depend on the accuracy of one single model anymore.

Secondly, the spread or variance of the different predictions can be used to derive a measure of confidence, called the model disagreement indicator. The idea is that in areas of high data density (information rich) the models in the ensemble will have the same behavior. However, in areas of low data density the various models will have more freedom and exhibit different behavior.

A third advantage is that the model disagreement indicator can be used for problem detection. Here the idea is that whenever something has changed in the process the models in the ensemble will show a high variance in their predictions whereas under normal circumstances the variance was much lower. This way drifting processes, faulty equipment or novelty can be detected easily.

Finally, ensembles enable redundancy by using models that depend on different input variables. In processing conditions it often occurs that one or more of the instruments measuring the input variables fail. The soft sensor can be made robust toward equipment failure since there will be at least one model available to predict in the absence of a certain input variable.

There are several methods to design ensembles [5, 6]. The key issue is to select models that are diverse enough to capture uncertainties, but similar enough to predict well. Designing such a robust ensemble can be very difficult, even for an experienced soft sensor developer. Utilizing the Pareto front, robust ensembles can be constructed with much more ease and objectivity.

### 2.3 Pareto front Method for Ensemble Model Selection

Since model selection is in principle a multi-objective problem (i.e. accuracy vs. complexity), the fundamentals of the Pareto-front can be applied. The Pareto front is defined by the dominant solutions satisfying both conflicting objectives. Using the Pareto front for GP-generated models has many advantages [1]. Firstly, the structural risk minimization principle [8], which finds the optimal balance between complexity and accuracy, can be easily applied to GP-generated models, see Fig. 1. Currently in GP the measure that is used for the complexity is the number of nodes needed to define the model.

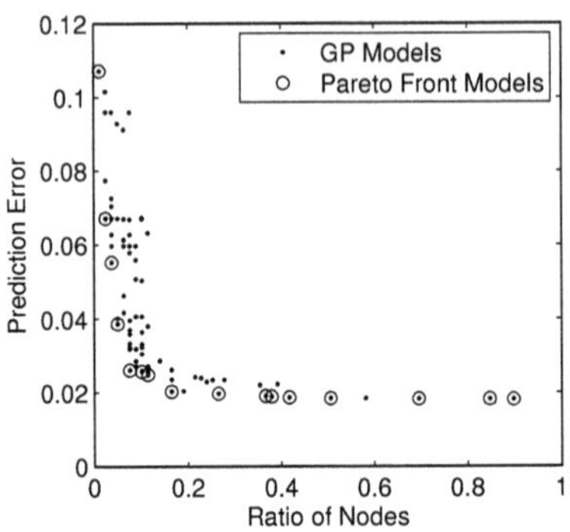


**Fig. 1.** Pareto front based on the performance of the training data

A second advantage is that the Pareto front effectively displays the trade-off between the measures, which enables the analyst to make an correct decision. The Pareto front models are models for which no improvement on one objective can be obtained without deteriorating the other. The best model as a result of multi-objective approach will therefore lie somewhere on the Pareto front. Typically, the most interesting models lie in the lower left corner.

The third advantage is that the number of models that needs to be inspected individually is decreased tremendously as only a small fraction of the generated models in GP will end up on the Pareto front. This is clearly seen in Fig. 1. Only 18 of the total of 88 models depicted in the figure lie on the Pareto front. Furthermore, as none of the models with a ratio of nodes higher than 0.3 significantly improve on the $R^2$, these may by omitted too.

## 3 Application

The described methodology for designing robust soft sensors based on an ensemble of GP-type predictors will be illustrated with an industrial application in a continuous chemical process. For a successful application of the approach to batch processes, see [2]. The symbolic regression-type models used in this application have been derived on an Dow-internally developed GP software.

The application is related to distillation column control. For better control of a distillation column, it is desirable to obtain an accurate and fast prediction of a process quality variable (in this case - propylene concentration). Current analytical technique allows measurement of propylene every 10 minutes, which is not sufficient for control purposes. The problem has been resolved by an ensemble-based soft sensor, developed by the described approach that can provide every minute a prediction of propylene.

As illustrated in the top graph of Fig. 2, the ensemble model performs well for the testing data ($R^2 = 0.985$; Root Mean Square Error Prediction (RMSEP) $= 0.0791$). The model disagreement indicator is the standard deviation of the three models and a critical limit was defined to quantify the effect. For this testing data, model disagreement indicators are below the critical limit for most of the data points. This suggests that the testing data and training data are similar and that the prediction is reliable.

The self-assessment capability of the ensemble is illustrated in the lower plot in Fig. 2, where the key inputs are increased by 15% in the testing data. The model disagreement indicator is now above the critical limit, which means that the simulated testing data are now outside the training range. The simulated abnormal process condition has been captured reliably by the model disagreement indicator.

## 4 Conclusions

In this paper we have shown a novel approach to improve the robustness of soft sensors. This approach involves the use of GP-generated models in an ensemble of predictors. We have described a mechanism which is based on the Pareto front to effectively construct ensemble. This mechanism enables us to compare the models produced by the various GP runs qualitatively through a performance-complexity trade-off. Furthermore, the number of interesting

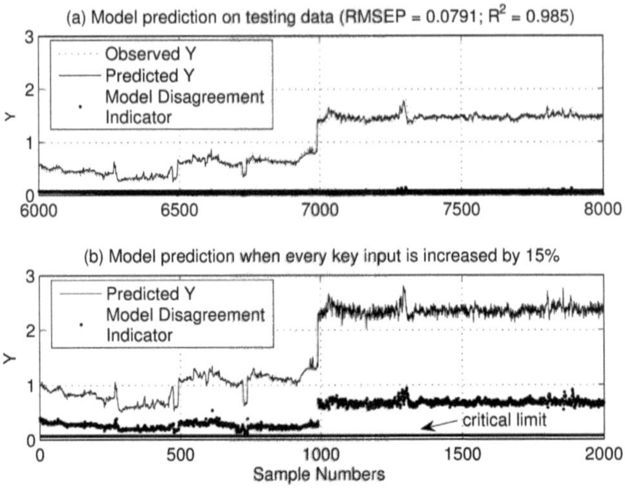


**Fig. 2.** Performance on the test data with normal and disturbed process conditions

models to inspect manually is decreased to a manageable number. The approach was successfully implemented in a chemical process.

## References


1. S. Bleuer, M. Brack, L. Thiele, and E. Zitzler. Multi-objective genetic programming. reducing bloat using SPEA-2. In *Proceedings of CEC 2001*, pages 536–543, 2001.
2. A. Kordon, E. Jordaan, L. Chew, G. Smits, T. Bruck, K. Haney, and A. Jenings. Biomass inferential sensor based on ensemble of models generated by genetic programming. In K. Deb, editor, *Proceedings of GECCO'2004 (Seattle, WA)*, pages 1078–1089, 2004.
3. A.K. Kordon and G.F Smits. Soft sensor development using genetic programming. In *Proceedings of GECCO'2001 (San Francisco,CA)*, pages 1346–1351, 2001.
4. J. Koza. *Genetic Programming. On the Programming of Computers by Means of Natural Selection.* MIT Press, Cambridge, MA, 1992.
5. Y. Liu, X. Yao, and T. Higuchi. Evolutionary ensembles with negative correlation learning. *IEEE Transactions on Evolutionary Computation*, 4(4):380–387, 2000.
6. A. Sharkey, editor. *Combining Neural Nets. Ensemble and Modular Multi-Net Systems.* Springer, London, UK, 1999.
7. G. Smits and M. Kotanchek. Pareto-front exploitation in symbolic regression. In R. Riolo and B. Worzel, editors, *Genetic Programming Theory and Practice.* Kluwer, (in press),2004.
8. V. Vapnik. *Statistical Learning Theory.* Wiley, New York, 1998.

# Two-Dimensional Patterns in High Frequency Plasma Discharges

D. Mackey[1] and M.M. Turner[2]

[1] School of Mathematics, Dublin City University, Ireland Dana.Mackey@dcu.ie
[2] School of Physics, Dublin City University, Ireland Miles.Turner@dcu.ie

**Summary.** Large area uniform plasmas are essential in microelectronics processing. Motivated by this application, a macroscopic model is proposed as a framework for investigating the occurrence of instabilities in high frequency plasma discharges for parallel plate geometries. This paper will concentrate on the formation of stationary, spatially inhomogeneous patterns.

**Key words:** plasma discharges, pattern formation.

## 1 Introduction

A key criterion for the utilisation of electrically driven discharges in semiconductor manufacturing or material surface modification is that the plasma be free of instabilities that produce stationary or slowly varying spatial structures since these may disastrously affect the quality of the process. Instabilities related to the balance of charged particle production and loss (which are usually observed experimentally as periodic spatial modulations of the plasma density and temperature) are a common occurrence in weakly ionized plasmas and have been most often studied in cylindrical geometries. In this context, a model was proposed in [3], consisting of two nonlinear partial differential equations (electron mass and energy balance) for which the existence of stationary, spatially periodic solutions was investigated using dynamical systems techniques.

In this paper, the model is generalised to parallel plate geometries, which are more relevant to plasma processing applications. The analytical study is quite different from that presented in [3] and consists of asymptotically reducing the model to a system of amplitude equations of Landau type and using a phase plane analysis for showing the existence of periodic attractors. Numerical simulations are also presented which illustrate the emergence of stable patterns from random electron temperature and density initial conditions .

## 2 Proposed Model

We consider a weakly ionized plasma which is sustained by a high-frequency current driven through a low pressure gas between two parallel electrodes. In industrial applications the electrode separation is usually small compared to the other two dimensions. In order to investigate the formation of instabilities in such discharges, we use the model proposed in [3] for a simpler one-dimensional geometry; the generalization to two spatial variables is straightforward. We give below the non-dimensional form of the model.

$$\frac{\partial n}{\partial t} = \Delta n + \alpha n \left[ \mathrm{e}^{\mathcal{E}\,(1-1/T)} - 1 \right] \tag{1}$$

$$\frac{\partial}{\partial t}(nT) + \nabla \cdot (\chi(T)\, \nabla n - \kappa n\, \nabla T) = P(n) - \frac{2}{3} \mathcal{E} \alpha \mathrm{e}^{\mathcal{E}\,(1-1/T)}\, n, \tag{2}$$

where $n(x, y; t)$ is the electron density (equal to the ion density, by the quasi-neutrality assumption), $T(x, y; t)$ is the electron temperature, $x$, $y \in [0, 2\pi]$ and $t \geq 0$. The first equation describes ambipolar diffusion of electrons, where particles are being created by ionization of neutrals and lost by transverse drift and recombination. The second equation describes electron energy transport, where the two terms on the right hand side represent power absorbed by the plasma (ohmic heating) and power dissipated by the electrons (electron-neutral collisions). Both equations have been averaged over the $z$ variable (measuring plate separation) and over the period of the driving current (which is small compared to the time scale over which instabilities develop). If we assume a simple $z$ dependence of the density $n$ (consistent with modelling the sheaths as capacitors), then an explicit formula can be found for the absorbed power, $P(n)$. The details of this calculation are not important for the subsequent analysis and will be omitted here. Equations (1) and (2) have a unique equilibrium point, which has been fixed to $n = 1$, $T = 1$ by the choice of non-dimensionalization (see [3]).

A similar expression for the energy flux in equation (2) was derived in [2]. It was argued there that the thermoelectric transport coefficient $\chi(T)$ should not be neglected in situations where the electron distribution function is non-Maxwellian. It is also one of the aims of our work to show that the inclusion of this transport coefficient in the model is essential for describing the onset of the plasma instabilities. From experimental data and kinetic simulations we find that $\chi$ (which is otherwise evaluated from velocity moments of the distribution function) can be approximated as $\chi(T) = \widetilde{\chi} - \Lambda\,(T-1)^2$. We choose $\widetilde{\chi}$ and $\Lambda$ as control parameters for the subsequent analysis.

## 3 Derivation and Analysis of Amplitude Equations

The linearization of (1) and (2) about the uniform steady state $n = 1$, $T = 1$ gives

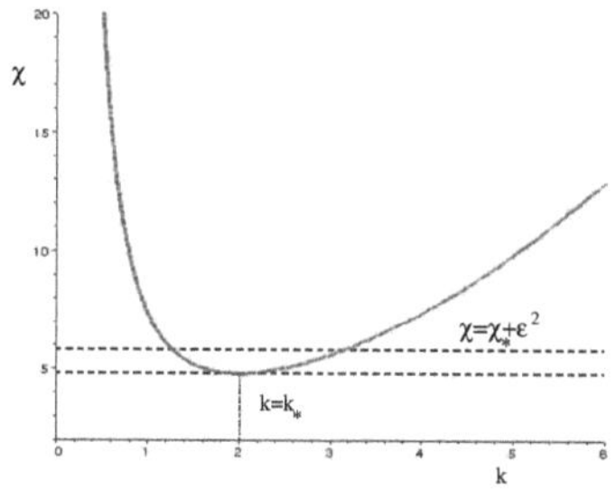


**Fig. 1.** Linear stability boundary

$$\frac{\partial \mathbf{U}}{\partial t} = \left(\mathbf{A}\,\nabla^2 + \mathbf{B}\right)\mathbf{U},$$

where

$$\mathbf{U} = \begin{pmatrix} n-1 \\ T-1 \end{pmatrix}, \quad \mathbf{A} = \begin{pmatrix} 1 & 0 \\ -(1+\widetilde{\chi}) & \kappa \end{pmatrix}, \; \mathbf{B} = \begin{pmatrix} 0 & p_1 \\ -q_1 & -(1+\gamma)p_1 \end{pmatrix},$$

and $p_1 = \alpha\mathcal{E} > 0$, $q_1 = P(1) - P'(1) > 0$. Imposing zero-flux boundary conditions (for mathematical convenience) leads to the following general solution

$$\mathbf{U}(x,y,t) = \sum_{k=1}^{\infty} \mathrm{e}^{\lambda(k)\,t}\,\widehat{\mathbf{U}}\,\cos(lx)\cos(my),$$

where $k^2 = l^2 + m^2$ ($l,\, m \in \mathbf{Z}$) and $\det\left(k^2\mathbf{A} - \mathbf{B} + \lambda\mathbf{I}\right) = 0$. Solutions grow or decay exponentially depending on whether $\lambda > 0$ or $\lambda < 0$ and the neutral stability boundary can be calculated as

$$\widetilde{\chi} = \mathcal{F}(k^2) \equiv \frac{\kappa}{p_1}\,k^2 + \frac{p_1 q_1}{k^2} + \gamma$$

which is (qualitatively) plotted in Figure 1. The linearly most unstable mode corresponds to the minimum of this curve, which is given by $k_*^2 = \sqrt{p_1 q_1/\kappa}$.

A weakly nonlinear analysis is carried out in order to determine the stability of a small periodic disturbance with fixed wave number $k = k_*$. For more details about this procedure see, for example, [1]. We assume we are close to the bifurcation point and let $\widetilde{\chi} = \widetilde{\chi}_* + \varepsilon^2$, where $\widetilde{\chi}_* = \mathcal{F}(k_*^2)$ and $0 < \varepsilon \ll 1$. We also introduce the slow time scale $\tau = \varepsilon^2 t$ and expand the density and temperature functions around the steady state as follows

$$n(x,y,t;\varepsilon) = 1 + \varepsilon n_1(x,y,t,\tau) + \varepsilon^2 n_2(x,y,t,\tau) + \cdots, \tag{3}$$

$$T(x,y,y;\varepsilon) = 1 + \varepsilon T_1(x,y,t,\tau) + \varepsilon^2 T_2(x,y,t,\tau) + \cdots. \tag{4}$$

Substitution of these expansions into the system (1)–(2) yields, at first order

$$\begin{bmatrix} \Delta & p_1 \\ \chi_*\,\Delta - q_1 & -\kappa\Delta + \gamma p_1 \end{bmatrix} \begin{bmatrix} n_1 \\ T_1 \end{bmatrix} = \begin{bmatrix} 0 \\ 0 \end{bmatrix}. \tag{5}$$

Without loss of generality we can assume the $\mathcal{O}(\varepsilon)$ solution to be of the form

$$\begin{pmatrix} n_1 \\ T_1 \end{pmatrix} = \frac{1}{\sqrt{k_*^4 + p_1^2}} \begin{pmatrix} p_1 \\ k_*^2 \end{pmatrix} [\mathcal{A}_1(\tau)\cos(k_* x) + \mathcal{A}_2(\tau)\cos(k_* y)] \,. \tag{6}$$

At order $\mathcal{O}(\varepsilon^2)$ we obtain the equations

$$\begin{bmatrix} \Delta & p_1 \\ \chi_* \,\Delta - q_1 & -\kappa\Delta + \gamma p_1 \end{bmatrix} \begin{bmatrix} n_2 \\ T_2 \end{bmatrix} = \begin{bmatrix} F_2 \\ G_2 \end{bmatrix} \tag{7}$$

where

$$F_2 = -p_1 n_1 T_1 - p_2\, T_1^2,\; G_2 = \kappa n_1 \Delta T_1 + \kappa\, \nabla n_1 \nabla T_1 + q_2 n_1^2 - \gamma\,(p_2 T_1 - p_1\, n_1)\, T_1$$

and $p_2 = \alpha\mathcal{E}\left(\frac{\mathcal{E}}{2} - 1\right)$, $q_2 = \frac{1}{2}P''(1)$. The solvability condition dictates that the inhomogeneous term of the linear equation (7) must be orthogonal to any solution $(\varphi, \psi)$ of the adjoint homogeneous problem, that is,

$$\int_0^{2\pi}\int_0^{2\pi} [F_2\varphi + G_2\psi]\; dxdy = 0. \tag{8}$$

It is easy to check that condition (8) is trivially satisfied. Finally, a similar solvability condition for the $\mathcal{O}(\varepsilon^3)$ problem yields evolution equations for the slowly-varying amplitudes,

$$\begin{aligned} \frac{\partial \mathcal{A}_1}{\partial \tau} &= m_1\, \mathcal{A}_1 + m_2\, \mathcal{A}_1^3 + m_3\, \mathcal{A}_1 \mathcal{A}_2^2, \\ \frac{\partial \mathcal{A}_2}{\partial \tau} &= m_1\, \mathcal{A}_2 + m_2\, \mathcal{A}_2^3 + m_3\, \mathcal{A}_2 \mathcal{A}_1^2, \end{aligned} \tag{9}$$

where $m_1$, $m_2$, $m_3$ are lengthy expressions of the physical parameters.

For the order 1 solution, (6), to converge to a stationary, spatially periodic pattern we need to show the existence, in the amplitude system (9), of at least one stable equilibrium point. A phase plane analysis reveals the following.

- The trivial equilibrium point $(0, 0)$ is always unstable.
- The equilibrium points $\left(\pm\sqrt{-\frac{m_1}{m_2+m_3}}, \pm\sqrt{-\frac{m_1}{m_2+m_3}}\right)$ exist for $m_2+m_3 < 0$ and are stable for $m_2 < m_3$. (This condition is equivalent to $\Lambda_{\min} < \Lambda < \Lambda_{\max}$, where $\Lambda_{\min}$ and $\Lambda_{\max}$ can be determined as functions of the system parameters.) This case corresponds to the formation of two-dimensional bounded patterns ("spots") in the original model.
- The equilibrium points $\left(0, \pm\sqrt{-\frac{m_1}{m_2}}\right)$ and $\left(\pm\sqrt{-\frac{m_1}{m_2}}, 0\right)$ exist for $m_2 < 0$ and are stable for $m_2 > m_3$ ($\Lambda > \Lambda_{\max}$). These correspond to one-dimensional bounded patterns ("stripes").

Note that no patterns are possible when $\widetilde{\chi} < \widetilde{\chi}_*$ since the equilibrium state $n = 1, T = 1$ is stable for these parameter values. Hence, as argued in Section 2, if we had neglected the thermoelectric transport coefficient $\chi$, the model would have failed to capture the formation of instabilities.

## 4 Numerical Results and Conclusions

The nonlinear system (1) – (2) with zero-flux boundary conditions was integrated using a finite difference ADI (Alternating Direction Implicit) scheme. All constants in the model were calculated from realistic physical parameters. The results shown in Figure 2 were obtained by choosing the control parameters so that $k_* = 1$ and $\Lambda$ corresponds to a 2-dimensional (spotted) pattern. After some transient behaviour, the temperature and density functions converged to steady-state spatially inhomogeneous solutions. In conclusion, the

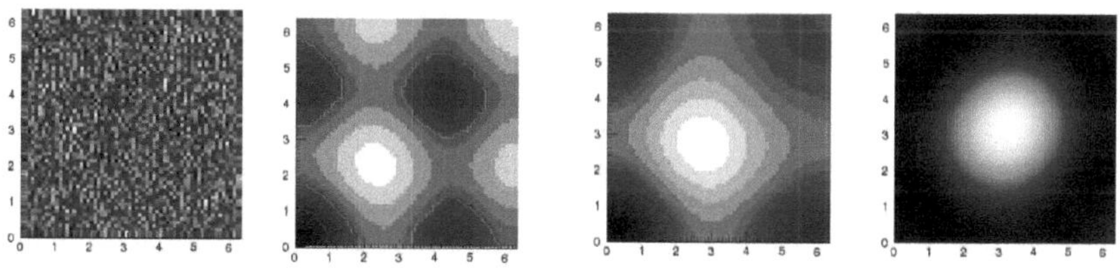

**Fig. 2.** Time evolution of the temperature function (bright colours denote maximum values). The spatial domain is $[0, 2\pi] \times [0, 2\pi]$ and the initial condition is a small random perturbation of the equilibrium state. Convergence to the final pattern is observed around $t \approx 100\mu s$.

proposed model predicts that stationary, spatially inhomogeneous patterns in parallel plate geometry discharges can occur as the equilibrium state loses stability. The type and stability of the resulting patterns can be determined by parameters relating to a certain thermoelectric transport coefficient. For more practical applicability, a similar analysis could be carried out in terms of physical control parameters such as the applied current density, gas pressure, etc. and this could also explain other interesting dynamical phenomena observed experimentally.

## References


1. M. C. Cross and P. C. Hohenberg. Pattern formation outside of equilibrium. *Rev. Mod. Phys.*, 65:851–1112, 1993.
2. J. H. Ingold. Nonequilibrium positive column. *Phys. Rev. E*, 56(5):5932–5944, 1997.
3. D. Mackey, L. Plantié, and M. M. Turner. Instabilities and pattern formation in low-temperature plasmas. *Applied Mathematics Letters*, to appear, 2004.

# A Mathematical Model for the Motion of a Towed Pipeline Bundle

N.W. Manson, S.K. Wilson, and B.R. Duffy

University of Strathclyde, 26 Richmond Street, Glasgow, G1 1XH, UK
rs.nman@maths.strath.ac.uk, s.k.wilson@strath.ac.uk,
b.r.duffy@strath.ac.uk

**Summary.** A simple mathematical model for the motion of a pipeline bundle being towed using the Controlled Depth Tow Method (CDTM) is constructed and analysed. When the forces exerted by the sea on the bundle are neglected the model predicts that the bundle is neutrally stable and that its motion involves two different timescales. When these forces are not neglected the model predicts that the bundle will always be stable if the tension in the bundle at its downstream end is sufficiently large.

**Key words:** Controlled Depth Tow Method, pipeline bundle

## 1 The Controlled Depth Tow Method (CDTM)

Pipeline bundles, consisting of a number of petroleum pipelines, control lines and umbilicals housed within a larger carrier pipe, are widely used in the North Sea offshore oil industry to carry oil between various underwater structures. In the North Sea, pipeline bundles are typically of around 1 metre in diameter but can be of up to 8 *kilometres* in length. Pipeline bundles are prefabricated on shore and towed into position using the Controlled Depth Tow Method (CDTM), which involves attaching a heavy towhead to each end of the bundle and suspending the whole system between two powerful tug boats. The front tug tows the bundle while the back tug maintains the tension in the bundle and assists with steering. A typical tow lasts two or three days, and during the tow the motion of the bundle is continuously monitored using an acoustic telemetry system. A variety of modes of oscillation are observed during the towing, and the tugs' velocities are continuously adjusted so as to control the motion of the pipeline bundle and, in particular, to keep it clear of the seabed and the free-surface waves, and to avoid excessive tension and distortion. As the final location is approached the tow speed is reduced and the bundle is lowered to within a few metres of the seabed before being carefully manoeuvred into its final position.

## 2 A Mathematical Model

In this simple first model for the motion of a pipeline bundle being towed using the CDTM, we assume that the bundle is neutrally buoyant, that all motion occurs in a vertical plane only, and that the displacement of the bundle is small.

Following Dowling [1] and Païdoussis [2, 3, 4, 5] the normal force balance equation for a bundle of length $l$ and radius $a$ is

$$\begin{aligned} m\frac{\partial^2 y}{\partial t^2} &= \frac{\partial}{\partial x}\left(T(x)\frac{\partial y}{\partial x}\right) + \rho\pi a^2\left(\frac{\partial}{\partial t} - U\frac{\partial}{\partial x}\right)^2 y \\ &- \rho\pi a U c_N\left(\frac{\partial y}{\partial t} + U\frac{\partial y}{\partial x}\right) + \rho\pi a U^2 c_T\frac{\partial y}{\partial x} - EI\frac{\partial^4 y}{\partial x^4}, \end{aligned} \tag{1}$$

where $x$ denotes position along the bundle from the front towhead, $t$ denotes time, $y(x,t)$ is the deflection of the centreline of the bundle, $U$ is the tow speed, $m$ is the mass per unit length of the bundle, $EI$ is the bending stiffness, $\rho$ is the density of sea water, and $c_N$ and $c_T$ are the skin friction coefficients in the normal and tangential directions, respectively. The tangential force balance equation determines the spatially varying tension in the bundle, $T(x)$, to be

$$T(x) = T(l) + \rho\pi a U^2 c_T(l - x), \tag{2}$$

where $T(l)$ is the tension at the downstream end.

We seek a Fourier-mode solution to (1) in the form

$$y(x,t) = \mathrm{Re}[\exp(i\omega t)\hat{Y}(x)], \tag{3}$$

where $\omega$ is the complex frequency. $\mathrm{Re}(\omega)$ determines the frequency of oscillation, while $\mathrm{Im}(\omega)$ determines the rate of temporal growth or decay. Specifically, if $\mathrm{Im}(\omega) > 0$ the bundle is stable, whereas if $\mathrm{Im}(\omega) < 0$ it is unstable; in the special case $\mathrm{Im}(\omega) = 0$ the bundle is neutrally stable.

Non-dimensionalising (1) in the obvious way leads to the fourth order ordinary differential equation

$$\varepsilon\frac{\mathrm{d}^4\hat{Y}}{\mathrm{d}X^4} - (X_c - X)\frac{\mathrm{d}^2\hat{Y}}{\mathrm{d}X^2} + b\frac{\mathrm{d}\hat{Y}}{\mathrm{d}X} + i\Omega b\hat{Y} = 0, \tag{4}$$

where we have written

$$\varepsilon = \frac{EI}{\rho\pi a U^2 c_T l^3}, \quad \Omega = \frac{\omega l}{U}, \quad L = \frac{l}{a}, \quad b = \frac{2i\Omega + c_N L}{c_T L}, \tag{5}$$

$$T = \frac{T(l)}{\rho\pi a^2 U^2}, \quad X_c = \frac{T - 1}{c_T L} + 1. \tag{6}$$

Here $X = X_c$ is the location of the so-called "critical point" at which the coefficient of the second derivative in (4) is zero. For a typical bundle being towed at a typical tow speed of 1.81 $\mathrm{ms}^{-1}$ typical values of $\varepsilon$, $L$, $T$ and $X_c$ are

$$\varepsilon \simeq 3.1 \times 10^{-5}, \quad L \simeq 18300, \quad T \simeq 112, \quad X_c \simeq 1.6. \tag{7}$$

In particular, typically $T$ is large and so we shall henceforth restrict our attention to the case $T > 1$ in which $X_c > 1$ and so the critical point does not lie on the bundle. (The case of a free downstream end, $T = 0$, in which the critical point can lie on the bundle, is treated by Dowling [1].) Moreover, typically $\varepsilon$ is small and so the fourth derivative term in (4) can be neglected everywhere. Thus we need to solve the second order ordinary differential equation

$$(X_c - X)\frac{\mathrm{d}^2\hat{Y}}{\mathrm{d}X^2} - b\frac{\mathrm{d}\hat{Y}}{\mathrm{d}X} - i\Omega b\hat{Y} = 0 \tag{8}$$

on $0 \leq X \leq 1$. Assuming that the towheads are fixed relative to the tugs the appropriate boundary conditions at the ends of the bundle are simply

$$\hat{Y}(0) = \hat{Y}(1) = 0. \tag{9}$$

The exact solution to the simplified equation (8) is

$$\begin{aligned} \hat{Y}(X) = &\, c_1(X - X_c)^{\frac{1}{2}(1-b)} I_{b-1}(2\sqrt{i\Omega b}\sqrt{X_c - X}) \\ &+ c_2(X - X_c)^{\frac{1}{2}(1-b)} K_{b-1}(2\sqrt{i\Omega b}\sqrt{X_c - X}), \end{aligned} \tag{10}$$

where $I_{b-1}$ and $K_{b-1}$ are modified Bessel functions of the first and second kind respectively of order $b-1$, and $c_1$ and $c_2$ are complex constants. Imposing the boundary conditions (9) leads to the determinantal equation

$$\begin{aligned} &K_{b-1}(-2i\sqrt{i\Omega b}\sqrt{1 - X_c}) I_{b-1}(-2i\sqrt{i\Omega b}\sqrt{-X_c}) \\ &= I_{b-1}(-2i\sqrt{i\Omega b}\sqrt{1 - X_c}) K_{b-1}(-2i\sqrt{i\Omega b}\sqrt{-X_c}). \end{aligned} \tag{11}$$

Equation (11) determines the complex frequency $\Omega$ and must, in general, be solved numerically. However, analytical solutions can be obtained in a number of special cases and asymptotic limits.

## 3 Analytical Solutions

### 3.1 Exact Solution in the Special Case $c_N = c_T = 0$

In the special case $c_N = c_T = 0$, in which the forces exerted by the sea on the bundle are neglected, the solutions for $\Omega$ and $\hat{Y}$ are simply

$$\Omega = \frac{\pi(T-1)}{\sqrt{2T-1}}, \tag{12}$$

$$\hat{Y}(X) = \sin(\pi X) \exp\left(\frac{i\pi X}{\sqrt{2T-1}}\right), \tag{13}$$

which means that the solution for the deflection of the bundle, $Y$, is

$$Y(X,t) = \sin(\pi X) \cos\left[\frac{\pi}{\sqrt{2T-1}}\left(X + (T-1)t\right)\right]. \tag{14}$$

This solution shows that in this case the bundle is neutrally stable and that its motion involves two different timescales, namely a short timescale of $(T-1)^{-1} \approx 0.009$ (corresponding to approximately 30 seconds) on which individual disturbances travel from the front to the back of the bundle, and a long timescale of $2\pi/\Omega = 2\sqrt{2T-1}/(T-1) \approx 0.269$ (corresponding to approximately 16 minutes) on which the entire bundle deforms significantly. The long timescale is consistent with the typical timescales observed during towing. While the short timescale is too short to be observed directly by the acoustic telemetry system, it is consistent with the observation that the bundle reacts rapidly if the tow is brought to a sudden halt.

### 3.2 Asymptotic Solution in the Limit $T \to \infty$

Typically $T$ is large, and so the behaviour of the solution in the limit of large tension, $T \to \infty$, is of particular interest. In this limit the solution for $\Omega$ is

$$\Omega = \frac{\pi}{\sqrt{2}}T^{\frac{1}{2}} + \frac{c_N L}{4}i + \frac{4\pi^2(Lc_T - 3) - (c_N L)^2}{16\sqrt{2}\pi}T^{-\frac{1}{2}} + O(T^{-1}), \tag{15}$$

and hence the pipeline bundle is neutrally stable and oscillates on a timescale of $2\sqrt{2}T^{\ 1/2}$.

### 3.3 General Stability Results

Triantafyllou and Chryssostomidis [6] derived stability results in the special case $c_N = c_T$. By extending some of their arguments to the general case $c_N \neq c_T$ we have been able to show that the bundle is *always stable* when $T > c_T/c_N$ and *always unstable* when $T < c_T/c_N - c_T L$. These analytical results are confirmed by Fig. 1, which shows that in the case $c_N = 0.0001$ and $c_T = 0.0439$ the numerically calculated neutral stability curve (i.e. the curve on which $\mathrm{Im}(\Omega) = 0$), which separates the unstable region of parameter space from the stable region, lies in the interval $c_T/c_N - c_T L < T < c_T/c_N$.

## 4 Summary

In this short paper we constructed and analysed a simple mathematical model for the motion of a pipeline bundle being towed using the CDTM. In the simplest case $c_N = c_T = 0$ the model predicts that the bundle is neutrally stable

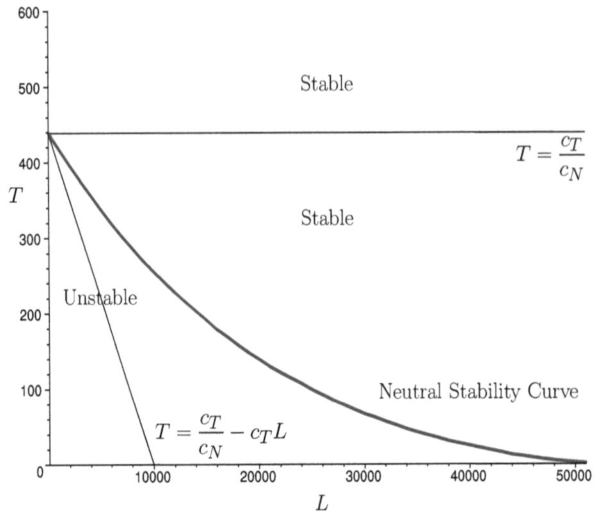


**Fig. 1.** Stability diagram in the case $c_N = 0.0001$ and $c_T = 0.0439$.

and that its motion involves two different timescales. In the more general case $c_N \neq 0$ and $c_T \neq 0$ the model predicts that the bundle will always be stable if $T > c_T/c_N$, i.e. if the tension in the bundle at its downstream end is sufficiently large.

## Acknowledgement

The first author (NWM) gratefully acknowledges the financial support of the United Kingdom Engineering and Physical Sciences Research Council (EPSRC) and Subsea 7, Aberdeen via the CASE Project Studentship Scheme under the auspices of the Faraday Partnership for Industrial Mathematics, managed by the Smith Institute for Industrial Mathematics and System Engineering.

## References

1. A. P. Dowling. The dynamics of towed flexible cylinders. Part 1. Neutrally buoyant elements. *J. Fluid Mech.*, 187:507–532, 1988.
2. M. P. Païdoussis. Dynamics of flexible slender cylinders in axial flow. Part 1. Theory. *J. Fluid Mech.*, 26:717–736, 1966.
3. M. P. Païdoussis. Dynamics of cylindrical structures subjected to axial flow. *J. Sound Vib.*, 29:365–385, 1973.

4. M. P. Païdoussis. *Fluid-Structure Interactions: Slender Structures and Axial Flow, Volume 1.* Academic, 1998.
5. M. P. Païdoussis. *Fluid-Structure Interactions: Slender Structures and Axial Flow, Volume 2.* Elsevier, 2004.
6. G. S. Triantafyllou and C. Chryssostomidis. Stability of a string in axial flow. *J. Eng. Res. Tech.*, 107:421–425, 1985.

# Operators and Criteria for Integrating FEA in the Design Workflow: Toward a Multi-Resolution Mechanical Model

J.-C. Léon[1], P.M. Marin[2], and G. Foucault[3]

Laboratoire Sols, Solides, Structures, INPG - UJF - UMR CNRS 5521, Domaine Universitaire BP 53, 38041 Grenoble Cedex 9, France
[1] jean-claude.leon@hmg.inpg.fr
[2] philippe.marin2@hmg.inpg.fr
[3] gilles.foucault.1@ens.etsmtl.ca

**Summary.** In the design workflow, CAD models of complex components include more and more details. A transformation of such models into Finite Element (F.E.) models often generates a much too large number of elements to be used directly. Generally, the removal of shape details or idealization operations are required to prepare F.E. models. These modifications must preserve the analysis result and the user must control the process in order to ensure sufficient accuracy of the F.E. results. In accordance to the analysis problems, the simplification process generates different appropriate F.E. models. In this paper, we present different operators and criteria to prepare analysis models from CAD models.

**Key words:** shape simplification, mesh, polyhedral model, mechanical criterion, finite element accuracy

## 1 Introduction

Design models are used by all the actors of the design process and therefore contain numerous details. These models are often too refined for mechanical analyses and their direct use would generate too many finite elements. The adaptation of the model shape needs the removal of its details when their presence has either no or limited effects on its mechanical behavior while requiring an important local mesh density. Examples of these details include fillets, but also detailed entities such as holes, small blocks, etc.

Various software make it possible to automate this step partially. Several categories of approaches have been proposed to solve the problems involved by the preparation of F.E. models from CAD data. A first one addresses configurations were small features must be removed to get the geometric model more compatible with the size of the F.E. required [7, 2, 5]. These approaches are strongly dependent on the modelling history of the part and work on the construction tree of the object and the removal of user-selected features. A second one starts with a polyhedral model of the part [8, 1, 3]. In order to simplify the model, different adaptation functions work on the initial polyhedral model. They combine decimation process and removal of topological details. Another category of approaches is characterized by idealization treatments. Such operations are often required to transform a volume into an open surface to model a plate behaviour. Similar operations hold for transforming a volume feature into a line to model a beam behaviour of the structure.

The accuracy of F.E. computations is one of the main concerns of the users. The sources of the errors are multiple, errors of discretization, uncertainty about the boundary conditions and the behaviour law of the constitutive material, simplification of the shape, ... The quality of F.E. computation can be strongly influenced by the simplifications carried out on the shape. Appropriately choosing and monitoring these simplifications is therefore of primary importance. When the preparation of the model is manual, its quality depends on the engineer's know-how. For an automatic simplification process, the monitoring process uses geometrical criteria, curve, size [9]... In a priori step, geometrical criteria related to the mechanical properties of the problem can be added, variation of mass, volume, sections, centre of inertia [4]... A posteriori indicators can be used also and adaptive simplification process can be performed to define the most suited simplified model for each analysis case.

Section 2 presents the existing polyhedral model simplification algorithms we use to automatically prepare F.E. models. In section 3, different implemented criteria are listed and some examples illustrate their efficiency.

## 2 Simplification operators

Our approach uses an intermediate polyhedral model of the object. Using such a representation, we can integrate data from CAD models, pre-existing F.E. meshes or 3D scans. algorithm.

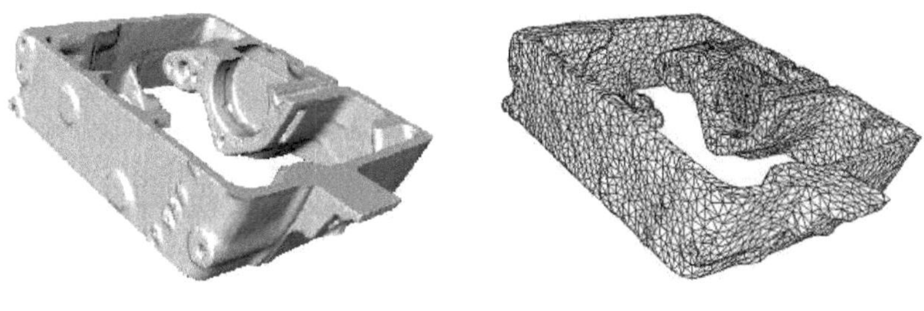

**Fig. 1.** Scan of a crankcase (courtesy Tomoadour-TurboMéca), generation of a F.E. mesh (total time ~2h), 427132 faces in the initial model and 11924 in the F.E. mesh.

The simplification process is based on an iterative vertex removal. Figure 1 shows the result of a simplification process.

## 3 Mechanical criteria

During a simplification process, a priori criteria are geometric ones but they can be related to mechanical property variations of mass, volume, section, centre of inertia. These indicators have been developed with our decimation process. We could provide the user either macro-scale information over the whole object or micro-scale information on the smallest possible entity of the geometrical model. Figure 2 shows two examples of such an indicator.

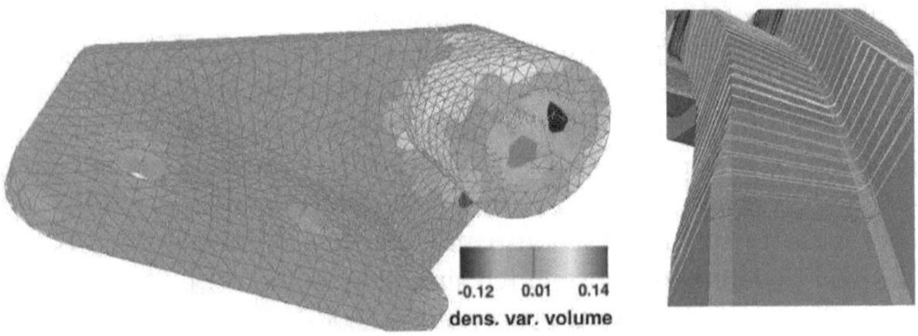


**Fig. 2.** Two examples of a priori criteria, variation of volume on the left part and variation of sections on the right part.

A priori criteria are an interesting information for the users but they cannot quantify directly the real influence of geometric simplifications on a F.E. simulation. For example, the errors generated by a hole removal will depend on the dimension of the hole but also of its location over the component. This form feature can be in an area involving either low or high stresses. To take into account its position, a mechanical criterion needs information about the highly stressed areas and hence a sketch of the F.E.A. results. To gain an idea of these results requires to set up an a posteriori process. After a simulation on the simplified part, mechanical criteria are computed for evaluate the influence of each suppressed feature. Based on criteria, the engineer can validate the quality of his (resp. her) F.E. results. These criteria can also used in an adaptive process of simplification to refine the simplified part by adding some of the suppressed features according to their mechanical influence.

Such a process of simplification was used in [10]. The criterion used was the error of discretization on the simplified problem and the map of sizes of an optimal mesh for this problem. The program removes details if the size of the detail is lower than that given in this area by the map of sizes. Another criterion can be estimation of the influence of each simplification on the strain energy variation [6]. In the framework of stationary linear problems, this variation is given by eq. (1) where $\sigma_2$ are the stresses of the simplified model, $U_1$ the displacements of the initial model and $n$ the outward unit normal. Obviously, these displacements are unknown and can be estimated by local computations around the removed feature.

$$\int_{feature\ boundary} \sigma_2.U_1.n.ds = \Delta\ strain\ energy\ variation \quad (1)$$

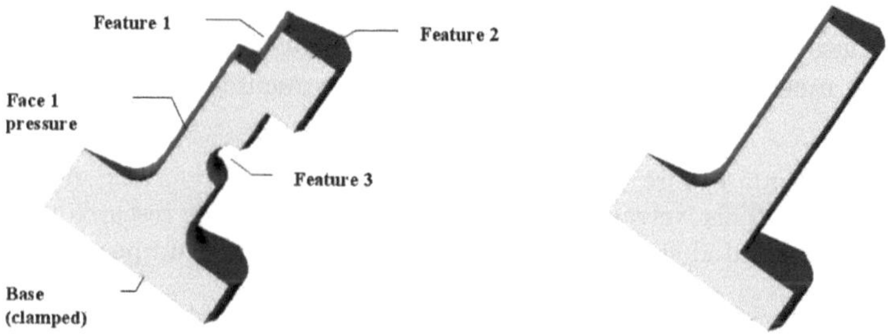


**Fig. 3.** Initial CAD model on the left and associated simplified model for analysis

On Fig. 3, we show an example of static problem that illustrates the efficiency of our criterion. The simplified model has been obtained by suppressing four features. In order to evaluate the efficiency of the criterion, we compute the solution and the strain energy on the initial part but also on the simplified one. We can see on table 1 that this criterion produces a good estimation of the influence of each simplification. During an adaptive process, the user computes the solution and the influence indicator on the simplified model. In this example, if the user wants to bound the accuracy to 1%, the simplified model must be refined by adding features 3 and 4.

**Table 1.** For each feature simplification, comparison between the real strain energy variation and our estimation of this variation.

| | A = real variation | B = Influence indicator | efficiency A/B |
|---|---|---|---|
| feature 1 | -0.014 % | -0.009 % | 1.56 % |
| feature 2 | 0.27 % | 0.26 % | 0.96 % |
| feature 3 | -8.4 % | -8.3 % | 1.01 % |
| feature 4 | 16 % | 18 % | 0.89 % |

## 4 Conclusion

The design process of complex structures needs the resolution of multiple mechanical analyses. Each analysis requires the generation of an adequate simulation model. To define easily such models and integrate F.E.A. in the design workflow, it is necessary to set up efficient simplification operators and criteria for evaluating and validating the simplification process. In this paper, we have presented a set of treatments contributing to this integration.

Developments are in progress to add other mechanical criteria, like a priori stiffness variation, a posteriori indicator for modal analysis, ... Work is also needed to automate the adaptive process using our a posteriori indicator.

All these treatments help transform CAD models from pure into multi-resolution mechanical models efficient for mechanical simulation preparation.

*Acknowledgement.* Future developments will be conducted in the AIM@SHAPE NoE framework. This Network of Excellence involves fourteen research teams about Shapes and associated semantic and it is supported by the European Community.

## References

1. J. Cohen, D. Manocha, A. Varshney, and G. Turk. Efficient model simplification with global error bounds. In *5th MSI Workshop on Computational Geometry*, Stony Brook, USA, 1995.
2. P. Dabke, V. Prabhakar, and S. Sheppard. Using features to support finite element idealisation. *Proc. of Int. ASME DETC*, 1:183–193, 1994.
3. L. Fine, L. Remondini, and J-C. Léon. Automated generation of FEA models though idealization operators. *Int. Journal of Numerical Methods in Engineering*, 49:83–108, 2000.
4. G. Foucault, P. Marin, and J-C Léon. Mechanical criteria for the preparation of finite element models. In *Proc. of 13th International Meshing Roundtable.* Sandia National Laboratories, 2004.
5. V. Fran¸cois and J-C. Cuillère. 3D automatic remeshing applied to model modification. *Computer Aided Design*, 33:377–388, 2000.
6. P. Marin. Influence of geometrical simplifications on the accuracy of finite element computation. In *Proc. of 4th International Conference on Engineering Computational technology*, Lisboa, 2004.
7. A. V. Mobley, M. P. Caroll, and S. A. Canann. An object oriented approach to geometry defeaturing for Finite Element Meshing. In *Proc. of 7th International Meshing Roundtable.* Sandia National Laboratories, 1998.
8. W.J. Schroeder, J. A. Zarge, and W. E. Lorensen. Decimation of triangle meshes. In *Computer Graphics SIGGRAPH'92*, pages 65–70, 1992.
9. P. Véron and J-C. Léon. Static polyhedron simplification using error measurements. *Computer Aided Design*, 29:287–298, 1997.
10. P. Véron and J-C. Léon. Shape preserving polyhedral simplification with bounded error. *Computers & Graphics*, 22:565–585, 1998.

# Wavelet Analysis of Sound Signal in Fluid-filled Viscoelastic Pipes

M. Prek

University of Ljubljana, Faculty of Mechanical Engineering, Askerceva 6, SI-1000 Ljubljana, Slovenia

**Summary.** In viscoelastic pipes, where the material properties depends on a complex bulk modulus as well as on a complex shear modulus, the sound field within the fluid is affected. Therefore, the dispersion of flexural waves occurs in the pipe, while the speed of flexural waves decreases due to the coupled fluid mass. Coupling between the pipe wall and the fluid also decreases the sound speed in the fluid. Likewise, the speed of sound in fluid is frequency-dependent, just as the group velocity of bending waves depends on the frequency. Wavelet transform of non-stationary sound signals was used to identify the frequency-dependent fluid sound speed. Measurement and analysis of non-stationary signals with the use of time-frequency method provides a view to frequency dependent transfer characteristics of fluid-pipe coupled system. The results also showed that, in the case of propagating small disturbances (such as acoustic waves), the pipe wall inertia has a minor influence on the wave propagation characteristics. The elastic reaction of the wall to expansion of the cross section greatly exceeds the inertial reactions.

**Key words:** wavelet transform, sound signal, viscoelastic pipe.

## 1 Introduction

In methods based on frequency response, the effect of viscoelastic properties is modelled through a frequency-dependent wave speed and a separate frequency-dependent damping factor. The impulse-response method has been utilized to calculate the water hammer [1]. Similarly, an impulse-response method applied to compute nonperiodic transients has been proposed [7]. The complex wave speed (complex-valued and frequency-dependent) is used in the standard impedance or transfer matrix method to analyse the oscillatory flow. A similar method has been proposed, which applied the concept of transmission loss instead of the concept of wall impedance [5]. An extended method uses the static mechanical properties and frequency-dependent mechanical properties of the pipe wall [8].

In this work the pulse method was used to measure propagation characteristics. An experimental study of axisymmetric propagation modes in fluid-filled viscoelastic pipes with emphasis on the two modes that exist down to the zero frequency limit is described. In our case the wavelet analysis was applied to the acoustic signal in order to analyse the structural features of the fluid-filled viscoelastic pipes, since the wavelet transform permits the characterization of a one-dimensional acoustic signal as a two-dimensional representation, evolving with time and period (frequency).

## 2 Experiment

For comparison, two different pipe materials with different properties were examined (steel, polypropylene (PP), and polybutylene (PB)). A 2.1 m long pipe of circular cross-section was filled with water and suspended vertically on a foam pad. One end of the pipe was tapped lightly with a hammer (Brüel-Kjær type 8202) and the output from the hammer was used for triggering. Measurement configuration consisted of a hydrophone (B-K type 8103) in conjunction with the charge amplifier (B-K type 2626), the FFT analyser (B-K type 2032) and a PC for additional computations. For the B-K type 2032 FFT analyser the sampling frequency was 65 kHz, while the chosen frequency span 12.8 kHz included all relevant frequency information. The sampling interval was 30.5 $\mu$s and record length 62.5 ms. Transient analysis of measurement data was done with a rectangular window as it provides equal weighting across the measurement period. The excited acoustic waves were registered and the input signal of 2048 samples has been processed.

## 3 Analysis and Results

In the investigation, the wavelet transforms were performed on the exponentially time-decaying frequency-dependent signals. The analysis was performed by using the Morlet wavelet function $\psi(\tau)$, which depends on a nondimensional 'time' parameter $\tau$ [2, 4, 3]. The Morlet wavelet $\psi(\tau)$ consists of a plane wave modulated by a Gaussian, such that:

$$\psi(\tau) = \pi^{-\frac{1}{4}} \cdot e^{i \cdot \omega_0 \cdot \tau} \cdot e^{-\frac{\tau^2}{2}} \tag{1}$$

where $\omega_0$ is the nondimensional frequency—in our case $\omega_0 = 6$ to satisfy the admissibility condition. The wavelet transform $W_n(x_n, s)$ of a discrete sequence $x_n$ (where $x_n = f(n \cdot \delta t)$) is defined as the convolution of $x_n$ by a scaled and translated version of wavelet function $\psi(\tau)$ as:

$$W_n(x_n, s) = \sum_{n=0}^{N-1} x_n \cdot \psi^* \left[ \frac{(n' - n) \cdot \delta t}{s} \right] A \tag{2}$$

The set of scales s for use in the nonorthogonal wavelet analysis were defined as fractional powers of two:

$$s_j = s_0 \cdot 2^{j \cdot \delta j} j = 0, 1, 2, \ldots, \quad J = \frac{1}{\delta j} \cdot \log_2 \left( \frac{N \cdot \delta t}{s_0} \right) \tag{3}$$

where $s_0$ is the smallest resolvable scale and $J$ determines the largest scale. The $s_0$ was chosen such that the equivalent Fourier period is approximately $2 \cdot \delta t$. Adequate sampling within the scale is provided by a $\delta j$ of about 0.5 for the Morlet wavelet. Smaller values of $\delta j$ give finer resolution; to provide a smooth picture of wavelet power, $\delta j = 0.25$ was used in the analysis. The time-frequency (or scale) representation of the energy concentration of the wavelet transform (WT) is called the ridge. The dominant features of each map are extracted by identifying correlation peaks [6]. Each peak in WT map represents the arrival time of a wave travelling with the group velocity. For a harmonic waves propagating in $L$ direction with small angular frequency difference $\Delta\omega$, the group velocity $c_g$ at the mean angular frequency $\omega_g$ can be defined as $c_g = \Delta\omega/\Delta k$, where $k$ is the wave number. The magnitude of WT takes its minimum value at $s = \omega_0/\omega_g$ and $x = (\Delta k/\Delta\omega) \cdot L = L/c_g$. Therefore, for a fixed distance $L$, a three-dimensional plot of $|W_n(x_n, s)|$ on the $(x, s)$ plane has a peak at $(x, s) = (L/c_g, \omega_0/\omega_g)$. In other words, the location of the peak on the WT map indicates the arrival time $x = L \cdot c_g$ of the wave having angular frequency $\omega_g = \omega_0/s$.

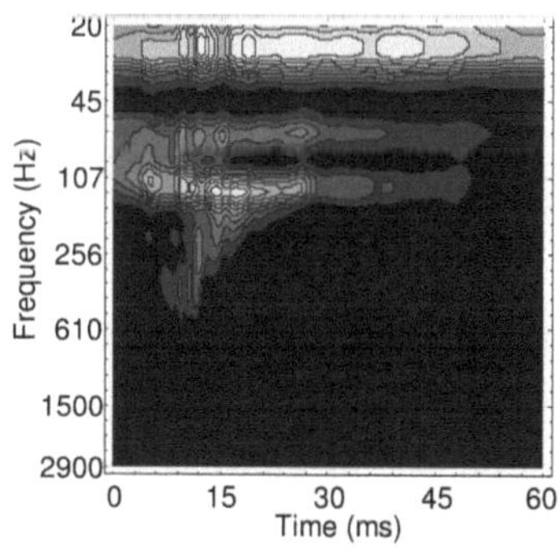


**Fig. 1.** Wavelet transform of sound signal in polybutilene pipe

In Fig. 1 is shown the wavelet transform of acoustic signal in polybutilene (PB) pipe. The influence of wall elasticity significantly alters the speed of sound (which is at zero frequency limit 470 $(+34/-25) m/s$ (at 95% confidence interval), as obtained from the map), while the structural damping causes wave attenuation. The confidence interval is defined as the probability that the true wavelet power at a certain time and scale lies within a certain interval about the estimated wavelet power. The confidence interval, defined as $Wt_n^2(x_n, s)$, is then

$$\frac{2}{\chi_2^2\left(\frac{p}{2}\right)} \cdot \left|W_n\left(x_n, s\right)\right|^2 \leqslant Wt_n^2\left(x_n, s\right) \leqslant \frac{2}{\chi_2^2\left(1-\frac{p}{2}\right)} \cdot \left|W_n\left(x_n, s\right)\right|^2 \qquad (4)$$

where $p$ is the desired significance ($p = 0.05$ for the 95% confidence interval) and $\chi_2^2\left(\frac{p}{2}\right)$ represents the value of $\chi^2$ at $p/2$.

There are two significant phenomena that appear when the pipe wall yields. With rigid walls, the lowest order mode is truly a plane wave mode, but with elastic walls, the lowest order mode exhibits a dispersive sound speed, which is at all frequencies slower than the free field value. With rigid walls, the higher order modes exhibit a cut-off frequency, that is, there is a frequency below which each higher order mode will not propagate, hence at low frequency, the plane wave mode exists alone. With elastic walls, this is no longer true. At least one higher order mode can exist down to zero frequency. The time-frequency distribution of the magnitude of the WT peaks depends on frequency of group velocity. Thus the changes in wavelets that correlate highly with the signal at different times indicate the changes in features of the signal at time progresses.

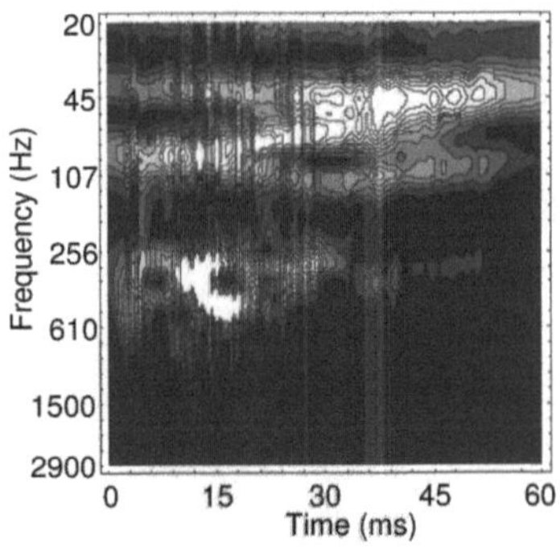


**Fig. 2.** Wavelet transform of sound signal in polypropylene pipe

Similar values were obtained for another viscoelastic pipe material (PP - Fig. 2), where the influence of wall elasticity also significantly alters the speed of sound (which is at zero frequency limit 380 $(+32/-28)m/s$ at 95% confidence interval). Since the speed of sound in fluid-filled pipes depends on the pipe wall material, this changes the effective bulk modulus.

## 4 Conclusions

The wavelet analysis was applied to the acoustic signal in order to analyse the structural features of the fluid-filled viscoelastic pipes. The main object was to reveal the influence of physical properties of pipe wall material on acoustic wave propagation. Wavelet transform of non-stationary acoustic signals was used to identify the frequency-dependent fluid sound speed. The dominant features of each map are extracted by identifying correlation peaks. Each peak in WT map represents the arrival time of a wave travelling with the group velocity. Using the time-frequency distribution of the WT, the dependency on frequency of group velocity and attenuation is evaluated. The influence of wall elasticity significantly alters the speed of sound, while the structural damping causes wave attenuation. The dispersive waves appear in he map as curved ridges that are asymptotic to the resonant frequency. Successive reflections appear in the map as families of high wavelet coefficients at constant period (or frequency). By measuring the time separation between ridges at a given frequency, the wave velocity at that frequency can be calculated. In the pipe with semi-rigid wall, the lowest order mode is truly a plane wave mode, but in the pipes with elastic walls, the lowest order mode exhibits a dispersive sound speed, which is at all frequencies significantly slower than the free field value.

## References


1. P.-G. Franke and F. Seyler. Computation of unsteady pipe flow with respect to visco-elastic material properties. *J. Hydraulic Res.*, 21:345–353, 1983.
2. A. Grossmann and J. Morlet. Decomposition of Hardy functions into square integrable wavelets of constant shape. *SIAM J. on Math. Analysis*, 4:723–736, 1984.
3. S. Mallat. *A Wavelet Tour of Signal Processing.* Academic Press, San Diego, 1999.
4. J. Morlet, G. Arens, I. Furgeau, and D. Giard. Wave propagation and sampling theory. *Geophysics*, 2:203–236, 1982.
5. M.L. Munjal and P.T. Thawani. Prediction of the vibro-acoustic transmission loss of planar hose-pipe systems. *J. Vib. Acoust*, 101:2524–2535, 1999.
6. D.E. Newland. Ridge and phase identification in frequency analysis of transient signals by harmonic wavelets. *J. Vib. Acoust*, 121:149–155, 1999.
7. L. Suo and E.B. Wylie. Complex wave speed and hydraulic transients in viscoelastic pipes. *J. Fluids Eng.*, 112:496–500, 1990.
8. J. Yu and E. Kojima. Wave propagation in fluids contained in finite-length anisotropic viscoelastic pipes. *J. Acoust. Soc. Am*, 104:3227–3235, 1998.

# Coarse-Grained Simulation and Bifurcation Analysis Using Microscopic Time-Steppers

P. Van Leemput[1], G. Samaey[1], K. Lust[1,2], D. Roose[1], and I.G. Kevrekidis[3]

[1] Department of Computer Science, K.U. Leuven, B-3001 Heverlee, Belgium
`{pieter.vanleemput, giovanni.samaey, dirk.roose}@cs.kuleuven.ac.be`

[2] Institute of Mathematics and Computing Science, University of Groningen, 9700 AV Groningen, the Netherlands `k.w.a.lust@math.rug.nl`

[3] Department of Chemical Engineering, PACM and Department of Mathematics, Princeton University, Princeton, USA `yannis@princeton.edu`

**Summary.** In many science and engineering problems, one observes smooth behaviour on macroscopic space and time scales. However, sometimes only a microscopic evolution law is known. In such cases, one can approximate the macroscopic time evolution by performing appropriately initialized simulations of the available microscopic model in small portions of the space-time domain. This coarse-grained time-stepper can be used to perform time-stepper based numerical bifurcation analysis. We discuss our recent results concerning the accuracy of the proposed methods.

**Key words:** coarse-graining, multiscale simulation, bifurcation analysis.

## 1 Introduction

Many systems for which only a microscopic evolution law is known, exhibit smooth behaviour on macroscopic space and time scales. In such systems, the evolution in the microscopic space takes place on a lower-dimensional manifold (sometimes called the slow manifold). For such time-dependent multiscale problems, a "coarse-graining" approach which exploits this property has been proposed [3]. It is assumed that one knows which variables determine evolution on macroscopic time and length scales, but one is unable to obtain an explicit closed macroscopic model. The key idea is to extract information on the evolution of these macroscopic variables through appropriately initialized simulations using the (given) microscopic evolution law. This information is then used to construct a "coarse-grained" time-stepper for the macroscopic variables, which consists of (1) construction of one or more microscopic initial states corresponding to the macroscopic initial condition (*lifting*); (2) simulation using the microscopic evolution law and (3) computation of a new macroscopic state (*restriction*). To reduce the computational cost, the *patch dynamics* scheme was proposed, which performs the simulations only in small portions of the space-time domain [3].

Once an accurate coarse-grained time-stepper is constructed, one can perform more general tasks, such as the computation and stability analysis of steady states and periodic solutions using *existing* time-stepper based numerical bifurcation analysis techniques, *e.g.*, [4].

Here we focus on two aspects of this coarse-graining approach, patch dynamics for diffusion problems and coarse-grained numerical bifurcation analysis of lattice Boltzmann models. In both cases the macroscopic model is known explicitly, which allows for a detailed study of the numerical accuracy.

## 2 Patch Dynamics

We consider the following homogenization problem for diffusion

$$\partial_t u = \partial_x \left( a\left( x/\varepsilon \right) \partial_x u \right), \qquad a(y) \text{ periodic in } y, \qquad \varepsilon << 1. \tag{1}$$

The solution of this partial differential equation (PDE) is highly oscillatory in space, but it is known from theory [1] that the macroscopic, averaged solution $U(x,t)$ satisfies a diffusion equation with *constant* diffusion coefficient $a^*$, which we assume to be unknown. The goal of patch dynamics is to perform simulations of this averaged equation, making only use of (1) in small portions of the space-time domain.

To obtain the averaged solution using an equidistant, macroscopic mesh of width $\Delta x$, we consider a small interval (box) of length $h$ around each mesh point, as well as a larger *buffer* box of size $H > h$ (Fig. 1, left). The coarse-grained time-stepper, called *gap-tooth scheme* [3, 7], is constructed as follows:

1. **Lifting.** Define the initial condition of (1) in each box as a Taylor expansion, based on the (given) box averages $U_i^n$, $i = 0, \ldots, N$, at $(x_i,t_n)$,

$$\tilde{u}_i(x,t_n) = D_i^0 + D_i^1(x-x_i) + D_i^2(x-x_i)^2/2, \; x \in [x_i - H/2, x_i + H/2], \tag{2}$$

with $D_i^k$, $k > 0$ a finite difference approximation for the $k$-th spatial derivative at $x_i$, and $D_i^0$ chosen to ensure $(1/h)\int_{x_i-h/2}^{x_i+h/2} \tilde{u}_i(\xi,t_n)\mathrm{d}\xi = U_i^n$.
2. **Simulation.** We compute the box solution $\tilde{u}_i(x,t)$, $t_n < t < t_n + \Delta t$, by solving equation (1) in the interval $[x_i - H/2, x_i + H/2]$ with the *built-in* boundary conditions of the microscopic code.
3. **Restriction.** We average $\tilde{u}_i$ over the *inner* box of size $h$ only

$$U_i^{n+\Delta} = (1/h) \int_{x_i-h/2}^{x_i+h/2} \tilde{u}_i(\xi, t_{n+\Delta})\mathrm{d}\xi.$$

We use the larger box of size $H$ to "shield" the effects of the artificial boundaries from the domain of interest (the box of size $h$). This works if the simulation is performed over *short enough* time intervals $\Delta t$, and $H$ is *large enough.*

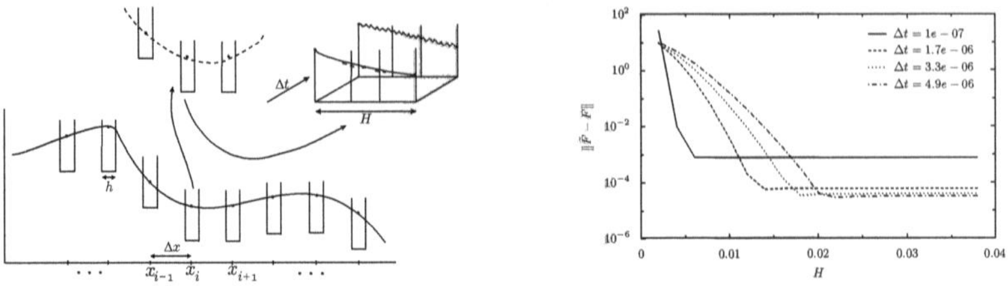


**Fig. 1.** Left: Schematic representation of the gap-tooth scheme with buffers. Right: $\|\tilde{F} - F\|$ as a function of $H$ for $\Delta t = 1 \cdot 10^{-7}$, $1.7 \cdot 10^{-6}$, $3.3 \cdot 10^{-6}$ and $4.9 \cdot 10^{-6}$.

To reduce the computational effort in time, we use the gap-tooth scheme as a time derivative estimator $\tilde{F}$ for the unavailable equation, which we then combine with a macroscopic forward Euler scheme. The resulting scheme is

$$U^{n+1} = U^n + M\Delta t\tilde{F}(U^n; H, h, \Delta t) = U^n + M(U^{n+\Delta} - U^n), M \gg 1. \quad (3)$$

We showed [6] that the error of $\tilde{F}$ is bounded by

$$\|\tilde{F} - F\| \leq C\|h^2 + \varepsilon/\Delta t + \Delta t^2 + E(\Delta t, H)\|, \quad (4)$$

where $F$ is the time derivative of the finite difference approximation of the macroscopic equation on the same mesh, and $E(\Delta t, H)$ is the error due to the boundary artefacts in each box. Figure 1 (right) shows the error for $a(y) = 1.1 + \sin(2\pi y)$, $\varepsilon = 1 \cdot 10^{-5}$, $h = 2 \cdot 10^{-3}$ and $\Delta x = 0.1$. Then $a^* \approx 0.45825686$. We see that $E(\Delta t, H)$ decreases faster than exponentially with $H$. The rate of decrease is higher for smaller $\Delta t$, but the optimal accuracy is lower, see (4).

## 3 Coarse-grained Numerical Bifurcation Analysis

As the macroscopic system, we consider the FitzHugh-Nagumo PDE [8]

$$\begin{cases} \partial_t \rho^{ac} = \partial_{xx}\rho^{ac} + \rho^{ac} - (\rho^{ac})^3 - \rho^{in}, \\ \partial_t \rho^{in} = d^{in}\, \partial_{xx}\rho^{in} + \lambda(\rho^{ac} - a_1\rho^{in} - a_0), \end{cases} \quad (5)$$

with homogeneous Neumann boundary conditions on the one-dimensional domain $[0, 20]$. The variables $\rho^{ac}(x,t)$ and $\rho^{in}(x,t)$ are the activator and inhibitor concentration. We set $d^{in} = 4$, $a_0 = -0.03$, $a_1 = 2$ and vary $\lambda \in [0,1]$. As the microscopic model, we used an equivalent lattice Boltzmann (LB) BGK model [5]. The LB variables are the distribution functions $f_j^s(x,t)$ (with $s \in \{ac, in\}$), defined on a space-time lattice with spacing $\delta x$ and $\delta t$ respectively, for "particle" velocities $v_j = j\,\delta x/\delta t$ with $j \in \{-1, 0, 1\}$. The concentration, our macroscopic variable, is then defined as

$$\rho^s(x,t) = \sum_{j=-1}^{1} f_j^s(x,t). \quad (6)$$

The coarse-grained time-stepper for this LB model is constructed as follows. Given the initial density $\rho^s(x,t)$, the lifting can be done in a number of ways. Good results were obtained with $f_j^s(x,t) = (1/3)\,\rho^s(x,t)$. This scheme initializes the LB model close to the correct point on the "slow manifold" [10]. Next, we run the LB model over a time interval $\Delta t$, after which we compute $\rho^s(x, t + \Delta t)$ from (6). This procedure is repeated within the time interval $[0, T]$. Due to errors in the lifting step, $\Delta t$ has to be large enough [10], but nonetheless small phase shifts remain. In [10] we showed that large errors can occur if the lifting is not performed properly. More sophisticated lifting schemes (as in [2]) are currently under investigation.

We performed a numerical bifurcation analysis of both the steady and periodic solutions using the Newton-Picard method [4, 9]. The solutions and their stability-determining (dominant) eigenvalues are computed through calls to the time-stepper only. In our case, the time-stepper is either *(a)* a Crank-Nicolson discretization of the PDE (5), *(b)* the coarse-grained LB time-stepper and *(c)* the full LB model [9].

Figure 2 shows the bifurcation diagram for periodic solutions. We use $\delta x = 0.1$, $\delta t = 0.001$ and $\Delta t \approx 5$. The periodic solution branch has a fold point at $\lambda \approx 0.00087$ and meets at $\lambda \approx 0.0183$ with a branch of steady states in a Hopf point. Although the unstable part of the branch after the fold has almost the same $(\lambda, T)$-projection as the stable one, the corresponding solutions are different. The solution curves for the coarse-grained LB *with appropriate lifting*, the full LB and the PDE model as wel as the position of the bifurcation points (determined by monitoring the dominant eigenvalues) correspond very well; their differences are of the order of the PDE discretization error.

# 4 Conclusions

We constructed different coarse-grained time-steppers to simulate the smooth macroscopic evolution of (reaction-)diffusion problems. First, we focused on increasing space-time efficiency using the patch dynamics scheme, where we introduced buffers to cope with artifical boundary effects. Secondly, we performed time-stepper based numerical bifurcation analysis of a coarse-grained lattice Boltzmann model and showed that the results correspond with those of the full lattice Boltzmann model and of the equivalent PDE. Further details can be found in [6, 7, 9, 10].

*Acknowledgement.* GS is a Research Assistant and KL was a Postdoctoral Fellow of the Fund for Scientific Research - Flanders which also provided further funding through project G.0130.03. This paper presents research results of the Belgian Programme on Interuniversity Attraction Poles, initiated by the Belgian Federal Science Policy Office. The work of IGK was partially supported by AFOSR and by an NSF/ITR grant.

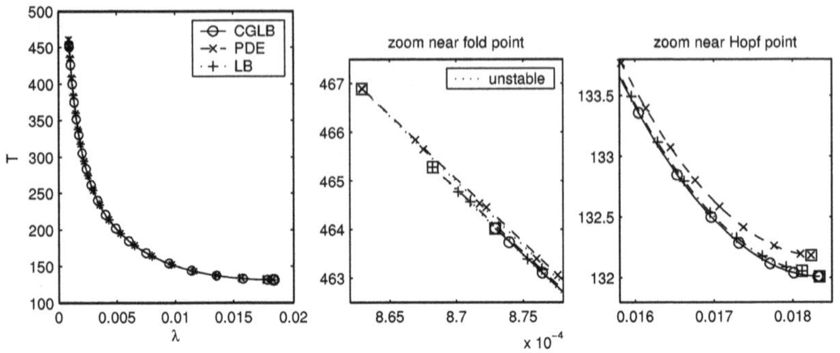


**Fig. 2.** Bifurcation diagram for periodic solutions. Unstable solutions are indicated by dotted lines and bifurcation points by boxed markers.

## References


1. A. Bensoussan, J.L. Lions, and G. Papanicolaou. *Asymptotic Analysis of Periodic Structures*, volume 5 of *Studies in Mathematics and its Applications*. North-Holland, Amsterdam, 1978.
2. C.W. Gear and I.G. Kevrekidis. Constraint-defined manifolds: a legacy code approach to low-dimensional computation. Technical Report physics/0312094, arXiv e-Print archive, 2003.
3. I.G. Kevrekidis, C.W. Gear, J.M. Hyman, P.G. Kevrekidis, O. Runborg, and C. Theodoropoulos. Equation-free, coarse-grained multiscale computation: Enabling microscopic simulators to perform system-level analysis. *Communications in Mathematical Sciences*, 1(4):715–762, 2003.
4. K. Lust, D. Roose, A. Spence, and A. Champneys. An adaptive Newton-Picard algorithm with subspace iteration for computing periodic solutions. *SIAM Journal on Scientific Computing*, 19(4):1188–1209, 1998.
5. Y.H. Qian and S.A. Orszag. Scalings in diffusion-driven reaction $A + B \rightarrow C$: Numerical simulations by Lattice BGK Models. *Journal of Statistical Physics*, 81(1/2):237–253, 1995.
6. G. Samaey, I.G. Kevrekidis, and D. Roose. Patch dynamics with buffers for homogenization problems. Technical Report physics/0412005, arXiv e-Print archive, 2004. Submitted to Journal of Computational Physics.
7. G. Samaey, D. Roose, and I.G. Kevrekidis. The gap-tooth scheme for homogenization problems. *SIAM Multiscale Modeling and Simulation*, 2004. In press.
8. C. Theodoropoulos, Y.H. Qian, and I.G. Kevrekidis. "Coarse" stability and bifurcation analysis using time-steppers: a reaction-diffusion example. *Proceedings of the National Academy of Sciences*, 97(18):9840–9843, 2000.
9. P. Van Leemput and K. Lust. Numerical bifurcation analysis of lattice Boltzmann models: a reaction-diffusion example. In M. Bubak, G.D. van Albada, P.M.A. Sloot, and J. Dongarra, editors, *Computational Science – ICCS 2004*, volume 3039 of *LNCS*, pages 572–579. Springer-Verlag, 2004.
10. P. Van Leemput, K. Lust, and I.G. Kevrekidis. Coarse-grained numerical bifurcation analysis of lattice Boltzmann models. Technical Report TW 410, Dept. of Computer Science, K.U.Leuven, 2004. Submitted to Physica D.

# Optimal Prediction in Molecular Dynamics

B. Seibold

University Kaiserslautern, Germany `seibold@mathematik.uni-kl.de`

**Summary.** Molecular dynamics simulations are typically very costly. We investigate whether optimal prediction, a method to approximate the mean solution of a large system of ordinary differential equations by a smaller system, can in principle be applied to speed up computations. A one-dimensional, solely classical model problem, describing some aspects of coating a copper layer onto a silicon crystal, is considered. Asymptotic methods are employed to approximate the high-dimensional conditional expectations, which arise in optimal prediction. Results of a comparison of the thus derived smaller system with the original system are shown.

**Key words:** optimal prediction, molecular dynamics, surface coating, hopping, Laplace's method, low temperature asymptotics.

## 1 Problem Description

Computations in molecular dynamics are typically costly due to a large number of atoms having to be computed over a large number of time steps. We introduce a one-dimensional model problem and investigate whether the method of optimal prediction can be applied to reduce the number of atoms being computed.

### 1.1 Industrial Problem

In the production of semiconductors a thin layer of copper has to be coated onto a silicon crystal. The crystal is bombarded by copper atoms, such that a copper layer forms on top. The time between two copper atoms hitting the crystal surface is about $10^{-4}$ seconds, while the system is out of its thermodynamical equilibrium for only $10^{-11}$ seconds after one copper atom has impacted. Hence, the system is in equilibrium nearly all the time.

However, even in equilibrium, single copper atoms can penetrate deep into the crystal due to *atomic hopping*: A copper atom gains by accident enough energy to overcome the potential barrier between two layers in the silicon crystal and hops to a neighboring cell. In practice, a clear separation between the copper layer and the silicon crystal is desired.

### 1.2 ITWM Project

The above described process is investigated in the *Institute for Industrial Mathematics* (ITWM), using molecular dynamics simulations [3]. The ITWM simulations include quantum mechanical effects. In order to speed up computations, it is desired to compute the 50 top layers of the crystal exactly, while the lower layers' dynamics should be replaced by something cheaper to compute, such as choosing fewer atoms (*coupling of length scales*) or introducing a continuum (partial differential equations). In the following, we show that optimal prediction can hold as another possibility to reduce the number of unknowns.

### 1.3 One Dimensional Model Problem

We introduce a one-dimensional model problem which preserves the properties of interest from three dimensions and also allows atomic hopping. We consider purely classical aspects of the dynamics. The pair potential between two silicon atoms is taken from the three-dimensional case (*Lennard-Jones* potential). The pair potential between silicon and copper, however, is changed to be finite at zero distance, *i.e.*, silicon and copper can interchange places (*hopping*), provided their energy is high enough.

## 2 Optimal Prediction

Optimal Prediction was introduced by Chorin, Kast, Kupferman [2], as a method for underresolved computation, *i.e.*, reducing computational effort by using prior statistical information. Sought is the mean solution of a system, where part of initial data is known and the rest is sampled from an underlying measure. Optimal prediction approximates the mean solution by a system smaller than the original system. Consider a $2n$-dimensional Hamiltonian system

$$\dot{\boldsymbol{q}} = \frac{\partial H}{\partial \boldsymbol{p}}, \quad \dot{\boldsymbol{p}} = -\frac{\partial H}{\partial \boldsymbol{q}} \tag{1}$$

with Hamiltonian function

$$H(\boldsymbol{q},\boldsymbol{p}) = \tfrac{1}{2}\sum_{i=1}^{n} \frac{p_i^2}{m_i} + V(\boldsymbol{q}). \tag{2}$$

The system may be distributed according to the canonical ensemble

$$f(\boldsymbol{q},\boldsymbol{p}) = Z^{-1}e^{-\beta H(\boldsymbol{q},\boldsymbol{p})}, \quad \beta = (k_B T)^{-1}. \tag{3}$$

Assume that only $m$ of the $n$ atoms are of interest, *i.e.*, $\hat{\boldsymbol{q}} = (q_1, \ldots, q_m)$ and $\hat{\boldsymbol{p}} = (p_1, \ldots, p_m)$ are sought, while $\tilde{\boldsymbol{q}} = (q_{m+1}, \ldots, q_n)$ and $\tilde{\boldsymbol{p}} = (p_{m+1}, \ldots, p_n)$ should be averaged out. Of the initial conditions only $\hat{\boldsymbol{q}}(0)$ and $\hat{\boldsymbol{p}}(0)$ are known, while $\tilde{\boldsymbol{q}}(0)$ and $\tilde{\boldsymbol{p}}(0)$ are sampled from the corresponding conditioned measure. Introducing the conditional expectation projection

$$Pu = \mathbb{E}[u|\hat{\boldsymbol{q}}, \hat{\boldsymbol{p}}] = \frac{\int\int u(\hat{\boldsymbol{q}}, \tilde{\boldsymbol{q}}, \hat{\boldsymbol{p}}, \tilde{\boldsymbol{p}}) e^{-\beta H(\hat{\boldsymbol{q}}, \tilde{\boldsymbol{q}}, \hat{\boldsymbol{p}}, \tilde{\boldsymbol{p}})} \mathrm{d}\tilde{\boldsymbol{q}} \mathrm{d}\tilde{\boldsymbol{p}}}{\int\int e^{-\beta H(\hat{\boldsymbol{q}}, \tilde{\boldsymbol{q}}, \hat{\boldsymbol{p}}, \tilde{\boldsymbol{p}})} \mathrm{d}\tilde{\boldsymbol{q}} \mathrm{d}\tilde{\boldsymbol{p}}}. \tag{4}$$

One can show that the mean solution is the projection $P$ applied to the solution. Optimal prediction sets up a $2m$-dimensional system which arises when applying $P$ to the right-hand side of the original system:

$$\dot{\boldsymbol{q}}_{op} = \mathbb{E}\left[\tfrac{\partial H}{\partial \boldsymbol{p}} |\hat{\boldsymbol{q}}, \hat{\boldsymbol{p}}\right], \quad \dot{\boldsymbol{p}}_{op} = -\mathbb{E}\left[\tfrac{\partial H}{\partial \boldsymbol{q}} |\hat{\boldsymbol{q}}, \hat{\boldsymbol{p}}\right]. \tag{5}$$

Hald showed in [1] that if a system is Hamiltonian, then its optimal prediction system is also Hamiltonian with Hamiltonian function

$$\mathfrak{H}(\hat{\boldsymbol{q}}, \hat{\boldsymbol{p}}) = -\tfrac{1}{\beta} \log \left( \int\int e^{-\beta H(\hat{\boldsymbol{q}}, \tilde{\boldsymbol{q}}, \hat{\boldsymbol{p}}, \tilde{\boldsymbol{p}})} \mathrm{d}\tilde{\boldsymbol{q}} \mathrm{d}\tilde{\boldsymbol{p}} \right) \tag{6}$$

$$= \tfrac{1}{2} \sum_{i=1}^{m} \frac{p_i^2}{m_i} \underbrace{- \tfrac{1}{\beta} \log \left( \int e^{-\beta V(\hat{\boldsymbol{q}}, \tilde{\boldsymbol{q}})} \mathrm{d}\tilde{\boldsymbol{q}} \right)}_{= \mathfrak{V}(\hat{\boldsymbol{q}})}. \tag{7}$$

## 2.1 Low Temperature Asymptotics

Note that (7) involves an $(n-m)$-dimensional integral, which for a general potential $V$ cannot be evaluated explicitly. We approximate expression (7) using *Laplace's method* [4]. Assuming that for any fixed $\hat{\boldsymbol{q}}$, the potential $V(\hat{\boldsymbol{q}}, \boldsymbol{r})$ has a unique minimizer $\boldsymbol{r}(\hat{\boldsymbol{q}})$ and that the Hessian $H_{\tilde{\boldsymbol{q}}} V = \frac{\partial^2 V}{\partial \tilde{\boldsymbol{q}}^2}(\hat{\boldsymbol{q}}, \boldsymbol{r}(\hat{\boldsymbol{q}}))$ is regular (see [5] on these assumptions), we can perform an asymptotic expansion for low temperature, *i.e.*, $\beta \to \infty$

$$\begin{aligned} \int e^{-\beta V(\hat{\boldsymbol{q}}, \tilde{\boldsymbol{q}})} \mathrm{d}\tilde{\boldsymbol{q}} &= \int e^{-\beta\left(V(\hat{\boldsymbol{q}}, \boldsymbol{r}) + \frac{1}{2}(\tilde{\boldsymbol{q}} - \boldsymbol{r})^T H_{\tilde{\boldsymbol{q}}} V (\tilde{\boldsymbol{q}} - \boldsymbol{r})\right)} \mathrm{d}\tilde{\boldsymbol{q}} \\ &= e^{-\beta V(\hat{\boldsymbol{q}}, \boldsymbol{r})} \int e^{-\frac{\beta}{2}(\tilde{\boldsymbol{q}} - \boldsymbol{r})^T H_{\tilde{\boldsymbol{q}}} V (\tilde{\boldsymbol{q}} - \boldsymbol{r})} \mathrm{d}\tilde{\boldsymbol{q}} \\ &= e^{-\beta V(\hat{\boldsymbol{q}}, \boldsymbol{r})} \, (2\pi/\beta)^{\frac{n-m}{2}} \, |\det H_{\tilde{\boldsymbol{q}}} V|^{-\frac{1}{2}}. \end{aligned} \tag{8}$$

The last expression is obtained using the transformation rule. Resubstitution yields the final result

$$\mathfrak{V}(\hat{\boldsymbol{q}}) = V(\hat{\boldsymbol{q}}, \boldsymbol{r}) + \tfrac{1}{2\beta} \log |\det H_{\tilde{\boldsymbol{q}}} V| + O(\tfrac{1}{\beta^2}). \tag{9}$$

In the sequel we will use the *zero temperature limit* only

$$\mathfrak{V}_0(\hat{\boldsymbol{q}}) = V(\hat{\boldsymbol{q}}, \boldsymbol{r}(\hat{\boldsymbol{q}})). \tag{10}$$

Note that the high-dimensional integration has been replaced by high-dimensional minimization in finding $\boldsymbol{r}(\hat{\boldsymbol{q}})$. This can be interpreted as: Given $m$ *real* atoms, place $n-m$ *virtual* atoms, such that the potential energy $V(\hat{\boldsymbol{q}}, \boldsymbol{r}(\hat{\boldsymbol{q}}))$ is minimized. Using the fact that $\frac{\partial \mathfrak{V}_0}{\partial \hat{q}}(\hat{\boldsymbol{q}}) = \frac{\partial V}{\partial \hat{q}}(\hat{\boldsymbol{q}}, \boldsymbol{r}(\hat{\boldsymbol{q}}))$, one can derive a new, and merely $(n+m)$-dimensional, system of equations of motions

$$\begin{aligned} \dot{\hat{\boldsymbol{q}}} &= \mathfrak{M}^{-1} \cdot \hat{\boldsymbol{p}} \\ \dot{\hat{\boldsymbol{p}}} &= \tfrac{\partial V}{\partial \hat{q}}(\hat{\boldsymbol{q}}, \boldsymbol{r}) \\ \dot{\boldsymbol{r}} &= \left(\tfrac{\partial^2 V}{\partial \bar{q}^2}(\hat{\boldsymbol{q}}, \boldsymbol{r})\right)^{-1} \cdot \tfrac{\partial^2 V}{\partial \hat{q} \partial \bar{q}}(\hat{\boldsymbol{q}}, \boldsymbol{r}) \cdot \mathfrak{M}^{-1} \cdot \hat{\boldsymbol{p}}. \end{aligned} \tag{11}$$

Here $\mathfrak{M}$ is a diagonal matrix containing the atoms' masses.

### 2.2 Boundary Layer Condition

In a crystal, atomic potentials typically reach only over $k \approx 10$ atomic distances. Consequently, in system (11) only the first $k$ *virtual* atoms have to be computed, the others align equidistantly while following the potential minimum. This allows to reduce (11) to a $(2m+k)$-dimensional system. The new system can be interpreted as a system of $m$ atoms with a *boundary layer condition* given by $k$ *virtual* atoms.

### 2.3 Computational Speed Up

In the model problem the optimal prediction system (11) with $n-m$ *virtual* atoms did not yield any speed up, but the system with only $k$ *virtual* atoms did. The choice $m = \frac{n}{2}$ resulted in a speed up factor of 2 for $n = 50$ and a speed up factor of 6 for $n = 100$.

## 3 Comparing Optimal Prediction to the Original System

The optimal prediction system described in Subsection 2.2 is compared to the original system in terms of statistical quantities, which are obtained by Monte Carlo sampling. Note that error bounds in phase space can be derived from equation (9), however, these are irrelevantly large. Two systems may significantly deviate in phase space, while exhibiting identical statistical mechanics. We investigate how well copper diffusion coefficients and energy fluctuations are preserved. In Fig. 1 one can observe that the diffusion behaviour of a copper atom in the silicon crystal is nonlinear, and this behaviour is reflected well by the optimal prediction system. In Fig. 2 the variance of the energy of the $m$ *real* atoms

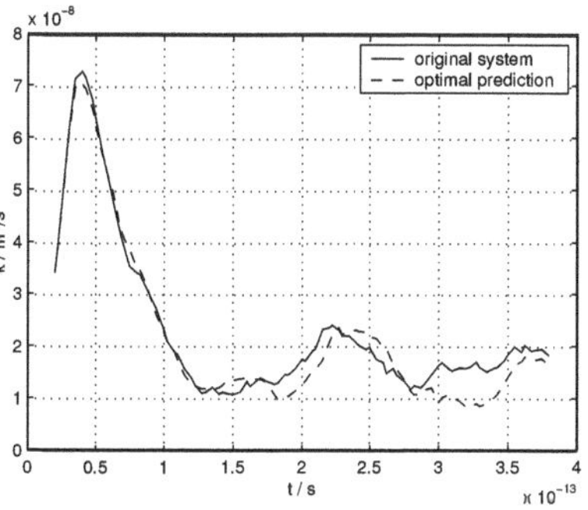


**Fig. 1.** Copper diffusion coefficient.

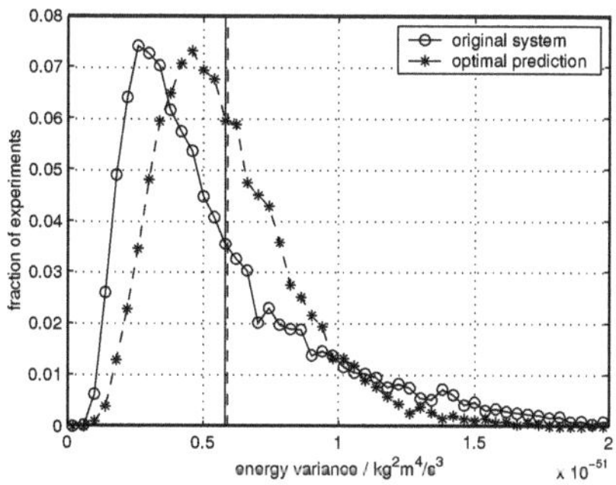


**Fig. 2.** Energy fluctuations.

$$\int_0^t |E_{\text{left}}(t) - E_{\text{left}}(0)|^2 dt \tag{12}$$

is plotted in a histogram plot. Both systems yield the same average energy fluctuation. However, optimal prediction yields fewer small and fewer high fluctuations. A possible reason for this effect could be the fact that the *virtual* atoms have no momentum and hence no free energy exchange is possible as it is with the full system.

Another important discrepancy is that the new system does not reproduce the correct behaviour in the presence of non-equilibrium effects. A sonic wave which normally travels through the crystal is instead reflected at the wall between *real* and *virtual* atoms.

# 4 Conclusions and Outlook

Optimal prediction, in combination with asymptotic methods, can in principle be applied in low temperature molecular dynamics to reduce the number of unknowns. In our one-dimensional model problem, the new system yields the correct diffusion behaviour, while it produces a discrepancy in the energy fluctuations. Also, non-equilibrium effects, such as sonic waves, are not reproduced correctly. Further research may on the one hand consider more complex three dimensional problems, on the other hand try to remedy the discrepancies, in particular consider non-equilibrium effects. A deeper discussion can be found in [5].

*Acknowledgement.* We would like to thank Alexandre Chorin, Thomas Götz, Peter Klein, Helmut Neunzert and Anja Streit for helpful discussions and comments.

## References


1. A.J. Chorin. Probability, Mechanics, and Irreversibility. Lecture notes, UC Berkeley Math. Dept., 2000.
2. A.J. Chorin, A.P. Kast, and R. Kupferman. Optimal prediction of underresolved dynamics. *Proc. Natl. Acad. Sci. USA*, 95(8):4094–4098, 1998.
3. T. Frauenheim, H. Hensel, P. Klein, and H.M. Urbassek. Comparison of classical and tight-binding molecular dynamics for silicon growth. *Phys. Rev. B*, 53:16497–16503, 1996.
4. J.D. Murray. *Asymptotic Analysis*, volume 48 of *Applied Mathematical Sciences*. Springer, New York, 1984.
5. B. Seibold. Optimal prediction in molecular dynamics. *Monte Carlo Methods and Applications*, 10(1):25–50, 2004.

# From CAD to CFD Meshes for Ship Geometries

V. Skytt[1]

SINTEF, Norway Vibeke.Skytt@sintef.no

**Summary.** The chart surface approach, a variational grid generation method for surface grids, is applied to CAD models describing ship hulls and propellers.

**Key words:** CAD, CFD, surface grid, variational design

## 1 Introduction

A CAD model and a CFD mesh both provide a geometric representation of some object, but these representations are very different in nature. In CAD the main objective is to describe with high accuracy the shape of the object to be produced. Central concepts are B-spline surfaces and boundary structures. A corresponding CFD mesh is often much rougher. The high level of detail represented in the CAD model is in general not required, and the processing time of the analysis will depend on the size of the geometry model. CFD requires an exact match between the various elements in the geometry model, while in CAD they normally match only within a tolerance. The main objective of this article is to look at how a surface grid for use in CFD can be obtained from a CAD model in the context of ship design.

The work presented here has been performed in the two EU projects Fantastic and Leading Edge. Fantastic was concerned with shape optimization of ship hull forms, see [7]. In Leading Edge an investigation of tip vortex cavitation on propellers is carried out. Both projects perform CFD calculations and the transition from a CAD model to a geometry model suitable for CFD computations is an important issue. This article focuses on the generation of surface grids from CAD models, and in particular on chart surfaces, which are a tool to assist in this process.

Grid generation is a significant part of a CFD application, and much of the time spent on grid generation is used for making a surface grid. [5] discusses this topic. Surface grids will then often be used to generate volume grids, where the quality and the density of the surface grid are important parameters regarding the reliability of the final CFD results.

Block structured grids are often found appropriate for CFD solvers, and are the type of grid to be addressed in this context. Adjacent blocks are required to have identical grid nodes at the common boundary, and the grid spacing should vary gradually over a block boundary. The chart surface method uses an approach based on variational design to make boundary fitted grids in one block. See also [1] for use of geometrical model tools in grid generation.

In general, high density of the grid points gives more accuracy, but the computations using this grid will take longer. Not all features in a CFD computation can be foreseen a priori. An initial grid may need to be modified for instance by refining the grid in critical areas. The outcome of the CFD computation may be a need to modify the initial geometry model. In this context a tight coupling between the CAD model and the CFD mesh is recommended.

## 2 Chart surfaces

One chart surface corresponds to one computational block, and a collection of chart surfaces provides a link between the CAD model and the CFD mesh. Each chart surface corresponds to a number of surfaces in the CAD model, which may be trimmed or not. The actual way of dividing the model into surfaces is normally of no interest for the meshing application. The aim is to replace the original parameter domains corresponding to a surface set with one domain that can serve as the computational block, where a mapping between the new and the original parameter domains is provided.

The following procedure defines a chart surface or computational block:

- The topology of the surface set is computed and a relation between surface boundaries and the parameter domain of the chart surface is defined.
- Sample points are fetched from the surfaces and parameterised.
- The parameterisations of the points from the individual patches are mapped to the composite parameter domain. The parameter domain of one surface maps to a closed polygon in the parameter domain of the chart surface. Keeping the relation between corner points of the original surface and the composed parameter domain provides a back-mapping.
- A tensor-product B-spline surface is defined on the new domain. This surface will approximate the sample points, but the surface also minimizes a smoothness functional, see expression( 1).

A chart surface is smooth, and has smooth iso-parametric curves. Evaluation in a uniform net gives a very regular grid which is not necessarily what we want. Along the leading edge of a propeller, for instance, there is a need for a tighter grid than in other areas. To allows a grid stretching independent from the parameterization of the chart surface, grid distribution functions are introduced to reparameterise the chart surface. This facilitates a tighter grid in some parts of the model. A chart surface has knowledge about its adjacent chart surfaces and in combination with the grid distribution functions this provides the means to get a smooth size distribution of grid elements across block boundaries.

A mesh over the entire surface set is generated by evaluating the set in a grid of points applying the following procedure:

- Apply the grid distribution function to compute the parameter value of the current grid point.
- Evaluate the approximating surface in this parameter value.
- Find the original surface corresponding to the parameter value. There is a mapping from the original surfaces to the parameter domain of the approximating surface defining a planar subdivision of this domain. Identifying the original surface corresponding to the current point is equivalent to locating a point in a planar graph.
- Compute the closest point in the original surface set to the current point in the approximating surface. A good starting point for this step is computed from the back-mapping information of the chart surface.

The approximating surface belonging to the chart surface is represented as a tensor product polynomial B-spline surface $F$ with surface coefficients $\vec{c}$. See [4] for background information on splines.

An initial surface is updated by approximating a set of points, $\{a\}_{r=1}^{R}$, evaluated from the original surface set. Simultaniously to the point approximation, smoothing is performed by applying the following functional:

$$\min_{\vec{c}} J(F) =$$

$$= \min_{\vec{c}} \Bigg[ \int\int_{\Omega} \int_{0}^{\pi} \sum_{l=1}^{3} \omega_l \left( \frac{\partial^l F(u + r\cos\varphi, v + r\sin\varphi)}{\partial r^l} \bigg|_{r=0} \right)^2 d\varphi \, du \, dv$$

$$+ \omega_4 \sum_{r=1}^{R} \Big( F(u_r, v_r) - a_r \Big)^2 \Bigg] \tag{1}$$

Some coefficients of $F$ situated at the surface boundary are defined due to the need of maintaining exact continuity between adjacent chart surfaces. The remaining coefficients are found by differentiating the functional $J$ with respect to the free coefficients and solving the resulting linear equation system.

The first three terms in the functional perform smoothing and are included to improve the quality of the surface and to make sure that a solvable linear equation system is created. Directional derivatives of the surface of varying orders are at all points in the parameter domain, $\Omega$, integrated around a circle. This expression is again integrated over $\Omega$. Minimizing 1st derivatives is an approximation to minimizing the area of the surface, 2nd order derivatives approximate minimization of curvature and 3rd order variation in curvature. For these approximations to be good, it is important for the points $\{a\}$ to have a close to iso-metric parameterisation. Some details are elaborated in [6]. More smoothing functionals can be found in [3].

## 3 Examples and Future Work

Often the CAD model of a ship hull will consist of a high number of surface patches. Each patch has a simple shape, but their number implies some complexity. Chart surfaces can be used to structure the patches into suitable computational blocks reducing the significance of the division of the CAD model into a lot surfaces.

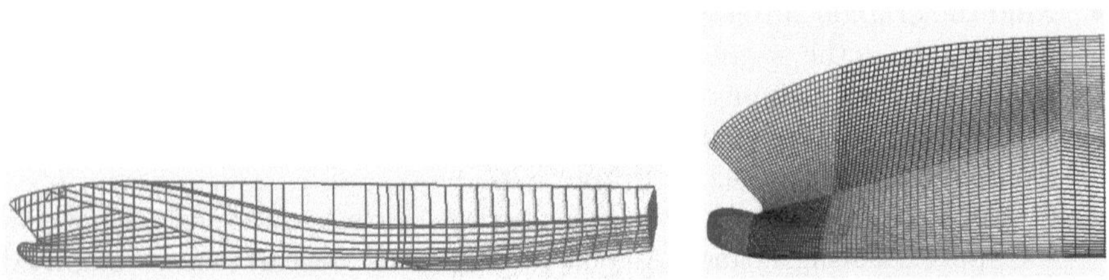

**Fig. 1.** Patch structure of a ship hull and a detail of the surface grid

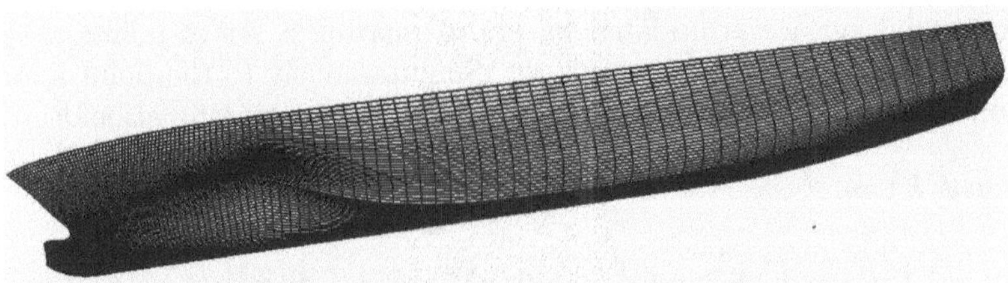

**Fig. 2.** Surface grid corresponding to the chart surfaces

Figure 1 shows the patch structure of a CAD model of a hull to the left and a detail of the corresponding surface grid to the right. The division into blocks is shown by different shades of colours. The nearly complete surface grid is shown in figure 2. The distribution of the grid cell sizes could be improved along the long blocks in the aft part of the ship, otherwise the distribution is good.

On the other hand, propellers have a complex geometrical shape, such as the conventional propeller of figure 3. It is rounded along the leading edge while the trailing edge is sharp. The tip is represented as a singular point in the surface model. The blade is trimmed towards the hub. The middle picture illustrates the parametrisation of the blade. The constant parameter curves join in one point at the tip. To the right is the surface grid. Each side is divided into 4 blocks: in the tip area, along the leading edge, along the trailing edge and in the inner of the blade side, see also [2].

One motivation for introducing the chart surface is to create a tight link between the CAD model and the CFD mesh. Such a tight connection will

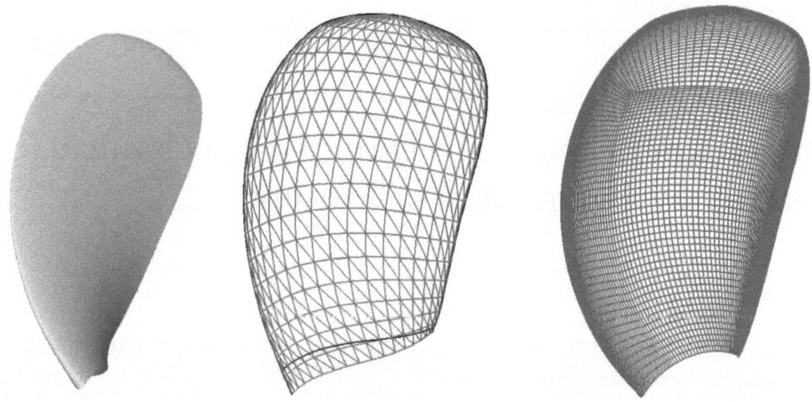

**Fig. 3.** Blade of conventional propeller

simplify mesh updates and refinement and also geometry updates of the CAD model based on simulation results. Currently, the path from the CAD model to a surface grid using the concept of chart surfaces is developed. The road back to perform geometry updates remains as future work.

*Acknowledgement.* Supperted in part by the EU projects Fantastic,GRD1-1999-10666, Leading Edge, G3RD-CT-2002-00818, and AIM@SHAPE, IST-NoE 506766.

## References

1. Khamayseh A. and Hamann B. Elliptic grid generation using nurbs surfaces. *Computer Aided Geometric Design*, 13:369–386, 1996.
2. M. Abdel-Maksoud, F. Menter, and H. Wuttke. Viscous flow simulations for conventional and high-skew marine propellers. *Schiffstechnik*, 45:64–71, 1998.
3. G. Greiner. Variational design and fairing of spline surfaces. *Computer Graphics Forum*, 13:143–154, 1994.
4. J. Hoschek and D. Lasser. *Fundamentals of Computer Aided Geometric Design.* Teubner, Stuttgart, 1989.
5. M. Sabin. *CAD and FEM: Theoretical Connections and Practical Links between two Design Tools.* Numerical Geometry Ltd., 2003.
6. V. Skytt. Parameterization of scattered data for surface generation. In H. Nowacki and P.D. Kaklis, editors, *Creating Fair and Shape-Preserving Curves and Surfaces*, pages 179–187. B.G. Teubner, Stuttgart, 1997.
7. F. Valdenazzi, S. Harries, C.-E. Janson, M. Leer-Andersen, Maisonneuve J.-J., Marzi J., and Raven H. The Fantastic Roro: CFD optimisation of the forebody and its experimental verification. In *NAV 2003 Conference Proceedings*, volume Vol 1, pages 3.7.1–3.7.14, 2003.

# Integration of Strongly Damped Mechanical Systems by Runge-Kutta Methods

T. Stumpp

Mathematical Institute Tübingen, Auf der Morgenstelle 10, D-72076 Tübingen, stumpp@na.uni-tuebingen.de

**Summary.** Strongly damped mechanical systems arise, for example, in vehicle dynamics and in modelling joints in biomechanics. Standard explicit integrators become unstable unless very small time steps are chosen. We are interested in the numerical solution of such systems with step sizes that are independent of the damping parameter. The smooth motion of the mechanical system is expanded in terms of solutions of differential-algebraic systems of index 2. These results hold for analytical solutions as well as for numerical solutions of suitable methods such as Radau collocation. In the border case of big damping constants it turns out that the error of numerical solutions of the strongly damped mechanical system is bounded by errors for the differential algebraic systems.

**Key words:** numerical integration, Runge-Kutta methods, strongly damped mechanical systems, error analysis.

## 1 Motivation

When strong damping forces arise in mechanical systems, special regard has to be paid on numerical simulations. An illustrative example for such a situation is a pendulum where a mass point is coupled to a spring-damper element via a massless interconnection (see Fig. 1). Simulating this test example with two different Runge-Kutta methods yields a result that highlights some basic difficulties. For the time integration, we used a MATLAB implementation [2] of the code RADAU5 [5], which is a RadauIIA method of order 5. The code ODE45 is an explicit Runge-Kutta method of order 4 [1]. Both implementations use an adaptive step size control. The numerical results are shown in Table 1. In the experiment, we varied the size of the damping constant and noted the needed numbers of steps and floating point operations (flops). Since RADAU5 is an implicit method, it is able to handle the different examples without choosing smaller steps, whereas the step size of ODE45 decreases proportionally to the size of the damping constants. The dashes in the column

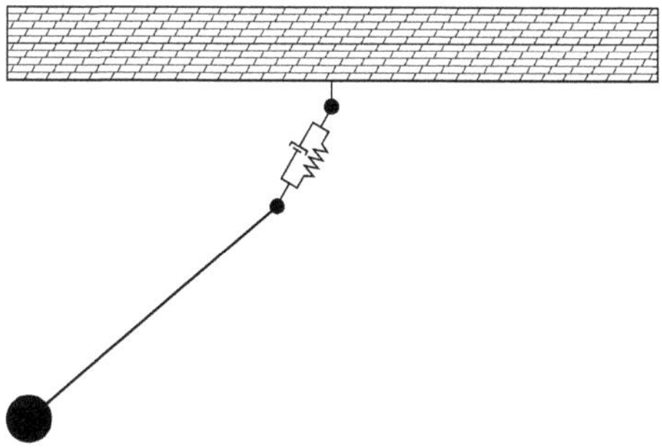

**Fig. 1.** Pendulum with spring-damper element.

of ODE45 mean that the numerical test was stopped since the iteration took already to long. (The iteration for the damping constant $10^2$ lasted 4 days on a workstation with an Athlon XP 1600+ and 1 GB memory). Of course, the explanation for this behaviour is to be found in the theory of stiff differential equations. With a growing damping constant, the equations of motion of the mechanical system get more and more stiff. But also among implicit Runge-Kutta methods, thereare differences in experiments. In [6] joints of human models are modelled by the same spring damper elements as in Fig. 1. In these more complicated biomechanical models it turns out that RadauIIA methods work significantly better than other implicit methods. We give an explanation for this numerical behaviour in the following sections.

**Table 1.** Simulation of the pendulum: numerical results

| damping constant in $Ns/m^2$ | RADAU5 | | ODE45 | |
|---|---|---|---|---|
| | flops | steps | flops | steps |
| $10^1$ | $6.51 \cdot 10^6$ | 209 | $3.39 \cdot 10^8$ | 75027 |
| $10^2$ | $1.68 \cdot 10^6$ | 46 | $3.4 \cdot 10^9$ | 753265 |
| $10^3$ | $1.67 \cdot 10^6$ | 37 | - | - |
| $10^5$ | $8.91 \cdot 10^6$ | 128 | - | - |
| $10^7$ | $2.18 \cdot 10^7$ | 291 | - | - |

A strongly damped mechanical system is represented by the second order differential equation

$$M(y)\ddot{y} = f(y, \dot{y}) - \frac{1}{\varepsilon} D(y)\dot{y} \tag{1}$$

with a small, positive parameter $\varepsilon$. We assume that, for all $y \in \mathbb{R}^d$, the mass matrix $M(y)$ is symmetric and positive definite and that the damping matrix $D(y)$ is symmetric, positive semidefinite and of constant rank $m$. Additionally, let $M$, $D$ and $f$ be bounded and sufficiently smooth. The matrix $D$ has a nontrivial kernel. We denote the dimension of this kernel by $l = d - m$. Since $D$ is symmetric and positive semidefinite, the term $D(y)\dot{y}$ vanishes on the $(d+l)$-dimensional submanifold

$$\mathcal{M}^0 = \{(y, v) \in \mathbb{R}^{2d} : D(y)v = 0\} \subseteq \mathbb{R}^{2d}.$$

## 2 Expansion of the Analytical Solution

Considering previous works to the related topics of singular perturbation problems [3] and stiff oscillatory mechanical systems [7] the first task is to show the existence of an $\varepsilon$-expansion for smooth solutions.

**Theorem 1.** *For $N \geq 1$ and $0 < \varepsilon \leq \varepsilon_0$ there is a $(d+l)$-dimensional manifold $\mathcal{M}^\varepsilon$, constructed by the bijection*

$$\Psi^\varepsilon : \mathcal{M}^0 \to \mathcal{M}^\varepsilon : (y^0, \dot{y}^0) \mapsto (y^\varepsilon, \dot{y}^\varepsilon)$$

*with $\Psi^\varepsilon = id + O(\varepsilon)$, and an $\varepsilon$-independent interval $[0, T]$, such that the following holds: if $(y(0), \dot{y}(0)) \in \mathcal{M}^\varepsilon$, then the solution $(y(t), \dot{y}(t))$ of (1) for this starting value fulfills*

$$(y(t), \dot{y}(t)) \in \mathcal{M}^\varepsilon + O(\varepsilon^N) \qquad \text{for } 0 \leq t \leq T,$$

*and there exist asymptotic $\varepsilon$-expansions*

$$\begin{aligned} y(t) &= y^0(t) + \varepsilon y^1(t) + \cdots + \varepsilon^N y^N(t) + O(\varepsilon^{N+1}), \\ \dot{y}(t) &= \dot{y}^0(t) + \varepsilon \dot{y}^1(t) + \cdots + \varepsilon^N \dot{y}^N(t) + O(\varepsilon^{N+1}). \end{aligned}$$

*Here $(y^0(t), \dot{y}^0(t)) \in \mathcal{M}^0$ and $(y^k(t), \dot{y}^k(t))$ are solutions of differential-algebraic equations for $0 \leq k \leq N$ that are stated below. The first $N$ derivatives of $y$ are bounded independently of $\varepsilon$.*

For $k = 0, \ldots, l$, the coefficients $(y^k(t), \dot{y}^k(t))$ are obtained from a differential-algebraic equation of index $2 + l$ ([8]).

## 3 RadauIIA Methods

Proving stability results yields further conditions which restrict the Runge-Kutta methods to a certain class of methods (see [8]), *e.g.*, RadauIIA methods. The main properties of these methods are that they are $A$-stable and stiffly accurate. Their classical order is $p = 2s - 1$ and the stage order equals the number of stages $s$. One first important result is that also the numerical solution $(y_n, \dot{y}_n)$ has an asymptotic $\varepsilon$-expansion.

**Theorem 2.** *Consider a RadauIIA method with stage order $q$. Assume that the starting value $(y_0^\varepsilon, \dot{y}_0^\varepsilon)$ is on $\mathcal{M}^\varepsilon$, i.e., that the solution of (1) with starting values $(y_0^\varepsilon, \dot{y}_0^\varepsilon)$ is smooth. For $0 < \varepsilon < h < h_0$, a unique solution $(y_n^\varepsilon, \dot{y}_n^\varepsilon)$ of the strongly damped mechanical system (1) exists. For $nh \in [0, T]$, approximations $(y_n^\varepsilon, \dot{y}_n^\varepsilon)$ can be represented as $\varepsilon$-expansions*

$$\begin{aligned} y_n^\varepsilon &= y_n^0 + \varepsilon y_n^1 + \cdots + \varepsilon^q y_n^q + O(\varepsilon^{q+1}), \\ \dot{y}_n^\varepsilon &= \dot{y}_n^0 + \varepsilon \dot{y}_n^1 + \cdots + \varepsilon^q \dot{y}_n^q + O(\varepsilon^{q+1}), \end{aligned}$$

*where $y_n^k$, $\dot{y}_n^k$ are Runge-Kutta solutions of the differential-algebraic systems. Their initial values $(y_0^k, \dot{y}_0^k)$ are chosen as the coefficients of $\varepsilon^k$ in the $\varepsilon$-expansion of $(y_0^\varepsilon, \dot{y}_0^\varepsilon)$.*

For the final error result it will be a deciding property that the numerical methods are suitable methods for the differential-algebraic systems. By using techniques from [4] we obtain that the error of the RadauIIA method for the DAE of index 2 is given by

$$y_n^0 - y^0(t_n) = O(h^p), \quad \dot{y}_n^0 - \dot{y}^0(t_n) = O(h^p), \quad \lambda_n^0 - \lambda^0(t_n) = O(h^{p-1})$$

uniformly for $0 \le t_n \le T$. Here, $p$ is the order of the method and consistent initial values $y_0^0$, $\dot{y}_0^0$, $\lambda_0^0$ are used.

A similar result can be derived for the error of the specified class of Runge-Kutta methods for the DAE of index 2+k. For consistent initial values $y_0^0$, $\dot{y}_0^0$, $\lambda_0^0$ we have

$$\begin{aligned} &y_n^k - y^k(t_n) = O(h^{q+1-k}), \quad \dot{y}_n^k - \dot{y}^k(t_n) = O(h^{q+1-k}), \\ &\lambda_n^k - \lambda^k(t_n) = O(h^{q-k}) \end{aligned}$$

uniformly for $0 \le t_n \le T$, where $q$ is again the stage order of the method.

## 4 Error Results

With the results of the Sections 2 and 3 we are now able to give an estimate of the global error.

**Theorem 3.** *Let a Runge-Kutta method according to Sect. 3 with stage order $q$ be given. Assume that the initial value $(y_0^\varepsilon, \dot{y}_0^\varepsilon)$ is on the manifold $\mathcal{M}^\varepsilon$, i.e., the exact solution of (1) with starting values $(y_0^\varepsilon, \dot{y}_0^\varepsilon)$ is smooth. For $0 < \varepsilon < h < h_0$, a unique Runge-Kutta solution of the strongly damped mechanical system (1) exists. The global error of this solution satisfies*

$$\begin{aligned} y_n^\varepsilon - y^\varepsilon(t_n) &= y_n^0 - y^0(t_n) + O(\varepsilon h^q), \\ \dot{y}_n^\varepsilon - \dot{y}^\varepsilon(t_n) &= \dot{y}_n^0 - \dot{y}^0(t_n) + O(\varepsilon h^q) \end{aligned}$$

*uniformly for* $\varepsilon \le h \le h_0$ *and* $0 \le t_n \le T$. *Here,* $y_n^0$, $\dot{y}_n^0$ *and* $y^0(t)$, $\dot{y}^0(t)$ *are Runge-Kutta solutions and exact solutions of the differential-algebraic equation of index 2, respectively. Their initial values* $(y_0^0, \dot{y}_0^0)$ *are the coefficients of* $\varepsilon^0$ *in the* $\varepsilon$*-expansion of* $(y_0^\varepsilon, \dot{y}_0^\varepsilon)$.

From this theorem we obtain the following final result for RadauIIA methods.

**Theorem 4.** *Assume that the initial value* $(y_0^\varepsilon, \dot{y}_0^\varepsilon)$ *is on the manifold* $\mathcal{M}^\varepsilon$, i.e., *the exact solution of (1) with starting values* $(y_0^\varepsilon, \dot{y}_0^\varepsilon)$ *is smooth. For* $0 < \varepsilon < h < h_0$ *there exists a unique solution according to the s-stage RadauIIA method of the strongly damped mechanical system (1) and the global error satisfies*

$$(y_n^\varepsilon, \dot{y}_n^\varepsilon) - (y^\varepsilon(t_n), \dot{y}^\varepsilon(t_n)) = O(h^{2s-1}) + O(\varepsilon h^s).$$

Since it is more exceptional that initial values with small perturbations are given, assuming $(y_0^\varepsilon, \dot{y}_0^\varepsilon) \in \mathcal{M}^\varepsilon$ is quite restrictive. But with the same techniques as in [7], we are able to derive the estimate

$$(y_n, \dot{y}_n) - (y_n^\varepsilon, \dot{y}_n^\varepsilon) = O(h\rho^n + \varepsilon^{q+1})$$

for starting values $(y_0, \dot{y}_0)$ that satisfy $D(y_0)\dot{y}_0 = O(h)$. The positive parameter $\rho$ is strictly smaller than one and does not depend on $h$, $\varepsilon$ and $n$.

These results show that RadauIIA methods are an excellent choice for the numerical time integration of strongly damped mechanical systems.

## References

1. J.R. Dormand and P.J. Prince. A family of embedded Runge-Kutta formulae. *J. Comp. Appl. Math.*, 6:19–26, 1980.
2. Ch. Engstler. Code for the MATLAB-implementation of RADAU5. Available from `http://na.uni-tuebingen.de/na/software.shtml`.
3. E. Hairer, Ch. Lubich, and M. Roche. Error of Runge-Kutta methods for stiff problems studied via differential algebraic equations. *BIT 29*, 1:77–90, 1989.
4. E. Hairer, Ch. Lubich, and M. Roche. *The Numerical Solution of Differential-Algebraic Systems by Runge-Kutta Methods*, volume 1409 of *Lecture Notes in Mathematics.* Springer, 1989.
5. E. Hairer and G. Wanner. *Solving Ordinary Differential equations II. Stiff and Differential-Algebraic Problems.* Springer-Verlag, Berlin, 1991.
6. T. Hans. *Interaktive Simulation Biomechanischer Bewegungsabläufe.* PhD thesis, University of Tübingen, Tübingen, 2004.
7. Ch. Lubich. Integration of stiff mechanical systems by Runge-Kutta methods. *Z. Angew. Math. Phys.*, 44(6):1022–1053, 1993.
8. T. Stumpp. *Integration stark gedämpfter mechanischer Systeme mit Runge-Kutta-Verfahren.* PhD thesis, University of Tübingen, Tübingen, 2004.

# Numerical Simulation of SMA Actuators

G. Teichelmann and B. Simeon

TU München, Zentrum Mathematik M2, Boltzmannstraße 3, 85748 Garching
email: teichelmann@ma.tum.de, simeon@ma.tum.de

**Summary.** This paper deals with Shape Memory Alloy (SMA) actuators for mechatronic applications. A mathematical model on the macroscopic level is discussed and a computational framework is introduced. The latter makes use of the method of lines and results in a system of differential-algebraic equations in time. Some first simulation results for a 1D wire in the isothermal case illustrate the approach.

**Key words:** Shape Memory Alloy (SMA), numerical simulation, Partial Differential-Algebraic Equation (PDAE).

## 1 Introduction

Shape Memory Alloy (SMA) materials have an enormous potential in technological applications. By focusing on the particular example of SMA actuators in mechatronics, we aim here at the development of computational methods to achieve improved understanding of how such materials perform their functions. SMA applications of industrial interest present major challenges since numerical methods need to be developed for heterogeneous coupled systems of partial differential and differential-algebraic equations (PDAEs).

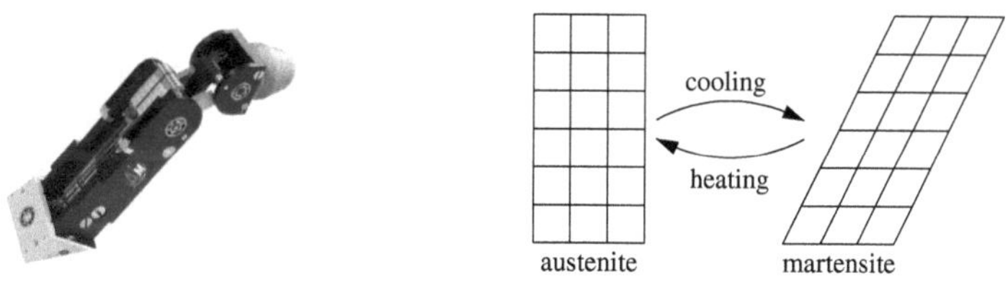


**Fig. 1.** SMA actuated robot finger (left); temperature action on SMA shape (right).

An example of an SMA actuator in mechatronics is given by the artificial finger shown in Fig. 1 on the left. It has been developed at the Institute of Applied Mechanics at the TU München and is driven by NiTi wires that are heated by electric current and act as flexor and extensor, similar to biological muscles. Along with theory and experiment, simulation represents an indispensable partner in the advance of such mechatronic applications. In case of the finger example, one of the key goals is to maximize the speed of the finger motion while minimizing the risk of failure and heat damage.

The paper is organized as follows. In Section 2, we discuss a mathematical model that is based on a macroscopic approach and that can be viewed as generalized viscoplastic constitutive law. Section 3 presents the computational treatment by the method of lines. A particularly challenging difficulty of the resulting differential-algebraic system are the discontinuities that mark the phase changes of the SMA material. Some first simulation results for a 1D wire in the isothermal case close the paper.

## 2 Mathematical Model

We discuss a model recently proposed by Helm [2]. Though not resolving the crystal grid, this model keeps track of the phase changes between austenite and martensite (Fig. 1 on the right) and is able to reproduce, depending on the temperature, the main effects of SMA structures.

Let $\Omega$ denote the domain in $d$-dimensional space that is occupied by the SMA material. As starting point, we formulate the momentum balance law

$$\rho\ddot{u}(t,x) = \operatorname{div}\, \sigma(u(t,x), \varepsilon_p(t,x), \theta(t,x)) + \beta(t,x)$$

for the displacement field $u : [0,T] \times \Omega \rightarrow \mathbb{R}^d$ and the stress tensor $\sigma : \mathbb{R}^d \times \mathbb{R}^{d\times d} \times \mathbb{R} \rightarrow \mathbb{R}^{d\times d}$, with $\beta$ standing for the density of body forces and $\rho$ for the mass density. The stress $\sigma$ depends on the displacement $u$, the so-called plastic strain $\varepsilon_p : [0,T] \times \Omega \rightarrow \mathbb{R}^{d\times d}$ and the temperature $\theta : [0,T] \times \Omega \rightarrow \mathbb{R}$. Furthermore, we have mixed boundary conditions $u(t,x) = u_D(t,x)$ on $\partial\Omega_D$ and $\sigma(t,x)\cdot\nu(x) = \tau(t,x)$ on $\partial\Omega_N$ with normal vector $\nu$. Proceeding like in viscoplasticity, we assume for the total strain

$$\varepsilon(t,x) = \frac{1}{2}\left(\nabla u(t,x) + \nabla u(t,x)^T\right), \quad \varepsilon(t,x) = \varepsilon_p(t,x) + \varepsilon_e(t,x),$$

where the first equation expresses the kinematics and the second an additive split into plastic and elastic strain. The relation between stress $\sigma$ and elastic strain $\varepsilon_e = \varepsilon - \varepsilon_p$ is then given by the generalized Hooke's law

$$\sigma = 2\mu(\theta)\varepsilon_e + \kappa(\theta)\,\mathrm{tr}\varepsilon_e I - 3\alpha(\theta)\kappa(\theta)(\theta - \theta_0)I \tag{1}$$

with temperature-depending material parameters $\mu$, $\alpha$, $\kappa$, reference temperature $\theta_0$, and identity tensor $I$.

Next, we go into the details of the material's evolution in a point $x$. There is a total of 6 possible phase changes between *temperature induced martensite* (TIM), *stress induced martensite* (SIM) and *austenite* (A). For example, the transition A $\to$ SIM takes place if the inequalities

$$\Delta\psi > 0, f > 0, z_{SIM} < 1, \tau_{eff} > \|\mathbf{X}_\theta\|, \varepsilon_p N \geq 0 \tag{2}$$

are satisfied. Herein, $N = (\sigma - X - X_\theta)/\|\sigma - X - X_\theta\|$ is the normal to the yield surface (see below) with the internal stress $X$ and the temperature dependent stress $X_\theta$, the quantity $\tau_{eff} = \|\sigma - X\|$ denotes the effective stress state, $z_{SIM}$ is the fraction of stress induced martensite, and $\Delta\psi$ is the free energy difference at temperature $\theta$, see [6] for more details.

The evolution equation for the plastic strain $\varepsilon_p$ reads

$$\dot{\varepsilon}_p = \lambda \frac{\partial f(\sigma,\theta,X)}{\partial \sigma} = \lambda N \text{ with yield function } f(\sigma,\theta,X) = \|\sigma - X\| - \sqrt{\frac{2}{3}} k(\theta)$$

and the inelastic multiplier $\lambda = \begin{cases} f^3/\nu & \text{if } \text{A} \leftrightarrow \text{SIM, TIM} \leftrightarrow \text{SIM} \\ 0 & \text{otherwise.} \end{cases}$

The fraction of temperature induced martensite behaves according to

$$\dot{z}_{TIM} = \begin{cases} -\frac{|\dot{\theta}|}{M_s - M_f} \frac{(\Delta\psi)}{|(\Delta\psi)|} & \text{A} \to \text{TIM} \\ -\frac{|\dot{\theta}|}{A_f - A_s} \frac{(\Delta\psi)}{|(\Delta\psi)|} \quad \text{if} & \text{TIM} \to \text{A} \\ -\dot{z}_{SIM} & \text{TIM} \leftrightarrow \text{SIM} \\ 0 & \text{otherwise} \end{cases} \quad \text{with} \quad \dot{z}_{SIM} = \sqrt{\frac{2}{3}} \frac{\varepsilon_p \cdot \dot{\varepsilon}_p}{\gamma_d \|\varepsilon_p\|}$$

The last two cases denote purely martensitic phase changes, where the sum $z_{TIM} + z_{SIM}$ of temperature induced and stress induced martensite is constant, while in the first two cases a phase transformation between martensite and austenite occurs. The internal stress $X$ finally satisfies the equation

$$\dot{X} = c_1 \dot{\varepsilon}_p - c_2 |\dot{\varepsilon}_p| X.$$

As some parameters depend on temperature, additionally the heat equation

$$c_0 \dot{\theta} + \frac{1}{\rho} \operatorname{div} Q - r = -\frac{3\alpha\kappa}{\rho} \theta \operatorname{tr}(\dot{\varepsilon} - \dot{\varepsilon}_p) - \Delta e_0 \dot{z}_{TIM}$$

is needed with heat flux $Q$, energy gap $\Delta e_0$, and reference heat capacity $c_0$.

In case of isothermal conditions, however, the heat equation can be omitted and, moreover, the above evolution equations simplify considerably. Assuming quasi-stationary deformation, we can write the model then as

$$0 = \operatorname{div} \sigma(u, \varepsilon_p) + \beta, \quad \dot{\varepsilon}_p = \gamma_1(u, \varepsilon_p, \xi), \quad \dot{\xi} = \gamma_2(u, \varepsilon_p, \xi) \tag{3}$$

where the internal variables $\xi$ comprise both $z_{TIM}$ and $X$. The coupled system (3) possesses the typical structure of viscoplastic materials [3], and accordingly, the theory of Alber [1] can be applied to analyse the existence of solutions. However, the phase changes introduce discontinuities into the right hand sides $\gamma_1$ and $\gamma_2$, which leads to both theoretical and numerical problems.

## 3 Numerical Treatment

We concentrate now on a computational framework to solve the model equations (3) in the isothermal case. Analogously to the discretization of viscoplastic materials, the weak form of the balance equation is formulated in the function space $V = \{v \in H^1(\Omega)^d : v|_{\partial\Omega_D} = 0\}$ and makes use of the relation $\sigma = C(\varepsilon(u) - \varepsilon_p)$ derived from (1). Assuming zero Dirichlet boundary conditions for brevity, the projection of the displacement field $u$ to its Galerkin approximation $u_S$ in some subspace $S$ of $V$ can be written as $u_S(t, x) = \Phi(x)q(t)$ with Ansatz functions $\Phi$ and unknown coefficients $q$. Correspondingly, the kinematic equation results in $\varepsilon(u_S) = B(x)q(t)$, and the discretized momentum balance law reads, cf. [4],

$$0 = K \cdot q + b(t) - \int_\Omega B^T C\varepsilon_p \, d\Omega \tag{4}$$

with stiffness matrix $K$ and force vector $b$.

Applying a quadrature rule to discretize the remaining integral in (4) by

$$\int_\Omega B^T C\varepsilon_p \, d\Omega \doteq G \cdot (\varepsilon_p(\zeta_1), \dots, \varepsilon_p(\zeta_k))^T = G\varepsilon_p$$

requires the knowledge of $\varepsilon_p$ in specific quadrature nodes $\zeta_i$. Correspondingly, all variables $\varepsilon_p$ and $\xi$ need to be evaluated in these nodes. In each node $\zeta_i$ it holds

$$\dot{\varepsilon}_p = \gamma_1(u, \varepsilon_p, \xi), \qquad \dot{\xi} = \gamma_2(u, \varepsilon_p, \xi). \tag{5}$$

Summing up, the differential equations (5) over all quadrature nodes and the discretized balance law (4) form the DAE

$$\dot{y} = \gamma(y, q), \qquad 0 = g(y, q, t) \tag{6}$$

with $y$ representing plastic strain and internal variables and $q$ the displacements. The function $g(y, q, t)$ is linear in $q$ with the invertible Jacobian $\partial g/\partial q = K$. Thus, the system (6) is of index 1, and a unique solution $q = q(y, t)$ of the algebraic part $g(y, q, t) = 0$ can be computed. Even more, an indirect approach $\dot{y} = \gamma(y, q(y, t))$ can be employed for the time integration in combination with standard ode solvers and step size control.

However, the discontinuities that arise from the multiple case distinctions for the phase changes still need special attention. The standard switching point technique is much too expensive here. A better choice is a regularization

$$m(x, x_0) = \begin{cases} 0 & x < x_0 \\ 1 & x > x_0 \end{cases} \quad \rightsquigarrow \quad m_S(x, x_0, c_S) = \frac{1}{1 + \exp(-(x - x_0)/c_S)},$$

where the jump at $x_0$ is approximated by a sigmoid function. The constant $c_S$ defines the transition slope and should be chosen carefully.

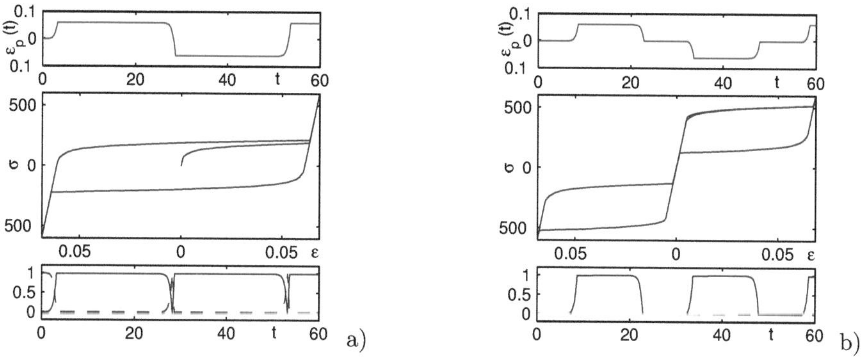


**Fig. 2.** Material behaviour at a) $240°K$, b) $330°K$.

**Simulation Example.** As final example, a one-dimensional wire [5] is discretized by linear finite elements under isothermal and quasistationary conditions. Using the smoothed phase transitions from above, we observe an increased stiffness whenever a transition takes place. So implicit integration schemes seem to be adequate for this problem, in contrast to the case where the discontinuities are not regularized. Our test runs with the implicit MATLAB codes `ode23tb/ode15s` gave good results, as can be seen in Fig. 2 showing the plastic strain $\varepsilon_p$, the stress-strain-diagram and the fractions $z_{SIM}$ (solid line) and $z_{TIM}$ (dashed line) for temperatures of $240°K$ and $330°K$. The wire was discretized by 10 equidistant finite element nodes and the results of Fig. 2 hold each for all quadrature nodes. A grid refinement in space does not show any perceptible change of the solution.

One of the next steps ahead is the coupling with heat conduction, which would allow to reproduce also phase changes from austenite to temperature induced martensite and vice versa.

## References


1. H.-D. Alber. *Materials with Memory.* Number 1682 in Lecture Notes in Mathematics. Springer, Berlin, 1998.
2. D. Helm and P. Haupt. Shape memory behaviour: modelling within continuum thermomechanics. *International J. of Solids and Structures*, 40:827–849, 2003.
3. T.J.R. Hughes. *Computational Inelasticity.* Springer, Berlin, 1996.
4. O. Scherf and B. Simeon. Differential-algebraic equations in elasto-viscoplasticity. volume 10 of *Lecture Notes in Mechanics*, pages 31–50. Springer, Berlin, 2003.
5. C. Schwarz. Modellierung und Simulation eines Formgedächtnisdrahtes. Master's thesis, Technische Universität München, 2003.
6. G. Teichelmann and B. Simeon. Numerical simulation of SMA actuators. Technical report, 2004.

# Color Plates

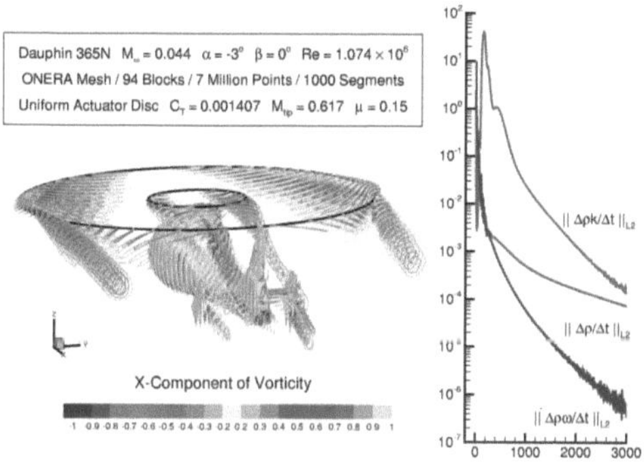


(a) Viscous calculation for Dauphin helicopter fuselage at $M_\infty = 0.044$, convergence behavior of mass and k-$\omega$ turbulence equations.

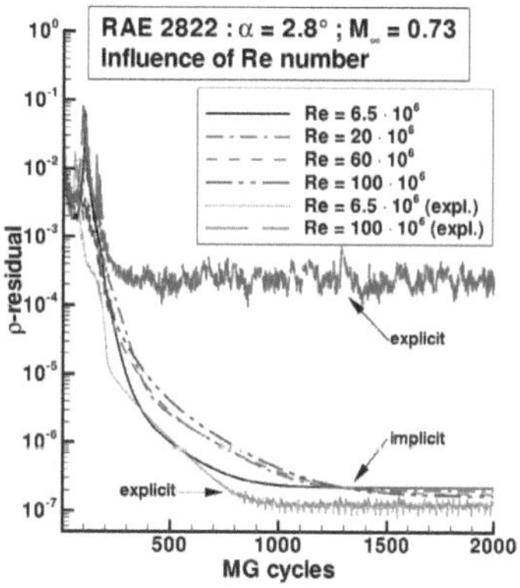


(b) Effect of Reynolds number on convergence for the RAE 2822 airfoil at $M_\infty = 0.73$, $\alpha = 2.8°$.

**(Rossow *et al.*, p. 7) Fig. 2.**

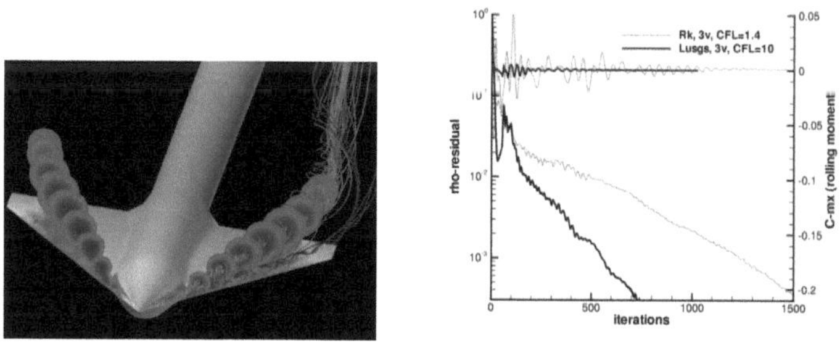


**(Rossow *et al.*, p. 11) Fig. 6.** Convergence behaviour of the hybrid TAU-Code for calculations of viscous flow around a delta wing at $M = 0.5$, $\alpha = 9°$. Comparison of the baseline Runge-Kutta scheme (RK) and the implicit LU-SGS scheme.

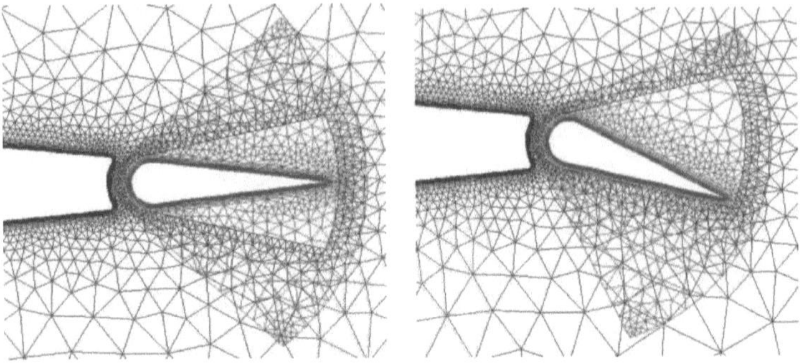

**(Rossow *et al.*, p. 12) Fig. 7.** Hybrid Chimera grid for delta wing with a movable flap.

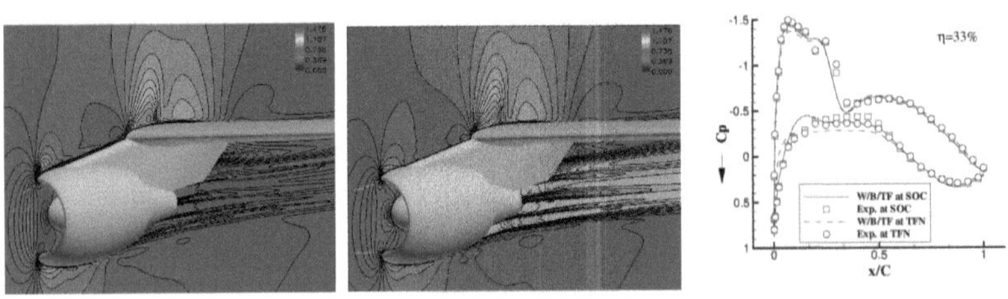


(a) TFN (b) SOC (c)

**(Rossow *et al.*, p. 18) Fig. 15.** Viscous calculation of DLR ALVAST configuration with FLOWer at $M_\infty = 0.75$, $C_L = 0.5$, influence of thrust condition of turbofan engine, (a) and (b) constant Mach number distribution for TFN and SOC, (c) surface pressure distribution at cross section $\eta = 33\%$.

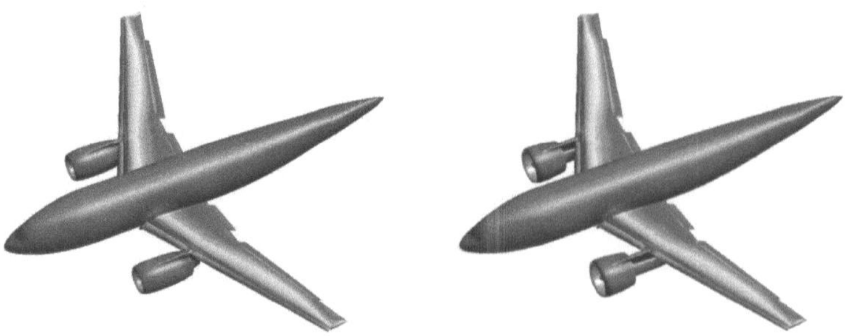

**(Rossow *et al.*, p. 19) Fig. 16.** Viscous simulation of the ALVAST high-lift configuration with VHBR (left) and UHBR (right) engine using the TAU-Code, surface pressure distribution, $M_\infty = 0.22$, $\alpha = 12°$, $Re = 2 \times 10^6$.

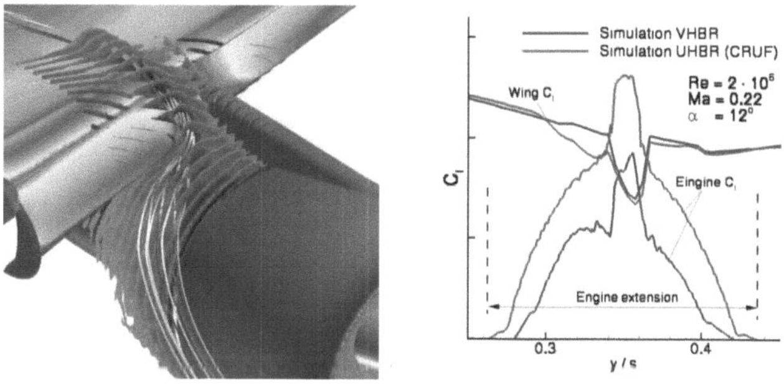


(a) Engine interference for ALVAST high-lift configuration with VHBR and UHBR engine $M_\infty = 0.22$, $\alpha = 12°, Re = 2 \times 10^6$ , left: nacelle vortex, right: lift distribution of wing and nacelle.

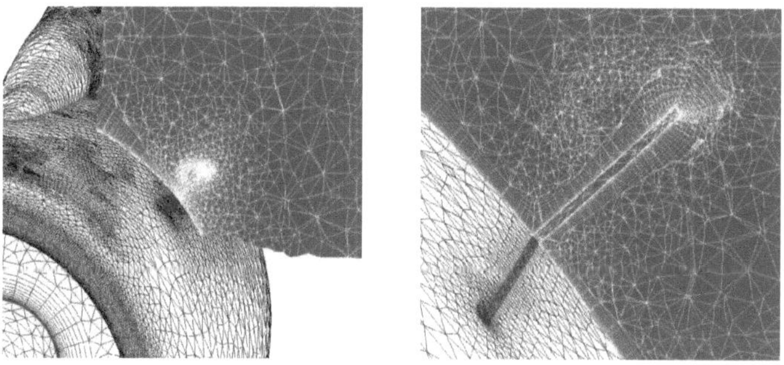

(b) Civil transport high-lift configuration with nacelle strakes, filled strake grid.

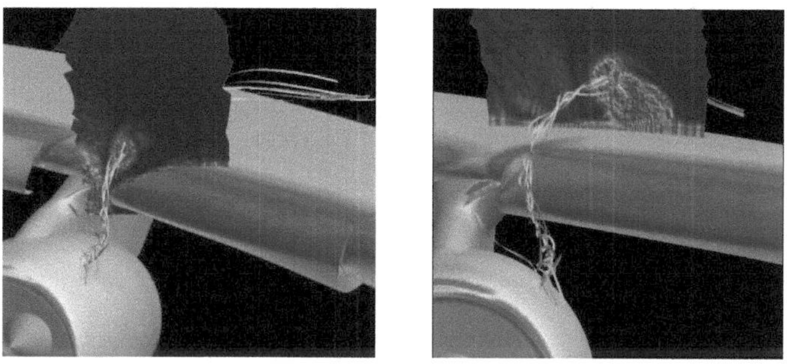

(c) Civil transport high-lift configuration with nacelle strakes, calculated streamlines and iso-vorticity cut planes.

**(Rossow *et al.*, p. 20) Fig. 17.**

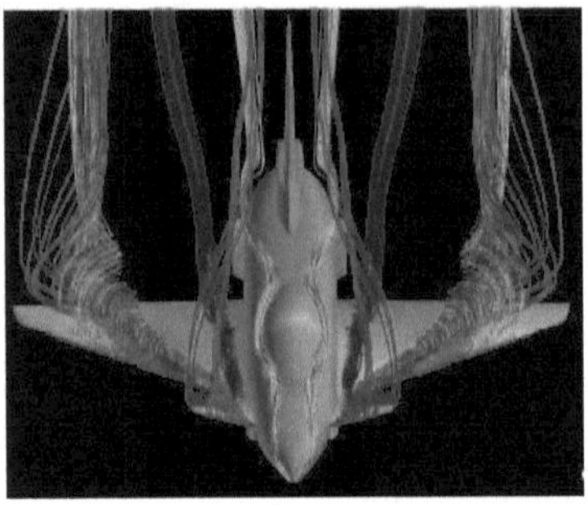

(a) 3D flow field of the X-31 configuration at 18° angle of attack, TAU-Code.

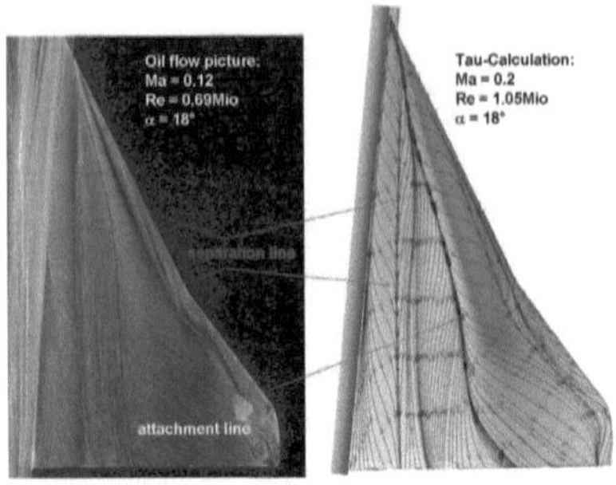


(b) X-31 clean wing, left: oil flow visualization, right: surface streamlines obtained with TAU-Code.

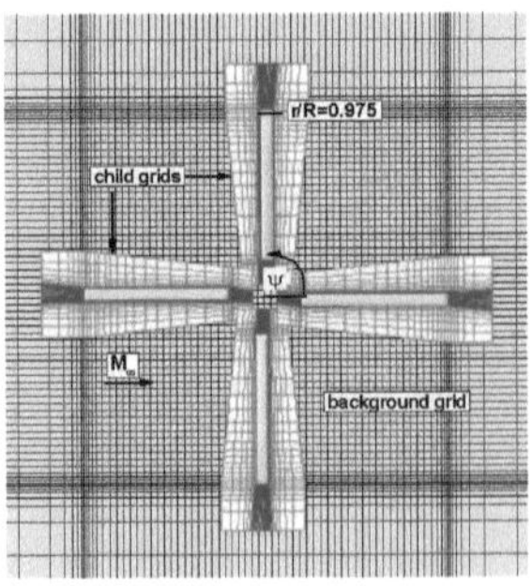


(c) Chimera grid system around 4-bladed 7A-rotor.

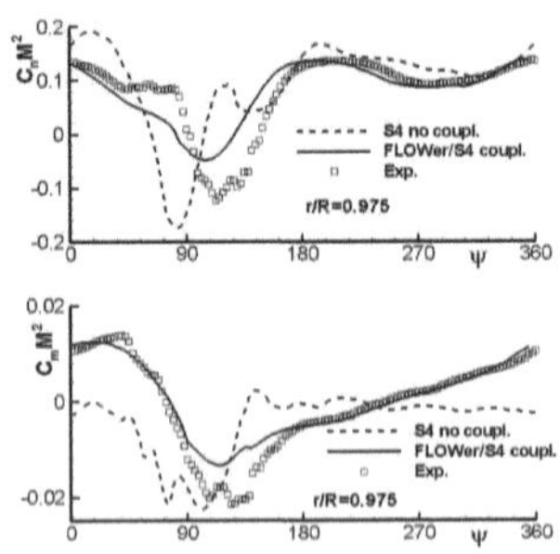


(d) Comparison of predicted and measured normal force and pitching moment coefficients versus azimuth for a high-speed forward flight test case of the 7A rotor.

**(Rossow *et al.*, p. 22) Fig. 18.**

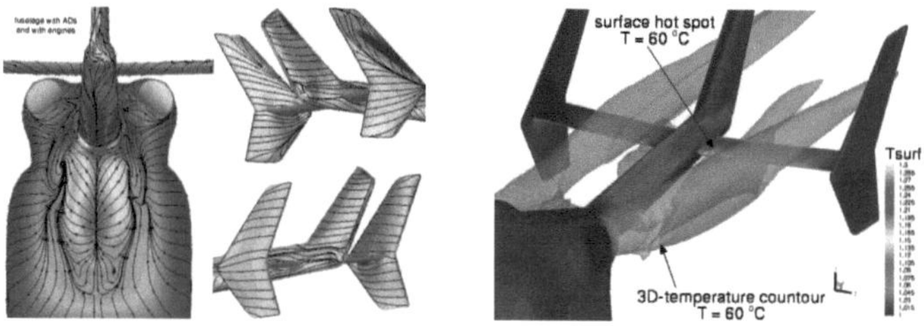


(a) (b)

**(Rossow *et al.*, p. 23) Fig. 19.** (a) $C_P$-distribution and friction lines on the EC145 fuselage, visualisation of separation areas on the boot and vertical stabilisers. (b) Temperature surface distribution and 3D-contour (T=60°C), visualisation of the impact of engine plumes on horizontal stabilisers.

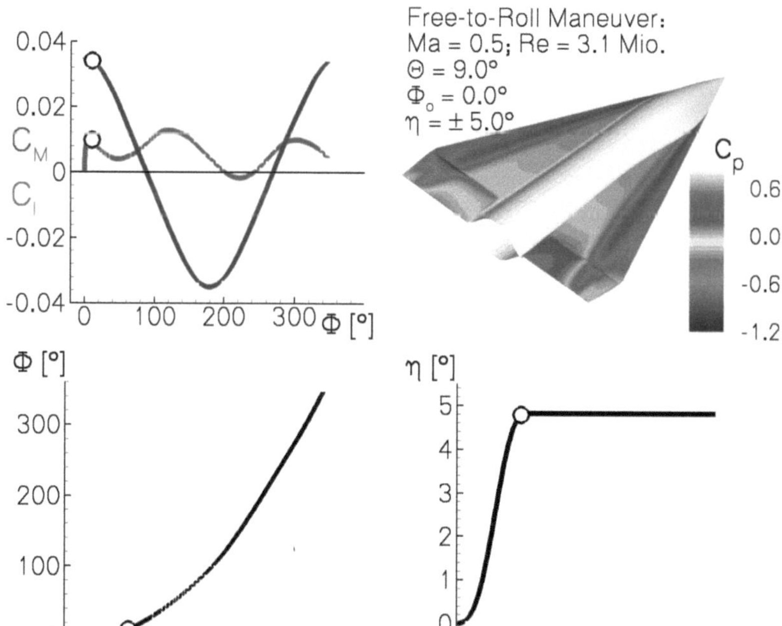


**(Rossow *et al.*, p. 25) Fig. 21.** Coupled aerodynamics and flight mechanics simulation for a rolling delta wing with trailing edge flaps using the TAU-Code.

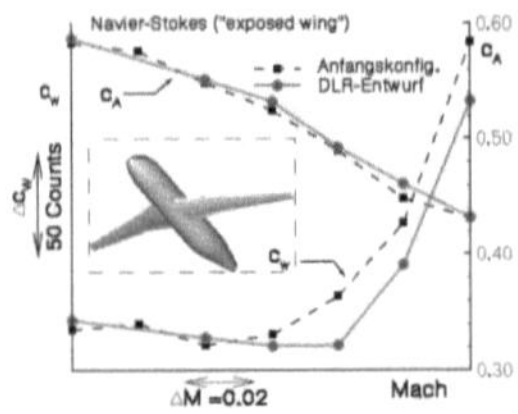


(a) Inverse wing design using FLOWer, drag rise lift as function of Mach number for baseline configuration and optimized configuration.

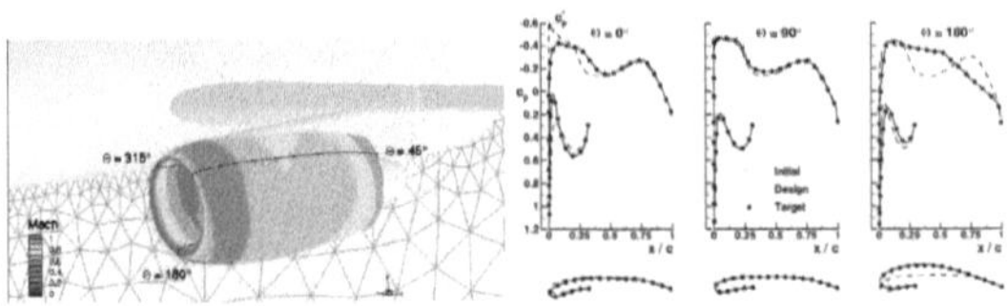

(b) Redesign of an installed nacelle using the TAU- Code, surface pressure distribution and nacelle profiles in three circumferential sections.

**(Rossow *et al.*, p. 26) Fig. 22.**

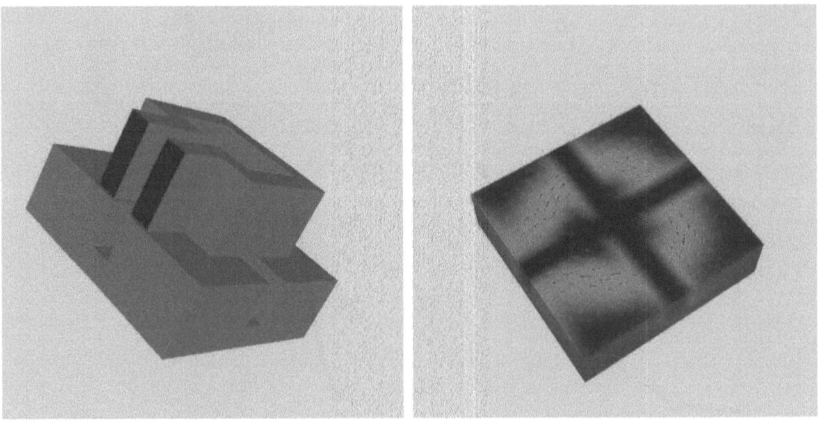

**(Schoenmaker *et al.*, p. 66) Fig. 3.** 3D view of a U-turn structure above a conductive substrate (left). The induced substrate current is shown (right).

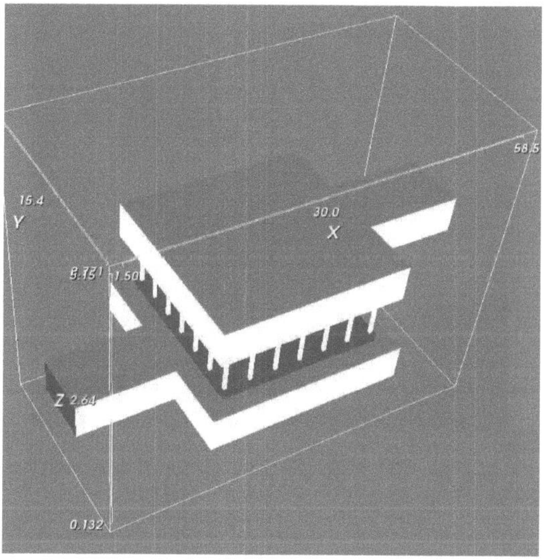


(**Schoenmaker *et al.*, p. 67) Fig. 4.** 3D view of the Metal-Insulator-Metal (MIM) capacitor.

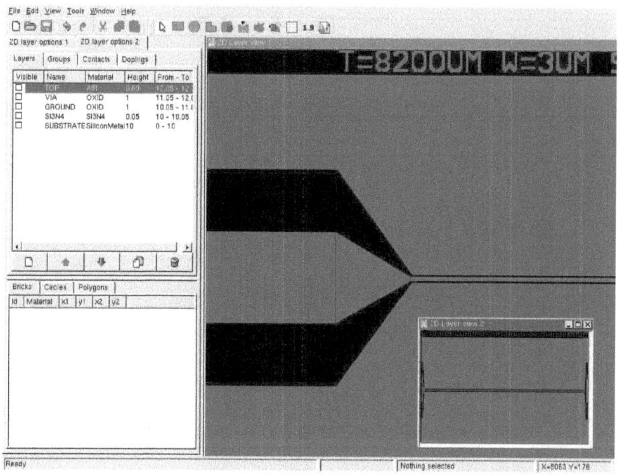


(**Schoenmaker *et al.*, p. 69) Fig. 6.** The layout of the coplanar line under study.

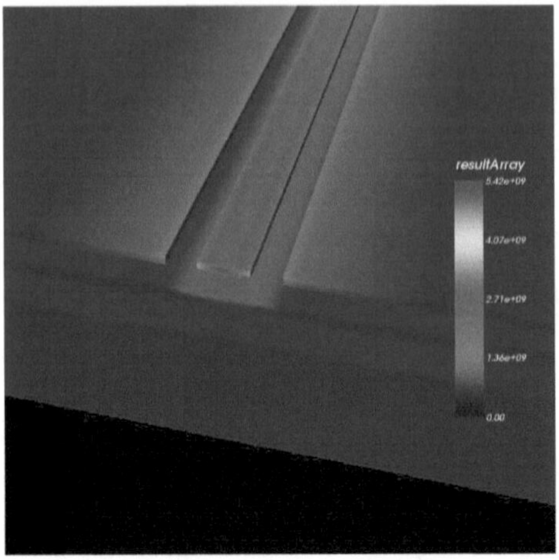


**(Schoenmaker *et al.*, p. 69) Fig. 7.** The current distribution at 30GHz in the coplanar line under study.

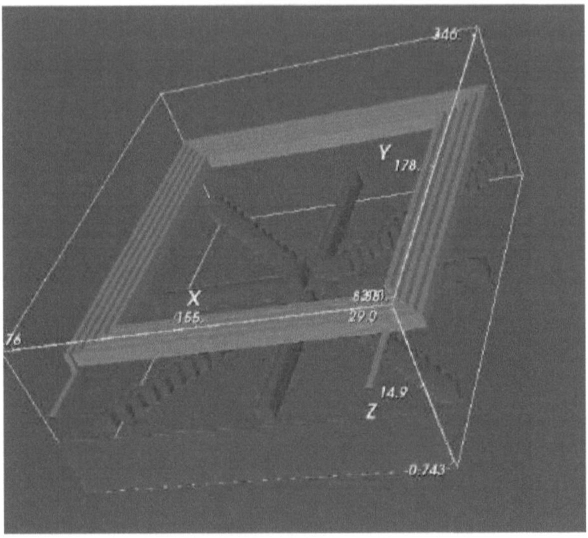


**(Schoenmaker *et al.*, p. 71) Fig. 12.** The geometry of the spiral inductor under study.

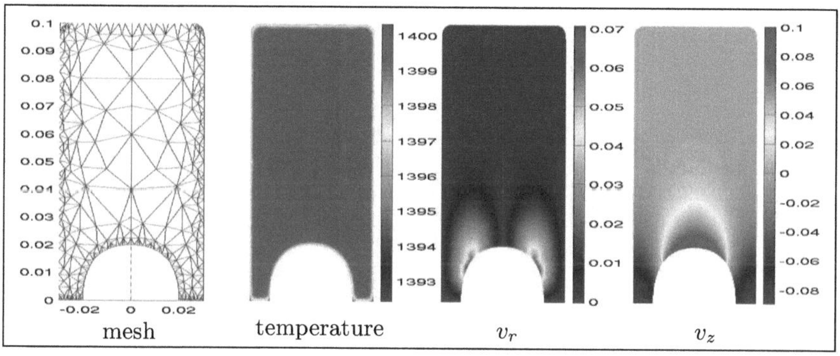


**(Kagan & Mattheij, p. 157) Fig. 1.** Mesh and velocity components at $t = 0$, temperature at $t = 0.03\,\mathrm{s}$

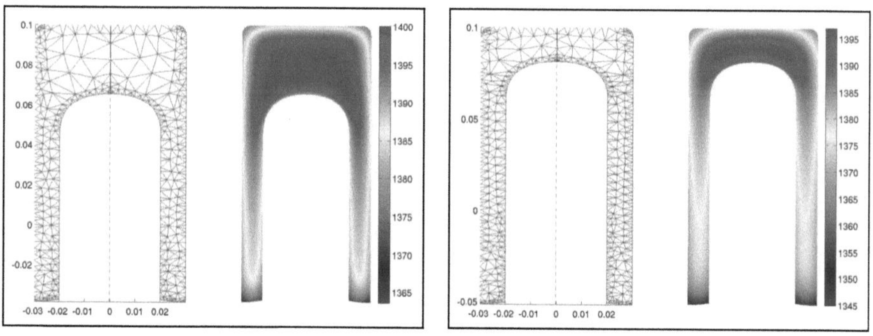

**(Kagan & Mattheij, p. 158) Fig. 2.** Mesh and temperature: left $t = 0.6$s, right $t = 1.2$s

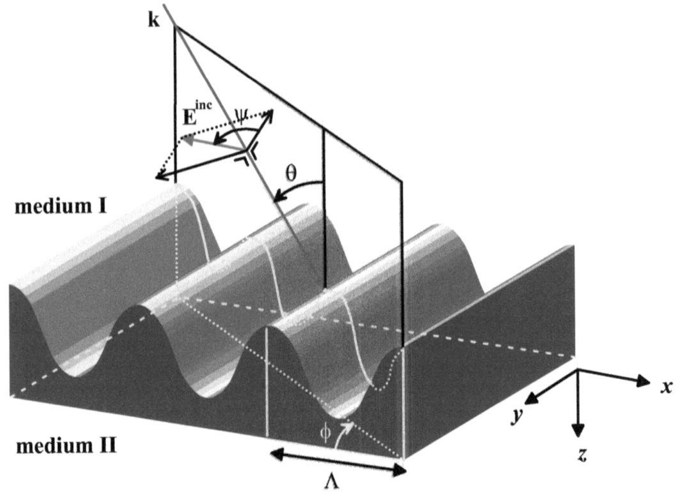


**(Van Kraaij & Maubach, p. 165) Fig. 1.** One-dimensional periodic grating in $\mathbb{R}^3$

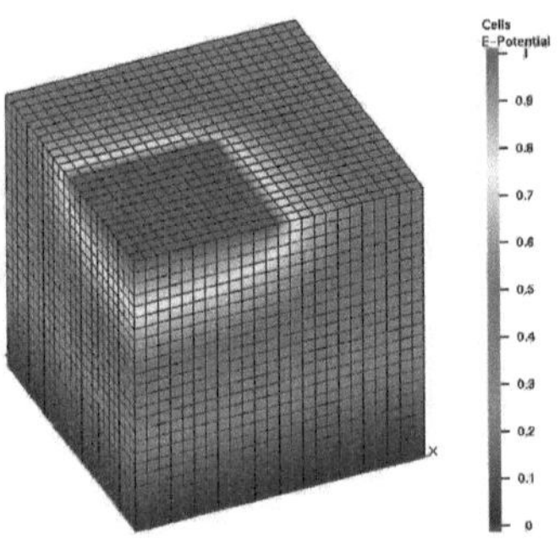


**(Mohr, p. 172) Fig. 1.** Electric potential distribution in the unit cube for boundary conditions (4).

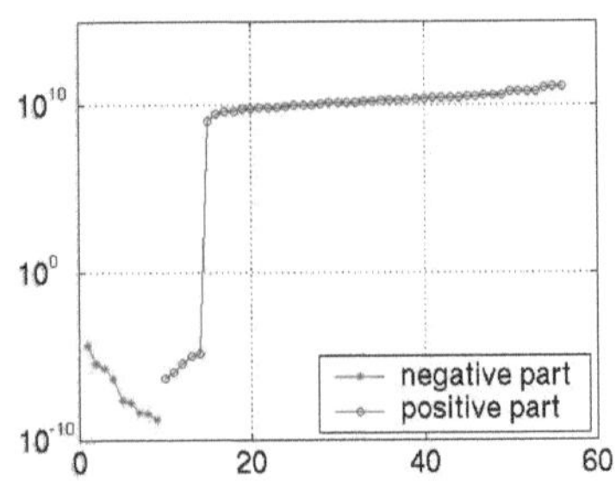


**(Mohr, p. 172) Fig. 2.** Logarithmic plot of magnitude of eigenvalues of the system matrix from (2) for test problem with 8 elements.

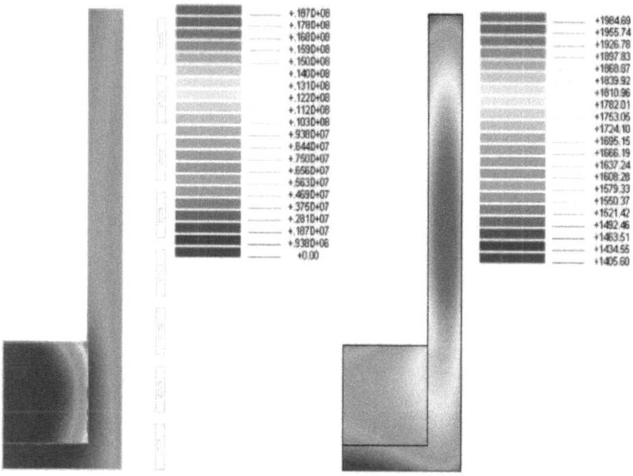


**(Bermúdez *et al.*, p. 236) Fig. 2.** Modulus of current density (left) and temperature (right) in the workpiece.

**(Graziadei & Ten Thije Boonkkamp, p. 246) Fig. 3.** Dimensionless temperature.

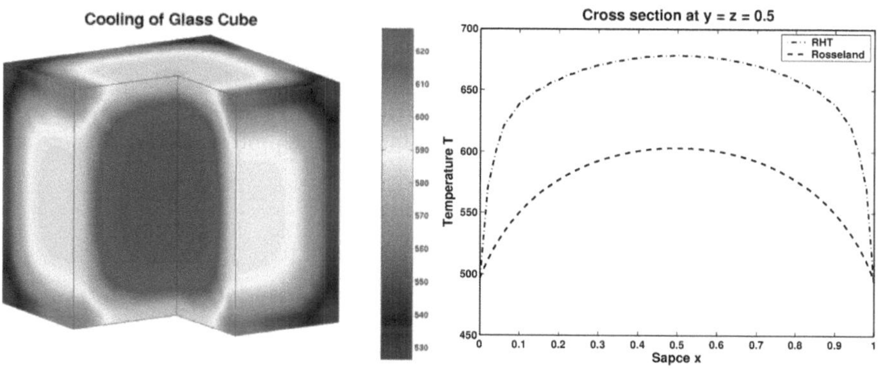


**(Seaïd & Klar, p. 286) Fig. 2.** Temperature distribution on the cube (left) and a section at $y = z = 0.5\,m$ for the computed solution by RHT equations and Rosseland approach (right).

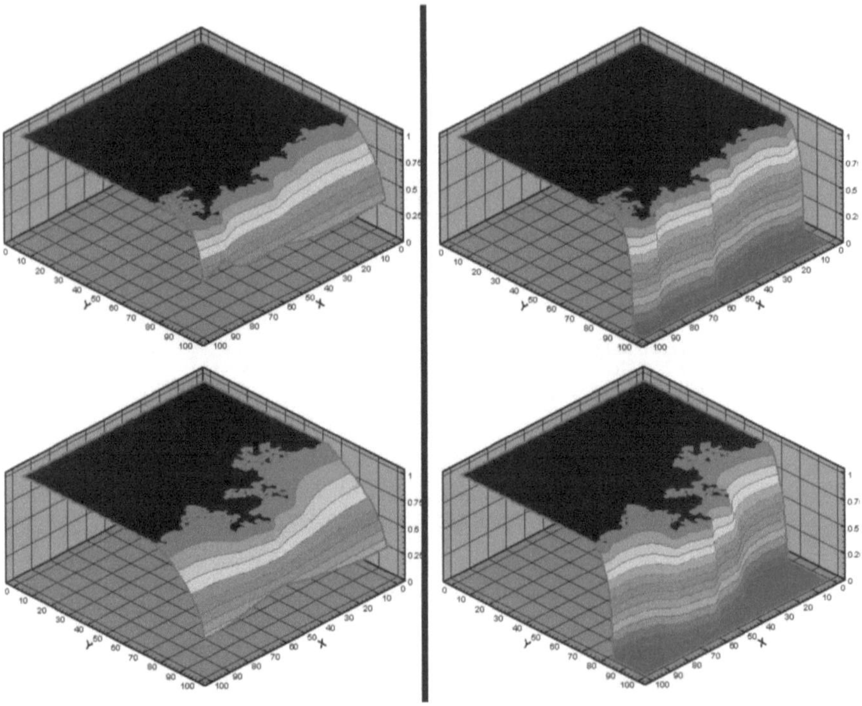

**(Yiotis *et al.*, p. 295) Fig. 2.** Profiles of the rescaled film radii for $Ca_F = 10^{-4}$ (left) and $Ca_F = 1$ (rigth) at two different stages of the process. Liquid clusters are in black, the fully dry region is in blue.

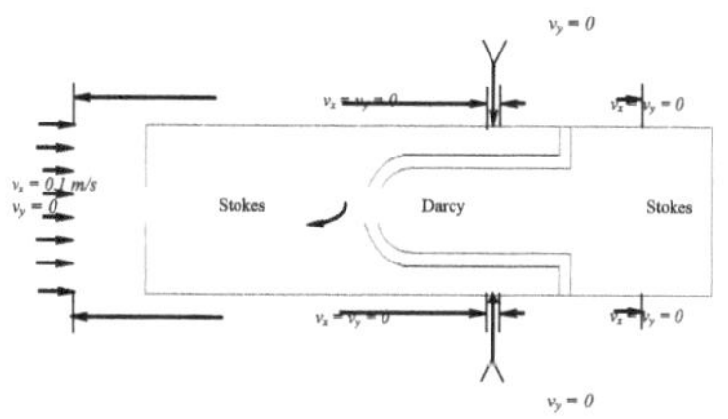


**(Nassehi *et al.*, p. 301) Fig. 1.** Schematic Representation of an Idealised Pleat of a Pleated Cartridge

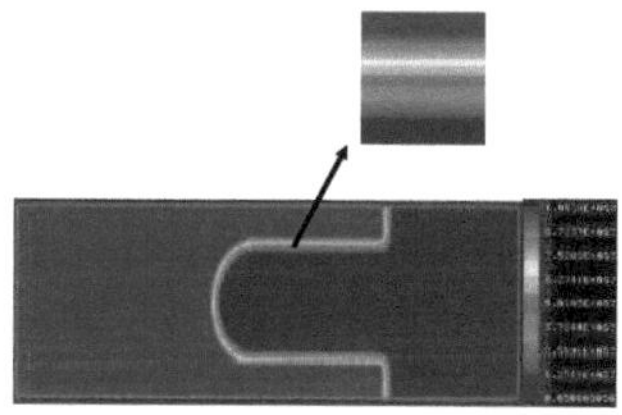

**(Nassehi *et al.*, p. 301) Fig. 2.** Predicted Pressure Distribution (Pa) in a Single Pleat Domain

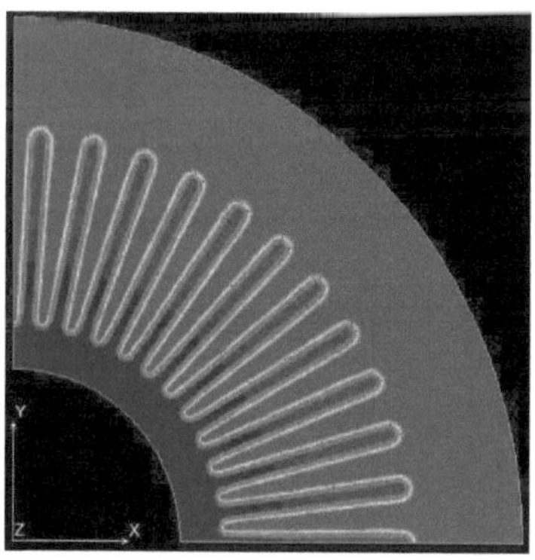


**(Nassehi *et al.*, p. 302) Fig. 5.** Predicted Pressure Field Distribution (Pa) in the Quarter Cartridge Domain of Pleated Cartridge Assembly

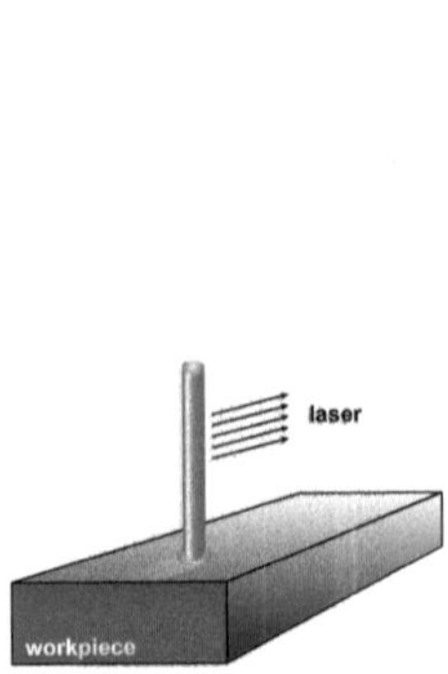


**(Anthonissen *et al.*, p. 357) Fig. 1.** Laser surface remelting

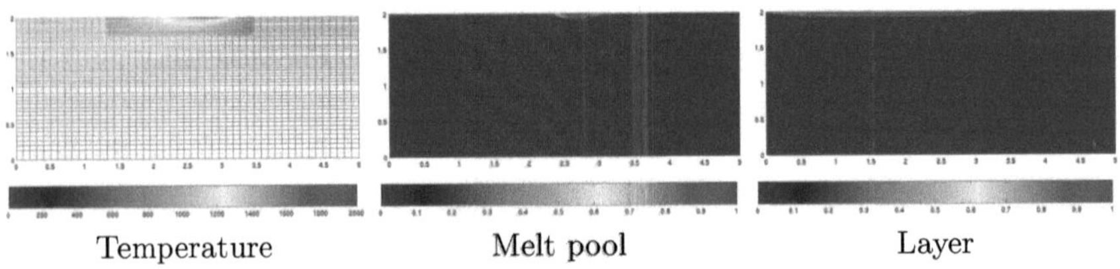


**(Anthonissen *et al.*, p. 359) Fig. 2.** Numerical results for the first simulation

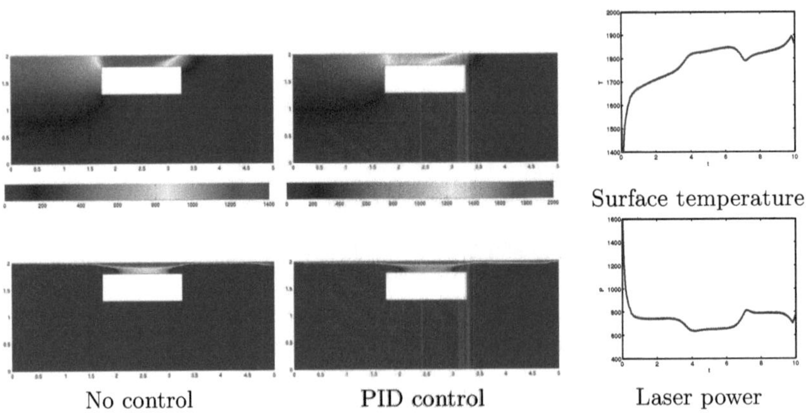


**(Anthonissen *et al.*, p. 360) Fig. 3.** Numerical results for the second simulation: temperature (top), hardening layer (bottom), surface temperature and laser power during PID control

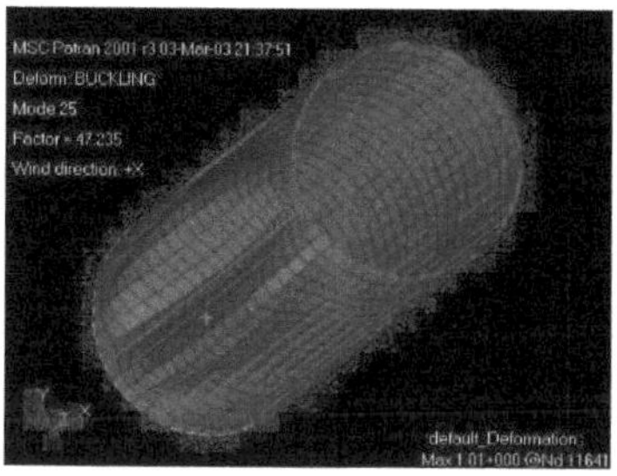


**(Morozov, p. 374) Fig. 3.** Buckling mode under the wind action.

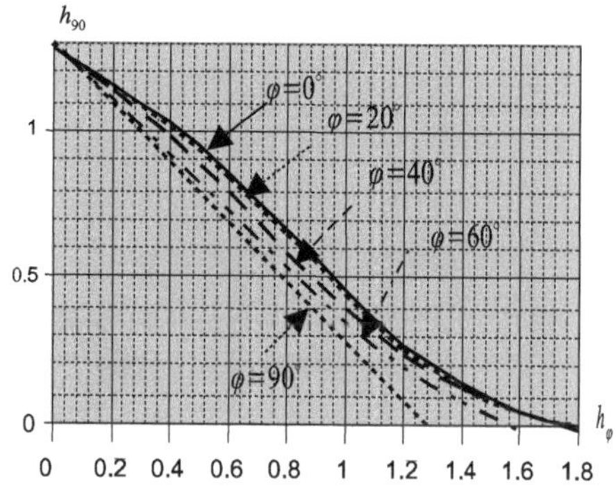


**(Morozov, p. 374) Fig. 4.** Critical thicknesses $h_{90}$ and $h_\phi$.

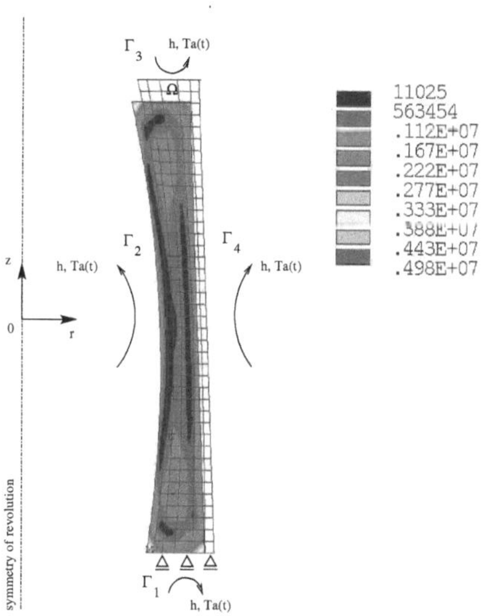


**(Sellier, p. 382) Fig. 1.** Initial glass piece geometry with corresponding Finite Element mesh and boundary conditions. The deformed glass geometry (after cooling) with the associated map of the residual Von Mises stresses is also shown.

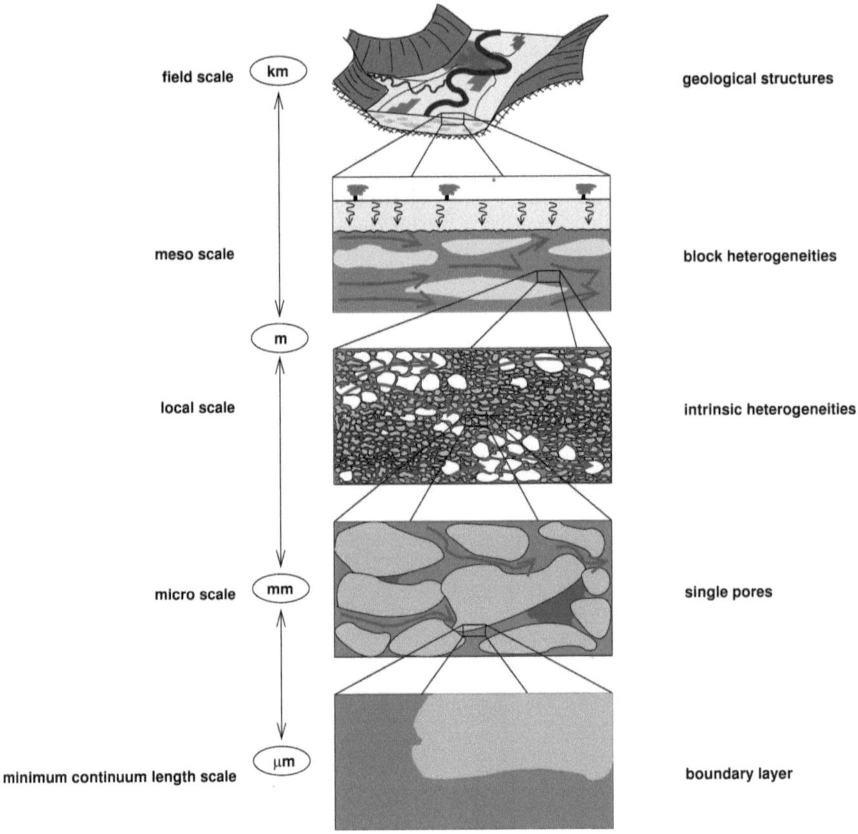


**(Helmig *et al.*, p. 451) Fig. 1.** Different scales for flow in porous media.

**(Helmig *et al.*, p. 457) Fig. 4.** TCE distribution for Subset A (TCE regions highlighted by red spots).

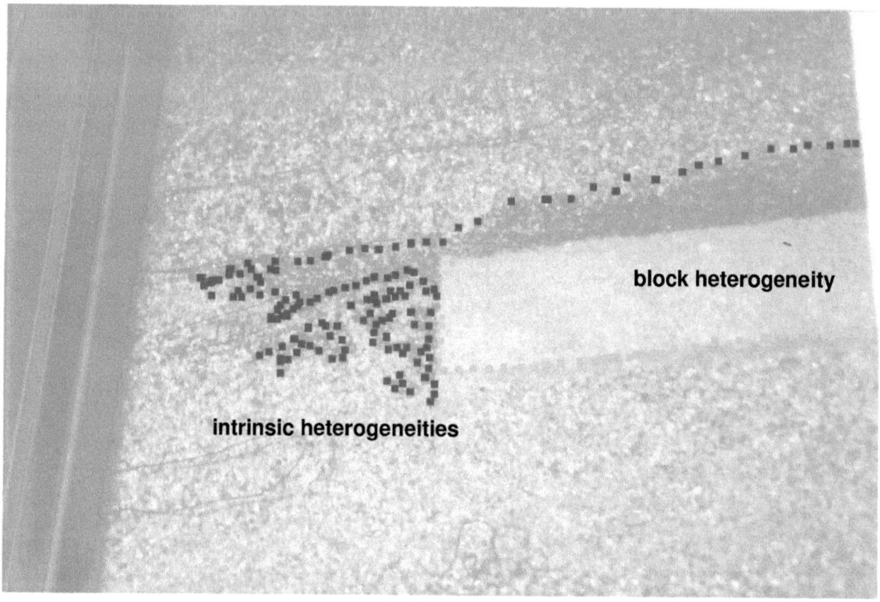


**(Helmig *et al.*, p. 458) Fig. 5.** TCE distribution for Subset B (TCE regions highlighted by red spots, and the fine sand lenses by yellow spots).

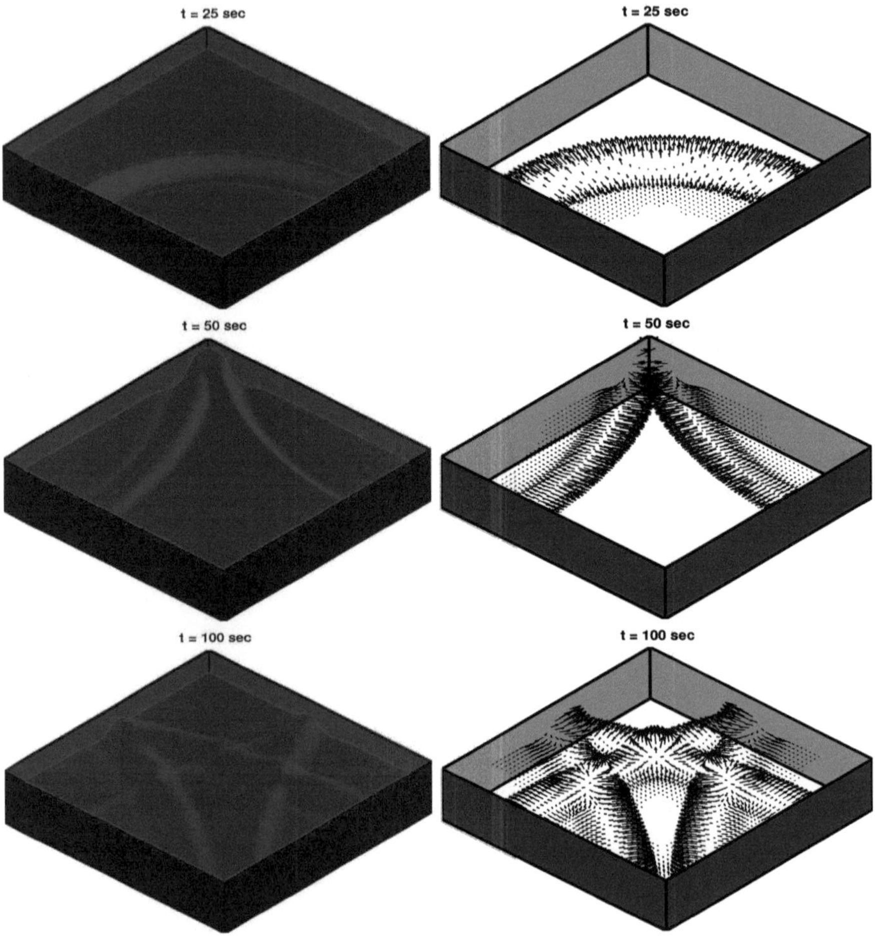


**(El Amrani & Seaïd, p. 497) Fig. 1.** Animating water waves in a squared pool.

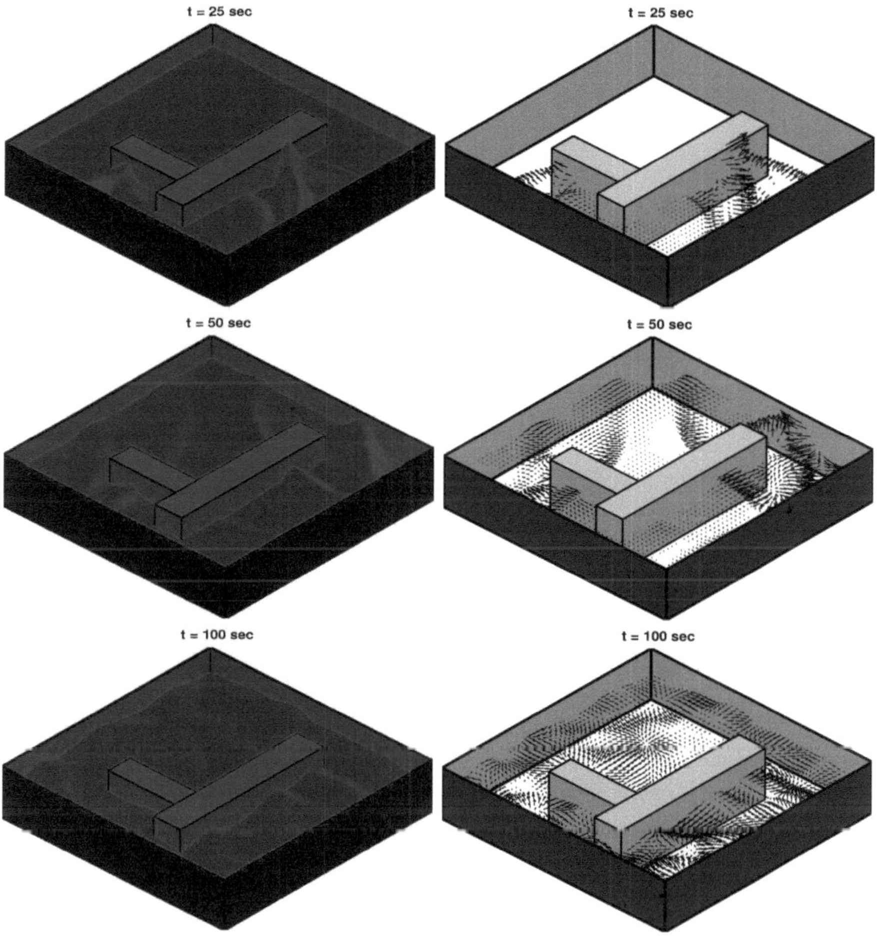


**(El Amrani & Seaïd, p. 498) Fig. 2.** Animating water waves in a squared pool with fixed obstacles.

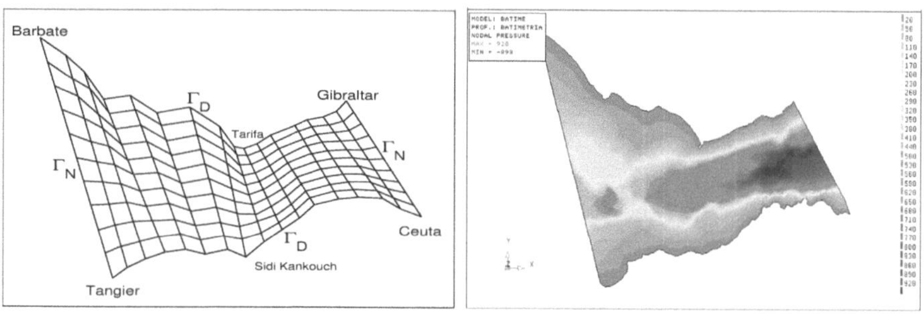


**(González & Seaïd, p. 520) Fig. 2.** Computational mesh and bathymetry of the strait.

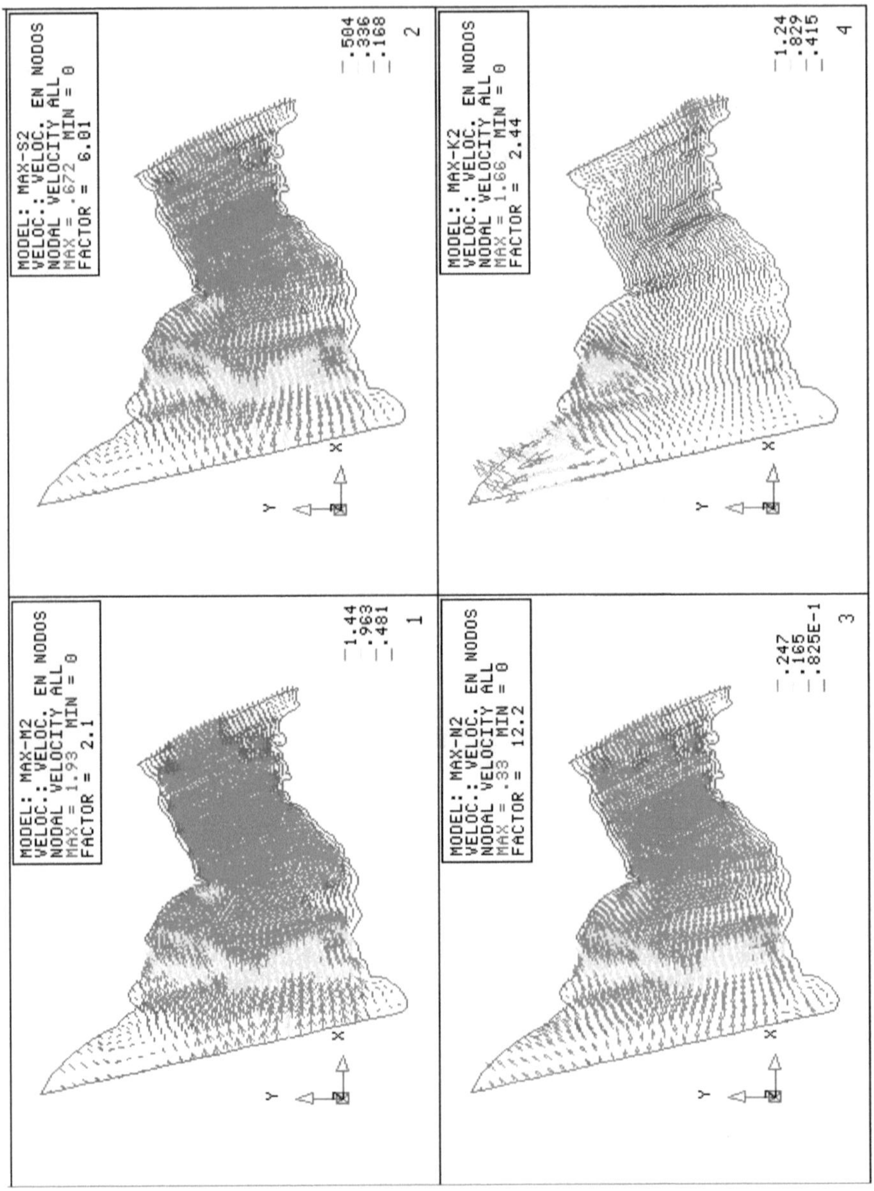


**(González & Seaïd, p. 522) Fig. 3.** Flow field for the main diurnal and semidiurnals K2, M2, N2 and S2.

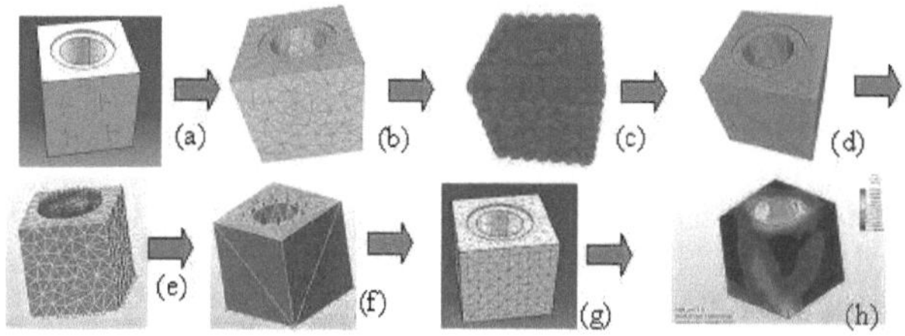


**(Hamri *et al.*, p. 586) Fig. 1.** The workflow of the analysis model generation: (a) case of study, (b) tessellated model, (c) envelope around the polyhedron, *i.e.* FE map of sizes, (d) adapted model (simulation model), (e) polyhedral model with boundary conditions, (f) transfert of boundary conditions on the adapted model, (g) mesh of the adapted model, (h) analysis result.

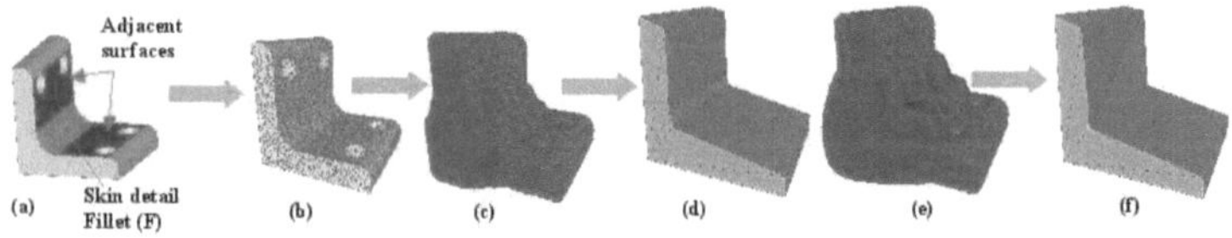


**(Hamri *et al.*, p. 588) Fig. 2.** The concept of simplification features: (a) initial CAD model(B-Rep), (b) polyhedral model, (c) small map of FE sizes attached to (b), (d) simplification result of (c), (e) large map of FE sizes attached to (b), (f) simplification result of (e)

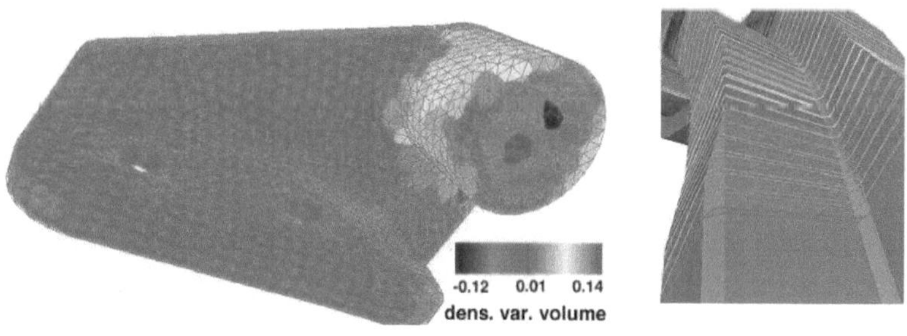


**(Léon *et al.*, p. 618) Fig. 2.** Two examples of a priori criteria, variation of volume on the left part and variation of sections on the right part.

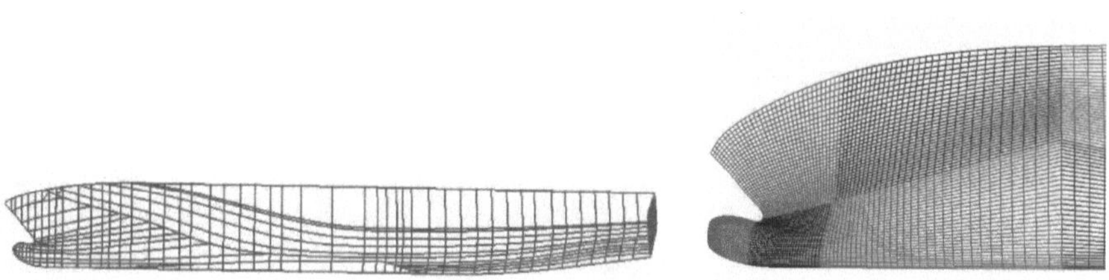

**(Skytt, p. 640) Fig. 1.** Patch structure of a ship hull and a detail of the surface grid

# Author index

PHILIPS

Eindhoven
leading in technology

MATHEMATICAL
RESEARCH
INSTITUTE
MRI
University of Nijmegen
P.O. Box 9010
6500 GL Nijmegen
The Netherlands
phone + 31 24 3652986
fax +31 24 3652140

J.M. Burgerscentrum

em
engineering
mechanics

ttec
THE EUROPEAN IT FACTORY

transtec
THE EUROPEAN IT FACTORY

eldim

THALES

Sodexho
Food and Management Services

TNO

DOW

océ

NWO

RS

ABN·AMRO

WL | delft hydraulics

CAMBRIDGE
UNIVERSITY PRESS

IBIS UvA BV
INSTITUUT VOOR BEDRIJFS- EN INDUSTRIËLE STATISTIEK

Springer

Koninklijke
Nederlandse
Akademie van
Wetenschappen

Can
Your partner in
Mathematics and
Statistics

Beta
Research School for Operations
Management and Logistics

EURANDOM

MATTeR

MIX
Papier aus verantwortungsvollen Quellen
Paper from responsible sources
FSC® C105338

If you have any concerns about our products,
you can contact us on
**ProductSafety@springernature.com**

In case Publisher is established outside the EU,
the EU authorized representative is:
**Springer Nature Customer Service Center GmbH**
**Europaplatz 3, 69115 Heidelberg, Germany**

Printed by Libri Plureos GmbH
in Hamburg, Germany